Lecture Notes in Artificial Intelligence 2035

Subseries of Lecture Notes in Computer Science
Edited by J. G. Carbonell and J. Siekmann

Lecture Notes in Computer Science
Edited by G. Goos, J. Hartmanis and J. van Leeuwen

Springer
Berlin
Heidelberg
New York
Barcelona
Hong Kong
London
Milan
Paris
Singapore
Tokyo

David Cheung Graham J. Williams
Qing Li (Eds.)

Advances in Knowledge Discovery and Data Mining

5th Pacific-Asia Conference, PAKDD 2001
Hong Kong, China, April 16-18, 2001
Proceedings

Series Editors

Jaime G. Carbonell, Carnegie Mellon University, Pittsburgh, PA, USA
Jörg Siekmann, University of Saarland, Saabrücken, Germany

Volume Editors

David Cheung
The University of Hong Kong, Dept. of Computer Science and Information Systems
Pokfulam, Hong Kong, China
E-mail: dcheung@csis.hku.hk

Graham J. Williams
CSIRO Mathematical and Information Sciences
GPO Box 664, Canberra, ACT 2601, Australia
E-mail: Graham.Williams@cmis.csiro.au

Qing Li
City University of Hong Kong, Department of Computer Science
83 Tat Chee Ave., Kowloon, Hong Kong, China
E-mail: csqli@cityu.edu.hk

Cataloging-in-Publication Data applied for

Die Deutsche Bibliothek - CIP-Einheitsaufnahme

Advances in knowledge discovery and data mining : 5th Pacific Asia conference ; proceedings / PAKDD 2001, Hong Kong, China, April 16 - 18, 2001. David Cheung ... (ed.). - Berlin ; Heidelberg ; New York ; Barcelona ; Hong Kong ; London ; Milan ; Paris ; Singapore ; Tokyo : Springer, 2001
(Lecture notes in computer science ; Vol. 2035 : Lecture notes in artificial intelligence)
ISBN 3-540-41910-1

CR Subject Classification (1998): I.2, H.3, H.5.1, G.3, J.1, K.4

ISBN 3-540-41910-1 Springer-Verlag Berlin Heidelberg New York

Springer-Verlag Berlin Heidelberg New York
a member of BertelsmannSpringer Science+Business Media GmbH

http://www.springer.de

Printed in Germany

Typesetting: Camera-ready by author, data conversion by PTP-Berlin, Stefan Sossna
Printed on acid-free paper SPIN: 10782484 06/3142 5 4 3 2 1 0

Preface

PAKDD 2001, Hong Kong, 16–18 April, was organized by the E-Business Technology Institute of The University of Hong Kong in cooperation with ACM Hong Kong, IEEE Hong Kong Chapter, and The Hong Kong Web Society. It was the Fifth Pacific-Asia Conference on Knowledge Discovery and Data Mining and the successor of earlier PAKDD conferences held in Singapore (1997), Melbourne, Australia (1998), Beijing, China (1999), and Kyoto, Japan (2000).

PAKDD 2001 brought together participants from universities, industry, and government to present, discuss, and address both current issues and novel approaches in the practise, deployment, theory, and methodology of Knowledge Discovery and Data Mining. The conference provides an international forum for the sharing of original research results and practical development experiences among researchers and application developers from the many KDD related areas including machine learning, databases, statistics, internet, e-commerce, knowledge acquisition, data visualization, knowledge-based systems, soft computing, and high performance computing.

The PAKDD 2001 conference included technical sessions organized around important subtopics such as: Web Mining; Text Mining; Applications and Tools; Interestingness; Feature Selection; Sequence Mining; Spatial and Temporal Mining; Concept Hierarchies; Association Mining; Classification and Rule Induction; Clustering; and Advanced Topics and New Methods.

Following careful review of the 152 submissions by members of the international program committee 38 regular papers and 22 short papers were selected for presentation at the conference and for publication in this volume.

The conference program also included invited keynote presentations from three international researchers and developers in data mining: H. V. Jagadish of the University of Michigan, Ronny Kohavi of Blue Martini, and Hongjun Lu of the University of Science and Technology, Hong Kong. Abstracts of their presentations are included in this volume.

The conference presented six tutorials from experts in their respective disciplines: An Introduction to MARS (Dan Steinberg); Static and Dynamic Data Mining Using Advanced Machine Learning Methods (Ryszard S. Michalski); Sequential Pattern Mining: From Shopping History Analysis to Weblog Mining and DNA Mining (Jiawei Han and Jian Pei); Recent Advances in Data Mining Algorithms for Large Databases (Rajeev Rastogi and Kyuseok Shim); Web Mining for E-Commerce (Jaideep Srivastava); and From Evolving Single Neural Networks to Evolving Ensembles (Xin Yao).

Associated workshops included: Spatial and Temporal Data; Statistical Techniques in Data Mining; and Data Mining and Electronic Business.

A conference such as this can only succeed as a team effort. We would like to thank the program committee members and reviewers for their efforts and the PAKDD steering committee members for their invaluable input and advice. Our sincere gratitude goes to all of the authors who submitted papers. We are grateful to our sponsors for their generous support. Special thanks go to Ms Winnie Yau, E-Business Technology Institute, The University of Hong Kong, for her considerable efforts, seamlessly keeping everything running smoothly and coordinating the many streams of the conference organization.

On behalf of the organizing and program committees of PAKDD 2001 we trust you found the conference a fruitful experience and hope you had an enjoyable stay in Hong Kong.

February 2001

Chung-Jun Tan, Jiawei Han
David Cheung, Graham J. Williams, Qing Li

PAKDD 2001 Organization

PAKDD 2001 was organized by the E-Business Technology Institute of The University of Hong Kong in cooperation with ACM Hong Kong, IEEE Hong Kong Chapter, and The Hong Kong Web Society.

Conference Committee

Conference Chairs:	Chung-Jen Tan (University of Hong Kong and IBM Watson)
	Jiawei Han (Simon Fraser University, Canada)
Program Chairs:	David Cheung (University of Hong Kong)
	Graham J. Williams (CSIRO, Australia)
	Qing Li (City University of Hong Kong)
Tutorial Chair:	Joshua Huang (University of Hong Kong)
Workshop Chair:	Michael Ng (University of Hong Kong)
Industrial Chair:	Joseph Fong (City University of Hong Kong)
Demonstration Chair:	Jiming Liu (Baptist University of Hong Kong)
Local Arrangements Chairs:	Ronnie Cheung (Hong Kong Polytechnic University)
	Ben Kao (University of Hong Kong)
Publicity Chairs:	Vincent Ng (Hong Kong Polytechnic University)
	Rohan Baxter (CSIRO, Australia)
	Hiroyuki Kawano (Kyoto University, Japan)
Treasurer:	Ada Fu (Chinese University of Hong Kong)

PAKDD Steering Committee

Xindong Wu	Colorado School of Mines, USA (Chair)
Hongjun Lu	Hong Kong Univ. of Science & Technology (Co-chair)
Ramamohanarao Kotagiri	University of Melbourne, Australia
Huan Liu	National University of Singapore
Hiroshi Motoda	Osaka University, Japan
Lizhu Zhou	Tsinhua University, China
Ning Zhong	Yamaguchi University, Japan
Masaru Kitsuregawa	University of Tokyo, Japan
Takao Terano	University of Tsukuba, Japan

Program Committee

Rohan Baxter	CSIRO, Australia
Keith Chan	Hong Kong Polytechnic University
Surajit Chaudhuri	Microsoft, USA
Ming-Syan Chen	National Taiwan University
David Cheung	University of Hong Kong
Umeshwar Dayal	Hewlett-Packard Labs, USA
Tharam Dillon	Hong Kong Polytechnic University
Guozhu Dong	Wright State University, USA
David Dowe	Monash University, Australia
Ling Feng	Tilburg University, The Netherlands
Ada Fu	Chinese University of Hong Kong
Chi Wai Fung	VTC-IVE, Hong Kong
Yike Guo	Imperial College, UK
Howard Hamilton	University of Regina, Canada
Jayant Haritysa	Indian Institute of Science, India
Simon Hawkins	CSIRO, Australia
Markus Hegland	Australian National University
Robert Hilderman	University of Regina, Canada
Howard Ho	IBM Almaden, USA
Joshua Z. Huang	University of Hong Kong
Moon Yul Huh	SKK University, Korea
Vijay Iyengar	IBM T.J. Watson Research Center, USA
Ben Kao	University of Hong Kong
Kamal Karlapalem	Hong Kong UST
Hiroyuki Kawano	University of Kyoto, Japan
Jinho Kim	Kangwon National U., Korea
Sang-Wook Kim	Kangwon National University, Korea
Masaru Kitsuregawa	University of Tokyo, Japan
Kevin Korb	Monash University, Australia
Laks V. S. Lakshmanan	Concordia U. and IIT, Bombay
Wai Lam	Chinese University of Hong Kong
Doheon Lee	Chonnam National University, Korea
Hong-Va Leong	Hong Kong Polytechnic University
Deyi Li	Electronic System Eng. Inst., Beijing
Qing Li	City University of Hong Kong
T. Y. Lin	San Jose State University, USA
Charles Ling	University of Western Ontario, Canada
Bing Liu	National University of Singapore
Huan Liu	Arizona State University, USA
Jiming Liu	Hong Kong Baptist University
Hongjun Lu	Hong Kong UST

Reviewers

David Albrecht
Lloyd Allison
A. Ammoura
Rohan Baxter
Keith Chan
Surajit Chaudhuri
Ming Syan Chen
David Cheung
W. Cheung
Yin Ling Cheung
Vishal Chitkara
Moon Young Chung
Woong Kyo Chung
Shi Chunyi
Hang Cui
Gautam Das
Manoranjan Dash
Umeshwar Dayal
Tharam Dillon
Guozhu Dong
David Dowe
Leigh Fitzgibbon
Ada Fu
Chi Wai Fung
Venkatesh Ganti
Moustafa Ghanem
Ishaan Goyal
Baohua Gu
Yike Guo
M. El Hajj
Howard Hamilton
Jayant Haritysa
Simon Hawkins
Markus Hegland
Chun Hung Heng
Robert Hilderman
Howard Ho
Kei Shiu Ho
Eui Kyeong Hong
Mintz Hsieh
Joshua Z Huang
Moon Yul Huh
Dongjoon Hyun
Vijay Iyengar
He Ji
Wenyun Ji
Ben Kao
Kamal Karlapalem
Hiroyuki Kawano
Jeong Ja Kim
Jinho Kim
Sang Wook Kim
Masaru Kitsuregawa
Kevin Korb
Laks V.S. Lakshmanan
Wai Lam
Yip Chi Lap
Doheon Lee
Raymond Lee
Hong Va Leong
Deyi Li
Qing Li
Seung Chun Li
Jong Hwa Lim
T Y Lin
Charles Ling
Feng Ling
Bing Liu
Huan Liu
James Liu
Jiming Liu
Hongjun Lu
Jinyoung Moon
Hiroshi Motoda
Ka Ka Ng
Michael Ng
Raymond Ng
Vincent Ng
Ann Nicholsons
Masato Oguchi
Tadashi Ohmori
Andrew Paplinski
Jong Soo Park
Sanghyun Park
Jon Patrick
Iko Pramudiono
Weining Qian
Kanagasabai Rajaraman
Kyuseok Shim
Ho Wai Shing
Ho Kei Shiu
Il Yeol Song
John Sum
Katsumi Takahashi
Ah Hwee Tan
Changjie Tang
Zhaohui Tang
Takao Terano
Bhavani Thurasingham
Masashi Toyoda
Dean Trower
Vassilios Verykios
Murli Viswanathan
Dianlui Wang
Haiyun Wang
Ke Wang
Wei Wang
Weinan Wang
Kyu Young Whang
Graham J. Williams
Ian Witten
Yong Gwan Won
Xindong Wu
Jim Yang
Jiong Yang
Yiyu Yao
Siu Ming Yiu
Ji Hee Yoon
Clement Yu
Jeffrey Yu
Philip Yu
Lui Yuchang
Osamar R. Zaiane
Mohammed Zaki
Shi Hui Zheng
Ning Zhong
Aoying Zhou
Lizhu Zhou

Sponsors

SAS

ETNet

Hong Kong Pei Hua Education Foundation

IEEE Computer Society, Hong Kong Section, Computer Chapter

ACM Hong Kong

Hong Kong Computer Society

Hong Kong Productivity Council

Hong Kong Web Society

E-Business Technology Institute, The University of Hong Kong

Department of Computer Science and Information Systems, The University of Hong Kong

Table of Contents

Keynote Presentations

Web Mining

Text Mining

Applications and Tools

Concept Hierarchies

Feature Selection

Interestingness

Sequence Mining

Spatial and Temporal Mining

Association Mining

Classification and Rule Induction

Clustering

Advanced Topics and New Methods

Incompleteness in Data Mining

Hosagrahar Visvesvaraya Jagadish*

University of Michigan,
Ann Arbor
jag@umich.edu

Abstract. Database technology, as well as the bulk of data mining technology, is founded upon logic, with absolute notions of truth and falsehood, at least with respect to the data set. Patterns are discovered exhaustively, with carefully engineered algorithms devised to determine all patterns in a data set that belong to a certain class. For large data sets, many such data mining techniques are extremely expensive, leading to considerable research towards solving these problems more cheaply.
We argue that the central goal of data mining is to find SOME interesting patterns, and not necessarily ALL of them. As such, techniques that can find most of the answers cheaply are clearly more valuable than computationally much more expensive techniques that can guarantee completeness. In fact, it is probably the case that patterns that can be found cheaply are indeed the most important ones.
Furthermore, knowledge discovery can be the most effective with the human analyst heavily involved in the endeavor. To engage a human analyst, it is important that data mining techniques be interactive, hopefully delivering (close to) real time responses and feedback. Clearly then, extreme accuracy and completeness (i.e., finding *all* patterns satisfying some specified criteria) would almost always be a luxury. Instead, incompleteness (i.e., finding only *some* patterns) and approximation would be essential.
We exemplify this discussion through the notion of *fascicles*. Often many records in a database share similar values for several attributes. If one is able to identify and group together records that share similar values for some – even if not all – attributes, one can both obtain a more parsimonious representation of the data, and gain useful insight into the data from a mining perspective. Such groupings are called fascicles. We explore the relationship of fascicle-finding to association rule mining, and experimentally demonstrate the benefit of incomplete but inexpensive algorithms. We also present analytical results demonstrating both the limits and the benefits of such incomplete algorithms.

* Supported in part by NSF grant IIS-0002356. Portions of the work joint with Cinda Heeren, Raymond Ng, and Lenny Pitt

D. Cheung, G.J. Williams, and Q. Li (Eds.): PAKDD 2001, LNAI 2035, p. 1, 2001.

Mining E-Commerce Data: The Good, the Bad, and the Ugly

Ronny Kohavi

Director of Data Mining,
Blue Martini Software
ronnyk@bluemartini.com

Abstract. Electronic commerce provides all the right ingredients for successful data mining (the Good). Web logs, however, are at a very low granularity level, and attempts to mine e-commerce data using only web logs often result in little interesting insight (the Bad). Getting the data into minable formats requires significant pre-processing and data transformations (the Ugly). In the ideal e-commerce architecture, high level events are logged, transformations are automated, and data mining results can easily be understood by business people who can take action quickly and efficiently. Lessons, stories, and challenges based on mining real data at Blue Martini Software will be presented.

D. Cheung, G.J. Williams, and Q. Li (Eds.): PAKDD 2001, LNAI 2035, p. 2, 2001.

Seamless Integration of Data Mining with DBMS and Applications

Hongjun Lu

The Hong Kong University of Science and Technology,
Hong Kong, China
luhj@cs.ust.hk

Abstract. Data mining has been widely recognized as a powerful tool for exploring added value from data accumulated in the daily operations of an organization. A large number of data mining algorithms have been developed during the past decade. Those algorithms can be roughly divided into two groups. The fist group of techniques, such as classification, clustering, prediction and deviation analysis, has been studied for a long time in machine learning, statistics, and other fields. The second group of techniques, such as association rule mining, mining in spatial-temporal databases and mining from the Web, addresses problems related to large amounts of data. Most classical algorithms in the first group assume that the data to be mined is somehow available in memory. Although initial effort in data mining has concentrated on making those algorithms scalable with respect to large volume of data, most of those scalable algorithms, even developed by database researchers, are still stand-alone. It is often assumed that data is available in desired forms, without considering the fact that most organizations store their data in databases managed by database management systems (DBMS). As such, most data mining algorithms can only be loosely coupled with data infrastructures in organizations and are difficult to infuse into existing mission-critical applications. Seamlessly integrating data mining techniques with database applications and database management systems remains an open problem.

In this paper, we propose to tackle the problem of seamless integration of data mining with DBMS and applications from three directions. First, with the recent development of database technology, most database management systems have extended their functionality in data analysis. Such capability should be fully explored to develop DBMS-awre data mining algorithms. Ideally, data mining algorithms can be fully implemented using DBMS supported functions so that they become database application themselves. Second, major difficulties in integrating data mining with applications are algorithm selection and parameter setting. Reducing or eliminating mining parameters as much as possible and developing automatic or semi-automatic mining algorithm selection techniques will greatly increase the application friendliness of data mining systems. Lastly, standardizing the interface among databases, data mining algorithms and applications can also facilitate the integration to certain extent.

D. Cheung, G.J. Williams, and Q. Li (Eds.): PAKDD 2001, LNAI 2035, p. 3, 2001.

Applying Pattern Mining to Web Information Extraction

Chia-Hui Chang, Shao-Chen Lui, and Yen-Chin Wu

Dept. of Computer Science and Information Engineering
National Central University, Chung-Li, 320, Taiwan
chia@csie.ncu.edu.tw, {anyway, trisdan}@db.csie.ncu.edu.tw

Abstract. Information extraction (IE) from semi-structured Web documents is a critical issue for information integration systems on the Internet. Previous work in *wrapper induction* aim to solve this problem by applying machine learning to automatically generate extractors. For example, WIEN, Stalker, Softmealy, etc. However, this approach still requires human intervention to provide training examples. In this paper, we propose a novel idea to IE, by repeated pattern mining and multiple pattern alignment. The discovery of repeated patterns are realized through a data structure call *PAT tree*. In addition, incomplete patterns are further revised by pattern alignment to comprehend all pattern instances. This new track to IE involves no human effort and content-dependent heuristics. Experimental results show that the constructed extraction rules can achieves 97 percent extraction over fourteen popular search engines.

Keywords: information extraction, semi-structured documents, wrapper generation, pattern discovery, multiple alignment

1 Introduction

Information extraction (IE) is concerned with extracting from a collection of documents the information relevant to a particular extraction task. For instance, the meta-search engine MetaCrawler extracts the search results from multiple search engines; and the shopping agent Junglee extracts the product information from several online stores for comparison. With the growth of the amount of online information, the availability of robust, flexible IE has become a stringent necessity.

Contrast to "traditional" information extraction which roots in natural language processing (NLP) techniques such as linguistic analysis, Internet information extraction rely on syntactic structures identification marked by HTML tags. The difference is due to the nature of Web such that the page contents have to be clear at glance. Thus, "itemized list" and "tabular format" have been the main presentation style for Web pages on the Internet. Such presentation styles together with the multiple records contained in one documents contribute the so called semi-structured Web pages.

D. Cheung, G.J. Williams, and Q. Li (Eds.): PAKDD 2001, LNAI 2035, pp. 4–15, 2001.

The major challenge of IE is the problem of scalability as the extraction rules must be tailored for each particular page collection, To automate the construction of extractors (or wrappers), recent research has identified important wrapper classes and induction algorithms. For example, Kushmerick et. al. identified a family of wrapper classes and the corresponding induction algorithms which generalize from labeled examples to extraction rules [9]. More expressive wrapper structure are introduced lately. *Softmealy* by Hsu and Dung [6] uses a wrapper induction algorithm to generate extractors that are expressed as finite-state transducers. Meanwhile, Muslea et al. [10] proposed *"STALKER"* that performs hierarchical information extraction to redeem Softmealy's inability to use delimiters that do not immediately precede and follow the relevant items with extra scans over the documents (see [11] for a complete survey).

In all this work, wrappers are induced from training examples such that landmarks or delimiters can be generalized from common prefixes or suffixes. However, labeling these training examples is sometimes time-consuming. Hence, another track of research is exploring new approaches to fully automate wrapper construction. For example, Embley et. al. describe a heuristic approach to discover record boundaries in Web documents by identifying *candidate separator tags* using five independent heuristics and selecting a consensus separator tag based on a heuristic combination [3]. However, one serious problem in this one-tag separator approach arises when the separator tag is used elsewhere among a record other than the boundary.

On the other hand, our work here attempts to eliminate human intervention by pattern mining. The motivation is from the observation that useful information in a Web page is often placed in a structure having a particular alignment and order. For example, Web pages produced by Web search engines generally have regular and repetitive patterns, which usually represent meaningful and useful data records. In the next section, we first give an example showing the repeated pattern formed by multiple aligned records.

2 Motivation

One observation from Web pages is that the information to be extracted is often placed in a structure having a particular alignment and forms *repetitive patterns*. For example, query-able or search-able Internet sites such as Web search engines often produce Web pages with large itemized match results which are displayed in a particular template format. The template can be recognized when the content of each match is ignored or replaced by some fixed-length string. Therefore, repetitive patterns are formed. For instance, in the example of Figure 1, the sequence "<LI>Text(_)<I>Text(_)</I>" is repeated four times, when all text strings between two tags such as "Congo", "Egypt", "Belize" etc. are replaced by token class *Text(_)*.

This is a simple example that demonstrates a repeated pattern formed by tag tokens in a Web page following a simple translation convention. In practice, many search-able Web sites also exhibit such repeated patterns since they

```
<H1>Country Code</H1><UL>
<LI>Congo<I>242</I>
<LI>Egypt<I>20</I>
<LI>Belize<I>501</I>
<LI>Spain<I>34</I>
</UL>
```

Fig. 1. Sample HTML page

usually extract data from relational database and produce dynamic Web pages with a predefined format style. Therefore, what we ought to do is kind of reverse engineering to discover the original format style and the content we need to extract. Meanwhile, we also find that extraction patterns of the desired information (called *main information block* as defined in [3]) often occur regularly and closely in a Web page. These observations motivate us to look for an approach to discover repeated patterns and validation criteria to filter desired repeats that are spaced regularly and closely.

Since HTML tags are the basic components for data presentation and the text string between tags are exactly what we see in the browsers. Hence, it is intuitive to regard the text string between two tags as one unit as well as each individual tag. This simple version of *HTML translation* will be used in the following paper where any text string between two tags is translated to one unit called Text(_) and every HTML tag is translated to a token Html(<tag>) according to its tag name.

Such translation convention enables the show-up of many repeated patterns. By *repeated patterns*, we mean any substring that occurs twice in the encoded token string. Thus, not only the sequence "Html(<LI>) Text(_) Html(<I>) Text(_) Html(<I>)" conforms to the definition of repeated pattern but also the subsequence "Html(<LI>) Text(_) Html(<I>)," "Text(_) Html (<I>) Text(_)", "HMLT(<I>)Text(_)Html(</I>)," etc. To distinguish from these repeats, we define *maximal repeats* to uniquely identify the longest pattern as follows.

Definition Given an input string S, we define maximal repeat α as a substring of S that occurs in k distinct positions $p_1, p_2, ..., p_k$ in S, such that the (p_i-1)th token in S is different from the (p_j-1)th token for at least one i, j pair, $1 \leq i < j \leq k$ (called left maximal), and the $(p_x + |\alpha|)$th token is different from the $(p_y + |\alpha|)$th token for at least one x, y pair, $1 \leq x < y \leq k$ (called right maximal).

The definition of maximal repeats is necessary for identifying the well-used and popular term, repeats. Besides, it also captures all interesting repetitive structures in a clear way and avoids generating overwhelming outputs. In the next section, we will describe how the problem of IE can be addressed by pattern discovery.

3 IE by Pattern Discovery

To discover patterns from an input Web page, first an encoding scheme is used to translate the Web page into a string of abstract representations, referred to here as tokens. Each token is represented by a binary code of length l. To enable pattern discovery, we utilizes a data structure called a *PAT tree* [4] in which repeated patterns in a given sequence can be efficiently identified. Using this data structure to index an input string, all possible repeats, including their occurrence counts and their positions in the original input string can be easily retrieved. Finally, the discovered maximal repeats are forwarded to the validator, which filters out undesired patterns and to produces a candidate pattern.

3.1 Translator

Since HTML tags are the basic components for document presentation and the tags themselves carry a certain structure information, it is intuitive to examine the tag token string formed by HTML tags and regard other non-tag text content between two tags as one single token called Text(_). Tokens seen in the translated token string include tag tokens and text tokens, denoted as Html(<tag_name>) and Text(_), respectively. For example, Html(</a>) is a tag token, where </a> is the tag. Text(_) is a text token, which includes a contiguous text string located between two HTML tags.

Tags tokens can be classified in many ways. The user can choose a classification depending on the desired level of information to be extracted. For example, tags in the BODY section of a document can be divided into two distinct groups: block-level tags and text-level tags. The former defines the structure of a document, and the latter defines the characteristics, such as format and style, of the contents of the text. Block level tags include categories such as headings, text containers, lists, and other classifications, such as tables and forms. Text-level tags are further divided into categories including logical markups, physical markups, and special markups for marking up texts in a text block.

The many different tag classifications allow different HTML translations to be generated. With these different abstraction mechanisms, different patterns can be produced. For example, skipping all text-level tags will result in higher abstraction from the input Web page than all tags are included. In addition, different patterns can be discovered and extracted when different encoding schemes are translated.

For example, when only block-level tags are considered, the corresponding translation of Fig. 1 is a token string:
"Html(<H1>)Text(_)Html(</H1>) Html(<UL>)Html(<LI>)
Text(_)Html(<LI>)Text(_)Html(<LI>)Text(_)Html(<LI>) Text(_) Html(</UL>)",
where each token is encoded as a binary strings of "0"s and "1"s with length l. For example, suppose three bits encode the tokens in the Congo code as shown in Fig. 2. The encoded binary string for the token string of the Congo code will be "100110 101000 010110 010110 010110 010110 001$" of 3*13 bits, where "$" represents the ending of the encoded string.

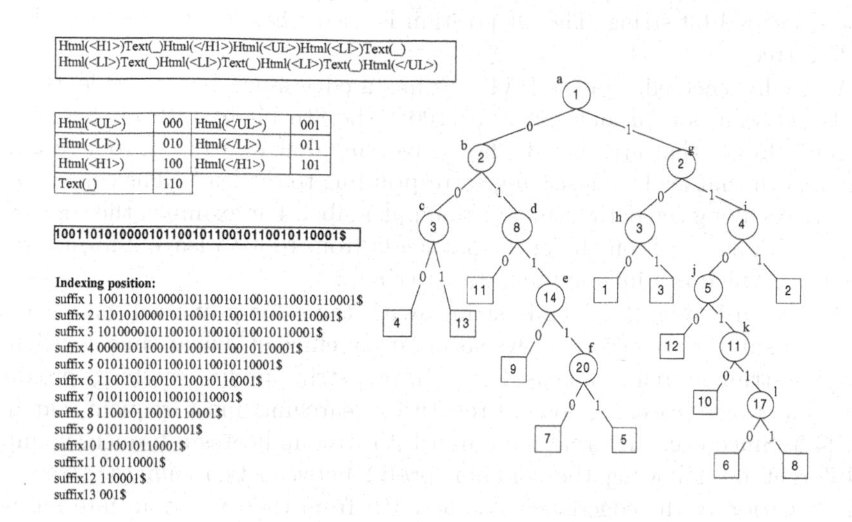

Fig. 2. The PAT tree for the Congo Code

3.2 The PAT Tree

Our approach for pattern discovery uses a PAT tree to discover repeated patterns in the encoded token string. A PAT tree is a Patricia tree (Practical Algorithm to Retrieve Information Coded in Alphanumeric [11]) constructed over all the possible suffix strings. A Patricia tree is a particular implementation of a compressed binary (0,1) digital tree such that each internal node in the tree has two branches: zero goes to left and one goes to right. Like a suffix tree [4], the Patricia tree stores all its data at the external nodes and keeps one integer, the bit-index, in each internal node as an indication of which bit is to be used for branching. For a character string with n indexing point (or n suffix), there will be n external nodes in the PAT tree and $n-1$ internal nodes. This makes the tree $O(n)$ in size.

When a PAT tree is to index a sequence of characters (or tokens here) not just 0 or 1, the binary codes for the characters can be used. For simplicity, each character is encoded as fixed-length binary code. Specifically, given a finite alphabet Σ of a fixed size, each character $x \in \Sigma$ is represented by a binary code of length $l = \lceil \log_2 |\Sigma| \rceil$. For a sequence S of n characters, the binary input B will have $n * l$ bits, but only the $[i * l + 1]$th bit has to be indexed for $i = 0, \ldots, n-1$.

Referring to Fig. 2, a PAT tree is constructed from the encoded binary string of the Congo example. The tree is constructed from thirteen sequences of bits, with each sequence of bits starting from each of the encoded tokens and extending to the end of the token string. Each sequence is called a "semi-infinite string" or "sistring" in short. Each leaf, or external node, is represented by a square labeled by a number that indicates the starting position of a sistring. For example, leaf 2 corresponds to sistring 2 that starts from the second token in the token string.

Each internal node is represented by a circle, which is labeled by a bit position in the encoded bit string. The bit position is used when locating a given sistring in PAT tree.

Virtually, each edge in the PAT tree has a edge label. For example, the edge labels between node d and e are "101100", the 8th bit to 13th bit for suffix 9, 7, and 5. Edges that are visited when traversing downward from root to a leave form a path that leads to a sistring corresponding to the leave. The concatenated edge labels along the path form a virtual path label. For example, the edge labels "1", "10", and "1..." on the path that leads from root to leave 2 form a prefix "1101...", which is a unique prefix for sistring 2.

As shown in Fig. 2, all suffix strings with the same prefix will be located in the same subtree. Hence, it allows surprisingly efficient, linear-time solutions to complex string search problems. For example, string prefix searching, proximity searching, range searching, longest repetition searching, most frequent searching, etc. [4,5] Since every internal node in a PAT tree indicates a branch, it implies a different bit following the common prefix between two suffixes. Hence, the concatenation of the edge-labels on the path from the root to an internal node represents one repeated string in the input string. However, not every path-label or repeated string represents a maximal repeat. Let's call the (p_k-1)th character of the binary string p_k the *left character*. For a path-label of an internal node v to be a maximal repeat, at least two leaves (suffixes) in the v's subtree should have different left characters. By recording the occurrence counts and the reference positions in the leaf nodes of a PAT tree, we can easily know how many times a pattern is repeated. Hence, given the pattern length, occurrence count, we can apply postorder traversal to the PAT tree to enumerate all repeats.

The essence of a PAT tree is a binary suffix tree, which has also been applied in several research field for pattern discovery. For example, Kurtz and Schleiermacher have used suffix trees in bioinformatics for finding repeated substring in genomes [8]. As for PAT trees, they have been applied for indexing in the field of information retrieval since a long time ago [4]. It has also been used in Chinese keyword extraction [1] for its simpler implementation than suffix trees and its great power for pattern discovery. However, in the application of information extraction, we are not only interested in repeats but also repeats that appear regularly in vicinity. Discovered maximal repeats have to be further validated or compared to find the best one that corresponds to the information to be extracted.

3.3 Pattern Validation Criteria

In the above section, we discussed how to find maximal repeats in a PAT tree. However, there may be over 60 maximal repeats discovered in an Web page. To classify these maximal repeats, we introduce two measures regularity, and compactness as described below. Let the suffixes of a maximal repeat α are ordered by its position such that suffix $p_1 < p_2 < p_3 \ldots < p_k$, where p_i denotes the position of each suffix in the encoded token sequence.

Regularity of a pattern is measured by computing the standard deviation of the interval between two adjacent occurrences $(p_{i+1}-p_i)$, that is, the sequence of spacing between two adjacent occurrences (p_2-p_1), (p_3-p_2), ..., (p_k-p_{k-1}). Regularity of the maximal repeat α is equal to the standard derivation of the sequence divided by the mean of the sequence.

Compactness is a measure of the density of a maximal repeat. It is used to eliminate maximal repeats that are scattered far apart beyond a given bound. Compactness is defined as $k * |\alpha| / \sum_{i=2}^{k} p_i - p_{i-1}$, where $|\alpha|$ is the length of α in number of tokens.

The value of regularity is located between 0 and 1 while the value of density is greater than 0. Ideally, the extraction pattern should have regularity equal to zero and compactness equal to one. To filter potentially good patterns, a simple approach will be to use a threshold for each of these measures above. Implicitly, good patterns have small regularity and density close to one. Therefore, only patterns with regularity less than the regularity threshold and density between the density thresholds are considered validated patterns.

4 Performance Evaluation

We first show the number of validated maximal repeats validated by our system using fourteen state-of-the-art search engines, each with ten Web pages. There are several control parameters which can affect the number of maximal repeats validated, including encoding scheme, minimum pattern length, occurrence count, and threshold values for regularity and compactness. Given the minimum length 3 and count 5, the effect of different encoding scheme is shown in Table 1. Conform to general expectation, higher-level encoding scheme often results in less patterns. From this table, we can also see how each control parameter filters patterns where the thresholds are decided by the following experiments. The value of density can be greater than one because maximal repeats may be overlapped. For example, suppose a maximal repeat α occurs ten times in a row. In such case, α will has regularity 0 and density 1. In addition, $\alpha\alpha$, $\alpha\alpha\alpha$, etc. are also qualified for regular maximal repeats, only with density greater than 1.

Table 1. No. of Patterns validated with different encoding scheme

Encoding Scheme	Maximal Repeat	Regularity < 0.5	Compactness > 1.5	< 0.25
All-tag	117	39	22	7.6
NoPhysical	88	41	25	6.5
NoSpecial	82	29	18	5.7
Block-level	66	32	17	3.9

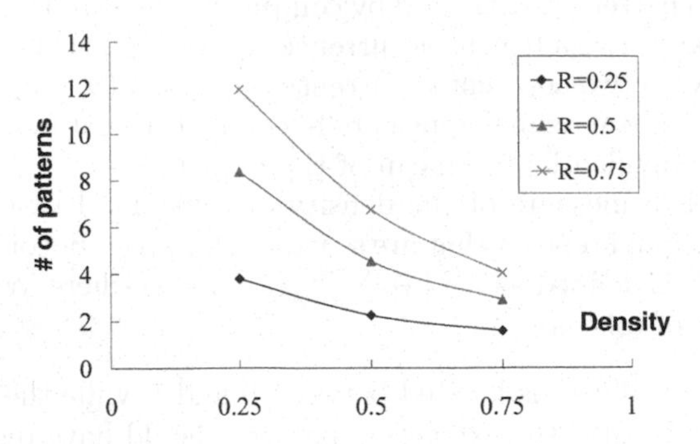

Fig. 3. # of patterns sucessfully validated

Fig. 3 shows the effect of various regularity and density thresholds using all-tag encoding scheme. Basically, low regularity threshold and high density threshold reduce the number of patterns, but could have missed good patterns. Therefore, the thresholds are chosen empirically to include as many good patterns as possible.

Table 2 shows the performance of different encoding scheme measured in retrieval rate, accuracy rate and matching percentage. Retrieval rate is defined as the ratio of the number of desired data records enumerated by a maximal repeat to the number of desired data records contained in the input text. Likewise, accuracy rate is defined as the ratio of the number of desired data records enumerated by a maximal repeat to the number of occurrence of the maximal repeat. A data record is said to be *enumerated* by a maximal repeat if the matching percentage is greater than a bound determined by the user. The matching percentage is used because the pattern may contain only a portion of the data record.

With the simple encoding scheme of using block-level tags, our approach could discover patterns which extract 86% records with matching percentage 78%. Nearly half the test Web sites are correctly extracted (with matching percentage greater than 90%). Among them, nine of the fourteen Web sites have retrieval rate and accuracy rate both greater than 0.9. However, examining other discovered patterns, many are incomplete due to exceptions. In the next section, we will further improve the performance by occurrence partition and multiple string alignment.

5 Constructing Extraction Pattern

Generally speaking, search engines utilize a "while loop" to output their results in some template. However, they may use "if clauses" inside the while loop to decorate the text content. For example, the keywords that are submitted

Table 2. Performance of different encoding scheme

Encoding Scheme	Retrieval Rate	Accuracy Rate	Matching Percentage
All-tag	0.73	0.82	0.60
NoPhysical	0.82	0.89	0.68
NoSpecial	0.84	0.88	0.70
Block-level	0.86	0.86	0.78

to search engines are shown in bold face for Infoseek and MetaCrawler, thus, breaking their "while loop" patterns.

From the statistics above, we summarize that "maximal repeat" and "regularity" are the two primary criteria we filter candidate patterns. However, we also found that the extraction pattern may not be maximal repeats and regular. For example, the regularity of the pattern for *Excite* is greater than default regularity threshold 0.5 because a banner is inserted among the search results, dividing the ten matches into two parts. Besides, the *"if-effect"* often hinders us from discovering complete patterns. These issues are what we would address in the following.

5.1 Occurrence Partition

To handle patterns with regularity greater than the specified threshold 0.5, these patterns are carefully segmented to see if any partition of the pattern's occurrences satisfies the requirement for regularity. By definition, the regularity of a pattern is computed through all occurrences of the pattern. For example, if there are k occurrences, the $k-1$ intervals (between two adjacent instances) are the statistics we use to compute the standard deviation and the mean. However, in examples such as *Lycos*, the search result is divided into three blocks. Such occurrences increase the regularity over all instances. Nonetheless, the regularity of the occurrences in each information block is still small. Therefore, the idea here is to segment the occurrences into partitions so that we can analyze each partition individually.

We don't really have to apply clustering algorithm on this matter, instead, a simple loop can accomplish the job if the occurrences are ordered by their position aforehand. Let $C_{i,j}$ denotes the set of occurrences $p_i, p_{i+1}, ..., p_j$ and initialize $s = 1, j = 1$. For instance p_{j+1}, if the regularity of $C_{s,j+1}$ is greater than θ then output $C_{s,j}$ as a partition and assign $j+1$ to s.

Once the partitions are separated, we can then compute the regularity for each individual partition. If a partition includes occurrences more the minimum count and has regularity less than threshold ϵ, the pattern as well as the occurrences in this partition are outputted. Note that the threshold ϵ is set to a small value much less than 0.5 to control the number of outputted patterns. With this modification, the performance is improved greatly. As shown in Table 3, the retrieval rate is increased to 93% and accuracy rate to 90%. The only tradeoff is the increased number of patterns from 3.9 to 8.9.

Table 3. Performance of advanced technique

Advanced Technique	Retrieval Rate	Accuracy Rate	Matching Percentage
Occurrence Partition	0.93	0.93	0.84
Multiple Alignment	0.97	0.94	0.90

5.2 Multiple String Alignment

For the tough work regarding incomplete pattern discovered, the technique for multiple string alignment is borrowed to find a good presentation of the critical common features of multiple strings. For example, suppose "adc" is the discovered pattern for token string "adcwbdadcxbadcxbcadc". If we have the following multiple alignment for strings "*adcwbd*", "*adcxb*" and "*adcxbd*":

$$\begin{array}{cccccc} a & d & c & w & b & d \\ a & d & c & x & b & - \\ a & d & c & x & b & d \end{array}$$

The extraction pattern can be *generalized* as "$adc[w|x]b[d|-]$" to cover these three instances. Specifically, suppose a validated maximal repeat has $k+1$ occurrence, p_1, p_2, ..., p_{k+1} in the encoded token string. Let string P_i denote the string starting at p_i and ending at $p_{i+1}-1$. The problem is to find the multiple alignment of the k strings $\mathcal{S} = \{P_1, P_2, ..., P_k\}$ so that the generalized pattern can be used to extract all records we need.

Multiple string comparison is a natural generalization of *alignment* for two strings which can be solved in $O(n*m)$ by *dynamic programming* to obtain optimal *edit distance*, where n and m are string lengths. As an example of *two string alignment*, consider the alignment of two strings *acwbd* and *adcxb* shown below:

$$\begin{array}{cccccc} a & - & c & w & b & d \\ a & d & c & x & b & - \end{array}$$

In this alignment, character w is mismatched with x, two ds are opposite hyphens (or called space), and all other characters match their counterparts in the opposite string. If we give each match a value of β, each mismatch a value of γ, and each space a value of δ, the *two string alignment problem* is to optimize the *weighted distance* $D(P_1, P_2) \equiv (nmatch * \beta + nmis * \gamma + nspace * \delta)$, where *nmatch*, *nmis*, and *nspace* denote the number of mismatch, match, and space, respectively ($nmatch = 3$, $nmis = 1$, and $nspace = 2$ here).

Extending dynamic programming to multiple string alignment yieds a $O(n^k)$ algorithm. Instead, an approximation algorithm is available such that the score of the multiple alignment is no greater than twice the score of optimal multiple alignment [5]. The approximation algorithm starts by computing the *center string* S_c in k strings $\mathcal{S}$ that minimizes *consensus error* $\sum_{P_i \in \mathcal{S}} D(S_c, P_i)$. Once the center string is found, each string is then iteratively aligned to the center string to construct multiple alignment, which is in turn used to construct the extraction pattern.

For each patterns with density less than one, the *center star* approximation algorithm for multiple string alignment is applied to generalize the extraction pattern. Suppose the generalized extraction pattern is expressed as "$c_1c_2c_3...c_n$", where each c_i is either a symbol or a subset of $\Sigma \cup \{-\}$ containing symbols that can appear at position i. An additional step is taken to generate pattern of this form '$c_jc_{j+1}c_{j+2}...c_nc_1c_2...c_{j-1}$" for position j with single symbol of the following special tags such as <DL>, <DT>, <TR> or <P>, <BR>, <HR>, because extraction patterns often begin or end up with them[1].

We adopt this additional step because the generated extraction pattern may not be the beginning of a record. The experimental results show that with the help of multiple string alignment and the additional step, the performance is improved to 97% retrieval rate, 94% accuracy rate and 0.90 matching percentage. The high percentage of retrieval rate is pretty encouraging. The ninety percent of matching percentage is actually higher in terms of the text content retrieved. For those Web sites with matching percentage greater than 85%, the text contents are all successfully extracted. What bothers is the accuracy rate, since the extraction pattern generalized from multiple alignment may comprehends more than the information we need. For example, the generalized rule for *Lycos* will extract information in all three blocks while only the information in one block is what we desired, causing lower accuracy rate.

6 Summary and Future Work

Information extraction from Web pages is a core technology for comparison-shopping agents [2], which Doorenbos et. al. regard as improvement in the axe of tolerating unstructured information. The characteristics of regularity, uniformity, and vertical separation enable the possibility of learning. In this paper, we have presented an unsupervised approach to semi-structured information extraction. We propose the application of PAT trees for pattern discovery in the encoded token string of Web pages. Once the PAT tree is constructed, we can easily traverse the tree to find all maximal repeats given the expected pattern frequency and length. The discovered maximal repeats are further filtered by three measures: regularity and compactness. The filtering criteria aim to keep the number of patterns as small as possible while at the same time have all interesting patterns. Furthermore, occurrence partition is applied to handle patterns with regularity greater than the default threshold. Finally, multiple string alignment is applied to patterns with density less than one to generalize extraction pattern. Thereby, the extraction module can simply adapt pattern matching algorithm to extract all records.

The extraction rule generalized from multiple string alignment has achieved 97% retrieval rate and 91% accuracy rate. The whole process requires no human intervention and training example. Comparing our algorithm to others, our approach is quick and expressive. It takes only three minutes to extract 140 Web

[1] Other tags include <TABLE>, <TD>, <UL>, <OL>, <LI>, <DD>.

pages. The extraction rule allowing alternative tokens and missing tokens, can tolerate exceptions and variance in the input.

We are currently applying this approach against more test data formatted in tabular form, which perform at the level of 80% retrieval rate. As more variances occur in input pages, it becomes even difficult to have good multiple string alignment. In such cases, the scoring of edit distance between two strings and the algorithm to construct multiple alignment become more important. In addition, filtering of the constructed patterns can also provide a reasonable number of patterns for user to choose.

Acknowledgements. This work is sponsored by National Science Council, Taiwan under grant NSC89-2213-E-008-056. Also, we would like to thank Lee-Feng Chien, Ming-Jer Lee and Jung-Liang Chen for providing their PAT tree code for us.

References

1. Chien, L.F. 1997. PAT-tree-based keyword extraction for Chinese information retrieval. In *Proceedings of the 20th annual international ACM SIGIR conference on Research and development in information retrieval.* pp.50–58. 1997.
2. Doorenbos, R.B., Etzioni, O. and Weld, D. S. A scalable comparison-shopping agent for the World-Wide Web. In *Proceedings of the first international conference on Autonomous Agents.* pp. 39–48, NewYork, NY, 1997, ACM Press.
3. Embley, D.; Jiang, Y.; and Ng. Y.-K. 1999. Record-boundary discovery in Web documents. In *Proceedings of the 1999 ACM SIGMOD International Conference on Management of Data (SIGMOD'99).* pp. 467–478, Philadelphia, Pennsylvania.
4. Gonnet, G.H.; Baeza-yates, R.A.; and Snider, T. 1992. New Indices for Text: Pat Trees and Pat Arrays. *Information Retrieval: Data Structures and Algorithms*, Prentice Hall.
5. Gusfield, D. 1997. *Algorithms on strings, trees, and sequences*, Cambridge. 1997.
6. Hsu, C.-N. and Dung, M.-T. 1998. Generating finite-state transducers for semi-structured data extraction from the Web. *Information Systems.* 23(8):521–538.
7. Knoblock, A. et al., ed., 1998. *Proc. 1998 Workshop on AI and Information Integration*, Menlo Park, California.: AAAI Press.
8. Kurtz, S. and Schleiermacher, C. 1999. REPuter: fast computation of maximal repeats in complete genomes. *Bioinformatics* 15(5):426–427.
9. Kushmerick, N.; Weld, D.; and Doorenbos, R. 1997 Wrapper induction for information extraction. In *Proceedings of the 15th International Joint Conference on Artificial Intelligence* (IJCAI).
10. Muslea, I.; Minton, S.; and Knoblock, C. 1999. A hierarchical approach to wrapper induction. In *Proceedings of the 3rd International Conference on Autonomous Agents* (Agents'99), Seattle, WA.
11. Muslea, I. 1999. Extraction patterns for information extraction tasks: a survey. In *Proceedings of AAAI'99: Workshop on Machine Learning for Information Extraction*

Empirical Study of Recommender Systems Using Linear Classifiers

Vijay S. Iyengar and Tong Zhang

IBM Research Division, T. J. Watson Research Center,
P.O. Box 218, Yorktown Heights, NY 10598, U.S.A.

Abstract. Recommender systems use historical data on user preferences and other available data on users (e.g., demographics) and items (e.g., taxonomy) to predict items a new user might like. Applications of these methods include recommending items for purchase and personalizing the browsing experience on a web-site. Collaborative filtering methods have focused on using just the history of user preferences to make the recommendations. These methods have been categorized as *memory-based* if they operate over the entire data to make predictions and as *model-based* if they use the data to build a model which is then used for predictions. In this paper, we propose the use of linear classifiers in a model-based recommender system. We compare our method with another model-based method using decision trees and with memory-based methods using data from various domains. Our experimental results indicate that these linear models are well suited for this application. They outperform the commonly proposed approach using a memory-based method in accuracy and also have a better tradeoff between off-line and on-line computational requirements.

1 Introduction

Recommender systems use historical data on user preferences and purchases and other available data on users and items to recommend items that might be interesting to a new user. One of the earliest techniques developed for recommendations was based on *nearest-neighbor collaborative filtering* algorithms [1,2] that used just the history of user preferences as input. Sometimes in the literature the term *collaborative filtering* is used to refer to just these methods. However, we will follow the taxonomy introduced by [3] in which collaborative filtering (*CF*) refers to a broader set of methods that use prior preferences to predict new ones. In this taxonomy, nearest-neighbor collaborative filtering algorithms are categorized as being memory-based *CF*. Nearest-neighbor methods use some notion of similarity between the user for whom predictions are being generated and users in the database. Variations on this notion of similarity and other aspects of memory-based algorithms are discussed in [3]. Scalability is an issue with nearest-neighbor methods. Proposed methods of addressing this issue range from the use of data structures like R-trees to the use of dimension reduction techniques like latent semantic indexing [4].

D. Cheung, G.J. Williams, and Q. Li (Eds.): PAKDD 2001, LNAI 2035, pp. 16–27, 2001.

In contrast, model-based *CF* methods use the historical data to build models which are then used for predicting new preferences. A model-based approach using Bayesian networks was found to be comparable to the memory-based approach in [3]. More recently, models based on a newer graphical representation called dependency networks [5] have been applied to this problem [6]. For this task, dependency network models seem to have slightly poorer accuracy but require significantly less computation when compared to Bayesian network models [6]. Another model based method is to use clustering to group users based on their past preferences. The parameters for this clustering model can be estimated by methods like Gibbs sampling and EM [7,8,3]. The clustering model explored in [3] was outperformed by the model-based approach using Bayesian networks and by the memory-based approach CR+ described in [3].

In this paper, we explore the use of various linear classifiers in a model-based approach to the recommendation task. Linear classifiers have been quite successful in the text classification domain [9]. Some of the characteristics shared between the text and CF domains include the high dimensionality and sparseness of the data in these domains. The main computational cost of using linear classifiers is in the model build phase, which is an off-line activity. The application of the models is very straightforward especially with sparse data. Our empirical study will use two data sets that reflect users' browsing behavior and one data set that captures their purchases. Because of its wider applicability, we focus on data that is implicitly gathered, e.g., boolean flag for each web page representing whether or not it was browsed as in the anonymous-msweb dataset in [10]. This is in contrast with explicitly collected data, e.g., ratings explicitly gotten for movies [11]. Section 3 presents results achieved on these datasets by various model-based approaches using linear classifiers. For comparison we also include results achieved by our implementation of the memory-based algorithm CR+ described in [3] and a model-based approach using decision trees. The linear models studied in this paper are described in the next section.

2 Model-Based Approaches

The problem of predicting whether a user (or a customer) will accept a specific recommendation can be modeled as a binary classification problem. However, in a recommender system, we are also interested in the likelihood that a customer will accept a recommendation. This information can be used to rank all of the potential choices according to their likelihoods, so that we can select the top choices to present to the customer. It is thus necessary that the classifier we use returns a score (or a confidence level), where a higher score corresponds to a higher possibility that the customer will accept the recommendation.

2.1 Linear Models

Formally, a two-class categorization problem is to determine a label $y \in \{-1, 1\}$ associated with a vector x of input variables. A useful method for solving this problem is by linear discriminant functions, which consist of linear combinations of the input variables. Various techniques have been proposed for determining the weight values for linear discriminant classifiers from a training set of labeled data $(x_1, y_1), \ldots, (x_n, y_n)$. Specifically, we seek a weight vector w and a threshold θ such that $w^T x < \theta$ if its label $y = -1$ and $w^T x \geq \theta$ if its label $y = 1$. A score of value $w^T x - \theta$ can be assigned to each data point to indicate the likelihood of x to be in class.

The problem just described may readily be converted into one in which the threshold θ is taken to be zero. One does this by converting a data point x in the original space into $\tilde{x} = [x, 1]$ in the enlarged space. Each hyperplane w in the original space with threshold θ can then be converted into $[w, -\theta]$ that passes through the origin in the enlarged space. Instead of searching for both an d-dimensional weight vector along with a threshold θ, we can search for an $(d+1)$-dimensional weight vector along with an anticipated threshold of zero. In the following, unless otherwise indicated, we assume that the vectors of input variables have been suitably transformed so that we may take $\theta = 0$. We also assume that x and w are d-dimensional vectors.

Many algorithms have been proposed for linear classification. We start our discussion with the least squares algorithm, which is based on the following formulation to compute a linear separator $\hat{w}$:

$$\hat{w} = \arg\min_{w} \frac{1}{n} \sum_{i=1}^{n} (w^T x_i - y_i)^2. \tag{1}$$

The least squares method is extensively used in engineering and statistics. Although the method has mainly been associated with regression problems, it can also be used in classification. Examples include use in text categorization [12] and uses in combination with neural networks [13].

The solution of (1) is given by

$$\hat{w} = (\sum_{i=1}^{n} x_i x_i^T)^{-1} (\sum_{i=1}^{n} x_i y_i).$$

One problem with the above formulation is that the matrix $\sum_{i=1}^{n} x_i x_i^T$ may be singular or ill-conditioned. This occurs, for example, when n is less than the dimension of x. Note that in this case, for any $\hat{w}$, there exists infinitely many solutions $\tilde{w}$ of $\tilde{w}^T x_i = \hat{w}^T x_i$ for $i = 1, \ldots, n$. This implies that (1) has infinitely many possible solutions $\hat{w}$.

A remedy of this problem is to use a pseudo-inverse [12]. However, one problem of the pseudo-inverse approach is its computational complexity. In order to handle large sparse systems, we need to use iterative algorithms which do not rely on matrix factorization techniques. Therefore in this paper, we use the standard ridge regression method [14] that adds a regularization term to (1):

$$\hat{w} = \arg\min_{w} \frac{1}{n} \sum_{i=1}^{n} (w^T x_i y_i - 1)^2 + \lambda w^2, \tag{2}$$

where λ is an appropriately chosen regularization parameter. The solution is given by

$$\hat{w} = (\sum_{i=1}^{n} x_i x_i^T + \lambda n I)^{-1} (\sum_{i=1}^{n} x_i y_i),$$

where I denotes the identity matrix. Note that $\sum_{i=1}^{n} x_i x_i^T + \lambda n I$ will always be non-singular, which solves the ill-condition problem. The regularized least squares formulation (2) can be solved by using a column relaxation method which is often called the Gauss-Seidel procedure in numerical optimization. The algorithm (see Algorithm 1 in Appendix A) cycles through components of w, and optimizes one component at a time (while keeping others fixed).

Another popular method is the support vector machine, which is a method originally proposed by Vapnik [15,16] that has nice properties from the sample complexity theory. Slightly different from our approach of forcing threshold $\theta = 0$, and then compensating by appending 1 to each data vector, the standard linear support vector machine (cf. [16]) explicitly includes θ in a quadratic formulation that can be transformed to:

$$(\hat{w}, \hat{\theta}) = \arg\inf_{w,\theta} \frac{1}{n} \sum_{i=1}^{n} g(y_i(w^T x_i - \theta)) + \lambda w^2, \tag{3}$$

where

$$g(z) = \begin{cases} 1 - z & \text{if } z \leq 1, \\ 0 & \text{if } z > 1. \end{cases} \tag{4}$$

It is interesting to compare the least squares approach and the support vector machine approach. In the least squares formulation, the loss function $(z-1)^2$ implies that we try to find a weight w such that $w^T x \approx 1$ for an in-class data point x, and $w^T x \approx -1$ for an out-of-class data point x. Although this means that the formulation attempts to separate the in-class data from the out-of-class data, it also penalizes a well behaved data point x such that $w^T xy > 1$. The support vector machine approach remedies this problem by choosing a loss function that does not penalize a well-behaved data point such that $w^T xy > 1$.

A popular method to obtain the numerical solution of an SVM is the SMO algorithm [17]. However, in general solving an SVM is rather complicated since g is not smooth. In this paper, we intentionally replace $g(z)$ by a smoother function to make the numerical solution simple. The following formulation modifies the least squares algorithm so that it does not penalize a data point with $w^T xy > 1$, and it has a loss function that is more smooth than that of an SVM:

$$\hat{w} = \arg\min_{w} \frac{1}{n} \sum_{i=1}^{n} h(w^T x_i y_i) + \lambda w^2, \tag{5}$$

where

$$h(z) = \begin{cases} (1 - z)^2 & \text{if } z \leq 1, \\ 0 & \text{if } z > 1. \end{cases} \tag{6}$$

This formulation is a mixture of the least squares method and a standard SVM. We thus call it modified least squares. Furthermore, a direct numerical optimization of (5) can be performed relatively efficiently. Similar to (2), the Algorithm 2 in Appendix A solves (5)

Another way to solve (5) is given in Algorithm 3 in Appendix A. It is derived by using convex duality. Because of space limitations, we skip the analysis.

2.2 Other Models

In the recommender system application, interpretability of the models used is an important characteristic to be considered in addition to the accuracy achieved and the computational requirements. We have included a decision tree based recommender system in this empirical study as an example of using an interpretable model. In this decision tree package, the splitting criteria during tree growth is a modified version of entropy and the tree pruning is done using a Bayesian model combination approach originated from data compression [18, 19]. A similar approach has appeared in [20].

We have also implemented a version of the nearest neighbor algorithm, CR+, described in [3] and included it in our study. As suggested in [3], inverse user frequency, case amplification and default voting heuristics are used in our implementation of CR+.

3 Experiments

The true value of a recommender system can only be measured by controlled experiments with actual users. Such an experiment could measure the relative lift achieved by a specific recommendation algorithm when compared to, say, recommending the most popular item. Experiments with historical data have been used to estimate the value of recommender algorithms in the absence of controlled live experiments [3,4]. In this paper we will follow experimental procedures similar to those introduced in [3].

3.1 Data Sets

Characteristics of the data sets used in our experiments are given in Table 1. The first dataset *msweb* was introduced in [3] and added to the UCI repository under the name *anonymous-msweb*. As described in [3], this dataset contains for each user the web page groups (called vroots) that were visited in a fixed time period. The total number of items is relatively small (around 300) for this dataset and this can be attributed to the fact that an item refers to a group of web pages.

The second dataset *pageweb* also captures visits by users to a different web site but at the individual page level (with about 6000 total items). Intuitively, one might expect the task of recommending specific pages to be more difficult than that of recommending page groups. But the other factor to be considered

Table 1. Description of the data sets.

Characteristics	Dataset		
	msweb	*pageweb*	*wine*
Training cases	32711	9195	13103
Total test cases	5000	1804	2610
Test cases with at least 2 items (*All But 1*)	3453	1243	1770
Test cases with at least 3 items (*Given 2*)	2213	932	1280
Test cases with at least 6 items (*Given 5*)	657	455	624
Test cases with at least 11 items(*Given 10*)	102	168	268
Total items	294	5781	663
Mean items per case in training set	3.02	4.36	4.60

is that we also have more fine-grained information at the individual page level about user preferences that can be used by the models. This dataset will be useful in evaluating how the various algorithms handle recommending from a large number of items.

The third dataset *wine* represents wine purchases made by customers of a leading supermarket chain store within a specified period. The dataset captures for each customer the wines purchased in this period as a binary value (purchased versus not purchased).

We have chosen to use a binary representation of the item/page variables in all the experiments. An alternative representation would be use more information like the number of visits to a web page or the time spent viewing a web page or the quantity of wine purchased.

3.2 Experimental Setup

Following the experimental setup introduced in [3], the datasets are split into two disjoint sets (training and test) of users. The entire set of visits (or purchases) for users in the training set is available for the model build process. The known visits (purchases) for users in the test set are split into two disjoint sets: given and hidden. The given set is used by the recommender methods to rank all the remaining items in the order of predicted preference to the user. This ranked list is evaluated by using the hidden set as the reference indicating what should have been predicted.

The evaluation metric, R, proposed in [3] is based on the assumption that each successive item in a list is less likely to be interesting to the user with an exponential decay. This metric uses a parameter, α, which is the position in the ranked list which has a 50-50 chance of being considered by the user. As in [3] we will set α such that the fifth position has a 50-50 chance of being considered.

The exponential decay in interest, which forms the basis for the R metric, may be a plausible behavior model for consumers. However, this metric is not easy to interpret. Also, the number of allowed recommendations can also be constrained by the environment. For example, the form factor of a hand-held device might restrict the number of recommendations on it to a small number. Hence, we will also report results using a simpler metric which measures the fraction of users for whom at least one valid recommendation (according to the hidden set as reference) was given in the top K items of the ranked list. In particular, we will report this metric for K values 1, 3 and 10.

The split of the test set data into the given and hidden sets is done as suggested in [3]. Three of these splits are denoted as *Given 2*, *Given 5*, and *Given 10*. These have 2, 5, and 10 items chosen into the given sets, respectively. The fourth split is denoted as *All But 1* because one item for each test user is randomly chosen to be hidden in this scenario. These scenarios can be used to assess how each recommender system handles different amounts of information being known and to be hidden (predicted) for each test user. Table 1 provides for each dataset the number of test users that are included in each of these scenarios.

Each scenario will be run five times with different random choices for the split between given and hidden subsets in the test data. Mean values and standard deviations are computed over these five experiments. We have adopted this approach to be compatible with the prior literature with regard to the training/test splits. A more traditional approach would have been to use n-fold cross validation where both training and test sets are different in the experiments. However, given the compatibility constraints, performing multiple experiments with the given/hidden splits provides some information on the experimental variability.

3.3 Results

The results achieved for all the four scenarios (*Given 2, Given 5, Given 10, All But 1*) are given in Tables 3, 4 and 5 for the datasets *msweb*, *pageweb* and *wine*, respectively. The format in which these results are provided in these tables for each combination of algorithm and scenario is explained in Table 2. The mean and the standard deviation for the R metric (expressed as percentage) are given on top. The three numbers below indicate the percentage of test users that had at least one successful recommendation in the top 1, 3 and 10 positions of the ranked list. The linear models were generated using 25 iterations with the regularization parameter λ set at 0.001.

The baseline approach of recommending popular items does significantly poorer when compared to the other algorithms on datasets with more items. The decision tree model also exhibits this pattern of not performing as well on datasets with more items.

As mentioned earlier, one advantage of model-based methods is that the model build is done off-line. The model build times for the dataset *msweb* were around 500 seconds for linear least squares and modified linear (primal) models and around 200 seconds for modified linear (dual) model. These times were

Table 2. Explanation of the entries in the Tables 3, 4 and 5.

R metric ± std. dev.		
Success within top 1	Success within top 3	Success within top 10

Table 3. Results on dataset *msweb*. For explanation on entries refer to Table 2. λ is set at 0.001 for the linear methods.

algorithm	Given2			Given5			Given10			AllBut1		
Popular	46.5 ± 0.2			43.7 ± 0.5			41.6 ± 2.0			46.5 ± 0.6		
	33.9	55.8	82.0	29.0	55.6	80.6	32.2	57.1	79.8	22.7	38.3	63.8
CR+	56.7 ± 0.1			54.2 ± 0.6			51.5 ± 1.9			60.8 ± 0.6		
	45.0	70.5	88.7	39.9	68.3	88.1	43.7	67.7	87.8	34.6	54.8	76.2
Decision Tree	53.4 ± 0.3			54.3 ± 0.7			53.0 ± 1.0			62.3 ± 0.5		
	46.6	71.3	87.4	48.0	73.9	88.5	51.6	72.9	87.8	38.4	58.4	74.9
Least Squares	55.7 ± 0.3			57.5 ± 0.7			57.0 ± 1.5			64.1 ± 0.5		
	46.9	72.4	89.6	49.9	75.0	90.9	55.5	77.1	91.0	38.5	58.8	79.2
Mod LS Primal	55.6 ± 0.3			57.7 ± 0.8			56.9 ± 1.4			64.4 ± 0.5		
	46.9	72.6	89.8	50.3	75.2	91.0	56.1	76.9	91.8	38.9	59.1	79.6
Mod LS Dual	55.2 ± 0.2			57.5 ± 0.9			56.7 ± 1.4			64.4 ± 0.6		
	46.5	72.9	89.7	50.5	75.6	90.5	57.1	77.3	91.8	39.0	59.0	79.4

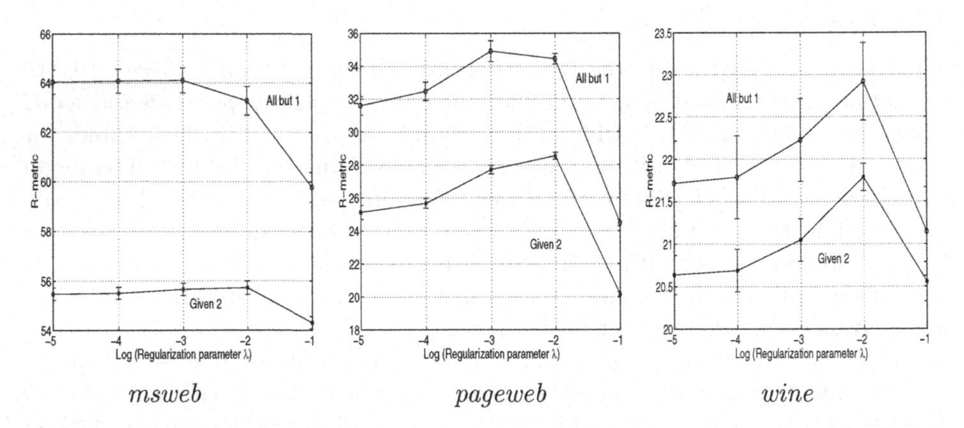

Fig. 1. Linear least squares classifier accuracy vs. regularization parameter λ.

Table 4. Results on dataset *pageweb*. For explanation on entries refer to Table 2. λ is set at 0.001 for the linear methods.

algorithm	Given2			Given5			Given10			AllBut1		
Popular	8.3 ± 0.3			7.0 ± 0.3			6.2 ± 0.7			7.6 ± 0.5		
	7.5	14.5	29.5	6.0	12.4	27.9	6.0	11.6	29.5	2.3	5.1	11.1
CR+	29.3 ± 0.8			31.9 ± 1.1			32.5 ± 1.4			33.3 ± 0.3		
	28.8	47.6	67.2	30.2	50.1	68.7	33.5	55.7	74.2	15.2	27.5	44.9
Decision	16.2 ± 0.1			19.9 ± 0.6			23.3 ± 1.2			22.7 ± 0.7		
Tree	23.3	33.5	47.0	28.9	41.6	52.7	36.9	50.7	63.2	13.5	20.2	28.4
Least	27.7 ± 0.3			32.5 ± 0.9			35.4 ± 0.8			34.9 ± 0.6		
Squares	30.1	48.4	66.8	33.7	53.2	71.6	41.7	62.9	77.6	17.1	29.8	46.6
Mod LS	28.3 ± 0.3			33.0 ± 0.9			35.7 ± 1.1			35.5 ± 0.5		
Primal	30.3	48.7	67.7	33.8	53.3	73.4	41.1	61.4	77.9	17.2	30.2	47.4
Mod LS	27.8 ± 0.3			32.9 ± 0.9			35.5 ± 1.2			35.2 ± 0.5		
Dual	29.9	48.7	67.1	34.6	53.4	73.1	41.8	62.3	77.6	17.4	30.0	46.7

Table 5. Results on dataset *wine*. For explanation on entries refer to Table 2. λ is set at 0.001 for the linear methods.

algorithm	Given2			Given5			Given10			AllBut1		
Popular	15.3 ± 0.2			14.5 ± 0.2			14.2 ± 0.4			13.6 ± 0.8		
	10.1	21.7	47.6	6.8	20.4	50.0	5.4	19.9	51.2	5.3	9.2	19.8
CR+	23.7 ± 0.3			24.6 ± 0.4			26.7 ± 0.8			21.4 ± 0.5		
	20.3	37.3	60.3	21.6	38.7	61.6	25.5	42.5	65.3	8.5	16.7	29.8
Decision	16.9 ± 0.4			18.6 ± 0.4			22.1 ± 0.5			17.9 ± 0.4		
Tree	16.8	27.8	49.0	18.9	33.9	52.8	21.4	41.3	60.1	7.6	14.3	24.1
Least	21.1 ± 0.3			24.7 ± 0.4			28.1 ± 0.3			22.2 ± 0.5		
Squares	19.4	35.8	57.1	23.5	43.2	63.8	25.9	46.3	68.1	8.8	17.5	31.3
Mod LS	20.8 ± 0.2			24.4 ± 0.4			27.8 ± 0.3			22.3 ± 0.4		
Primal	19.2	35.7	56.9	23.9	42.8	63.6	25.7	46.9	67.2	8.7	17.5	31.2
Mod LS	19.7 ± 0.2			23.4 ± 0.4			27.1 ± 0.5			21.8 ± 0.4		
Dual	18.0	34.0	55.6	23.1	41.8	62.9	26.6	46.2	66.9	8.6	17.3	30.2

recorded using our prototype implementation on an IBM RISC System/6000, Model 43P-140 using a 332 MHz PowerPC 604e processor. These model build times can be compared to those reported in [6] for Bayesian networks (144.65 seconds) and for dependency networks (98.31 seconds) on a 300 MHz Pentium system. However, our linear models are orders of magnitude more efficient than both Bayesian networks and decision networks [6] when generating recommendations because of the simplicity in computing the scores for a test user.

The accuracy achieved on the public dataset *msweb* cannot be directly compared with the results in [3] and [6] because of random choices made in the given/hidden sets. The results in Tables 3, 4 and 5 suggest that linear least squares and the primal and dual forms of the modified version fare well in comparison with our implementation of CR+. For example, the linear (dual) model is more accurate than CR+ in 11 out of the 12 experimental setups (3 datasets with 4 scenarios in each) using the success in the top 3 metric. If we use the R metric the linear (dual) model beats CR+ in 8 out of the 12 experimental setups.

The impact of the regularization parameter λ on the accuracy of one of the linear models (least squares) is shown in Figure 1. Similar behavior has been observed for the other linear models. The choice of λ makes a non-negligible difference for all algorithms. The value for this parameter could be chosen using cross-validation experiments with the training data, though this was not done in our study. The figures also suggest that in practice, one may choose a fixed λ with reasonable performance across a number of datasets, without any cross-validation λ selection. We would like to mention that for all algorithms, the value of λ should be same for every potential recommendation item. Otherwise, the computed scores $w^T x$ will not be comparable for different items. A side effect is that we only have a single λ to determine for each algorithm. Therefore a cross-validation procedure can be used to determine this value stably.

It is interesting to observe that in this application, the standard least squares method does as well as the more complicated modified least squares method. This phenomenon is not true for many other classification problems. One explanation could be that we use the ranking of the items, rather than a form of classification error, to measure the quality of a recommendation system. The impact of penalizing a point $w^T x > 1$ in the formulation for this task is clearly different when compared to a standard classification task.

4 Conclusion

This paper presents a model-based approach to recommender systems using linear classification models. We focus on three different linear formulations and the corresponding algorithms for building the models. Experiments are performed with three datasets and recommendation accuracies compared using two different metrics. The experiments indicate that the linear models are more accurate than a memory-based collaborative filtering approach reported earlier. This improved accuracy in combination with the better computational characteristics makes these linear models very attractive for this application.

Acknowledgments. We would like to thank Murray Campbell, Richard Lawrence and George Almasi for their help during this work.

Appendix A. Details of Algorithms

Algorithm 1 *(Least Squares Primal)*

let $w = 0$ and $r_i = 0$
for $k = 1, 2, \ldots$
  for $j = 1, \ldots, d$
    $\Delta w_j = -(\sum_i (r_i - 1) x_{ij} y_i + \lambda n w_j)/(\sum_i x_{ij}^2 + \lambda n)$
    update r: $r_i = r_i + \Delta w_j x_{ij} y_i$ $\quad (i = 1, \ldots, n)$
    update w: $w_j = w_j + \Delta w_j$
  end
end

Algorithm 2 *(Mod Least Squares Primal)*

let $w = 0$ and $r_i = 0$
pick a decreasing sequence of $1 = c_1 \geq c_2 \geq \cdots \geq c_K = 0$
for $k = 1, 2, \ldots, K$
  define function $C_k(r_i) = 1$ if $r_i \leq 1$ and $C_k(r_i) = c_k$ otherwise
  for $j = 1, \ldots, d$
    $\Delta w_j = -0.5[\sum_i C_k(r_i)(r_i - 1) x_{ij} y_i + \lambda n w_j]/[\sum_i C_k(r_i) x_{ij}^2 + \lambda n]$
    update r: $r_i = r_i + \Delta w_j x_{ij} y_i$ $\quad (i = 1, \ldots, n)$
    update w: $w_j = w_j + \Delta w_j$
    end
  end

Algorithm 3 *(Mod Least Squares Dual)*

let $\zeta = 0$ and $v_j = 0$ for $j = 1, \ldots, d$
for $k = 1, 2, \ldots$
  for $i = 1, \ldots, n$
    $\Delta \zeta_i = \min(-\zeta_i, -\frac{(2+\zeta_i)\lambda n + v^T x_i y_i}{\lambda n + x_i^2})$
    update v: $v_j = v_j + \Delta \zeta_i x_{ij} y_i$ $\quad (j = 1, \ldots, d)$
    update ζ: $\zeta_i = \zeta_i + \Delta \zeta_i$
  end
end
let $w = -\frac{1}{2\lambda n} v$.

References

1. P. Resnick, N. Iacovou, M. Suchak, P. Bergstrom, and J. Riedl. Grouplens: An open architecture for collaborative filtering of netnews. In *Proceedings of CSCW*, 1994.
2. U. Shardanand and P. Maes. Social information filtering: Algorithms for automating word of mouth. In *Proceedings of CHI'95*, 1995.
3. J.S. Breese, D. Heckerman, and C. Kadie. Emperical analysis of predictive algorithms for collaborative filtering. In *Proceedings of Fourteenth Conference on Uncertainty in Artificial Intelligence*. Morgan Kaufmann, 1998.

4. B. Sarwar, G. Karypis, J. Konstan, and J. Riedl. Analysis of recommendation algorithms for e-commerce. In *Proceedings ACM E-Commerce 2000 Conference*, 2000.
5. R. Hofmann and V. Tresp. Nonlinear markov networks for continuous variables. In *Advances in Neural Information Processing Systems 9, editors: M. Mozer, M. Jordan and T. Petsche*. MIT Press, 1997.
6. D. Heckerman, D. Chickering, C. Meek, R. Rounthwaite, and C. Kadie. Dependency networks for collaborative filtering and data visualization. In *Proc. of 16th Conf. on Uncertainty in Artificial Intelligence*. Morgan Kaufmann, 2000.
7. L. Ungar and D. Foster. Clustering methods for collaborative filtering. In *Workshop on Recommendation Systems at the Fifteenth National Conference on AI*, 1998.
8. L. Ungar and D. Foster. A formal statistical approach to collaborative filtering. In *CONALD'98*, 1998.
9. T. Joachims. Text categorization with support vector machines: Learning with many relevant features. In *Proceedings of the Tenth European Conference on Machine Learning*, pages 137–142, 1998.
10. C. Blake, E. Keogh, and C.J. Merz. UCI repository of machine learning databases. University of California, Irvine, Dept. of Information and Computer Science, URL=http://www.ics.uci.edu/~mlearn/MLRespository.html, 1998.
11. S. Glassman. Eachmovie data set. URL=http://research.compaq.com/SRC-/eachmovie/.
12. Yiming Yang and Christopher G. Chute. An example-based mapping method for text categorization and retrieval. *ACM Transactions on Information Systems*, 12:252–277, 1994.
13. B.D. Ripley. *Pattern recognition and neural networks*. Cambridge university press, 1996.
14. A. E. Hoerl and R. W. Kennard. Ridge regression: Biased estimation for nonorthogonal problems. *Technometrics*, 12(1):55–67, 1970.
15. C. Cortes and V.N. Vapnik. Support vector networks. *Machine Learning*, 20:273–297, 1995.
16. V.N. Vapnik. *Statistical learning theory*. John Wiley & Sons, New York, 1998.
17. John C. Platt. Fast training of support vector machines using sequential minimal optimization. In Bernhard Scholkopf, Christopher J. C. Burges, and Alex J. Smola, editors, *Advances in Kernel Methods: Support Vector Learning*, chapter 12. The MIT press, 1999.
18. F.M.J. Willems, Y.M. Shtarkov, and T.J. Tjalkens. The context tree weighting method: basic properties. *IEEE Trans. on Inform. Theory*, 41(3):653–664, 1995.
19. Tong Zhang. Compression by model combination. In *Proceedings of IEEE Data Compression Conference, DCC'98*, pages 319–328, 1998.
20. M. Kearns and Y. Mansour. A fast, bottom-up decision tree pruning algorithm with near optimal generalization. In *Proceedings of the 15th International Conference on Machine Learning*, pages 269–277, 1998.

iJADE eMiner - A Web-Based Mining Agent Based on Intelligent Java Agent Development Environment (iJADE) on Internet Shopping

Raymond S.T. Lee and James N.K. Liu

Department of Computing
Hong Kong Polytechnic University
Hung Hom, Hong Kong
{csstlee, csnkliu}@comp.polyu.edu.hk

Abstract. With the rapid growth of e-commerce applications, Internet shopping is becoming part of our daily lives. Traditional Web-based product searching based on keywords searching seems insufficient and inefficient in the 'sea' of information. In this paper, we propose an innovative intelligent multi-agent based environment, namely (iJADE) - intelligent **J**ava **A**gent **D**evelopment **E**nvironment - to provide an integrated and intelligent agent-based platform in the e-commerce environment. In addition to contemporary agent development platforms, which focus on the autonomy and mobility of the multi-agents, iJADE provides an intelligent layer (known as the 'conscious layer') to implement various AI functionalities in order to produce 'smart' agents. From the implementation point of view, iJADE eMiner consists of two main modules: 1) a visual data mining and visualization system for automatic facial authentication based on the FAgent model, and 2) a fuzzy-neural based shopping agent (FShopper) to facilitate Web-mining on Internet shopping in cyberspace.

1 Introduction

Owing to the rapid development in e-commerce, ranging from C2C e-commerce applications such as e-auction to sophisticated B2B e-commerce activities such as e-Supply Chain Management (eSCM), the Internet is becoming a common virtual marketplace for us to do our business, search for information and communicate with each other.

However, owing to these ever-increasing tons of information in cyberspace, information searching, or more precisely knowledge discovery and Web-mining is becoming the critical key to success at doing business in the cyberworld. Moreover, with the advance of PC computing technology in terms of computational speed and popularity, intelligent software applications known as agents, with their autonomous properties, automatic delegation of jobs, and highly mobile and adaptive behavior in the Internet environment, are becoming a potential area of development for e-business in the new millennium [12].

In a typical e-shopping scenario, there are two fundamental aspects of functionality in which Web-mining and visual data mining might help. The first one is customer authentication. Traditional authentication, based on username and password over a security transport layer such as SSL protocol, although providing a secured user authentication scheme, requires customer pro-active login to grant the access right, which may discourage the customer from his or her shopping intention. Other authentication schemes based on digital certificate with smart card technology [22], or

D. Cheung, G.J. Williams, and Q. Li (Eds.): PAKDD 2001, LNAI 2035, pp. 28-40, 2001.

biometric authentication techniques based on iris or palm recognition, might provide some alternatives of automatic authentication scheme. However, they all need special authentication equipment which limits usability in the e-commerce environment, not to mention the legal implications of accessing personal privacy data such as iris and palm patterns. In contrast, automatic authentication based on human face recognition can get rid of all these limitations. In terms of visual processing equipment, the standard Web-camera is already good enough for facial pattern extraction, which is nowadays more or less standard equipment for Web browsing. Moreover, this kind of authentication scheme can provide a truly automatic scheme in which the customer does not need to provide any special identity information, and more importantly it does not need to explore any 'confidential' or 'sensitive' data such as fingerprints and iris patterns.

The other area is the automation of the online shopping process via agent technology. Traditional shopping models include Consumer Buying Behavior Models such as the Blackwell model [7], the Bettman model [3] and the Howard-Sheth model [11], which all share a similar list of six fundamental stages of consumer buying behavior: 1) consumer requirement definition, 2) product brokering, 3) merchant brokering, 4) negotiation, 5) purchase and delivery, and 6) after-sale services and evaluation. In reality, the first three stages in the consumer buying behavior model involve a wide range of uncertainty and possibilities - or what we called 'fuzziness' - ranging from the setting of buying criteria and provision of products by the merchant, to the selection of goods. So far these are all 'grey areas' that we need to thoroughly explore in order to apply agent technology to the e-commerce environment.

In this paper, we propose an integrated intelligent agent-based framework, known as iJADE (pronounced as 'IJ') - **I**ntelligent **J**ava **A**gent-based **D**evelopment **E**nvironment. To accommodate the deficiency of contemporary agent software platforms such as IBM Aglets [1] and ObjectSpace Voyager Agents [25], which mainly focus on multi-agent mobility and communication, iJADE provides an ingenious layer called the 'Conscious (Intelligent) Layer', which supports different AI functionalities to the multi-agent applications. From the implementation point of view, we will demonstrate one of the most important applications of iJADE in the e-commerce environment - **iJADE eMiner**. iJADE eMiner is a truly intelligent agent-based Web-mining application which consists of two major modules: 1) an agent-based facial pattern authentication scheme - FAgent, an innovative visual data mining and visualization scheme using EGDLM technique [14]; 2) a Web-mining tool based on fuzzy shopping agents for product selection and brokering - FShopper. This paper is organized as follows. Section 2 presents an overview on Web-mining, a vital extension of data mining in cyberspace. Section 3 presents the model framework of iJADE, and the two major components of iJADE eMiner: FAgent and FShopper. System implementation will be discussed in section 4, which is followed by a brief conclusion.

2 Web Mining - A Perspective

As an important extension of data mining [8], Web-mining is an integrated technology of various research fields including computational linguistics, statistics, informatics, artificial intelligence (AI) and knowledge discovery [25]. It can also be interpreted as the discovery and analysis of useful information from the Web.

Although there is no definite principle of the Web-mining model, basically it can be categorized into two main areas: Content Mining and Structural Mining [6][25]. The taxonomy of Web-mining is depicted in Fig. 1.

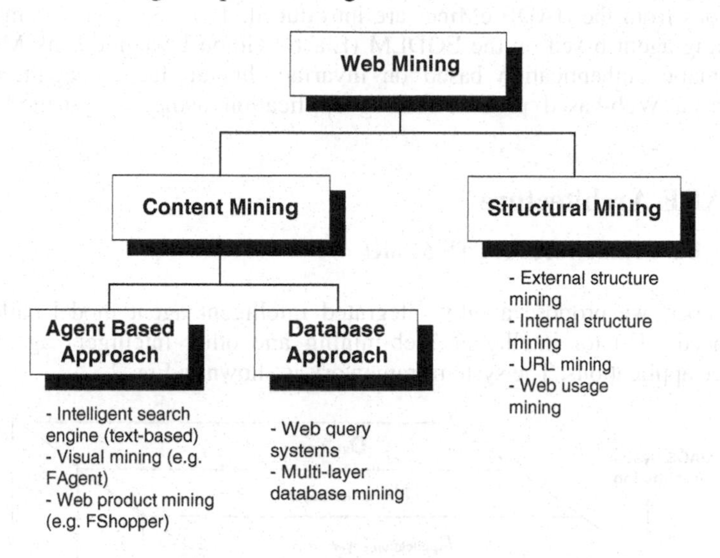

Fig. 1. A taxonomy of Web-mining

'Content Mining' focuses on the extraction and knowledge discovery (mining) of the Web content, ranging from the HTML, XML documents found in the Web servers to the mining of data and knowledge from the data source (e.g. databases) attached to the backend of Web systems. On the other hand, 'Structural Mining' focuses on knowledge discovery for the structure of the Web system, including the mining of the user preferences on Web browsing (Web usage mining), the usage of the different URLs in a particular Website (URL mining), external structure mining (for hyperlinks between different Web pages) and internal structure mining (within a particular Web page). Active research includes Spertus et al. [24] on internal structure mining and Pitkow [21] on Web usage mining.

With the concern of 'Content Mining', classical search engines such as Lycos, WebCrawler, Infoseek and Alta Vista do provide some sort of searching aid. However, they failed to provide concrete and structural information [19]. In recent years, interest has been focused on how to provide a higher level of organization for semi-structured and even unstructured data on the Web using Web-mining techniques. Basically, there are two main research areas: the Database Approach and the Agent-based Approach. The Database Approach Web-mining tries to focus on the techniques for organizing the semi-structured / unstructured data on the Web into more structured information and resources, such that traditional query tools and data mining can be applied for data analysis and knowledge discovery. Typical examples can be found in the ARANEUS system [2] using a multi-level database for Web-mining, and Dunren et al. [5] using a complex Web query system for the Web-mining of G-Protein Coupled Receptors (GPCRs).

The Agent-based Approach focuses on the provision of 'intelligent' and 'autonomous' Web-mining tools based on agent technology. Typical examples can be found in FAQFinder [9] for intelligent search engines and Firefly [23] for personalized Web agents.

In our proposed iJADE intelligent agent model, two innovative Web-mining applications from the iJADE eMiner are introduced: 1) FAgent, an automatic visual data mining agent based on the EGDLM (Elastic Graph Dynamic Link Model [16]) for automatic authentication based on invariant human face recognition, and 2) FShopper, a Web-based product mining application using fuzzy-neural shopping agents.

3 iJADE Architecture

3.1 iJADE Framework: ACTS Model

In this paper, we propose a fully integrated intelligent agent model called iJADE (pronounced 'IJ') for intelligent Web-mining and other intelligent agent-based e-commerce applications. The system framework is shown in Fig. 2.

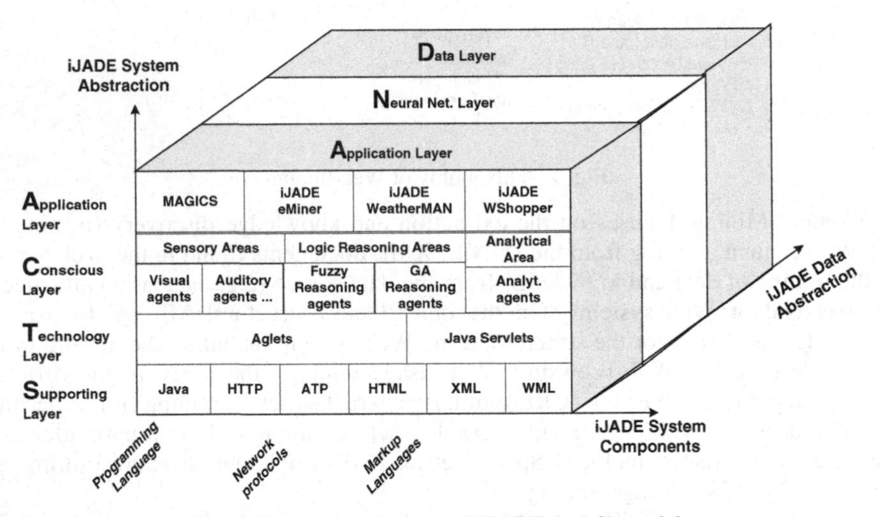

Fig. 2. System architecture of iJADE (v 1.0) model

Unlike contemporary agent systems and APIs such as IBM Aglets [1] and ObjectSpace Voyager [25], which focus on the multi-agent communication and autonomous operations, the aim of iJADE is to provide comprehensive 'intelligent' agent-based APIs and applications for future e-commerce and Web-mining applications.

Fig. 2 depicts the two level abstraction in the iJADE system: a) iJADE system level - **ACTS** model, and b) iJADE data level - **DNA** model. The ACTS model consists of 1) the **A**pplication Layer, 2) the **C**onscious (Intelligent) Layer, 3) the **T**echnology Layer, and the 4) **S**upporting Layer. The DNA model is composed of 1) the **D**ata Layer, 2) the **N**eural Network Layer, and 3) the **A**pplication Layer.

Compared with contemporary agent systems which provide minimal and elementary data management schemes, the iJADE DNA model provides a comprehensive data manipulation framework based on neural network technology. The 'Data Layer' corresponds to the raw data and input 'stimulates' (such as the facial images captured from the Web camera and the product information in the cyberstore) from the environment. The 'Neural Network Layer' provides the 'clustering' of different types of neural networks for the purpose of 'organization', 'interpretation', 'analysis' and 'forecasting' operations based on the inputs from the 'Data Layer', which are used by the iJADE applications in the 'Application Layer'.

Another innovative feature of the iJADE system is the ACTS mode, which provides a comprehensive layering architecture for the implementation of intelligent agent systems, will explain in the next sections.

3.2 Application Layer Including iJADE eMiner

This is the uppermost layer that consists of different intelligent agent-based applications. These iJADE applications are developed by the integration of intelligent agent components from the 'Conscious Layer' and the data 'knowledge fields' from the DNA model.

Concurrent applications (iJADE v1.0) implemented in this layer include:

- MAGICS (Mobile Agent-based Internet Commerce System), a collection of intelligent agent-based e-commerce applications including the MAGICS shopper (Internet shopping agent), and the MAGICS auction (intelligent auction agent) [4].
- iJADE eMiner, the intelligent Web-mining agent system proposed in this paper. It consists of the implementation of 1) FAgent, an automatic authentication system based on human face recognition, and 2) FShopper, a fuzzy agent-based Internet shopping agent.
- iJADE WeatherMAN, an intelligent weather forecasting agent which is the extension of previous research on multi-station weather forecasting using fuzzy neural networks [20]. Unlike traditional Web-mining agents, which focus on the automatic extraction and provision of the latest weather information, iJADE WeatherMAN possesses neural network-based weather forecasting capability (AI services provided by the 'Conscious Layer' of the iJADE model) to act as a 'virtual' weather reporter as well as an 'intelligent' weather forecaster for weather prediction.
- iJADE WShopper, an integrated intelligent fuzzy shopping agent with WAP technology for intelligent mobile shopping on the Internet.

3.3 Conscious (Intelligent) Layer

This layer provides the intelligent basis of the iJADE system, using the agent components provided by the 'Technology Layer'. The 'Conscious Layer' consists of the following three main intelligent functional areas:

1) 'Sensory Area' - for the recognition and interpretation of incoming stimulates. It includes a) visual sensory agents using EGDLM (Elastic Graph Dynamic Link Model) for invariant visual object recognition [16][17][18], and b) auditory sensory agents based on wavelet-based feature extraction and interpretation technique [10].
2) 'Logic Reasoning Area" - conscious area providing different AI tools for logical 'thinking' and rule-based reasoning, such as fuzzy and GA (Genetic Algorithms) rule-based systems [15].

3) 'Analytical Area' - consists of various AI tools for analytical calculation, such as recurrent neural network-based analysis for real-time prediction and data mining [13].

3.4 Technology Layer Using IBM Aglets and Java Servlets

This layer provides all the necessary mobile agent implementation APIs for the development of intelligent agent components in the 'Conscious Layer'.

In the current version (v1.0) of the iJADE model, IBM Aglets [1] are used as the agent 'backbone'. The basic functionality and runtime properties of Aglets are defined by the Java Aglet, AgletProxy and AgletContext classes. The abstract class Aglet defines the fundamental methods that control the mobility and lifecycle of an aglet. It also provides access to the inherent attributes of an aglet, such as creation time, owner, codebase and trust level, as well as dynamic attributes, such as the arrival time at a site and address of the current context.

The main function of the AgletProxy class is to provide a handle that is used to access the aglet. It also provides location transparency by forwarding requests to remote hosts and returning results to the local host. Actually, all communication with an aglet occurs through its aglet proxy. The AgletContext class provides the runtime execution environment for aglets within the Tahiti server. Thus when an aglet is dispatched to a remote site, it is detached from the current AgletContext object, serialized into a message bytestream, sent across the network, and reconstructed in a new AgletContext, which in turn provides the execution environment at the remote site. The other critical component of the Aglet environment is the security issue. Aglets provide a security model in the form of an AgletSecurityManager, which is a subclass of the "standard" Java SecurityManager.

In this layer, server-side computing using Java Servlet technology is also adopted due to the fact that for certain intelligent agent-based applications, such as the WShopper, in which limited resources (in terms of memory and computational speed) are provided by the WAP devices (e.g. WAP phones), all the iJADE agents interactions are invoked in the 'backend' WAP server using Java Servlet technology.

3.5 Supporting Layer

This layer provides all the necessary system supports to the 'Technology Layer'. It includes 1) Programming language support based on Java, 2) Network protocols support such as HTTP, HTTPS, ATP, etc., and 3) Markup languages support such as HTML, XML, WML, etc.

4 Implementation

4.1 iJADE eMiner: Overview

In this paper, an iJADE eMiner for intelligent e-shopping is introduced. This system consists of two modules: 1) Visual data mining for intelligent user authentication module based on FAgent, and 2) Web product mining using neuro-fuzzy shopping agent based on FShopper.

Intelligent Visual Data Mining for User Authentication Using FAgent

In short, there are three kinds of intelligent agents operating within the FAgent system. They are 1) the FAgent Feature Extractor, a stationary agent situated within the client machine to extract the facial features from a facial image which is captured from the client's digital camera, 2) the FAgent Messenger, a mobile agent which acts as a messenger which on one hand "carries" the facial image to the server-side agent and on the other hand "reports" the latest recognition results back to the client machine, and 3) the FAgent Recognizer, a stationary agent situated within the server (e.g. virtual shopping mall) using EGDLM (Elastic Graph Dynamic Link Model) for invariant facial recognition - an innovative object recognition scheme based on neural networks and elastic graph matching techniques, which have promising results in human face recognition [16], scene analysis [18] and tropical cyclone identification [17]. The main duty is to perform automatic and invariant facial pattern matching against the server-side facial database. A schematic diagram of the FAgent is shown in Fig. 3.

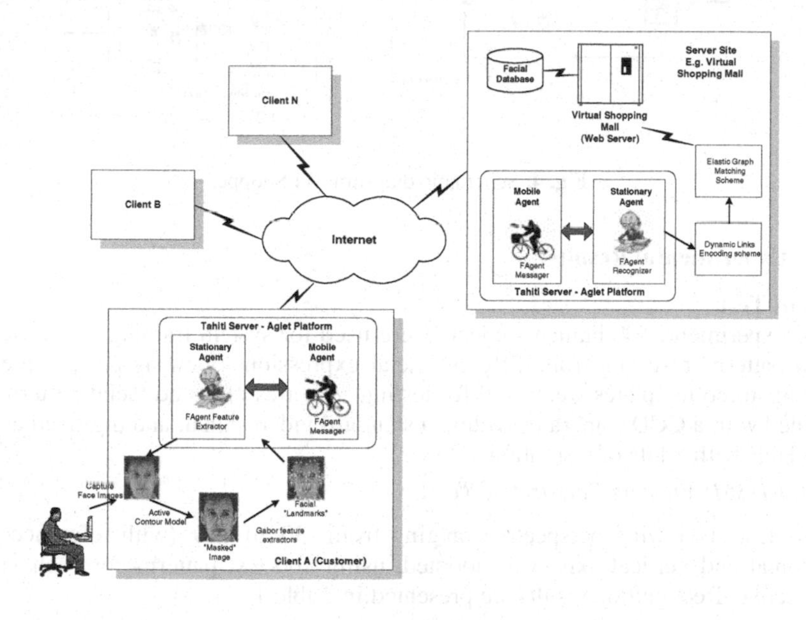

Fig. 3. Schematic diagram of FAgent System

Neural-Fuzzy Agent-Based Shopping Using FShopper

The system framework of FShopper consists of the following six main modules:

1. Customer requirement definition (CRD)
2. Requirement fuzzification scheme (RFS)
3. Fuzzy agents negotiation scheme (FANS)
4. Fuzzy product selection scheme (FPSS)
5. Product defuzzification scheme (PDS)
6. Product evaluation scheme (PES)

A schematic diagram of the Fuzzy Shopper is shown in Fig. 4.

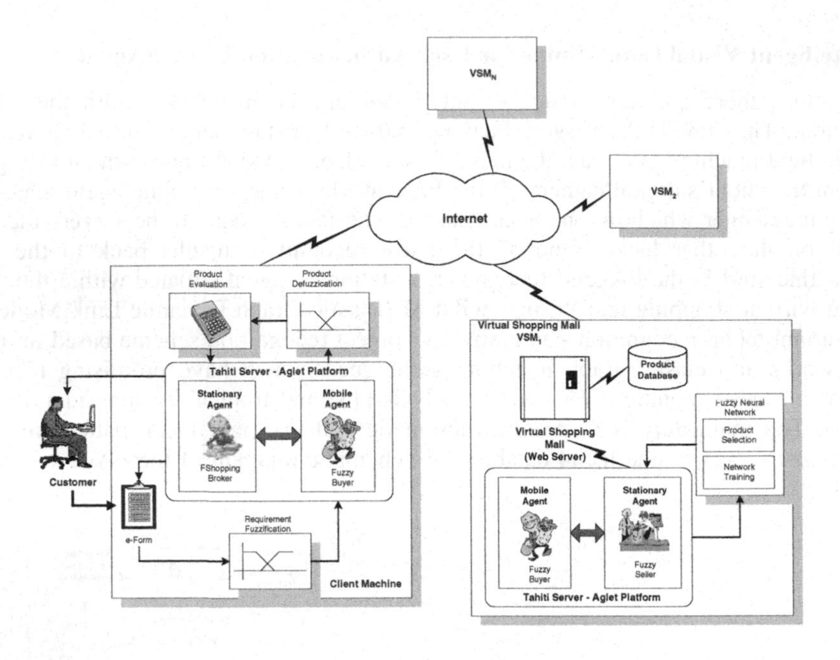

Fig. 4. Schematic diagram of FShopper

4.2 Experimental Results

FAgent Test

In the experiment, 100 human subjects were used for system training. A set of 1,020 tested patterns resulting from different facial expressions, viewing perspectives, and sizes of stored templates were used for testing. A series of tested facial patterns were obtained with a CCD camera providing a standard video signal, and digitized at 512 x 384 pixels with 8 bits of resolution.

FAgent Test I: Viewing Perspective Test

In this test, a viewing perspective ranging from -30° to +30° (with reference to the horizontal and vertical axis) was adopted, using 100 test patterns for each viewing perspective. Recognition results are presented in Table 1.

Table 1. Results for viewing perspective test

Viewing perspective (from horiz. axis)	Correct Classification	Viewing perspectives (from vertical axis)	Correct Classification
+30°	84%	+30°	86%
+20°	90%	+20°	88%
+10°	92%	+10°	91%
-10°	91%	-10°	92%
-20°	89%	-20°	87%
-30°	85%	-30°	82%

According to the "Rotation Invariant" property of the EGDLM model [16][18], the FAgent possesses the same characteristic in the "contour maps elastic graph matching" process. An overall correct recognition rate of over 86% was achieved.

FAgent Test II: the Facial Pattern Occlusion and Distortion Test

In this test, the 120 test patterns are basically divided into three categories:

- Wearing spectacles or other accessories
- Partial occlusion of the face by obstacles such as cups / books (in reading and drinking processes)
- Various facial expressions (such as laughing, angry and gimmicky faces).

Pattern recognition results are shown in Table 2.

Table 2. Recognition results for occlusion /distortion test

Pattern Occlusion & Distortion Test	Correct Classification
Wearing spectacles (or other accessoires)	87%
Face partially hidden by obstacle (e.g. books, cups)	72%
Facial expressions (e.g. laughing, angry and gimmicky faces)	83%

Compared with the three different categories of facial occlusion, "wearing spectacles" provides the least negative effect to facial recognition, owing to the fact that all the main facial contours are still preserved in this situation. In the second situation, the influence to the recognition rate depends on the proportion and which portion of the face is being obscured. Nevertheless, the average correct recognition rate was found to be over 73%.

Facial expressions and gimmicky faces gave the most striking results. Owing to the "Elastic Graph" characteristic of the model, the recognition engine "inherited" the "Distortion Invariant" property and an overall correct recognition rate of 83% are attained.

The FShopper Test

For the product database, over 200 items under eight categories were being used to construct the e-catalog. These categories were: T-shirt, shirt, shoes, trousers, skirt, sweater, tablecloth, napkins. We deliberately choose soft-good items instead of hard-goods such as books or music CDs (as commonly found in most e-shopping agent systems), so that it would allow more room for fuzzy user requirement definition and product selection. For neural network training, all the e-catalog items were 'pre-trained' in the sense that we had pre-defined the attribute descriptions for all these items to be 'fed' into the fuzzy neural network for product training (for each category). Thus totally, eight different neural networks were constructed according to each different category of product.

From the experimental point of view, two sets of tests were conducted: the Round Trip Time (RTT) test and the Product Selection (PS) test. The RTT test aims at an evaluation of the 'efficiency' of the FShopper in the sense that it will calculate the whole round trip time of the agents, instead of calculating the difference between the arrival and departure time to/from the server. The RTT test will calculate the entire "component" time fragments starting from the collection of the user requirement, fuzzification process, to the product selection and evaluation steps, so that a total picture of the performance efficiency can be deduced. In the PS test, since there was no definite answer to whether a product would 'fit' the taste of the customer or not, a sample group of 40 candidates was used to judge the 'effectiveness' of the FShopper. Details are illustrated in the following sections.

FShopper Test I: The Round Trip Time (RTT) test

In this test, two iJADE Servers were being used: the T1server and the T2server. T1server was situated within the same LAN as the client machine, while the T2server was in a remote site (within the campus). Results of the mean RTT after 100 trials for each server are shown in Table 3.

As shown in Table 3, the total RTT is dominated by the Fuzzy Product Selection Scheme (FPSS), but the time spent is still within an acceptable timeframe: 5 to 7 seconds. Besides, the difference of RTT between the server situated in the same LAN and the remote site was not significant except in the FANS, whereas the Fuzzy Buyer needs to take a slightly longer "trip" than the other. Of course, in reality, it depends heavily on the network traffic.

Table 3. Mean RTT summary after 100 trials

Time (msec)	T1server	T2sever
Server location	Same LAN as client	Remote site (within campus)
A. Client machine		
CRD	-	
RFS	310	305
B. Server machine		
FANS	320	2015
FPSS	4260	4133
C. Client machine		
PDS	320	330
PES	251	223
TOTAL RTT	5461	7006

FShopper Test II: The Product Selection (PS) test

Unlike the RTT test, in which objective figures can be easily obtained, the PS test results rely heavily on user preference. In order to get a more objective result, a sample group of 40 candidates was invited for system evaluation. In the test, each candidate would 'buy' one product from each category according to his or her own requirements. For evaluation, they would browse around the e-catalog to choose a list of the 'best five choices' (L) which 'fit his/her taste'. In comparison with the 'top five' recommended product items (i) given by the fuzzy shopper, the 'Fitness Value (FV)' is calculated as follows:

$$FV = \frac{\sum_{n=1}^{5} n \times i}{15} \quad \text{where } i = \begin{cases} 1 & \text{if } i \in L \\ 0 & \text{otherwise} \end{cases}$$

In the calculation, scores of 5 to 1 were given to 'correct matches' of the candidate's first to fifth 'Best five' choices with the FShopper's suggestion. For example, if out of the five 'best choices' selected by the customer, products of rank no. 1, 2, 3 and 5 appear in the fuzzy shopper recommended list, the fitness value will be 73%, which is the sum of 1, 2, 3 and 5 divided by 15. Results under the eight different categories are shown in Table 4.

It is not difficult to predict that the performance of the FShopper is highly dependent on the 'variability' (or 'fuzziness') of the merchandise. The higher the fuzziness (which means more variety), the lower the score. As shown in Table 4, skirts and shoes are typical examples in which skirts scores 65% and shoes scores 89%. Nevertheless, the average score is still over 81%. Note that these figures are only for illustration purposes, as human justification and product variety in actual scenarios do vary case by case.

Table 4. Fitness value for the eight different product categories

Product category	Fitness Value % (FV)
T-shirt	81
Shirt	78
Shoes	89
Trousers	88
Skirts	65
Sweater	81
Tablecloth	85
Napkins	86
Average score	**81.6%**

5 Conclusion

In this paper, an innovative intelligent agent-based Web-mining application, iJADE eMiner, is proposed. Based on the integration of neuro-fuzzy based Web-mining technology (FShopper) and intelligent visual data-mining technology (FAgent) for automatic user authentication. It will hopefully provide a new era of Web-based data mining and knowledge discovery using intelligent agent-based systems.

Acknowledgement. The authors acknowledge partial supports from the Central Research Grants G-T142 and G-YC50 of the Hong Kong Polytechnic University.

References

1. Aglets. URL: http://www.trl.ibm.co.jp/aglets/.
2. Atzeni, P., Masci, A., Mecca, G., Merialdo, P. and Tabet, E. (1997): ULIXES: Building Relational Views over the Web. In Proceedings of 13th Int'l Conference on Data Engineering 576.
3. Bettman J. (1979): An Information Processing Theory to Consumer Choice. Addison-Wesley.

4. Chan, H. C. B, Lee, R. S. T. and Lam, B. (May 2001): Mobile AGent based Internet Commerce System (MAGICS): An Overview. Submitted to IEEE Trans. Systems, Man and Cybernetics.
5. Che, D., Chen, Y., Aberer, K. and Eisner, H. (1999): The Advanced Web Query System of GPCRDB. In Proceedings of the 11th Int'l Conference on Scientific and Statistical Database Management 281.
6. Cooley, R., Mobasher, B. and Srivastava, J. (1997): Web Mining: Information and Pattern Discovery on the World Wide Web. In Proceedings of the 9th Int'l Conference on Tools with A. I. 97 558-567.
7. Engel, J. and Blackwell, R. (1982): Consumer Behavior. CBS College Publishing.
8. Fayyad, U. M., Piatetsky-Shapiro, G., Smyth, P. and Uthurusamy, R. (eds) (1996): Advances in Knowledge Discovery and Data Mining. AAAI Press/ The MIT Press.
9. Hammond, K., Burke, R., Martin, C. and Lytinen, S. (1995): FAQ-Finder: A Case-based Approach to Knowledge Navigation. In Working Notes of the AAAI Spring Symposium: Information Gathering from Heterogeneous, Distributed Environments. AAAI Press.
10. Hossain, I., Liu, J. and Lee, R. (1999): A Study of Multilingual Speech Feature: Perspective Scalogram Based on Wavelet Analysis. In Proceedings of IEEE International Conference on Systems, Man, and Cybernetics (SMC'99), vol. II 178-183, Tokyo, Japan.
11. Howard J. and Sheth J. (1969): The Theory of Buyer Behavior. John Wiley and Sons.
12. Klusch M. (eds) (1999): Intelligent Information Agents: Agent-Based Information Discovery and Management on the Internet. Springer-Verlag Berlin Heidelberg.
13. Lee, R.S.T. and Liu, J.N.K. (2000): Tropical Cyclone Identification and Tracking System using integrated Neural Oscillatory Elastic Graph Matching and hybrid RBF Network Track mining techniques. IEEE Transaction on Neural Network **11**(3) 680-689.
14. Lee, R. S. T. and Liu, J. N.K. (2000): FAgent - An Innovative E-Shopping Authentication Scheme using Invariant Intelligent Face Recognition Agent. In Proceedings of International Conference on Electronic Commerce (ICEC2000) 47-53, Seoul, Korea.
15. Lee, R. S. T. and Liu, J. N. K. (2000): Teaching and Learning the A. I. Modeling. In Innovative Teaching Tools: Knowledge-Based Paradigms (Studies in Fuzziness and Soft Computing **36**) 31-86, L. C. Jain (eds). Physica-Verlag, Springer.
16. Lee, R. S. T. and Liu, J. N. K. (1999): An Integrated Elastic Contour Fitting and Attribute Graph Matching Model for Automatic Face Coding and Recognition. In Proceedings of the Third International Conference on Knowledge-Based Intelligent Information Engineering Systems (KES'99) 292-295, Adelaide, Australia.
17. Lee, R.S.T. and Liu, J.N.K. (1999): An Automatic Satellite Interpretation of Tropical Cyclone Patterns Using Elastic Graph Dynamic Link Model. International Journal of Pattern Recognition and Artificial Intelligence **13**(8) 1251-1270.
18. Lee, R.S.T. and Liu, J.N.K. (1999): An Oscillatory Elastic Graph Matching Model for Scene Analysis. In Proceedings of International Conference on Imaging Science, Systems, and Technology (CISST'99) 42-45, Las Vegas, Nevada, USA.
19. Leighton, H. V. and Srivastava, J. (1997): Precision among WWW Search Services (Search Engines): Alta Vista, Excite, Hotbot, Infoseek, Lycos. http://www.winona.musus.edu/is-f/library-f/webind2/webind2.htm.
20. Liu, J. N. K. and Lee, R. S. T. (1999): Rainfall Forecasting from Multiple Point Source Using Neural Networks. In Proceedings of IEEE International Conference on Systems, Man, and Cybernetics (SMC'99), vol. II 429-434, Tokyo, Japan.
21. Pitkow, J. (1997): In Search of Reliable Usage Data on the WWW. In Proceedings of the Sixth Int'l WWW Conference 451-463.
22. Rankl, W. and Effing, W. (1997): Smart card handbook. Wiley.

23. Shardanand, U. and Maes, P. (1995): Social Information Filtering: Algorithms for Automating 'word of mouth'. In Proceedings of 1995 Conference on Human Factors in Computing Systems (CHI-95) 210-217.
24. Spertus, E. (1997): ParaSite: Mining Structural Information on the Web. In Proceedings of the Sixth Int'l WWW Conference 485-492.
25. Voyager. URL: http://www.objectspace.com/voyager/.
26. Wang, J., Huang, Y., Wu, G. and Zhang, F. (1999): Web Mining: Knowledge Discovery on the Web. In Proceedings of Int'l Conference on Systems, Man and Cybernetics (SMC'99) **2** 137-141.

A Characterized Rating Recommend System[1]

Yao-Tsung Lin and Shian-Shyong Tseng

Department of Computer and Information Science, National Chiao Tung University, Hsinchu 300, Taiwan, R.O.C.
{gis88801, sstseng}@cis.nctu.edu.tw

Abstract. In recent years, due to the rapid growth of Internet usage, the problem of how to avoid inappropriate Internet contents accessing becomes more and more important. To solve the problem, a *Collaborative Rating System* [3, 4] based upon PICS protocol has been proposed. However, since the users usually would like to consult the opinions of the user group with similar rating tendency rather than the common opinions from the majority, it means the opinion of second majority with sufficient number of voters should also be considered. So does third majority, and so on. In order to provide a characterized rating service, a *Characterized Rating Recommend System* is designed to provide more precise and proper rating service for each user. Also, in this work, a questionnaire is designed to get users' opinions, and some experimental results show that the system can provide acceptable rating service.

1 Introduction

In recent years, due to the rapid growth of Internet usage, the problem of how to avoid inappropriate Internet contents accessing becomes more and more important. The concept of content selection is proposed to solve this problem, and there are many previous researches about content selection; e.g., PICS [5] protocol, which is proposed by W3C [8]. But there are still some problems; e.g., problem of rating information collecting. To solve this problem, a *Collaborative Rating System* [3, 4] has been proposed. However, users' rating tendencies can not be considered in the system. In order to provide a *Characterized Rating Service* which take care of this problem, in this work, the opinions of participants will be first represented in well-structured data, *Rating Vectors*, and these *Rating Vectors* will be clustered into *Rating Groups*, corresponding to different rating opinions. Then the properties of each *Rating Groups* will be mined by using *Rating Decision Tree Constructing Algorithm.* To prevent the problem of over-fitting in a decision tree, a *Precision and Support Based Decision Pruning Algorithm* will be applied. Finally, the rules about *Rating Groups* generated will be used to provide users characterized rating services. Based on these concepts, a *Characterized Rating Recommend System* is designed. In the experiment, a question naire is designed to efficiently get opinions about content rating. 700 participants are

[1] This work was supported by Ministry of Education and National Science Council of the Republic of China under Grand No. 89-E-FA04-1-4, High Confidence Information Systems.

D. Cheung, G.J. Williams, and Q. Li (Eds.): PAKDD 2001, LNAI 2035, pp. 41-46, 2001.

asked to answer the questionnaire, 616 of the filled questionnaires are useable in the clustering without missing data. Some experimental results with cross validation will also be shown in this paper.

2 Related Works

In order to solve the problem of how to avoid inappropriate Internet content, many researches are proposed [1, 3, 4, 7], and among these researches, *Collaborative Rating System* [3, 4] based upon *PICS* [5] protocol, provides practical solution. And *Characterized Rating Recommend System* is proposed to make *Collaborative Rating System* more adaptive to different user requirements.

2.1 PICS, Platform for Internet Content Selection [5]

To solve the problems of selecting appropriate or desired content via the Internet, many researches have been proposed. In these researches, the PICS[5] protocol was proposed by W3C[8] organization and provided a systematic architecture for document rating system, and it also provides the methods of rating information collecting. In PICS protocol, the rating information is provided by two methods, self-labeling and third-party labeling. In self-labeling method, the rating information is provided by the content providers of each web page. In third-party labeling, the rating information is provided by specific groups or organizations instead of the content provider. However, these rating information collecting methods seem to be too weak, since there is no obligation for content providers to provide the rating, it is impossible to rate all documents by few voluntary or non-profit organizations, and it is hard to design an acceptable automatic rating system.

2.2 Collaborative Rating System [3, 4]

To solve the issues on rating information collecting, a Collaborative Rating System [3, 4] was proposed, which collects rating information by the help of huge amount of volunteers. In *Collaborative Rating System*, participants are asked to rate web contents as they browse web pages according to a selected rating category, and their ratings will be collected and used to conclude a more objective result. The attributes of collected rating data consist of:

Category	*Web Page*	*User*	*Level*

In the collected rating data, *Category* represents the selected rating category of this rating, and *Web Page* indicates the address of target web page. The information in *User* attribute records who had made the rating, and *Level* attribute is the rated level in the selecting rating category the user thought. The rating data of the same web page collected from huge amount of users will be used to conclude result rating levels for each web page. *Collaborative Rating System* provides a more practical method for rating information collecting, and reduces the effort of each volunteer and organization to construct real non-profit rating system.

3 Characterized Rating Recommend System

Since different users may have other opinions than the majority; it means a unique rating result may not satisfy the needs of all users. In this section, a Characterized Rating Recommend System is designed to provide recommendation on document rating according to the characters of users in a collaborative rating system. The architecture of the Characterized Rating Recommend System can be shown in Fig. 1.

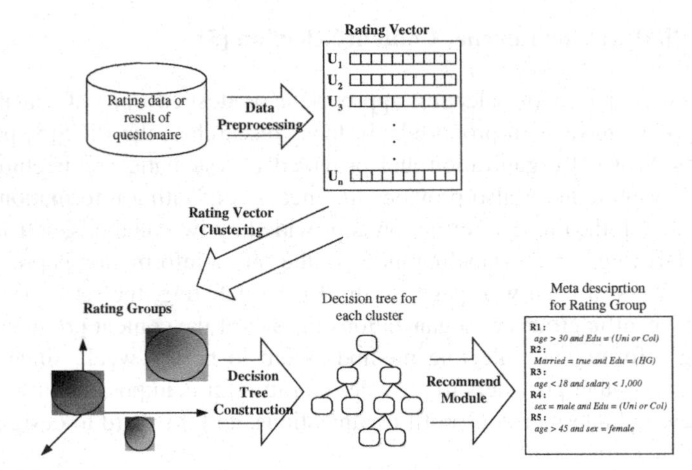

Fig. 1. The architecture of Characterized Rating Recommend System

3.1 Rating Data Preprocessing

In a *Collaborative Rating System* [3,4], volunteers provide ratings to web pages they browsed. Besides, the ratings a user made can be structured into well-formatted *Rating Vector* to represent his/her rating opinion. A *Rating Vector* consists of ratings of a user to specific web pages which are browsed by many participants. The difference in rating vectors of each participant will be mapped to the difference of their rating opinions, and this property can be used to cluster different user groups. For k given popular web pages, the participants who have rate all these pages will be selected and their information will be analyzed in the system. The rating vectors of these participants can be constructed by arranging each participant's ratings for these k web pages in a specific order.

3.2 Rating Group Clustering and Selecting

As we have defined, each element of rating vector of a participant is the rating he/she has made for a specific web page, and assume the rating options provided are orderly arranged according to the difference of meaning, for example, from *harmless* to *dangerous*. It means the numeric difference in the same element of *Rating Vectors* of different users will correspond to the difference of opinions to the web page. And then the difference of *Rating Vector* can be easily defined by this concept.

For *Rating Vectors*, we can then apply clustering algorithm to find the rating groups representing different rating opinions. K-means algorithm [2] will be used here to cluster *Rating Vectors* into clusters, and only these clusters including participants more than a threshold will be selected. After clustering and selecting, *Rating Vectors* will be clustered into several different *Rating Groups*.

3.3 Rating Group Character Analyzing

Each *Rating Vector* corresponds to a unique participant, and a *Rating Group* can be thought as a group of users with similar rating tendency. It seems that some common properties of the users' characters in *Rating Groups* can be analyzed. A symbolic learning approach will try to conclude some common characters form the characters of users in the *Rating Groups*. Among symbolic learning algorithms, decision tree algorithm [6] is used due to the flexibility of handling the training data containing both symbolic and numeric data. To apply the decision tree algorithm, the participants of different Rating Group are treated as different kinds of samples used in decision tree algorithm, and the characters of participants are used for decision tree learning. After applying the algorithm, the constructed model can be further used to analyze the properties of users for different *Rating Groups*.

3.4 Precision and Support Based Decision Tree Pruning

In order to prevent the problem of over-fitting and find the overall trend in users' characters to their rating tendencies, not only precision is concerned, but also the support of rules should also be considered. In order to generate rules with more support, a *Precision and Support Based Decision Tree Pruning Algorithm* is proposed in this section, and the detail of the algorithm is shown in Algorithm 1.

Notations:

$E(n)$	The formula to evaluate the expected error ratio below node n in a tree.
$Su(n)$	The formula to evaluate the expected support ratio below node n in a tree.
G	Gain value of a node, the sum of E( ) and Su( ) of a node.
G_{sum}	Sum of Gain value of all child nodes.

3.5 Rating Recommend Module

From the pruned decision tree, the rules about *Rating Groups* will be generated [6], and the rules will be used for rating recommending. After all the rules of *Rating Groups* are generated, the users of rating services can be partitioned into different groups, so does the rating information. For the users using rating service, a recommend can be made based on his/her characters, and corresponding rating information by evaluating ratings of the same group. With this mechanism, users may get rating information which is more adapted to their opinions, since the rating information comes from users who have similar characters to him/her.

Algorithm 1: Precision and Support Based Decision Tree Pruning Algorithm

```
For each non-leaf node p in Decision Tree from the bottom of Decision Tree
        For each sub node b of n
                G = Su(b) + E(b) ;
                G_Sum = G_Sum + G ;
        endFor
        If G_Sum > Su(p) + E(p) then
                Prune the node p from T_i;
        endIf
endFor
```

4 Experiments and Implementation

In the experiment, a questionnaire was designed to efficiently collect the users' opinions to analyze. In out experiment, 700 users are asked to answer the questionnaire, and 616 of the filled questionnaires are usable in the clustering without missing data. The rating vectors of each questionnaire of the same category are clustered by k-means algorithm. Then the decision tree algorithm and corresponding pruning algorithm are applied to find rules about the attributes of users in this cluster, and the rules are generated from the constructed decision tree model. The results of experiment are shown in following tables:

Table 1. Experiment results of both training set and test set

		Training Set		Test Set	
	Number of Rules	Avg. Precision	Avg. Support	Avg. Precision	Avg. Support
1st set	11	64.15 %	39.18 / 431	62.13 %	16.82 / 185
2nd set	8	58.79 %	53.88 / 431	54.75 %	23.13 / 185
3rd set	10	62.56 %	43.1 / 431	57.56 %	18.5 / 185
4th set	15	65.47 %	28.73 / 431	60.4 %	12.33 / 185

5 Conclusion

In this paper, users' rating opinions can be represented in *Rating Vectors* using the rating data collected from *Collaborative Rating System,* and clustered into several *Rating Groups* which correspond to different rating tendencies. And then a decision tree algorithm is proposed to find common characters of participants of *Rating Groups*. Besides, a *Precision and Support Based Decision Pruning Algorithm* will be applied to prune tree branches by considering both precision and support. Finally, the generated rules from pruned decision tree will be used to provide users characterized rating services and construct a *Characterized Rating Recommend System.* In the experiment, a questionnaire was designed to get users' rating opinions efficiently. Some experimental results of proposed system showed that the proposed system can provide acceptable precision and rating recommending service.

Table 2. Experiment results before and after pruning

	Before Pruning		After Pruning	
	Avg. Precision	No. of rules	Avg. Precision	No. of rule
1st set	85.16 %	15	64.15 %	11
2nd set	82.75 %	13	58.79 %	8
3rd set	88.13 %	17	62.56 %	10
4th set	84.27 %	23	65.47 %	15

Reference

1. ICRA, Internet Content Rating Association, *http://www.icra.org*, 2000
2. A. K. Jain and R. C. Dubes, *Algorithms for clustering data*, Prentice-Hall Inc., pp. 58-89, 1988.
3. Mon-Fong Jiang, Shian-Shyong Tseng, and Yao-Tsung Lin, "Collaborative Rating System for Web Page Labeling," World Conference of the WWW and Internet, Honolulu, Hawaii, 1999
4. Yao-Tsung Lin, Shian-Shyong Tseng, and Mon-Fong Jiang, "Voting Based Collaborative Platform for Internet Content Selection", The 4th Global Chinese Conference on Computers in Education Proceeding, 2000
5. PICS, Platform for Internet Content Selection, *http://www.w3.org/PICS*, 2000
6. J. R. Quinlan , *C4.5: programs for machine learning*, Morgan Kaufmann Publishers, Inc., 1988
7. RSAC, Recreational Software Advisory Council, http://www.rsac.org, 1999
8. W3C, World Wide Web Consortium, *http://www.w3.org*, 2000

Discovery of Frequent Tree Structured Patterns in Semistructured Web Documents

Tetsuhiro Miyahara[1], Takayoshi Shoudai[2], Tomoyuki Uchida[1],
Kenichi Takahashi[1], and Hiroaki Ueda[1]

[1] Faculty of Information Sciences,
Hiroshima City University, Hiroshima 731-3194, Japan
{miyahara@its, uchida@cs, takahasi@its, ueda@its}.hiroshima-cu.ac.jp
[2] Department of Informatics, Kyushu University, Kasuga 816-8580, Japan
shoudai@i.kyushu-u.ac.jp

Abstract. Many documents such as Web documents or XML files have no rigid structure. Such semistructured documents have been rapidly increasing. We propose a new method for discovering frequent tree structured patterns in semistructured Web documents. We consider the data mining problem of finding all maximally frequent tag tree patterns in semistructured data such as Web documents. A tag tree pattern is an edge labeled tree which has hyperedges as variables. An edge label is a tag or a keyword in Web documents, and a variable can be substituted by any tree. So a tag tree pattern is suited for representing tree structured patterns in semistructured Web documents. We present an algorithm for finding all maximally frequent tag tree patterns. Also we report some experimental results on XML documents by using our algorithm.

1 Introduction

Web documents have been rapidly increasing as the Information Technologies develop. Our target for knowledge discovery is the Web documents which have tree structures such as documents on World Wide Web or XML/SGML files. Such Web documents are called semistructured data [1]. In order to extract meaningful and hidden knowledge from semistructured Web documents, we need first to discover frequent tree structured patterns from them.

In this paper, we adopt a variant of Object Exchange Model (OEM, for short) in [1] for representing semistructured data. For example, we give an XML file *xml_sample* and a labeled tree T as its OEM data in Fig. 1. Many real semistructured data have no absolute schema fixed in advance, and their structures may be irregular or incomplete. As knowledge representations for semistructured data, for example, the type of objects [7], tree-expression pattern [10] and regular path expression [3] were proposed. In [4], we presented the concept of term trees as graph patterns suited for representing tree-like semistructured data. A term tree is pattern consisting of variables and tree-like structures. A term tree is different from other representations proposed in [3,7,10] in that a term tree has structured variables which can be substituted by arbitrary trees.

D. Cheung, G.J. Williams, and Q. Li (Eds.): PAKDD 2001, LNAI 2035, pp. 47–52, 2001.

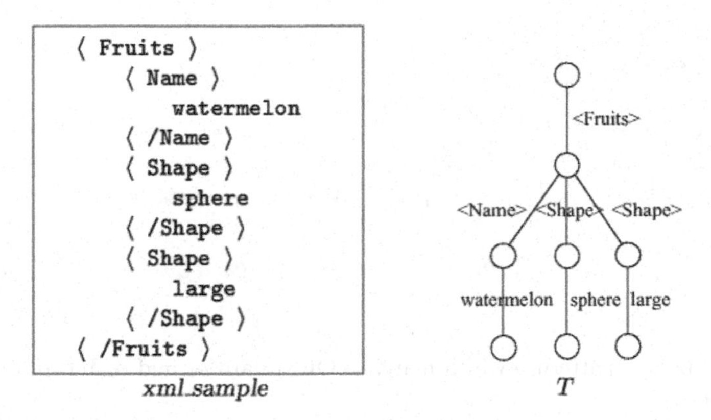

Fig. 1. An XML file *xml_sample* and a labeled tree T as its OEM data.

In [5], we gave the knowledge discovery system KD-FGS which receives graph structured data and produces a hypothesis by using Formal Graph System [9] as a knowledge representation language. In [8], we designed an efficient knowledge discovery system having polynomial time matching algorithms and a polynomial time inductive inference algorithm from tree-like semistructured data. The above systems find a hypothesis consistent with all input data or a term tree which can explain a minimal language containing all input data, respectively. These systems work correctly and effectively for complete data. However, for irregular or incomplete data, the systems may output obvious or meaningless knowledge. In this paper, in order to obtain knowledge efficiently from irregular or incomplete semistructured data, we define a tag tree pattern which is a special type of a term tree. In Fig. 2, for example, the tag tree pattern p matches the OEM data o_1 and o_2, but p does not match the OEM data o_3.

The purpose of this work is extraction of tree structured patterns from tree structured data which are regarded as positive samples. So, overgeneralized patterns explaining the given data are meaningless. Finding least generalized patterns explaining the given data, which are called maximally frequent tag tree patterns, is reasonable. We propose a method for discovering all maximally frequent tag tree patterns. To do this, we present an algorithm which generates all maximally frequent tag tree patterns by employing the algorithm in [2] which generates the canonical representations of all rooted trees. By using this algorithm, we can exclude meaningless tag tree patterns and avoid missing useful tag tree patterns. And we report some experimental results on XML documents by using our algorithm.

2 Tag Tree Patterns and Data Mining Problems

Let $T = (V_T, E_T)$ be a rooted unordered tree (or simply tree) with an edge labeling. A *variable* in V_T is a list $[u, u']$ of two distinct vertices u and u' in V_T.

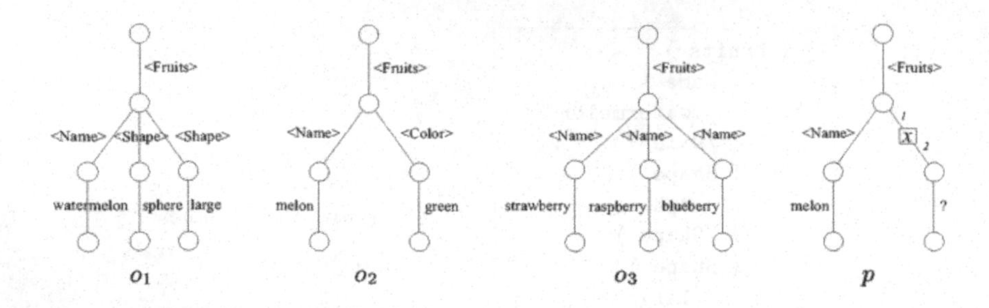

Fig. 2. A tag tree pattern p which matches OEM data o_1 and o_2 but does not match OEM data o_3.

A label of a variable is called a *variable label.* Λ and X denote a set of edge labels and a set of variable labels, respectively, where $\Lambda \cap X = \phi$. A triplet $g = (V_g, E_g, H_g)$ is called *rooted term tree* (or simply *term tree*) if (V_g, E_g) is a tree, H_g is a finite set of variables, and the graph $(V_g, E_g \cup E'_g)$ is a tree where $E'_g = \{\{u, v\} \mid [u, v] \in H_g\}$. A term tree g is called *regular* if all variables have mutually distinct variable labels in X. Let f and g be term trees with at least two vertices. Let $\sigma = [u, u']$ be a list of two distinct vertices in g. The form $x := [g, \sigma]$ is called a *binding* for x. A new term tree $f\{x := [g, \sigma]\}$ is obtained by applying the binding $x := [g, \sigma]$ to f in the following way: Let $e_1 = [v_1, v'_1], \ldots, e_m = [v_m, v'_m]$ be the variables in f with the variable label x. Let $g_1, \ldots, g_m$ be m copies of g and u_i, u'_i the vertices of g_i corresponding to u, u' of g. For each variable e_i, we attach g_i to f by removing the variable $e_i = [v_i, v'_i]$ from H_f and by identifying the vertices v_i, v'_i with the vertices u_i, u'_i of g_i. Let the root of the resulting term tree be the root of f. A *substitution* θ is a finite collection of bindings $\{x_1 := [g_1, \sigma_1], \cdots, x_n := [g_n, \sigma_n]\}$, where x_i's are mutually distinct variable labels in X.

Tag Tree Patterns. Let Λ_{Tag} and Λ_{KW} be two languages which contain infinitely many words where $\Lambda_{Tag} \cap \Lambda_{KW} = \emptyset$. We call words in Λ_{Tag} and Λ_{KW} a *tag* and a *keyword*, respectively. A *tag tree pattern* is a regular term tree such that each edge label on it is any of a tag, a keyword, and a special symbol "?". A tag tree pattern with no variable is called a *ground tag tree pattern.* For an edge $\{v, v'\}$ of a tag tree pattern and an edge $\{u, u'\}$ of a tree, we say that $\{v, v'\}$ *matches* $\{u, u'\}$ if the following conditions (1)-(3) hold: (1) If the edge label of $\{v, v'\}$ is a tag then the edge label of $\{u, u'\}$ is the same tag or a tag which is considered to be identical under an equality relation on tags. (2) If the edge label of $\{v, v'\}$ is a keyword then the edge label of $\{u, u'\}$ is a keyword and the label of $\{v, v'\}$ appears as a substring in the edge label of $\{u, u'\}$. (3) If the edge label of $\{v, v'\}$ is "?" then we don't care the edge label of $\{u, u'\}$.

A ground tag tree pattern $\pi = (V_\pi, E_\pi, \emptyset)$ *matches* a tree $T = (V_T, E_T)$ if there exists a bijection φ from V_π to V_T such that (i) the root of π is mapped to the root of T by φ, (ii) $\{v, v'\} \in E_\pi$ if and only if $\{\varphi(v), \varphi(v')\} \in E_T$, and (iii) for all $\{v, v'\} \in E_\pi$, $\{v, v'\}$ matches $\{\varphi(v), \varphi(v')\}$. A tag tree pattern π *matches* a tree T if there exists a substitution θ such that $\pi\theta$ is a ground tag term tree and $\pi\theta$ matches T. Then *language* $L(\pi)$, which is the descriptive power of a tag tree pattern π, is defined as $L(\pi) = \{\text{a tree } T \mid \pi \text{ matches } T\}$.

Data Mining Problems. A *set of semistructured data* $\mathcal{D} = \{T_1, T_2, \dots, T_m\}$ is a set of trees. The *matching count* of a given tag tree pattern π w.r.t. $\mathcal{D}$, denoted by $match_{\mathcal{D}}(\pi)$, is the number of trees $T_i \in \mathcal{D}$ $(1 \le i \le m)$ such that π matches T_i. Then the *frequency* of π w.r.t. $\mathcal{D}$ is defined by $supp_{\mathcal{D}}(\pi) = match_{\mathcal{D}}(\pi)/m$. Let σ be a real number where $0 \le \sigma \le 1$. A tag tree pattern π is *σ-frequent* w.r.t. $\mathcal{D}$ if $supp_{\mathcal{D}}(\pi) \ge \sigma$. We denote by $\Pi(L)$ the set of all tag tree patterns such that all edge labels are in L. Let Tag be a finite subset of Λ_{Tag} and KW a finite subset of Λ_{KW}. A tag tree pattern $\pi \in \Pi(Tag \cup KW \cup \{?\})$ is *maximally σ-frequent* w.r.t. $\mathcal{D}$ if (1) π is σ-frequent, and (2) if $L(\pi') \subsetneq L(\pi)$ then π' is not σ-frequent for any tag tree pattern $\pi' \in \Pi(Tag \cup KW \cup \{?\})$. In Fig. 2, for example, we give a maximally $\frac{2}{3}$-frequent tag tree pattern p in $\Pi(\{\langle\text{Fruits}\rangle, \langle\text{Name}\rangle, \langle\text{Shape}\rangle\} \cup \{\text{melon}\} \cup \{?\})$ with respect to OEM data o_1, o_2 and o_3. The tag tree pattern p matches o_1 and o_2, but p does not match o_3.

All Maximally Frequent Tag Tree Patterns Problem
Input: A set of semistructured data $\mathcal{D}$, a threshold $0 \le \sigma \le 1$, and finite sets of edge labels Tag and KW.
Problem: Generate all maximally σ-frequent tag tree patterns w.r.t. $\mathcal{D}$ in $\Pi(Tag \cup KW \cup \{?\})$.

We gave a polynomial time algorithm for finding *one* of maximally σ-frequent tag tree patterns [6]. Here we propose an algorithm for generating *all* maximally σ-frequent tag tree patterns with at most n vertices by generating all canonical level sequences of trees with n vertices [2]. Let T be a tree with n vertices. A *level sequence* $\ell(T) = [\ell_1\ \ell_2\ \cdots\ \ell_n]$ is obtained by traversing T in preorder, and recording the level (=depth+1) of each vertex as it is visited. The *canonical level sequence of* T, denoted by $\ell(T)^*$, is the lexicographically last level sequence of T. In order to prune the hypothesis space, we define a function r_p for reducing a given level sequence $\ell(T)$ with n elements to a level sequence with $n-1$ elements as follows: Let q' be the leftmost position following p such that $\ell_{q'} \le \ell_p$. If there is not such a position, let q' be $n+1$ for convenience. If the pth vertex has only one child, we define $r_p(\ell(T))$ to be $[\ell_1\ \cdots\ \ell_{p-1}\ \ell_{p+1}-1\ \cdots\ \ell_{q'-1}-1\ \ell_{q'}\ \cdots\ \ell_n]$, if the pth vertex is a leaf such that $\ell_p > \ell_{p+1}$ or $p = n$, then $r_p(\ell(T)) = [\ell_1\ \cdots\ \ell_{p-1}\ \ell_{p+1}\ \cdots\ \ell_n]$, otherwise $r_p(\ell(T))$ is undefined.

Given a set of semistructured data $\mathcal{D}$, let n be the maximum number of vertices of trees in $\mathcal{D}$. We repeat the following three steps for $k = 1, \dots, n$: Let Π_k^σ be the set of all σ-frequent tag tree patterns with at most k vertices and no edge. Let $\Pi_k^\sigma(L)$ be the set of all σ-frequent tag tree patterns with at most k vertices and edge labels in L.

Step 1. Generate the canonical level sequences of all tag tree patterns with k vertices. For each canonical level sequence of length k, we determine whether or not $(r_p(\ell(\pi)))^*$ is in Π_{k-1}^σ for each $p = 2, \dots, k$. If there is p such that $(r_p(\ell(\pi)))^*$

Experiment (frequency σ)	Exp.1 (σ=0.3)						Exp.2 (σ=0.5)					
max # of vertices in TTPs	2	3	4	5	6	7	2	3	4	5	6	7
# of max freq TTPs	1	2	4	9	15	34	1	2	2	4	3	5
run time (secs)	7	32	159	630	1948	6162	9	39	107	312	627	2721

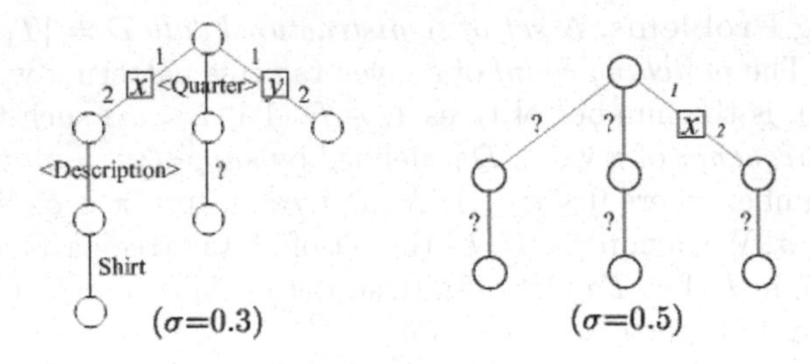

Fig. 3. Experimental results for generating all maximally frequent tag tree patterns and maximally σ-frequent tag tree patterns obtained in the experiments.

is not in Π_{k-1}, π is not σ-frequent, otherwise we compute the frequency of π and if the frequency is greater than or equals to σ then we add π to Π_k^σ.

Step 2. For each $\pi \in \Pi_k^\sigma$, we try to substitute variables of π with edges labeled with "?" as many as possible so that all σ-frequent tag tree patterns in $\Pi_k^\sigma(\{?\})$ are generated. This work can be done in a backtracking way. Then for each $\pi \in \Pi_k^\sigma(\{?\})$, we try to replace ?'s with labels in $Tag \cup KW$ as many as possible so that all σ-frequent tag tree patterns in $\Pi_k^\sigma(Tag \cup KW \cup \{?\})$ are generated. This work can be done in a backtracking way.

Step 3. Finally we check whether or not $\pi \in \Pi_k^\sigma(Tag \cup KW \cup \{?\})$ is maximally σ-frequent. Let g be a tag tree pattern $(\{u_1, u_2, u_3\}, \emptyset, \{[u_1, u_2], [u_2, u_3]\})$. Let $\theta_1^x = \{x := [g, [u_1, u_2]]\}$, $\theta_2^x = \{x := [g, [u_2, u_3]]\}$, and $\theta_3^x = \{x := [g, [u_1, u_3]]\}$, for each variable x appearing in π. If there exists a variable labeled with x such that $\pi\theta_i^x$ is σ-frequent, then π is not maximal, otherwise we can conclude that π is maximally σ-frequent.

3 Implementation and Experimental Results

We have implemented the algorithm for generating all maximally frequent tag tree patterns on a SUN workstation Ultra-10 with clock 333 MHz. We report some experiments on a sample file of semistructured data. The sample file is converted from a sample XML file about garment sales data such as *xml_sample* in Fig. 1. The sample file consists of 32 tree structured data. The maximum number of vertices in a tree in the file is 58, the maximum depth is 3 and the maximum number of children of a vertex is 6. In the experiments described in Fig. 3, we gave the algorithm "<Quarter>" and "<Description>" as tags, and "Summer" and "Shirt" as keywords. The algorithm generated all maximally σ-frequent tag tree patterns w.r.t. the sample file for a specified minimum frequency

σ. We can set the maximum number ("max # of vertices in TTPs") of vertices of tag tree patterns in the hypothesis space.

We explain the results of Fig. 3 by taking the last column of Experiment Exp. 1 as an example. The specified minimum frequency σ is 0.3. The total number ("# of max freq TTPs") of all maximally σ-frequent tag tree patterns with at most 7 vertices is 34. The total run time is 6162 secs. One of such maximally frequent patterns is shown in Fig. 3.

4 Conclusions

In this paper, we have considered knowledge discovery from semistructured Web documents such as XML files. We have proposed a tag tree pattern which is suited for representing tree structured pattern in such semistructured data. We have given an algorithm for solving All Maximally Frequent Tag Tree Patterns Problem. We have reported some experimental results by applying our algorithm for a sample file of semistructured Web documents.

Acknowledgement. This work is partly supported by Grant-in-Aid for Scientific Research (11780279) from the Ministry of Education, Culture, Sports, Science and Technology, Japan and Grant for Special Academic Research (0066,0070) from Hiroshima City University.

References

1. S. Abiteboul, P. Buneman, and D. Suciu. *Data on the Web: From Relations to Semistructured Data and XML*. Morgan Kaufmann, 2000.
2. T. Beyer and S. Hedetniemi. Constant time generation of rooted trees. *SIAM J. Comput.*, 9:706–712, 1980.
3. M. Fernandez and Suciu D. Optimizing regular path expressions using graph schemas. *Proc. Intl. Conf. on Data Engineering (ICDE-98)*, pages 14–23, 1998.
4. T. Miyahara, T. Shoudai, T. Uchida, K. Takahashi, and H. Ueda. Polynomial time matching algorithms for tree-like structured patterns in knowledge discovery. *Proc. PAKDD-2000, Springer-Verlag, LNAI 1805*, pages 5–16, 2000.
5. T. Miyahara, T. Uchida, T. Kuboyama, T. Yamamoto, K. Takahashi, and H. Ueda. KD-FGS: a knowledge discovery system from graph data using formal graph system. *Proc. PAKDD-99, Springer-Verlag, LNAI 1574*, pages 438–442, 1999.
6. T. Miyahara, T. Shoudai and T. Uchida. Discovery of maximally frequent tag tree patterns in semistructured data. *Proc. LA Winter Symposium, Kyoto*, pages 15–1 – 15–10, 2001.
7. S. Nestorov, S. Abiteboul, and R. Motwani. Extracting schema from semistructured data. *Proc. ACM SIGMOD Conf.*, pages 295–306, 1998.
8. T. Shoudai, T. Miyahara, T. Uchida, and S. Matsumoto. Inductive inference of regular term tree languages and its application to knowledge discovery. *Information Modelling and Knowledge Base XI, IOS Press*, pages 85–102, 2000.
9. T. Uchida, T. Shoudai, and S. Miyano. Parallel algorithm for refutation tree problem on formal graph systems. *IEICE Trans. Inf. Syst.*, E78-D(2):99–112, 1995.
10. K. Wang and H. Liu. Discovering structural association of semistructured data. *IEEE Trans. Knowledge and Data Engineering*, 12:353–371, 2000.

Text Categorization Using Weight Adjusted k-Nearest Neighbor Classification*

Eui-Hong (Sam) Han, George Karypis, and Vipin Kumar

Department of Computer Science & Engineering, University of Minnesota
4-192 EE/CSci Building
200 Union Street SE, Minneapolis, MN 55455
{han, karypis, kumar}@cs.umn.edu

Abstract. Text categorization presents unique challenges due to the large number of attributes present in the data set, large number of training samples, attribute dependency, and multi-modality of categories. Existing classification techniques have limited applicability in the data sets of these natures. In this paper, we present a Weight Adjusted k-Nearest Neighbor (WAKNN) classification that learns feature weights based on a greedy hill climbing technique. We also present two performance optimizations of WAKNN that improve the computational performance by a few orders of magnitude, but do not compromise on the classification quality. We experimentally evaluated WAKNN on 52 document data sets from a variety of domains and compared its performance against several classification algorithms, such as C4.5, RIPPER, Naive-Bayesian, PEBLS and VSM. Experimental results on these data sets confirm that WAKNN consistently outperforms other existing classification algorithms.

Keywords: text categorization, k-NN classification, weight adjustments

1 Introduction

We have seen a tremendous growth in the volume of online text documents available on the Internet, digital libraries, news sources, and company-wide intranet. Automatic text categorization [24,16,10], which is the task of assigning text documents to prespecified classes (topics or themes) of documents, is an important task that can help people finding information on these huge resources.

Text categorization presents unique challenges due to the large number of attributes present in the data set, large number of training samples, attribute dependency, and multi-modality of categories. Existing classification algorithms [18,2,6,10,24] address these challenges to varying degrees [7].

* This work was supported by NSF ACI-9982274, by Army Research Office contract DA/DAAG55-98-1-0441, by the DOE grant LLNL/DOE B347714, and by Army High Performance Computing Research Center contract number DAAH04-95-C-0008. Access to computing facilities was provided by AHPCRC, Minnesota Supercomputer Institute. Related papers are available via WWW at URL: http://www.cs.umn.edu/~han

D. Cheung, G.J. Williams, and Q. Li (Eds.): PAKDD 2001, LNAI 2035, pp. 53–65, 2001.

k-nearest neighbor (k-NN) classification is an instance-based learning algorithm that has shown to be very effective for a variety of problem domains in which underlying densities are not known [6]. In particular, this classification paradigm works well in the data sets with multi-modality. It has been applied to text categorization since the early days of research [23], and has been shown to produce better results when compared against other machine learning algorithms such as C4.5 [18] and RIPPER [2]. In text categorization task, the class of a new text document is determined by computing the similarity between the test document and individual instances of the training documents, and determining the class based on the class distribution of the nearest instances. A major drawback of this algorithm is that it uses all the features while computing the similarity between a test document and training documents. In many text data sets, relatively small number of features (or words) maybe useful in categorizing documents, and using all the features may affect performance. A possible approach to overcome this problem is to learn weights for different features (or words).

In this paper, we present a Weight Adjusted k-Nearest Neighbor (WAKNN) classification that learns weights for words. WAKNN finds the optimal weight vector using an optimization function based on the leave-one-out cross validation and a greedy hill climbing technique. WAKNN has better classification results than many other classifiers, but it has a high computational cost. One of the key challenges of the weight adjustment algorithm is how to reduce the high computational cost. We present two performance optimizations of WAKNN in this paper. The first optimization intelligently selects words used for weight adjustment. The second optimization reduces the computational cost involved with the evaluation of weight adjustments by clustering documents within each class. Experimental results show that these two optimizations do not compromise the quality of the classification, but improve the computational performance by a few orders of magnitude. We experimentally evaluated WAKNN-C, which is the performance-improved version of WAKNN, on 52 document data sets from a variety of domains and compared its performance against several classification algorithms, such as C4.5 [18], RIPPER [2], Naive-Bayesian [16], PEBLS [3], and Variable-Kernel Simulation Metric (VSM) [15]. Experimental results on these data sets show that WAKNN-C consistently outperforms other existing classification algorithms.

Section 2 describes challenges of text categorization and provides a brief overview of existing algorithms for text categorization. Section 3 presents the weight adjustment algorithm WAKNN and its performance improvements. Section 4 shows experimental evaluation of the WAKNN. Finally, Section 5 provides conclusions and directions for future research.

2 Previous Work

Text categorization is essentially a classification problem [22,18,2,6]. The words occurring in the document sets become variables or features for the classification problem. The class profile or model is described based on these words. There are several classification schemes that can potentially be used for text categorization.

A classification decision tree, such as C4.5 [18], is a widely used classification paradigm that has been shown to produce good classification results, primarily on low dimensional data sets. Decision tree based schemes like C4.5 or rule induction algorithms such as C4.5rules [18] and RIPPER [2] are not very effective in text data sets due to overfitting [7]. The overfitting occurs, because the number of samples is relatively small with respect to the distinguishing words, which leads to very large trees with limited generalization ability.

The Naive-Bayesian (NB) classification algorithm has been widely used for document classification, and has been shown to produce very good performance [13,16]. Even though Naive-Bayes classification techniques, such as Rainbow [16], are commonly used in text categorization [13,16], the independence assumption limits their applicability in the document classes and unimodal density assumptions might not work well in document data sets with multi-modal densities [6].

There have been several approaches to learn feature weights for k-NN. PEBLS [3] and k-NN with mutual information [22,16] are approaches that compute weights of features in prior to the classification learning. These approaches compute the importance of a feature independent of all the other features. Variable-Kernel Simulation Metric (VSM) [15] learns the feature weights using non-linear conjugate gradient optimization. VSM has a very structured approach to find weights, but requires optimization functions to be differentiable and does not have the convergence guarantees like the linear conjugate gradient optimization. RELIEF-F [12] is another weight adjustment technique that learns weights based on the nearest neighbors in each class.

Support Vector Machines (SVM) is a new learning algorithm that was introduced to solve two-class pattern recognition problem using the Structural Risk Minimization principle [10]. An efficient implementation of SVM and its application in text categorization of Reuters-21578 corpus is reported in [10].

3 Weight Adjusted k-Nearest Neighbor Classification (WAKNN)

The Weight Adjusted k-Nearest Neighbor Classification (WAKNN) tries to learn the best weight vector for classification. Given the weight vector W, the similarity between two documents X and Y is measured using the weighted cosine measure [19]:

$$cos(X,Y,W) = \frac{\sum_{t\in T}(X_t \times W_t) \times (Y_t \times W_t)}{\sqrt{\sum_{t\in T}(X_t \times W_t)^2} \times \sqrt{\sum_{t\in T}(Y_t \times W_t)^2}},$$

where T is the set of terms (or words), X_t and Y_t are normalized text frequency (TF) of word t for X and Y, respectively, and W_t is the weight of word t.

Figure 1 illustrates the objective of weight adjustment in the classification task. In the original data without weights, the test sample of class A is equally close to the training samples in class A and class B. After weight adjustment, it is much closer to the training samples in class A and thus is correctly classified.

We can visually see the effect of weight adjustment on real data sets using the class-preserving projection [5]. This projection tries to find the optimal 2-dimensional display of high-dimensional data sets with class labels. This projection works the best when there are 3 classes in the data sets, because 3 class means automatically determine a plane for 2-dimensional display. Hence we selected 3 classes from the training data set of *west1*, which is described in Section 4, and learned weights using WAKNN. Figure 2 shows the original data set on the left and the weight-adjusted data set on the right. Figure 2 shows that WAKNN was able to find weights that can separate data points in different classes. For instance, data points in class "010" (depicted as "x" in the figure) are almost completed separated from the other two classes. Classes "008" (depicted as "+") and "054" (depicted as "*") are also well separated except several points mixed up around the coordinate (-0.1, -0.2).

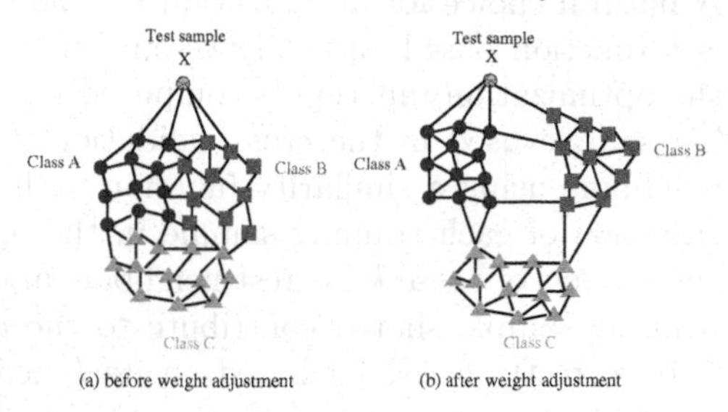

Fig. 1. Weight Adjustment in k-NN Classification.

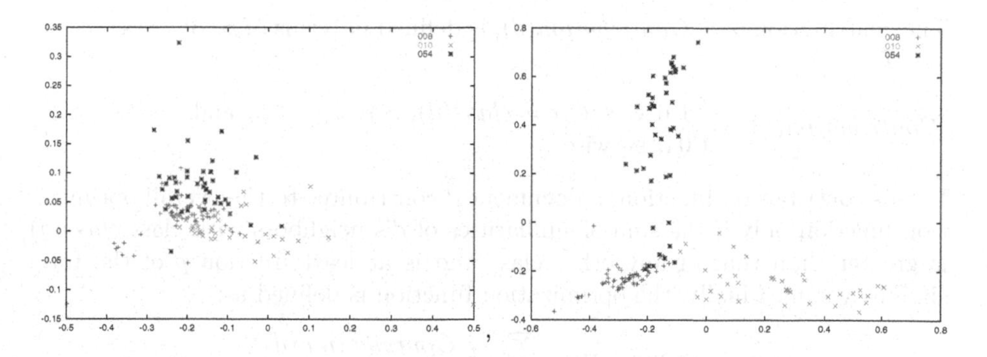

Fig. 2. 2-dimensional display of west1 data set before and after weight adjustments.

The weight adjustment process is a search process in the word weight vector space. The key design decisions in this process include which optimization function to use, which search method to use, and how to make the algorithm efficient in terms of run time and memory requirements. In the rest of this section, we will discuss these design decisions and present the algorithm.

3.1 Optimization Function and Search Strategy

A natural choice for optimization function in the k-NN classification paradigm is a function based on the leave-one-out cross validation. The overall value for the optimization function is computed based on the contribution of each training sample used in the cross validation. Each training sample finds k-nearest neighbors using a similarity function with the current weight vector. The contribution of each training sample in the optimization function depends on the class labels of these k-nearest neighbors and similarities to these neighbors. The training sample should contribute to the overall optimization function only if it is correctly classified based on its k-nearest neighbors and its classification is clear cut. One way of achieving this goal is to use the following contribution function. Let D be a set of training documents and W be a weight vector. Let C be a set of classes of D and $class(x)$ be a function that returns the class of document x. For a training sample $d \in D$, let $N_d = \{n_1, n_2, \ldots, n_k\}$ be the set of k-nearest neighbors of d. Given N_d, the similarity sum of d's neighbors that belong to class c is:

$$S_c = \sum_{n_i \in N_d, class(n_i)=c} cos(d, n_i, W) \tag{1}$$

Then the total similarity sum of d is:

$$T = \sum_{c \in C} S_c$$

The contribution of d, $Contribution(d)$, is defined in terms of S_c of classes $c \in C$ and T:

$$Contribution(d) = \begin{cases} 1 \text{ if } \forall c \in C, c \neq class(d),\ S_{class(d)} > S_c \text{ and } \frac{S_{class(d)}}{T} \geq p \\ 0 \text{ otherwise} \end{cases}$$

In this contribution function, a document d contributes to the overall optimization function only if the sum of similarities of d's neighbors with class $class(d)$ is greater than that of any other class and is at least fraction p of the total similarity sum. Finally, the optimization function is defined as:

$$OPT(D, W) = \frac{\sum_{d \in D} Contribution(d)}{|D|}$$

Different $Contribution(d)$ functions describe different optimization functions. We have experimented with several contribution functions [7] and chosen the above contribution function for WAKNN. The optimization function based on this contribution function allows the weight adjustment to be flexible in terms of local minima and overfitting. If the optimization function uses a low p (close to 0.0), then the weight adjustment will tend to overfit the training documents but will move out of possible local minima more easily. On the other hand, if the optimization function uses high p (close to 1.0), then the weight adjustment will tend to avoid overfitting but have tendency to be trapped in a local minima.

Several space search techniques are available including hill-climbing (greedy) search [11], best-first search [11], beam search [11], or bidirectional search [11]. WAKNN searches for the best weight vector using hill-climbing (greedy) method. We chose a hill-climbing (greedy) search strategy, because this approach is computationally fast and memory efficient. In the greedy search of WAKNN, the weight for each word is perturbed slightly and its effect on the optimization function is evaluated. The perturbation of the weight is proposed by multiplying the current weight by different factors. The weight perturbation with the most improvement in the optimization function is chosen as the winner and its weight is updated. This process is repeated until no further progress is obtained.

3.2 Performance Improvements of WAKNN

The major drawback of WAKNN is its high computational cost. After each possible weight adjustment of a word in the greedy search, the merit of the possible weight adjustment is evaluated using the optimization function $OPT(D, W)$. The cost of this evaluation is $O(n^2)$ where n is the number of training documents, because finding the k nearest neighbors of each training document requires comparison to the rest of the training documents. Hence the computational cost of each iteration of the WAKNN algorithm is $O(mn^2)$ where m is the number of words in the document data sets. We propose two computational performance improvements of WAKNN. The first method (WAKNN-F) improves the performance by reducing the number of words used for the weight adjustment, i.e, reducing m in $O(mn^2)$. The second method (WAKNN-C) improves the performance by reducing the cost of evaluating $OPT(D, W)$, i.e., reducing n in $O(mn^2)$.

WAKNN-F. In order to select words for weight adjustment, we need a ranking method and a selection method. We use mutual information [22,16] to rank words. The mutual information can be computed in $O(nl)$ where n is the number of training documents and l is the average length of the document. In the single scan of training documents, we can compute class distributions of each word. Based on this information, the mutual information of each word can be computed. The preliminary experiments show that only 10 to 70 weights are changed under WAKNN and most of these words ranked in the top 1000 according to the mutual information of words. These results show that the ranking method based on mutual information could be effective in finding words that are important in weight adjustment. We also consider pairwise word dependency in the ranking by computing mutual information of the pair of words. The mutual information of a word is then determined as the maximum of the mutual information of the single word and any pair of words with this word in it.

There are three desired goals of the selection method. We want to include as many top ranked words as possible. In the same time, we want to make sure that the selected words cover as many documents as possible. It is possible that a small set of documents contain many words that have high mutual information.

If we select words based purely on the mutual information, we will not be able to improve the classification accuracies of documents not in this set. Finally, we want to select as small number of words as necessary to control the computational cost.

We propose the following scheme to achieve these three goals. The overall steps of the scheme is shown in Figure 3. We rank words globally according to the mutual information. For each document, we sort its words according to the global ranking. We have chosen an incremental approach to determine the right level of coverage of documents. We start collecting top 10% (MinCov = 0.1) of the words according to the global ranking in each document. The union of all these words constitute the selected words. By doing this, we guarantee that words selected cover at least 10% of the words in each document. The k-NN classification accuracy of training documents using only these selected words is calculated. This accuracy is compared to the baseline classification accuracy that is computed by using all the words. If this ratio is less than a user specified minimum ratio (MinRatio), the selection process is repeated with the minimum coverage (MinCov) incremented by 0.05 (5%). The selection mechanism stops as soon as the classification power of the selected words is at least MinRatio of the classification power of all the words. The MinRatio controls the total number of words selected.

```
SelectWord(D, MinRatio)
  1. Rank words (W) in training document D according to mutual information.
  2. For each d ∈ D, sort words in d according to the global word ranking.
  3. BaseAcc = Accuracy(D , W)
  4. AccRatio = 0.0; MinCov = 0.1; Selected = {};
  5. While (AccRatio < MinRatio)
      5.1 for each d ∈ D
              Selected = Selected ∪ {top (MinCov * 100.0) % of words in d}
      5.2 Acc = Accuracy(D , Selected)
      5.3 AccRatio = Acc / BaseAcc
      5.4 MinCov = MinCov + 0.05
  6. return Selected
```

Fig. 3. Major steps of word selection scheme.

The proposed scheme guarantees that distinguishing or important words are selected, all the documents have minimum coverage, and selected words have classification capacity of at least MinRatio of the full set of words. However, when the classification using all the words is very poor, this scheme will select very few words and stop. We fixed this problem by selecting at least 500 words. We call this optimized version of WAKNN as WAKNN-F.

WAKNN-C. One obvious choice for reducing the computational cost of $O(n^2)$ part of the overall computational cost is by sampling training data. However, sampling in the presence of large variations of class size is troublesome. Furthermore, this might mitigate the advantage of k-NN classification. k-NN performs well when there is multi-modal class distribution in which summarization of the whole class does not accurately depict the class distribution. In this multi-modal distribution, k-NN works well because each data point can find a small set of neighbors that are in the same class and are very close. Random sampling could thin out these neighbors such that each data point can be pulled toward other classes in k-NN classification.

We have chosen to cluster training documents. We cluster similar documents within each class and represent each cluster by a centroid. Then in the optimization function, we find k-nearest centroids instead of k-nearest neighbors. If we have c clusters, then we will have computational complexity of $O(cn)$ instead of $O(n^2)$. Assuming that we can find $c \ll n$, the computational cost would decrease dramatically.

In the top down clustering of documents within each class, we start with each class being a single cluster. We select the next cluster to divide based on the recall of each cluster. The recall of each cluster is defined as follows. For each cluster, we first find training samples (regardless of their class labels) that have the centroid of this cluster as the closest centroid. The recall for the cluster is defined as the percentage of these training samples that have the same class as the cluster.

We pick the cluster with the worst recall and divide this cluster into two clusters, and then perform the refinement similar to k-means clustering algorithm [9] with the weighted cosine similarity measure for assigning documents to clusters. Note that refinement of clusters is performed among clusters of the same class. This approach gives the natural stopping point: we stop clustering when all the clusters have perfect recalls. We augment this stopping criterion by forcing each class to have at least k clusters.

As the weight adjustment progresses, the weight changes can cause the cluster to change. After each weight change, we refine the clusters with new weights and repeat the recall-based clustering process to enforce that each cluster has a perfect recall. In general, the number of clusters increases as the weight adjustment progresses. It is entirely possible that better weights can allow clusters to be merged and yet have a perfect recall. In the current implementation this possibility has not been explored. We call this optimization of WAKNN-F as WAKNN-C.

The major steps of WAKNN-C are shown in Figure 4. In steps 1, 2 and 3, training document matrix is constructed and the matrix is normalized to mitigate the effect of different document lengths. In step 8.3, the weight for each word is perturbed slightly and its effect on the optimization function is evaluated. The perturbation of the weight is proposed by multiplying the current weight by different factors. The weight perturbation with the most improvement in the optimization function is chosen as the winner and its weight is updated. This process is repeated until no further progress is obtained.

The differences between WAKNN-F and WAKNN are steps 5 and 8.3. Instead of checking all the words for weight changes, WAKNN-F only checks the words selected from the SelectWord function. The difference between WAKNN-C and

```
1. Construct a training matrix D.
   - each row corresponds to a training document
   - each column corresponds to a word
   - value in the matrix D(i, j) corresponds frequency of word j in document i.
2. Let F be a multiplication factors and W be a weight vector.
3. Normalize word frequencies in each document such that they add up to 1.0.
4. For each j, W_j = 1.0.
5. Selected = SelectWord(D)
6. C = FindClusters(D, W)
7. bestopt = OPT(D, W, C); oldopt = 0;
8. While (bestopt > oldopt)
     8.1 oldopt = bestopt
     8.2 bestword = -1; newval = -1
     8.3 For each word j ∈ Selected
           For each f ∈ F
             W' = W
             W'_j = W_j × f
             if (OPT(D, W', C) > bestopt)
                   bestword = j
                   bestval = W'_j
                   bestopt = OPT(D, W', C)
     8.4 W_bestword = bestval
     8.5 RefineClusters(D, C, W)
```

Fig. 4. Major steps of WAKNN-C.

WAKNN-F are step 6 in the algorithm for computing initial clusters and step 8.5 for refining clusters after a weight change. Another difference is the definition of neighbors in the computation of similarity sum in Eq 1. Now the N_d' is the set of k-nearest clusters of d. k-nearest clusters of d is determined by computing similarity to the centroids of clusters. S_c is computed based on the similarities between d and the centroids of k-nearest clusters.

4 Experimental Results

In all of the data sets, we have used stop words to remove common words and stemmed words using Porter's suffix-stripping algorithm [17]. For the data representation, we have followed the vector space model commonly used in Information Retrieval systems [19].

We have used 52 total data sets for experiments. More detailed information on these data sets is available in [7].[1] Data sets *west1, west2, ..., west7* are from the statutory collections of the legal document publishing division of West Group described in [4]. Data sets *tr11, tr12, ..., tr45, fbis, trec6*, are derived from TREC-5 [21], TREC-6 [21], and TREC-7 [21] collections. The classes of the various TREC data sets were generated from the relevance judgment provided in these collections. Data sets *oh0, oh1, ..., oh19* are from OHSUMED collection [8] subset of MEDLINE database. We took different subsets of categories to construct these data sets. Data sets *re0* and *re1* are from Reuters-21578 text

[1] These data sets are available from *http://www.cs.umn.edu/~han/data/tmdata.tar.gz*.

categorization test collection Distribution 1.0 [14]. We removed dominant classes such as “earn” and “acq” that have been shown to be relatively easy to classify. We then divided the remaining classes into 2 sets. Data set *wap* is from the WebACE project (WAP) [1]. Each document corresponds to a web page listed in the subject hierarchy of Yahoo!.

We compare WAKNN against C4.5, RIPPER, Rainbow, k-NN, PEBLS, k-NN with mutual information (MI), RELIEF-F, and VSM. We have not included SVM in the comparison, because the SVM code [10] available was designed for two class problem. We implemented k-NN, MI, and RELIEF-F; we used publicly available implementations of C4.5, RIPPER, Rainbow, PEBLS, and VSM. In all the k-NN based algorithms (WAKNN-C, kNN, MI, RELIEF-F, and VSM) except PEBLS, the number of neighbors $k = 10$ is used. For PEBLS, we used the nearest neighbor (i.e. $k = 1$), because the results using more than one neighbor were significantly worse. For Rainbow, we used the multinomial event model as the document representation model, because results from other works [16,24] indicate that this event model is better than the multi-variate Bernoulli event model and our preliminary experiments also confirmed this claim. For other parameters in these algorithms, we used default parameter values suggested by the developers of the algorithms.

In the first experiment, we compared the classification accuracy of WAKNN to other classifiers on a subset of documents described above. In these experiments, we used $p = 0.5$ in the optimization function and $\{0.2, 0.5, 0.8, 1.5, 2.0, 4.0, 9.0, 15.0, 30.0, 50.0\}$ as multiplication factors for weight perturbation. Out of 19 data sets, WAKNN has the best classification accuracy on 13 data sets. None of the other classifiers has the best classification accuracy on more than one data set. Even for 6 data sets that WAKNN did worse than some other classification algorithms, the classification accuracy of WAKNN was within less than 2% of the best results.

Even though WAKNN provides better classification accuracies over other classifiers, the computational cost of WAKNN is very high. The runtime of WAKNN varied from a few hours on 300 to 500 training sample size (west1 and tr11) to a few days on 1000 training sample size (oh8, oh12, and oh18). WAKNN-F and WAKNN-C significantly reduced the runtime of WAKNN. For instance, WAKNN-F reduced the runtime of WAKNN by a factor of 2 to 6. WAKNN-C further reduced runtime dramatically in large data sets (oh8, oh12, and oh18) compared to WAKNN-F. The runtime of WAKNN-C on these data sets ranged from 19 to 25 minutes, whereas the runtime of WAKNN-F on the same data set ranged from 5 to 8 hours. While WAKNN-F and WAKNN-C reduced the runtime of WAKNN significantly, WAKNN-F and WAKNN-C have statistically equal quality classification according to the sign test [20,7] and two sample significance test [20,7].

We now present the classification accuracy of WAKNN-C compared to other classifiers on all 52 data sets in Table 1. We used MinRatio=0.75 for selecting words in WAKNN-C. The number of words selected ranged from one half to one tenth of the original number of words in the data sets. The minimum coverage

Table 1. Classification accuracies of different classifiers. Note that the highest accuracy for each data set is highlighted with bold font.

	C4.5	RIPPER	Rainbow	kNN	PEBS	MI	RELIEF-F	VSM	WAKNN-C
west1	82.40	84.87	84.40	76.73	78.50	86.80	76.87	85.27	**87.40**
west2	74.20	72.22	72.11	68.33	67.80	75.89	68.44	76.11	**81.78**
west3	75.40	78.48	79.92	67.83	67.00	79.71	68.03	79.92	**85.86**
west4	80.50	73.88	79.61	67.26	70.70	74.96	67.80	75.49	**83.54**
west5	84.20	80.35	88.73	77.78	83.40	90.02	78.10	87.12	**94.04**
west6	76.60	78.14	85.66	72.27	74.30	82.92	72.40	85.11	**86.89**
west7	79.90	76.05	74.35	67.46	68.40	78.08	67.68	78.64	**81.69**
tr11	79.70	79.23	**85.99**	85.02	81.20	85.51	85.02	83.58	85.99
tr12	86.00	82.17	80.25	82.17	79.60	86.62	82.17	**87.90**	87.90
tr13	88.90	88.36	87.69	**93.10**	87.70	92.02	93.10	92.56	92.69
tr14	88.90	90.40	91.64	92.88	79.30	93.19	92.88	90.71	**95.05**
tr15	94.90	93.57	97.75	94.86	94.20	97.43	94.86	**98.07**	97.43
tr21	80.50	82.25	60.95	81.66	75.70	80.47	81.66	80.47	**90.53**
tr22	88.60	87.40	91.06	93.09	84.60	**93.50**	93.09	91.06	92.28
tr23	**93.20**	93.20	73.79	82.52	79.60	89.32	82.52	76.70	85.44
tr24	92.50	85.63	73.12	88.75	90.00	86.25	88.75	86.25	**96.25**
tr25	83.40	**91.72**	85.80	80.47	72.80	81.07	80.47	84.02	88.76
tr31	90.50	87.93	92.46	91.59	86.90	**95.69**	91.59	92.46	94.61
tr32	86.80	83.72	78.68	75.97	71.70	75.58	75.97	80.62	**90.31**
tr33	89.50	90.83	67.69	86.46	86.50	90.39	86.46	89.52	**95.63**
tr34	**94.00**	87.99	77.74	88.34	85.20	85.51	88.34	86.57	93.99
tr35	87.90	90.00	87.50	88.93	78.90	91.79	88.93	90.00	**96.07**
tr41	90.70	**95.67**	94.08	89.07	86.30	89.07	89.07	92.71	94.99
tr42	89.70	92.31	87.91	85.35	79.50	87.55	85.35	91.58	**94.87**
tr43	**95.30**	89.67	88.73	83.57	83.60	84.04	83.57	88.73	92.96
tr44	86.60	79.77	**89.31**	83.97	80.50	82.06	83.97	80.92	85.88
tr45	87.90	83.53	84.68	86.99	70.80	85.55	87.28	88.73	**96.53**
fbis	57.10	74.84	76.38	78.49	69.80	76.54	78.49	74.67	**83.52**
trec6	67.50	82.79	92.16	91.99	84.30	88.42	91.99	87.56	**94.21**
oh0	84.70	84.06	90.21	87.08	55.60	85.73	86.98	77.19	**90.73**
oh1	81.20	79.82	84.37	86.03	37.50	81.93	86.03	82.48	**87.36**
oh2	88.90	85.31	91.57	90.57	76.10	87.49	90.48	86.40	**92.38**
oh3	85.60	81.58	88.69	86.03	55.10	87.40	86.03	84.83	**91.35**
oh4	82.00	86.33	92.07	90.52	58.20	86.87	90.43	82.04	**93.53**
oh5	76.70	72.73	83.83	84.06	53.20	81.94	84.06	80.87	**88.78**
oh6	83.90	84.43	88.58	**90.71**	53.30	82.64	90.59	81.86	89.92
oh7	77.90	76.33	87.46	85.80	34.40	81.30	85.56	73.49	**87.81**
oh8	76.90	71.60	85.44	82.40	55.00	80.95	82.52	79.85	**87.99**
oh9	84.50	76.90	90.48	88.81	56.10	83.10	88.81	83.33	**91.19**
oh10	74.00	70.36	79.69	77.26	39.90	76.09	77.26	69.00	**84.55**
oh11	83.30	84.32	82.81	80.93	65.80	85.74	80.93	82.91	**91.31**
oh12	82.80	81.24	86.63	85.17	73.90	80.45	85.28	81.46	**88.88**
oh13	80.00	80.55	91.97	89.64	63.50	86.36	89.75	86.05	**92.92**
oh14	82.10	79.79	88.51	86.88	43.50	86.88	86.97	83.53	**90.13**
oh15	73.00	76.67	82.81	80.25	49.80	75.11	80.13	77.12	**84.26**
oh16	87.70	86.63	92.15	90.60	74.80	90.34	90.68	87.49	**93.70**
oh17	85.20	80.16	90.72	89.04	56.30	86.77	89.04	85.59	**92.20**
oh18	78.60	77.63	82.44	78.30	56.60	76.17	78.41	77.63	**86.02**
oh19	80.60	77.49	86.81	85.70	42.90	80.82	85.70	81.49	**87.25**
re0	66.90	67.27	76.45	81.04	67.90	79.44	**81.44**	81.04	78.44
re1	78.60	77.27	77.73	74.35	67.00	76.19	74.35	80.49	**84.49**
wap	71.30	72.95	83.97	81.92	41.80	79.23	81.92	68.08	**84.23**

of each document with selected words ranged from 10% to 85%. The number of clusters found ranged from one half to one tenth of the original number of training documents. Detailed parameter studies of WAKNN-C are available in [7]. Out of 52 data sets, WAKNN-C has the best classification accuracy on 38 data sets. None of the other classifiers has the best classification accuracy on more than three data sets. These results show that WAKNN-C consistently outperforms other classifiers in most of the data sets. In order to verify whether these differences are statistically significant, we performed the sign test [20,7] and the two sample significance test [20,7]. The two sample significance test shows that WAKNN-C is not statistically worse than any of the classifiers on any data set. The results also show that WAKNN-C is statistically better (P-value ≤ 0.05) than C4.5, RIPPER, Rainbow, k-NN, PEBLS, MI, RELIEF-F, and VSM in 43, 39, 22, 34, 49, 41, 34 and 35 data sets respectively. According to the sign test, WAKNN-C is statistically better (with P-value ≤ 0.05) than all the other classifiers.[2]

5 Conclusions and Directions for Future Research

In this paper, we presented a Weight Adjusted k-Nearest Neighbor (WAKNN) classification that retains the power of the k-NN while further enhancing its ability by learning feature weights. Experimental results show that WAKNN-C outperforms existing state of the art classification algorithms quite consistently on the document data sets from a variety of domains. Even though the focus of this paper has been the application of WAKNN in the text categorization task, WAKNN is applicable in any domain with large number of attributes.

Several issues that we identified in the course of this work deserve further attention. In the greedy search technique used in WAKNN, only one weight change is selected at a time. It will be rewarding to explore methods for changing multiple words at a time in the greedy search process. One possibility is to use association rules to identify set of words that are strongly related, and then use this knowledge in determining which sets of words to select for simultaneous weight adjustments. Another research focus is the incremental weight adjustment as new training samples are available.

References

1. D. Boley, M. Gini, R. Gross, E.H. Han, K. Hastings, G. Karypis, V. Kumar, B. Mobasher, and J. Moore. Document categorization and query generation on the world wide web using WebACE. *AI Review*, 13(5-6), 1999.

---

[2] We also performed a set of experiments in which we kept only the top few hundred discriminating words according to mutual information or the same words that WAKNN-C chose for weight adjustment. According to the sign test and two sample significance test, WAKNN-C consistently outperformed other classifiers on these data sets as well [7].

2. W.W. Cohen. Fast effective rule induction. In *Proc. of the Twelfth International Conference on Machine Learning*, 1995.
3. S. Cost and S. Salzberg. A weighted nearest neighbor algorithm for learning with symbolic features. *Machine Learning*, 10(1):57–78, 1993.
4. T. Curran and P. Thompson. Automatic categorization of statute documents. In *Proc. of the 8th ASIS SIG/CR Classification Research Workshop*, Tucson, Arizona, 1997.
5. I.S. Dhillon and D.M. Modha. Visualizing class structure of multi-dimensional data. In *Proc. of the 30th Symposium of the Interface: Computing Science and Statistics*, pages 488–493, 1998.
6. R.O. Duda and P.E. Hart. *Pattern Classification and Scene Analysis.* John Wiley & Sons, 1973.
7. E.H. Han. *Text Categorization Using Weight Adjusted k-Nearest Neighbor Classification.* PhD thesis, University of Minnesota, October 1999.
8. W. Hersh, C. Buckley, T.J. Leone, and D. Hickam. OHSUMED: An interactive retrieval evaluation and new large test collection for research. In *SIGIR-94*, pages 192–201, 1994.
9. A.K. Jain and R. C. Dubes. *Algorithms for Clustering Data.* Prentice Hall, 1988.
10. T. Joachims. Text categorization with support vector machines: Learning with many relevant features. In *Proc. of the European Conference on Machine Learning*, 1998.
11. L. N. Kanal and Vipin Kumar, editors. *Search in Artificial Intelligence.* Springer-Verlag, New York, NY, 1988.
12. I. Kononenko. Estimating attributes: Analysis and extensions of relief. In *Proc. of the 1994 European Conference on Machine Learning*, 1994.
13. D. Lewis and M. Ringuette. Comparison of two learning algorithms for text categorization. In *Proc. of the Third Annual Symposium on Document Analysis and Information Retrieval*, 1994.
14. D. D. Lewis. Reuters-21578 text categorization test collection distribution 1.0. *http://www.research.att.com/ lewis*, 1999.
15. D.G. Lowe. Similarity metric learning for a variable-kernel classifier. *Neural Computation*, pages 72–85, January 1995.
16. A. McCallum and K. Nigam. A comparison of event models for naive bayes text classification. In *AAAI-98 Workshop on Learning for Text Categorization*, 1998.
17. M. F. Porter. An algorithm for suffix stripping. *Program*, 14(3):130–137, 1980.
18. J. Ross Quinlan. *C4.5: Programs for Machine Learning.* Morgan Kaufmann, San Mateo, CA, 1993.
19. G. Salton. *Automatic Text Processing: The Transformation, Analysis, and Retrieval of Information by Computer.* Addison-Wesley, 1989.
20. G.W. Snedecor and W.G. Cochran. *Statistical Methods.* Iowa State University Press, 1989.
21. TREC. Text REtrieval conference.
22. D. Wettschereck, D.W. Aha, and T. Mohri. A review and empirical evaluation of feature-weighting methods for a class of lazy learning algorithms. *AI Review*, 11, 1997.
23. Y. Yang. Expert network: Effective and efficient learning from human decisions in text categorization and retrieval. In *SIGIR-94*, 1994.
24. Y. Yang and X. Liu. A re-examination of text categorization methods. In *SIGIR-99*, 1999.

Predictive Self-Organizing Networks for Text Categorization

Ah-Hwee Tan

Kent Ridge Digital Labs, 21 Heng Mui Keng Terrace, Singapore 119613
ahhwee@krdl.org.sg

Abstract. This paper introduces a class of predictive self-organizing neural networks known as Adaptive Resonance Associative Map (ARAM) for classification of free-text documents. Whereas most statistical approaches to text categorization derive classification knowledge based on training examples alone, ARAM performs supervised learning and integrates user-defined classification knowledge in the form of IF-THEN rules. Through our experiments on the Reuters-21578 news database, we showed that ARAM performed reasonably well in mining categorization knowledge from sparse and high dimensional document feature space. In addition, ARAM predictive accuracy and learning efficiency can be improved by incorporating a set of rules derived from the Reuters category description. The impact of rule insertion is most significant for categories with a small number of relevant documents.

1 Introduction

Text categorization refers to the task of automatically assigning documents into one or more predefined classes or categories. It can be considered as the simplest form of text mining in the sense that it abstracts the key content of a free-text documents into a single class label. In recent years, there has been an increasing number of statistical and machine learning techniques that automatically generate text categorization knowledge based on training examples. Such techniques include decision trees [2], K-nearest-neighbor system (KNN) [7,19], rule induction [8], gradient descent neural networks [9,16], regression models [18], Linear Least Square Fit (LLSF) [17], and support vector machines (SVM) [6, 7]. All these statistical methods adopt a supervised learning paradigm. During the learning phase, a classifier derives categorization knowledge from a set of prelabeled or tagged documents. During the testing phase, the classifier makes prediction or classification on a separate set of unseen test cases. Supervised learning paradigm assumes the availability of a large pre-labeled or tagged training corpus. In specific domains, such corpora may not be readily available. In a personalized information filtering application, for example, few users would have the patience to provide feedback to a large number of documents for training the classifier. On the other hand, most users are willing to specify what they want explicitly. In such cases, it is desirable to have the flexibility of building a text

D. Cheung, G.J. Williams, and Q. Li (Eds.): PAKDD 2001, LNAI 2035, pp. 66–77, 2001.

classifier from examples as well as obtaining categorization knowledge directly from the users.

In machine learning literatures, hybrid models have been studied to integrate multiple knowledge sources for pattern classification. For example, Knowledge Based Artificial Neural Network (KBANN) refines imperfect domain knowledge using backpropagation neural networks [15]; Predictive self-organizing neural networks [4,12] allow rule insertion at any point of the incremental learning process. Benchmark studies on several databases have shown that initializing such hybrid learning systems with prior knowledge not only improves predictive accuracy, but also produces better learning efficiency, in terms of the learning time as well as the final size of the classifiers [13]. In addition, promising results have been obtained by applying KBANN to build intelligent agents for web page classification [11].

This paper reports our evaluation of a class of predictive self-organizing neural networks known as Adaptive Resonance Associative Map (ARAM) [12] for text classification based on a popular public domain document database, namely Reuters-21578. The objectives of our experiments are twofold. First, we study ARAM's capability in mining categorization rules from sparse and high dimensional document feature vectors. Second, we investigate if ARAM's predictive accuracy and learning efficiency can be enhanced by incorporating a set of rules derived from the Reuters category description.

The rest of this article is organized as follows. Section 2 describes our choice of feature selection and extraction methods. Section 3 presents the ARAM learning, classification, and rule insertion algorithms. Section 4 reports the experimental results. The final section summarizes and concludes.

2 Features Selection/Extraction

As in statistical text categorization systems, we adopt a bag-of-words approach to representing documents in the sense that each document is represented by a set of keyword features. The keyword features can be obtained from two sources. Through rule insertion, a keyword feature can be specified explicitly by a user as an antecedent in a rule. Features can also be selected from the words in the training documents based on certain feature ranking metric. Some popularly used measures for feature ranking include keyword frequency, χ^2 statistics, and information gain. In our experiments, we only use χ^2 statistics which has been reported to be one of the most effective measures [20].

During rule insertion and keyword selection, we use an in-house morphological analyzer to identify the part-of-speech and the root form of each word. To reduce complexity, only the root forms of the noun and verb terms are extracted for further processing.

During keyword extraction, the document is first segmented and converted into a keyword feature vector

$$\mathbf{v} = (v_1, v_2, \ldots, v_M). \tag{1}$$

where M is the number of keyword features selected. We experiment with three different document representation schemes described below.
tf* encoding:** This is the simplest and the first method that we used in earlier experiments [14]. The feature vector $\mathbf{v}$ simply equals the term frequency vector ***tf such that the value of feature j

$$v_j = tf_j \tag{2}$$

where tf_j is the in-document frequency of the keyword w_j.
***tf*idf* encoding:** A term weighting method based on *inverse document frequency* [10] is combined with the term frequency to produce the feature vector $\mathbf{v}$ such that

$$v_j = tf_j \log_2 \frac{N}{df_j} \tag{3}$$

where N is the total number of documents in the collection and df_j is the number of documents containing the keyword w_j.
***log-tf*idf* encoding:** This is a variant of the *tf*idf* scheme. The feature vector $\mathbf{v}$ is computed by

$$v_j = (1 + \log_2 tf_j) \log_2 \frac{N}{df_j}. \tag{4}$$

After encoding using one of the three feature representation schemes, the feature vector $\mathbf{v}$ is normalized to produce the final feature vector

$$\mathbf{a} = \mathbf{v}/v_m \quad \text{where} \quad v_m >= v_i \ \ \forall i \neq m. \tag{5}$$

before presentation to the neural network classifier.

3 ARAM Algorithms

ARAM belongs to a family of predictive self-organizing neural networks known as predictive Adaptive Resonance Theory (predictive ART) that performs incremental supervised learning of recognition categories (pattern classes) and multidimensional maps of patterns. An ARAM system can be visualized as two overlapping Adaptive Resonance Theory (ART) [3] modules consisting of two input fields F_1^a and F_1^b with an F_2 category field. For classification problems, the F_1^a field serves as the input field containing the input activity vector and the F_1^b field servers as the output field containing the output class vector. The F_2 field contains the activities of the recognition categories that are used to encode the patterns.

In an ARAM network (Figure 1), the unit for recruiting an F_2 category node is a complete pattern pair. Given a pair of input patterns, the category field F_2 selects a winner that receives the largest overall input from the feature fields F_1^a and F_1^b. The winning node selected in F_2 then triggers a top-down priming on F_1^a and F_1^b, monitored by separate reset mechanisms. Code stabilization is ensured by restricting encoding to states where resonances are reached in both modules. By synchronizing the unsupervised categorization of two pattern sets, ARAM learns supervised mapping between the pattern sets. Due to the code stabilization mechanism, fast learning in a real-time environment is feasible.

In addition, the knowledge that ARAM discovers during learning, is compatible with IF-THEN rule-based representation. Specifically, each node in the F_2 field represents a recognition category associating the F_1^a input patterns with the F_1^b input vectors. Learned weight vectors, one for each F_2 node, constitute a set of rules that link antecedents to consequents. At any point during the incremental learning process, the system architecture can be translated into a compact set of rules. Similarly, domain knowledge in the form of IF-THEN rules can be inserted into ARAM architecture.

3.1 Learning

The ART modules used in ARAM can be ART 1 [3], which categorizes binary patterns, or analog ART modules such as ART 2, ART 2-A, and fuzzy ART [5], which categorize both binary and analog patterns. The fuzzy ARAM model, that is composed of two overlapping fuzzy ART modules (Figure 1), is described below.

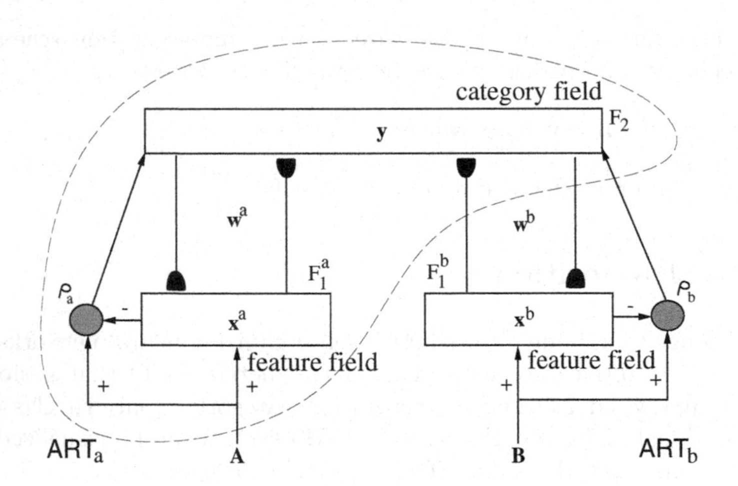

Fig. 1. The Adaptive Resonance Associative Map architecture.

Input vectors: Normalization of fuzzy ART inputs prevents category proliferation. The F_1^a and F_1^b input vectors are normalized by complement coding that preserves amplitude information. Complement coding represents both the on-response and the off-response to an input vector $\mathbf{a}$. The complement coded F_1^a input vector $\mathbf{A}$ is a 2M-dimensional vector

$$\mathbf{A} = (\mathbf{a}, \mathbf{a}^c) \equiv (a_1, \ldots, a_M, a_1^c, \ldots, a_M^c) \tag{6}$$

where $a_i^c \equiv 1 - a_i$.

Similarly, the complement coded F_1^b input vector $\mathbf{B}$ is a 2N-dimensional vector

$$\mathbf{B} = (\mathbf{b}, \mathbf{b}^c) \equiv (b_1, \ldots, b_N, b_1^c, \ldots, b_N^c) \tag{7}$$

where $b_i^c \equiv 1 - b_i$.
Activity vectors: Let $\mathbf{x}^a$ and $\mathbf{x}^b$ denote the F_1^a and F_1^b activity vectors respectively. Let $\mathbf{y}$ denote the F_2 activity vector. Upon input presentation, $\mathbf{x}^a = \mathbf{A}$ and $\mathbf{x}^b = \mathbf{B}$.
Weight vectors: Each F_2 category node j is associated with two adaptive weight templates $\mathbf{w}_j^a$ and $\mathbf{w}_j^b$. Initially, all category nodes are uncommitted and all weights equal ones. After a category node is selected for encoding, it becomes *committed.*
Parameters: Fuzzy ARAM dynamics are determined by the choice parameters $\alpha_a > 0$ and $\alpha_b > 0$; the learning rates $\beta_a \in [0, 1]$ and $\beta_b \in [0, 1]$; the vigilance parameters $\rho_a \in [0, 1]$ and $\rho_b \in [0, 1]$; and the contribution parameter $\gamma \in [0, 1]$.
Category choice: Given a pair of F_1^a and F_1^b input vectors $\mathbf{A}$ and $\mathbf{B}$, for each F_2 node j, the choice function T_j is defined by

$$T_j = \gamma \frac{|\mathbf{A} \wedge \mathbf{w}_j^a|}{\alpha_a + |\mathbf{w}_j^a|} + (1 - \gamma) \frac{|\mathbf{B} \wedge \mathbf{w}_j^b|}{\alpha_b + |\mathbf{w}_j^b|}, \tag{8}$$

where the fuzzy AND operation $\wedge$ is defined by

$$(\mathbf{p} \wedge \mathbf{q})_i \equiv min(p_i, q_i), \tag{9}$$

and where the norm $|.|$ is defined by

$$|\mathbf{p}| \equiv \sum_i p_i \tag{10}$$

for vectors $\mathbf{p}$ and $\mathbf{q}$.

The system is said to make a choice when at most one F_2 node can become active. The choice is indexed at J where

$$T_J = \max\{T_j : \text{for all } F_2 \text{ node } j\}. \tag{11}$$

When a category choice is made at node J, $y_J = 1$; and $y_j = 0$ for all $j \neq J$.
Resonance or reset: Resonance occurs if the *match functions*, m_J^a and m_J^b, meet the vigilance criteria in their respective modules:

$$m_J^a = \frac{|\mathbf{A} \wedge \mathbf{w}_J^a|}{|\mathbf{A}|} \geq \rho_a \quad \text{and} \quad m_J^b = \frac{|\mathbf{B} \wedge \mathbf{w}_J^b|}{|\mathbf{B}|} \geq \rho_b. \tag{12}$$

Learning then ensues, as defined below. If any of the vigilance constraints is violated, mismatch reset occurs in which the value of the choice function T_J is set to 0 for the duration of the input presentation. The search process repeats to select another new index J until resonance is achieved.

Learning: Once the search ends, the weight vectors $\mathbf{w}_J^a$ and $\mathbf{w}_J^b$ are updated according to the equations

$$\mathbf{w}_J^{a(\text{new})} = (1 - \beta_a)\mathbf{w}_J^{a(\text{old})} + \beta_a(\mathbf{A} \wedge \mathbf{w}_J^{a(\text{old})}) \tag{13}$$

and

$$\mathbf{w}_J^{b(\text{new})} = (1 - \beta_b)\mathbf{w}_J^{b(\text{old})} + \beta_b(\mathbf{B} \wedge \mathbf{w}_J^{b(\text{old})}) \tag{14}$$

respectively. For efficient coding of noisy input sets, it is useful to set $\beta_a = \beta_b = 1$ when J is an uncommitted node, and then take $\beta_a < 1$ and $\beta_b < 1$ after the category node is committed. *Fast learning* corresponds to setting $\beta_a = \beta_b = 1$ for committed nodes.
Match tracking: Match tracking rule as used in the ARTMAP search and prediction process [4] is useful in maximizing code compression. At the start of each input presentation, the vigilance parameter ρ_a equals a baseline vigilance $\overline{\rho_a}$. If a reset occurs in the category field F_2, ρ_a is increased until it is slightly larger than the match function m_J^a. The search process then selects another F_2 node J under the revised vigilance criterion. With the match tracking rule and setting the contribution parameter $\gamma = 1$, ARAM emulates the search and test dynamics of ARTMAP.

3.2 Classification

In ARAM systems with category choice, only the F_2 node J that receives maximal $F_1^a \rightarrow F_2$ input T_j predicts ART$_b$ output. In simulations,

$$y_j = \begin{cases} 1 \text{ if } j = J \text{ where } T_J > T_k \text{ for all } k \neq J \\ 0 \text{ otherwise.} \end{cases} \tag{15}$$

The F_1^b activity vector $\mathbf{x}^b$ is given by

$$\mathbf{x}^b = \sum_j \mathbf{w}_j^b y_j = \mathbf{w}_J^b. \tag{16}$$

The output prediction vector $\mathbf{B}$ is then given by

$$\mathbf{B} = (b_1, b_2, \ldots b_{2N}) = \mathbf{x}^b \tag{17}$$

where b_i indicates the confidence of assigning a pattern to category i.

3.3 Rule Insertion

ARAM incorporates a class of if-then rules that maps a set of input attributes (antecedents) to a disjoint set of output attributes (consequents). The rules are conjunctive in the sense that the attributes in the IF clause and in the THEN clause have an *AND* relationship. Conjunctive rules has limited expressive power but is intuitive and adequate for representing simple heuristic for categorization.

ARAM rule insertion proceeds in two phases. The first phase parses the rules for keyword features. When a new keyword is encountered, it is added to a keyword feature table containing keywords obtained through automatic feature selection from training documents. Based on the keyword feature table, the second phase of rule insertion translates each rule into a 2M-dimensional vector **A** and a 2N-dimensional vector **B**, where M is the total number of features in the keyword feature table and N is the number of categories. Given a rule of the following format,

$$\text{IF} \quad x_1, x_2, \ldots, x_m, \neg y_1, \neg y_2, \ldots, \neg y_n \qquad \text{THEN} \quad z_1, z_2, \ldots, z_p$$

where $x_1, \ldots, x_m$ are the positive antecedents, $y_1, \ldots, y_n$ are the negative antecedents, and $z_1, \ldots, z_p$ are the consequents, the algorithm derives a pair of vectors **A** and **B** such that for each index $i = 1, \ldots, M$,

$$(a_i, a_i^c) = \begin{cases} (1,0) \text{ if } w_i = x_j \text{ for some } j \in \{1, \ldots, m\} \\ (0,1) \text{ if } w_i = y_j \text{ for some } j \in \{1, \ldots, n\} \\ (1,1) \text{ otherwise} \end{cases} \quad (18)$$

where w_i is the ith entry in the keyword feature table; and for each index $i = 1, \ldots, N$,

$$(b_i, b_i^c) = \begin{cases} (1,0) \text{ if } w_i = z_j \text{ for some } j \in \{1, \ldots, p\} \\ (0,0) \text{ otherwise} \end{cases} \quad (19)$$

where w_i is the class label of the category i.

The vector pairs derived from the rules are then used as training patterns to initialize a ARAM network. During rule insertion, the vigilance parameters ρ_a and ρ_b are each set to 1 to ensure that only identical attribute vectors are grouped into one recognition category. Contradictory symbolic rules are detected during rule insertion when a perfect match in F_1^a ($m_J^a = 1$) is assciated with a mismatch in F_1^b ($m_J^b < 1$).

4 Experiments: Reuters-21578

Reuters-21578 is chosen as the benchmark domain for a number of reasons. First, it is reasonably large, consisting of tens of thousands of pre-classified documents. Second, there is a good mix of large and small categories (in terms of the number of documents in the category). It enables us to examine ARAM learning capability and the effect of rule insertion using different data characteristics. The last but not the least, Reuters-21578 has been studied extensively in statistical text categorization literatures, allowing us to compare ARAM performance with prior arts.

To facilitate comparison, we used the recommended *ModApte* split (Reuters version 3) [1,19] to partition the database into training and testing data. By selecting the 90 (out of a total of 135) categories that contain at least one training and one testing documents, there were 7770 training documents and 3019 testing documents.

4.1 Performance Measures

ARAM experiments adopt the most commonly used performance measures, namely *recall*, *precision*, and the F_1 measure. *Recall* (r) is the percentage of the documents for a given category (i.e. topic) that are classified correctly. *Precision* (p) is the percentage of the predicted documents for a given category that are classified correctly. It is a normal practice to combine *recall* and *precision* in some way, so that classifiers can be compared in terms of a single rating. Two common ratings are the *break-even point* and the F_1 measure. *Break-even point* is the value at which *recall* equals *precision*. F_1 measure is defined as

$$F_1(r,p) = \frac{2rp}{r+p}. \tag{20}$$

These scores can be calculated for a series of binary classification experiments, one for each category, and then averaged across the experiments. Two types of averaging methods are commonly used: (1) *micro-averaging* technique that gives equal weight to each document; and (2) *macro-averaging* technique that gives equal weight to each category [19]. As micro-averaging F_1 scores are computed on a per-document basis, they tend to be dominated by the classifier's performance on large categories. Macro-averaging F_1 scores, computed on a per-category basis, are more likely to be influenced by the classifier's performance on small categories.

4.2 Learning and Classification

ARAM experiments used the following parameter values: choice parameters $\alpha_a = 0.1$, $\alpha_b = 0.1$; learning rates $\beta_a = \beta_b = 1.0$ for fast learning; contribution parameter $\gamma = 1.0$, and vigilance parameters $\bar{\rho_a} = 0.8$, $\rho_b = 1.0$. Using a voting strategy, 10 voting ARAM produced a probabilistic score between 0 and 1. The score was then thresholded at a specific cut off point to produce a binary class prediction.

We fixed the number of keyword features at 100 determined empirically through earlier experiments. Null feature vectors and contradictory feature vectors were first removed from the training set before training. We experimented with all the three feature encoding schemes, namely *tf*, *tf*idf*, and *log-tf*idf*. Table 1 summarizes the performance of ARAM averaged across 90 categories of Reuters-21578. Among the three encoding schemes, *log-tf*idf* produced the best performance in terms of micro-averaged F_1. *tf*idf* however performed better in terms of macro-averaged F_1.

4.3 Rule Insertion

A set of IF-THEN rules was crafted based on a description of the Reuters categories provided in the Reuters-21578 documentation (cat-descriptions_120396.txt). The rules simply linked the keywords mentioned in the description to their respective category labels. Creation of such rules was

Table 1. Performance of ARAM using the three feature encoding schemes in terms of micro-averaged recall, micro-averaged precision, micro-averaged F_1 and macro-averaged F_1 across all 90 categories of Reuters-21578.

Encoding Method	*miR*	*miP*	miF_1	maF_1
tf	0.8251	0.8376	0.8313	0.5497
*tf*idf*	0.8368	0.8381	0.8375	0.5691
*log-tf*idf*	0.8387	0.8439	0.8413	0.5423

rather straight-forward. A total of 150 rules was created without any help from domain experts. They are generally short rules containing one to two keywords extracted from the category description. A partial set of rules is provided in Table 2 for illustration.

Table 2. An illustrative set of rules generated based on the Reuters category description.

acq	:- acquire acquisition
acq	:- merge merger
crude	:- crude oil
grain	:- grain
interest	:- interest
interest	:- rate
money-fx	:- foreign exchange
money-fx	:- money exchange

In the rule insertion experiments, rules were parsed and inserted into the ARAM networks before learning and classification. Table 3 compares the results obtained by ARAM (using *log-tf*idf*) with and without rule insertion on the 10 most populated Reuters categories. The micro-averaged F_1 and the macro-averaged F_1 scores across the top 10 and all the 90 categories are also given. Eight out of the top 10 categories, namely *acq*, *money-fx*, *grain*, *crude*, *interest*, *ship*, *wheat*, and *corn*, showed noticeable improvement in F_1 measures by incorporating rules. Interestingly, one category, namely *trade*, produced worse results. No improvement was obtained for *earn*, the largest category. The overall improvement on the micro-averaged F_1 scores across the top 10 and all the 90 categories were 0.004 and 0.011 respectively. The improvement obtained on the macro-averaged F_1 scores, recorded at 0.006 for the top 10 and 0.055 for the 90 categories, were much more significant. This suggests that rule insertion is most effective for categories with a smaller number of documents. The results are encouraging as even a simple set of rules is able to produce a noticeable improvement.

Table 3. Predictive performance of ARAM with and without rule insertion on Reuters-21578. *Epochs* refers to the number of learning iterations for ARAM to achieve 100% accuracy on the training set. *Nodes* refers to the number of ARAM recognition categories created. The last two rows show the *micro-averaged* F_1 and the *macro-averaged* F_1 across the top 10 and the 90 categories respectively. Boldfaced figures highlighted improvement obtained by rule insertion.

	Number of	ARAM			ARAM w/rules		
Category	Documents	*Epochs*	*Nodes*	F_1	*Epochs*	*Nodes*	F_1
earn	1087	4.5	717.3	0.984	5.3	722.4	0.984
acq	719	5.1	732.0	0.930	6.1	732.9	**0.938**
money-fx	179	4.2	334.4	0.750	5.5	336.0	**0.763**
grain	149	4.7	104.5	0.895	5.0	108.3	**0.906**
crude	189	6.7	86.4	0.802	6.9	86.7	**0.813**
trade	117	6.0	257.2	0.689	6.8	260.9	0.661
interest	131	4.7	281.3	0.727	5.6	287.4	**0.740**
ship	89	4.6	46.8	0.793	4.8	48.7	**0.796**
wheat	71	4.2	87.7	0.789	8.1	89.0	**0.803**
corn	56	3.6	89.6	0.748	8.0	90.2	**0.765**
Top 10 (miF_1,maF_1)			(0.897,0.811)			(**0.901**,**0.817**)	
All 90 (miF_1,maF_1)			(0.841,0.542)			(**0.852**,**0.597**)	

Table 4 compares ARAM results with top performing classification systems on Reuters-21578 [19]. ARAM performed noticeably better than the gradient descent neural networks and the Native Bayes classifiers. Its miF_1 scores were comparable with those of SVM, KNN, and LLSF, but the maF_1 scores were significantly higher. As miF_1 scores are predominantly determined by the largest categories and miF_1 scores are dominated by the large number of small categories, the results indicate that ARAM performs fairly well for large categories and outperforms in small categories.

Table 4. Performance of ARAM compared with other top performing text classification systems across all 90 categories of Reuters-21578.

Classifiers	*miR*	*miP*	miF_1	maF_1
ARAM w/rules	0.8909	0.8155	0.8515	0.5967
ARAM	0.8961	0.7922	0.8409	0.5422
SVM	0.8120	0.9147	0.8599	0.5251
KNN	0.8339	0.8807	0.8567	0.5242
LLSF	0.8507	0.8489	0.8498	0.5008
Gradient descent NNet	0.7842	0.8785	0.8287	0.3765
Native Bayes	0.7688	0.8245	0.7956	0.3886

5 Conclusion

This paper has presented a novel approach to incorporate domain knowledge into a learning text categorization system. ARAM can be considered as a scaled down version of KNN. Increasing the ART_a baseline vigilance parameter $\bar{\rho_a}$ to 1.0 would cause ARAM's performance to converge to that of KNN with the price of storing all unique training examples. ARAM, therefore, is more scalable than KNN and is useful in situations when it is not practical to store all the training examples in the memory. Comparing with SVM, ARAM has the advantage of on-line incremental learning in the sense that learning of new examples does not require re-computation of recognition nodes using previously learned examples.

The most distinctive feature of ARAM, however, is its rule-based domain knowledge integration capability. The performance of ARAM is expected to improve further as good rules are added. The rule insertion capability is especially important when few training examples are available. This suggests that ARAM could be suitable for on-line text classification applications such as document filtering and personalized content management.

References

1. C. Apte, F. Damerau, and S.M. Weiss. Automated learning of decision rules for text categorization. *ACM Transactions on Information Systems*, 12(3):233–251, 1994.
2. C. Apte, F. Damerau, and S.M. Weiss. Text mining with decision rules and decision trees. In *Proceedings of the Conference on Automated Learning and Discovery, Workshop 6: Learning from Text and the Web*, 1998.
3. G. A. Carpenter and S. Grossberg. A massively parallel architecture for a self-organizing neural pattern recognition machine. *Computer Vision, Graphics, and Image Processing*, 37:54–115, 1987.
4. G. A. Carpenter, S. Grossberg, N. Markuzon, J. H. Reynolds, and D. B. Rosen. Fuzzy ARTMAP: A neural network architecture for incremental supervised learning of analog multidimensional maps. *IEEE Transactions on Neural Networks*, 3:698–713, 1992.
5. G. A. Carpenter, S. Grossberg, and D. B. Rosen. Fuzzy ART: Fast stable learning and categorization of analog patterns by an adaptive resonance system. *Neural Networks*, 4:759–771, 1991.
6. S. Dumais, J. Platt, D. Heckerman, and M. Sahami. Inductive learning algorithms and representation for text categorization. In *Proceedings, ACM 7th International Conference on Information and Knowledge Management*, pages 148–155, 1998.
7. T. Joachims. Text categorization with support vector machines: Learning with many relevant features. In *Proceedings, 10th European Conference on Machine Learning (ECML'98)*, pages –, 1998.
8. D.D. Lewis and M. Ringuette. A comparison of two learning algorithms for text categorization. In *Proceedings, Third Annual Symposium on Document Analysis and Information Retrieval (SDAIR'94), Las Vegas*, pages 81–93, 1994.
9. H.T. Ng, W.B. Goh, and K.L. Low. Feature selection, perceptron learning, and a usability case study for text categorization. In *Proceedings, 20th Annual International ACM SIGIR Conference on Research and Development in Information Retrieval (SIGIR'97)*, pages 67–73, 1997.

10. G. Salton and C Buckley. Term weighting approaches in automatic text retrieval. *Information Processing and Management*, 24(5):513–523, 1988.
11. J.W. Shavlik and T. Eliassi-Rad. Building intelligent agents for web-based tasks: A theory-refinement approach. In *Working Notes of the Conf on Automated Learning and Discovery Workshop on Learning from Text and the Web, Pittsburgh, PA*, 1998.
12. A.-H. Tan. Adaptive Resonance Associative Map. *Neural Networks*, 8(3):437–446, 1995.
13. A.-H. Tan. Cascade ARTMAP: Integrating neural computation and symbolic knowledge processing. *IEEE Transactions on Neural Networks*, 8(2):237–250, 1997.
14. A-H. Tan and Lai F-L. Text categorization, supervised learning, and domain knowledge integration. In *Proceedings, KDD-2000 International Workshop on Text Mining, Boston*, pages 113–114, 2000.
15. G. G. Towell, J. W. Shavlik, and M. O. Noordewier. Refinement of approximately correct domain theories by knowledge-based neural nctworks. In *Proceedings, 8th National Conference on AI, Boston, MA*, pages 861–866. AAAI Press/The MIT Press, 1990.
16. E. Wiener, J. O. Pedersen, and A. S. Weigend. A neural network approach to topic spotting. In *Proceedings of the Fourth Annual Symposium on Document Analysis and Information Retrieval (SDAIR'95)*, 1995.
17. Y. Yang. Expert network: Effective and efficient learning from human decisions in text categorization and retrieval. In *Proceedings, 17th Annual International ACM SIGIR Conference on Research and Development in Information Retrieval (SIGIR'94)*, 1994.
18. Y. Yang and C. G. Chute. An exampled-based mapping method for text categorization and retrieval. *ACM Transactions on Information Systems*, 12(3):252–277, 1994.
19. Y. Yang and X. Liu. A re-examination of text categorization methods. In *Proceedings, 22th Annual International ACM SIGIR Conference on Research and Development in Information Retrieval (SIGIR'99)*, pages 42–49, 1999.
20. Y. Yang and J. P. Pedersen. Feature selection in statistical learning for text categorization. In *Proceedings, Fourteehth International Conference on Machine Learning*, pages 412–420, 1997.

Meta-learning Models for Automatic Textual Document Categorization

Kwok-Yin Lai* and Wai Lam

Department of Systems Engineering and Engineering Management,
Ho Sin Hang Engineering Building,
The Chinese University of Hong Kong,
Shatin,
Hong Kong
{kylai,wlam}@se.cuhk.edu.hk

Abstract. We investigate two meta-model approaches for the task of automatic textual document categorization. The first approach is the linear combination approach. Based on the idea of distilling the characteristics of how we estimate the merits of each component algorithm, we propose three different strategies for the linear combination approach. The linear combination approach makes use of limited knowledge in the training document set. To address this limitation, we propose the second meta-model approach, called Meta-learning Using Document Feature characteristics (MUDOF), which employs a meta-learning phase using document feature characteristics. Document feature characteristics, derived from the training document set, capture some inherent properties of a particular category. Extensive experiments have been conducted on a real-world document collection and satisfactory performance is obtained.

Keywords: Text Categorization, Text Mining, Meta-Learning

1 Introduction

Textual document categorization aims to assign none or any number of appropriate categories to a document. The goal of automatic text categorization is to construct a classification scheme, or called the classifier, from a training set. A training set contains sample documents and their corresponding categories. Specifically, there is a classification scheme for each category. During the training phase, documents in training set are used to learn a classification scheme for each category by using a learning algorithm. After completing the whole training phase, each category will have a different learned classification scheme. After the training phase, the learned classification scheme for each category will be used to categorize unseen documents.

There has been some research conducted for automatic text categorization. Yang and Chute [12] proposed a statistical approach known as Linear Least

* Correspondence Author

D. Cheung, G.J. Williams, and Q. Li (Eds.): PAKDD 2001, LNAI 2035, pp. 78–89, 2001.

Squares Fit (LLSF) which estimates the likelihood of the associations between document terms and categories via a linear parametric model. Lewis et al. [8] explored linear classifiers for the text categorization problem. Yang [10] developed an algorithm known as ExpNet which derives from the k-nearest neighbor technique. Lam et al. [6] attempted to tackle this problem using Bayesian networks. Lam and Ho [5] proposed the generalized instance set approach for text categorization. Joachims [4], as well as, Yang and Liu [13] recently compared support vector machines with k-NN. Dumais et al. [2] compared support vector machines, decision trees and Bayesian approaches on the Reuters collection. All the above approaches developed a single paradigm to solve the categorization problem.

In the literature, several methods on multi-strategy learning or combination of classifiers have been proposed. Chan and Stolfo [1] presented their evaluation of simple voting and meta-learning on partitioned data, through inductive learning. Recently, several meta-model methods have been proposed for text domains. Yang et al. [11] proposed the Best Overall Results Generator (BORG) system which combined classification methods linearly for each classifier in the Topic Detection and Tracking (TDT) domain. Larkey et al. [7] reported improved performance, by using new query formulation and weighting methods, in the context of text categorization by combining different classifiers. Hull at al. [3] examined various combination strategies in the context of document filtering.

Instead of using only one algorithm, meta-model learning involves more than one categorization algorithm. Under the approach, classification schemes that have been separately learned by different algorithms for a category, are combined together in a certain way, to yield one single meta-model classification scheme. Given a document to be categorized, the meta-model classification scheme can be used for deciding the document membership for the category. As a result, each meta-model classifier for a category is the combined contributions of all the involved algorithms.

All existing meta-model approaches for text categorization are based on linear combination of several basic algorithms. In this paper, we investigate the linear combination approach by distilling the characteristic of how we estimate the relative merit of each component algorithm for different categories. Based on this idea, we propose three different strategies for the linear combination approach. The linear combination approach makes use of limited knowledge in the training document set. To address this limitation, we propose a second meta-model approach, called Meta-learning Using Document Feature characteristics (MUDOF), which employs a meta-learning phase using document feature characteristics. Document feature characteristics, derived from the training document set, capture some inherent properties of a particular category. This approach aims at recommending algorithms automatically for different categories.

We have conducted extensive experiments on a real-world document collection The results demonstrate that our new approaches of meta-learning models for text categorization outperforms all other component algorithms under various measures.

2 Linear Combination Approach

The first approach we investigate is based on the weighted sum of linear combination of classifiers. Under this approach, the contribution of each individual component algorithm j to the final meta-model classification scheme for a category i, is represented by a weight factor w_{ij}. Consider a document m which is to be categorized. Instead of using the score calculated from a single classification scheme of a particular category, the linear combination approach calculates a combined score which is the weighted sum of contributions of all component algorithms in a linear fashion. Suppose there are n component algorithms. The combined score for m is computed by Equation 1.

$$S^m_{i,comb} = \sum_{j=1}^{n} \omega^m_{ij} * S^m_{ij} \tag{1}$$

where $S^m_{i,comb}$ is the final combined score for m for the category i. S^m_{ij} is the score calculated between m and the classifier learned by algorithm j for category i. The value of ω^m_{ij} is the weight factor, or the contribution, of the classifier to the score S^m_{ij}, and $\sum_{j=1}^{n} \omega^m_{ij}$ equals to 1.

If the final combined score for m is larger than the threshold value of a category, that category is assigned to m. To reflect the significance of contribution by different classifiers for a category, various methods can be employed to determine the weight (ω^m_{ij} in Equation 1). We have implemented three strategies under this linear combination approach, to study the categorization performance differences due to the use of different weight determination for the combination.

2.1 Equal Weighting Strategy

The first strategy, called LC1, is an equal weighting scheme. Under this scheme, the weight of all classifiers are the same, as indicated in Equation 2. As a result, the contribution of each classification algorithm to the final combined score for m is equal.

$$\omega^m_{i1} = \omega^m_{i2} = \cdots = \omega^m_{ij} = \cdots = \omega^m_{in} = \frac{1}{n} \qquad \text{for all } i \tag{2}$$

2.2 Weighting Strategy Based on Utility Measure

The second strategy, called LC2, determines the weighting scheme based on utility measure from training. Under this strategy, the relative contribution, w^m_{ij}, of a classification scheme, which is constructed by algorithm j for category i, to the final combined score for document m, depends on the performance, u_{ij}, of the learned classifier in the training phase. The relationship between the contribution of the classifier and its categorization performance, is represented as a function indicated in Equation 3.

$$\omega^m_{ij} = f(u_{ij}) \tag{3}$$

where function f is expressed in terms of u_{ij}, which is the utility score obtained by the classifier.

The function f is a transformation function from certain utility scores to corresponding contribution weights. The transformation is restricted by the condition of $\sum_{j=1}^{n} \omega_{ij}^{m}$ equals to 1. Conceptually, a well-performed classification scheme constructed by an algorithm should be given a heavier weight than the others during the score combination. In our investigation, we adopt the function f as shown in Equation 4.

$$\omega_{ij}^{m} = f(u_{ij}) = \frac{u_{ij}}{\sum_{k=1}^{n} u_{ik}} \qquad \text{for } 1 \leq j \leq n \tag{4}$$

We make use of a set of documents, called tuning set, obtained from a subset of the training set to calculate u_{ij}. Specifically, u_{ij} is the classification performance of the tuning set using the classification scheme constructed by algorithm j for category i.

2.3 Weighting Strategy Based on Document Rank

Our third strategy, called LC3, determines the contribution weights of the involved component algorithms, based on the rank of scores, S_{ij}^{m} for document m. The scores are first ranked. By mapping from the rank R to a set of predetermined weight factors using the function g, a particular weight, say P_d, is assigned to the corresponding algorithm as its contribution in the final combined score for m. The idea of this strategy is illustrated in Equation 5.

$$\omega_{ij}^{m} = P_d = g(R_{ij}^{m}) \qquad \text{for } P_d \in \{P_1, P_2, \ldots, P_n\} \text{ and } 1 \leq j \leq n \tag{5}$$

where P_d is one of the n pre-determined weights, and R_{ij}^{m} is the rank of score of m by algorithm j under category i. g is a mapping function from the rank R_{ij}^{m} to the assignment of the weight P_d for the document m.

3 Meta-learning Using Document Feature Characteristics (MUDOF)

We propose our second approach of the meta-learning framework for text categorization, based on multivariate regression analysis, by capturing category specific feature characteristics. In MUDOF, there is a separate meta-learning phase using document feature characteristics. Document feature characteristics, derived from the training set of a particular category, can capture some inherent properties of that category. Different from existing categorization methods, instead of applying a single method for all categories during classification, this new meta-learning approach can automatically recommend a suitable algorithm during training and tuning steps, from an algorithm pool, for each category based on the category specific statistical characteristics and multivariate regression analysis. The problem of predicting the expected classification error of an algorithm for a category can be interpreted as a function of feature characteristics.

In particular, we wish to predict the classification error for a category based on the feature characteristics. This is achieved by a learning approach based on regression model, in which, the document feature characteristics are the independent variables, while the classification error of an algorithm is the dependent variable. Feature characteristics are derived from the categories. We further divide the training collection into two sets, namely the training set and the tuning set. Two sets of feature characteristics are collected separately from training set and tuning set. Statistics from training set are for parameter estimations. Together with the estimated parameters, the statistics from tuning set are used for predicting the classification error of an algorithm for a category. The algorithm with the minimum estimated classification error for a category will be recommended for that category during the testing, or validation, phase. Classification errors need to undergo a logistic transformation to yield the response variable, or the dependent variable, for the meta-model. The transformation ensures that the fitted error to be in the range from 0 to 1. Consider the ith category and the jth algorithm. The response variable, y_{ij} is related to the feature characteristics by the regression model, as shown in Equation 6.

$$y_{ij} = \ln \frac{e_{ij}}{1 - e_{ij}} = \beta_j^0 + \sum_{k=1}^{p} \beta_j^k * F_i^k + \epsilon_{ij}, \tag{6}$$

where e_{ij} is the classification error, obtained for the ith category by using the jth algorithm. F_i^k is the kth feature characteristic in the ith category. The number of feature characteristics used in the meta-model is p. β_j^k is the parameter estimate for the kth feature, by using algorithm j. ϵ_{ij} is assumed to follow an $N(0,var(\epsilon_{ij}))$. Based on the regression model above, the outline of meta-model for text categorization is given in Figure 1.

Step 1 to 9, in Figure 1, aims to estimate a set of betas ($\widehat{\beta_j^k}$), the parameter estimates of the feature characteristics in the regression model, for each individual algorithm. In Step 2, an algorithm, with optimized parameter settings, is picked from the algorithm pool. By repeating Step 3 to 7, the algorithm is applied on training and tuning examples to yield classification errors of the classifier for all categories. Documents in tuning set, as shown in Step 5, are used for obtaining the classification performance, and so the classification error, of a trained classifier for each category. A set of betas, belonging to the algorithm being considered, can be obtained by fitting all classification errors of the categories, and their corresponding feature characteristics in the training set, into the regression model. After Step 9, there will be j sets of estimated parameters, the betas, which are then used for the subsequent steps.

The predictions on the classification errors of the involved algorithms are made from Step 10 to Step 16. In Step 12, one algorithm with the same optimized parameter settings as in Step 2, is picked from the algorithm pool. The corresponding set of betas of the selected algorithm, together with the feature characteristics of a category in the tuning set, will be fitted into the regression model, in Step 13, to give the estimated classification errors of the algorithm on the category. Decisions, about which algorithm will be applied on the cate-

Input: The training set TR and tuning set TU
An algorithm pool A and categories set C

1) Repeat
2) Pick one algorithm ALG_j from A.
3) For each category C_i in C
4) Apply ALG_j on TR for C_i to yield a classifier CF_{ij}.
5) Apply CF_{ij} on TU for C_i to yield classification error e_{ij}.
6) Take logistic transformation on e_{ij} to yield y_{ij} for later parameter estimation.
7) End For
8) Estimate $\widehat{\beta}_j^k$ (k=0,1,2,...,p) for ALG_j by fitting y_{ij} and F_i^k (in TR) into the regression model.
9) Until no more algorithms in A.
10) For each category C_i in C
11) Repeat
12) Pick one algorithm ALG_j from A.
13) Estimate the classification error $\widehat{e}_{ij}$ by fitting $\widehat{\beta}_j^k$ and corresponding F_i^k (in TU) into the regression model.
14) If $\widehat{e}_{ij}$ is minimum, recommend ALG_j for C_i as the output.
15) Until no more algorithms in A.
16) End For

Fig. 1. The Meta-Model algorithm.

gory, are based on the predicted minimum classification errors in Step 14. After Step 16, classification algorithms are recommended for categories, and the recommended algorithm will be applied to each category during the validation step. The whole process, from parameter estimation to recommending algorithms for categories, of our proposed meta-model approach is fully automatic.

4 Experiments and Empirical Results

4.1 Document Collection and Experimental Setup

Extensive experiments have been conducted on the Reuters-21578 corpus, which contains news articles from Reuters in 1987. 90 categories are used in our experiments. We divided the 21,578 documents in the Reuters-21578 document collection according to the "ModApte" split into one training collection of 9603 documents, and one testing collection of 3299 documents. The remaining 8,676 documents are not used in the experiments as the documents are not classified by human indexer. For those meta-models requiring a tuning set, we further divided the training collection into training set of 6000 documents and 3603 tuning documents. For each category, we used the training document collection to learn a classification scheme. The testing collection is used for evaluating the classification performance.

Six component classification algorithms have been used in our meta-model approaches. They are Rocchio, WH, k-NN, SVM, GIS-R and GIS-W, with optimized parameter settings. These are six recent algorithms, each of which exhibits certain distinctive nature: Rocchio and WH are linear classifiers, k-NN is instance-based learning algorithm, SVM is based on Structural Risk Minimization Principle [9] and both GIS-R and GIS-W [5] are based on generalized instance approach.

In MUDOF, seven feature characteristics are used in our regression model as independent variables:

1. ***PosTr***: The number of positive training examples of a category.
2. ***PosTu***: The number of positive tuning examples of a category.
3. ***AvgDocLen***: The average document length of a category. Document length refers to the number of indexed terms within a document. The average is taken across all the positive examples of a category.
4. ***AvgTermVal***: The average term weight of documents across a category. Average term weight is taken for individual documents first. Then, the average is taken across all the positive examples of a category.
5. ***AvgMaxTermVal***: The average maximum term weight of documents across a category. Maximum term weight of individual documents are summed, and the average is taken across all the positive examples of a category.
6. ***AvgMinTermVal***: The average minimum term weight of documents across a category. Minimum term weight of individual documents are summed, and the average is taken across all the positive examples of a category.
7. ***AvgTermThre***: The average number of terms above a term weight threshold. The term weight threshold is optimized and set globally. Based on the preset threshold, the number of terms with term weight above the threshold within a category are summed. The average is then taken across all the positive examples of the category.

Two sets of normalized feature characteristics are collected separately from training set and tuning set. As illustrated in Step 8 and Step 13 in Figure 1, the feature characteristics from these two data sets serve different purposes in the meta-model: feature characteristics from training set are combined for parameters estimation, while feature characteristics from tuning set are used for predicting classification errors, base on which algorithms are recommended.

To measure the performance, we use both micro-averaged recall and precision break-even point measure (MBE) [8], as well as the macro-averaged recall and precision break-even point measure (ABE). In micro-averaged recall and precision break-even point measure, the total number of false positive, false negative, true positive, and true negative are computed across all categories. These totals are used to compute the micro-recall and micro-precision. Then we use the interpolation to find the break-even point. In macro-averaged recall and precision break-even point measure, break-even point for individual category is calculated first, and the simple average of all those break-even points is taken across all the categories to obtain the final score.

4.2 Empirical Results

After conducting extensive experiments for each component algorithm in order to search for the most optimized parameters setting, we further conduct experiments for the linear combination approach and MUDOF.

Table 1 shows the micro-recall and precision break-even point measure of our proposed meta-learning models. Table 2 shows the classification performance improvement, based on the micro-recall and precision break-even point measure, obtained by the meta-learning models under linear combination and MUDOF approach. Meta-learning models demonstrate improvement over all component algorithms in various extent. The improvement over Rocchio is the largest for all approaches. In particular, the LC2 strategy obtains the best improvement than the other strategies under linear combination approach.

Table 1. Micro-recall and precision break-even point measures over 90 categories for the meta-learning models.

	LC1	LC2	LC3	MUDOF
MBE	0.860	0.862	0.858	0.858

Table 2. Classification improvement by meta-model approaches over component algorithms based on micro-recall and precision break-even point measures over 90 categories.

ALG	MBE	LC1+(%)	LC2+(%)	LC3+(%)	MUDOF+(%)
RO	0.776	10.825	11.082	10.567	10.567
WH	0.820	4.878	5.122	4.634	4.634
KNN	0.802	7.232	7.481	6.983	6.983
SVM	0.841	2.259	2.497	2.021	2.021
GISR	0.830	3.614	3.855	3.373	3.373
GISW	0.845	1.775	2.012	1.538	1.538

Table 3 shows the parameter estimates for the document feature characteristics of different component algorithms in MUDOF approach. Based on these parameter estimates and the corresponding feature characteristics, on category basis, the estimated classification errors of different algorithms on the categories can be obtained. It should be noted that, a negative parameter estimate will contribute to a smaller estimated classification error for an algorithm on a category. As a result, a feature characteristic with a large negative parameter estimate, will make itself a more distinctive feature in voting for the algorithm than others. For example, as shown in the table, *PosTr* has a more favourable impact for Rocchio, k-NN and SVM than other features.

Table 3. Parameter estimates for document feature characteristics of different algorithms.

Features	RO	WH	KNN	SVM	GISR	GISW
PosTr	-4.75	9.24	-1.21	-0.46	5.98	9.84
PosTu	-0.82	-17.05	-5.88	-8.68	-17.26	-20.87
AvgDocLen	2028.77	3199.54	1514.31	2275.34	2567.84	2676.94
AvgTermVal	103.51	154.21	81.74	117.62	151.54	155.67
AvgMaxTermVal	14.04	23.21	12.37	15.59	16.95	21.39
AvgMinTermVal	-69.79	-104.37	-68.77	-92.12	-124.89	-138.73
AvgTermThre	-2006.50	-3164.39	-1495.49	-2250.07	-2534.19	-2640.33

Table 4 shows the performance of all component algorithms and our proposed meta-learning models, based on the macro-recall and precision break-even point measures of the ten most frequent categories, which are those categories with top-ten number of positive training documents. It shows that, while linear combination approach outperforms Rocchio, WH, k-NN as well as SVM, MUDOF can show even better performance over all the component algorithms. Such observation attributes to the fact that, under MUDOF approach, feature characteristics derived from those categories, with a larger number of positive training examples, have better predictive power for the classification errors of individual component algorithms.

Table 4. Macro-recall and precision break-even point measures of the 10 most frequent categories.

	RO	WH	KNN	SVM	GISR	GISW	LC1	LC2	LC3	MUDOF
Top 10 ABE	0.730	0.851	0.781	0.859	0.814	0.871	0.867	0.865	0.868	0.874

Besides of comparing the performance of the MUDOF approach over individual component algorithms, we set up the ideal combination of algorithms as another benchmark for our MUDOF approach. The ideal combination of algorithms is set up manually and is composed of the best algorithms, which are the true algorithms that MUDOF should recommend for each category accordingly. Table 5 depicts the selected algorithms and their performance within a category by MUDOF and the ideal combination (IDEAL), for the ten most frequent categories. Meta-model can estimate the ideal algorithms (in bold) correctly for 6 categories out of the 10 most frequent categories. For the remaining 4 categories, our meta-model can estimate the second best algorithms. Our results, not shown in the table, show that the meta-model can identify the ideal algorithms for 60 categories out of the total 90 categories.

Since the ideal combination consists of the most appropriate algorithm for each category, it sets an upper bound for the amount of improvement that can

Table 5. Macro-recall and precision break-even point measures of the 10 most frequent categories, for individual classifiers, meta-model approach (MUDOF) and the ideal combination (IDEAL).

Category	RO	WH	KNN	SVM	GISR	GISW	MUDOF		IDEAL	
acq	0.829	0.870	0.859	0.931	0.932	0.909	0.932	**GISR**	0.932	GISR
corn	0.614	0.867	0.690	0.832	0.867	0.885	0.885	**GISW**	0.885	GISW
crude	0.793	0.853	0.823	0.871	0.813	0.869	0.869	GISW	0.871	SVM
earn	0.956	0.969	0.956	0.980	0.959	0.962	0.980	**SVM**	0.980	SVM
grain	0.803	0.887	0.820	0.917	0.804	0.910	0.910	GISW	0.917	SVM
interest	0.481	0.881	0.721	0.962	0.721	0.881	0.881	GISW	0.962	SVM
money-fx	0.582	0.718	0.674	0.717	0.681	0.756	0.756	**GISW**	0.756	GISW
ship	0.800	0.860	0.800	0.845	0.825	0.872	0.860	WH	0.872	GISW
trade	0.732	0.763	0.740	0.715	0.714	0.788	0.788	**GISW**	0.788	GISW
wheat	0.713	0.839	0.727	0.820	0.825	0.875	0.875	**GISW**	0.875	GISW
Top 10 ABE	0.730	0.851	0.781	0.859	0.814	0.871	0.874		0.884	

be made under our meta-model. Table 6 shows the comparison of performances, under different aspects of measures, between MUDOF and the ideal combination (IDEAL). Based on the utility measures as shown in the table, our results, not shown in this paper due to space limit, show that both MUDOF or the ideal combination have more than 10% improvement over Rocchio in both aspects. Improvement made by the meta-model over k-NN and GIS-R is more than 5% and 3% in all aspects respectively. When considering the improvement bound set by the ideal combination, our approach has attained more than 90% of the improvement bound for both Rocchio and k-NN under the Top 10 ABE measure.

Table 6. Classification performances of meta-model (MUDOF) and the ideal combination (IDEAL) under different groups of categories.

Utility Measure	MUDOF	IDEAL
Top 10 ABE	0.874	0.884
All 90 MBE	0.858	0.868

Table 7 shows that incremental improvement can be obtained as more robust and more classifiers are included in the algorithm pool of the meta-model. Performance obtained after adding GIS-R to Combination 1 is increased. After replacing GIS-W with GIS-R, the improvement over Combination 1 is more significant. After adding the robust SVM to Combination 3, the MBE performance is further increased, as indicated in Combination 4. Combination 5 is actually the whole algorithm pool of our meta-model. The results demonstrate that our meta-model under MUDOF approach is not limited to combining a fixed number of classifiers, or combining classifiers of same type, instead, it allows flexible additions or substitutions of different classifiers in its algorithm pool.

Table 7. Performance with different combinations of classifiers based on our meta-model under micro-recall and precision break-even point measure over 90 categories.

Algorithm Combination	MBE
1) KNN+WH+RO	0.820
2) GIS-R+KNN+WH+RO	0.842
3) GIS-W+KNN+WH+RO	0.848
4) SVM+GIS-W+KNN+WH+RO	0.854
5) GIS-R+SVM+GIS-W+KNN+WH+RO	0.858

5 Conclusions

We have investigated two meta-model approaches for the task of automatic textual document categorization. The first approach is the linear combination approach. Under the approach, we propose three different strategies to combine the contributions of component algorithms. We have also proposed a second meta-model approach, called Meta-learning Using Document Feature characteristics (MUDOF), which employs a meta-learning phase using document feature characteristics. Different from existing categorization methods, MUDOF can automatically recommend a suitable algorithm for each category based on the category-specific statistical characteristics. Moreover, MUDOF allows flexible additions or replacement of different classification algorithms, resulting in the improved overall classification performance. Extensive experiments have been conducted on the Reuters-21578 corpus for both approaches. Satisfactory performance is obtained for the meta-learning approaches.

References

1. P. K. Chan and S. J. Stolfo. Comparative evaluation of voting and meta-learning on partitioned data. In *Proceedings of the International Conference on Machine Learning (ICML'95)*, pages 90–98, 1995.
2. S. Dumais, J. Platt, D. Heckerman, and M. Sahami. Inductive learning algorithms and representations for text categorization. In *Proceedings of the Seventh International Conference on Information and Knowledge Management*, pages 148–155, 1998.
3. D. A. Hull, J. O. Pedersen, and H. Schutze. Method combination for document filtering. In *Proceedings of the Nineteenth International ACM SIGIR Conference on Research and Development in Information Retrieval*, pages 279–287, 1996.
4. T. Joachims. Text categorization with support vector machines: Learning with many relevant features. In *European Conference on Machine Learning (ECML'98)*, pages 137–142, 1998.
5. W. Lam and C. Y. Ho. Using a generalized instance set for automatic text categorization. In *Proceedings of the Twenty-First International ACM SIGIR Conference on Research and Development in Information Retrieval*, pages 81–89, 1998.
6. W. Lam, K. F. Low, and C. Y. Ho. Using a Bayesian network induction approach for text categorization. In *Proceedings of the Fifteenth International Joint Conference on Artificial Intelligence, (IJCAI), Nagoya, Japan*, pages 745–750, 1997.

7. L. S. Larkey and W. B. Croft. Combining classifiers in text categorization. In *Proceedings of the Nineteenth International ACM SIGIR Conference on Research and Development in Information Retrieval*, pages 289–297, 1996.
8. D. D. Lewis, R. E. Schapore, J. P. Call, and R. Papka. Training algorithms for linear text classifiers. In *Proceedings of the Nineteenth International ACM SIGIR Conference on Research and Development in Information Retrieval*, pages 298–306, 1996.
9. V. Vapnic. *The Nature of Statistical Learning Theory.* Springer, New York, 1995.
10. Y. Yang. Expert network: Effective and efficient learning from human decisions in text categorization and retrieval. In *Proceedings of the Seventeenth International ACM SIGIR Conference on Research and Development in Information Retrieval*, pages 13–22, 1994.
11. Y. Yang, T. Ault, and T. Pierce. Combining multiple learning strategies for effective cross validation. In *Proceedings of the International Conference on Machine Learning (ICML 2000)*, pages 1167–1174, 2000.
12. Y. Yang and C. D. Chute. An example-based mapping method for text categorization and retrieval. *ACM Transactions on Information Systems*, 12(3):252–277, 1994.
13. Y. Yang and X. Liu. A re-examination of text categorization methods. In *Proceedings of the Twenty-First International ACM SIGIR Conference on Research and Development in Information Retrieval*, pages 42–49, 1999.

Efficient Algorithms for Concept Space Construction

Chi Yuen Ng, Joseph Lee, Felix Cheung, Ben Kao, and David Cheung

Department of Computer Science and Information Systems
The University of Hong Kong
{cyng, jkwlee, kmcheung, kao, dcheung}@csis.hku.hk

Abstract. The vocabulary problem in information retrieval arises because authors and indexers often use different terms for the same concept. A thesaurus defines mappings between different but related terms. It is widely used in modern information retrieval systems to solve the vocabulary problem. Chen et al. proposed the *concept space* approach to automatic thesaurus construction. A concept space contains the associations between every pair of terms. Previous research studies show that concept space is a useful tool for helping information searchers in revising their queries in order to get better results from information retrieval systems. The construction of a concept space, however, is very computationally intensive. In this paper, we propose and evaluate efficient algorithms for constructing concept spaces that include only *strong* associations. Since *weak* associations are not useful in thesauri construction, our algorithms use various prunning techniques to avoid computing weak associations to achieve efficiency.

Keywords: concept space, thesaurus, information retrieval, text mining

1 Introduction

The vocabulary problem has been studied for many years [5,3]. It refers to the failure of a system caused by the variety of terms used by its users during human-system communication. Furnes et al. studied the tendency of using different terms among different users to describe a similar concept. For example, they discovered that for spontaneous word choice for concepts, in certain domain, the probability that two people choose the same term is less than 20% [5]. In an information retrieval system, if the keywords that a user specifies in his query are not used by the indexer, the retrieval fails. To solve the vocabulary problem, a *thesaurus* is often used. A thesaurus contains a list of terms along with the relationships between them. During searching, a user can make use of the thesaurus to design the most appropriate search strategy. For example, if a search retrieves too few documents, a user can expand his query by consulting the thesaurus for similar terms. On the other hand, if a search retrieves too many documents, a user can use a more specific term suggested by the thesaurus. Manual construction of thesauri is a very complex process and often involves human experts.

D. Cheung, G.J. Williams, and Q. Li (Eds.): PAKDD 2001, LNAI 2035, pp. 90–101, 2001.

Previous research works have been done on automatic thesaurus construction [4].

In [6], Chen et. al. proposed the *concept space approach* to automatic thesaurus generation. A concept space is a network of terms and their weighted associations. The association between two terms is a quantity between 0 and 1, computed from the co-occurrence of the terms from a given document collection. Its value represents the strength of similarity between the terms. When the association between two terms is zero, the terms have no similarity. It is because the terms never co-exist in a document. When the association from a term a to another term b is near to 1, term a is highly related to term b in the document collection. Based on the idea of concept space, Schatz et al. constructed a prototype system to provide interactive term suggestion to searchers of the University of Illinois Digital Library Initiative test-bed [7]. Given a term, the system retrieves all the terms from a concept space that has non-zero associations to the given term. The associated terms are presented to the user in a list, sorted in decreasing order of association value. The user then selects new terms from the list to refine his queries interactively. Schatz showed that users could make use of the terms suggested by the concept space to improve the *recall* of their queries.

The construction of a concept space involves two phases: (1) an automatic indexing phase in which a document collection is processed to build inverted lists [4], and (2) a co-occurrence analysis phase in which the associations of every term pair are computed. Since there could be tens of thousands of terms in a document collection, computing the associations of all term pairs is very time consuming. In order to apply the concept space approach in large-scale document collections, efficient methods are needed.

We observe that in many applications, a complete concept space is not needed. In typical document collections, most of the associations are zero, i.e., most term pairs are not associated at all. Also, among the non-zero associations, only a very small fraction of them have significant values. In typical applications, small-valued associations (or weak associations) are not useful. For example, in query augmentation, recommending weakly-associated terms to those keywords specified by a user query lowers the precision of the retrieval result.

In this paper, we propose and evaluate a number of efficient algorithms for constructing concept spaces that only contain strong associations. The challenge is how one could deduce that a certain term pair association is weak without actually computing it. We consider a number of pruning techniques that efficiently and effectively make such deductions.

The rest of the paper is organized as follows. In Section 2 we give a formal definition of concept space and discuss how term-pair associations are calculated. In Section 3 we consider three algorithms for the efficient construction of concept spaces that only contain strong associations. Experiment results comparing the performance of the algorithms are shown in Section 4. Finally, Section 5 concludes the paper.

2 Concept Space Construction

A concept space contains the associations, W_{jk} and W_{kj}, between any two terms j and k found in a document collection. Note that the associations are asymmetric, that is, $W_{jk} \neq W_{kj}$. According to Chen and Lynch [6], W_{jk} is computed by the following formula:

$$W_{jk} = \frac{\sum_{i=1}^{N} d_{ijk}}{\sum_{i=1}^{N} d_{ij}} \times \mathit{WeightingFactor}(k). \tag{1}$$

The symbol d_{ij} represents the weight of term j in document i based on the term-frequency-inverse-document-frequent measure [2]:

$$d_{ij} = \mathit{tf}_{ij} \times \log(\frac{N}{\mathit{df}_j} \times w_j)$$

where

$$\begin{aligned} \mathit{tf}_{ij} &= \text{number of occurrences of term } j \text{ in document } i, \\ \mathit{df}_j &= \text{number of documents in which term } j \text{ occurs}, \\ w_j &= \text{number of words in term } j, \\ N &= \text{number of documents.} \end{aligned}$$

The symbol d_{ijk} represents the combined weight of both terms j and k in document i. It is defined as:

$$d_{ijk} = \mathit{tf}_{ijk} \times \log(\frac{N}{\mathit{df}_{jk}} \times w_j) \tag{2}$$

where

$$\begin{aligned} \mathit{tf}_{ijk} &= \text{number of occurrences of both terms } j \text{ and } k \text{ in document } i, \\ &\quad \text{i.e., } \min(\mathit{tf}_{ij}, \mathit{tf}_{ik}), \\ \mathit{df}_{jk} &= \text{number of documents in which both terms } j \text{ and } k \text{ occur.} \end{aligned}$$

Finally, $\mathit{WeightingFactor}(k)$ is defined as:

$$\mathit{WeightingFactor}(k) = \frac{\log(N/\mathit{df}_k)}{\log N}.$$

$\mathit{WeightingFactor}(k)$ is used as a weighting scheme (similar to the concept of inverse document frequency) to penalize general terms (terms that appear in many documents). Terms with a high df_k value has a small weighting factor, which results in a small association value. Chen showed that this asymmetric similarity function (W_{jk}) gives a better association than the popular cosine function [6].

In the following discussion, for simplicity, we assume that $w_j = 1$ for all j (i.e., all terms are single-word ones). We thus remove the term w_j from the formula of d_{ij} and d_{ijk}.

As we have mentioned, concept space construction is a two-phase process. In the first phase (automatic indexing), a term-document matrix, TF is constructed. Given a document i and a term j, the matrix TF returns the term frequency, tf_{ij}. In practice, TF is implemented using inverted lists. That is, for each term j, a linked list of [document-id,term-frequency] tuples is maintained. Each tuple records the occurrence frequency of term j in the document with the corresponding id. Documents that do not contain the term j are not included in the inverted list of j.

Besides the matrix, TF, the automatic indexing phase also calculates the quantity df_j (the number of documents containing term j) and well as $\sum_{i=1}^{N} tf_{ij}$ (the sum of the term frequency of term j over the whole document collection) for each term j. These numbers are stored in arrays for fast retrieval during the second phase (co-occurrence analysis).

In the co-occurrence analysis phase, associations of every term pair are calculated. According to Equation 1 (page 92), to compute W_{jk}, we need to compute the values of three factors, namely, $\sum_{i=1}^{N} d_{ijk}$, $\sum_{i=1}^{N} d_{ij}$, and $WeightingFactor(k)$. Note that

$$\sum_{i=1}^{N} d_{ij} = \sum_{i=1}^{N} [tf_{ij} \times \log\left(\frac{N}{df_j}\right)] = \log\left(\frac{N}{df_j}\right) \times \sum_{i=1}^{N} tf_{ij}.$$

Since both df_j and $\sum_{i=1}^{N} tf_{ij}$ are already computed and stored during the automatic indexing phase, $\sum_{i=1}^{N} d_{ij}$ can be computed in constant time. Similarly, $WeightingFactor(k)$ can be computed in constant time as well.

Computing $\sum_{i=1}^{N} d_{ijk}$, however, requires much more work. From Equation 2, one needs to compute df_{jk} (i.e., the number of documents containing both terms j and k) and $\sum_{i=1}^{N} tf_{ijk}$ in order to find $\sum_{i=1}^{N} d_{ijk}$. Figure 1 shows an algorithm for computing W_{jk}.

The execution time of Weight is dominated by the for-loop in line 3. Basically, most of the work is spent on scanning the inverted lists of terms j and k.

2.1 Algorithm A

As we have mentioned, most of the associations have zero or very small values. Our goal is to construct a concept space that contains only *strong* associations. Given a user-specified threshold λ, an association W_{jk} is *strong* if $W_{jk} \geq \lambda$; otherwise, the association is *weak*.

To construct a concept space with only strong associations, our base algorithm first identifies all term pairs that have non-zero associations and then applies the function Weight on each pair. In particular, during the automatic indexing phase, a two-dimensional triangular bit matrix C is built. The entry $C(j,k)$ is set to 1 if there exists a document that contains both terms j and k (i.e., $W_{jk} > 0$); otherwise, $C(j,k)$ is set to 0.

During the co-occurrence analysis phase, the matrix C is consulted. The associations, W_{jk} and W_{kj}, are computed only if $C(j,k)$ is set. Associations that are less than λ are discarded. We call this base Algorithm A(Figure 2).

```
WEIGHT(j, k)
 1  df_jk ← 0
 2  sum_tf_ijk ← 0
 3  for each (i, tf_ij) in the adjacency list of j
 4  do
 5      if there exists (i, tf_ik) in the adjacency list of k
 6         then
 7              df_jk ← df_jk + 1
 8              if tf_ij < tf_ik
 9                 then
10                      sum_tf_ijk ← sum_tf_ijk + tf_ij
11                 else
12                      sum_tf_ijk ← sum_tf_ijk + tf_ik
13  sum_d_ijk ← sum_tf_ijk × log(N/df_jk)
14  sum_d_ij ← sum_tf_ij × log(N/df_j)
15  weighting_factor_k ← log(N/df_k)/ log N
16  return sum_d_ijk × weighting_factor_k/sum_d_ij
```

Fig. 1. Function Weight

```
ALGORITHM-A(λ)
 1  (* Automatic indexing phase *)
 2  C ← 0
 3  for each document i
 4  do
 5      for each term j in document i
 6      do
 7          tf_ij ← no. of occurrences of j in i
 8          append (i, tf_ij) to j's inverted list
 9          sum_tf_ij ← sum_tf_ij + tf_ij ; df_j ← df_j + 1
10      for each term pair j and k (j < k) in document i
11      do
12          C(j, k) ← 1 (* j, k have non-zero associations *)
13  (* Co-occurrence analysis phase *)
14  for each C(j, k) = 1
15  do
16      if Weight(j, k)(or Weight(k, j)) ≥ λ
17         then output Weight(j, k)(or Weight(k, j))
```

Fig. 2. Algorithm A

3 Pruning Algorithms

Algorithm A is not particularly efficient. It basically computes all possible non-zero associations before filtering out those that are weak. As an example, we ran Algorithm A on a collection of 23,000 documents. It took 100 minutes for the algorithm to terminate. The major source of inefficiency lies in the Weight function, which scans two inverted lists for every non-zero association. In the collection, there are about 9 millions non-zero associations, and hence Algorithm A had to scan about 18 millions inverted lists. In this section, we consider a few algorithms for improving the efficiency of Algorithm A. All of these algorithms share the following feature. Before computing an association W_{jk}, each one first computes an *easy-to-compute estimate* W'_{jk}. If the estimate suggests that W_{jk} is likely to be strong, the algorithm will execute the more expensive Weight function. The algorithms gain efficiency by avoiding the computation of weak associations.

3.1 Algorithm B

Our first efficient algorithm, B, computes an estimate W'_{jk} that is always an *upper bound* of W_{jk}[1]. If W'_{jk} is smaller than the threshold λ, we have $W_{jk} \leq W'_{jk} < \lambda$. Hence, the association W_{jk} must be weak and needs not be computed.

Recall that

$$W_{jk} = \frac{\sum_{i=1}^{N} d_{ijk}}{\sum_{i=1}^{N} d_{ij}} \times WeightingFactor(k)$$

and

$$\sum_{i=1}^{N} d_{ijk} = \sum_{i=1}^{N} \left[tf_{ijk} \times \log(\frac{N}{df_{jk}}) \right] = \log(\frac{N}{df_{jk}}) \times \sum_{i=1}^{N} tf_{ijk}.$$

By definition, $tf_{ijk} = \min(tf_{ij}, tf_{ik})$. Note that

$$\sum_{i=1}^{N} tf_{ijk} = \sum_{i=1}^{N} \min(tf_{ij}, tf_{ik}) \leq \min\left(\sum_{i=1}^{N} tf_{ij}, \sum_{i=1}^{N} tf_{ik} \right).$$

Also, unless the terms j and k are negatively co-related, we have

$$\frac{df_{jk}}{N} \geq \frac{df_j}{N} \times \frac{df_k}{N}.$$

That is, the probability that a random document contains both terms j and k is at least as large as that probability when j and k are independent.

Now, consider W'_{jk}, defined as:

$$W'_{jk} = \frac{\min\left(\sum_{i=1}^{N} tf_{ij}, \sum_{i=1}^{N} tf_{ik}\right) \times \log(\frac{N}{df_j} \times \frac{N}{df_k})}{\sum_{i=1}^{N} d_{ij}} \times WeightingFactor(k).$$

[1] This condition holds unless terms j and k are *negatively* co-related.

We have $W'_{jk} \geq W_{jk}$. Note that all the quantities that are needed to compute W'_{jk} are made available in the automatic indexing phase. Hence, W'_{jk} can be computed in constant time. Figure 3 shows the function Weight1 for computing W'_{jk} and Figure 4 shows Algorithm B, which uses Weight1 as a pruning test.

WEIGHT1(j,k)
1 $sum\_d_{ijk} \leftarrow \min(sum\_tf_{ij}, sum\_tf_{ik}) \times \log((N \times N)/(df_j \times df_k))$
2 $sum\_d_{ij} \leftarrow sum\_tf_{ij} \times \log(N/df_j)$
3 $weighting\_factor\_k \leftarrow \log(N/df_k)/\log N$
4 **return** $sum\_d_{ijk} \times weighting\_factor\_k / sum\_d_{ij}$

Fig. 3. Function Weight1

ALGORITHM-B(λ)
1 (* Automatic indexing phase *)
2 (* same as Algorithm A *)
3 (* Co-occurrence analysis phase *)
4 **for** each $C(j,k) = 1$
5 **do**
6 **if** Weight1(j,k)(*or Weight1*(k,j)) $\geq \lambda$
7 **then** output Weight(j,k)(*or Weight*(k,j))

Fig. 4. Algorithm B

3.2 Algorithm C

With W'_{jk}, we replace the term N/df_{jk} by the bound $N^2/(df_j \cdot df_k)$. This bound can be very loose. Many weak associations W_{jk} may have the estimate W'_{jk} exceeds λ and hence the expensive Weight function is called. To improve the effectiveness of pruning, we consider another association estimate.

We first consider how the quantity df_{jk} can be estimated efficiently. For each term j, we compute a *signature*, S_j. Each signature is an array of Q bits (with indices from 0 to $Q-1$). Let $H()$ be a hash function such that, given a document i, $H(i)$ returns an index in $[0..Q-1]$. For each document i, $S_j[H(i)]$ is set to 1 if document i contains term j; otherwise $S_j[H(i)]$ is 0.

Given two terms j and k, we estimate df_{jk} by counting the number of '1' bits in the result of applying the bit-wise AND operation on the signatures S_j and S_k. We denote this estimate by $\hat{df}_{jk}$. Note that the estimate $\hat{df}_{jk}$ could be incorrect if Q is too small (which leads to many hash collisions). However, as we have observed in our experimental study, setting Q to 5% of the total number of documents in a collection is sufficient to reduce the error probability to a negligible level. Also, we note that computing $\hat{df}_{jk}$ using the small bit-vector signatures is much faster than computing the exact value of df_{jk} by scanning the inverted lists.

Another estimate we consider is to approximate the value $\sum_{i=1}^{N} tf_{ijk}$. First, if we use $\max_p(tf_{pj})$ to denote the largest term frequency of a term j over any document, then clearly, $tf_{ij} \leq \max_p(tf_{pj})$ for any document i. Hence,

$$tf_{ijk} = \min(tf_{ij}, tf_{ik}) \leq \min(\max_p(tf_{pj}), \max_p(tf_{pk})).$$

Thus,

$$\begin{aligned}\sum_{i=1}^{N} tf_{ijk} &= \sum_{j,k\in \text{ doc } i} tf_{ijk} \leq \sum_{j,k\in \text{ doc } i} \min(\max_p(tf_{ij}), \max_p(tf_{ik})) \\ &= df_{jk} \times \min(\max_p(tf_{pj}), \max_p(tf_{pk})).\end{aligned}$$

Substituting these estimates to the association formula, we obtain

$$W''_{jk} = \frac{\min\left(\max(tf_{ij}), \max(tf_{ik})\right) \times \hat{df}_{jk} \times \log(\frac{N}{\hat{df}_{jk}})}{\sum_{i=1}^{N} d_{ij}} \times \mathit{WeightingFactor}(k).$$

Note that W''_{jk} can be computed in constant time.

Figure 5 shows the function Weight2 for computing W''_{jk}. Comparing Weight1 and Weight2, we note that Weight1 is a bit more efficient than Weight2 since no bit vector processing is needed. However, the estimate computed by Weight1 is less tight. Our algorithm (Algorithm C) thus uses Weight1 as the first pruning test, then applies Weight2 if necessary. Figure 6 shows Algorithm C.

WEIGHT2(j, k)
1 $df_{jk} \leftarrow$ number of '1' bits in $S_j \wedge S_k$
2 $sum\_d_{ijk} \leftarrow \min(\max\_tf_{pj}, \max\_tf_{pk}) \times df_{jk} \times \log(N/df_{jk})$
3 $sum\_d_{ij} \leftarrow sum\_tf_{ij} \times \log(N/df_j)$
4 $weighting\_factor\_k \leftarrow \log(N/df_k)/\log N$
5 **return** $sum\_d_{ijk} \times weighting\_factor\_k / sum\_d_{ij}$

Fig. 5. Function Weight2

3.3 Algorithm D

In Algorithm C, the quantity df_{jk} is estimated by processing the signatures S_j and S_k. We note that df_{jk} appears only in a log term (see Equation 2). A mild mis-estimation of df_{jk} should not affect the value of the association W_{jk} by much. Moreover, by inspecting a few document collections, we discovered that

```
ALGORITHM-C(λ)
 1   (* Automatic indexing phase *)
 2   C ← 0
 3   for each document i
 4   do
 5       for each term j in document i
 6       do
 7           tf_ij ← no. of occurrence of j in i
 8           append (i, tf_ij) to j's inverted list
 9           S_j[H(i)] ← 1
10           sum_tf_ij ← sum_tf_ij + tf_ij
11           max_tf_pj ← max(max_tf_pj, tf_ij) ; df_j ← df_j + 1
12       for each term pair j and k (j < k) in document i
13       do
14           C(j,k) ← 1 (* j,k have non-zero associations *)
15   (* Co-occurrence analysis phase *)
16   for each C(j,k) = 1
17   do
18       if Weight1(j,k)(or Weight1(k,j)) ≥ λ
19         then if Weight2(j,k)(or Weight2(k,j)) ≥ λ
20                then output or Weight(j,k)(or Weight(k,j))
```

Fig. 6. Algorithm C

for terms that have a not-too-small association (e.g., > 0.6), it is almost always the case that $df_{jk} = \min(df_j, df_k)$. Our next algorithm uses this observation to compute a quick estimate, W'''_{jk}, for the association between terms j and k:

$$W'''_{jk} = \frac{\min\left(\sum_{i=1}^{N} tf_{ij}, \sum_{i=1}^{N} tf_{ik}\right) \times \log(\frac{N}{\min(df_j, df_k)})}{\sum_{i=1}^{N} d_{ij}} \times WeightingFactor(k).$$

Same as Weight1, we estimate $\sum_{i=1}^{N} tf_{ijk}$ by an upper bound $\min\left(\sum_{i=1}^{N} tf_{ij}, \sum_{i=1}^{N} tf_{ik}\right)$. Also, we use $\min(df_j, df_k)$ to approximate df_{jk}. Figure 7 shows the function Weight3 for computing W'''_{jk}. Figure 8 shows Algorithm D which uses Weight3 as the pruning test.

```
WEIGHT3(j,k)
1  sum_d_ijk ← min(sum_tf_ij, sum_tf_ik) × log(N/min(df_j, df_k))
2  sum_d_ij ← sum_tf_ij × log(N/df_j)
3  weighting_factor_k ← log(N/df_k)/log N
4  return sum_d_ijk × weighting_factor_k/sum_d_ij
```

Fig. 7. Function Weight3

```
ALGORITHM-D(λ)
1   (* Automatic indexing phase *)
2   (* same as Algorithm A *)
3   (* Co-occurrence analysis phase *)
4  for each C(j, k) = 1
5  do
6      if Weight3(j, k)(or Weight3(k, j)) ≥ λ
7         then output Weight(j, k)(or Weight(k, j))
```

Fig. 8. Algorithm D

4 Performance Evaluation

In this section we compare the performance of the four algorithms. Due to space limitation, only some representative results are shown. We applied the algorithms on a few document collections. For example, we had a medical document collection (Medlars) taken from the SMART project [1], and a collection of web pages taken from a news web site. For illustration, we show the performance results using the collection of news documents.

The document collection consists of 22,613 documents with 55,772 terms. The document database is 20.9 MBytes large (after stop-word removal and stemming). There are about 1.2×10^8 non-zero associations in the dataset. The bar graph in Figure 9 shows the frequency distributions of the non-zero associations under 10 ranges of association values. This distribution shows that only a very small fraction of the associations have large values.

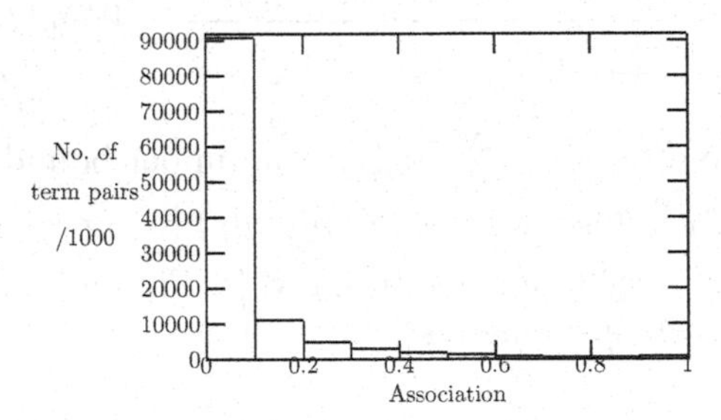

Fig. 9. Distribution of non-zero associations

We ran the algorithms A, B, C, and D on a 250 MHz UltraSparc machine. Figure 10 shows the runtime of the algorithms. For Algorithm B, the number of bits for a signature vector is set to 5% of the total number of documents in the collection (i.e., 2,789 bits).

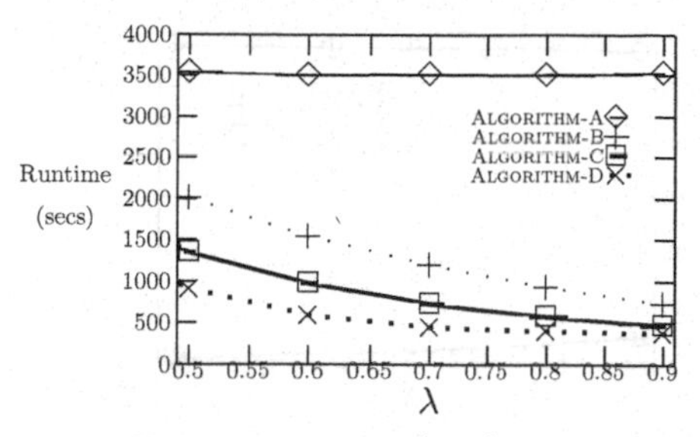

Fig. 10. Runtime (secs) of algorithms

From the figure, we see that Algorithm A is the slowest, taking about 3,500 seconds to complete. The runtime of Algorithm A is independent of λ, since it simply computes all non-zero associations. Algorithms B, C, and D are much faster than A. For example, when $\lambda = 0.9$, Algorithm D only takes 380 seconds – a speedup of 9.2 times over Algorithm A.

Among the three pruning algorithms, Algorithm D is the most efficient, followed by C, and then by B. Moreover, their speed increases with λ. This is because a larger λ value means that more associations are considered weak. This allows the algorithms to prune more weak associations.

It is interesting to see how much *unnecessary* work that each algorithm has done. Figure 11 shows the number of weak associations that each algorithm has computed (using the Weight function). The figure shows, for example, that when $\lambda = 0.9$, Algorithm B computed about 20 million associations whose values are less than 0.9. Algorithm A, on the other hand, computed about 110 million of such associations, or 5.5 times as many as that of Algorithm B. Algorithms C and D are even more effective in pruning weak associations. For example, at $\lambda = 0.9$, Algorithm D only computed 349638 weak associations, or 3% of that of Algorithm A. Finally, we note that even though algorithms C and D are equally effective in pruning weak associations, Algorithm C is less efficient than Algorithm D. This is because Algorithm C uses signature vectors S_j and S_k to estimate df_{jk}. Processing these signatures requires $O(Q)$ time, where Q is the number of bits in a signature vector. Computing the function Weight2 thus takes $O(Q)$ time. Algorithm D, on the other hand, computes the function Weight3 in constant time. Hence, Algorithm D is faster.

As we have discussed, the efficiency of the pruning algorithms come from the use of estimates that act as pruning tests to filter out weak associations (without computing them). Although very unlikely, it is possible that these pruning tests fail. That is to say, it is possible (with a very small probability) that a strong association is mistakenly indicated by a pruning test to be weak. In such a case, the association is (erroneously) not computed. We compare the output of Algorithm A with those of the three pruning algorithms. Fortunately, we discover that only a very small fraction of the strong associations are discarded. In particular, Algorithm B never misses any strong associations; also, for $\lambda \geq 0.7$, algorithms C and D missed at most 8 and 7 strong associations (out of 1844552).

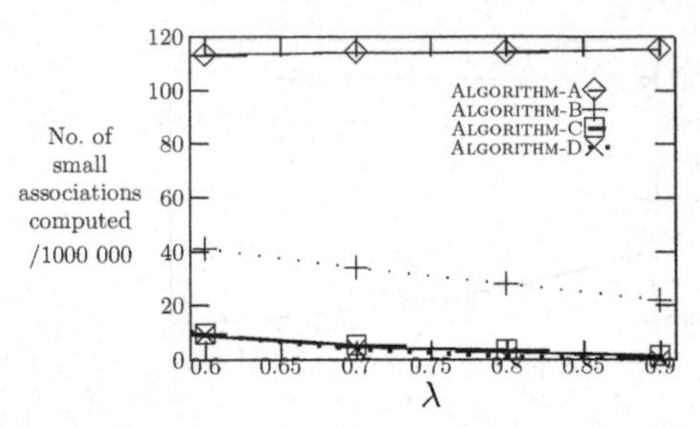

Fig. 11. Number of weak associations computed by the algorithms

5 Conclusion

This paper studied the problem of concept space construction. Previous studies have shown that the concept space approach to automatic thesaurus construction is a useful tool for information retrieval. The construction of concept spaces, however, is very time consuming. In many applications, a full concept space is not needed, in particular, only strong associations are used. We proposed three pruning algorithms for constructing concept spaces containing only strong associations. We evaluated these algorithms using a number of document collections. We found that the three pruning algorithms are very effective in avoiding the computation of unneeded weak associations. A 10-time speedup of the construction process can be achieved.

References

1. *The SMART retrieval system.* ftp://ftp.cs.cornell.edu/pub/smart/med.
2. R. Baeza-Yates and Berthier Ribeiro-Neto. *Modern Information Retrieval.* Addison Wesley, 1999.
3. Hsinchun Chen, Joanne Martinez, Tobun D. Ng, and Bruce R. Schatz. A concept space approach to addressing the vocabulary problem in scientific information retrieval: an experiment on the worm community system. *Journal of American Society for information Science*, 48(1):17–31, 1997.
4. W.B. Frakes and R. Baeza-Yates. *Information Retreival: Data Structures and Algorithms.* Prentice Hall, 1992.
5. G.W. Furnas et al. The vocabulary problem in human-system communicaiton. *Comm. ACM*, 30(11):964–971, 1987.
6. H.Chen and K.J. Lynch. Automatic construction of networks of concepts characterizing document databases. *IEEE Transaction of Systems, Man, and Cybernetics*, 22(5):885–902, Sep/Oct 1992.
7. B.R. Schatz, E. Johnson, P. Cochrane, and H. Chen. Interactive term suggestion for users of digital libraries: using subject thesauri and co-occurrence lists for information retrieval. In *Digital Library 96*, Bethesda MD, 1996.

Topic Detection, Tracking, and Trend Analysis Using Self-Organizing Neural Networks

Kanagasabi Rajaraman and Ah-Hwee Tan

Kent Ridge Digital Labs
21 Heng Mui Keng Terrace
Singapore
{kanagasa,ahhwee}@krdl.org.sg
http://textmining.krdl.org.sg

Abstract. We address the problem of Topic Detection and Tracking (TDT) and subsequently detecting trends from a stream of text documents. Formulating TDT as a clustering problem in a class of self-organizing neural networks, we propose an incremental clustering algorithm. On this setup we show how trends can be identified. Through experimental studies, we observe that our method enables discovering interesting trends that are deducible only from reading all relevant documents.

Keywords: Topic detection, topic tracking, trend analysis, text mining, document clustering

1 Introduction

In this paper, we address the problem of analyzing trends from a stream of text documents, using an approach based on the Topic Detection and Tracking initiative. Topic Detection and Tracking (TDT) [1] research is a DARPA-sponsored effort that has been pursued since 1997. TDT refers to tasks on analyzing time-ordered information sources, e.g news wires. Topic detection is the task of detecting topics that are previously unknown to the system[8]. Topic here is an abstraction of a cluster of stories that discuss the same event. Tracking refers to associating incoming stories with topics (i.e. respective clusters) known to the system[8]. The topic detection and tracking formalism together with the time ordering of the documents provides a nice setup for tracing the evloution of a topic. In this paper, we show how this setup can be exploited for analyzing trends.

Topic detection, tracking and trend analysis, the three tasks being performed on incoming stream of documents, necessitate solutions based on incremental algorithms. A class of models that enable incremental solutions are the Adaptive Resonance Theory (ART) networks[2], which we shall adopt in this paper.

D. Cheung, G.J. Williams, and Q. Li (Eds.): PAKDD 2001, LNAI 2035, pp. 102–107, 2001.

2 Document Representation

We adopt the traditional vector space model[6] for representing the documents, i.e. each document is represented by a set of keyword features. We employ a simple feature selection method whereby all words appearing in less than 5% of the collection are removed and, from each document, only the top n number of features based on $tf.idf$ ranking are picked. Let M be the number of keyword features selected through this process. With these features, each document is converted into a keyword weight vector

$$\mathbf{a} = (a_1, a_2, \ldots, a_M) \quad (1)$$

where a_j is the normalized word frequency of the keyword w_j in the keyword feature list. The normalization is done by dividing each word frequency with the maximum word frequency.

We assume that text streams are provided as document collections ordered over time. The collections must be disjoint sets but could have been collected over unequal time periods. We shall call these time-ordered collections as *segments*.

3 ART Networks

Adaptive Resonance Theory (ART) networks are a class of self-organizing neural networks. Of the several varieties of ART networks proposed in the literature, we shall adopt the fuzzy ART networks[2] .

Fuzzy ART incorporates computations from fuzzy set theory into ART networks. The crisp (nonfuzzy) intersection operator ($\cap$) that describes ART 1 dynamics is replaced by the fuzzy AND operator ($\wedge$) of fuzzy set theory in the choice, search, and learning laws of ART 1. By replacing the crisp logical operations of ART 1 with their fuzzy counterparts, fuzzy ART can learn stable categories in response to either analog or binary patterns.

Each fuzzy ART system includes a field, F_0, of nodes that represents a current input vector; a field F_1 that receives both bottom-up input from F_0 and top-down input from a field, F_2, that represents the active code or category. The F_0 activity vector is denoted $\mathbf{I}$. The F_1 activity vector is denoted $\mathbf{x}$. The F_2 activity vector is denoted $\mathbf{y}$.

Due to space constraints, we skip the description of fuzzy ART learning algorithm. The interested reader may refer to [2] for details.

4 Topic Detection, Tracking, and Trend Analysis

In this section we present our topic detection, tracking and trend analysis methods.

4.1 Topic Detection Algorithm

As described in Section 3, ART formulates recognition categories of input patterns by encoding each input pattern into a category node in an unsupervised manner. Thus each category node in F_2 field encodes a cluster of patterns. In other words, each node represents a topic. Hence, identification of new topics translates to the method of creation of new categories in the F_2 field as more patterns are presented. Using this idea, we derive the topic detection algorithm in Table I.

Table 1. Topic Detection Algorithm.

Step 1. Initialize network and parameters.
Step 2. Load previous network and cluster structure, if any.
Step 3. Repeat
- present the document vectors
- train the net using fuzzy ART Learning Algorithm

until convergence
Step 4. Prune the network to remove low confidence category nodes
Step 5. Save the net and cluster structure.

4.2 Topic Tracking Algorithm

For tracking new documents, the latest topic structure is loaded before processing the documents. For an incoming document, the activities at the F_2 field are checked to select the winning node, i.e. the one receiving maximum input. The document is then assigned to the corresponding topic. This is the idea behind the tracking algorithm presented in Table II.

Table 2. Topic Tracking Algorithm.

Step 1. Initialize network and parameters.
Step 2. Load previous network and cluster structure, if any.
Step 3. Present the document to be tracked, to the net
Step 4. Assign the document to the topic corresponding to the winning category node, i.e. category node that receives maximum input.

4.3 Trend Analysis

The topic detection and tracking setup together with the time ordering of the documents provides a natural way for topic-wise focussed trend analysis. In particular, for every topic, suppose we plot the number of documents per segment versus time. This plot can be thought of as a trace of the evolution of a topic. The 'ups' and 'downs' in the graph can be used to deduce the trends for this topic. For more specific details on the trends, one can zoom in and view documents on this topic segment-wise. This process is illustrated in the following section.

5 Experiments

For our experiments, we have grabbed daily news articles from CNET and ZD-Net and grouped the articles into weekly segments. Starting from 1st week of September 2000 up till 4th week of October 2000, we collected 8 segments in all. Totally there were 1468 documents at an average of about 180 documents in each segment. Documents in each segment are converted into weight vectors as described in Section 2. We then applied our topic detection and tracking and performed trend analysis. Some qualitative results are presented below:

5.1 Topic Detection and Tracking

Typically we observed 10 to 15 new topics being identified per segment when choice parameter $\alpha = 0.1$ and vigilance parameter $\rho = 0.01$ (ignoring small clusters with 1 or 2 documents only).

A list of some of the hot topics that have been identified by the topic detection algorithm can be viewed at http://textmining.krdl.org.sg/people/kanagasa/tdt. The tracking results can also be viewed at the same URL. We skip the details due to space constraints.

5.2 Trend Analysis

The evolution graphs for some selected topics are shown below. Time is represented through the segment ID which takes values $1,\cdots,8$. ID=1 corresponds to Sep 1st week, ID=2 corresponds to Sep 2nd week and so on.

The topics 'MS Case' (i.e. Microsoft Case), 'Linux' and 'Windows ME' have been plotted in Fig 1. The 'MS Case' topic shows an initial up trend early September. An examination of the documents under this topic reveals the reason to be Bristol Technology ruling against Microsoft. Similarly the topic on 'Linux' shows a peak for early October when the Open source conference was held. 'Windows ME' graph peaks during September 2nd week coinciding with Win ME release.

The topics 'Apple' and 'Hackings' have been plotted in Fig 2. The 'Apple' topic shows an up trend during mid September when Apple Expo was on. The Microsoft hack-in can be seen to have lead to the sudden peak in 'Hackings' topic around late October.

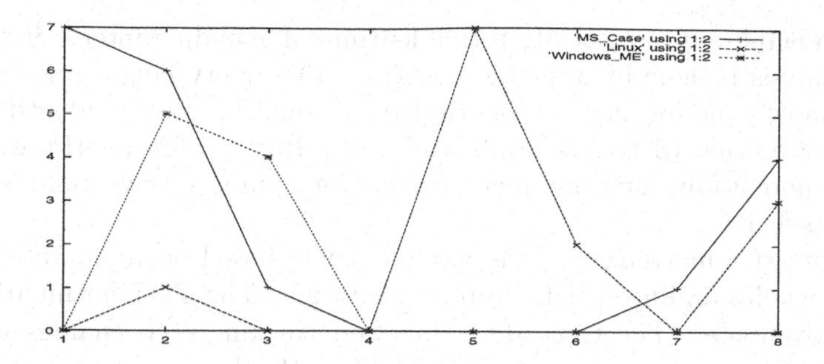

Fig. 1. Trends for 'MS Case', 'Linux' and 'Windows ME'.

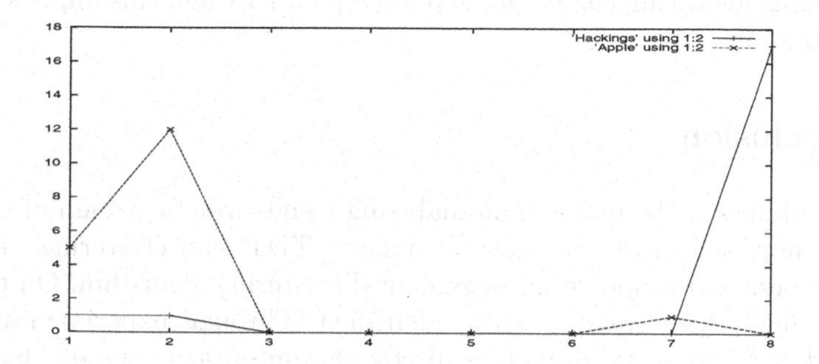

Fig. 2. Trends for 'Apple' and 'Hackings'.

The above study thus shows that our method can be used to detect hot topics automatically and track the evolution of detected topics. The method also serves to spot emerging trends (with respect to the timescale defined) on topics of interest.

6 Related Work

TDT research has been predominantly 'pure IR' based and can be categorized as based on either incremental clustering (e.g. [7]) or routing-queries (e.g. [5]). (One notable exception is the tracking method by Dragon Systems which is based on language-modelling techniques.) Incremental clustering based methods come the closest to our work, but we use ART networks for document processing in contrast to the traditional document similarity measures. Our main motivation is that ART networks enable truly incremental algorithms.

Trend analysis for numerical data has been well investigated. For free-text, where the challenge is tough, we are aware of only very few papers. [3] defines concept distributions and propose a trend analysis method by comparing distributions of old and new data. Typically, the trends discovered are of the type "keyword 'napster' appeared x% more now than in old data", "keyword 'divx' appeared y% less now than in old data", etc. [4] uses the popular a-priori algorithm employed in association-rule learning, for finding interesting phrases. Trend analysis is done by applying a shape based query language on the identified phrases. Queries like 'Up' or 'BigDown' could be used to identify upward and strong downward trends respectively, in terms of phrases. However, there could be potentially large number of candidate phrases that could make this method inefficient.

In contrast, our trend analysis method being based on topic detection and tracking enables finding specific, topic-wise trends. The TDT formulation offers several advantages. The topic detection and tracking step enables the trend analysis be focussed and more meaningful. Since the documents under each topic are relatively small, the analysis can be done efficiently. (On a related note, the ART learning algorithm can be implemented parallelly and this implies potential further speedup.)

7 Conclusion

We have addressed the problem of analyzing trends from a stream of text using the TDT approach. First we have formulated TDT as a clustering problem in ART networks and proposed an incremental clustering algorithm. On this setup we have shown how trends can be identified. Through experimental studies, we have found our method enables discovering interesting trends that are not directly mentioned in the documents but deducible only from reading all relevant documents.

References

1. TDT homepage. `http://www.itl.nist.gov/iad/894.01/tests/tdt/index.htm`, 2000.
2. G. A. Carpenter, S. Grossberg, and D. B. Rosen. Fuzzy ART: Fast stable learning and categorization of analog patterns by an adaptive resonance system. *Neural Networks*, 4:759–771, 1991.
3. R. Feldman and I. Dagan. Knowledge discovery in textual databases (KDT). In *Proceedings of KDD-95*, 1995.
4. Brian Lent, Rakesh Agrawal, and R. Srikant. Discovering trends in text databases. In *Proceedings of KDD-97*, 1997.
5. Ron Papka, James Allan, and Victor Lavrenko. UMASS approaches to detection and tracking at TDT2. In *Proceedings of the TDT-99 workshop*. NIST, 1999.
6. G. Salton and M. J. McGill. *Introduction to Modern Information Retrieval*. McGraw-Hill, New York, 1983.
7. Fredrick Walls, Hubert Jin, Sreenivasa Sista, and Richard Schwartz. Topic detection in broadcast news. In *Proceedings of the TDT-99 workshop*. NIST, 1999.
8. Charles Wayne. Overview of TDT. `http://www.itl.nist.gov/iaui/894.01/tdt98/doc/tdtslides/sld001.htm`, 1998.

Automatic Hypertext Construction through a Text Mining Approach by Self-Organizing Maps

Hsin-Chang Yang and Chung-Hong Lee

Department of Information Management, Chang Jung University, Tainan, 711, Taiwan
{hcyang, leechung}@mail.cju.edu.tw

Abstract. In this work we developed a new automatic hypertext construction method based on a proposed text mining approach. Our method applies the self-organizing map algorithm to cluster some flat text documents in a training corpus and generate two maps. We then use these maps to identify the sources and destinations of some important hyperlinks within these training documents. The constructed hyperlinks are then inserted into the training documents to translate them into hypertext form. Such translated documents form the new corpus. Incoming documents can also be translated into hypertext form and added to the corpus through the same approach. Our method had been tested on a set of flat text documents collecting from several newswire sites. Although we only used Chinese text documents, our approach can be applied to any document that can be transformed to a set of indexed terms.

1 Introduction

The use of hypertexts for information representation has been widely recognized and accepted because they provide a feasible mechanism to retrieve related documents. Unfortunately, most hypertext documents were created manually using some kind of authoring tools. Although such authoring tools are easy to use and provide sufficient functionality for individual users, manual construction of hypertexts is still, if not impossible, a very hard work. Moreover, manual construction is always unstructured and unmethodological. To remedy such inefficiency, we need a method to automatically transform a 'flat' text into a hypertext rather than creating a hypertext from the ground up. Thus research of automatic hypertext construction arises rapidly in recent years.

To transform a flat text to a hypertext we need to decide where to insert a hyperlink in the text. A hyperlink connects between two documents where at one end of the hyperlink is the source text which may be an individual term or a sentence and at the other end is the destination text which may be another document or a different location of the same document. Different types of hyperlinks may be used depending to the kinds of functionality that need to be implemented by the hypertext. For example, according to Agosti et al. [1], there are three types of hyperlinks, namely structural links, referential links, and associative links. The first two types of hyperlinks are usually explicit and may be

D. Cheung, G.J. Williams, and Q. Li (Eds.): PAKDD 2001, LNAI 2035, pp. 108–113, 2001.

easily created manually or automatically. The associative hyperlinks, however, require a understanding of the semantics of the connecting documents. Such understanding also requires much human effort. Nowadays, automatic creating of associative hyperlinks, or semantic hyperlinks, plays a central role in the development of automatic hypertext construction methods because these hyperlinks may provide the most effective exploring paths to fulfill the users' information need. The critical point in creating associative hyperlinks is to find the documents which are semantically relevant to the sources of these hyperlinks. Such semantic relevance could be revealed by a text mining process.

Text mining concerns of discovering knowledge from a textual database and attracts much attention from both researchers and practitioners. The problem is not easy to tackle due to the semi-structured or unstructured nature of the text documents. Many approaches have been devised in recent years(for example, [2]). In this work, we apply the self-organizing map model to perform the text mining process. Since the self-organizing process could reveal the relationship among documents as well as indexed terms, such text mining process may also be used to find the associative hyperlinks.

2 Related Work

Research on automatic construction of hypertext arose mostly from the information retrieval field. A survey of the use of information retrieval techniques for the automatic hypertext construction can be found in [1]. There was no neural network based methods had made significant contribution in this field according to their survey.

Text mining has received lots of attention in recent few years. Many researchers and practitioners have involved in this field using various approaches [7]. Among these approaches, the self-organizing map (SOM) [3] models played an important role. Lots of works had used SOM to cluster large collection of text documents. However, there is few works had applied text mining approaches, particularly the SOM approach, to automatic hypertext construction. One close work by Rizzo et al. [5] used the SOM to cluster hypertext documents. However, their work was used for interactive browsing and document searching, rather than hypertext authoring.

3 Text Mining by Self-Organizing Maps

Before we can create hyperlinks, we first perform a text mining process, which is similar to the work described in [4], on the corpus. The popular self-organizing map (SOM) algorithm is applied to the corpus to cluster documents. We adopt the vector space model [6] to transform each document in the corpus into a binary vector. These document vectors are used as input to train the map. We then apply two kinds of labeling process to the trained map and obtain two feature maps, namely the document cluster map (DCM) and the word cluster map (WCM). In the document cluster map each neuron represents a document cluster which contains several similar documents with high word co-occurrence. In the word cluster map each neuron represents a cluster of words which reveal the general concept of the corresponding document cluster associated to the same neuron in the document cluster map.

Since the DCM and the WCM use the same neuron map, a neuron represents a document cluster as well as a word cluster simultaneously. By linking the DCM and the WCM according to the neuron locations we may discover the underlying ideas of a set of related documents. This is essentially a text mining process because we can discover related terms among a set of related documents through their co-occurrence patterns which are hard to extract directly. By virtue of the SOM algorithm, terms that are often co-occurred will tend to be labeled to the same neuron, or neighboring neurons because the neurons are willing to learn all of them simultaneously. Thus the co-occurrence patterns of indexed terms may be revealed. Moreover, the terms that associated to the same neuron also reveal the common themes of the associated documents of the neuron. Essentially these related terms construct a thesaurus that is derived from the context of the underlying documents rather than their grammatical meaning. Thus the indexed terms associated to the same neuron in the WCM compose a pseudo-document which represents the general concept of the documents associated to that neuron. Therefore our approach may also generate a thesaurus automatically, another text mining application.

4 Automatic Hypertext Construction

We divide the hypertext construction process into two parts. In the first part we concern about finding the *source* of a hyperlink. In the second part we will try to find the *destination* of a hyperlink. The entire process will be described in the following subsections.

4.1 Finding Sources

To find the source of a hyperlink within a document, we should first decide the important terms that are worth further exploration. In this work, two kinds of terms are used as sources. The first kind of terms include the terms that are the themes of other documents but not of this document. Such terms are generally recognized as necessary sources of hyperlinks because they fulfill users' need during browsing this document. We call these hyperlinks the *inter-cluster hyperlinks* because they often connect documents which locate on different document clusters in the DCM. The reason of such disparity is that a document cluster in the DCM contains documents that have common terms often co-occur. That is, the corresponding word cluster will contain these common terms. Therefore, the first kind of terms are used in creating inter-cluster hyperlinks.

The second kind of terms include terms that are the themes of this document. This kind of terms are used to include documents that are related to this document for referential purpose. Such hyperlinks may be created by adding links between each pair of documents associated to the same document cluster in the DCM. Since these documents share some common concepts after the text mining process, we may consider them related and use them to create the *intra-cluster hyperlinks.*

In the following we will describe how to obtain the sources of these two kinds of hyperlinks. To create a inter-cluster hyperlink in a document D_j associated to a word cluster W_c, we may find its source by selecting a term that is associated to other word clusters but not W_c. That is, a term k_i is selected as a source if:

$$k_i \notin W_c \text{ and } k_i \in W_m \text{ for } m \neq c, \tag{1}$$

where W_m is the set of words associated to neuron m in the WCM and W_c is the word cluster associated to the document cluster that contains D_j. To find the sources of the intra-cluster hyperlinks in document D_j, we simply find the terms in the word cluster W_c which is associated to the document cluster that contains D_j. That is, we select all k_i if:

$$k_i \in W_c. \tag{2}$$

4.2 Finding Destinations

It is straightforward to find the destinations of hyperlinks after determining the sources as described in Sec. 4.1. For an inter-cluster hyperlink in document D_j with source k_i, we assign a document D_i as its destination if it fulfills the following requirements:

1. D_j and D_i belong to different document clusters and $k_i \notin W_c$.
2. $k_i \in W_m, m \neq c$ and
$$w_{im} = \max_{l \neq c, 1 \leq l \leq M} w_{il}, \tag{3}$$
where c is the neuron index of the word cluster that contains D_j.
3. The distance between D_i and D_j is minimum, i.e.
$$||\mathbf{d}_i - \mathbf{d}_j|| = \min_{1 \leq l \leq N} ||\mathbf{d}_l - \mathbf{d}_j||. \tag{4}$$

The first requirement states that the destination document should be reasonably differed from the source document and the source document should not have k_i as its theme. The second requirement selects the word cluster that is the most relevant to k_i. Since a word cluster may have several documents associated to it, we need the third requirement to choose the most similar one.

To find the destination of a intra-cluster hyperlink starting from $k_i \in W_c$, we simply connect it to a document associated to neuron c in the DCM because this document cluster contains the most related documents. Since a document cluster may contain multiple documents, the document which has minimum distance to the source document (the document containing the source) will be selected as the destination of this hyperlink. This is also formulated in Eq. 4 where $\mathbf{d}_j$ and $\mathbf{d}_i$ are the source and destination documents, respectively.

5 Experimental Results

The experiments were based on a corpus collected by the authors. The test corpus contains 3268 news articles which were posted by the CNA (Central News Agency[1]) during Oct. 1, 1996 to Oct. 10, 1996. Each article contains a subject line starting with a '@'. The test documents were written in Chinese, so we applied a Chinese term extraction program to every document in the corpus and obtained the indexed terms for each document. The overall vocabulary contains 10937 terms for these 3268 documents. To reduce the number of terms, we manually discarded those terms that occur only once in a document. We also discarded those terms that occur in a manually constructed stoplist which contains 259 Chinese terms. This reduces the number of indexed terms to 1976.

Each document is transformed to an 1976-dimensional binary vector and fed into a SOM network for training. The network contains a 20×20 grid of neurons which each is represented by a 1976-dimensional synaptic weight vector. All documents were used to train the network. We set the maximum training epoch to 500 and the training gain to 0.4. After the training process, we labeled the neurons with documents and terms respectively and obtained the DCM and WCM. We started creating hypertext documents after obtaining the DCM and the WCM. The sources and destinations were determined by the method described in Sec. 4. The spanning factors σ_1 and σ_2 were set to 10 and 5 respectively. For each document we also generated an aggregate link which contains hyperlinks to all relevant documents. The flat text documents were then converted to their corresponding hypertext form by a text conversion program. We adopted the standard HTML format to represent the hypertexts for easy access via Internet. An example flat text document and its hypertext form are shown in Figure 1.[2] The underlined terms in the figure were the obtained sources where the italic ones depicted the sources for intra-cluster hyperlinks.

In Figure 1 a intra-cluster hyperlink connects the source document to a randomly selected document in the same document cluster. A hyperlink was created on every occurrence of each source term. The aggregate links were appended to the converted hypertexts beneath the "Relevant links:" line. The number of relevant documents is equal to the number of documents in the document cluster to which the source document was associated minus one.

6 Conclusions

In this work we devised a novel method for automatic hypertext construction. To construct hypertexts from flat texts we first developed a text mining approach which adopted the self-organizing map algorithm to cluster these flat texts and generate two feature maps. The construction of hypertexts were achieved by analyzing these maps. Two types of hyperlinks, namely the intra-cluster hyperlinks

[1] `http://www.cna.com.tw`

[2] Interested readers may access these hypertext documents at the author's web site `http://www.im.cju.edu.tw/~hcyang/ahc`.

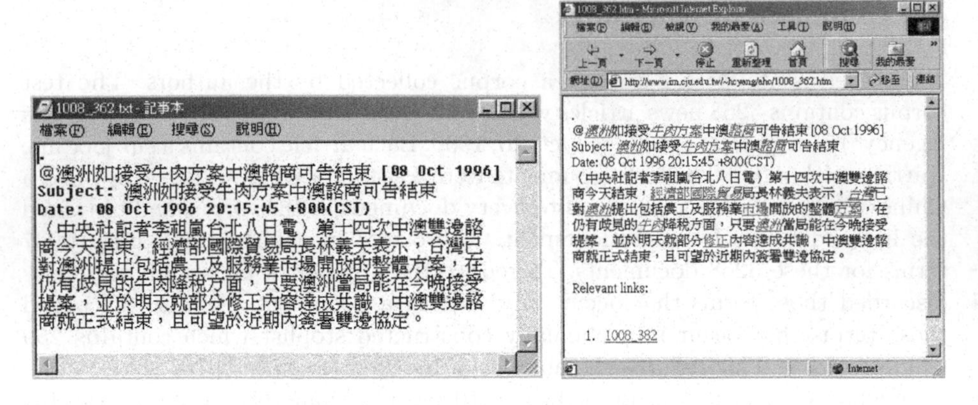

Fig. 1. An example flat text(left) and its hypertext form(right)

and the inter-cluster hyperlinks, were created. The intra-cluster hyperlinks created connections between a document and its relevant documents while the inter-cluster hyperlinks connected a document to some irrelevant documents which reveal some keywords occurred in the source document. Experiments showed that not only the text mining approach successfully revealed the co-occurrence patterns of the underlying texts, but also the devised hypertext construction process effectively constructed semantic hyperlinks among these texts.

References

1. M. Agosti, F. Crestani, and M. Melucci. On the use of information retrieval techniques for the automatic construction of hypertext. *Information Processing & Management*, 33:133–144, 1997.
2. R. Feldman and I. Dagan. Knowledge discovery in textual databases (kdt). In *Proceedings of the First International Conference on Knowledge Discovery and Data Mining (KDD-95)*, pages 112–117, Montreal, Canada, 1995.
3. T. Kohonen. *Self-Organizing Maps*. Springer-Verlag, Berlin, 1997.
4. C. H. Lee and H. C. Yang. A web text mining approach based on self-organizing map. In *Proc. ACM CIKM'99 2nd Workshop on Web Information and Data Management*, pages 59–62, Kansas City, MI, 1999.
5. R. Rizzo, M. Allegra, and G. Fulantelli. Developing hypertext through a self-organizing map. In *Proc. WebNet 98*, pages 768–772, Orlando, USA, 1998.
6. G. Salton and M. J. McGill. *Introduction to Modern Information Retrieval.* McGraw-Hill, New York, 1983.
7. A. H. Tan. Text mining: The state of the art and the challenges. In *Proceedings of PAKDD'99 Workshop on Knowledge discovery from Advanced Databases (KDAD'99)*, pages 65–70, Beijing, China, 1999.

Semantic Expectation-Based Causation Knowledge Extraction: A Study on Hong Kong Stock Movement Analysis

Boon-Toh Low, Ki Chan, Lei-Lei Choi, Man-Yee Chin, and Sin-Ling Lay

Department of Systems Engineering and Engineering Management
Chinese University of Hong Kong
Shatin, N.T., Hong Kong
{btlow, kchan}@se.cuhk.edu.hk

Abstract. Human beings generally analyze information with some kinds of semantic expectations. This not only speeds up the processing time, it also helps to put the analysis in the correct context and perspective. To capitalize on this type of intelligent human behavior, this paper proposes a semantic expectation-based knowledge extraction methodology (SEKE) for extracting causation relations from text. In particular, we study the application of a causation semantic template on the Hong Kong Stock market movement (Hang Seng Index) with English financial news from Reuters, South China Morning Post and Hong Kong Standard. With one-month data input and over a two-month testing period, the system shows that it can correctly analyzes single reason sentences with about 76% precision and 74% recall rates. If partial reason extraction (two out of one reason) is included and weighted by a factor of 0.5, the performance is improved to about 83% and 81% respectively. As the proposed framework is language independent, we expect cross lingual knowledge extraction can work better with this semantic expectation-based framework.

Keywords: knowledge extraction, semantic-based natural language processing, expectation-based information extraction.

1 Introduction

With rapid advances of technologies and the availability of vast information in the World Wide Web, information is easily accessible. This overwhelming load of information, especially in the form of articles (text), calls for good solution(s) in knowledge extraction technology to help understand the contents. The success of this important area will help us handle textual information more efficiently, such as indexing and relevant article searching, and more effectively, such as direct knowledge understanding and gathering.

In the information rich financial sector, active research has been carried out with the information needs of banks and financial companies [1]. Typical examples include information extraction projects about take over activities from financial news (e.g. [2], [3]). In a broader view, the movements of major financial markets around the world such as New York, Tokyo, Europe, Hong Kong. etc., have closer links and greater interests with the general public. Many comments and analytical articles are readily available on the electronic newspapers as well as the information providers.

D. Cheung, G.J. Williams, and Q. Li (Eds.): PAKDD 2001, LNAI 2035, pp. 114-123, 2001.

The articles usually analyze movements of a particular market, typically in some indices (e.g. Dow Jones), with the reason(s) affecting it. These reasons can range from the recent movements of other influencing financial markets, the states of the current world economy and the outlook, possibility of outbreak of war in certain regions, to some micro factors such as policies of local government.

Human can extract the knowledge rather easily but due to the rapid development of the financial situation, very few ordinary people can keep herself up-to-date with the large volume of news articles available. One alternative solution is to develop a knowledge extraction system whose main purpose is to analyze the news articles and provide us with the summaries.

Full natural language understanding and processing has long been recognized as an important topic. However, we are still far away from a truly versatile and general system. By looking at a narrower domain, some researchers have report successes. Most Information Extraction Systems to-date, such as the current NYU proteus system [4] and the SRI FASTUS system [5], use syntactic parsing as the main technology, while some use syntactic parsing with the aid of semantic analysis for solving their information extraction problems.

In this paper, we explore a semantic expectation-based approach for the extraction of financial movement knowledge from news articles. Hong Kong stock market news is readily available to our research team over the internet and it is chosen as the topic of our study. In addition, we have on-line access to Reuters financial information service including financial news. It therefore becomes a part of the information sources.

We first describe the semantic expectation-based knowledge extraction methodology (SEKE) for extracting market movements and the associated reasons in the following section. In section 3, we outline the studies carried out on Hong Kong Stock market and the findings are discussed in section 4.

2 Semantic Expectation-Based Knowledge Extraction (SEKE) for Causation Knowledge

The main purpose of an article is to convey information. In news articles about financial markets, the context of the information is more focused. Readers of these articles usually have certain semantic expectations in mind. For example, some may be interested in knowing the latest market movements, some may want to read about the analysis of cross-market influences, and some may want to know what causes the recent market to move up or down.

To design an effective knowledge extraction system, we can learn from the relevant human behavior (as those developed in the field of Artificial Intelligence). Two main characteristics are being observed:

1. There is always an expected semantic in mind; and
2. The expected semantic is used to guide the search and understanding.

Based on these characteristics, we can design the *semantic expectation-based knowledge extraction system* (SEKE) for causation financial market movement knowledge analysis with the following assumptions:

(a) There are some semantic concepts associated with market movements. In the most general cases, there are three types of movements: *upward*, *downward* and *no movement*.
(b) There are analyses about recent movements and the influencing factor(s) is discussed. The *reason(s)* and *consequence* (market movement) are presented together, most likely in the *same sentence*, or, close to each other.
(c) Although there are many different ways to present the information (sentence styles), the same semantics are usually preserved. *Semantic templates* can be used to extract the encapsulated knowledge base on the expected semantics, and different sentence styles can be associated with the corresponding templates.
(d) The reasons are usually restricted to those associated to the particular market and its movements from recent information sources. It is possible to generate a *set of expected reasons* at the beginning of the knowledge extraction process based on recent information. The same is applicable to the expected consequences.

A causation relation typically has two entities, *reason(s)* and *consequence*, linked by a directional causation indicator. It is expected that market movement analysis also has these two main concepts and its semantic template can be illustrated using the following figure:

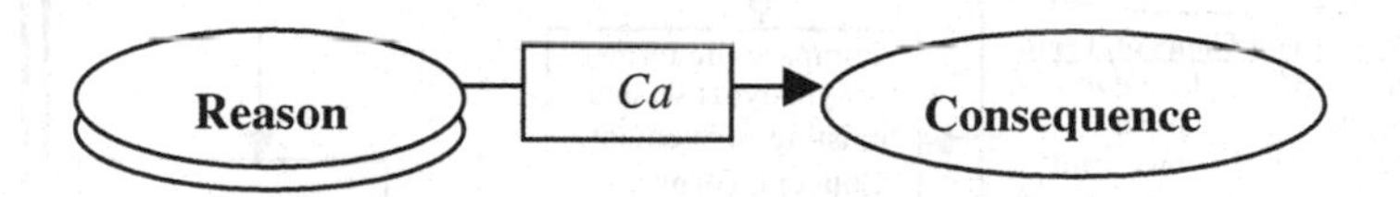

Fig. 1. The causation semantic template

This semantic template states that one or more reasons cause the occurrence of the consequence. As the semantic template is language independent but information is expressed in some languages, we have to associate the sentence styles (usually there are more than one sentence style to express a causality in a given natural language) pertaining to a particular language to it. These sentence styles are called the *sentence templates*.

For a given market, SEKE requires a preliminary study on the expected semantics for the knowledge to be extracted. A set of example/historical texts from the same (or similar) sources will be useful for this purpose. From this set of training data, three groups of information are gathered: the ***sentence templates*** with possibly different styles matching the *semantic template*, the set of concepts matching the ***Reasons*** and the set of concepts matching the ***Consequence***. Some examples of the sentence template, consequences and reasons are given below.

Sentence template:
The movement of stock   is caused by   a factor with its movement
"*Hang Seng Index rose*   *as*   *Wall Street gains*" *(an example sentence)*
where "*Hang Seng Index rose*" is the consequence, "*Wall Street gains*" is the reason, and the "*as*" is the English Language causation expression which links the reason to the consequence in the reversed order as compared to the causation semantic template in Figure 1. This example is also a simplified version of Figure 1 since it only has one reason.

Reasons: *Wall Street gains, US interest rate down, Nasdaq sinks.*
Consequences:
Hang Seng Index rise, Hong Kong Stocks gains (upward movement)
Hang Seng Index drops, Hong Kong Stocks sink (downward movement)
Hang Seng Index unchanged, Hong Kong Stocks barely changed (no movement)

Since most training data set cannot cover the full spectrum of the interested domain, SEKE introduces a way to conglomerate terms expressing similar concepts in future encounters. It makes use of an electronic thesaurus (e.g. WordNet [6], Roget's Thesaurus [7]) to group unseen terms with the same category into the pre-defined concepts. The following figure shows an example on how an unseen work "*rose*" is absorbed into the upward movement concept using WordNet as the thesaurus in SEKE.

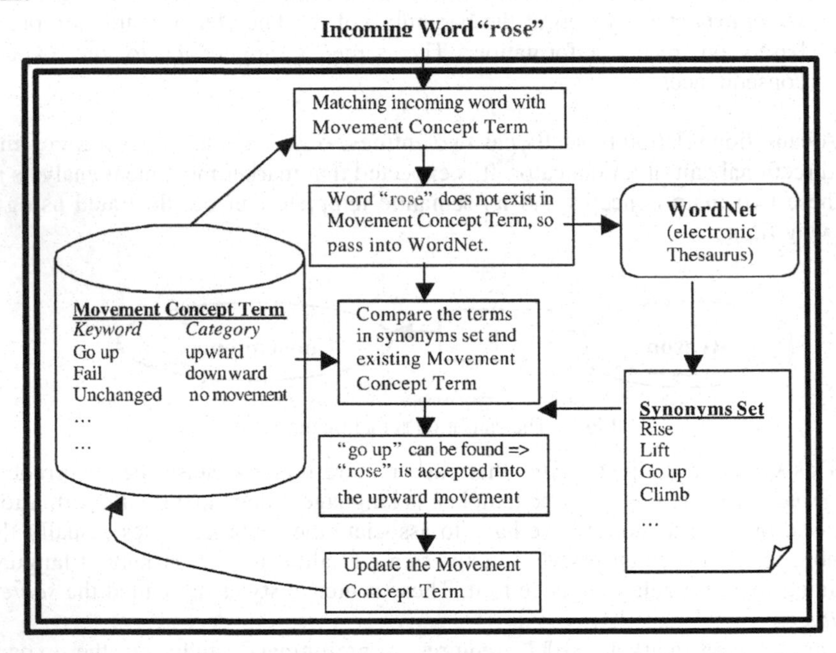

Fig. 2. Movement Extraction through WordNet with an example movement concept term

Due to the limited coverage of any given thesaurus, there is no guarantee that a new term can be found in the thesaurus and it is correctly classified in the context concerned. Thus, it is impossible to fully automate the concept grouping. For example, "*Down Jones*" and "*Wall Street*" denote the same entity to us but they may not be captured in the thesaurus used. Human identification of similar concepts has to be carried out from time to time to resolve the newly encountered terms.

The procedure for the semantic expectation-based knowledge extraction (SEKE) system for causation knowledge is outlined below:

(1) Determine the domain area and scope of the study.
(2) Define the expected semantic template. If there are more than one thing to be extracted, multiple semantic templates have to be defined.

(3) Collect a set of training data set.
(4) From the data, generate the reason and consequence sets, and the set of sentences matching the template.
(5) Analyze and match the concepts and template to the sentences. Collect this set of styles with their correspondences to the semantic templates.
(6) When a new article is fed to the system, use the semantic template with the collected sentence, reason and consequence sets to filter out unrelated information and zoom in to the possibly relevant sentence(s) that can be fully or partially matched.
(7) If the sentence can match the expected semantic template, collect it.
(8) Otherwise, use a thesaurus to check whether the unmatched terms has similar meaning to our collected concepts. If all terms are matched with the semantic template, then collect it.
(9) For those terms that cannot be matched by the thesaurus, human inspection is required.
(10) It is possible that a sentence matches the concept terms but fails on the sentence level. This may mean a possible unseen sentence style and human inspection has to be carried out to decide if it is the required knowledge. Collect it into the sentence concept knowledge base for future use if it expresses the intended knowledge.

3 A Study on Hong Kong Market Movements and the Influencing Reasons

A study to extract the reasons affecting the movements of Hang Seng Index (HSI) in Hong Kong Stock Market based on the SEKE framework is carried out. English news articles for the study were solicited from the most reliable and relevant sources available in Hong Kong, namely Reuters news, South China Morning Post and Hong Kong Standard, the two local English newspapers.

Relevant news articles in December 1999 were collected as the training data. From this set of text, we analyze the single sentence knowledge expressing Hong Kong Stock movements with their influencing reasons. Our observations are summarized below:

1) Both the consequences and reasons have some movements. The causality relation in this study is about how the movements of some factors affect the Hong Kong market's movements (measured by HSI). Therefore, the concept of *movement* is common to both the reasons and the consequences. The corresponding semantic template becomes:

    Some samples sentences are :

    "The Hang Seng Index (HSI) rebounded as the interest rate rise."

    "Hong Kong's Hang Seng Index led the charge downwards, tumbling 7.18 per cent to close at 15,846.72 points due to overseas market drop."

    "Hong Kong stocks opened moderately higher on Thursday led by technology-related companies after the U.S. Federal Reserve increased interest rates by a quarter point as predicted."

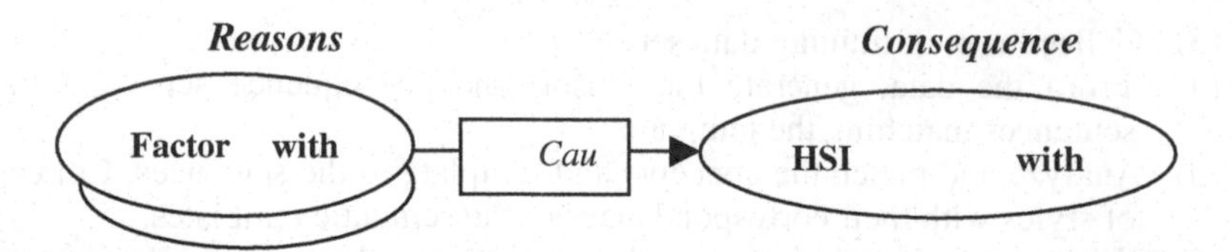

Fig. 3. The causation semantic template for the Hong Kong market movements

2) The Hang Seng Index (HSI) movements can be divided into three categories and each is assigned a symbol. The categories are *upward* (+), *downward* (-) and *no movement* (0). The concept seeding words are {*up, rise, soar, gain, boost, high, rebound, strong, recover*}, {*down, fall, sink, lose, tumble, low, weak, sell-off, retreat*} and {*no change, consolidate, barely changed, easing*} respectively.
3) The most common factors in the reasons are:
Wall Street/ Dow Jones, Nasdaq, US interest rate, Overseas market,

   Internal factors, e.g. technical stocks.

   They are collected from the training data set as the fundamental concepts for the reasons in the causation semantic template. They will have to be combined with the three categories of movements to form the complete semantic descriptions of a reason. For example: interest rate fall, overseas market rises.
4) The set of causes for linking consequence to reasons is {*as, ","*} and the set of conjunction terms amongst the factors in the reasons is {*and, ",", but*}.

Based on the SEKE framework described in the previous section, a system for this study is implemented in Java and it is illustrated in Figure 3 below. In this SEKE ystem, we highlight the human intervention steps with dotted lines. If these human assisted steps are included, we can achieve almost 100% accuracy in the analysis, they are therefore excluded from the evaluation of the experiment. Those news articles that cannot be processed automatically by the SEKE system (represented by solid lines in Figure 4) will be treated as incorrectly extracted (fail).

The SEKE system for Hong Kong market movement analysis takes in raw news articles either from the Reuters news data feed, or electronic news articles from the web sites of the other two newspapers: South China Morning Post and Hong Kong Standard. Preprocessing is carried out to filter the irrelevant information and zoom in to the relevant sentences. This is achieved with the expected sentence templates and the semantics of the set of possible consequences generated from the initial training data set (e.g. HSI rose). Those sentences identified as relevant will be semantically parsed (concept term matching without syntactical knowledge) using the domain knowledge captured in the reason factors concept, consequence (HSI) concept and movement concept. Successful sentences will be collected in the form of the causation semantic template (Figure 3).

Some sentences may not be parsed semantically. Since irrelevant sentences have been filtered earlier, there are only two possibilities here: either some movement terms, or, reason factors cannot be recognized at this stage. Each unseen movement concept term is passed to the module on movement extraction using WordNet (double outlined box in Figures 4 and 2). If the unseen term can be classified, then it is updated to the movement concept knowledge and the parsing of the original sentence continues. If the term cannot be recognized by the module then it will be kept for human recognition and the sentence is considered failed in this study (marked as **F2** in Figure 4). For unseen reason factors, since they are compounded and specialized concepts (e.g. "*World Bank support*", "*European Community's intervention*") and most of them are not captured by the current thesaurus, they have to be decided by human recognition. Hence sentences containing this type of factors are considered failed in this study (**F1** in Figure 4).

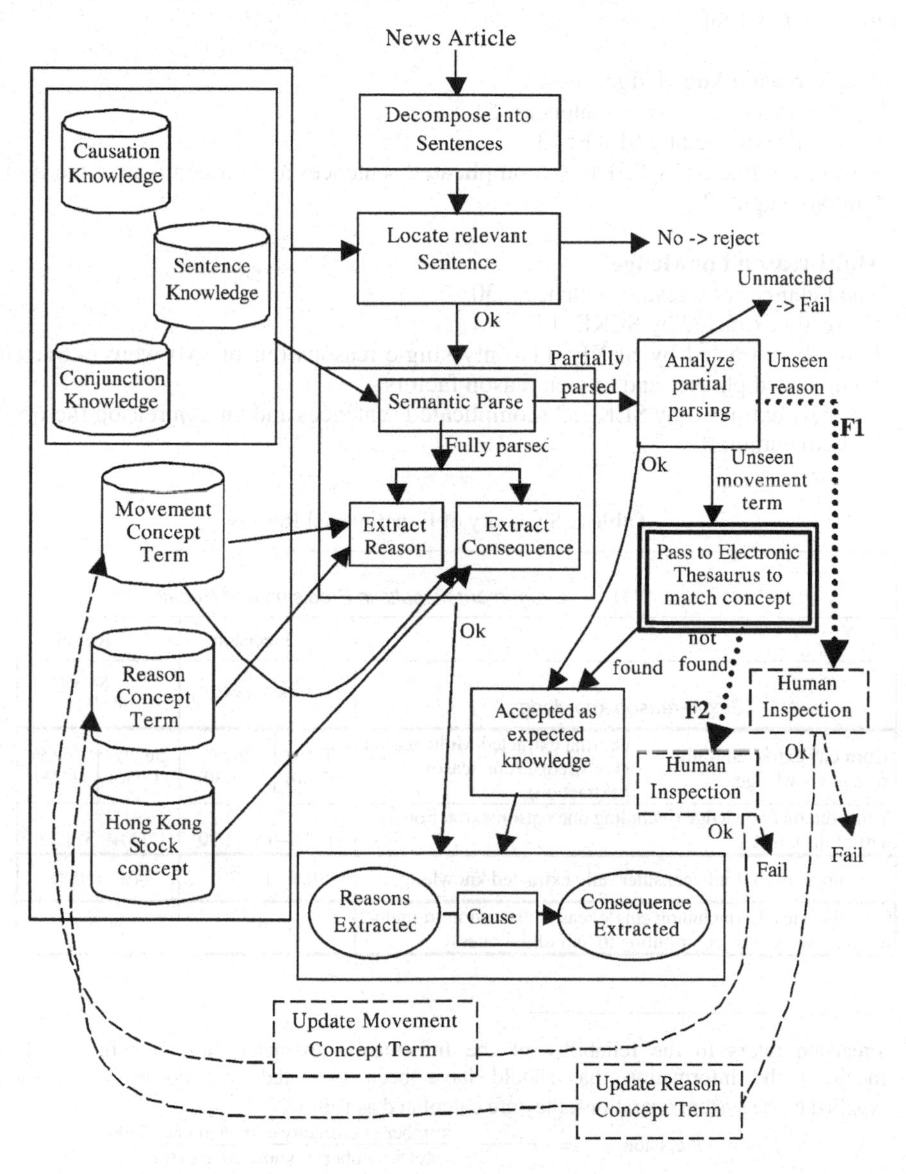

Fig. 4. The SEKE system for Hong Kong Stock Movement Analysis

News articles of the three information sources from January to February 2000 are used as the testing data set in the experiment. These news articles were classified manually and they serve as the correct answers. A total of 81 relevant news sentences (only single sentence information is studied) were collected (a 100% systems success rate compared to human classifications). Among them, 30 have multiple reasons associated with a Hong Kong market movement while the other 51 have only single reasons. The results are given below and the precision and recall[1] rates are summarized in Table 1.

Single-reason knowledge
Total number of relevant sentences: 51
Correctly extracted by SEKE: 43
Wrongly extracted by SEKE: 6 (complicated sentences and unseen reason factors)
Fail/No output: 2

Multi-reason knowledge
Total number of relevant sentences: 30
Correctly extracted by SEKE: 17
Partially extracted by SEKE: 11 (only single reasons out of two were extracted, complicated phrases and unseen reason factors)
Wrongly extracted by SEKE: 2 (complicated sentences and unseen reason factors)
Fail/No output: 0

Table 1. Summary of Experimental Results

<table>
<tr><td colspan="6">(11) Experiment Results in Precision and Recall</td></tr>
<tr><td colspan="2"></td><td colspan="2">Precision</td><td colspan="2">Recall</td></tr>
<tr><td colspan="2">Single-reason knowledge</td><td colspan="2">87.8% (43/49)</td><td colspan="2">84.3% (43/51)</td></tr>
<tr><td>Correctly extracted Multi-reason knowledge</td><td>Partial extracted Multi-reason knowledge (one reason extraction)</td><td>56.7% (17/30)</td><td>36.6% (11/30)</td><td>56.7% (17/30)</td><td>36.6% (11/30)</td></tr>
<tr><td colspan="2">Multi-reason knowledge (including one reason extraction as half of the rates)</td><td colspan="2">75% (17/30+0.5*11/30)</td><td colspan="2">75% (17/30+0.5*11/30)</td></tr>
<tr><td colspan="2">Combined result (only consider fully extracted knowledge)</td><td colspan="2">76% (60/79)</td><td colspan="2">74% (60/81)</td></tr>
<tr><td colspan="2">Combined result (including single reason extraction in multi-reason sentences as contributing to half of the rates)</td><td colspan="2">82.9%</td><td colspan="2">80.9%</td></tr>
</table>

---

[1] Precision refers to the reliability of the information extracted. Recall refers to how much of the information that should have been extracted was correctly extracted. Applied to the system operations, they are calculated as follow.

$$\text{Precision} = \frac{\text{number of correctly extracted knowledge}}{\text{total number of knowledge extracted}}$$

$$\text{Recall} = \frac{\text{number of correctly extracted knowledge}}{\text{total number of knowledge}}$$

4 Discussion and Future Work

This paper proposes a semantic expectation-based knowledge extraction system SEKE for causation knowledge and a study using SEKE for extracting Hong Kong stocks market movement knowledge was carried out. By excluding human recognition in the SEKE framework, the experiment only measures the performance of the fully automated SEKE system.

For single-reason sentences, an 88% precision and 84% recall rates were achieve. For multi-reason sentences (all test sentences only have two reasons), full knowledge extraction with all reasons and movements discovered is about 57% for both the precision and recall; those have half of the reasons discovered constitute another 37% for both performance measurements. None of the multi-reason sentences has no knowledge extracted. The overall precision and recall are 76% and 74 % if we only consider full discovery of knowledge, and if we include partially discovered knowledge with a 0.5 factor (since one out of two reasons is extracted) then the rates improve to about 83% and 81% respectively. Errors contributing to wrong knowledge extraction of single-reason sentences are mainly due to the following three problems:

1. The movement of the reason cannot be extracted which causes an incomplete reason. For example:
   "*Hong Kong stocks were lower in lackluster afternoon trade on Thursday with investors seeking out Cable & Wireless HKT <0008.HK> (C&W HKT) and some other technology stocks but shunning much of the rest of the market.*"
2. The movements for the reason are wrongly extracted due to complex sentence expression. Example:
   "Hong Kong stocks are expected to ease further on Thursday under ***steady*** *pressure from interest rate concerns. "*
   In this example, *steady* were wrongly recognized as the movement for the reason factors where the correct one should be the increase of interest rate.
3. Unseen reason factor and/or movement. Example:
   "Hong Kong stocks are expected to mirror Wall Street overnight action on Tuesday, with blue chips easing on interest rate fears but with technology stocks defying the trend."
   Here "*overnight action*" was unrecognizable to the SEKE system.

For multiple reason knowledge, there are a number of cases in which the SEKE system can only extract one out of two reasons. Although problems encountered by the single-reason sentence extraction may occur, our analysis reveals that the main problem is due to unseen reason factors. Some examples are shown below:

> *"Hong Kong stocks finished strongly higher on Wednesday morning following a rebound on Wall Street and after a Hutchison Whampoa Ltd <0013.HK> Internet deal inspired fresh interest in technology stocks."*
>
> *"Hong Kong stocks finished Friday morning flat as a combination of interest rate concerns, covered warrants and fund outflows pared early gains."*

The third problem listed under single-reason sentence can be addressed by human recognition of the unseen concepts but when the sentence structure becomes too complex and there is no way to semantically match the expected concepts, the current

SEKE system fails to process the knowledge extraction task. Two cases were discovered for the single-reason sentences.

By using the expected semantic knowledge to extract information from financial news articles, similar to how human beings understand texts, we have shown that it is a workable approach. It is simple and computationally less demanding than the syntactical, statistical and/or probabilistic approaches. The unresolved problem of complex sentences may not be easily tackled by these approaches as well. In our context, we believe that more complex semantic templates have to be defined and captured for this purpose, and we will look into this possible direction.

SEKE system is an evolving system with the capability of incrementally improving itself with more input examples, we will have to try out different sequence of tests to check its average extraction capability. In terms of the granularity of the degree of modifier for the movement knowledge, we would also like to explore more detailed information. Distinctions such as *"HSI rises"*, *"HSI rises sharply"*, and *"HSI rises moderately"* will be incorporated. In addition, the causation semantic template generated from for the analysis of Hong Kong stocks is believed to be equally applicable to other market movement analysis, if the reason and consequence concepts are collected from the relevant training articles. Since the semantic template is language independent, the same process can be ported over to another language. This also sets our future direction.

Reference

1. Yorick Wilks, Information Extraction as a Core Language Technology, International Summer School on Information Extraction, SCIE'97, Fracscati.
2. Rau, L.F., Conceptual information extraction from financial news. System Sciences, 1988. Vol.III. Decision Support and Knowledge Based Systems Track, Proceedings of the Twenty-First Annual Hawaii International Conference, 1988 Pg. 501–509.
3. Costantino, M.; Morgan, R.G.; Collingham, R.J.; Carigliano, R., Natural language processing and information extraction: qualitative analysis of financial news articles . Computational Intelligence for Financial Engineering (CIFEr), 1997., Proceedings of the IEEE/IAFE 1997, 1997 Pg. 116–122.
4. Grishman, R.; Sterling, J., New York University: Description of the Proteus System as used for MUC-5, Proceedings of the Fifth Message Understanding Conference, 1993 Pgs. 181–194.
5. Appelt, D.; Hobbs, J.; Bear, J.; Israel, D.; Kameyama, M.; Tyson, M., SRI: Description of the JV-Fastus System Used for MUC-5, Proceedings of the Fifth Message Understanding Conference, 1993, Pg. 221–236.
6. WORDNET is an on-line lexical reference system whose design is inspired by current psycholinguistic theories of human lexical memory. English nouns, verbs, adjectives and adverbs are organized into synonym sets, each representing one underlying lexical concept. Different relations link the synonym sets. WordNet was developed by the Cognitive Science Laboratory at Princeton University under the direction of Professor George A. Miller (Principal Investigator). http://www.cogsci.princeton.edu/~wn/
7. Roget's Thesaurus, a Project for American and French Research on the Treasury of the French Language (ARTFL), University of Chicago http://humanities.uchicago.edu/ARTFL/forms_unrest/ROGET.html

A Toolbox Approach to Flexible and Efficient Data Mining

Ole M. Nielsen[1]*, Peter Christen[1], Markus Hegland[1], Tatiana Semenova[1], and Timothy Hancock[2]

[1] Australian National University, Canberra, ACT 0200, Australia
URL: http://csl.anu.edu.au/ml/dm/
[2] James Cook University, Townsville, QLD 4811, Australia

Abstract. This paper describes a flexible and efficient toolbox based on the scripting language Python, capable of handling common tasks in data mining. Using either a relational database or flat files the toolbox gives the user a uniform view of a data collection. Two core features of the toolbox are caching of database queries and parallelism within a collection of independent queries. Our toolbox provides a number of routines for basic data mining tasks on top of which the user can add more functions – mainly domain and data collection dependent – for complex and time consuming data mining tasks.

Keywords: Python, Relational Database, SQL, Caching, Health Data

1 Introduction

Due to the availability of cheap disk space and automatic data collection mechanisms huge amounts of data in the Terabyte range are becoming common in business and science [7]. Examples include the customer databases of health and car insurance companies, financial and business transactions, chemistry and bioinformatics databases, and remote sensing data sets. Besides being used to assist in daily transactions, such data may also contain a wealth of information which traditionally has been gathered independently at great expense. The aim of data mining is to extract useful information out of such large data collections [3].

There is much ongoing research in sophisticated algorithms for data mining purposes. Examples include predictive modelling, genetic algorithms, neural networks, decision trees, association rules, and many more. However, it is generally accepted that it is not possible to apply such algorithms without careful data understanding and preparation, which may often dominate the actual data mining activity [5,11]. It is also rarely feasible to use off-the-shelf data mining software and expect useful results without a substantial amount of data insight. In addition, data miners working as consultants are often presented with data sets from an unfamiliar domain and need to get a good feel for the data and the

* Corresponding author, E-Mail: Ole.Nielsen@anu.edu.au

D. Cheung, G.J. Williams, and Q. Li (Eds.): PAKDD 2001, LNAI 2035, pp. 124–135, 2001.

domain prior to any "real" data mining. The ease of initial data exploration and preprocessing may well hold the key to successful data mining results later in a project.

Using a portable, flexible, and easy to use toolbox can not only facilitate the data exploration phase of a data mining project, it can also help to unify data access through a middleware library and integrate different data mining applications through a common interface. Thus it forms the framework for the application of a suite of more sophisticated data mining algorithms. This paper describes the design, implementation, and application of such a toolbox in real-life data mining consultancies.

1.1 Requirements of a Data Mining Toolbox

It has been suggested that the size of databases in an average company doubles every 18 months [2] which is akin to the growth of hardware performance according to Moore's law. Yet, results from a data mining process should be readily available if one wants to use them to steer a business in a timely fashion. Consequently, *data mining software has to be able to handle large amounts of data efficiently and fast.*

On the other hand, data mining is as much an art as a science, and real-life data mining activities involve a great deal of experimentation and exploration of the data. Often one wants to "let the data speak for itself". In these cases one needs to conduct experiments where each outcome leads to new ideas and questions which in turn require more experiments. Therefore, it is mandatory that *data mining software facilitates easy querying of the data.*

Furthermore, data comes in many disguises and different formats. Examples are databases, variants of text files, compact but possibly non-portable binary formats, computed results, data downloaded from the Web and so forth. Data will usually change over time – both with respect to content and representation – as will the demands of the data miner. It is desirable to be able to access and combine all these variants uniformly. *Data mining software should therefore be as flexible as possible.*

Finally, data mining is often carried out by a group of collaborating researchers working on different aspects of the same dataset. A suitable software library providing shared facilities for access and execution of common operations leads to safer, more robust and more efficient code because the modules are tested first by the developer and then later by the group. A shared toolbox also tends to evolve towards efficiency because the best ideas and most useful routines will be chosen among all tools developed by the group.

This paper describes such a toolbox – called *DMtools* – developed by and aimed at a small data mining research group for *fast*, *easy*, and *flexible* access to large amounts of data.

The toolbox is currently under development and a predecessor has successfully been applied in health data mining projects under the ACSys CRC[1]. It assists our research group in all stages of data mining projects, starting from data preprocessing, analysis and simple summaries up to visualisation and report generation.

2 Related Work

Database and data mining research are two overlapping fields and there are many publications dealing with their intersection. An overview of database mining is given in [6]. According to the authors the efficient handling of data stored in relational databases is crucial because most available data is in a relational form. Scalable and efficient algorithms are one of the challenges, as is the development of high-level query languages and user interfaces. Another key requirement is interactivity.

A classification of frameworks for integrating data mining applications and database systems is presented in [4]. Three classes are presented: (1) Conventional – also called loosely coupled – where there is no integration between the database system and the data mining applications. Data is read tuple by tuple from a database, which is very time consuming. The advantage of this method is that any application previously running on data stored in a file system can easily be changed, but the disadvantage is that no database facilities like optimised data access or parallelism are used. (2) In the tightly coupled class data intensive and time-consuming operations are mapped to appropriate SQL queries and executed by the database management system. All applications that use SQL extensions or propose such extensions to improve data mining algorithms are within this class. (3) In the black box approach complete data mining algorithms are integrated into the database system. The main disadvantage of such an approach is its lack of flexibility. Following this classification, our *DMtools* belong to the tightly-coupled approach, as we generate simple SQL queries and retrieve the results for further processing in the toolbox. As the results are often aggregated data or statistical summaries, communication between the database and data mining contexts can be reduced significantly.

Several research papers address data mining based on SQL databases and propose extensions to the SQL standard to simplify data mining and make it more efficient. In [8] the authors propose a new SQL operator that enables efficient extraction of statistical information which is required for several classification algorithms. The problem of mining general association rules and sequential patterns with SQL queries is addressed in [13], where it is shown that it is possible to express complex mining computations using standard SQL. Our data mining toolbox is currently based on relational databases, but can also integrate

[1] ACSys CRC stands for 'Advanced Computational Systems Collaborative Research Centre' and the data mining consultancies were conducted at the Australian National University (ANU) in collaboration with the Commonwealth Scientific and Industrial Research Organisation (CSIRO).

flat files. No SQL extension is needed, instead we put a layer on top of SQL where most of the "intelligent" data processing is done. Database queries are cached to improve performance and re-usability.

Other toolbox approaches to data analysis include the IDEA (Interactive Data Exploration and Analysis) system [12], where the authors identify five general user requirements for data exploration: Querying (the selection of a subset of data according to the values of one or more attributes), segmenting (splitting the data into non-overlapping sub-sets), summary information (like counts or averages), integration of external tools and applications, and history mechanisms. The IDEA framework allows quick data analysis on a sampled sub-set of the data with the possibility to re-run the same analysis later on the complete data set. IDEA runs on a PC, with the user interacting on a graphical interface.

Yet another approach used in the Control [9] project is to trade quality and accuracy for interactive response times, in a way that the system quickly returns a rough approximation of a result that is refined continuously. The user can therefore get a glimpse at the final result very quickly and use this information to change the ongoing process. The Control system, among others, includes tools for interactive data aggregation, visualisation and data mining.

An object-oriented framework for data mining is presented in [14]. The described Data Miner's Arcade provides a collection of APIs for data access, plug'n'play type tool integration with graphical user interfaces, and for communication of results. Access to analysis tools is provided without requiring the user to become proficient with the different user interfaces. The framework is implemented in Java.

3 Choice of Software

The *DMtools* are based on the scripting language *Python* [1], an excellent tool for rapid code development that meets all of the requirements listed in Section 1.1 very well. Python handles large amounts of data efficiently, it is very easy to write scripts as well as general functions, it can be run interactively (interpretable) and it is flexible with regards to data types because it is based on general lists and dictionaries (associative arrays), of which the latter are implemented as very efficient hash-tables. Functions and routines can be used as templates which can be changed and extended as needed by the user to do more customised analysis tasks. Having a new data exploration idea in mind the data miner can implement a rapid prototype very easily by writing a script using the functions provided by our toolbox.

Databases using SQL are a standardised tool for storing and accessing transactional data in a safe and well-defined manner. The *DMtools* are accessing a relational database using the Python database API [2]. Currently, we are using MySQL [15] for the underlying database engine, but modules for other database servers are available as well. Both MySQL and Python are freely available, li-

[2] Available from the Python homepage at `http://www.python.org/topics/database/`

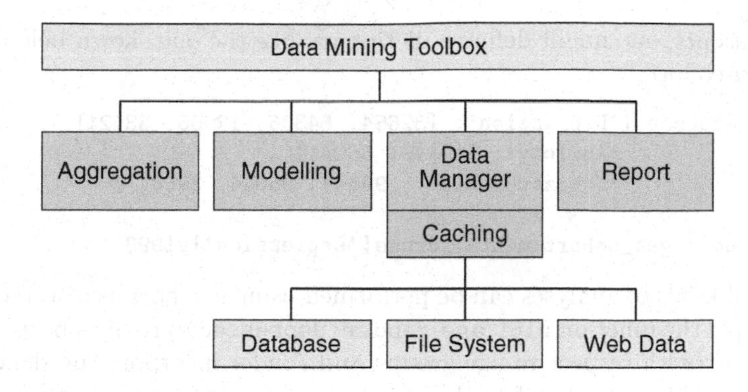

Fig. 1. Architecture of DMtools

censed as free software and enjoy very good support from a large user community. In addition, both products are very efficient and robust.

3.1 Toolbox Architecture

In our toolbox the ease of SQL queries and the safety of relational databases are combined with the efficiency of flat file access and the flexibility of object-oriented programming languages in an architecture as shown in Figure 1. Based on relational databases, flat files, the Web, or any other data source a *Data Manager* deals with retrieval, caching and storage of data. It provides routines to execute an arbitrary SQL query and to read and write binary and text files. The two important core components of this layer are its transparent caching mechanism and its parallel database interface which intercepts SQL queries and parallelises them on-the-fly. The *Aggregation* module implements a library of Python routines taking care of simple data exploration, statistical computations, and aggregation of raw data. The *Modelling* module contains functions for parallel predictive modelling, clustering, and generation of association rules. Finally, the *Report* module provides visualisation and allows facilities for simple automatic report generation.

Functions defined in the toolbox layer are designed to deal with issues specifically for a given data mining project, which means they use knowledge about a given database structure and return customised results and plots. This layer contains routines that are not available in standard data analysis or data mining packages.

Example 1. Dictionary of Mental Health Medications
A central object within the domain of health statistics is a *cohort*, defined here as a Python dictionary of *entities* like customers or patients fulfilling a given criterion. As one task in a data mining project might be the analysis of a group of entities (e.g. all patients taking certain medication), one can use the function `get_cohort` to extract such a cohort once and cache the resulting dictionary so it is readily available for subsequent invocations. Being interested in mental

health patients, one might define a dictionary like the one shown below and use it to get a cohort.

```
mental_drugs = {'Depression': [32654, 54306, 12005, 33421],
                'Anxiety': [10249, 66241],
                'Schizophrenia': [99641, 96044, 39561]}

depressed = get_cohort(mental_drugs['Depression'],1998)
```

Several kinds of analyses can be performed using a cohort as a starting point. For example the function `plot_age_gender(depressed)` provides barplots of the given cohort with respect to age groups and gender incorporating denominator data if available. Another function `list_drug_usage(depressed)` gives a list of all medication prescribed to patients in the given cohort. This list includes description of the drug, number of patients using it, the total number of prescriptions and the total cost of each drug. Routines from all modules can either be used interactively or added to other Python scripts to build more complex analysis tasks.

4 Caching

Caching of function results is a core technology used throughout *DMtools* in order to render the database approach feasible. We have developed a methodology for *supervised* caching of function results as opposed to the more common (and also very useful) *automatic* disk caching provided by most operating systems and Web browsers.

Like automatic disk caching, supervised caching trades space for time, but the approach we use is one where time consuming operations such as database queries or complex functions are intercepted, evaluated and the resulting objects are made persistent for rapid retrieval at a later stage. We have observed that many of these time consuming functions tend to be called repetitively with the same arguments. Thus, instead of computing them every time, the cached results are returned when available, leading to substantial time savings. The repetitiveness is even more pronounced when the toolbox cache is shared among many users, a feature we use extensively. This type of caching is particularly useful for computationally intensive functions with few frequently used combinations of input arguments. Supervised caching is invoked in the toolbox by explicitly applying it to chosen functions. Given a Python function of the form `T = func(arg1,...,argn)` caching in its simplest form is invoked by replacing the function call with `T = cache(func,(arg1,...,argn))`.

Example 2. Function Caching
Caching of a simple SQL query using the toolbox function `execquery` can be done as follows:

```
database = 'CustomerData'
query = 'select distinct CustomerID, count(*) from %s;' %database
customer_list = cache(execquery,(query))
```

Table 1. Function Caching Statistics

Function Name	Hits	Time (sec)		Gain(%)	Size (MB)
		Exec	Cache		
execquery	4,149	130	6	91.43	4.53
get_mbs_patients	172	1,281	76	93.92	48.53
get_selected_transactions	420	1,507	5	99.33	6.67
multiquery	46	133	0	99.69	0.76
simplequery	5	50	0	99.86	0.08
get_cohort	168	489	0	99.92	0.20
get_drug_usage	95	1,388	0	99.99	0.02

The user can take advantage of this caching technique by applying it to arbitrary Python functions. However, this technique has already been employed extensively in the *Data Manager* module so using the high level toolbox routines will utilise caching completely transparently with no user intervention – the caching supervision has been done in the toolbox design. For example, most of the SQL queries that are automatically generated by the toolbox are cached in this fashion. Generating queries automatically increases the chance of cache hits as opposed to queries written by the end user because of their inherent uniformity. In addition to this, caching can be used in code development for quick retrieval of precomputed results. For example if a result is obtained by automatically crawling the Web and parsing HTML or XML pages, caching will help in retrieving the same information later – even if the Web server is unserviceable at that point.

The function `get_cohort` used in a particular project required on the average 489 seconds worth of CPU time on a Sun Enterprise server and the result took up about 200 Kilobytes of memory. Subsequent loading takes 0.22 seconds – more than 2,000 times faster than the computing time. This particular function was hit 168 times in a fortnight saving four users a total of 23 hours of waiting. Table 4 shows some caching statistics from a real-life data mining consultancy in health services obtained from four users over two weeks. The table has one entry for each Python function that was cached. The second column shows how many times a particular instance of that function was hit, i.e. how many times results were retrieved from the cache rather than being computed. The third column shows the average CPU time which was required by instances of each function when they were originally executed, and the fourth column shows the average time it took to retrieve cached results for each function. The fifth column then shows the average percentile gain $((Exec - Cache)/Exec * 100)$ achieved by caching instances of each function, and the sixth column shows the average size of the cached results for each function. The table is sorted by average gain.

If the function definition changes after a result has been cached or if the result depends on other files wrong results may occur when using caching in it simplest form. The caching utility therefore supports specification of explicit dependencies in the form of a file list, which, if modified, triggers a recomputation.

Other options include forced recomputation of functions, statistics regarding time savings, sharing of cached results, clean-up routines and data compression. Note that if the inputs or outputs are very large, caching might not save time because disk access may dominate the execution time. This is due to overheads consisting mainly of input checks, hashing and comparisons of argument list, as well as writing and reading cache files. If caching does not lead to any time savings, a warning is given. Very large datasets are dealt with through blocking into manageable chunks and separate caching of these.

Example 3. Caching of XML Documents
Supervised caching is used extensively in the toolbox for database querying but is by no means restricted to this. Caching has proven to be useful in other aspects of the data mining toolbox as well. An example is a Web application built on top of the toolbox which allows managers to explore and rank branches according to one or more user-defined features such as *Annual revenue*, *Number of customers serviced relative to total population*, or *Average sales per customer*. The underlying data is historical sales transaction data which is updated monthly, so features need only be computed once for new data when it is added to the database. Because the data is static, cached features are never recomputed and the application can therefore make heavy use of the cached database queries. Moreover, no matter how complicated a feature is, it can be retrieved as quickly as any other feature once it has been cached. In addition, the Web application is configured through an XML document defining the data model and describing how to compute the features. The XML document must be read by the toolbox, parsed and converted into appropriate Python structures prior to any computations. Because response time is paramount in an interactive application, parsing and interpretation of XML is prohibitive, but by using the caching module, the resulting Python structures are cached and retrieved quickly enough for the interactive application. The caching function was made dependent on the XML file itself, so that all structures are recomputed whenever the XML file has been edited – for example to modify an existing feature-definition, add a new month, or change the data description. Below is a code snippet from the Web application. The XML configuration file is assumed to reside in `sales.xml`. The parser which builds Python structures from XML is called `parse_config` and it takes the XML filename as input. To cache this function, instead of the call `(feature_list, branch_list) = parse_config(filename)` we write:

```
filename = 'sales.xml'
(feature_list, branch_list) = cache(parse_config, (filename),
                                    dependencies = filename)
```

5 Database Access and Parallelism

The toolbox provides powerful and easy-to-use access to an SQL database using the Python database API. We are using MySQL but any SQL database known to Python will do. In its simplest form it allows execution of any valid SQL query. If a *list* of queries is given, they are executed in parallel by the database server if a multiprocessor architecture is available and the results are returned in a list.

The achievable speedup of this procedure will naturally depend on factors such as the amount of *communication* and the *load balancing*. For large results communication time will dominate the execution time thus reducing the speedup. In addition, the total execution time of a parallel query is limited by the slowest query in the list, so if the queries are very different in their complexity, speedup will only be modest. However, a well balanced parallel query where results are of a reasonable size can make very good use of a parallel architecture. For example, executing a parallel query over five tables of size varying from 250 thousand to 13 million transactions took 2,280 seconds when run sequentially and 843 seconds when run in parallel. This translates into a speedup of 2.7 on five processors or an parallel efficiency of 0.54.

The database interface makes use of supervised caching technology and caches the results of queries as described in the previous section. This can be enabled or disabled through a keyword argument in the function `execquery`. The data_manager module also contains a number of functions to perform standard queries across several tables. One example is the function `standardquery` which takes as input two attribute names, A_1 and A_2, a list of (conforming) database tables, and an optional list of criteria to impose simple restrictions on the query. The function returns all occurrences of attribute A_2 for each distinct value of A_1 from all tables where all the given criteria are met (using conjunction). For example, the call

```
standardquery('Company', 'Customer', tables,
              [('Year', 1997), ('Qtr', 1)])
```

yields a count of customers for each company in the first quarter of 1997. Another example is the function `joinquery` which improves the performance of of normal SQL joins. It takes as arguments a list of fields, a list of tables, a list of joins of the form '`table1_name.field = table2_name.field`' and a list of conditions, and returns a dictionary of results.

6 Applications

To illustrate the application of the *DMtools* we give two examples designed and used for a health services data mining consultancy. The data collection we had to our disposal consisted of two tables, one containing medication prescriptions and the other containing doctor consultancies by patients. In addition, we had specialty information about doctors and geographical information that associated each post code with one of seven larger area codes (like capital, metropolitan or rural). All patient and doctor identifiers were coded to protect individual privacy. Finally, we had data describing different drugs and different treatments obtained from the Web.

Example 4. Doctors Prescription Behaviour
In this example, we describe how we analysed prescription patterns of specialist doctors. The aim was to find unusual behaviour such as over-prescribing. In particular, we wanted to know for each specialist how many patients he or she serviced and how many of these were prescribed a particular type of medication.

For this task, we linked medication prescription information with patient information for a user given doctor specialty. The domain specific function

```
get_doctor_behaviour(items, years, specialty_code)
```

takes as input a list of drug code items, a list of years (as we are interested in the temporal changes in prescription behaviour over a time period) and a doctor specialty code.

The toolbox is used as follows: First, the cohort of all patients taking medications in the given items list is extracted from the medication prescription database and the cohort of all patients seeing doctors with the particular specialty code is extracted from the doctor consultancy database. These lists are then matched resulting in the desired table (see example below). Sorting this table gives the highest ratio of prescriptions per patients which can lead to the detection procedure for over-prescribing.

The first run of `get_doctor_behaviour` with `items` being psycho tropic drugs and `specialty_code` being psychiatrists, over a five years period, required a run time of about two hours, extracting about 115 psychiatrists, almost 10,000 patients and more than half a million transactions to analyse. Subsequent studies with different medication groups were each processed in less than a minute thanks to caching.

```
Doctor Code |                        1995  |  1996  |  1997  |  1998
---------------------------------------------------------------------
x42r19$     | Total  Patients:       424   |  450   |  241   |   199
            | Mental Patients:       167   |  198   |  142   |   131
            |            Ratio:      39%   |  44%   |  59%   |   66%
---------------------------------------------------------------------
7%w#t0q     | Total  Patients:       372   |  336   |  335   |   389
            | Mental Patients:       101   |  115   |  121   |   156
            |            Ratio:      27%   |  34%   |  36%   |   40%
```

Example 5. Episode Extraction
Episodes are units of health care related to a particular type of treatment for a particular problem. An episode may last anywhere from one day to several years. Analysing temporal episodic data from a transactional database is a hard task not only because there are very many episodes within a large database but also because episodes are complex objects with different lengths and contents. To facilitate better understanding and manipulation, the *DMtools* contain routines

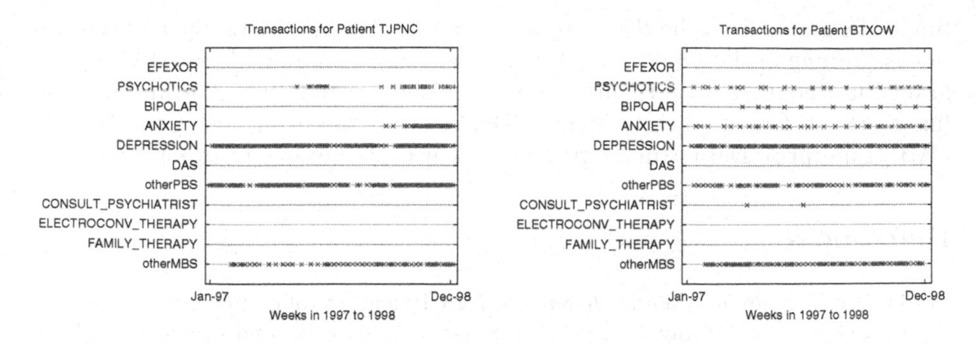

Fig. 2. Timelines: Medical services for two patients

to examine their basic characteristics, like length, number of transactions, average cost, etc. One example is the *timelines* diagram which displays all medical transactions for a single patient split up for different groups of items as is shown in Figure 2. Detecting and extracting episodes from a transactional database is very time consuming and caching of such a functions is very helpful. It is feasible to cache several hundred thousand episodes – even if the resulting cache file has a size of hundred Megabytes – because the access time to get all these episodes is reduced from hours to minutes.

7 Outlook and Future Work

The *DMtools* is a project driven by the needs of a group of researches who are doing consultancies in health data mining. With this toolbox we try to improve and facilitate routine tasks in data analysis, with an emphasis on the exploration phase of a data mining project. It is important to have tools at hand that help to analyse and get a "feel" for the data interactively in the early stages of a data mining project, especially if the data is provided from external sources. This is in contrast to many data mining and knowledge discovery algorithms that aim at extracting information automatically from the data without any guidance from the user.

Ongoing work on the *DMtools* includes the extension of the toolbox with more analysis routines and the integration of algorithms like clustering, predictive modelling and association rules. As these processes are time consuming we are exploring methods to integrate external parallel applications (optimised C code using communication libraries like MPI [10]). Building graphical user interfaces (GUI) on top of our toolbox, Web enabling interfaces and exporting results via XML are on our wish list as is the publication of the *DMtools* as a package under a free software licence.

Acknowledgements. This research was partially supported by the Australian Advanced Computational Systems CRC (ACSys CRC). Peter Christen was funded by grants from the *Swiss National Science Foundation* (SNF) and the *Novartis Stiftung*, Switzerland; and Tatiana Semenova is funded by a ARC SPIRT grant with the Australian Health Insurance Commission (HIC). The authors also like to thank Christopher Kelman (CDHAC) and the members of the CSIRO CMIS Enterprise Data Mining group for their contributions to this project.

References

1. D. Beazly, *Python Essential Reference*, New Riders, October 1999.
2. G. Bell and J. N. Gray, *The revolution yet to happen*, Beyond Calculation (P. J. Denning and R. M. Metcalfe, eds.), Springer Verlag, 1997.
3. M.J. A. Berry and G. Linoff, *Data Mining, Techniques for Marketing, Sales and Customer Support.* John Wiley & Sons, 1997.
4. E. Bezzaro, M. Mattoso and G. Xexeo, *An Analysis of the Integration between Data Mining Applications and Database Systems*, Proceedings of the Data Mining 2000 conference, Cambridge, 2000.
5. P. Chapman, R. Kerber, J. Clinton, T. Khabaza, T. Reinartz and R. Wirth, *The CRISP-DM Process Model*, Discussion Paper, March 1999. `www.crisp.org`
6. M.-S. Chen, J. Han and P.S. Yu, *Data Mining: An Overview from a Database Perspective*, IEEE Transactions on Knowledge Discovery and Data Engineering, Vol. 8, No. 6, December 1996.
7. D. Düllmann, *Petabyte databases.* Proceedings of the ACM SIGMOD International Conference on Management of Data (SIGMOD-99), ACM Press, July 1999.
8. G. Graefe, U. Fayyad and S. Chaudhuri, *On the Efficient Gathering of Sufficient Statistics for Classification from Large SQL Databases*, Proceedings of KDD-98, 4th International Conference on Knowledge Discovery and Data Mining, AAAI Press, 1998.
9. J.M. Hellerstein, R. Avnur, A. Chou, C. Hidber, C. Olston, V. Raman,T. Roth and P. Haas, *Interactive Data Analysis: The Control Project*, IEEE Computer, Vol. 32, August 1999.
10. W. Gropp, E. Lusk and A. Skjellum, *Using MPI – Portable Parallel Programming with the Message-Passing Interface.* MIT Press, Cambridge, Massachusetts, 1994.
11. D. Pyle, *Data Preparation for Data Mining.* Morgan Kaufmann Publishers, Inc., 1999.
12. P.G. Selfridge, D. Srivastava and L.O. Wilson, *IDEA: Interactive Data Exploration and Analysis*, Proceedings of the ACM SIGMOD International Conference on Management of Data, 1996.
13. S. Thomas and S. Sarawagi, *Mining Generalized Association Rules and Sequential Patterns Using SQL Queries*, Proceedings of KDD-98, 4th International Conference on Knowledge Discovery and Data Mining, AAAI Press, 1998.
14. G. Williams, I. Altas, S. Barkin, P. Christen, M. Hegland, A. Marquez, P. Milne, R. Nagappan and S. Roberts, *The Integrated Delivery of Large-Scale Data Mining: The ACSys Data Mining Project*, In Large-Scale Parallel Data Mining, M. J. Zaki and C.-T. Ho (Eds.), Springer Lecture Note in Artificial Intelligence 1759, 1999.
15. R.J. Yarger, G. Reese and T. King, *MySQL & mSQL*, O'Reilly, July 1999.

Determining Progression in Glaucoma Using Visual Fields

Andrew Turpin[1], Eibe Frank[2], Mark Hall[2],
Ian H. Witten[2], and Chris A. Johnson[1]

[1] Discoveries in Sight, Devers Eye Institute, Portland OR USA
{aturpin, cjohnson}@discoveriesinsight.org

[2] Department of Computer Science
University of Waikato
Hamilton, New Zealand
{eibe, mhall, ihw}@cs.waikato.ac.nz

Abstract. The standardized visual field assessment, which measures visual function in 76 locations of the central visual area, is an important diagnostic tool in the treatment of the eye disease glaucoma. It helps determine whether the disease is stable or progressing towards blindness, with important implications for treatment. Automatic techniques to classify patients based on this assessment have had limited success, primarily due to the high variability of individual visual field measurements.
The purpose of this paper is to describe the problem of visual field classification to the data mining community, and assess the success of data mining techniques on it. Preliminary results show that machine learning methods rival existing techniques for predicting whether glaucoma is progressing—though we have not yet been able to demonstrate improvements that are statistically significant. It is likely that further improvement is possible, and we encourage others to work on this important practical data mining problem.

1 Introduction

Glaucoma, a disease that affects the optic nerve, is one of the leading causes of blindness in the developed world. Its prevalence in populations over the age of 40 of European extraction is about 1 to 2%; it occurs less frequently in Chinese populations, and much more often in people descending from West African nations [9]. There are several different types of glaucoma, but all share the characteristic that structural damage to the optic nerve tissue eventually leads to loss of visual function. If left untreated, or treated ineffectively, blindness ultimately occurs.

The cause of glaucoma involves many factors, most of which are poorly understood. If diagnosed early, however, suitable treatment can usually delay the loss of vision. If a patient continues to lose visual function after the initial diagnosis, their glaucoma is said to be *progressing*.

D. Cheung, G.J. Williams, and Q. Li (Eds.): PAKDD 2001, LNAI 2035, pp. 136–147, 2001.

Determining progression is of vital importance to patients for two reasons. First, if a patient's vision continues to deteriorate under treatment, the treatment must be altered to save their remaining visual function. Second, progression, or rate of progression, is used as the main outcome measure of clinical research trials involving new glaucoma drugs. If effective medication is to be developed and made available for widespread use, accurate detection of progression is essential. Patient care is as dependent on monitoring the progression of glaucoma as it is on the original diagnosis of the disease.

Unfortunately, there is no universally accepted definition of glaucomatous progression. Various clinical drug trials use different definitions, and criteria used by practicing ophthomologists in vision clinics also differ widely. All agree, however, that *visual field* measurements are an essential tool for detecting progression.

The next section explains the standard technique for measuring visual fields. Following that, we describe standard methods that have been used to determine progression. In Section 4 we introduce a dataset that has been assembled for the purpose of testing different progression metrics, and describe the data mining methods that we have applied. Then we summarize the results obtained, using the standard technique of pointwise linear regression as a benchmark. Surprising results were obtained using a simple 1R classifier, while support vector machines equaled and often outperformed the benchmark. We also investigated many other techniques, including both preprocessing the data by smoothing, taking gradients, using *t*-statistics, and physiologically clustering the data; and different learning schemes such as decision trees, nearest-neighbor classifiers, naive Bayes, boosting, bagging, model trees, locally weighted regression, and higher-order support vector machines. Although we were able to obtain little further improvement from these techniques, the importance of the problem merits further study.

2 Visual Field Measurement

The standard visual field assessment currently employed in glaucoma management requires the patient to sit facing a half sphere with a white background that encompasses their entire field of vision. Typically only the central 30° of the visual field is tested. Subjects are instructed to fixate in the center of the hemisphere and press a button whenever they see a small white light flash in the "bowl." Lights of varying intensities are flashed in 76 locations of the visual field, and the minimum intensity at which the patient can see the target is recorded as their *threshold*. It is not feasible to spend more than about 10 seconds determining patient thresholds at each location because fatigue influences the results, but consistent threshold measurements are possible—particularly in reliable patients. Thresholds are scored on a logarithmic decibel scale. At any particular location a score of 0 dB indicates that the brightest light could not be seen (blindness), while 35 dB indicates exceptional vision at that point.

Fig. 1 shows visual field measurements taken on the right eye of a progressing glaucoma patient at an interval of five years. Each number is a dB threshold value. There are 76 locations in each field, separated by 6° of visual angle. Note the blind spot (readings ≤ 1 dB) at $(15, -3)$ where the optic nerve exits the eye. In a left eye this blind spot occurs at $(-15, -3)$—assuming the patient remains fixated on the center spot for the duration of the test. For convenience in data processing, the asymmetry between eyes is removed by negating left eye x-coordinates, so that all data is in a right eye format.

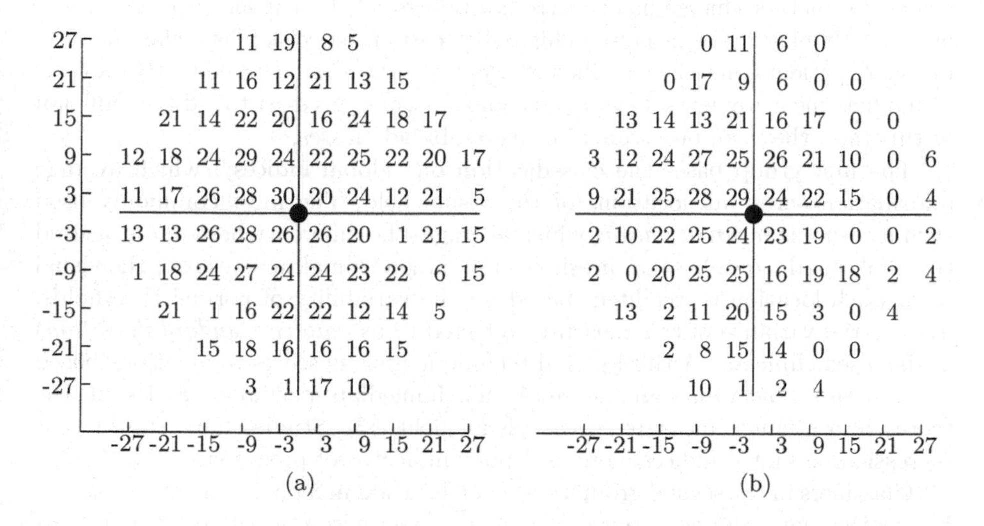

Fig. 1. Visual fields for a progressing right eye. (a) Baseline measurement, (b) the same eye five years later. The axes measure degrees of visual angle from the center fixation mark.

To assist the diagnosis of glaucoma, it is useful to compare the thresholds recorded in a visual field to a database of normal thresholds. To monitor progression, visual field measurements must be compared from one visit to the next, seeking locations whose thresholds have decreased. In Fig. 1 it is obvious that visual sensitivity around $(15, 15)$ and $(9, -15)$ has decreased from Fig. 1(a) to Fig. 1(b). Visual fields are quite noisy, however, so thresholds from a large series of visits are usually required to distinguish true progression from measurement noise. For example, the threshold at location $(-9, -27)$ in Fig. 1 increases from 3 dB to 10 dB, which in this case is probably because the original estimate of 3 dB is low. The entire field can fluctuate from visit to visit ("long term variability" in the ophthomology literature) depending on factors such as the patient's mood and alertness, as well as physiological factors like blood pressure and heart rate. Moreover, each location can vary during the test procedure ("short term variability") depending on a patient's criterion for pressing the response button, fatigue, learning effects, and mistakes.

The challenge is to detect progression as early as possible, using the smallest number of sequential visual fields.

3 Determining Progression

Glaucoma patients are usually tested at yearly or half-yearly intervals. When presented with visual fields from successive visits, the ophthomologist's task is to decide whether change has occurred, and if so whether it indicates glaucoma or is merely measurement noise. This paper casts the decision as a classification problem: patients must be classified as *stable* or *progressing* based solely on their visual field measurements. Several automatic techniques exist to aid the clinician in this task; they can be divided into three broad classes.

The first group bases the classification on "global indices," which average information across all locations of the visual field. The most commonly used such measure is *mean deviation*, which averages the difference between measured thresholds and a database of thresholds in non-glaucomatous eyes over the visual field. Each location is weighted, based on the variability of normal thresholds. The spatial variance of this measure (referred to as *pattern standard deviation*) is also used clinically. A third global technique assigns scores to locations based on their threshold values and those of their immediate neighbors, and sums the scores into a single measure of the visual field [13]. Studies have shown that regression on global indices alone is a poor indicator of progression [2,6,20,22].

Classifiers in the second group treat each location independently of the others. By far the most common approach is to use pointwise univariate linear regression (PWLR) to detect trends in individual locations [2,8,20,22,23,27]. Typically this is applied to each individual location, and if the slope of the fitted line is significantly less than zero the patient is classified as *progressing*. Several variations on this theme have been investigated, the most notable being to correct for natural age-related decline in thresholds [27]. Performing PWLR on each of the 76 locations introduces a multiple-comparison problem. This is solved previously using multiple t-tests with a Bonferroni correction, ignoring the fact that locations in the visual field are not strictly independent.

Other pointwise techniques have been investigated, including multivariate regression [20] and using high-order polynomials to fit threshold trends [26]. The *glaucoma change probability* [19] calculates the likelihood that the difference between a threshold and a baseline measure falls outside the 95% confidence limits established by a database of stable glaucomatous visual fields.

All these pointwise techniques can be refined by requiring that a cluster of points show progression, rather than a single location. Alternatively, progression can be confirmed on subsequent tests by requiring that points not only show progression after examining n visual fields, but also when $n+1$ fields are examined. These techniques are currently employed in combination with the glaucoma change probability as outcome measures in several drug trials [17].

The final group of classifiers falls between the two extremes of global indices and pointwise modeling. These classifiers attempt to take account of the spatial relationship between neighboring locations. Some approaches cluster the locations based on known physiological retinal cell maps [18,24], and apply regression to the mean of the clusters [20,22]. Others use neural networks as classifiers, relying on the network to learn any spatial relations [3,11,16, and references therein].

It is difficult to compare different approaches, because most studies do not report standard classification metrics. Moreover, patient groups used in experimental studies differ in size, stage of glaucoma, and type of glaucoma. To provide a baseline for comparison, we include results from one of the best-performing statistical methods, pointwise univariate regression, which is specifically tailored to this application [2]. This technique is detailed in Section 4.2 below.

4 Experiments

Diagnosing glaucoma progression is a prime example of a medical classification task that is ideally suited for data mining approaches because pre-classified training data, although arduous to collect, is available. However, apart from neural networks [16] no standard data mining paradigms have been applied to this problem, and there is no evidence that neural networks outperform well-known special-purpose statistical algorithms designed for this application.

We present an empirical evaluation of two standard data mining algorithms, support vector machines [7] and 1R decision stumps [12], and show that they detect progressing glaucoma more accurately than pointwise univariate regression.

4.1 The Data

Data was collected retrospectively from patient charts of 113 glaucoma patients of the Devers Eye Institute, each having at least 8 visual field measurements over at least a 4 year period as part of their regular ophthomologic examination. Unlike many previous studies, no special efforts were made to ensure patient reliability, nor was the quality of the visual fields evaluated. These are typical patient records from a typical clinical situation.

The patients were classed as *progressing*, *stable*, or *unknown* by an expert (author CAJ), based on optic disk appearance and their visual field measurements. The final data set consisted of 64 progressing eyes and 66 stable eyes, each with eight visual field measurements at different points in time. The visual field threshold values were adjusted for age at measurement by 1 dB per decade, so in effect all eyes were from a 45 year old patient. Left eyes were transformed into right eye coordinates.

4.2 The Methods

Pointwise univariate linear regression analysis (PWLR) is a well-established diagnostic tool in the ophthomology literature. In an empirical comparison of several statistical methods for detecting glaucoma progression—pointwise univariate regression, univariate regression on global indices, pointwise and cluster-wise multivariate regression, and glaucoma change analysis—it emerged as the most accurate predictor of clinician diagnosis (Table 4 in [20].) It consists of three steps. First, a linear regression function is fitted to each of the 76 locations in a visual field, using all available measurements for the patient (8 in the data we used.) Second, a significance test is performed on those locations with a negative slope to ascertain whether the decrease in sensitivity is statistically significant. Third, a series of visual field measurements is declared to be *progressing* if at least two adjacent test locations are deemed significant, and *stable* otherwise.

In our experiments, we used a one-sided t-test to detect progression [25]. We experimented with different significance levels α and report results for both $\alpha = 0.01$ and $\alpha = 0.05$. We did not adjust for multiple significance tests using the Bonferroni correction (which would result in significantly smaller significance levels) because, like Nouri-Madhavi [20] we have noticed that it fails to detect many cases of progressing glaucoma.

In contrast to this statistical approach, data mining algorithms exploit training data to build a classification model that can diagnose progressing glaucoma. In our experiments two very different data mining approaches turned out to perform well: linear support vector machines and 1R decision stumps. We describe these next. In Section 4.4 we mention other approaches that failed to improve upon them.

Linear support vector machines (LSVM) construct a hyperplane in the input space to classify new data. All data on one side of the hyperplane is assigned to one class; all that on the other side to the other class. Unlike hyperplanes constructed by other learning algorithms—for example, the perceptron—support vector machine hyperplanes have the property that they are maximally distant from the convex hulls that surround each class (if the classes are linearly separable.) Such a hyperplane is called a "maximum-margin" hyperplane, and is defined by those feature vectors from the training data that are closest to it. These are called "support vectors."

Support vector machines are very resilient to overfitting, even if the feature space is large—as it is in this application—because the maximum-margin hyperplane is very stable. Only support vectors influence its orientation and position, and there are usually only a few of them. If the data is not linearly separable, for example, because noise is present in the domain, learning algorithms for support vector machines apply a "soft" instead of a "hard" margin, allowing training instances to be misclassified by ending up on the "wrong" side of the hyperplane. This is achieved by introducing an upper bound on the weight with which each support vector can contribute to the position and orientation of the hyperplane.

The glaucoma data is very noisy: it is often hard even for experts to agree on the classification of a particular patient. Thus our experiments impose a low upper bound on the support vectors' weights, namely 0.05. To learn the support vector machines we employed the sequential minimal optimization algorithm [21] with the modifications described in [14]. An implementation of this algorithm is included in the WEKA machine learning workbench [28].

The most straightforward way to apply support vector machines to this problem is to use each individual visual field measurement as an input feature. With 8 visual field measurements and 76 locations for each one, this produces 608 features. However, proceeding this way discards valuable information—namely the fact that there is a one-to-one correspondence between the 76 locations for different visual field measurements. A more promising approach is to use the per-location degradation in sensitivity as an input feature, yielding 76 features in total. This produced consistently more accurate results in our experiments. We experimented with three different ways of measuring degradation: (a) using the slope of a univariate regression function constructed from the 8 measurements corresponding to a particular location, (b) using the value of the *t*-statistic for that slope, and (c) simply taking the difference between the first and the last measurement. Surprisingly, methods (a) and (b) did not result in more accurate classifications than (c). Thus the experimental results presented in Section 4.3 are based on method (c).

A slightly more sophisticated approach is to sort the per-location differences into ascending order before applying the learning scheme. This is motivated by the fact that different patients exhibit progressing glaucoma in different locations of the visual field. A disadvantage is that it prevents the learning scheme from exploiting correlations between neighboring pixels. With this approach, the learning scheme's first input feature is the location exhibiting the smallest decrease in sensitivity, and its 76th feature is the location exhibiting the largest decrease. The median is represented by the 38th feature. In our experiments, the sorted differences produced more accurate predictions. All the results presented in Section 4.3 are based on this approach.

Compared to a support vector machine, a 1R decision stump is a very elementary classification scheme that simply determines the single most predictive feature and uses it to classify unknown feature vectors. Numeric features are discretized into intervals before applying the scheme. If the value of the chosen feature of a test instance falls into a particular interval, that instance is assigned the majority class of the training examples in this interval. The discretization intervals for a particular feature are constructed by sorting the training data according to the value for that feature and merging adjacent feature vectors of the same class into one interval. To prevent overfitting, one additional constraint is employed: the majority class in a particular interval must be represented by a certain minimum number of feature vectors in the training data. Holte [12] recommends a minimum of 6 as the threshold: this is what we used in our experiments.

As with support vector machines, we used the sorted differences in sensitivity between the first and last measurement for each location as a set of 76 input features for 1R. Applying this single-attribute scheme on the unsorted differences is not appropriate: classifications would be based on the single test location in the series of visual field measurements that is the most predictive one across all patients in the training data. It is unlikely that the same location is affected by progressing glaucoma in every patient. As we show in the next section, 1R tends to choose a feature for prediction that is close to the median per-location difference.

4.3 The Results

Tables 1 and 2 summarize the experimental results that we obtained by applying the three different techniques to our dataset consisting of 130 visual field sequences. For PWLR we show results for two different significance levels α, namely 0.05 and 0.01. All performance statistics are estimated using stratified 10-fold cross-validation [28]. The same folds are used for each method. Standard deviations for the 10 folds are also shown. Note that PWLR does not involve any training. Thus cross-validation is not strictly necessary to estimate its performance. However, it enables us to compare PWLR to the other methods and lets us test potential performance differences for statistical significance.

Table 1 shows the estimated percentage of correct classifications. Note that 50.8% is the accuracy achieved by a classifier that always predicts the majority class (i.e. "stable glaucoma"). The five different columns correspond to different numbers of visual fields: for the column labeled "8 VFs" we used all 8 visual field measurements corresponding to a particular patient to derive a classification (and for training if applicable), for the column labeled "7 VFs" we used the first 7, etc. The classification problem gets harder as fewer visual field measurements become available, and corresponds to an earlier diagnosis.

Table 2 shows estimated performance according to the kappa statistic [1]. This statistic measures how much a classifier improves on a chance classifier that assigns class labels randomly in the same proportions. It is defined by:

$$\kappa = \frac{p_c - p_r}{1 - p_r}, \tag{1}$$

where p_c is the percentage of correct classifications made by the classifier under investigation, and p_r is the corresponding expected value for a chance classifier. Following the convention established by Landis and Koch [15], values of kappa above 0.8 represent excellent agreement, values between 0.4 and 0.8 indicate moderate-to-good agreement, and values less than 0.4 represent poor agreement.

The first observation is that the accuracy of the different methods depends on the number of visual field measurements that are available. With the exception of PWLR given a setting of $\alpha = 0.05$ (for reasons explained below), all methods achieve kappa values greater than 0.4 for 7 and 8 VFs, and exhibit a decline in performance as fewer VFs become available. For 6 and fewer VFs all estimates of kappa are below 0.4.

The second observation is that the performance of PWLR depends on an appropriately chosen significance level. According to a paired two-sided t-test,[1] PWLR with $\alpha = 0.01$ performs significantly better than PWLR with $\alpha = 0.05$

[1] Significant at the 0.05%-level.

Table 1. Percent correct, and standard deviation, for different numbers of visual field measurements (estimated using stratified 10-fold cross-validation).

Method	8 VFs	7 VFs	6 VFs	5 VFs	4 VFs
PWLR (α=0.01)	75.4±8.7	72.3±10.4	62.3±9.2	60.8±9.2	53.9±6.3
PWLR (α=0.05)	61.5±8.1	60.8±8.5	61.5±8.9	60.8±8.5	63.1±8.7
1R	80.0±9.0	78.5±12.5	64.6±13.2	61.5±11.5	56.2±10.3
LSVM	75.4±9.5	72.3±8.3	68.5±6.7	61.5±8.9	63.9±12.1

Table 2. Value of kappa statistic, and standard deviation, for different numbers of visual field measurements (estimated using stratified 10-fold cross-validation).

Method	8 VFs	7 VFs	6 VFs	5 VFs	4 VFs
PWLR (α=0.01)	0.51±0.18	0.45±0.2	0.24±0.18	0.2±0.17	0.06±0.13
PWLR (α=0.05)	0.24±0.16	0.23±0.15	0.23±0.19	0.23±0.15	0.25±0.19
1R	0.59±0.19	0.56±0.25	0.28±0.27	0.23±0.22	0.12±0.2
LSVM	0.50±0.19	0.44±0.17	0.36±0.14	0.23±0.18	0.28±0.24

for 8 and 7 VFs (according to both percent correct and kappa); for 6 and 5 VFs there is no significant difference between the two parameter settings; and for 4 VFs $\alpha = 0.05$ significantly outperforms $\alpha = 0.01$ according to the percent correct measure.

The reason for this result is that $\alpha = 0.05$ is too liberal a significance level if 7 or 8 VFs are available: it detects too many decreasing slopes in the series of visual field measurements, consequently classifying too many glaucoma patients as progressing. On the other hand, $\alpha = 0.01$ is too conservative if only a few VF measurements are present: with 4 VFs it almost never succeeds in diagnosing progressing glaucoma.

The third observation is that LSVM does not share this disadvantage of PWLR. It does not require parameter adjustment to cope with different numbers of visual field measurements. For 7 and 8 VFs it performs as well as PWLR with $\alpha = 0.01$ and significantly better than PWLR with $\alpha = 0.05$; for 6 VFs it performs better than both; and for 4 VFs it performs as well as PWLR with $\alpha = 0.05$ and significantly better than PWLR with $\alpha = 0.01$.

The fourth observation is that the 1R-based method is the best-performing one for 8 and 7 VFs. For 6 and less VFs it appears to be less accurate than LSVM. However, the only differences that are statistically significant occur for 7 and 8 VFs between 1R and PWLR using $\alpha = 0.05$. Interestingly, if 7 or 8 VFs are available, 1R consistently bases its predictions on the 33rd-largest per-location difference. Forcing the scheme to use the median per-location difference—the 38th-largest difference—slightly decreases performance. Unfortunately, due to lack of additional independently sampled data, we could not test whether this is a genuine feature of the domain or just an artifact of the particular dataset we used.

4.4 Other Approaches

During the course of our experiments we tried many other learning schemes and data pre-processing regimes:

- Replacing the above classifiers with decision trees, nearest-neighbor rules, naive Bayes, model trees, and locally weighted regression [28]. We also tried higher-order support vector machines, which are able to represent non-linear class boundaries [7].
- Performance enhancing wrappers applied to the above classifiers such as boosting [28], bagging [28], and additive regression [10].
- A custom stacking approach where several classifiers are built using the difference between the ith and the $(i-1)$th visual field and a meta-classifier to arbitrate between the predictions of these base classifiers.
- Smoothing the measurements for each location using a rectangular filter before applying the learning scheme (with varying filter sizes.)
- Using physiologically clustered data where the 76 test locations are reduced into several cluster based on physiological criteria [24].
- Using simulated data [23] in addition to data based on real patients.
- Focusing on the N locations with the largest decrease in sensitivity and using only those points, along with some of the surrounding ones, as the input to the learning scheme.
- Using the number of per-location-decreases as feature value instead of the difference between the first and the last measurement.
- Adding the difference between the mean of the per-location measurements for the first and the last visual field as an extra attribute.

None were able to improve on the results reported above.

5 Discussion

Determining whether a glaucoma patient has deteriorating vision on the basis of visual field information is a challenging, but important, task. Each visual field assessment provides a large number of attributes, and the task lends itself to data mining approaches. We hope that exposing the problem to the data mining community will stimulate new approaches that reduce the time required to detect progression.

The results of our experiments show that both 1R and support vector machines appear to improve on pointwise univariate regression, a method that is commonly used for glaucoma analysis. Among all methods tested, 1R produced the best results if a series of eight or seven visual field assessments is available for each patient. If six or fewer measurements are available, a linear support vector machine appears to be the better choice, and we found it to be the most reliable method across different numbers of measurements. Compared to pointwise univariate regression analysis, whose performance depends strongly on an appropriately chosen significance level, this stability is an outstanding advantage.

We have focused on white-on-white visual field data measured in 76 locations. However, many other attributes are available that assist in progression analysis, and these may help learning schemes to yield more accurate results. For example, there are other visual field tests that detect glaucoma earlier and signal progression more rapidly, such as blue-on-yellow perimetry and Frequency Doubling Technology perimetry. The latter is particularly promising because it appears less variable than white-on-white perimetry [5].

Supplementary data is often available from measures of the structure of the optic nerve head. For example, confocal scanning laser tomography obtains 32 optical sections of the optic nerve with a laser and uses them to generate a 3-D topographic map [4]. This allows changes in nerve head shape, which may signal progression of glaucoma, to be quantified. However, these measures are very noisy, and it remains to be seen whether they can increase accuracy.

A serious impediment to research is the effort required to gather clean data sets. Not only is it arduous to collect clinical data, but significant expert time is needed to classify each patient. Computer simulation of visual field progression, which generates data that closely models reality, may offer an alternative source of training data [23]. An added advantage is that simulated visual fields are known to be progressing or stable by design, so the classification operation introduces no noise into the data.

References

1. P. Armitage and G. Berry. *Statistical Methods in Medical Research.* Blackwell Scientific Pulications, Oxford, third edition, 1994.
2. M.K. Birch, P.K. Wishart, and N.P. O'Donnell. Determining progressive visual field loss in serial humphrey visual fields. *Ophthomology*, 102(8):1227–1235, 1995.
3. L. Brigatti, K. Nouri-Mahdavi, M. Weitzman, and J. Caprioli. Automatic detection of glaucomatous visual field progression with neural networks. *Archives of Ophthalmol.*, 115:725–728, 1997.
4. R.O. Burk, A. Tuulonen, and P.J. Airaksinen. Laser scanning tomography of localised nerve fibre layer defects. *British J. of Ophthalmol*, 82(10):1112–1117, 1998.
5. B.C. Chauhan and C.A. Johnson. Test-retest variability of frequency-doubling perimetry and conventional perimetry in glaucoma patients and normal subjects. *Investigative Ophthomology and Vision Science*, 40(3):648–656, 1999.
6. B.C. Chauhan, S.M.Drance, and G.R. Douglas. The use of visual field indices in detecting changes in the visual field in glaucoma. *Investigative Ophthomology and Vision Science*, 31(3):512–520, 1990.
7. C. Cortes and V. Vapnik. Support vector networks. *Machine Learning*, 20:273–297, 1995.
8. D.P. Crabb, F.W. Fitzke, A.I. McNaught, D.F. Edgar, and R.A. Hitchings. Improving the prediction of visual field progression in glaucoma using spatial processing. *Ophthomology*, 104(3):517–524, 1997.
9. M. Fingeret and T.L. Lewis. *Primary care of the glaucomas.* Mc Graw Hill, New York, second edition, 2001.
10. J.H. Friedman. Greedy function approximation: A gradient boosting machine. Technical report, Department of Statistics, Stanford University, CA, 1999.

11. D.B. Henson, S.E. Spenceley, and D.R. Bull. Artificial neural network analysis of noisy visual field data in glaucoma. *Art. Int. in Medicine*, 10:99–113, 1997.
12. R.C. Holte. Very simple classification rules perform well on most commonly used datasets. *Machine Learning*, 11:63–91, 1993.
13. J. Katz. Scoring systems for measuring progression of visual field loss in clinical trials of glaucoma treatment. *Ophthomology*, 106(2):391–395, 1999.
14. S. Keerthi, S. Shevade, C. Bhattacharyya, and K. Murthy. Improvements to platt's SMO algorithm for SVM classifier design. Technical report, Dept. of CSA, Banglore, India, 1999.
15. J.R. Landis and G.G. Koch. An application of hierarchical kappa-type statistics in the assessment of majority agreement among multiple observers. *Biometrics*, 33(2):363–374, 1977.
16. T. Leitman, J. Eng, J. Katz, and H.A. Quigley. Neural networks for visual field analysis: how do they compare with other algorithms. *J Glaucoma*, 8:77–80, 1999.
17. M.C. Leske, A. Heijl, L. Hyman, and B. Bengtsson. Early manifest glaucoma trial: design and baseline data. *Ophthomology*, 106(11):2144–2153, 1999.
18. S. Mandava, M. Zulauf, T. Zeyen, and J. Caprioli. An evaluation of clusters in the glaucomatous visual field. *American J. of Ophthalmol.*, 116(6):684–691, 1993.
19. R.K. Morgan, W.J. Feuer, and D.R. Anderson. Statpac 2 glaucoma change probability. *Archive of Ophthalmol.*, 109:1690–1692, 1991.
20. K. Nouri-Mahdavi, L. Brigatti, M. Weitzman, and J. Caprioli. Comparison of methods to detect visual field progression in glaucoma. *Ophthomology*, 104(8):1228–1236, 1997.
21. J. Platt. Fast training of support vector machines using sequential minimal optimization. In *Advances in Kernel Methods—Support Vector Learning.* MIT Press, Cambridge, MA, 1998.
22. S.D. Smith, J. Katz, and H.A. Quigly. Analysis of progressive change in automated visual fields in glaucoma. *Investigative Ophthomology and Vision Science*, 37(7):1419–1428, 1996.
23. P.G.D. Spry, A.B. Bates, C.A. Johnson, and B.C. Chauhan. Simulation of longitudinal threshold visual field data. *Investigative Ophthomology and Vision Science*, 41(8):2192–2200, 2000.
24. J. Weber and H. Ulrich. A perimetric nerve fiber bundle map. *International Ophthalmology*, 15:193–200, 1991.
25. C.J. Wild and G.A.F. Weber. *Introduction to probability and statistics.* Department of Statistics, University of Auckland, New Zealand, 1995.
26. J.M. Wild, M.K. Hussey, J.G.Flanagan, and G.E. Trope. Pointwise topographical and longitudinal modeling of the visual field in glaucoma. *Investigative Ophthomology and Vision Science*, 34(6):1907–1916, 1993.
27. J.M. Wild, N. Hutchings, M.K. Hussey, J.G.Flanagan, and G.E. Trope. Pointwise univariate linear regression of perimetric sensitivity against follow-up time in glaucoma. *Ophthomology*, 104(5):808–815, 1997.
28. Ian H. Witten and Eibe Frank. *Data Mining: Practical Machine Learning Tools and Techniques with Java Implementations.* Morgan Kaufmann, San Francisco, CA, 2000.

Seabreeze Prediction Using Bayesian Networks

Russell J. Kennett, Kevin B. Korb, and Ann E. Nicholson

School of Computer Science and Software Engineering
Monash University, VIC 3800, Australia,
russk88@hotmail.com,{korb,annn}@csse.monash.edu.au

Abstract. In this paper we examine the use of Bayesian networks (BNs) for improving weather prediction, applying them to the problem of predicting sea breezes. We compare a pre-existing Bureau of Meteorology rule-based system with an elicited BN and others learned by two data mining programs, TETRAD II [Spirtes et al., 1993] and Causal MML [Wallace and Korb, 1999]. These Bayesian nets are shown to significantly outperform the rule-based system in predictive accuracy.

1 Introduction

Bayesian networks have rapidly become one of the leading technologies for applying artificial intelligence to real-world problems, as a decision support tool for reasoning under uncertainty. The usual approach is one of knowledge engineering: elicit the causal structure and conditional probabilities from domain experts. This approach, however, suffers from the same headaches that accompanied early expert systems; frequently, for example, experts are unavailable or they generate inconsistent probabilities [Wallsten and Zwick, 1993]. This has led to a recent upsurge in interest in the automated learning of Bayesian networks. Here we examine the use of Bayesian networks (BNs) in improving weather prediction. In particular, we compare a pre-existing rule-based system for the prediction of seas breezes provided by the Australian Bureau of Meteorology (BOM) with BNs developed by expert elicitation and BNs learned by two machine learning programs, TETRAD II (from [Spirtes et al., 1993]) and Causal MML (CaMML; [Wallace and Korb, 1999]). We first describe the domain, BN methodology and data mining methods and then examine predictive accuracy.

2 The Seabreeze Prediction Problem

Sea breezes occur because of the unequal heating and cooling of neighbouring sea and land areas. As warmed air rises over the land, cool air is drawn in from the sea. The ascending air returns seaward in the upper current, building a cycle and spreading the effect over a large area. If wind currents are weak, a sea breeze will usually commence soon after the temperature of the land exceeds that of the sea, peaking in mid-afternoon. A moderate to strong prevailing offshore wind will delay or prevent a sea breeze from developing, while a light to moderate prevailing offshore wind at 900 metres (known as the gradient level) will reinforce

D. Cheung, G.J. Williams, and Q. Li (Eds.): PAKDD 2001, LNAI 2035, pp. 148–153, 2001.

a developing sea breeze. The sea breeze process is also affected by time of day, prevailing weather, seasonal changes and geography [Batt, 1995,Bethwaite, 1996, Houghton, 1992].

BOM provided a database of meteorological information from three types of sensor sites in the Sydney area. We used 30MB of data from October 1997 to October 1999, with about 7% of cases having missing attribute values. Automatic weather stations (AWS) provided ground level wind speed (ws) and direction (wd) readings at 30 minute intervals (date and time stamped). Olympic sites provided ground level wind speed (ws), direction (wd), gust strength, temperature, dew temperature and rainfall. Weather balloon data from Sydney airport (collected at 5am and 11pm daily) provided vertical readings for gradient-level wind speed (gws), direction (gwd), temperature and rainfall. (Predicted variables arc wind speed [wsp] and wind direction [wdp] below.)

Seabreeze forecasting is currently done using a simple rule-based system, which predicts them by applying several conditions: if the wind is offshore and is less than 23 knots, and if part of the forecast timeslice falls in the afternoon, then a sea breeze is likely to occur. Its predictions are generated from wind forecasts produced from large-scale weather models. According to BOM, this rule-based system is the best they have been able to produce and correctly predicts approximately two thirds of the time.

3 Bayesian Network Methodology

Bayesian methods provide a formalism for reasoning under conditions of uncertainty. A Bayesian network is a directed acyclic graph representing a probability distribution [Pearl, 1988]. Network nodes represent random variables and arcs represent the direct dependencies between variables. Each node has a conditional probability table (CPT) which indicates the probability of each possible state of the node given each combination of parent node states. The tables of root nodes contain unconditional prior probabilities.

A major benefit of BNs is that they allow a probability distribution to be decomposed into a set of local distributions. The network topology indicates how these local distributions should be combined to produce the joint distribution over all nodes in the network. This allows the separation of the quantification of influence strengths from the qualitative representation of the causal influences between variables, making the knowledge engineering and/or interpretation of BNs significantly easier. The task of building a Bayesian model can therefore be split in two: the specification of the structure of the domain, and the quantification of the causal influences. Various tools for efficient inference in BNs have been developed; we used Netica [Norsys, 2000]. In the remainder of this section we describe the tasks involved in applying BNs to the seabreeze prediction problem.

Netica learns BN CPTs by counting combinations of variable occurrences, a method developed by [Spiegelhalter and Lauritzen, 1990]. We applied this technique to all networks, but only after the qualitative causal structure was fixed, either by expert elicitation or the causal discovery methods described below.

The first method we used for network construction was expert elicitation, with meteorologists at the BOM (Figure 1(a). The links between network nodes described causal relationships between the wind to be predicted and the current wind, the time of day, and the month of the year. Arc direction was selected to reflect the temporal relationship between variables.

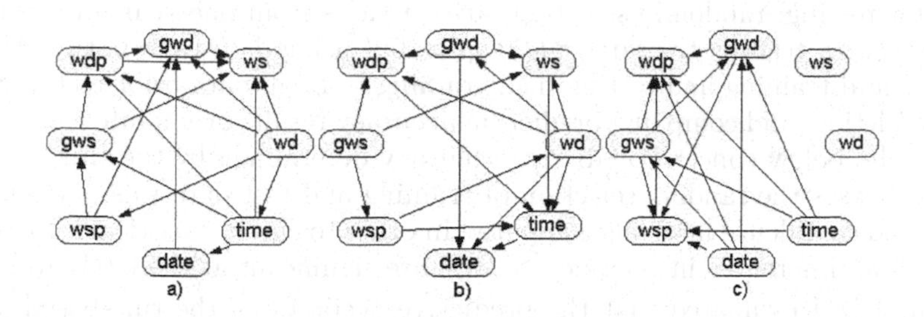

Fig. 1. Airport data networks - a) CaMML, b) TETRAD II with prior temporal ordering, c) Expert elicitation

Learning causal structure by testing data for conditional independencies (CI learning) was introduced by [Verma and Pearl, 1991]. The basic algorithm presumes the existence of an 'oracle' which can provide a true or false answer to any question of type $X \amalg Y|S$ (i.e., X and Y are independent given the variables in set S). The algorithm is not guaranteed to discover the original Bayesian network, but will find all direct arcs between nodes and orient many of them. TETRAD II [Spirtes et al., 1993] provided the first practical implementation of this algorithm, replacing an oracle with significance tests for vanishing conditional dependencies. TETRAD II asymptotically obtains the causal structure of a distribution to within the statistical equivalence class of the true model. Unfortunately, as a CI learner, TETRAD II does not always specify the direction of a link between nodes. To compensate for this, it is possible to specify a partial temporal ordering of variables, should the user have such prior information. Therefore, we generated two networks with TETRAD II for each data set: the first simply from the data, and the second using a full temporal ordering of variables (Figure 1b). The first run was performed to allow a fair comparison of TETRAD II's performance with CaMML, given no extra domain information.

An alternative type of causal learning employs a scoring metric to rank each network model and searches through the model space, attempting to maximize its metric. MML (Minimum Message Length) [Wallace and Boulton, 1968] uses information theory to develop a Bayesian metric. MML metrics for causal models have been developed by [Wallace et al., 1996,Wallace and Korb, 1999]. Here, we applied CaMML (Causal MML; [Wallace and Korb, 1999]), which conducts the MML search through the space of causal models using stochastic sampling.

4 Experimental Results

Here we consider the performance of the different tools in sea breeze prediction. First we compare the BNs with the existing predictor provided by BOM, and then we compare the different BNs against each other.

All of the BNs were trained on weather data provided by BOM. The elicited BN was parameterized by Netica using those data; the TETRAD II and MML BN structures were learned from the data and then parameterized by Netica. We examined four different testing regimes: using 1997 data for training and 1998 data for testing; randomly selecting 80% of cases from one year and using the remainder for testing; the same 80-20 split, but using data from both years; and incremental training and testing, i.e., training from all data prior to the date of the prediction, and compiling prediction accuracy results over a full year. Most of the results below concern predictive accuracy determined by the third method. In these cases the random selection of training and test data was performed 15 times and confidence intervals computed in order to check statistical significance; in general, differences in accuracy > 10% are significant at the 0.05 level.

The BOM estimated that the predictive accuracy of the rule-based system (RB) was approximately 60 to 70 per cent (more detailed statistics were unavailable). A sea breeze is defined as occurring when and only when the gradient wind direction is offshore and the ground level wind onshore. Weather balloons providing usable data were launched from Sydney airport twice a day, at about 5am and 11pm. Predictions were produced over a period of two years, from both sets of data, with the lookahead time of the prediction in increments of three hours. Predictions of both seabreeze existence and (more interesting and difficult) wind direction were generated for each AWS site (see [Kennett et al., 2001] for full results). The predictive accuracy of the system varied in a rough sine pattern, with a maximum of 0.8 reached at about 4pm. Wind direction accuracy was approximately 10% lower, in the same pattern.

We tested four Bayesian networks for predictive accuracy (see Figure 1), one discovered by MML, two by TETRAD II and one elicited from experts. Since TETRAD II without the aid of a prior temporal ordering produced undirected arcs and cycles, its network was modified by resolving inconsistencies and ambiguities to TETRAD II's advantage. The accuracy results for these four BNs, together with the BOM RB system, are given in Figure 2. It is clear that, with the exception of the earliest 3 hour prediction, the differences between the BNs are not statistically significant, while the simple BOM RB system is clearly (statistically significantly) underperforming all of the Bayesian networks. This demonstrates that the automated data mining methods are capable of improving on the causal relations encoded in the rule-based system. In this problem, the difference between CaMML and TETRAD II was largely in usability: TETRAD II either required additional prior information (temporal constraints) or else hand-made posterior edits to reach the level of performance of CaMML.

In general, network performance was maximal (up to 80% accuracy) at look ahead times which were multiples of 12 hours, corresponding to late afternoon or early morning. Clearly, there is a strong periodicity to this prediction problem. In the future, such periodicity could be explicitly incorporated into models using MML techniques (e.g., in selecting parameters for a sine function).

The training method examined thus far has the drawback that both the structure of the model and its parameters are learned in batch mode, with predictions generated from a fixed, fully specified model. We speculated that since weather systems change over time a better approach would be to learn the causal

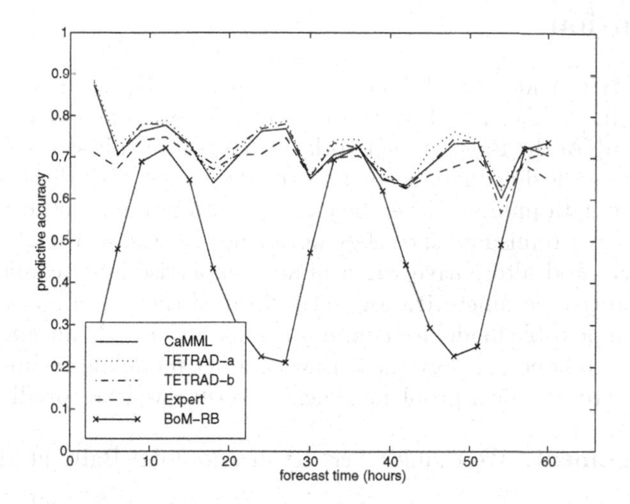

Fig. 2. Comparison of Airport site type network versions

structure in batch mode, but to reparameterize the Bayesian network incrementally, applying a time decay factor so as to favor more recent over old data. The (unnormalized) weight applied to data for incremental updating of the network parameters was optimized by a greedy search. Figure 3 shows the average performance of the MML Bayesian networks when incrementally reparameterized over the 1998 data. The improvement in predictive accuracy is statistically significant at the 0.05 level, despite the fact that the average scores reported here are themselves presumably suboptimal, since the predictions made early in the year use parameters estimated from small data sets.

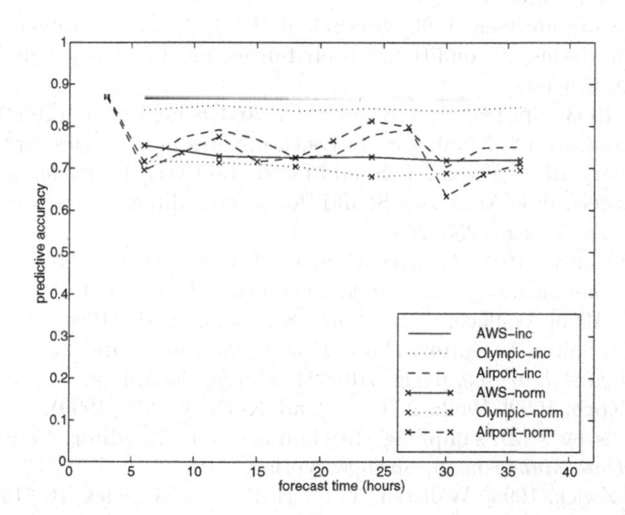

Fig. 3. Comparison of incremental and normal training method by BN type

5 Conclusion

This case study provides a useful model for employing Bayesian network technology in data mining. The initial rule-based predictive system was demonstrated to be inferior to all of the Bayesian networks developed in this study. The Bayesian net elicited from the domain experts performed on a par with those generated automatically by data mining. Nevertheless, the data mining methods show themselves to be very promising, since they performed as well as the elicited network and so offer a good alternative when human expertise is unavailable. Furthermore, the adaptive parameterization outperformed the static Bayesian networks and provides a possible model for combining elicitation with automated learning. In the future, we hope to apply these Bayesian net modeling techniques to more challenging meteorological problems, such as severe weather prediction.

Acknowledgement. We thank Chris Ryan and John Bally at the Bureau of Meteorology.

References

[Batt, 1995] Batt, K. (1995). Sea breezes on the NSW coast. *Offshore Yachting*, Oct/Nov.

[Bethwaite, 1996] Bethwaite, F. (1996). *High Performance Sailing*. Waterline Books.

[Houghton, 1992] Houghton, D. (1992). *Wind Strategy*. Fernhurst Books.

[Kennett et al., 2001] Kennett, R., Korb, K., and Nicholson, A. (2001). Seabreeze prediction using Bayesian networks: A case study. Technical Report TR 2001/86, School of Computer Science and Software Engineering, Monash University.

[Norsys, 2000] Norsys (2000). Netica. http://www.norsys.com.

[Pearl, 1988] Pearl, J. (1988). *Probabilistic Reasoning in Intelligent Systems*. Morgan Kaufmann, San Mateo, Ca.

[Spiegelhalter and Lauritzen, 1990] Spiegelhalter, D. J. and Lauritzen, S. L. (1990). Sequential updating of conditional probabilities on directed graphical structures. *Networks*, 20:579–605.

[Spirtes et al., 1993] Spirtes, P., Glymour, C., and Scheines, R. (1993). *Causation, Prediction and Search*. Number 81 in Lecture Notes in Statistics. Springer Verlag.

[Verma and Pearl, 1991] Verma, T. S. and Pearl, J. (1991). Equivalence and synthesis of causal models. In D'Ambrosio, S. and Bonissone, editors, *Uncertainty in Artificial Intelligence Vol 6*, pages 255–268.

[Wallace and Boulton, 1968] Wallace, C. S. and Boulton, D. M. (1968). An information measure for classification. *The Computer Journal*, 11:185–194.

[Wallace et al., 1996] Wallace, C. S., Korb, K., and Dai, H. (1996). Causal discovery via MML. In Saitta, L., editor, *Proceedings of the Thirteenth International Conference on Machine Learning*, pages 516–524. Morgan Kaufman.

[Wallace and Korb, 1999] Wallace, C. S. and Korb, K. B. (1999). Learning linear causal models by MML samplling. In Gammerman, A., editor, *Causal Models and Intelligent Data Management*. Springer-Verlag.

[Wallsten and Zwick, 1993] Wallsten, T. S., B. D. V. and Zwick, R. (1993). Comparing the calibration and coherence of numerical and verbal probability judgments. *Management Science*, 39(2):176–190.

Semi-supervised Learning in Medical Image Database

Chun Hung Li and Pong Chi Yuen

Department of Computer Science, Hong Kong Baptist University, Hong Kong
{chli, pcyuen}@comp.hkbu.edu.hk

Abstract. This paper presents a novel graph-based algorithm for solving the semi-supervised learning problem. The graph-based algorithm makes use of the recent advances in stochastic graph sampling technqiue and a modeling of the labeling consistency in semi-supervised learning. The quality of the algorithm is empirically evaluated on a synthetic clustering problem. The semi-supervised clustering is also applied to the problem of symptoms classification in medical image database and shows promising results.

1 Introduction

In machine learning for classification problems, there are two distinct approaches to learning or classifying data: the supervised learning and un-supervised learning. The supervised learning deals with problem where a set of data are labeled for training and another set of data would be used for testing. The un-supervised learning deals with problem where none of the labels of the data are available. In recent years, important classification tasks have emerged with enormous volume of data. The labeling of a significant portions of the data for training is either infeasible or impossible. Sufficient labeled data for training are often unavailable in data mining, text categorization and web page classification. A number of approaches have been proposed to combine a set of labeled data with unlabeled data for improving the classification rate. The naive Bayes classifier and the EM algorithm have been combined for classifying text using labeled and unlabeled data [1]. The support vector machines have been extended with transductive inference to classify text [2]. A modified support vector machine and non-convex quadratic optimization approaches have been studied for optimizing semi-supervised learning [3]. Graph based clustering has received a lot of attention recently. A factorization approach has been proposed for clustering[4], the normalized cuts have been proposed as a generalized method for clustering [5] and [6]. In this paper, we investigated the use of stochastic graph-based sampling approach for solving the semi-clustering problem. Graph-based clustering is also shown to be closely related to similarity-based clustering [7].

D. Cheung, G.J. Williams, and Q. Li (Eds.): PAKDD 2001, LNAI 2035, pp. 154–160, 2001.

2 Graph Based Clustering

The cluster data is modeled using an undirected weighted graph. The data to be clustered are represented by vertices in the graph. The weights of the edge of vertices represents the similarity between the object indexed by the vertices. The similarity matrix $S \in R^{N^2}$ where S_{ij} represents the similarity between the object o_i and the object o_j. A popular choice of the similarity matrix is of the form

$$S_{ij} = \exp\left(-\frac{d^2(i,j)}{\sigma^2}\right) \tag{1}$$

where $d(i,j)$ is a distance measure for the object i and the object j. The exponential similarity function have been used for clustering in [4], [5] and [6]. The graph G is simply formed by using the objects in the data as nodes and using the similiarity S_{ij} as the values(weights) of the edges between nodes i and j. The minimum cut algorithm partitions the graph G into two disjoint sets (A, B), that minimizes the following objective function,

$$f(A,B) = \sum_{i \in A} \sum_{j \in B} S_{ij}. \tag{2}$$

The minimum cut problem has been a widely studied problem. The classical approach for solving the minimum cuts is via solving the complementary problem of maximum flows. Recently, Karger introduced a randomized algorithm which solve the minimum cuts problem in $O(n^2 \log^3 n)$ [8]. The randomized algorithm makes use of a contraction algorithm for evaluating the minimum cuts of the graph and we will show in later section how we modify the contraction algorithm for semi-supervised learning.

3 Graph Contraction Algorithm

The contraction algorithm for unsupervised clustering consists of an iterative procedure that contract two connected vertex i and j:

- Randomly select an edge (i,j) from G with probability proportional to S_{ij}
- Contract the edge (i,j) into a meta node ij
- Connect all edges incident on i and j to the meta node ij while removing the edge (i,j)

This contraction is repeated until all nodes are contracted into a specified number of clusters. The semi-supervised algorithm assumes a set of given labeled samples. Initially, assign empty label to all nodes,i.e. $L_i = \phi$ for all nodes i. Then assign the labels of given labeled nodes to their respective labels. After this initialization, the following contraction algorithm is applied:

1. Randomly select an edge (i,j) from G with probability proportional to S_{ij}, depending on the labels of i and j, do one of the following:

a) If $L_i = \phi$ and $L_j = \phi$, then $L_{ij} = \phi$
b) If $L_i = \phi$ and $L_j \neq \phi$, then assign $L_{ij} = L_j$
c) If $L_j = \phi$ and $L_i \neq \phi$, then assign $L_{ij} = L_i$
d) If $L_i \neq L_j$ and $L_i \neq \phi$ and $L_j \neq \phi$, then remove the edge (i, j) from G, and return to step 1

2. Contract the edge (i, j) into a meta node ij
3. Connect all edges incident on i and j to the meta-node ij while removing the edge (i, j)

This contraction is repeated until all nodes are contracted into a specified number of clusters or until all edges are removed. This semi-supervised contraction guarantees the consistency in the labeling outcome in merging the meta-nodes. Furthermore, both of the algorithms above can be repeated as separate randomized trials on the data. The results of each trial can be considered as giving the probability of connectivity between individual nodes and then be combined to give more accurate estimation of the probability of connectivity.

4 Testing on Synthetic Data

A synthetic dataset is used for testing the semi-supervised clustering algorithm. Figure 1 shows the data for a two cluster clustering problem. The two clusters are seperated with a sinusoidal boundary. The minimum cut algorithm with randomized contraction is applied to the synthetic data. The result of a typical trial is shown in 2. The cross on the right middle part of the figure is the sigeleton that is separated from the rest of the nodes. The algorithm fails to recover the two clusters as the minimum cut is often given by splitting the data into a big cluster and an outlying point which have a large distance from its neighbors. To

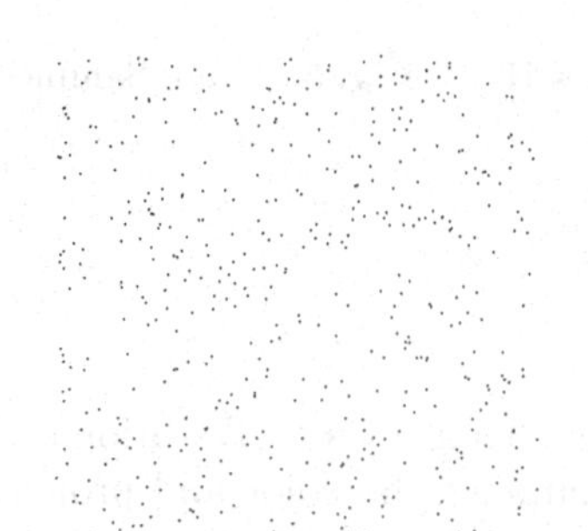

Fig. 1. Two clusters

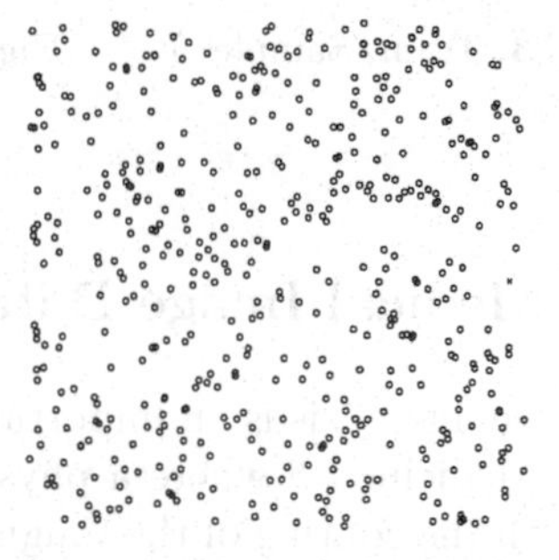

Fig. 2. One solution of randomized contraction

test the performance of the semi-supervised clustering algorithm on the synthetic data, three test situations are considered:

1. one sample from the top half and one sample from the bottom half is randomly chosen as training samples
2. two samples from top cluster and two samples from the bottom cluster
3. ten samples from top cluster and ten samples from the bottom cluster,

the training samples in the above three cases are shown in Figure 3, Figure 4 and Figure 5 respectively. The training samples for the top cluster are shown with plus sign and the training samples for the lower cluster are shown with circle sign. With such a small number of training samples, one can judge from the figures that inference based/decision surface based classifier will not be able to determine accurately the sinusoidal boundary between the cluster. The solution for one trial of the semi-supervised clustering has been shown in 7. There are a few misclassified data points on the middle left portion of the graph. The average error percentage of the semi-supervised clustering is shown in Table 6. From the table, we can see that as the number of training samples is increased the average error percentage of classification drops, which is consistent with expectation. Furthermore, if we considered a single trial as given the probability of a pixel belonging to a cluster, we can average out the probabilities obtained from different trials and obtain the error of the combined estimation. From Table 6, we can see that the combined estimation is very accurate, having errors of less than one percent.

Fig. 3. Train. Samples I **Fig. 4.** Train. Samples II **Fig. 5.** Train. Samples III

5 Medical Image Database

Tongue diagnosis is an important part of the Four Diagnosis in Traditional Chinese Medicine [9] where a physician visually examines the color and properties of both the coating of the tongue and the tongue proper [10],[11]. Sample tongue images from nine patients are shown in Figure 8. The images show that there are large variations in the color of tongue proper and the color of surface coating. In the center image of the middle row, the tongue is covered with yellowish coating. In the right image of the bottom row, the tongue is covered with dense white-gray coating. The left image in the top column shows a pink tongue and the center image of the bottom row shows a deep red tongue. In western medicine, the

	Average error	Error of combined est.
1	16.58%	0.2%
2	9.24%	0.6%
3	2.82%	0.4%

Fig. 6. Performance of semi-supervised clustering: (1) Average error of single trials (2) Error of combined estimation

Fig. 7. One solution of semi-supervised randomized contraction

visual examination of tongue also reveals important information on the patient. The glossitis and the geographic tongue can be diagnosed by visually examining the tongue. Glossitis maybe caused by local bacterial or viral infections on the tongue or be caused by systemic origin such as: iron deficiency anemia, pernicious anemia and vitamin deficiencies.

Images drawn from sections of tongue images from different patients are used for comparison. The tongue image is then segmented into square blocks of 36x36 and blocks which cover the tongue are then selected from the image. Figure 9 shows the partitioned blocks from a sample image. Tongue images are taken from 64 patients and a total of 6788 blocks are extracted. The color mean and variance of a color block is used as the representative features of the block. Thus, a color block is represented by its color means and variances as six attributes $(\mu_1, \mu_2, \mu_3, \sigma_1, \sigma_2, \sigma_3)$. The color attributes are color values in RGB color space or color values in CIE $L^*u^*v^*$ color space.

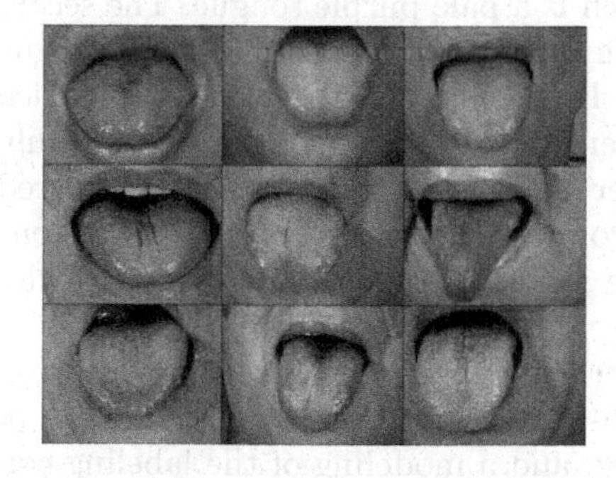

Fig. 8. Sample Tongue Images

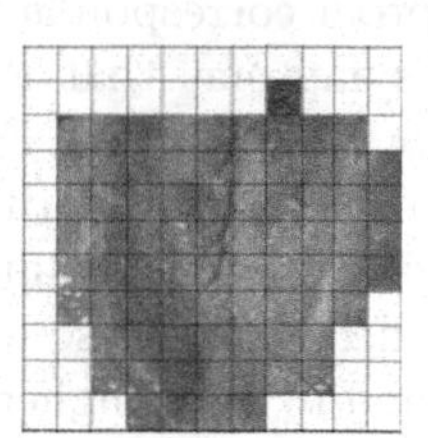

Fig. 9. Partition of blocks in a sample image

We have previously designed algorithms for color cluster analysis on the tongue image database [12]. In this section, we extended the analysis by incorporating labeled samples corresponding to typical symptoms shown in various tongue diagnosis. Twenty color blocks are selected from different patients corresponding to typical symptoms. The following table shows the diagnosis of the 20 samples selected. Each color block is indexed by three coordinates (n, bx, by), where n is the the patient number in the database, bx is the horizontal block number in the image and by is the vertical block number. The semi-supervised

Table 1. Training samples for tongue image database

	TBC	Symptoms		TBC	Symptoms
1	(8,8,7)	pale purple	2	(11,5,9)	pale purple
3	(43,11,9)	red	4	(31,8,6)	red
5	(27,12,4)	light pale	6	(51,10,9)	thin white coat (pale red)
7	(56,8,9)	thin white coat (light pale)	8	(59,8,6)	thin white coat (pale)
9	(42,7,5)	thin white coat (dark red)	10	(49,7,12)	thin white coat (red)
11	(31,5,9)	thin while coat (red)	12	(32,4,5)	pale yellow coat (pale red)
13	(52,3,4)	pale yellow coat (light pale)	14	(40,3,6)	dark yellow (thick)
15	(27,4,11)	dark yellow (light pale)	16	(29,3,3)	dark yellow (thick)
17	(17,4,8)	thick white (light pale)	18	(18,13,6)	"kong"
19	(61,5,5)	thick white	20	(22,4,4)	light yellow (pale purple)

clustering is then applied to the clustering of tongue image blocks in the medical image database. The semi-supervised clustering is repeated ten times for the complete image database. Each block will be classified into one of the 20 sample classes in each trail. To find the major symptoms associated with a patient, we first calculated the matches of all the image blocks of the patient in a single trial. The sum of the number of matches of all image blocks in the 10 trials are accumulated. The five highest matches are taken as the major symptoms for the patients. To show the typical performance of the algorithm, the results on the first seven patients are tabulated here. Table 3 shows the results of the semi-supervised clustering and the results for the nearest neighbor classifier are shown in Table 2 for comparison. For example, for patient 1 in Table 2, the first symptom corresponds to class 2, which is a pale purple tongue. The second symptom corresponds to class 11 which is a thin white coating on red tongue. The symptoms that is not consistent with the judgement of Chinese medical doctor is underlined. For most of the patient cases, the differences in the results between the two algorithms lies in the order of the first five symptoms discovered. Judging from the symptoms discovered from the algorithms and the symptoms classified by Chinese medical doctors, the semi-supervised clustering algorithm is found to have higher consistency.

To conclude, a novel graph-based algorithm for solving the semi-supervised learning problem is introduced. The graph-based algorithm makes use of the recent advances in stochastic graph sampling and a modeling of the labeling consistency in semi-supervised learning. The quality of the algorithm is empirically evaluated on a synthetic clustering problem. The semi-supervised clustering is also applied to the problem of symptoms classification in medical image database and promising results have been obtained.

Table 2. Major symptoms discovered by the nearest neighbour classifier

Patient no.	Sym I	II	III	IV	V
1	2	11	5	1	10
2	10	1	6	2	5
3	10	5	6	20	15
4	19	8	12	11	13
5	12	15	8	11	19
6	19	5	1	6	9
7	8	20	2	11	15

Table 3. Major symptoms discovered by the semi-supervised clustering

Patient no.	Sym I	II	III	IV	V
1	11	2	5	10	1
2	10	6	11	5	1
3	20	10	5	6	19
4	8	12	20	19	13
5	12	13	8	19	20
6	9	7	19	5	6
7	8	20	19	12	2

References

1. K. Nigam, A. McCallum, Sebastian Thrun, and Tom Mitchell. Text classification from labeled and unlabeled documents using em. *Machine Learning*, 34(1), 1999.
2. Thorsten Joachims. Transductive inference for text classification using support vector machines. In *International Conference on Machine Learning (ICML)*, 1999.
3. T. S. Chiang and Y. Chow. Optimization approaches to semi-supervised learning. In M. C. Ferris, O. L. Mangasarian, and J. S. Pang, editors, *Applications and Algorithms of Complementarity*. Kluwer Academic Publishers, 2000.
4. P. Perona and W. T. Freeman. A factorization approach to grouping. In *Proceedings of European Conference on Computer Vision*, page ?, 1998.
5. J. Shi and J. Malik. Normalized cuts and image segmentation. *IEEE Trans. Pattern Analysis and Machine Intelligence*, 21, 2000.
6. Y. Gdalyahu, D. Weinshall, and M. Werman. Stochastic image segmentation by typical cuts. In *Proceedings of the IEEE CVPR 1999*, volume 2, pages 596–601, 1999.
7. J. Puzicha, T. Hofmann, and J. M. Buhmann. A theory of proximity based clustering: structure detection by optimization. *Pattern Recognition*, 33:617–634, 2000.
8. D. R. Karger and C. Stein. A new approach to the minimum cut problem. *Journal of the ACM*, 43(4):601–640, 1996.
9. X.Z. Shuai. *Fundamentals of Traditional Chinese Medicine*. Foreign Languages Press, Beijing, China, second edition, 1995.
10. C. C. Chiu, H.S. Lin, and S.L. Lin. A structural texture recognition approach for medical diagnosis through tongue. *Biomedical Engineering, Application, Basis, Communication*, 7(2):143–148, 1995.
11. C. C. Chiu. The development of a computerized tongue diagnosis system. *Biomedical Engineering, Application, Basis, Communication*, 8(4):342–350, 1996.
12. C. H. Li and P. C. Yuen. Regularized color clustering for medical image database. *IEEE Trans. on Medical Imaging*, 19(11), 2000.

On Application of Rough Data Mining Methods to Automatic Construction of Student Models

Feng-Hsu Wang and Shiou-Wen Hung

Graduate School of Information Management, Ming Chuan University,
5 Teh-Ming Rd., Gwei Shan District, Taoyuan County 333, Taiwan
{fhwang@mcu.edu.tw, jess2939@ms26.hinet.net}

Abstract. Student modeling has been an active research area in the field of intelligent tutoring systems. In this paper, we propose a rough data mining approach to the student modeling problems. The problem is modeled as a knowledge discovery process in which a student's domain knowledge (classification rules) was discovered and rebuilt using rough set data mining techniques. We design two knowledge extraction modules based on the lower approximation set and upper approximation set of the rough set theory, respectively. To verify the effectiveness of the knowledge extraction modules, two similarity metrics are presented. A set of experiments is conducted to evaluate the capability of the knowledge extraction modules. At last, based on the experimental results some suggestions about a future knowledge extraction module are outlined.

1 Introduction

Building student models [9] to effectively represent a learner state has been an active research area in the field of intelligent tutoring systems. There has been much effort in constructing student models in the literature, including the overlay model [3], debug model [4], dynamic model [8], and model tracing [1]. In this paper, we propose a rough data mining approach based on rough set theory [5] [6] to the student-modeling problem. Rough set theory has found many applications in artificial intelligence and cognitive science, and it is applied successfully to many practical real-world problems. Not like the fuzzy set theory [10], rough set theory does not need a well-predefined membership functions to proceed successfully. Instead, it relies only on the available data and attributes to work on the generation of rules. In particular rough set theory is a deterministic data mining methodology useful not only for large data sets but also for small amount of data, for which statistical methods may not be adequate. This is especially important for the context of this research in which a student's answering records may not be a large record set.

In this paper the student-modeling problem is represented as a knowledge discovery process in which a student's domain knowledge (classification rules) was discovered and rebuilt using rough data mining techniques. Two knowledge extraction modules were designed based on the lower approximation set and upper approximation set of

D. Cheung, G.J. Williams, and Q. Li (Eds.): PAKDD 2001, LNAI 2035, pp. 161-166, 2001.

the rough set theory, respectively. To verify the effectiveness of the knowledge extraction modules, we present two similarity metrics. Based on these metrics, a set of simulation experiments is conducted to evaluate the capability of the knowledge extraction modules. At last following the experimental results some suggestions about a future knowledge extraction module are outlined.

2 Rough Set Theory

A rough set [7] is represented by two sets, a lower approximation set and an upper approximation set. Let U be the closed universe set of objects, $Q=C \cup D$ be the attribute set where C is the condition attribute set and D is the decision attribute set, and A be any subset of Q, then define $I(A)$ as a binary relation on U, called "indiscernibility relation", that is,

$$x\, I(A)\, y \text{ iff } a(x) = a(y) \text{ for every } a \in A,\ x,\ y \in U \quad (1)$$

where $a(x)$ denotes the value of attribute a for object x. Note that $I(A)$ is an equivalence relation. Besides, let $U/I(A)$ be the partition determined by the relation $I(A)$, denoted as U/A, and let $[x]_A$ be the equivalent class containing object x. Then we say that (x, y) is A-indiscernible if (x, y) belongs to $I(A)$ and the set of equivalent classes of the relation $I(A)$ is called the A-elementary sets. Now, the lower approximation set and upper approximation set of X based on A are defined as follows:

(1) $\underline{A}(X) = \{x \in U: [x]_A \subseteq X\}$ is the set of objects whose equivalence class is included in X. It is called the A-lower approximation set of X.

(2) $\overline{A}(X) = \{x \in U: [x]_A \cap X \neq \phi\}$ is the set of objects whose equivalence class is overlapped with X. It is called the A-upper approximation set of X.

For an equivalence class $[x]_A \in \overline{A}(X)$, define the confidence degree of $[x]_A$ in $\overline{A}(X)$ as:

$$\mu_X([x]_A) = \frac{|[x]_A \cap X|}{|[x]_A|} \quad (2)$$

3 Rough Set Data Mining

In this section our rough data mining approach for building student models is outlined. The system first generates testing examples for the student. The student tries to classify the examples correctly using whatever he/she knows. All the answering records are kept in a database for further analysis to construct the student model (classification rules). Some form of interface was designed to allow the student to visually edit the classification rules. With these facilities, the student and the system can collaboratively work out the real student model, which can then be used for further more accurate diagnosis.

As to the application of rough set theory in data mining, refer to the work of [2], which is based only on lower-approximations. Let $|U|$ be the total number of the testing examples, and $|C|$ be the number of condition attributes, then the complexity order of the *Lower_Set_Rule_Extraction* algorithm in [4] is $O(|C|^2|U|^2)$. Note that every lower set found in this algorithm is a potential rule for output if its support degree is higher than some threshold value. The support degree for a specific rule is defined as follows:

$$spt_A(W) = \frac{|S_A(W)|}{|U|}, \text{ where } S_A(W) = \underline{A}(W) = \bigcup_{V \in U/A, V \subseteq W} V, \tag{3}$$

where W is the set of examples that match the conditions of the rule.

Since the *Lower_Set_Rule_Extraction* algorithm is based on lower set approximations, all the classification rules it generate is deterministic. We would like to investigate how things are going on when the rules are generated based on upper set approximations, so we devise another *Upper_Set_Rule_Extraction* algorithm, which is outlined as follows:

```
Algorithm Upper_Set_Rule_Extraction
Step1. Use the Lower_Set_Rule_Extraction algorithm to
generate all classification rules.
Step2. For those upper sets X of W∈U/D generated in the
last iteration of the Lower_Set_Rule_Extraction algo-
rithm, compute the support degree of X and the confi-
dence degree of X. If the support degree is greater
than a user-specified threshold θs, and the confidence
degree of X is greater than a user-specified threshold
θc, and X hasn't been included in previously output
rules, then output the corresponding rule.
```

The *Upper_Set_Rule_Extraction* algorithm generates those *deterministic* rules generated by the *Lower_Set_Rule_Extraction* algorithm as well as those *non-deterministic* rules with high support degree and confidence degree. Its time complexity is also $O(|C|^2|U|^2)$.

4 Effectiveness Metrics

In this section we propose two similarity metrics: M_1 and M_2. Let S be the set of student classification rules, E be the set of the extracted rules, and E_S be the similarity degree of E with respect to S, and S_E be the similarity degree of S with respect to E. Besides, let r_c be the set of conditions in rule $r \in S$, and t_c be the set of conditions in rule $t \in E$. Then

$$M_1 = \left(\frac{|S \cap E|}{|S|}\right) \times w_1 + \left(\frac{|S \cap E|}{|E|}\right) \times w_2, \quad (4)$$

where $w_1 + w_2 = 1$, $0 <= w_1, w_2 <= 1$. That is, M_1 measures the degree of *exact matches* between the rules in E and S. Since this measurement might somewhat too strict, we define another metric M_2. Let $K_{|S|\times|E|}$ be a rule-by-rule similarity matrix in which $K(r,t) = \left(\frac{|r_c \cap t_c|}{|r_c|}\right) \times w_3 + \left(\frac{|r_c \cap t_c|}{|t_c|}\right) \times w_4$, if rule r and rule t are classification rules for the same decision pattern; otherwise $K(r,t)=0$, where $w_3 + w_4 = 1$, $0 <= w_3, w_4 <= 1$. Then

$$M_2 = \left(\frac{S_E}{|S|}\right) \times w_1 + \left(\frac{E_S}{|E|}\right) \times w_2, \quad (5)$$

where $E_S = \sum_{r \in S} \left(\underset{t \in E}{Max}\ K(t,r)\right)$, and $S_E = \sum_{t \in E} \left(\underset{r \in S}{Max}\ K(t,r)\right)$. That is, in the condition level of similarity, we keep the strict-match principle, but we take a softer measurement for the similarity between rule sets. This is inspired by the fact that partial matches between rule sets also contribute to the collaborative construction of the student models.

5 Simulation Experiments

5.1 Experiment Design

The experiment process is as follows. First, expert classification rules are obtained through a knowledge engineering process. The expert rules are then *transformed* in specific manners to simulate various kinds of students' misconceptions. The transformed rule sets ($R_1, R_2, \ldots, R_n$) are called *Student Rule Sets*. These rule sets are then applied to a sample database (with more than 1500 samples) and the classification results are stored as the student answering records. Now we can apply the rule extraction modules to get the *Extracted Rule Sets* ($R_1', R_2', \ldots, R_m'$), respectively. Finally, the similarity between each R_i and set and R_i' set is computed using M_1 and M_2, respectively.

Some typical kinds of students' misconceptions are considered in the experiment sets. For example, in case of over-generalized/over-specialized concepts some negative/positive samples are misclassified as positive/negative samples. In case of not sufficient knowledge, students are unable to classify some samples. In case of redundant concepts, students would be able to classify all samples correctly if the redundant concepts were removed. At last, the case of lack of knowledge is aimed to describe a new beginner who holds little domain knowledge. To simulate these kinds of misconceptions, we design six kinds of rule transformations. For each kind of the misconceptions aforementioned, three sub-experiments are conducted and analyzed using the M_1 and M_2 similarity metrics for the lower-set extraction algorithm and the two upper-set extraction algorithms with $\theta_c = 0.5$ and $\theta_c = 0.8$, respectively. All the three algorithms are executed with $\theta_s = 0.5$.

5.2 Experimental Results

The Delete-Some-Conditional-Attributes Experiment. This experiment is planned to simulate misconceptions of over-generalized concepts. It includes sub-experiments on sets of student rules, each of which was gained by deleting one conditional attribute in specific location from each expert rule. In average the upper-set rule extraction algorithms perform better than the lower ones.

The Delete-Some-Rules Experiment. This experiment is planned to simulate misconceptions of insufficient knowledge by deleting a portion of the expert rules. It includes six sub-experiments for which the deleted rules are, respectively, (1) the even-number indexed rules, (2) the rules whose indices are multiples of 3, (3) the rules whose indices are multiples of 4, (4) the first one-third of the rules, (5) the second one-third of the rules and (6) the last one-third of the rules. The results show that the upper algorithms outperform the lower one with M2 > 0.53.

The Change-Conditional-Attribute-Values Experiment. This experiment is planned to simulate misconceptions of mixing over-generalized and over-specialized concepts. It includes seven sub-experiments. The first five experiments change the conditional attribute values, while the last two change the decision attribute values. In these cases the upper extraction algorithms perform much better and more stable than the lower one.

The Add-Conditional-Attributes Experiment. This experiment is planned to simulate misconceptions of over-specialized concepts. It includes five sub-experiments, each of which adds specific conditional attribute and values to the expert rules. The M2 values of the three algorithms lie between 0.62 and 0.79, which show the reconstruction capability of the rule extraction algorithms is acceptable.

The Add-Some-Rules Experiment. This experiment is planned to simulate misconceptions of redundant concepts. It includes five sub-experiments, each of which adds one specific rule into the expert rules. Again the upper rule extraction algorithms outperform the lower one and are more stable.

The Delete-Large-Amount-of-Rules Experiment. This experiment is planned to simulate misconceptions of a new beginner with very few concepts. It includes six sub-experiments, each of which deletes randomly five-sixth of the expert rules. The result is quite satisfying.

As a summary, the lower algorithm performs better in M_1 metric, while the upper algorithms outperform the lower one in M_2 metric significantly. The confidence threshold θ_c did not show significant difference in the M_2 metric for the upper algorithms. Nevertheless, our results show that the upper algorithm with $\theta_c = 0.5$ performs a little better than the one with $\theta_c = 0.8$.

6 Conclusive Remarks

In this paper, we investigate the feasibility of applying rough data mining methods to automatic construction of student models. The simulation results show that it is hard to extract "*just-the-same*" rules using current rough data mining methods. Nevertheless, The rough data mining methods, especially the upper algorithms, perform significantly well to extract "*almost-the-same*" rules. Besides, when inconsistency exists in the student rules, the upper algorithms can deal with the inconsistency problem very well. Finally, to improve the rule extraction effectiveness, the extraction algorithms should use only those attributes that students had really adopted to give the answers. This will contribute much especially to the extraction of *inconsistent* student rules.

References

1. Anderson, J. R., Boyle, C. F., Corbett, A. T., Lewis, M. W: Cognitive modeling and intelligent tutoring. Artificial Intelligence **42** (1990) 7-49.
2. Bell,D. A., Guan J. W.: Computational Methods for Rough Classification and Discovery. Journal of the American Society for Information Science. **49**(5) (1998) 403-414.
3. Carr, B., Goldstein, I.: Overlays: A theory of modeling for computer-aided instruction. Technical Report A.I. Memo 406, Cambridge, MA: MIT (1977).
4. Finin,T.: GUMS:A general user modeling system. In: Kobsa, A. Wahlster, W. (eds.): User models in Dialog Systems. (1989) 411-430.
5. Lingras, P. J., Yao, Y. Y.: Data Mining Using Extensions of the Rough Set Model. Journal of the American Society for Information Science. **49**(5) (1998) 415-422.
6. Pawlak, Z.: Rough Sets. Intern. J. of Computer and Information Sciences, **11**(5) (1982) 341-356.
7. Pawlak, Z. (1996). Why Rough Sets? Institute of Electrical and Electronics Engineers.
8. Sleeman, D., Hirsh, H., Ellery, I., Kim, I.: Extending domain theories: Two case studies in student modeling. Machine Learning **5** (1990) 11-37.
9. Ohlsson, S. (1992). Constraint-Based Student Modeling. Journal of Artificial intelligence in Education, **3**(4), 429-447.
10. Zadeh, L. A.: Fuzzy Sets. Information and Control, **8**(3) (1965) 338-353.

Concept Approximation in Concept Lattice

Keyun Hu[1], Yuefei Sui[2], Yuchang Lu[1], Ju Wang[2], and Chunyi Shi[1]

[1]Department of Computer Science, Tsinghua University, Beijing 100084, P.R.China
[2]Institute of Software, Academia Sinica, Beijing 100080, P.R.China
hky@s1000e.cs.tsinghua.edu.cn

Abstract. In this paper we present a novel approach to the concept approximations in concept lattice. Using the similar idea of rough set theory and unique properties of concept lattice, upper and lower approximations of any object or attribute set can be found by exploiting meet-(union-)irreducible elements in concept lattice, the approximations can be performed on the fly. We show that our approach is more natural and effective than existing approach.

1 Introduction

Concept lattice, also called Galois lattice, was first proposed by Wille[3]. A node of concept lattice is an objects/attributes pair, called a formal concept, consisting of two parts: the extension (objects the concept covers) and intension (attributes describing the concept). Concept lattice gives a vivid and concise account of relations (generalization /specialization) among those concepts through Hasse Diagram. Concept lattice is useful for data mining[4,5,9], information retrieval[7], and soft engineering[8], etc.

Not every pair of objects and attributes defines a formal concept. Only those maximally extended ones are included in concept lattice, i.e., attributes in intension are maximal common attributes of objects in extension and vice versa. This brings forward a problem that how to best approximate a set of objects or attributes if there is no exact match in the concept lattice. Furthermore, is it possible to approximate them without knowing the whole concept lattice? It is an expensive operation to generate the whole lattice after all.

This kind of approximation may be useful in many situations. For example, assuming that we have a concept lattice describing a set of documents and a set of keywords, when given a query (a set of keywords), we could find out the best approximate result if an exact match failed.

Rough set theory[6] efficiently approximates a given concept by using a pair of concepts, namely the upper and lower approximations. In this paper, using the similar idea we propose a concept approximation method in concept lattice. However, there is something different. First, our approach is not based on the equivalence classes. Second, our approach makes use of the properties of lattice, that is, the existence of meet-irreducible elements, which greatly simplifies the computation of approximation.

There are some similar works [1,2]. We argue that our method is more natural and effective than existing approach, and by exploiting meet-irreducible elements in concept lattice, we could generate concept approximations on the fly.

D. Cheung, G.J. Williams, and Q. Li (Eds.): PAKDD 2001, LNAI 2035, pp. 167-173, 2001.

The rest of the paper is organized as follows. Section 2 recalls necessary notions used in this paper. Section 3 introduces related work. Our approach is presented in section 4 and an illustrating example is given in section 5. Section 6 concludes the paper.

2 Basic Notions

In this section we recall necessary basic notions of concept lattice and rough set briefly. The detail description can be found in [3,6].

First, we begin with some notions from concept lattice.

Suppose given the context (***O***, ***D***, ***R***) describing a set ***O*** of objects, a set ***D*** of descriptors and a binary relation ***R***, there is a unique corresponding lattice structure ***L***, which is known as *concept lattice*. Each node in lattice ***L*** is a pair, noted (X, Y), where $X \in P(O)$ is called *extension* of the concept, $Y \in P(D)$ is called *intension* of concept. Each pair must be complete with respect to ***R***. i.e.:

(1) $X = \beta(Y) = \{x \in O \mid \forall y \in Y, yRx\}$ •(2) $Y = \alpha(X) = \{y \in D \mid \forall x \in X, yRx\}$

A partial order relation can be built on all concept lattice nodes. Given $H_1=(X_1, Y_1)$ and $H_2=(X_2, Y_2)$•let $H_1 \leq H_2 \Leftrightarrow X_1 \subseteq X_2$, the precedent order means H_1 is a direct parent of H_2. The Hasse diagram of the lattice can be generated using the partial order relation. If $H_1 \leq H_2$ and there is no other node H_3 such that $H_1 \leq H_3 \leq H_2$, there is an edge from H_1 to H_2.

In rough set theory, information system plays similar role as context in concept lattice. An information system is a ordered pair S=(***O***, ***D***), where ***O*** is a non-empty, finite set called the universe, ***D*** is a non-empty, finite set of attributes. The elements of the universe are called objects.

Let S=(***O***, ***D***) be an information system, every subset $B \subseteq D$ defines an equivalence relation IND(B), called an indiscernibility relation, defined as $IND(B)=\{(x,y) \in O \times O: a(x)=a(y) \text{ for every } a \in B\}$.

Given an information system S=(***O***, ***D***), let $X \subseteq U$ be a set of objects and $B \subseteq D$ a selected set of attributes. The lower approximation of X with respect to B is $B_*(X)=\{x \in O: [x]_B \subseteq X\}$. The upper approximation of X with respect to B is $B^*(X)=\{x \in O: [x]_B \cap X \neq \Phi\}$, where $[x]_B=\{y \in O: (x,y) \in IND(B)\}$.

Upper approximate consists of all objects possibly belonging to X and lower approximate consists of all objects definitely belonging to X. Obviously we have $B_*(X) \subseteq X \subseteq B^*(X)$.

Concept lattice includes all concepts in context, and assembles them in a visual concept hierarchy while rough set provides a powerful concept approximation mechanism.

3 Related Work

To the best of the author's knowledge, there are two existing works on concept approximation, that is, Kent's work[1] and Saquer's work[2].

Kent used an equivalence relation $\boldsymbol{E}$ on the set of objects $\boldsymbol{O}$ provided by an expert. A pair $(\boldsymbol{O}, \boldsymbol{E})$, where $\boldsymbol{E}$ is an equivalence relation on $\boldsymbol{O}$, is called an approximation space. An $\boldsymbol{E}$-definable formal context of $\boldsymbol{O}$-objects and $\boldsymbol{D}$-attributes is a formal context $(\boldsymbol{O}, \boldsymbol{D}, \boldsymbol{R})$ whose elementary extents $\{g \in \boldsymbol{O} | g\boldsymbol{R}m, m \in \boldsymbol{D}\}$ are $\boldsymbol{E}$-definable subsets of $\boldsymbol{O}$-objects. Two new concept lattices--the lower and upper $\boldsymbol{E}$-approximation of $\boldsymbol{R}$ with respect to $(\boldsymbol{O}, \boldsymbol{E})$, are then defined. Undefined concepts are approximated by finding closest elements in the two new approximating lattices.

Kent's work is not natural because the upper and lower approximation of the concept lattice have to be found first and the resulting approximations depend on the equivalence relation chosen.

Saquer and Deogun defined a natural equivalence relation on $\boldsymbol{O}$ as $g_1 I g_2$ iff $g_1\boldsymbol{R} = g_2\boldsymbol{R}$ where $g\boldsymbol{R} = \{m \in \boldsymbol{D} | g\boldsymbol{R}m, g \in \boldsymbol{O}\}$. If a set $A \subseteq \boldsymbol{O}$ is an extent of concept lattice, it is called feasible; if a set $A \subseteq \boldsymbol{O}$ is the union of feasible set, it is called definable. By declaring each equivalence class $\boldsymbol{O}/I$ is a feasible set, a non-definable set $A \subseteq \boldsymbol{O}$ is approximated by $A_* = \{g \in O | [g] \subseteq A\}$ and $A^* = \{g \in O | [g] \cap A \neq \Phi\}$.

Saquer's idea is to approximate a new concept using the union of existing concepts in concept lattice. That is similar to the idea of rough set, but things in concept lattice are different. We could observe that all possible concepts in $\boldsymbol{O}$ are included in concept lattice and the definable sets may be not in concept lattice. This is to say, the definable sets that may be not a concept can be expressed by the concept lattice. In such a situation the approximation become meaningless. Another thing, it is not so easy to know whether a set is definable according to the theory given.

What is the best approximation of a given concept? We argue that the best approximation of a given concept is two existing concepts in concept lattice, which approximate the concept in both directions. This is because existing concepts in concept lattice are all possible concepts we "know" from original context.

4 Our Approach

First we introduce the following definition and theorem from the lattice theory.

Definition 4.1 An element a is **meet-irreducible** in a lattice $\boldsymbol{L}$ if for any $b,c \in \boldsymbol{L}$, $a=b \wedge c$ implies $a=b$ or $a=c$; dually, an element a is **union-irreducible** in a lattice $\boldsymbol{L}$ if for any $b,c \in \boldsymbol{L}$, $a=b \vee c$ implies $a=b$ or $a=c$.

Theorem 4.1 Every element is the meet(union) of the meet-irreducible (union-irreducible) elements.

Given context $(\boldsymbol{O}, \boldsymbol{D}, \boldsymbol{R})$ and corresponding concept lattice $\boldsymbol{L}$, let $g\boldsymbol{R} = \{m \in \boldsymbol{D} | g\boldsymbol{R}m, g \in \boldsymbol{O}\}$.

Definition 4.2 Given context $(\boldsymbol{O}, \boldsymbol{D}, \boldsymbol{R})$, a binary relation J on $\boldsymbol{O}$ is defined as

$$g_1 J g_2 \text{ iff } g_1\boldsymbol{R} \subseteq g_2\boldsymbol{R}, \text{ where } g_1, g_2 \in \boldsymbol{O}$$

Clearly, J is reflexive, anti-symmetric and transitive. Thus, J is a partial relation on $\boldsymbol{O}$. We denote partial class of g as [g], namely, $[g] = \{g' \in \boldsymbol{O} : gJg'\}$.

Theorem 4.2 Every pair $([g], \beta([g]))$ is a union-irreducible element of $\boldsymbol{L}$.

Proof. First we prove $([g], \beta([g]))$ is an element of $\boldsymbol{L}$. We need only show [g] is maximally extended. By the definition of [g], all elements have maximal common attribute with g are in [g], so [g] must be maximally extended. Therefore the pair must be in the *L*.

Assume ([g], β([g])) is not a union-irreducible element of *L*, there are two pairs (A, β(A)), (B, β(B)⊆O such that [g]=A∨B and [g]≠A or B. So [g] ⊃A, [g] ⊃B, and g∈A or g∈B. we assume g∈A. From the duality of concept lattice, we have β([g]) ⊈β (A). This is to say, g must have a larger attribute set in pair (A, β(A)), but this is a contradiction, because from the definition of [g], we know that the attribute set processed by g should be smallest. Therefor ([g], β([g])) must be a union-irreducible elements of *L*.

Let UI be the set of all the union-irreducible elements in *L*, P be the set of all pair ([g], β([g])).

Theorem 4.3 UI=P

proof. From theorem 4.2, we know every element in P is also an element in UI. We only need prove that the union-irreducible elements must be elements in P.

Assume (A, β(A)) is one of the union-irreducible elements. If there is an object g∈A such that g***R***=β(A), so ([g], β(A)) must be in P; or for every object g∈A, there is g***R***⊅ (A), so (A, β(A)) can be decomposed into $\vee(g_i, \beta(g_1))$, $g_i \in A$. This contradicts to that (A, β(A)) is a union-irreducible element. Therefore, there must be an object g∈A such that g***R***=β(A). i.e, the union-irreducible elements must be elements in P.

From theorem 4.3 and 4.1, we know all elements in concept lattice can be expressed in term of union of elements in P.

Similarly, Let ***R***m={g∈***O***|g***R***m}, we have:

Definition 4.3 Given context (***O***, ***D***, ***R***), a binary relation J' on ***D*** is defined as

$$m_1 J' m_2 \text{ iff } \boldsymbol{R}m_1 \subseteq Rm_2, \text{ where } m_1, m_2 \in \boldsymbol{D}$$

We denote partial class of m as [m], namely, [m]={m'∈***O*** : m J'm'}.

Theorem 4.4 Every pair (α[m], [m]) is a meet-irreducible element of ***L***.

Let MI be set of all the meet-irreducible elements, Q be the set of all pair (α[m], [m]).

Theorem 4.5 MI=Q

The proof is similar to theorem 4.2 and 4.3.

From theorem 4.5 and 4.1, we know all elements in concept lattice can also be expressed in term of meet of elements in Q.

Definition 4.4 Given X⊆***O***, the lower and upper approximations of X with respect to *L* are

$$X_* = \cup\{X' \subseteq X | (X', \beta(X')) \in UI\}$$

and

$$X^* = \cap\{X \subseteq X' | (X', \beta(X')) \in MI\}$$

Theorem 4.6 Given X, X'⊆***O***, The lower approximation satisfies the following properties:

(1) $U_* = U$

(2) $X_* \subseteq X$

(3) $X \subseteq X' \rightarrow X_* \subseteq X'_*$

(4) $(X \cup X')_* = X_* \cup X'_*$

(5) $X_* = X_{**}$

and the upper approximation satisfies the following properties:

(6) $\varnothing^* = \varnothing$

(7) $X \subseteq X^*$

(8) $X \subseteq X' \rightarrow X^* \subseteq X'^*$
(9) $(X \cap X')^* = X^* \cap X'^*$
(10) $X^* = X^{**}$
Theorem 4.7 Given $X \subseteq \boldsymbol{O}$, the best lower and upper approximations are given by $(X_*, \beta(X_*))$ and $(X^*, \beta(X^*))$, respectively.
proof. We prove that X_* is the best lower approximation and the proof of upper approximation is similar. Assume there is another concept (A, β(A) such that $X_* \subset A \subseteq X$. So A can be expressed as union of X_* and some union-irreducible elements. This is $A = X_* \vee (\vee ui_i) \subseteq X$, where $ui_i \in UI$. Thus $ui_i \subseteq X$. According the definition of X_*, $ui_i \subseteq X_*$. Then we have $A = X_*$. Therefore X_* is the best approximation of X.
Dually, we have the following definition and theorems for attribute set approximation.
Definition 4.5 Given $Y \subseteq \boldsymbol{D}$, the lower and upper approximations of Y with respect to ***L*** are

$$Y_* = \cap\{Y \subseteq Y' \mid (\alpha(Y'), Y') \in MI\}$$

and

$$Y^* = \cup\{Y' \subseteq Y \mid (\alpha(Y'), Y') \in UI\}$$

Theorem 4.8 Given $Y \subseteq \boldsymbol{D}$, the best lower and upper approximations are given by $(\alpha(Y_*), Y_*)$ and $(\alpha(Y^*), Y^*)$, respectively.

Note we do not find upper and lower approximations for any pair(X, Y), $X \subseteq \boldsymbol{O}$, $Y \subseteq \boldsymbol{D}$, because this kind of combination may be meaningless in the context, i.e., it does not represent any meaningful concepts. For example, finding approximation for (***O***, ***R***) is meaningless and impossible.

5 An Illustrating Example

In this section we illustrate our idea using an example from [3]. Figure 1 is a simple context, and figure 2 is the corresponding concept lattice. We have ***O***={1, 2, 3, 4, 5, 6, 7, 8}, ***D***={a, b, c, d, e, f, g, h, i}, and R describing objects in ***O*** processing some attributes in ***D***.

		a	b	c	d	e	f	g	h	i
1	Leech	×	×					×		
2	Bream	×	×					×	×	
3	Frog	×	×	×				×	×	
4	Dog	×		×				×	×	×
5	Spike-weed	×	×		×		×			
6	Reed	×	×	×	×		×			
7	Bean	×		×	×	×				
8	Maize	×		×	×		×			

Fig. 1 A context excerpted from [3] p18. a=needs water; b=lives in water; c=lives on land; d=needs chlorophyll; e=two seeds leaf; f=one seed leaf; g=can move around; h=has limbs; i=suckles it offspring.
From definition 4.2 and 4.3, partial class of objects and attributes can be computed as follows:
[1]={1,2, 3}, corresponding union-irreducible elements is ({1,2,3},{a,b,g})
[2]={2,3}, corresponding union-irreducible elements is ({2,3},{a,b,g,h})

[3]={3}, corresponding union-irreducible elements is ({3},{a,b,c,g,h})
[4]={4}, corresponding union-irreducible elements is ({4},{a, c, g,h,i})

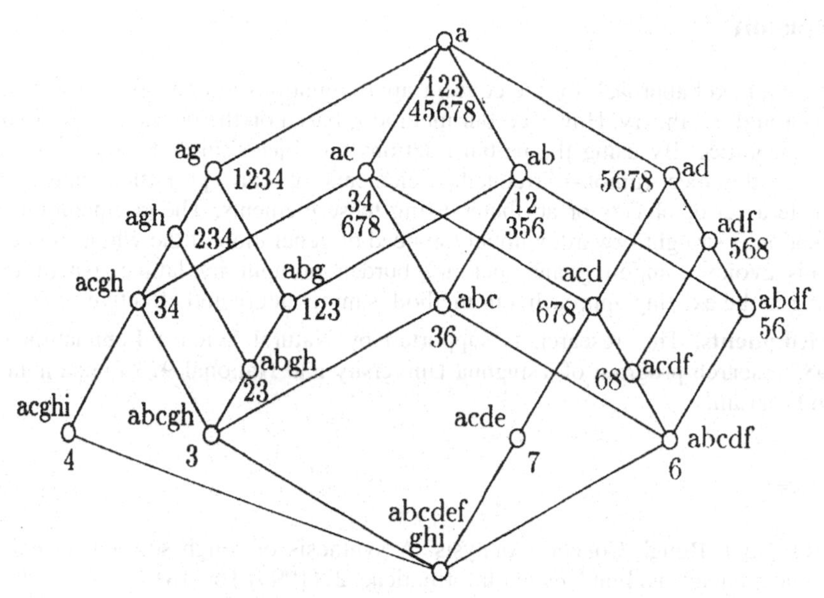

Fig. 2. Concept lattice for the context of figure 1.

[5]={5,6}, corresponding union-irreducible elements is ({5,6},{a,b,d,f})
[6]={6}, corresponding union-irreducible elements is ({6},{a,b,c,d,f})
[7]={7}, corresponding union-irreducible elements is ({7},{a,c,d,e})
[8]={6,8}, corresponding union-irreducible elements is ({6,8},{a,c,d,f})

[a]={a}, corresponding meet-irreducible elements is ({1,2,3,4,5,6,7,8},{a})
[b]={a,b}, corresponding meet-irreducible elements is ({1,2,3,5,6},{a,b})
[c]={a,c}, corresponding meet-irreducible elements is ({3,4,6,7,8},{a,c})
[d]={a,d}, corresponding meet-irreducible elements is ({5,6,7,8},{a,d})
[e]={a,c,d,e}, corresponding meet-irreducible elements is ({7},{a,c,d,e})
[f]={a,d,f}, corresponding meet-irreducible elements is ({5,6,8},{a,d,f})
[g]={a,g}, corresponding meet-irreducible elements is ({1,2,3,4},{a,g})
[h]={a,g,h}, corresponding meet-irreducible elements is ({2,3,4},{a,g,h})
[i]={a,c,g,h,i}, corresponding meet-irreducible elements is ({4},{a,c,g,h,i})

We can verify in figure 2 that they are meet-irreducible and union-irreducible elements respectively.
Given a set X={2,4}, its lower and upper approximation can be computed as:

$X_* = \cup\{X' \subseteq X \mid (X', \beta(X')) \in UI\} = \cup\{4\} = \{4\}$

$X^* = \cap\{X \subseteq X' \mid (X', \beta(X')) \in MI\} = \cap\{\{1,2,3,4,5,6,7,8\},\{1,2,3,4\},\{2,3,4\}\} = \{2,3,4\}$.

We can verify that X_* and X^* are the best approximations in the concept lattice.

Using method in [2], we can get $X_*=\{2,4\}$, $X^*=\{2,4\}$. And $\{2,4\}$ is not an expressible concept by the concept lattice.

6 Conclusion

We propose a novel approach for the concept approximations in concept lattice using the idea of rough set theory. However, our method is based on the partial relations and properties of lattice. By using the partial relations on objects and attributes, we are able to find the meet-(union-) irreducible elements of concept lattice, and then approximate a set of objects or attributes using these elements. The computation of approximation is straightforward, without the need of generation of the whole concept lattice, this avoids computing and memory burden brought by lattice generation. Comparing to the existing approach, our method is more nature and effective.

Acknowledgments. This research is supported by Natural Science Foundation of China, 985 research program of Tsinghua University and National 973 Fundamental Research Program.

References

[1] R.E.Kent, Rough Concept Analysis: A synthesis of rough set and formal concept analysis, Fundamenta Informaticae, 27(1996):169-181

[2] J.Saquer, J.S.Deogun, Formal rough concept analysis, N.Zhong, A.Skowron and S.Ohsuga (eds.) Proceedings of RSFDGrC'99, Japan, Springer, 1999, 91-99

[3] B.Ganter, R.Wille, Formal concept analysis: mathematical foundations. Berlin: Springer, 1999.

[4] K.Hu, Y.Lu, L.Zhou, C.Shi, Integrated classification and association rule mining based on concept lattice, N.Zhong, A.Skowron, and S.Ohsuga (eds.) Proceedings of RSFDGrC'99, Japan, Springer, 1999.443-447

[5] K.Hu, Y.Lu, C.Shi. Incremental association rule mining: a concept lattice approach[A]. in Ning Z, Lizhu Z. Eds. Proceedings of PAKDD'99. Beijing, Springer, 1999.109-113

[6] Z.Pawlak, Rough sets – theoretical aspects of reasoning about data, Kluwer publishers, Boston, Dordrecht, 1991

[7] Eklund P W, Martin P. WWW indexation and document navigation using conceptual structures. 2nd IEEE Conference on Intelligent Information Processing Systems (ICIPS '98), IEEE Press, 1998, 217-221

[8] Godin R, Mineau G, Missaoui R, St-Germain M, Faraj N. Applying concept formation methods to software reuse. International Journal of Software Engineering and Knowledge Engineering, 5(1), 119-142.

[9] Godin R. Incremental concept formation algorithm based on Galois (concept) lattices. Computational Intelligence, 11(2) (1995) 246-267

Generating Concept Hierarchies/Networks: Mining Additional Semantics in Relational Data

T.Y. Lin[1,2]

[1] Department of Mathematics and Computer Science
San Jose State University, San Jose, California 95192
[2] Berkeley Initiative in Soft Computing
Department of Electrical Engineering and Computer Science
University of California, Berkeley, California 94720
tylin@cs.sjsu.edu, tylin@cs.berkely.edu

Abstract. In relational theory, attribute domains are classical sets; no interactions among attribute values are modeled. So the concept hierarchies, which are additional semantics, used in data mining have to be input by users. In this paper, "real world" data model - relational model with *additional semantics* specified by binary relational structures (adopt from first order logic) - are explored; in such model, concept hierarchies/networks can be generated automatically. In fact, there are two families of concepts. One family forms a traditional hierarchy. Another forms a hierarchy syntactically, but semantically the hierarchy is a network; this is due to the fact that distinct concepts may be semantically related. A simple example is illustrated.

Keywords: attribute oriented generalization, binary relation, concept hierarchies/networks, data mining, partition, granulation, neighborhood systems

1 Introduction

In [4] we stated to the effect that concept hierarchy are supplied by human, and "some automations of building such semantic relations are needed for large scaled applications. We will report our exploration in future papers." This paper is to honor the statement.

In traditional *attribute-oriented generalization*(AOG) approaches [1], concept hierarchies are specified as trees, with the attribute values (known as *base concepts*) at the leaves, and higher level concepts as the interior nodes. These hierarchies embody certain implicit assumption on data structures of the active attribute domains [3]. The primary assumption is that there is a nested sequence of equivalence relations among these leaf concepts. The first level parent nodes represent the equivalence classes of the inner most equivalence relation. Such assumption restricts hierarchies to have tree structures, precluding other types of relationships among concepts. In [8], the notion has been successfully extended to binary relations, in other words, instead, there is a nested sequence of binary

D. Cheung, G.J. Williams, and Q. Li (Eds.): PAKDD 2001, LNAI 2035, pp. 174–185, 2001.

relations among the leaf concepts. We call it the concept hierarchy/network, since its structure is more than a tree.

In this paper, "real world database - relational database with additional semantics - are investigated. Based on the additional semantics, concept hierarchies/networks can be automatically generated. In fact, somewhat surprisingly, we find there are two concept "hierarchies;" one is the same as traditional one, another one is syntactically a hierarchy, but semantically forms a network

2 Relational and "Real World" Model

The intent of database is to model the universe of real world entities by tuples of elementary concepts (attribute values). For simplicity relational theory tactically assumes everything, such as the universe and attribute domains, are classical Cantor sets. In other words, the interactions among elements in real world sets are "forgotten" in relational modeling. However, in practical data processing, some additional semantics derived from "real world" attributes are often processed. For example, in numerical attributes, the order of numbers is often used in SQL statement. In multimedia databases, the geographical relationships, such as "near," "in the same area" are often used in data processing by human operators. Therefore these "real world" semantics implicitly *exist* in the stored data. To capture such additional semantics in data mining, we need the "real world" structure to replace Cantor set theory.

What would be the "correct" mathematical structure for "real world"? This is a question that may have many add hoc answers. We decide to consult the history. In the model theory of first order logic, relational structures have been used to model the "real world." So we will assume the universe of discourse has relational structures. As a first step, we will confine ourselves to the simplest kind of relational structure namely, *binary* relations. So we will assume that all attribute domains are embedded with binary relations; see Section 3.1.

2.1 "Real World" Structure-Additional Semantics

To illustrate the fundamental idea, let us modify a relation (Table 1) from a popular text [2] by giving some additional semantics using binary relations:

1. order relation, $<$, is defined on $Dom(STATUS)$ naturally,
2. "near" semantics in the attribute domain $Dom(CITY)$ is defined algebraically by Table 2 or geometrically by Table 3.

2.2 "Real World" Dependencies - Strong and Weak Dependencies

Note that Table 1 has a binary relation VC on $CITY$ attribute, an order relation $<$ on $STATUS$, and a discrete (identity) relation on each of the rest of attributes. If we had forgotten the semantic relations in $CITY$ and $STATUS$ attributes,

Table 1. The Supplier Table

SNUM	*SNAME*	*STATUS*	*CITY*
S_1	Smith	*TWENTY*	*B*
S_2	Jones	*TEN*	*C*
S_3	Blake	*TEN*	*C*
S_4	Clark	*TWENTY*	*B*
S_5	Adams	*THIRTY*	*D*
S_6	Peterson	*FORTY*	*E*
S_7	Ewing	*EIGHTY*	*F*
S_8	Johnson	*EIGHTY*	*F*
S_9	Pike	*FORTY*	*E*
S_{10}	Meyers	*NINTY*	*G*

Table 2. *VC*-Binary Relation

CITY	*CITY*
B	*B*
B	*C*
C	*C*
C	*B*
C	*D*
D	*D*
D	*E*
E	*E*
E	*D*
F	*F*
F	*G*
G	*G*

Table 3. Binary Granulation and Neighborhood System

CITY		Elementary neighborhood (granule)		Elementary concept (granule name)
B	$\longrightarrow$	$\{C, B\}$	=	*J*
C	$\longrightarrow$	$\{C, B, D\}$	=	*K*
D	$\longrightarrow$	$\{D, E\}$	=	*L*
E	$\longrightarrow$	$\{D, E\}$	=	*L*
F	$\longrightarrow$	$\{F, G\}$	=	*M*
G	$\longrightarrow$	$\{G\}$	=	*N*

then one would think that there were an extensional, (two directions) functional dependencies, namely, $CITY \longrightarrow STATUS$ and $STATUS \longrightarrow CITY$. However, for example, B and C are symmetrically VC-related ($B\ VC\ C$ and $C\ VC\ B$), but their images, $TWENTY$ and TEN, are not symmetrically related ($TWENTY \not< TEN$ and $TEN < TWENTY$). So only the map $STATUS \longrightarrow CITY$ respects the semantic relations. In such a semantically richer situations, a map should not be treated as a functional dependency unless it respects the semantics; see Sction 3.1 The only functional dependency, called weak dependency, in Table 1 is $STATUS \longrightarrow CITY$.

3 Binary Relations and The Induced Equivalence Relations

We will recall some theory of binary relations from [6].

3.1 Binary Relations and Binary Neighborhood System Spaces

Binary Relation. Let $B \subseteq V \times V$ be a *binary relation* on V. For each object $p \in V$, we associate a subset

$$NEIGH_B(p) = \{u \mid p\ B\ u\},$$

called elementary (binary) B-neighborhood $NEIGH_B(p)$ (Note: *elementary* neighborhood is used as a generalization of *elementary* set (= equivalence class) in rough set theory, and *binary* neighborhood is used to reminding binary relation). Note that each point has only one elementary neighborhood (intuitively it can be viewed as the nearest neighborhood) of p. This association defines a map, called binary B-granulation:

$$B : V \longrightarrow 2^V : p \longrightarrow NEIGH_B(p).$$

Next we gather all the B-neighborhoods together, i.e., we set

$$NEIGH_B(V) = \{NEIGH_B(p) \mid \forall\, p \in V\}$$

and call the collection a *binary B-neighborhood system (BNS).*

Conversely suppose $NEIGH_B(V) = \{NEIGH_B(p)\}$ is given, we can get the binary relation back by defining

$$B = \{(p, u) \mid u \in NEIGH_B(p)\}.$$

We summarize the discussion to a

Proposition. There is a one-to-one correspondence between binary neighborhood systems, binary granulations and binary relations.

So from now on, we will treat them as synonyms and use them interchangeably; we will use *the same notation* B for all of them. In particular, we will write

$$NEIGH_B(V) = B(V) \text{ and } NEIGH_B(p) = B(p)$$

If the binary relation B is an equivalence relation E, then the binary granulation is a partition. An elementary set is the elementary E-neighborhood of its members (in rough set theory an equivalence class is called an elementary set). We will use *elementary granule* to denote both elementary neighborhood and elementary set. For $B = VC$, it is illustrated in Table 3.

Binary Neighborhood System Space. The pair (V, B) has been called a binary neighborhood system space (BNS-space) [7], [6]. A BNS-space is a pre-topological space; it is a variant of Frechet(V)-space [13]. In the case that B is an equivalence relation E, (V, E) is a clopen topological space and is the focus of rough set theory [9]. We will collect few simple properties of BNS-space. Let B' be another binary relation.

Definition

1. A subset X is called a definable B-neighborhood, if X is a union of elementary (binary) neighborhoods of B;
2. The set of all definable B-neighborhoods *at* p is denoted by $BS(p)$; The set of all definable B-neighborhoods *in* U is denoted by $BS(U)$.
3. Let S be a subset of V. $NEIGH(S) = \bigcup_{p|inX} B_(p)$ is called the elementary B-neighborhood of S; note that it is a definable neighborhood.
4. B' weakly (or continuously) depends on B iff every elementary B-neighborhood at p is contained in B'-neighborhood; in other words, the identity map is continuous
5. B' strongly (or definably) depends on B, denoted by $B \Longrightarrow B'$, iff every elementary B'-neighborhood is a definable B-neighborhood; we say B is definably finer than B' or B' is definably coarser than B.

Strongly dependence is an elaborate extension of refinement of equivalence relations. The obvious extension, weak dependence, has no desirable properties of "functional" dependency (or knowledge dependency).

3.2 The Induced Equivalence Relations

Classification is often an important knowledge for human. Its mathematical term is partition or equivalence relation. Though a binary neighborhood system (binary relation) is not a partition(equivalence relation), we will show that it does induce a partition(equivalence relation).

Given a map $f : X \longrightarrow Y : x \longrightarrow y$, its family of complete inverse image $f^{-1}(y)$ forms a partition, e.g., see [6]. Here we take the map, binary granulation,

$$B : V \longrightarrow 2^V : p \longrightarrow B(p).$$

The inverse image $B^{-1}(B(p))$ is called *the center* of $B(p)$. The family of all centers forms a partition, called *induced partition* of a binary relation:

$$E_B = \{B^{-1}(B(p)) \mid \text{for all elementary neighborhood } B(p) \text{ in } V\}$$

By abuse of language, we may also call each member of the center *a center*.

1. The binary granulations or binary neighborhood systems (Section 4.2)
   a) $VC : Dom(A) \longrightarrow 2^{Dom(A)}$;
   b) $VC \circ VC : Dom(A) \longrightarrow 2^{Dom(A)}$;
   c) $VC \circ VC \circ VC : Dom(A) \longrightarrow 2^{Dom(A)}$;
   induces 3 partitions; see Table 4, 5 and 6
2. Each elementary set (=equivalence class) of the 3 induced partitions is the center of the corresponding elementary neighborhood.
3. Each distinct center of an induced partition is labeled by a distinct elementary concept. Since all centers of a induced partitions are mutually disjoints, they are independent to each other (within an attribute doamin).
4. Each distinct elementary neighborhood of a binary granulation is labeled by a distinct elementary concept; However, elementary neighborhoods may overlap, so these elementary concepts are *only syntactically distinct, but semantically may be related.* For example, K and J are distinct, but semantically related; however, the label of the centers, K_P and J_P are distinct and semantically independent; see Table 4, 5, 6.
5. If VC-binary relation is an equivalence relation, then $VC = E_{VC}$ and hence, these elementary concepts are *both syntactically and semantically independent.*
6. In databases, elementary concepts are referred to as attribute values.

Table 4. VC-Binary Granulation and Induced Partition

Center(Elementary set of E_VC)		Elementary neighborhood of VC
$\{B\} = J_P$	$\longrightarrow$	$\{C, B\} = J$
$\{C\} = K_P$	$\longrightarrow$	$\{C, B, D\} = K$
$\{D, E\} = L_P$	$\longrightarrow$	$\{D, E\} = L$
$\{F\} = M_P$	$\longrightarrow$	$\{F, G\} = M$
$\{G\} = N_P$	$\longrightarrow$	$\{G\} = N$

4 Generating Concept Hierarchies/Networks

4.1 Concept Hierarchies/Networks

In [8], a *concept* is recursively defined to be:

1. a label (base concept) for each distinct attribute value, for example, the first column of Table 8
2. a label (parent concept) for a set of concepts (referred to as *sibling concepts*), which are not mutually recursive. For example, see the second or third columns of Table 8

Table 5. $VC \circ VC$-Binary Granulation and Induced Partition

Center(Elementary set of $E_{VC \circ VC}$)		Elementary neighborhood of $VC \circ VC$
$\{B\} = K_P$	$\longrightarrow$	$\{C, B, D\} = K$
$\{C\} = O_P$	$\longrightarrow$	$\{C, B, D, E\} = O$
$\{D, E\} = L_P$	$\longrightarrow$	$\{D, E\} = L$
$\{F\} = M_P$	$\longrightarrow$	$\{F, G\} = M$
$\{G\} = N_P$	$\longrightarrow$	$\{G\} = N$

Table 6. $VC \circ VC \circ VC$-Binary Granulation and Induced Partition

Center(Elementary set of $E_{VC \circ VC \circ VC}$)		Elementary neighborhood of $VC \circ VC \circ VC$
$\{B, C\} = O_P$	$\longrightarrow$	$\{C, B, D, E\} = O$
$\{D, E\} = L_P$	$\longrightarrow$	$\{D, E\} = L$
$\{F\} = M_P$	$\longrightarrow$	$\{F, G\} = M$
$\{G\} = N_P$	$\longrightarrow$	$\{G\} = N$

Higher level concepts derived by part 2 are typically specified by an outside source. In this paper, the set of higher level concepts will be generated automatically, only the interpretation of labels may be provided by an outside source.

A traditional concept hierarchy is *deterministic.* Every concept is the child of at most one higher level concept. In such a hierarchy, the base concepts are grouped into a nested sequence of partitions: A level zero concept is a base concept(distinct attribute value). A level one concept is an equivalence class of the *innermost* equivalence relation;a level two concept is that of the second innermost relation, and etc. In contrast, the new approach relax the equivalence relations to general binary relations. The new concept hierarchy/network is *syntactically deterministic*, but may be *semantically non-deterministic.* In such a hierarchy/network, the base concepts are grouped into a nested sequence of binary neighborhood systems(binary relations); here the nesting is in the sense of strong dependencies Section 4.2. Note that this nested sequence induces a nested sequence of *induced equivalence relations.* So we have two nested sequences: Both sequences have the same level zero concept; they are attribute values. In level one, each sequence has its own concept: one is an elementary neighborhood of the *innermost* binary relation, the other one is the corresponding center. Next proceed to second innermost, and etc. So we have a hierarchy and a network. In the first sequence the siblings no longer form a partition. Nevertheless each point (attribute value) still only have a unique elementary neighborhood. So each lower concept is syntactically group into a unique higher level concept. However, a concept may be covered by several elementary neighborhoods, so semantically, it has several "parents."

4.2 Generating Hierarchy for Strong Dependencies

Let X, B, and $\circ$ be a finite set, a binary relation and composition of binary relations respectively.

Propositions

1. if B is reflective, then

$$B \longrightarrow B^2 = B \circ B \longrightarrow B^3 = B \circ B \circ B, \ldots, \longrightarrow B^m \qquad (*)$$

   is a strongly dependent sequence ; see Section 3.1.
2. if B is reflective and symmetric, then there is the smallest m such that B^m is an equivalence relation (transitive closure).
3. the generated sequence (*) is a concept hierarchy/network ; see Table 7.
4. (*) induces a nested induced equivalence relations

$$E_B \longrightarrow E_{B^2} \longrightarrow E_{B^3}, \ldots, \longrightarrow E_{B^m} \qquad (E_*)$$

   Comments on Table 7

1. Table 7 illustrates the nested sequence (*) for the binary relation in Table 2.
2. Note that syntactically, Table 7 is a hierarchy. For example, O and L are syntactically independent, but semantically $L \subset O$.
3. Also note that C syntactically has a unique parent K (elementary neighborhood), but semantically it has another "parent" J (sine C is an element of J). Note that $\{C, B\} = J$ is the unique elementary neighborhood of B, not for C. C's unique elementary neighborhood is $\{C, B, D\} = K$.

   Comments on Table 8

4. Table 8 illustrates the nested induced partitions (E_*); a hierarchy in the traditional sense.
5. Table 7 and Table 8 are syntactically two way isomorphic, but semantically there is only a one way map from Table 8 to Table 7. Note that semantically $L \subset O$, but L_P and O_P are independent.

5 Mining High Level Data

5.1 Real World Interpretations of VC-Binary Relation

1. We interpret VC-binary relation as "one hour drive." Since some public roads may be "one way street", so VC, is reflexive, but not necessarily symmetric nor transitive.
2. VC-elementary neighborhood is the "nearest neighborhood". We will define it by the measurement "one hour drive;" see Table 3.

Table 7. Concept Hierarchies/Networks Based on Binary Granulations

$CITY$	MeaningfulName of VC - Elementary neighborhood	MeaningfulName of $VC \circ VC$ - Elementary neighborhood	MeaningfulName of $VC \circ VC \circ VC$ - Elementary neighborhood
B	$\longrightarrow J$	$\longrightarrow K$	$\longrightarrow O$
C	$\longrightarrow K$	$\longrightarrow O$	$\longrightarrow O$
D	$\longrightarrow L$	$\longrightarrow L$	$\longrightarrow L$
E	$\longrightarrow L$	$\longrightarrow L$	$\longrightarrow L$
F	$\longrightarrow M$	$\longrightarrow M$	$\longrightarrow M$
G	$\longrightarrow N$	$\longrightarrow N$	$\longrightarrow N$

Table 8. Concept Hierarchies by Induced Partitions

$CITY$	MeaningfulName of E_{VC} - Elementary set (the center)	MeaningfulName of $E_{VC \circ VC}$ - Elementary set (the center)	MeaningfulName of $E_{VC \circ VC \circ VC}$ - Elementary set (the center)
B	$\longrightarrow J_P$	$\longrightarrow K_P$	$\longrightarrow O_P$
C	$\longrightarrow K_P$	$\longrightarrow O_P$	$\longrightarrow O_P$
D	$\longrightarrow L_P$	$\longrightarrow L_P$	$\longrightarrow L_P$
E	$\longrightarrow L_P$	$\longrightarrow L_P$	$\longrightarrow L_P$
F	$\longrightarrow M_P$	$\longrightarrow M_P$	$\longrightarrow M_P$
G	$\longrightarrow N_P$	$\longrightarrow N_P$	$\longrightarrow N_P$

3. Each VC-elementary neighborhood assigned a symbol that represents a meaningful name. For example, L is the name of an "informal city L" (for example Greater San Francisco) that includes two distinct cities, D and E (San Francisco and Berkeley); and D and E are the centers of L.

   Next items are about $VC \circ VC$-binary relation.

4. $VC \circ VC$-binary relation is "two hours drive."
5. $VC \circ VC$-nearest neighborhood is the "informal county" which contains "informal city." For example, K is the name of a "informal county K" (for example, Silicon Valley) that includes informal city J (Greater San Jose) and real city C (Palo Alto); and J and C are part of the centers of K.

   Next items are about $VC \circ VC \circ VC$-binary relation.

6. $VC \circ VC \circ VC$-binary relation is "three hours drive."

7. $VC \circ VC \circ VC$–nearest neighborhood is the "informal state" which contains "informal counties."

   Next items are about the binary granulations.

8. Table 9 is the proposed concept hierarchy/network defined by the nested binary relations VC, $VC \circ VC$, $VC \circ VC \circ VC$.
9. All symbols in this proposed concept hierarchy/network are syntactically distinct and semantically may be related.

   Next items are about the induced partitions.

10. Table 10 is the traditional concept hierarchy defined by the nested equivalence relations $E_{VC}, E_{VC \circ VC}, E_{VC \circ VC \circ VC}$.
11. All symbols in this traditional concept hierarchy are syntactically and semantically independent.
12. Each induce equivalence class represents the center cities of respective elementary neighborhoods of nested binary relations.

    Next items are about the relationship between granulations and partitions.

13. There is a semantic preserving map from the concept hierarchies/networks of Table 9 to Table 10; and an inverse map on syntactical level.

5.2 High Level Rules

To mining high level rule, we create high level tables Table 9, and Table 10 from Table 1, using new concept hierarchy/network induced by VC (Table 7), and traditional concept hierarchy induced by $E_V C$ (Table 8). Here we give verbally typical AOG results from the two hierarchies, namely

Rule 1: From Table 10, we conclude that the suppliers' status is high on the real city $M_P = \{F\}$ and low on the real city $O_P = \{B\}$

Rule 2: From Table 9, we conclude that the suppliers' status is high on the informal city $M = \{F, G\}$ and low on "informal state " $O = \{C, B, D, E\}$

These two Rules are related. Let us interpret them as follow. People live in an informal area (city, county, state) and work at its center. The Rule 1 is saying about the pay in centers, while Rule 2 is about the income of individuals resided in informal areas. So to discuss cooperate pay, Table 10 is better, to discuss family average income, Table 9 is better, (couple may work at different cities in the same informal area).

6 Conclusion

In practical databases, attribute values often carry additional semantics specified by binary relations, such as numerical orders, physical distances and etc.. In this paper, we demonstrate that one can use such semantics to generate concept hierarchy/network and to mine high level rules. We believe this paper provides an initial foundation for a new promising data mining technique.

Table 9. The Supplier Table Granulation Level 2

SNUM	*SNAME*	STATUS	$VC \circ VC$ - Elementary neighborhood
S_1	Smith	*TWENTY*	*K*
S_2	Jones	*TEN*	*O*
S_3	Blake	*TEN*	*O*
S_4	Clark	*TWENTY*	*K*
S_5	Adams	*THIRTY*	*L*
S_6	Peterson	*FORTY*	*L*
S_7	Ewing	*EIGHTY*	*M*
S_8	Johnson	*EIGHTY*	*M*
S_9	Pike	*FORTY*	*L*
S_{10}	Meyers	*NINTY*	*N*

Table 10. The Supplier Table Partition Level 2

SNUM	*SNAME*	STATUS	$E_{VC \circ VC}$- Elementary neighborhood
S_1	Smith	*TWENTY*	K_P
S_2	Jones	*TEN*	O_P
S_3	Blake	*TEN*	O_P
S_4	Clark	*TWENTY*	K_P
S_5	Adams	*THIRTY*	L_P
S_6	Peterson	*FORTY*	L_P
S_7	Ewing	*EIGHTY*	M_P
S_8	Johnson	*EIGHTY*	M_P
S_9	Pike	*FORTY*	L_P
S_{10}	Meyers	*NINTY*	N_P

References

1. Y.D. Cai, N. Cercone, and J. Han. "Attribute-oriented induction in relational databases," in Knowledge Discovery in Databases, pages 213-228. AAAI/MIT Press, Cambridge, MA, 1991.
2. Date, C., An Introduction to Database Systems - Vol I, 7th ed., Addison-Wesley, 2000.
3. Maier,D.: The Theory of Relational Databases. Computer Science Press, 1983 (6th printing 1988).
4. T. Y. Lin, "Data Mining and Machine Oriented Modeling: A Granular Computing Approach," Journal of Applied Intelligence, Vol 13, N. 2, September/October, Kluwer,2000, pp.113-124.

5. Eric Louie and T.Y. Lin, "Finding Association Rules using Fast Bit Computation: Machine-Oriented Modeling." In: Proceeding of 12th International Symposium ISMIS2000, Charlotte, North Carolina, Oct 11-14, 2000. Lecture Notes in AI 1932. 486-494.
6. T.Y. Lin, "Granular Computing on Binary Relations I: Data Mining and Neighborhood Systems." In: Rough Sets In Knowledge Discovery, A. Skoworn and L. Polkowski (eds), Springer-Verlag, 1998, 107-121
7. An Overview of Rough Set Theory from the Point of View of Relational Databases, Bulletin of International Rough Set Society, Vol I, No1, March , 1997, 30-34.
8. M. Hadjimichael and T. Y. Lin, "Non-classificatory Generalization in Data Mining." In: the Proceedings of The Fourth Workshop on Rough Sets, Fuzzy Sets and Machine Discovery, Tokyo, Japan, November 8-10,1996, 404-411.
9. T. Y. Lin, "Topological and Fuzzy Rough Sets." In: Decision Support by Experience - Application of the Rough Sets Theory, R. Slowinski (ed.), Kluwer Academic Publishers, 1992, 287-304.
10. Pawlak, Z.: Rough Sets: Theoretical Aspects of Reasoning about Data, Kluwer Academic, Dordrecht (1991)
11. W. Ziarko, "Variable Precision Rough Set Model." Journal of Computer and Systems Science Vol 46,No1, February, Academic Press, 1993, pp.38-59.
12. Y.Y. Yao, "Information tables with neighborhood semantics." In: Proceeding of 14th Annual International Symposium Aerospace/Defense Sensing, Simulation, and Controls , SPIE Vol 4057, pp. 108- 116, Orlando, April 24-28, 2000
13. W. Sierpenski and C. Krieger, General Topology, University of Torranto Press 1952.

Representing Large Concept Hierarchies Using Lattice Data Structure

Yanee Kachai[1] and Kitsana Waiyamai[2]

Department of Computer Engineering, Faculty of Engineering, Kasetsart University, Chatuchak, Bangkok, Thailand
[1]g4265084@ku.ac.th, [2]wkitsana@master.cpe.ku.ac.th

Abstract. With the rapid growth in size and number of available databases, the manipulation of large concept hierarchies that cannot be fit in main memory becomes more and more frequent. Several representations of concept hierarchies are possible, for example tree, lattice, table, linked list, etc. In this paper, we propose an efficient implementation technique to manipulate large concept hierarchies. We use a lattice data structure to represent concept hierarchies and encode such a lattice into a boolean transitive closure matrix. A set of lattice operators are defined and implemented as abstract data types on the top of an object-relational database management system, and are used to perform generalization and specialization operations. We show the efficiency of the lattice operators to perform generalization and specialization in large concept hierarchies and compare their performance with the START WITH and CONNECT BY clauses of SQL.

1 Introduction

A concept hierarchy defines a sequence of mappings from a set of low-level concepts to higher level, more general concepts [1]. Such mappings may organize the set of concepts in partial order, such as in the shape of a tree (a hierarchy, a taxonomy), a lattice, a directed graph, etc. While in a strict concept hierarchy such as a tree or taxonomy, each concept has exactly one parent (super-concept) but in a lattice or directed graph hierarchy, there are many paths to a particular concept. We believe that such a lattice or directed graph data structure has advantage over the tree data structure to represent real world concept hierarchies. In this paper, we use a lattice data structure to represent concept hierarchies

Concept hierarchies are very important in data mining and data warehousing. In data mining, they allow knowledge discovery at different conceptual levels [2]. In data warehousing, concept hierarchies are necessary for operations such as drill-down and roll-up fact dimension [3]. With the rapid growth in size and number of available databases, the manipulation of large concept hierarchies that cannot be fit in main memory becomes more and more frequent. In order to improve the efficiency of a knowledge discovery process, effective implementation techniques to manage large concept hierarchies have to be proposed.

In a relational database management system, large concept hierarchies are represented as a collection of tables. Each table has two columns, which respectively represents a node and its direct parents. Generalization and specialization are then performed using the START WITH and CONNECT BY clauses of SQL [4]. Operating in a depth first search manner, such operators are inefficient with respect to

D. Cheung, G.J. Williams, and Q. Li (Eds.): PAKDD 2001, LNAI 2035, pp. 186-197, 2001.

the response time in presence of large concept hierarchies. Furthermore, SQL lacks essential features for generalization and specialization of n input where $n \geq 1$.

In this paper we propose a new approach to manipulate large concept hierarchies. Instead of representing them as a traditional collection of tables, we use a lattice data structure to represent concept hierarchies. We propose a new mechanism to manipulate large lattice data structures using efficient implementation techniques. We use a transitive closure encoding method to represent the lattice in the form of boolean transitive closure matrix. Encoding method leads in decreasing of almost constant time in performing generalized and specialized operations in the hierarchies. We propose an efficient method for representing such a matrix in an object-relational database where only the position of all elements that equals to 1 in the matrix is stored. We show that such a data structure is very efficient in terms of storage space used. Generalization and specialization of concept hierarchies are performed using the proposed lattice operators, which are UB, LB, Meet and Join. We show their performances in terms of efficiency of the generalization and specialization operations, and the ability to handle large concept hierarchies, which cannot fit in memory. Lattice operators have been tested against the NCBI/Genbank molecular databases [5]. NCBI database contains relationships between more than 80,000 organisms spanning up to 35 levels. The experimental results show that our new operators are more effective with respect to the response time of performing generalized and specialized operations compared to the results obtained using the START WITH and CONNECT BY operators.

The paper is organized as follows: Section 2 gives basic definitions of concept hierarchy, lattice and its operators. Section 3 describes how concept hierarchies are represented and queried in an object-relational database using START WITH and CONNECT BY clauses of SQL. Section 4, presents how lattice can be encoded into a boolean matrix of binary words, and how such a matrix can be efficiently represented in object-relational databases. Section 5 explains the implementation of lattice operators and describes how generalization and specialization can be performed using lattice operators. Section 6 compares performance between the new lattice operators and the START WITH-CONNECT BY clauses of SQL in term of efficiency of the generalization and specialization operations against large concept hierarchies. Section 7 concludes the paper and presents future works.

2 Basic Definitions

The basic notions *concept hierarchy, lattice and its operators* are given in this section. Readers are referred to [6] for further information on this theory.

2.1 Concept Hierarchy

A concept hierarchy is a sequence of mapping from a set of lower-level concepts to their high-level correspondences [1]. Such mapping may organize the set of concepts in *Partial Order*, which the most general concepts is the null description whereas the most specific concepts corresponds to the specific values of attribute in the database. Let H be the hierarchy defined on a set of domains $\bullet_i, \ldots, \bullet_k$. Formally we have

$$H_l : \bullet_i \times \ldots \times \bullet_k \rightarrow H_{l-1} : \bullet \ldots \rightarrow H_0$$

Where H_l represents the set of concepts at the primitive level, H_{l-1} represents the concepts at one level higher than those at H_l etc., and H_0 represents the highest level of the hierarchy.

2.2 Lattice and Its Operators

A non-empty partial ordered set $< L, \bullet >$ is a lattice if $\forall x, y \in L$, each pair {x, y} always has the least upper bound denoted as Join ({x, y}) and the greatest lower bound denoted as Meet ({x, y}). In the following, lattice operators are given. These operators allow generalized and specialized operations to be performed within a hierarchy.

UB/LB. Let P be a partial ordered set, and $S \subseteq P$. An element $u \in P$ is an *upper bound* of S if $s \bullet u$ for all $s \in S$. A *lower bound* is defined dually. The set of all upper bounds of elements in S is denoted by $UB\ (S)$ and the set of all lower bounds is denoted by $LB\ (S)$.

$UB\ (S) = \{u \in P \mid (\forall s \in S)\ s \bullet u\}$ $\qquad$ $LB\ (S) = \{l \in P \mid (\forall s \in S)\ s \bullet l\}$

Join/Meet. Let P be a partial ordered set, and $S \subseteq P$. The *join* of elements in S when it exists is the least element of the set of common upper bounds of elements in S. In the same way, the *meet* of elements in S when it exists is the greatest element of the set of common lower bounds of elements in S. Hence the join and meet of elements in S can be, respectively determined by

$Join\ (S) = Min(UB(S))$ $\qquad$ $Meet(S) = Max(LB(S))$

The best way to understand how the concept hierarchy can be organized in the form of a lattice is to consider an example. In Figure 1, a concept hierarchy of organisms A, B, C, D, E, F and G is organized into a lattice, which represents the relationships between organisms.

Once a concept hierarchy is represented using a lattice data structure, generalization and specialization of such a hierarchy can be performed using lattice operators, which are UB, LB, Meet and Join. According to the given example, we are able to answer the following queries:

- Find all organisms that are more general than F? The answers are {A, B, C, D and E} by using UB operator.
- Find all organisms that are more specific than A? The answers are {B, C, D, E, F and G} by using LB operator.
- Find the least common super-concept of D and E? The answer is {B} by using Join operator.
- Find the greatest common sub-concept of D and E? The answer is {G} by using Meet operator.

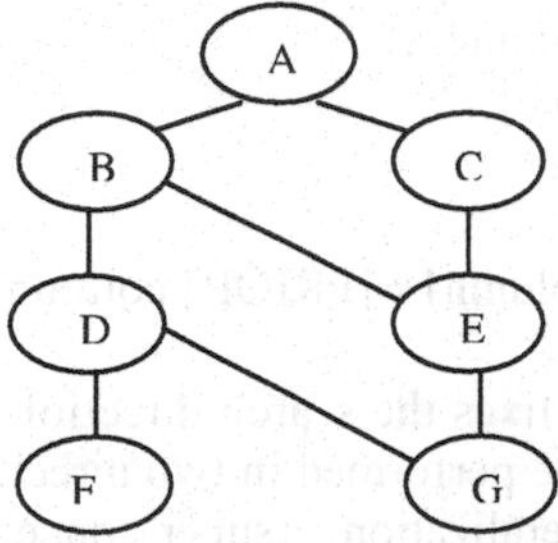

Fig. 1. Representation of a concept hierarchy using lattice data structure

3 Querying Concept Hierarchies in Relational Databases

In a relational database management system, we are able to represent concept hierarchies in a simple form of a collection of tables. Each table consists of two columns, which respectively represents a node and its direct parents. Relational table representing our concept hierarchy example is given in figure 2. The first column is used to store the identifier of a given organism and the second one of the same type is used to store its direct parent(s). It must be noticed that each node_id can have more than one parent_node_id that is the case when a node has multiple parents. From our concept hierarchy example in figure 1, organism D is a direct child of organism B and C. Thus, we store the identifier of its direct parent organisms that are D-B in one row and D-C in another.

Node_id	Parent_node_id
A	
B	A
C	A
D	B
E	B
E	C
F	D
G	D
G	E

Fig. 2. Relational database table representing concept hierarchy

3.1 Generalization and Specialization Operators in Relational Databases

In a relational database system, the START WITH and CONNECT BY PRIOR clauses can be used to perform generalization and specialization against concept hierarchies in a dept first search manner. The START WITH clause specifies the root of the hierarchy while the CONNECT BY PRIOR clause sets up the relationships between elements in the hierarchy. The level function gives the distance between the root and the current node, starting with 1 for the root. The number of levels returned by a generalization/specialization query may be limited according to available resource (main memory).

General syntax of a generalized/specialized query.

```
SELECT      column1, column2, …
FROM        tables [variable]…
[WHERE      condition]
START WITH condition
CONNECT BY [PRIOR] column1 =[PRIOR] column2
[ORDER BY LEVEL]
```

The CONNECT BY clause fixes the search direction in a hierarchy. Searching for elements in the hierarchy can be performed in two directions, which are specialization (sub-concept search) and generalization (super-concept search). In the following, examples of generalization and specialization queries using the START WITH and CONNECT BY clauses are given.

Query 1: Find all organisms that are more specific than organism A.
Select node_id from Table
Start with node_id = 'A ' Connect by parent_node_id = prior node_id;
Query 2: Find all organisms that are more general than organism D.
Select node_id from Table
Start with node_id = 'D' Connect by node_id = prior parent_node_id;

4 Encoding Lattice Data Structure

In this section, we present an efficient implementation technique to represent large concept hierarchies. This is archived by plunging the hierarchy into a boolean lattice matrix of binary words. In [7], methods are described for implementing lattice operators for implementing type hierarchies in the context of object-oriented languages. Here, we use a transitive closure encoding method for effective implementation of our lattice operators. In the following, we start by presenting how a concept hierarchy can be encoded into a boolean transitive closure matrix of binary words. Then, an efficient method for representing such a matrix in an object-relational database is proposed.

4.1 Boolean Transitive Closure Encoding

The boolean transitive closure encoding is a mapping of a given lattice into a boolean lattice matrix of binary words. A 1 in the $(i,j)^{th}$ element in the matrix means that element i is an ancestor of element j. Row i represents descendants of element i. Column j represents the ancestors of element j. The size of matrix is equal to n x n where n is the number of elements in the lattice. We name each row and each column of the matrix with the name of each element in the lattice.

To encode the concept hierarchy into a matrix of binary words, we use the "immediate greater than" relation covered by the ordering in that hierarchy. This relation is obtained by computing all descendants of every element in the concept hierarchy. Therefore each row of the matrix contain 1's only in those columns headed by elements which are immediately less than the element heading the row; and it contains 0's otherwise. Thus, each row is a characteristic representation of the set of all its' sub-concepts and each column is a characteristic representation of the set of all its' super-concepts dually. For example, the boolean transitive closure matrix as shown in figure 3 can represent the concept hierarchy of figure 1. The row bit string code 0001011 for element D represents the sub-concepts of D = {D, F and G}. The column bit string code 1101010 for element F represents the super-concepts of F = {A, B, D and F}. Using these codes, it is possible to compute lattice operations using only logical AND operator on bit string, which we will describe in section 5.

4.2 Boolean Transitive Closure Matrix Representation

We have seen how a concept hierarchy can be encoded into a boolean transitive closure matrix of binary words. In the following, we explain how such a matrix can be efficiently represented in an object-relational database. Given a concept hierarchy of n elements and each element is allocated its own code, then the concept hierarchy is encoded with n^2 bits. In the presence of sparse matrix, such representation is inefficient with respect to the amount of storage space. In order to reduce the storage space of the boolean transitive closure matrix in an object-relational database, we propose an efficient representation by storing only the position of elements that equals to 1.

	A	B	C	D	E	F	G
A	1	1	1	1	1	1	1
B	0	1	0	1	1	1	1
C	0	0	1	0	1	0	1
D	0	0	0	1	0	1	1
E	0	0	0	0	1	0	1
F	0	0	0	0	0	1	0
G	0	0	0	0	0	0	1

Fig. 3. Boolean transitive closure matrix

If the $(i,j)^{th}$ element in the boolean transitive closure matrix is equal to 1 then we store value i in the RowInd column and value j in the ColInd column. With this tabular representation, storage space of a matrix in an object-relational database system is by far reduced. From our example of boolean transitive closure matrix in figure 3, we are able to transform such a matrix into an object-relational database table using this technique as shown in figure 4. With this representation, we are able to retrieve row and column of the matrix using simple SQL statements. In the following algorithms, we describe the steps for retrieving row and column of boolean transitive closure matrix from the table stored in an object-relational database.

RowInd	1	1	1	1	1	1	1	2	2	2	2	2	3	3	3	4	4	4	5	5	6	7
ColInd	1	2	3	4	5	6	7	2	4	5	6	7	3	5	7	4	6	7	5	7	6	7

Fig. 4. Table representing boolean transitive closure matrix

In algorithm 1 of figure 5, we propose a method for retrieving row bit string of the matrix from an object-relational database table. To retrieve row X of the matrix, we first select the ColInd from the table where the RowInd is equal to X. Then, we set value of bit string at the returned column position to 1 and the other positions to 0.

To retrieve column bit string of the matrix, similar operations can be performed as described in algorithm 2. Getting column X of the matrix we first select the RowInd from the table where the ColInd is equal to X. Then we set value of bit string at the returned row position to 1 and the other positions to 0.

5 Implementation of Lattice Operators

In this section we discuss how lattice operators can be used to perform generalization and specialization of a concept hierarchy. Previously, we have seen how a concept hierarchy can be encoded into a boolean transitive closure matrix and how such a matrix is represented in an object relational database. Using these codes, it is possible to compute lattice operations using logical AND on bit string. We use column bit string of the matrix to compute UB and Join operations and use row bit string to compute LB and Meet operations. From our matrix example in figure 3, we are able to compute UB, LB, Meet and Join operations of element D and E as shown in the following:

UB $\{D, E\} = 1101000 \cap 1110100 = 1100000 = \{A, B\}$
LB $\{D, E\} = 0001011 \cap 0000101 = 0000001 = \{G\}$
Join $\{D, E\} = 1101000 \cap 1110100 = 1100000 = \{B\}$
Meet $\{D, E\} = 0001011 \cap 0000101 = 0000001 = \{G\}$

Algorithm 1. Get a row bit string of the matrix	**Algorithm 2.** Get a column bit string X of the matrix
Procedure GetRowCode(X:element); Returns : bit string code begin CodeX = 0; S = Select ColInd from matrix_table Where RowInd = X; For each i ∈ S CodeX(i) = 1; Return(CodeX); End	Procedure GetColCode(X:element); Returns : bit string code Begin CodeX = 0; S = Select RowInd from matrix_table Where ColInd = X; For each i ∈ S CodeX(i) = 1; Return(CodeX); End

Fig. 5. Algorithms for retrieving row (X) and column (X) of a given transitive closure matrix

Steps for computing UB and Join operations are given in algorithm 3 and 4 of figure 6. To perform generalization of elements X and Y, we first retrieve the column bit string code of element X and Y from the matrix stored in a database using *GetColCode* procedure given in algorithm2 of figure 5. Then, we apply the logical AND operator to the retrieved codes. All elements that equal to 1 of the resulting code are super-concepts of X and Y.

Computation of the Join operation is performed in a similar way, with the only difference is that the answer is the element in the matrix where its code matches the result of logical AND operator. In the case that the resulting code is the code of none element in the matrix, the minimal common super-concepts of X and Y is returned instead.

Steps for computing the other lattice operations, which are LB and Meet, are given in algorithm 5 and 6 of figure 7. To perform specialization of elements X and Y, we first retrieve the row bit string code of element X and Y from the matrix stored in database using *GetRowCode* procedure in algorithm 1 of figure 5. Then we apply logical AND operator to the retrieved codes. All elements that equal to 1 of the resulting code are sub-concepts of X and Y.

Computation of the Meet operation is performed in a similar way, with the only difference is that the answer is the element in the matrix where its code matches the resulting of logical AND operator. In the case that the resulting code is the code of none element in the matrix, the maximum common sub-concepts of X and Y is returned instead.

<table>
<tr>
<td>

Algorithm 3. UB Operator

```
Procedure UB (X, Y:element);
 Returns : element
 begin
  While X, Y is not null
  begin
   CodeX = GetColCode(X);
   CodeY = GetColCode(Y);
   Result = CodeX•CodeY;
  end
  for i = 0 to CodeLength
   if Result(i) = 1 then
    return {element(i)};
   end if
end
```

</td>
<td>

Algorithm 4. Join Operator

```
Procedure Join (X, Y:element);
 Returns : element
 begin
  While X, Y is not null
  begin
        CodeX = GetRowCode(X);
        CodeY = GetRowCode(Y);
        Result = CodeX•CodeY;
  end
  if ∀element ∈ Matrix|
   code(element)=Result then
        return(element)
  else
   return minimum
     {element|Result(i)=1};
end
```

</td>
</tr>
</table>

Fig. 6. Algorithms for UB and Join operators

<table>
<tr>
<td>

Algorithm 5. LB Operator

```
Procedure LB (X, Y:element);
 Returns : element
 begin
  While X, Y is not null
  begin
        CodeX = GetRowCode(X);
        CodeY = GetRowCode(Y);
        Result = CodeX•CodeY;
  end
  for i = 0 to CodeLength
   if Result(i) = 1 then
    return {element(i)};
   end if
end
```

</td>
<td>

Algorithm 6. Meet Operator

```
Procedure Meet (X, Y:element);
 Returns : element
 begin
  While X, Y is not null
  begin
        CodeX = GetColCode(X);
        CodeY = GetColCode(Y);
        Result = CodeX•CodeY;
  end
  if ∀element ∈ Matrix|
   code(element)=Result then
        return(element)
  else
   return maximum
     {element|Result(i)=1};
end
```

</td>
</tr>
</table>

Fig. 7. Algorithms for LB and Meet operators

6 Performance Evaluation

To access their performance, lattice operators are implemented in C/C++ programming language provided by the Oracle ORDBMS, and compared against START WITH and CONNECT BY clauses of SQL. The platform we used was a SUN E3500 running SunOS version 2.5.1 with CPU clock rate of 366 MHz, 6144 MB of main memory and 9GB disk. Data is stored in Oracle database version 8.1.5 [8]

We ran our experiments using the NCBI/Gen bank molecular databases. NCBI database contains relationships between more than 80,000 organisms spanning up to 35 levels. Indexes have been created on the two columns in order to have the best performance when using the START WITH and CONNECT BY clauses. It has been noticed that generalization and specialization with n input parameters where $n \geq 1$ are not supported by the START WITH and CONNECT BY clauses. In order to compare these operators with our lattice operators, with n input parameters, we have developed SQL-embedded procedures using PL/SQL to perform this task.

We encoded the NCBI data into boolean transitive closure matrix and then stored such a matrix using the proposed matrix representation as described in subsection 4.2. The total size of the encoded hierarchy is 15MB where only position of elements equals to 1 is stored compared against the total size of 7.5GB if all elements of the matrix are stored.

Figure 8, 9, 10 and 11 show the average CPU time (including disk access) of generalized and specialized operations given in microsecond. In order to study the behaviors of the proposed lattice operators and the START WITH-CONNECT BY clauses in various sizes of concept hierarchies, experiments were conducted on the database using different numbers of records ranging from 20,000 to 90,000 records. The numbers of input concepts for each operator are chosen randomly form the total number of existing concepts in the concept hierarchy. Figure 8 shows the average CPU time used to perform generalization using UB operator compared with START WITH and CONNECT BY clauses, with respect to the different size of concepts. We have found that the UB operator outperforms the START WITH-CONNECT BY clauses at every data size. Furthermore, the result shows that increasing number of records in database also increase average execution time of operation to be longer for both new and existing operators.

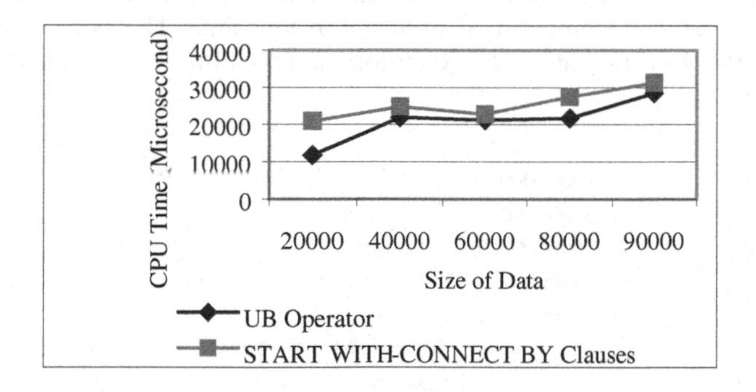

Fig. 8. Average CPU time of generalization in various data size

The average CPU time used to perform specialization using LB operator compared against START WITH-CONNECT BY clauses in different data size was shown in figure 9. The result shows that the average CPU time of new LB operator is faster than the START WITH and CONNECT BY clauses. Increasing number of records in database also increase average execution time of operation to be longer the same as UB operator.

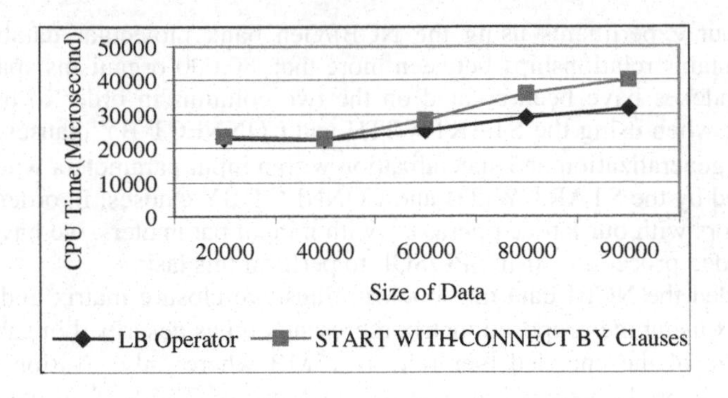

Fig. 9. Average CPU time of specialization in various data size

Figure 10 shows the average CPU time of Meet operator compared to START WITH-CONNECT BY clauses. For the existing operators, we observe that the more number of concept hierarchies is the longer execution time searching for the greatest common sub-concept. Furthermore, the execution time of START WITH-CONNECT BY clauses exponentially increases when the size of the concepts (number of record) is larger. For our Meet operators, we observe that its execution time is almost instantaneous (less than 6-7 milliseconds). The important difference between the execution time of both operators can be explained by the method used for searching the greatest common sub-concept in the hierarchy. Indeed, START WITH-CONNECT BY clauses operate in a depth first search manner, they have to move toward to every element in the hierarchy start at the root node searching for the greatest common sub-concept. Whereas, new Meet operator uses logical AND operator apply on the boolean transitive closure matrix to perform the same operation. It can find the searched element in an almost constant time. The same observation can be made for the Join operator. Its execution time is almost constant (less than 6-7 milliseconds).

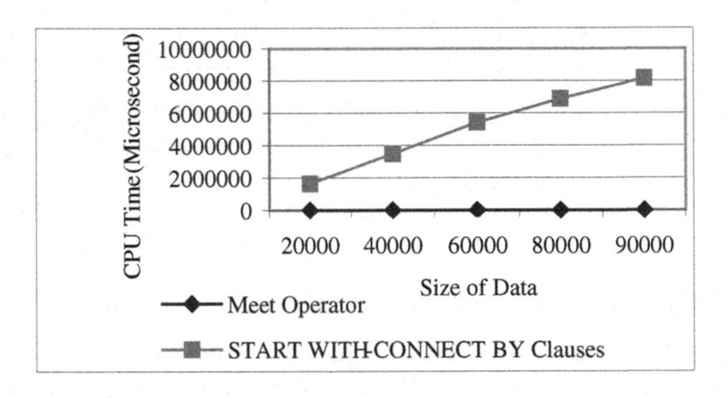

Fig. 10. Average CPU time of searching the greatest common sub-concept in various data size

Figure 11 shows the average CPU time searching for the least common super-concept using Join operator and the START WITH-CONNECT BY clauses in different data size.

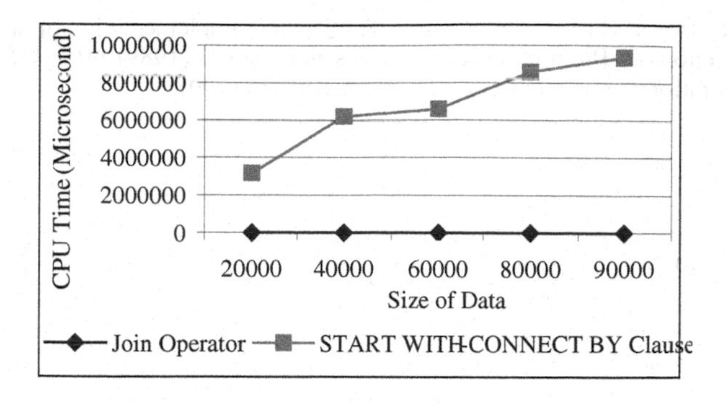

Fig. 11. Average CPU time of searching the least super-concept in various data size

7 Conclusion

In this paper, we use a lattice data structure to represent concept hierarchies and improve response time of generalization and specialization operations comparing against the START WITH and CONNECT BY clauses of SQL. The concept hierarchy is encoded into $n \times n$ boolean transitive closure matrix, where n is the number of elements in the hierarchy. Each element is encoded into binary words. We use only logical AND operator to apply on those binary words to perform lattice operators during generalization and specialization operations. In order to reduce storage space of the underlying $n \times n$ boolean transitive closure, we store only position of elements equal to 1. With this new representation, we are able to retrieve row and column of the matrix using simple SQL statements. The benefit of the new representation of concept hierarchies is to reduce time needed to perform generalization and specialization operations in the object-relational databases. This paper essentially presents an efficient method for manipulating large concept hierarchies. In the future, we plan to study how the proposed method effect efficiency in the mining of KDD process on specific data mining systems.

References

1. Han, J., Fu, Y.: Dynamic Generation and Refinement of Concept Hierarchies for Knowledge Discovery in Databases. AAAI'94 Workshop on Knowledge Discovery in Databases (KDD'94), Seattle, WA, July. (1994) 157-168
2. Han, J.: Mining Knowledge at Multiple Concept Levels. Proc. 4th Int'l Conf. on Information and Knowledge Management (CIKM'95), Baltimore, Maryland, Nov. (1995) 19-24
3. Han, J., Kamber, M.: Data Mining: Concepts and Techniques. Morgan Kufmann Publisher (2000)
4. Oracle Coporation.: Oracle 7 Server SQL Language Reference Manual (1992)
5. URL: http://www.ncbi.nlm.nih.gov/Taxonomy/taxonomyhome.html
6. Davey, B.A., Priestley, H.A.: Introduction to Lattices and Order. Cambridge University Press (1990)

7. Aït-Kaci ,H., Boyer ,R., Lincoln ,P., Nasr, R.: Efficient implementation of lattice operations. ACM transactions on Programming Languages and systems (1989) 11(1) 115-146
8. Oracle Coporation.: Oracle 8i Concepts. Release 8.1.5 (1999)

Feature Selection for Temporal Health Records

Rohan A. Baxter, Graham J. Williams, and Hongxing He

CSIRO Mathematical and Information Sciences,
GPO Box 664, Canberra 2602, Australia,
{Rohan.Baxter,Graham.Williams,Hongxing.He}@cmis.csiro.au

Abstract. In this paper we consider three alternative feature vector representations of patient health records. The longitudinal (temporal), irregular character of patient episode history, an integral part of a health record, provides some challenges in applying data mining techniques. The present application involves episode history of monitoring services for elderly patients with diabetes. The application task was to examine patterns of monitoring services for patients. This was approached by clustering patients into groups receiving similar patterns of care and visualising the features devised to highlight interesting patterns of care.

1 Introduction

We are interested in the problem of clustering individuals given observed data about the individuals where the observed data does not naturally occur in vector form. Clustering algorithms are typically applied to data in vector form. For example, we may have k-measurements on a set of patients and so the measurements on each individual i are represented as a k-dimensional vector. For vector-form data well-known and widely-applied clustering techniques can be applied. Such techniques are generally model-based methods include mixture modelling [6], or distance-based methods [3].

Much real world data is actually in non-vector form consisting of observations of an individual, recording information at particular time points. Such variable-length event sequence data is described in Sect. 2, but examples include a patient's usage of medical services and an individual's stock trading behaviour. The data is characterised as irregular events where each event may encapsulate a different type of action.

The data mining practitioner wishing to cluster event sequence data appears to have three options. The first option is to convert the event sequence data into feature vectors [4]. A problem with this approach is that information is inevitably lost in the vectorisation process. The second option is to use a distance-based clustering method which allows for non-vector data. An edit-distance metric [5] which uses insert, delete and replace operations to turn one sequence into another is an example of this approach. A difficulty here is in defining an effective distance metric. A suitable distance metric needs to be created for each new application. The third option is the use of mixtures of a generative probabilistic model [2,1]. This is an attractive approach but not further explored here.

D. Cheung, G.J. Williams, and Q. Li (Eds.): PAKDD 2001, LNAI 2035, pp. 198–209, 2001.

We chose the first option for the application described in this paper. An aim was to minimise the loss of information relevant to the data mining objectives in choosing the feature vectors. We present three alternative feature vectors for representing medical event sequence data. Our exploration provides insights into the process of developing alternative feature sets. We identify feature sets that are useful for clustering event sequence data.

Sect. 2 describes the patient health record data and Sect. 3 describes the objectives for investigating patterns of care received by patients. Sect. 4 describes the feature vectors we have used in looking for patterns of care. To the best of our knowledge two of the three feature vectors we use here are novel. Clustering results and their visualisations are presented in Sect. 5.

2 Health Care Data

Medicare is the Australian Government's universal health care system. Each visit to a medical practitioner or hospital is covered by Medicare and recorded as a transaction in the Medicare Benefits Scheme (MBS) database. This data has been collected in Australia since the inception of Medicare in 1975. Such a massive collection of data provides an extremely rich resource that has not been fully utilised in the exploration of health care delivery in Australia.

For this current exploration we use a subset of de-identified data (to protect privacy) based on Medicare transactions from Western Australia (WA) for the period 1994 to 1998. Our particular focus is on patterns of care related to diabetes for elderly patients (over 65 years of age). We have only limited demographic information about each patient, such as age, gender and location. For each patient we also have the sequence of diabetes-related monitoring tests they have received over this time interval.

The four monitoring tests included in our dataset are given in Table 1. Glycated hemoglobin measurements (*Gl*) provide information about the accumulated effect of glucose levels. Ophthalmologic examinations (*Op*) are important in the early identification of complications related to eye sight. Cholesterol measurements via lipid studies (*Ch*) help identify possible complications relating to heart conditions. Microalbuminuria tests (*Al*) provide early indications of possible future kidney function loss.

Table 1. Types of services received by Patients and indicative guidelines.

Abbrev	Description	Guidelines
Gl	Quantitation of glycosylated hemoglobin.	2–4 times per year
Op	Ophthalmologic examination.	Every 1-2 years
Ch	Cholestorol measurement via lipid studies.	Every year
Al	Microalbuminuria test	Every year

A sample patient record is illustrated in Fig. 1. The event sequence data can be augmented with any available vector based data.

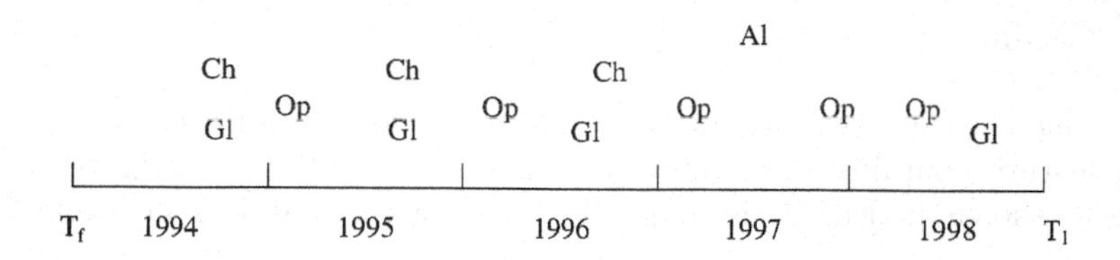

Fig. 1. A sample patient's health record, showing the four types of tests received over five years. The tests are: glycated hemoglobin (Gl); ophthalmology (Op); cholesterol (Ch); and micro-albuminuria (Al).

3 Patterns of Care in the Management of Diabetes

An important area in health population research is the investigation of patterns of care received by patients. Are there distinct patterns of care for these diabetes patients? Are there groups of patients receiving similar patterns of care? Are the patterns of care related to their doctor? Do patients of different age or gender or location receive differing patterns of care to other patients?

We have some prior expectations about the desired patterns of care for elderly patients with diabetes. The Australian National Health and Medical Research Council (NHMRC) publishes clinical guidelines for looking after patients. Patients with diabetes are at risk of developing complications such as eye problems, loss of kidney function and circulatory problems. The clinical guidelines recommend monitoring services, such as those in Table 1, be carried out at certain regular intervals. There is no compulsion for general practitioners to adhere to these guidelines and the guidelines cannot be expected to be appropriate for everyone.

To complicate matters, published clinical guidelines can differ in their details from state to state, and from country to country. We use the NHMRC guidelines as our starting point, but refer to other guidelines where they differ and where they may have an effect on clinical practice in Western Australia.

For example, according to NHMRC guidelines, glycated hemoglobin measurement should be done every six months (or every four months for some guidelines). Ophthalmologic examinations should be done every two years (or annually for some guidelines). The cholesterol measurement via lipid studies should be performed once a year. The microalbuminuria test should be done annually.

4 Selecting Features

We now present three methods for mapping the non-vector sequence data onto feature vectors. The first is the obvious *count* approach of having one feature for each type of service representing the number of times the service was used. The other two methods, which we call *average-residual-deviance* and the *gap*, are less obvious and overcome some shortcomings of the *count* method.

4.1 Count

In the *count* feature vector approach we have one feature for each type of service. Each feature contains the number of services received. The original sequence data, as shown in Fig. 1, is mapped to the features shown in Table 2. This

Table 2. Count Feature Vectors.

Patient	*Gl*	*Op*	*Ch*	*Al*
1	4	5	3	1
2	5	0	0	0
2	1	1	1	1
3	16	20	17	17

feature representation has the advantage of being easily interpretable. However the obvious loss of information is a concern for the goals of our project. We have lost information relating to the time between successive services and also to the overall coverage of the services across the five years. For example, an individual with a count of 15 for a service appears to be well-monitored, but if those 15 services all occurred in 1994 and none occurred in 1995, 1996, 1997 and 1998, then that is a pattern we would like to identify.

4.2 Average, Residual, Deviance

We have devised the *average, residual, deviance* feature vector for capturing the required temporal information missed by the *count feature vector* approach.

Let $t_{i,j}^{k}(i = 1, 2, ...n_j)$ be the date of the ith service for service type j on patient k and n_j be the total number of type j services received by patient k. Define T_f and T_l to be the beginning and ending dates of the time interval covered by the study.

Definition 1 (Mean Interval). *The Mean Interval, $MI_{j,k}$, for the patient k on service j when $n_j > 0$ is defined as :*

$$MI_{j,k} = \frac{\sum_{i=1}^{n_j-1}(t_{i+1,j}^{k} - t_{i,j}^{k})}{n_j - 1} \tag{1}$$

If $n_j = 0$ then $MI_{j,k} = T_f - T_l$.

For example, we can calculate the Mean Interval for a patient who had thirteen tests for Quantitation of glycosylated hemoglobin on the following dates:

$$9044, 9272, 9377, 9527, 9592, 9766, 9875, 9985, 10101, 10154, 10334, 10413, 10510 \tag{2}$$

where, for computational convenience, these dates are expressed as the number of days since January 1st, 1970. The interval (in days) between two consecutive tests are then

$$228, 105, 150, 65, 174, 109, 110, 116, 53, 180, 79, 97 \quad (3)$$

For this patient we have $MI_{j,k} = \frac{228+105+150+65+...+116+53+180+79+97}{12} = 122.2$.

Definition 2 (Deviation Interval). *The Deviation Interval,$DI_{j,k}$, for patient k on service j for $n_j > 0$ is defined as:*

$$DI_{j,k} = \sqrt{\frac{\sum_{i=1}^{n_j-1}(t_{i+1,j}^k - t_{i,j}^k - MI_{j,k})^2}{n_j - 1}} \quad (4)$$

If $n_j = 0$ then $DI_{j,k} = T_f - T_l$.

For example, using the patient with the health record for a single test given in Eqn. (2), the Deviation Interval is

$$DI_{j,k} = \sqrt{\frac{(228 - 122.2)^2 + (105 - 122.2)^2 + (150 - 122.2)^2 + ...}{12}} = 47.9 \quad (5)$$

Definition 3 (Residual Time). *The Residual Time, $RT_{j,k}$, for patient k on service j is defined as:*

$$DI_{j,k} = t_{1,j}^k - T_f + T_l - t_{n_j,j}^k \quad (6)$$

For example, using the patient with the health record given in Eqn. (2), we have $T_f = 8765$ (January 1st, 1994) and $T_l = 10592$ (December 31st, 1997), so that the Residual Time is $DI_{j,k} = 9044 - 8765 + 10592 - 10510 = 361$

The feature *Mean Interval* measures the average interval between receiving the same service. The *Interval Deviation* measures whether the service intervals are regular or irregular. The third feature provides a way of accounting for the windowing effects of having data for 5 years only. The time interval from the window boundary to the time of the first service and from the last service to the window boundary are not considered in the definition of the first two feature definitions. The third feature is used to account for these boundary effects.

The feature vector for a service should have reasonably small values for all three features if the patient is treated according to the clinical guidelines. Typically, some patients do need more frequent services as their diabetic condition is serious. We do not consider the possibility of over-servicing by medical practitioners, where more services than are clinically necessary are provided.

Patterns of care contrary to the clinical guidelines can arise from insufficient numbers of services provided over the five years. This type of pattern is detected by the count feature vector and by a large Mean Interval value. The *average, residual, deviance* feature vector also represents patterns of care where the services provided are clustered in time, or are absent near the boundaries of the time window.

The features are still relatively easy to interpret. However, we now need twelve features in our present application instead of the four for the *count* feature vector.

4.3 Gap

Our third feature vector representation is the most specific to the task of investigating service patterns with reference to service clinical guidelines. The motivation is to describe the total length of time when the regular required tests are not carried out.

Once again, let $t^k_{i,j} (i = 1, 2, ...n_j)$ be the date of the ith service for service type j on patient k and n_j be the total number of type j services received by patient k. Define T_f and T_l to be the beginning and ending dates of the time interval covered by the study.

We require that service type j have a *desirable gap*, DG_j, as given by some clinical guidelines.

Definition 4 (Gap). *If patient k has $n_j = 0$ (the patient has no services) then the gap, $G_{j,k}$, is defined as:*

$$G_{j,k} = T_l - T_f - DG_j \tag{7}$$

If patient k has $n_j > 0$ (the patient has one or more services)

$$G^{initial}_{j,k} = \begin{cases} t^k_{1,j} - T_f - DG_j & \textit{if } t^k_{1,j} - T_f > 0 \\ 0 & \textit{otherwise} \end{cases} \tag{8}$$

The following counts the time intervals between services received:

$$G^i_{j,k} = \begin{cases} t^k_{i+1,j} - t^k_{i,j} - DG_j & \textit{if } t^k_{i+1,j} - t^k_{i,j} > DG_j \\ 0 & \textit{otherwise} \end{cases} \tag{9}$$

We then include the final service interval:

$$G_{final} = \begin{cases} T_l - t^k_{n_j,j} - DG_j & \textit{if } T_l - t^k_{n_j,j} > DG_j \\ 0 & \textit{otherwise} \end{cases} \tag{10}$$

The three Gap sub-parts are now summed:

$$G_{j,k} = G^{initial}_{j,k} + \sum_{i=1}^{n_j-1} G^i_{j,k} + G^{final}_{j,k} \tag{11}$$

For example, using the patient with the health record given in Eqn. (2), and assuming that the $DG_1 = 120$ for the Quantitation of glycosylated hemoglobin test ($j = 1$). As shown previously, $T_f = 8765$ and $T_l = 10592$. The Gap between T_f and the first test is $9044 - 8765 = 179$ which is greater than DG_1, so it contributes $179 - 120 = 59$ to the sum. There are four time intervals exceeding DG_1 among the 12 time intevals. They are 228, 150, 174 and 180 days respectively. Their contribution to the sum is 108, 30, 54 and 60 days respectively. The last test was done on day 10510 and the gap between T_l and the last test is $10592 - 10510 = 82$, which is less than DG_1 and therefore contributes nothing to the sum. Therefore we have $G_{j,k} = 59 + 108 + 30 + 54 + 60 = 311$.

The advantage of this feature vector is that it is low-dimensional and easy to interpret. This feature is particularly useful when there is an expectation of regularity in the events and this regularity is to be explored.

5 Results

We used a model-based clustering program called Snob[7,8] using a Bayesian mixture-modelling method with a Poisson distribution for the *count* feature vectors and a log-normal distribution for the *average, residual, deviance* feature vectors.

5.1 Clustering Using *Count*

Fig. 2 gives the means and membership size of the 23 clusters found using Poisson mixture models. The Poisson distribution was suitable for these features because the counts are positive integers. Note that the counts are only approximately Poissson, because very large counts of services do not occur at all in practice. We now interpret these clusters. First recall that to receive care conforming to clinical guidelines for the *Gl* test over five years you would need between 10 and 15 tests. The population mean is 5 tests. It is apparent from Fig. 2 that most individuals do not receive conforming care because the large membership clusters (e.g., 4, 5, 6, 7) have mean counts below 6. Only two clusters (e.g., 2 and 20) have means of 10 or more. Cluster 2 individuals do not conform on the other two tests because they have means less than 3 for *Op* and *Cl*. In contrast, Cluster 20 individuals receive better than conforming care for all three tests. The 70 individuals in that group are apparently better looked after than all the others. The next best groups for all three tests are clusters 15, 16 and 17.

In follow-up work we plan to examine the characteristics (e.g., number of GP consultations, whether they are in the community or a nursing home, number of days in hospital) of individuals in the very 'good' and very 'bad' clusters to see if they can be distinguished from those receiving other patterns of care.

5.2 Clustering Using *Mean, Residual, Deviance*

Fig. 3 gives the means and standard deviations of the 17 clusters found using log-normal mixture models. The log-normal distribution was suitable for these features because the mean, residual and standard deviation have positive continuous values. A mean interval of six months or so is indicative of care conforming to clinical guidelines for the *Gl* test. Looking at Fig. 3 we see that clusters 3, 6, 7, 8, 9, 10, 15 and 17 have mean intervals less than 10. Cluster 17 has less than 10 members and so we ignore it for the moment. Individuals in clusters 8 and 9 receive the best patterns of care for this population. The 700 cluster 3 individuals receive regular conforming *Cl* tests, but very infrequent *Op* tests. In follow-up work we hope to characterise these individuals further. It may be possible to devise a policy to improve their quality of ophthalmology care. Opposite to cluster 3, clusters 6, 7 and 9 receive frequent *Op* tests, but infrequent *Cl* tests.

At the other end of the quality of care, the 850 individuals in cluster 16 receive less care than the other individuals in the population.

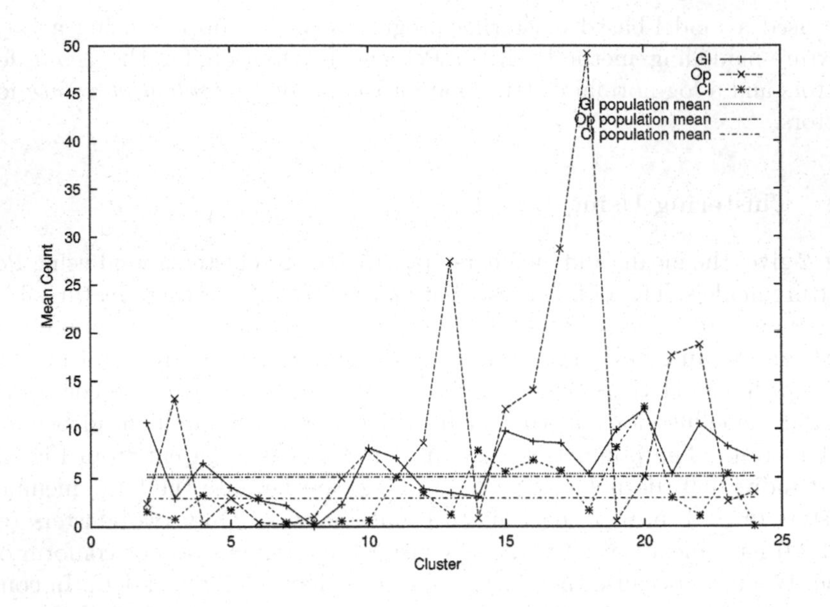

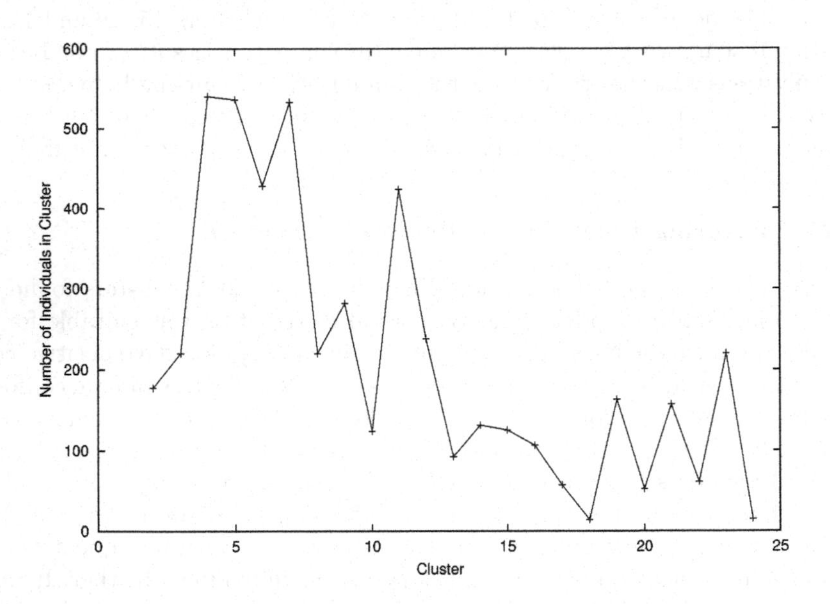

Fig. 2. Clustering results based on the *count* features.

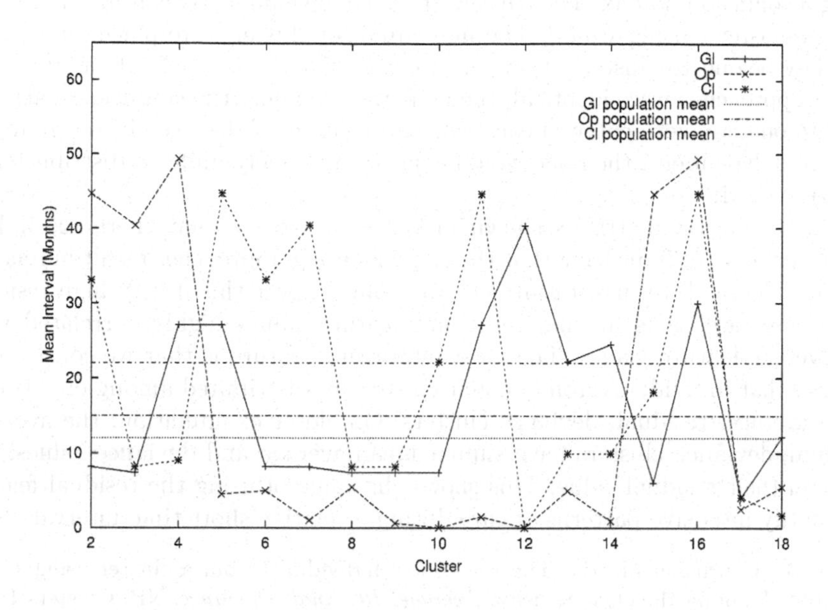

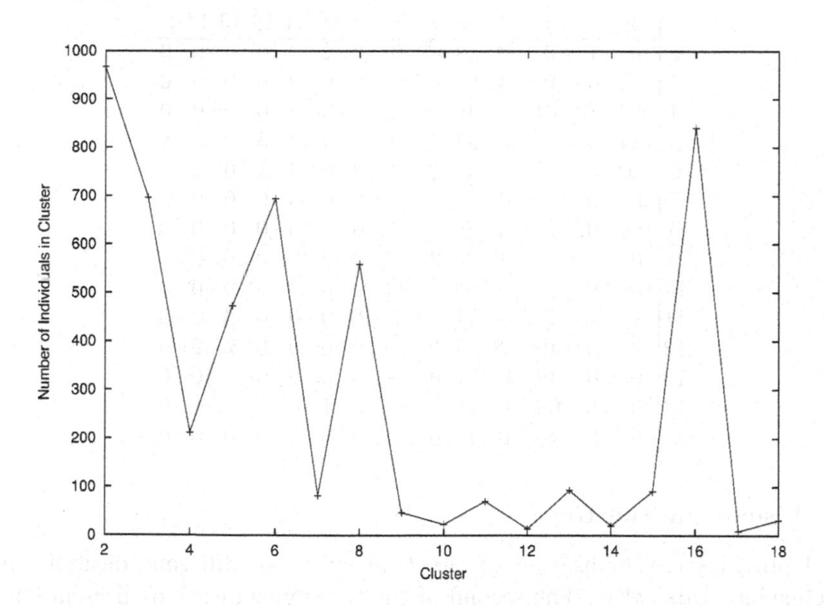

Fig. 3. Clustering based on /textitaverage, residual, deviance features. Residual and Deviance features were used in the clustering but are not shown.

We now compare the clustering results from Fig. 3 with those from Fig. 2 using a confusion matrix. The ith row of the confusion matrix contains the members of cluster i using *Count*. The individuals of cluster i are placed in column j if they belong to cluster j using *Average, Residual, Deviance.* If the two clustering approaches were identical, then one would expect the confusion matrix to contain one non-zero entry in each row and column. If the two clustering methods are independent, then one would expect a relatively uniform distribution of non-zero entries.

The confusion matrix is shown in Table 3. We see that there are indeed many zero entries indicating that the two clustering approaches result in related, but not identical, results. Intuitively, we would expect this if very high residual values are rare, thus making the count feature values highly correlated with the average feature values. The most interesting feature is that we consistently observe that individuals from a count cluster are distributed among one, two or three average, residual, deviance clusters. On closer examination, the average, residual, deviance clusters have similar mean average and deviance values, but differ in their residual value. This shows the value of using the residual feature to identify intensive patterns of care during a relative short time interval.

Table 3. Confusion Matrix. The rows show individuals from a cluster using *Count* distributed among the clusters using *Average, Residual, Deviance.* NB: Clusters 16–23 for *Count* and clusters 16–17 for *Average, Residual, Deviance* have been omitted.

	2	3	4	5	6	7	8	9	10	11	12	13	14	15
2	67	1	0	0	18	15	0	0	2	0	0	0	0	0
3	0	0	0	137	51	0	0	0	0	0	0	0	0	0
4	257	183	10	0	0	0	0	0	0	0	0	0	0	0
5	114	0	4	38	38	25	0	0	4	22	3	2	3	8
6	21	0	107	0	0	0	0	0	0	0	1	0	3	27
7	37	0	0	0	0	0	0	0	0	0	0	0	0	13
8	109	0	0	0	0	5	0	0	4	0	0	0	0	0
9	0	0	0	147	10	2	0	0	0	21	0	0	0	1
10	0	0	0	0	80	1	0	0	0	0	0	0	0	0
11	1	77	2	0	24	1	179	21	0	0	0	9	2	0
12	0	0	0	23	36	0	29	0	0	0	0	52	2	0
13	0	0	0	42	37	0	0	0	0	0	0	1	0	0
14	0	27	63	0	0	0	0	0	1	0	0	0	1	0
15	0	0	0	0	17	0	74	0	0	0	0	0	0	0

5.3 Visualising the *Gap*

Fig. 4 provides a visualisation of the *Gap* for three different desired clinical guideline intervals (*DG*). The second of the three visualisations presents of Fig. 4 sets $DG = 6$ months. The distinct mode at zero indicates good conformance with the guidelines. In all three visualisations there is a mode around 20–24 months, worthy of further investigation: Is there some structural feature in the health system that has patients receiving this test every two years, rather than not at all (in the worst case). In general we note a peak at $Gap = 0$ and another

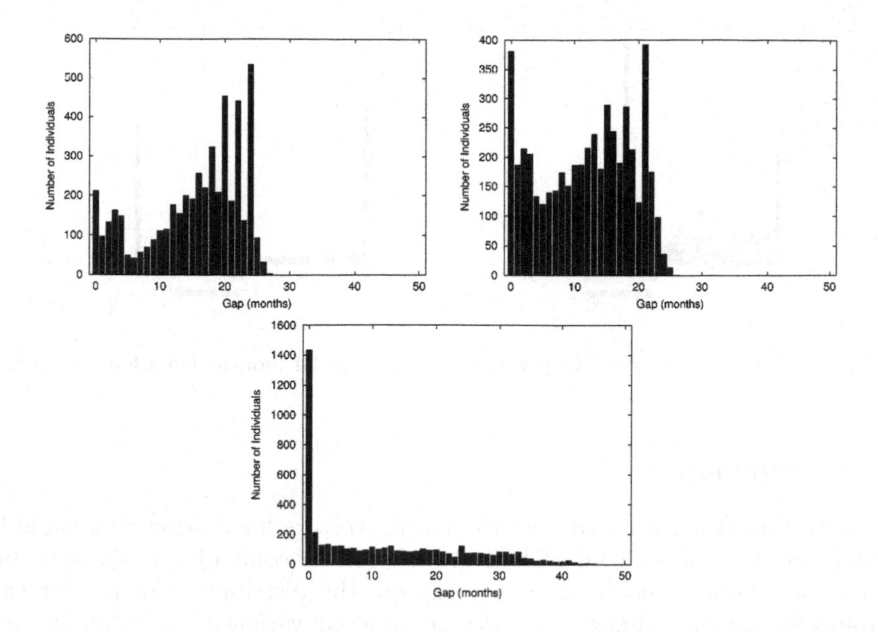

Fig. 4. Visualisation of *Gl* Gap with *DG* = 3, 6 *and* 12 months from left to right.

peak at the other end of the scale. These two peaks represent extremes: the first peak corresponds to conformance while the other peak corresponds to non-conformance to the guidelines.

Fig. 5 and Fig. 6 provide a visualisation of the *Op* and *Cl* tests for two different published clinical guideline intervals. At the right extreme are the individuals who did not receive any tests and so do not conform to the guidelines. At the left extreme are those individuals who conform to the guidelines. In between we see how patterns of care slowly degrade in terms of conformity. Note the mode at 12 months on the left-hand side panels (where *DG* = 12 months). This is indicative of the population of individuals receiving a precisely conforming pattern of care.

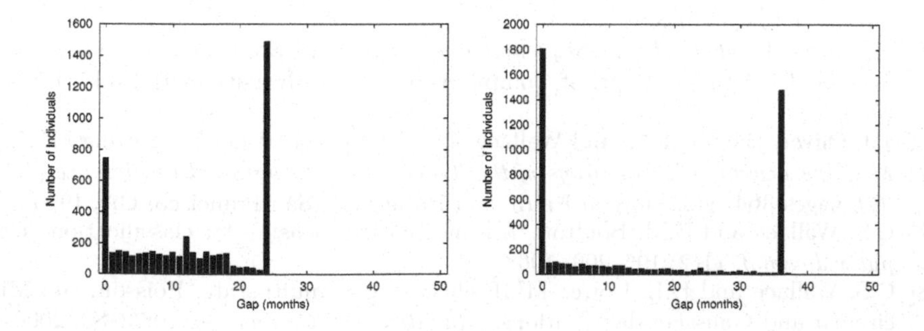

Fig. 5. Visualisation of *Op* Gap with *DG* = 12 *and* 24 months from left to right.

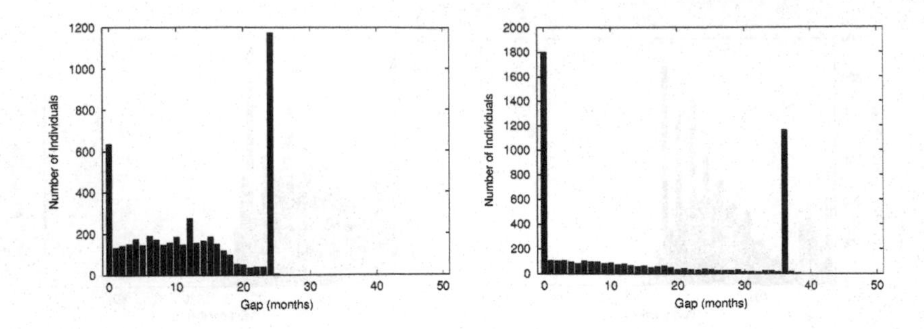

Fig. 6. Visualisation of *Cl* Gap with $DG = 12$ *and* 24 months from left to right.

6 Conclusion

We have considered three alternative feature vectors for representing variable-length patient health records. The feature vector of counts is the simplest, but can be misleading since it does not capture the distribution of patient care throughout the data window. The average, residual, variance feature vector overcomes this problem. For the specific task of characterising relationships to clinical guidelines, the gap feature vector most directly represents the required information. We expect the features created here for event sequence data for this health application will be applicable to other event sequence data such as trading and web log data.

References

[1] E. Arjas, H. Mannila, M. Salmenkivi, R. Suramo, and H Toivonen. Bass: Bayesian analyzer of event sequences. In *Proceedings in Computational Statistics (COMPSTAT'96)*, pages 199–204. Barcelona, Spain, Physica-Verlag, 1996.

[2] I. Cadez and P. Smyth. Probabilistic clustering using hierarchical models. Technical Report 99-16, Department of Information and Computer Science, University of California, Irvine, March 1999.

[3] A. Jain and R. Dubes. *Algorithms for Clustering.* Prentice-Hall, Englewood Cliffs, NJ, 1988.

[4] Huan Liu and Hiroshi Motoda. *Feature Selection for knowledge discovery and data mining.* Kluwer, 1998.

[5] P. Moen. *Attribute, Event Sequence, and Event Type Similarity Notions for Data Mining.* PhD thesis, Dept. of Computer Science, University of Helsinki, Finland, 2000.

[6] J.J. Oliver, Baxter R.A., and Wallace C.S. Unsupervised Learning using MML. In *Machine Learning: Proceedings of the Thirteenth International Conference (ICML 96)*, pages 364–372. Morgan Kaufmann Publishers, San Francisco, CA, 1996.

[7] C.S. Wallace and D.M. Boulton. An information measure for classification. *Computer Journal*, 11(2):195–209, 1968.

[8] C.S. Wallace and D.L. Dowe. MML clustering of multi-state, Poisson, von Mises circular and Gaussian distributions. *Statistics and Computing*, 10:73–83, 2000.

Boosting the Performance of Nearest Neighbour Methods with Feature Selection

Shlomo Geva

Smart Devices Laboratory
Queensland University of Technology
GPO Box 2434
Brisbane Q 4001
Australia
s.geva@qut.edu.au

Abstract. This paper describes a Nearest Neighbour procedure for variable selection in function approximation, pattern classification, and time series prediction. Given a training set of input/output vector pairs the procedure identifies a subset of input vector components that effectively capture the input-output relationship implicit in the training set. The utility of this procedure is demonstrated with numerous data sets from the UCI repository of machine learning databases and the Mackey-Glass time series prediction. A comprehensive set of benchmark problems is used to demonstrate comparable performance to that of much more complex boosted C4.5 decision trees.

1 Introduction

Subset selection is a special case of feature extraction. Given a training set of input/output vector pairs, one assumes that the observed output is a function of some subset (not necessarily unique) of input vector components. The objective is then to identify such a subset. A related problem, which was more extensively studied in relation to pattern classification problems, involves the weighting of input variables. Each variable is assigned a weight so as to improve classification accuracy. For instance, with methods that are based on distance metric, such as Nearest Neighbour classifiers or Radial Basis Functions networks, it is often useful to apply a transformation to the input vectors in distance calculations. The Mahalanobis distance (D_ψ) supports an arbitrary linear transformation of the input vectors:

$$D_\psi(\mathbf{x},\mathbf{y}) = \| \mathbf{x} - \mathbf{y} \|_\psi = \sqrt{(\mathbf{x}-\mathbf{y})^T \psi(\mathbf{x}-\mathbf{y})}$$

A diagonal matrix ψ is often useful when the components of the input vector are independent. The use of D_ψ can sometimes produce better results than the use of Euclidean Distance, which is a special case of D_ψ with $\psi = \mathbf{I}$. In practice however, it is usually difficult to determine ψ on the basis of a-priori knowledge, and therefore this is done implicitly by an adaptive optimisation procedure.

D. Cheung, G.J. Williams, and Q. Li (Eds.): PAKDD 2001, LNAI 2035, pp. 210-221, 2001.

Improvements in classification accuracy by several weight selection procedures was previously reported [2, 6, 9].

The Wrapper method [5] was also studied in the context of pattern classification. In that approach variables are assigned weights which are optimised through a search in weight space (ie, D_ψ is used and a diagonal ψ is determined empirically.) The fitness of a particular set of weights is measured by the accuracy that is achieved by an underlying induction algorithm. The subset selection is 'wrapped' around the particular induction method, hence the name Wrapper. A comprehensive coverage of feature selection approaches can be found in [1].

A different subset selection problem arises in relation to time series prediction. A frequently used method for subset selection is based on *Takens Theorem* [10]. A subset of time delayed co-ordinates is used, where d is the embedding dimension:

$$\mathbf{x} = (x(t), x(t-\tau), x(t-2\tau), \ldots, x(t-d\tau))$$

This approach assumes that the future dynamical behaviour of the whole system is largely dependent upon the time series itself, up to the present time; this is imperative if the procedure is to be effective. Reference [14] describes the False Nearest Neighbour method that describes how to discover an appropriate embedding dimension for phase space reconstruction. In this paper we describe a related, by very different method for the determination of a suitable subset of attribute that is based on nearest neighbour analysis of the data. The method is applicable to classification as well as to function approximation and time series prediction problems.

2 Estimating the Nearest Neighbour Error

The quantity that we will be looking to minimise, in searching for an appropriate subset of variables, is the Nearest Neighbour **leave-one-out** error E. It is closely related to the often-used leave-one-out cross validation error measure. Here we define it a little more broadly than usual to also cover function approximation tasks with vector, rather than scalar, output. The error is calculated as follows:

1. Leave out one of the input/output pairs in the training set, say, $\{\mathbf{x}_i, \mathbf{y}_i\}$
2. Find $\mathbf{x}_n$ the nearest neighbour vector of $\mathbf{x}_i$ in the training set. The approximation error of $\mathbf{y}_i$ is now defined as

$$e_i = (\mathbf{y}_i - \mathbf{y}_n)^T (\mathbf{y}_i - \mathbf{y}_n),$$

where e_i is the error of approximating $\mathbf{y}_i$ at $\mathbf{x}_i$ by the function value $\mathbf{y}_n$ observed at $\mathbf{x}_n$

3. Repeat steps 1-3, leaving out each of the N training set examples in turn, to accumulate the global quantity E:

$$E = \sum_{i=1}^{N} e_i$$

The error E is an approximation of the nearest neighbour classification, or function approximation, error of the data set. E has a useful property that we can exploit in subset selection. Suppose that we have a training set derived with the **actual** subspace of M significant components in R^M. We compute the error E^M in this subspace. Now consider the addition of an irrelevant component to the input vectors, so that they lie in R^{M+1}. Computing the error E^{M+1} we observe a desirable property of E - one would generally expect to find that $E^{M+1} > E^M$. This is because the irrelevant component will cause the neighbourhood relationship among the vectors in R^{M+1} to be different to that in R^M. This means in **some** cases that the approximated function values will not be derived from the nearest neighbours in R^M but rather from other more distant vectors. On average we would expect the approximation error to increase. Even in the presence of noisy training examples this should hold on average, as confirmed by our experiments

3 Subset Selection

Two different approaches are commonly used in subset selection, Backward Sequential Selection and Forward Sequential Selection [1]. The BSS procedure is somewhat more exhaustive and slower but sometimes produces more reliable results.

3.1 Backward Sequential Selection (BSS)

Starting from a training set with input vectors in R^M we compute E^M in all M subspaces R^{M-1}, each of which has one of the original M components left out. Following from our argument in the preceding section, we expect that when an irrelevant component is dropped we will find that $E^M > E^{M-1}$. It is possible that more than one of the components can be left out with that result. We eliminate that component which when dropped leaves a subspace having the **lowest** value of E^M. The process is then repeated in the selected subspace until a stopping criterion is met. Our experiments reveal that the error can sometimes increase when a component is eliminated, only to continue and decrease as more components are eliminated. Therefore the process is repeated until all but one component is left. The subset along the path that gives the lowest leave-one-out nearest neighbour error E is selected.

In practice it may be that one is interested in dimensionality reduction to the extent that one is prepared to accept some decrease in approximation accuracy. In that situation it may be possible to identify a smaller than optimal subset which still provides an acceptable error rate.

This procedure gradually eliminates components in reverse order of significance and its time complexity is quadratic in the number of attributes and quadratic in the number of training instances. Our experiments show that it can be done in a computationally effective manner with data sets having hundreds of components and thousands of training instances. The ideal process of computing E in an exhaustive manner for all possible subsets of components is computationally prohibitive because of the size of the power set involved.

3.2 Forward Sequential Selection (FSS)

This procedure starts with an empty subset. Each of the M components are tested to find out which produces the lowest error, when used in isolation. That component is then added to the subset and each of the remaining $M-1$ components are tested, in conjunction with the already selected component, to discover the pair that leads to the lowest error. This process is repeated, adding one component at a time, until the approximation error starts to increase. The FSS procedure is more economical than BSS with data sets having a large number of variables and where the actual subset of significant components is relatively small. However, it is generally unknown in advance whether this would be the case with a given data set. Although FSS is usually more economical it is often argued that FSS might be less effective in discovering *combinations* of attributes, which BSS is more likely to preserve.

4 Experimental Results

To test the effectiveness of the procedure we experimented with many benchmark problems from the UCI repository of machine learning databases and with the Mackey-Glass time series prediction problem in a function approximation scenario. Often with real world problems, one encounters the problem of missing values and the choice of a metric for nearest neighbour calculations. We have adopted several strategies to encounter these difficulties. In the case of missing values the distance calculations are performed in the available subspace. This can lead to some anomalies, but our experiments show that despite this the procedure is robust and performs well. All ordered numeric variables (discrete or continuous) were normalized to the range 0..1. The Euclidean distance calculations were then carried out as usual. Symbolic variables are often encoded as discrete numeric, but have no ordering associated with the actual values. With symbolic domains the distance calculation was based on the *hamming distance* (the number of different symbols). In problems where a mixture of numeric and symbolic variables were involved the distance calculation was also mixed. With this we were able to deal with all of the data sets in the UCI repository.

4.1 Pattern Classification

We have experimented with 18 data sets from the UCI repository. These data sets cover many different types of problems having discrete, continuous, and symbolic variables. Some data sets have missing values, and some have a mixture of all the above. The 18 data sets are listed in Table 1. The rightmost column lists the size of the selected subset, averaged over 10 runs of BSS.

Significant reduction in the number of attributes was obtained in all cases. The important question though is whether this reduction also leads to improved accuracy. We compare the classification accuracy results, of a nearest neighbour classifier – in the selected subspace - with results obtained with **C4.5** (Quinlan's implementation **See5**). It should be noted we used default training parameters to test **C4.5**. However, the error rates we obtained (Table 2) agree with results previously reported in [11]. We have also used the boosting option with a committee of 10 classifiers. Boosted **C4.5** classifiers produced better results than **C4.5** without boosting. **C4.5** [12] is a well-understood tree building procedure and it is used in numerous publications as a yardstick to performance comparisons. All tests were conducted using 10-fold cross validation (10-fold-XV). The data was partitioned into 10 disjoint subsets having similar statistics with respect to class membership proportions. Using the training data set alone we then carried out subset selection. Only the variables that were selected as relevant were then used to test the training set as a Nearest Neighbour classifier (in the reduced subspace) against the held-out data. The results, averaged over all 10 experiments are depicted in Figure 1. The performance of Nearest Neighbour classifier, in a suitably reduced subspace, compares well with boosted C4.5.
The error rates of both procedures usually fall within one standard deviation from each other. This is a remarkably consistent result that seems to have been overlooked in the past. It demonstrates that a nearest neighbour classifier can often perform as well as a sophisticated committee of decision trees – provided that a suitable subset selection procedure is applied beforehand. The results depicted here correspond to the BSS procedure, but these are similar to the results obtained with the FSS procedure. In none of the data sets did we observe BSS to discover a subset of attributes that was not also discovered by FSS. Therefore, the conjecture that BSS is superior in discovering combinations of attributes is not confirmed by our results.

To appreciate the improvement in Nearest Neighbour classification due to subset selection we also tested the accuracy of a classifier that is based on the full training set (i.e., in the full input space available with no subset selection). Again, we used the same 10-fold cross validation partitions to classify each partition by the remaining 9 partitions. In addition we also used LVQ [13] to obtain a Nearest Neighbour classifier having a single prototype vector from each class. This of course represents a very limited classifier whereby linear discriminant functions separate the classes. We refer to this classifier as a Linear Machine. We used the error rate of the linear machine as a base-line error rate (it is the simplest nearest neighbour classifier possible). The error rates of all the methods were normalized by the error rate of the Linear Machine and are depicted in Figure 2.

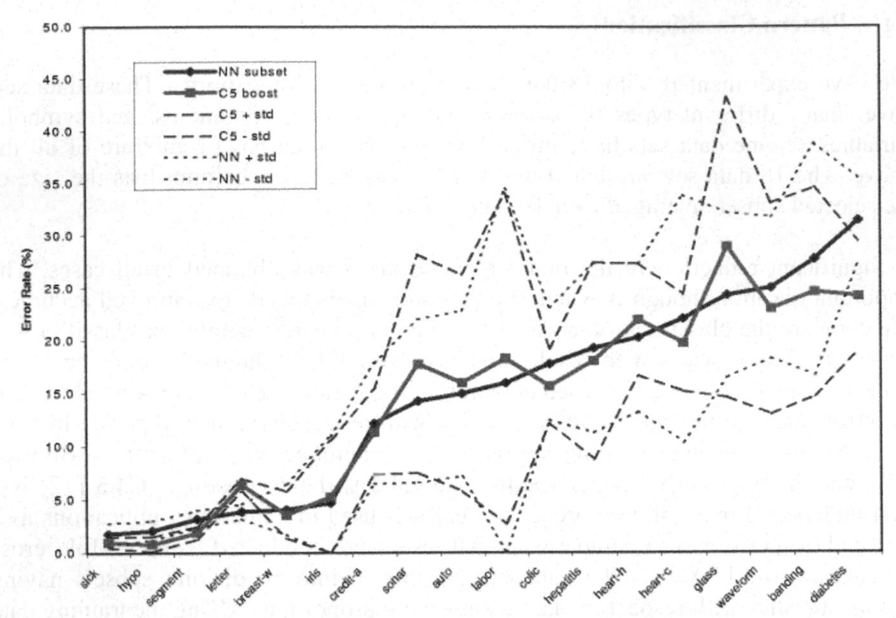

Fig. 1. 10-fold-XV error rates ± 1 standard deviation

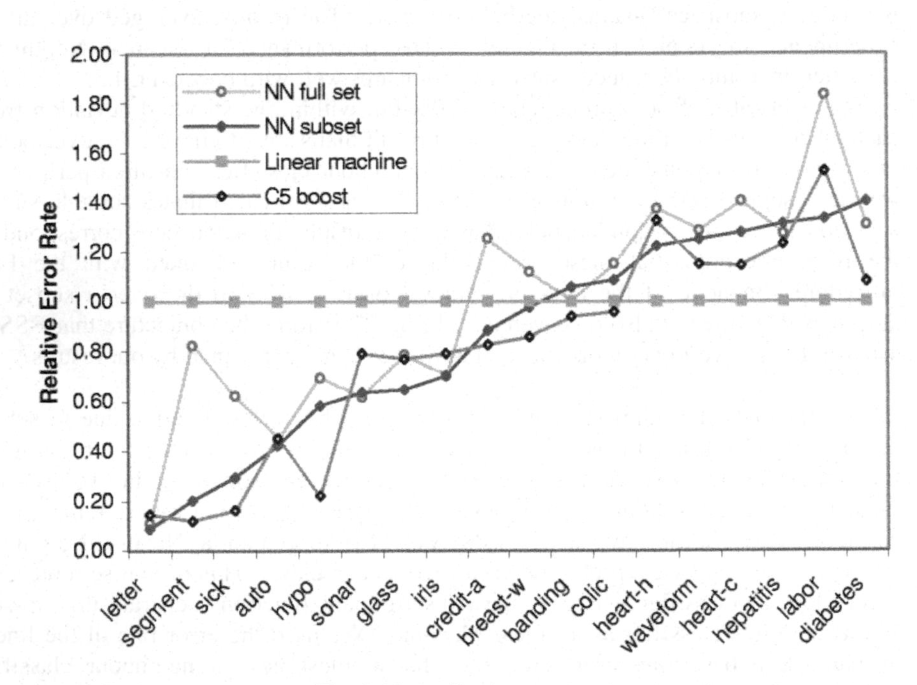

Fig. 2. Normalized error rates

Table 1. Data sets used in subset selection for pattern classification

Domain	Number of Cases	Number of classes	Number of attributes	Subset Size
Auto insurance	205	6	25	7.3
Breast cancer (Wisc)	699	2	9	5.7
horse colic	368	2	22	9.0
Credit screening (Aus)	490	2	15	9.1
Pima diabetes	768	2	8	4.9
Glass identification	214	6	9	5.7
Heart disease (Clev)	303	2	13	7.4
Heart disease (Hun)	294	2	13	2.7
Hepatitis prognosis	155	2	19	9.1
Hypothyroid diagnosis	3772	5	25	7.5
Iris classification	150	3	4	2.1
Labor negotiations	57	2	16	7.5
Image segmentation	2310	7	19	6.2
Sick euthyroid	3772	2	29	9.1
Sonar classification	208	2	60	12.2
waveform differentiation	300	3	21	13.6
Letter identification	20000	26	16	11.0
Banding	138	2	29	13.5

The complete results of 10-fold-XV tests are given in Table 2 where the best figures for each data set are shaded. It is evident that subset selection significantly improves the performance of a nearest neighbour classifier, which is otherwise significantly outperformed by a boosted C4.5 classifier. Perhaps more interesting is the fact that there are several data sets, on the right-hand side of the Figure 2, for which not only there is no advantage in using a non-linear classifier, but there is actually a distinct disadvantage to doing so. It also shows that boosting was not effective, in our experiments (using ensembles of 10 classifiers), in avoiding over-fitting of training data.

4.2 Function Approximation

This Mackey-Glass benchmark problem has been widely used to test time series prediction methods. Lapedes and Farber [7] have used a multilayer perceptron trained on a short sequence of 500 points to predict future values of the series. Farmer and Sidorowich [3] have used a form of local regression to tackle the same problem. Given a segment of the time series sequence their procedure finds a set of similar segments in a long stored sequence. Prediction from the current sequence into the future then follows by linear regression on those similar past segments and the way ***they*** evolved.

Table 2. Pattern classification error rates (%)

Domain	NN full set	NN subset	C5 boost	C5	Linear machine
Sonar	13.9	14.3	17.8	22.0	22.5
Letter	4.9	3.9	6.4	13.0	43.9
Glass	29.5	24.3	28.9	30.0	37.6
Auto	16.0	15.0	16.0	17.6	35.5
Iris	4.7	4.7	5.3	5.3	6.7
Segment	13.0	3.2	1.9	2.9	15.8
Sick	3.8	1.8	1.0	1.3	6.1
Hypo	2.5	2.1	0.8	0.9	3.6
Credit-a	17.3	12.2	11.4	12.3	13.8
Breast-w	4.7	4.1	3.6	4.4	4.2
Banding	26.2	27.7	24.6	27.6	26.2
Colic	18.9	17.8	15.7	16.3	16.4
Heart-h	22.8	20.3	22.0	21.7	16.6
Waveform	25.7	25.0	23.0	25.7	20.0
Heart-c	24.3	22.1	19.8	25.1	17.3
Hepatitis	18.7	19.3	18.1	18.7	14.7
Labor	22.0	16.0	18.3	22.0	12.0
Diabetes	29.3	31.4	24.2	25.8	22.4

The Mackey-Glass differential delay equation is defined as:

$$\frac{dx}{dt} = -bx(t) + a\frac{x(t-\tau)}{1+x(t-\tau)^{10}}$$

We used the values $a=0.2$ and $b=0.1$ As the value of τ is increased the series exhibits a more chaotic behaviour. We have experimented with a value of $\tau=30$. The usual approach to the solution of this problem involves the prediction of $x(t+T)$) from a vector of $m+1$ components: $\{x(t), x(t-\delta), x(t-2\delta),..., x(t-m\delta)\}$. The values that are typically used for the prediction of the Mackey-glass equation with the choice $\tau=30$ are $\delta = 6$ and $m = 5$.

To demonstrate the utility of subset selection we tackled the problem *without* the knowledge of a suitable subset. Starting from a past sequence, we discover a suitable set of attributes to use in function approximation. We created a training sequence of 500 50-dimensional vectors by sliding a window of size 50 along the given training sequence, with the value to predict being the next point, i.e. one step ahead. The input to the subset selection procedure was $\{x(t), x(t-1), x(t-2),..., x(t-49)\}$ with the function value sought being $x(t+1)$. We conducted many experiments with statistically independent sequences of 500 points, using both BSS and FSS. The FSS procedure proved superior on this data set always converging on the subset $x(t)$ and $x(t-30)$. The BSS results varied slightly between experiments. We consistently obtained 5 or fewer attributes, which always included values from the sets $\{x(t), x(t-1), x(t-2)\}$ and $\{x(t-29), x(t-30), x(t-31)\}$ It was usually, but not always, the case that $x(t)$ and $x(t-30)$ were the most significant - as one would expect - given the form of the Mackey-Glass equation (with longer sequences it is easy to obtain this result). A plot of the nearest neighbour error as subset selection progressed with BSS is depicted in Figure 3. It demonstrates how the prediction error decreases while irrelevant attributes are removed, and how a sharp rise is observed when significant attributes are removed.

The False Nearest Neigbours (FNN) procedure is a related method that was designed to obtain the optimum embedding dimension for phase space reconstruction. [14]. Unlike our procedure, FNN requires a pre-determined delay constant (spacing between points), and it then discovers the number of delays that are required. Therefor, FNN is a more restricted procedure in that it requires prior analysis to determine the delay constant.

To test the result obtained with BSS, which did not always converge on the ideal subset, we intentionally selected the input sequence $x(t-2)$ and $x(t-31)$ which appeared as the result in one of the subset selection experiments. This is *not* the set of inputs that appear in the Mackey-Glass differential equation for $x(t+1)$. The prediction accuracy for this subset was tested with 3 different methods of function approximation in predicting the time series from 1 to 400 steps ahead.

A multilayer perceptron with two hidden layers (2:10:10:1) was trained to predict one step ahead using a PC with a MATLAB toolbox implementation of the Levenberg-Marquardt training algorithm (a solution was obtained in less than 1 minute on a PC). The neural network was tested in predicting the future evolution of the series by using the iterated prediction method. Starting from an initial sequence, the next (future) value is predicted by the network. This predicted value is then appended to the initial sequence, which is then used to predict the next value in the sequence. This process is repeated to obtain 400 time steps ahead prediction. We conducted a set of 100 independent prediction experiments. Each experiment was performed with a statistically independent training sequence of 500 points and test sequence of 1000 points. Figure 4 depicts the results obtained showing the mean error rates over 100 independent experiments. The Error Index is plotted against the prediction time step. The Error Index is the root mean squared error, divided by the standard deviation of the sequence.

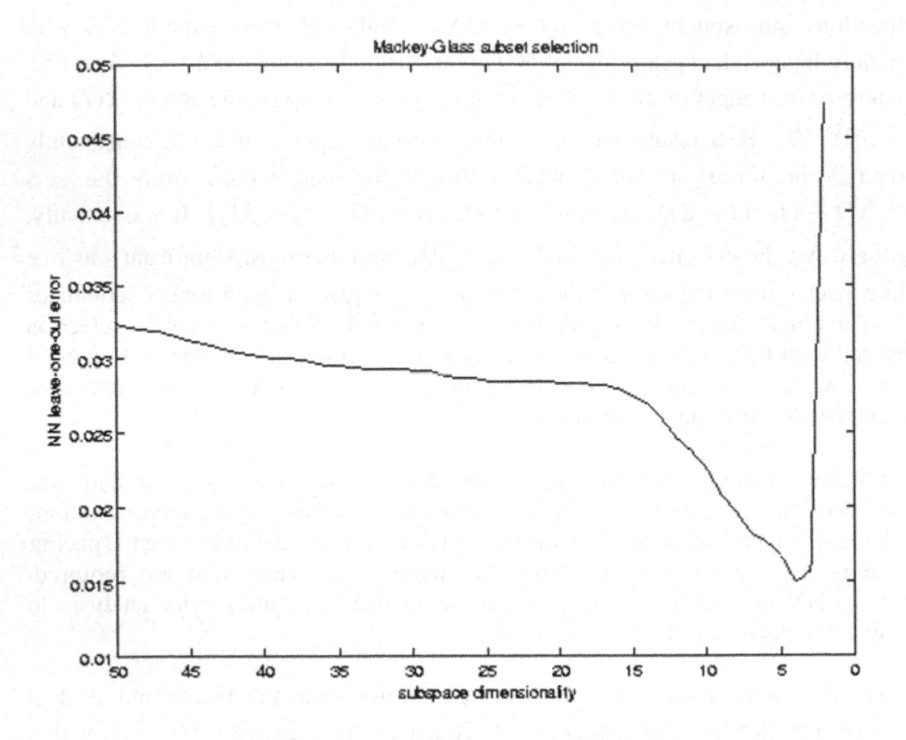

Fig. 3. Backward Sequential Selection for Mackey-Glass time series

The results obtained with the neural network are very similar to the results obtained by Lapedes and Farber using the same network architecture [11]. However, our results were obtained with the input sequence { $x(t-2), x(t-31)$ } obtained empirically by DSS. The results of Lapedes and Farber were obtained with the extended sequence of input vector components identified by analysis of the properties of the Mackey-Glass chaotic time series properties and using *Takens Theorem* Using { $x(t), x(t-6), x(t-12), x(t-18), x(t-24), x(t-30)$ } -the sequence contains the actual variables that were used to generate the series.

Also depicted in Figure 4 is the result of iterated nearest-neighbour approximation. In this approach we simply predict the next value in the sequence by using the value observed to follow the nearest neighbour vector in the training sequence. Iterated prediction is then used to predict multiple time steps ahead. Although less accurate than the neural network - as one would expect from a piecewise constant approximation of a continuous function - the approximation is still rather useful in selecting relevant variables.

The third approach that we used is the regression approach of Farmer and Sidorowich [3]. Like the Nearest Neighbour approach it involves identification of a subset of past sequences that are most similar to the current sequence. The future value is then computed by the application of linear regression using the past sequences and *their* future values. Farmer and Sidorowich used much longer sequences than were used by

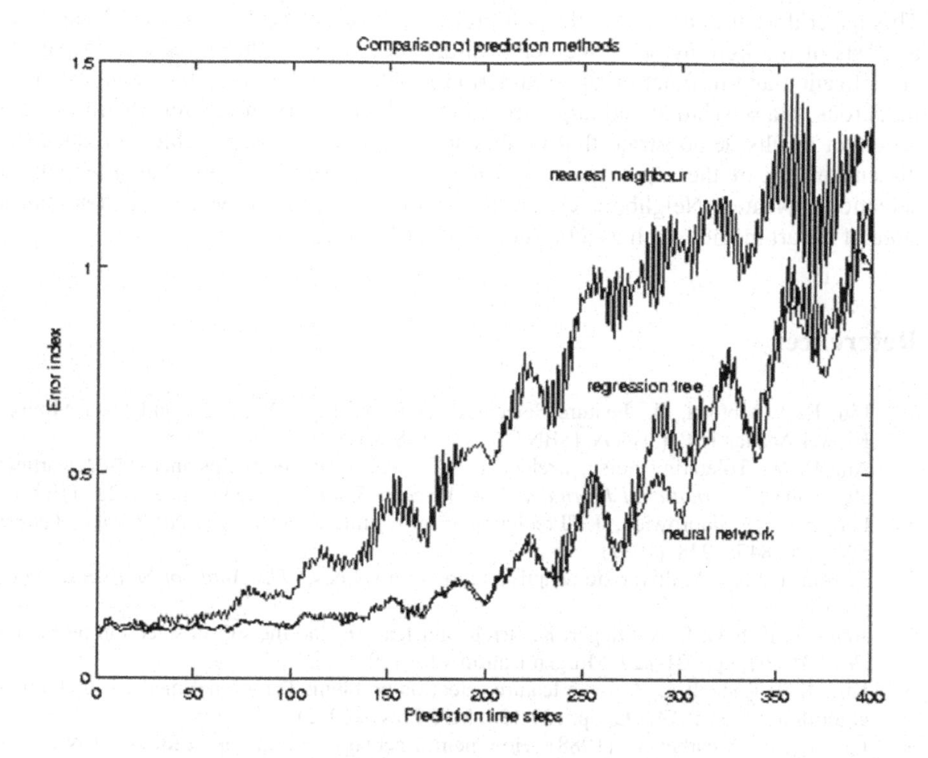

Fig. 4. The Error Index averaged over 100 independent segments predictions

Lapedes and Farber, and they did not use the iterated prediction method. To obtain a more meaningful comparison we used the same short training sequences of only 500 points in implementing the regression method. The relatively large number of points, presumably required for the regression method to work, was one criticism leveled at this approach, and we wanted to test the method with the same length sequence as the neural network. A search tree was used to locate nearest neighbour sequences so that searching for past sequences required very few distance calculations at tree nodes. To predict up to 400 steps ahead we again used the iterated prediction method. The results are depicted in figure 4. The regression method is only slightly less effective than the neural network. Lapedes and Farber indicated that both methods seem to perform similarly and our experiments confirm it by this direct comparison. However, by keeping all conditions equal, our results demonstrate that the regression method does not require larger training sets and longer sequences than the neural network does in order to perform as accurately – provided that a suitable subset of variables is used.

5 Conclusion

This paper describes a subset selection procedure that is based on Nearest Neighbour analysis of the training set. We demonstrate the utility of the procedure in pattern classification and in function approximation problems. The procedure was tested on numerous data sets, small and large, from the UCI repository of machine learning data sets. Our results demonstrate that in almost every case it was possible to reduce the dimensionality of the input space. Furthermore, our results show that after subset selection a Nearest Neighbour classifier can often perform as well or better than a state of the art method such as a boosted C4.5 classifier.

References

1. Liu, H. & Motoda, H., Feature Selection for Knowledge Discovery and Data Mining, Kluwer Academic Publishers, ISBN 0-7923-8198-X (1998)
2. Aha D. W. Tolerating noisy, irrelevant and novel attributes in instance-based learning algorithms," *International Journal of Man-Machine Studies*}," **36**(1), pp. 267-287 (1992)
3. Farmer, j. & Sidorowich, J. "Predicting chaotic time series," *Physical Review Letters* **59**(8), pp. 845—948 (1987)
4. Friedman, J.H., "Multivariate adaptive regression splines," *The Anals of Statistics*, **19**(1), pp. 1-82 (1991)
5. John, G., Kohavi, R. & Pfleger, K. "Irrelevant features and the subset selection problem," *Proc. ML-94*, pp. 121-129, Morgan Kaufmann. (1994)
6. Kira, K. & Randell, L. A. "The feature selection problem: Traditional methods and a new algorithm," *Proc. AAAI-92*, pp. 129-134, MIT Press (1992)
7. Lapedes, A. & Farber, R. (1988) "How neural networks work," proceedings of NIPS ed. D. Anderson *Neural Information Processing Systems,* pp. 442-456.
8. Mackey, M. & Glass, L. "Oscillation and Chaos in Physiological Control Systems," *Science* **197**, pp. 287-289 (1977)
9. Scott, S. & Salzberg S. "A weighted nearest neighbour algorithm for learning with symbolic features," *Machine Learning* **10**(1), pp. 57-78 (1993)
10. Takens, F. "Detecting strange attractors in turbulence," in *Dynamical Systems and Turbulence*. Eds. D. Rand. and L. Young (Springer, Berlin) pp. 366-381 (1981)
11. Quinlan, J.R. "Improved Use of Continuous Attributes in C4.5," *Journal of Artificial Intelligence Research*, vol 4, pp. 77-90 (1996)
12. Quinlan, J. R.. C4.5: Programs for Machine Learning. Morgan Kaufmann. San Mateo, CA. (1993)
13. Kohonen T. Self-Organization and Associative Memory, 2nd Ed, Springer-Verlag, ISBN 3-540-18314-0 2 (1988)
14. Kennel M.B, Brown R., and Abarbanel H., Determining embedding dimension for phase-space reconstruction using a geometrical construction. Phys. Rev. A, 45(6), pages 3403-3411 (1992)

Feature Selection for Meta-learning

Alexandros Kalousis and Melanie Hilario

CSD University of Geneva,
CH-1211 Geneva 4, Switzerland
{kalousis,hilario}@cui.unige.ch

Abstract. The selection of an appropriate inducer is crucial for performing effective classification. In previous work we presented a system called NOEMON which relied on a mapping between dataset characteristics and inducer performance to propose inducers for specific datasets. Instance-based learning was applied to meta-learning problems, each one associated with a specific pair of inducers. The generated models were used to provide a ranking of inducers on new datasets.
Instance-based learning assumes that all the attributes have the same importance. We discovered that the best set of discriminating attributes is different for every pair of inducers. We applied a feature selection method on the meta-learning problems, to get the best set of attributes for each problem. The performance of the system is significantly improved.

1 Introduction

One of the most difficult problems in the machine learning field is that of the appropriate selection of a classification algorithm for a specific classification task. As it is known from the various NFL theorems [31], there is no classification algorithm that is superior over all the others for all the possible classification problems.

In the past the problem was tackled mainly through the use of measures that describe properties of the dataset on which the classification task is to be performed. These dataset measures, along with performance measures of the classification algorithms, are used as data to (meta-)learn models that guide the application of the classification algorithms.

There has been a substantial amount of research effort that follows the previous mentioned approach, using different sets of dataset characteristics, and different ways of constructing the meta-level problems and the meta-models. Nevertheless little attention has been given to the usefulness and the discriminating power of the dataset characteristics used, with the only notable exception of [28]. In this paper we examine the meta-feature selection problem in the context of pairwise comparisons between base-level learning algorithms. We believe this setup gives us a better insight into the usefulness of each characteristic. Our goal is twofold: first to improve the overall performance, and second to acquire a better understanding of the way the dataset characteristics affect the relative performance of the algorithms.

D. Cheung, G.J. Williams, and Q. Li (Eds.): PAKDD 2001, LNAI 2035, pp. 222–233, 2001.

2 Background

There is a huge amount of literature -both theoretical and empirical- on model / algorithm selection for a specific classification task. We restrict our attention to empirical work, and more specifically to approaches that handle the problem of model selection as a learning problem on the meta-level (meta-learning). Here the goal of meta-learning is the construction of meta-models that associate dataset descriptions with algorithm performance.

The idea of meta-learning appears for the first time, in a very simple form, in the work of Rendell et al,[22]. Meta-learning regains attention as a byproduct of the STATLOG project [17]. STATLOG's main goal was the study of the performance of a variety of classification algorithms on different problems and the acquisition of knowledge on what works well and where. To this end, they tried to explain the performance of the algorithms with respect to dataset properties. These properties included mainly statistical and information based measures taken on the datasets. They went one step further and automated the process by constructing meta-learning problems, one for each classification algorithm. Algorithms were characterized as applicable or non-applicable for each dataset, and it was that characterization that they tried to predict and explain using the dataset properties, (see [3], for a thorough description).

The approach followed in STATLOG had two main limitations. First a dataset's properties were mainly means of statistical and information theoretical measures computed overall its attributes, due to the restrictions imposed by propositional learning. Second for a given dataset, algorithms were characterized only as applicable or non-applicable, i.e. they do not provide a way to rank the algorithms; furthermore, that characterization was based on a simple comparison of accuracies devoid of any statistical significance test.

Kalousis and Theoharis in [10] tried to overcome those limitations. To alleviate the problem of the constraints imposed by propositional learning they introduced histograms to describe the distributions of the measures computed for each attribute. To provide a ranking of the algorithms and to statistically control it, they created $\binom{n}{2}$ meta-learning problems, n being the number of algorithms. Each meta-learning problem concerns a specific pair of algorithms. For every dataset in order to determine which algorithm from a pair was better, a statistical test was employed. This formulation of meta-learning problems provides great flexibility. It allows the analyst to focus on specific pairs of algorithms and handle each pair in a different way, e.g. by applying a different learning algorithm on the meta-level, or even by using a different set of discriminating features. In short the power of the approach comes from the possibility of combining different meta-learning models.

A very elegant way to make full use of the information contained in the statistical and information theoretical measures is inductive logic programming, which overcomes the representational limitations of propositional learning [27].

Soares and Brazdil [25] provide a method for ranking, not only in terms of accuracy, but in terms of the accuracy-time tradeoff. The ranking is produced from a ratio of accuracy over time estimated from the performance of the classifiers averaged on a number of datasets. Their approach is similar to the DEA approach (Data Envelopment Analysis) [18] which also takes into account various performance measures of the algorithms. Both approaches are very effective rankers, but when they are used in a meta-learning context their flexibility diminishes. To produce a ranking for a new dataset, a set of similar datasets has to be examined (similar in terms of the dataset properties). This imposes the use of a single and global meta-learning model, which may only be based on a k-nearest neighbor algorithm. Furthermore as it has been shown in a previous study [12], a nearest neighbor algorithm is not necessarily the best choise, as the inducer of the meta-level.

A new approach to dataset characterization appeared in Phahringer et al [19]. Instead of using statistical and information based measures to describe the datasets, they use the performance of very simple learning algorithms, which they call landmarkers. A crucial factor here is the appropriate selection of the landmarkers, which should cover a wide range of learning biases, and at least be representative of the learning biases of the base learners. The results presented in that paper seem quite promising.

3 Meta-feature Selection

A dimension that has received little attention, if any, in the meta-learning field, is the explanation and the understanding of the factors that affect inducer performance. All previous efforts have aimed at maximizing the predictive capabilitics of the meta-learner without understanding the factors (i.e. properties of the datasets) that affect the performance of the algorithms. Applying feature selection to the meta-level can cover this gap and at the same time improve meta-learning performance. Using feature selection we can have a better idea of the factors that affect the performance of the learners. This is especially true when the meta-learning algorithm used is an instance based learner, which gives no insight into the relevance of the atributes used for learning.

The first attempt at meta-feature selection appeared in the meta-learning framework of zooming-ranking [28]. As it was mentioned previously the main limitation of this framework, is the mandatory use of a single and global meta-learning model (an instance based model). However the factors that determine the relative performance of a group of algorithms, may be quite different from the factors that determine the relative performance of another group of algorithms. It is exactly this case that calls for use and combination of different meta-learning models. Here the diversity of the meta-models comes from the use of different sets of meta-features (i.e. dataset characteristics). This is where we can make full use of the flexibility of the meta-learning framework of [10]. By applying feature selection to pairwise comparisons of learning algorithms we get different sets of meta-features which will give rise to different meta-learning problems and finally to different meta-learned models.

4 Feature Selection

Two are the main ways that feature selection is performed in machine learning, the *filter* and the *wrapper* approach [16]. In the filter approach, solely properties of the datasets are used in order to perform the selection of the features. These properties could be measures of association between features, measures of distance or dependence.

In the wrapper approach [14], the driving force is the accuracy of the learning algorithm that is going to be applied on the dataset. An extensive and systematic search is performed in the state space of all the possible feature subsets using heuristic search methods, like hill climbing, simulated annealing or best first. The search can begin either from the full set of features (*backward elimination*) or from the empty set (*forward selection*). The feature selection algorithm conducts the search using the estimated accuracy of the induction algorithm as the evaluation function. At the end, the feature set achieving the highest accuracy is selected.

In this study we have chosen to use the wrapper approach to perform the feature selection on the meta-level. Although it requires a substantial amount of computational time, in the case of meta-learning this factor is not so important, since it will only be perfrormed once.

5 Experimental Setup and Results

We used a variety of learning algorithms as the base learners whose relative perfromance we try to predict. An orthogonal decision tree inducer from Quinlan's C5.0 (c50tree), an oblique decision tree inducer Ltree [8], two rule inducers–Ripper [5] and the rule version of C5.0 (c50rules), a linear discriminant (lindiscr), a boosting algorithm from C5.0 (c50boost), an instance-based learner (IBL), and Naive Bayes (NB) (the last two from the MLC++ library [13]). To perform feature selection we used MLC++ feature selection capabilities. We started the search from the full set of characteristics, thus using backward elimination. The search strategy used was best first search. The evaluation function for the quality of a state in the search space (i.e. a subset of features) was the accuracy of the instance-based inducer as it is estimated by 10-fold cross validation.

We used 1082 datasets, including benchmarks from the UCI repository [30] as well as artificial datasets generated by modifying the former. The modified datasets were produced in the context of two big scale studies designed to explore the behavior of the learning method used in response to two additional dataset deficiencies, namely missing values and irrelevant attributes. For a complete description of the way that those datasets were created see [11,9].

Since one of the main goals is to predict which inducer(s) to use, we measure performance in terms of predictive accuracy. Accuracy is estimated not only for the final suggestions, but also for the individual meta-models extracted from the meta-learning problems, since these critically affect the performance of the whole system. We give results both for meta-models that have been created with feature selection and for meta-models that have been created with no feature selection, and compare the two approaches.

5.1 Dataset Characteristics

By *dataset characteristics* we mean a set of structural characteristics of a dataset that jointly determine the performance of an inducer when applied to it. They constitute the attributes of the meta-learning problems. The full set of characteristics is presented in Table 1. Characteristics 8 to 11 give the maximum, minimum, mean and standard deviation of distinct values for the nominal attributes. The *concentration coefficient* is a measure of association between nominal attributes [1]. The non computable features are associated with pathological cases where for various reasons the corresponding measure cannot be computed. Binary attributes is the number of binary attributes obtained, when all nominal attributes are represented with binary encoding. *Canonical correlation* is the first canonical correlation between a linear combination of the class variable and a linear combination of the attributes. *Fract* is the proportion of total variation that is explained by the first canonical discriminant. *Equivalent number of attributes* is the number of attributes required to describe the class, assuming that they all have the same mutual information with the class, that equals to the mean mutual information. *Noise to signal ratio* is a rough indication of the noise contained in the dataset. *Mean multiple correlation* denotes the average correlation between each attribute and a linear combination of all the other attributes. Finally *SDratio* is a measure of the homogeneity of the covariance matrices of the different classes. A complete description of the characteristics can be found in [10,17].

Table 1. Dataset Characteristics.

1	# classes	45	non comput. correl. hist.
2	# attributes	46..55	missing values histogram
3	# instances	56,57	$\frac{\#\text{ continuous}}{\#\text{ attributes}}, \frac{\#\text{ nominal}}{\#\text{ attributes}}$
4	$\frac{\#\text{ attributes}}{\#\text{instances}}$	58	Binary Attributes
5	# unknown values	59	Fract
6	$\frac{\#\text{ unknown values}}{\#\text{ attributes} * \#\text{ instances}}$	60	Canonical Correlation
7	# nominal attributes	61	Mean Skew
8..11	max, min, mean, stdv of nominal attribute values	62	Mean Kurtosis
12..21	concentration histogram	63	Class Entropy
22	non computable conc. histogram	64	Mean Attribute Entropy
23..32	concentration histogram with class	65	Mean Mutual Information
33	non comput. conc. hist. with class	66	Equivalent number of attributes
34	# continuous attributes	67	Noise to signal ratio
35..44	correlation histogram	68	Multi Attribute Correlation
		69	SDratio

Table 2. Class Distributions for each of the meta-learning problems.

(algo–x, algo–y) pairs	algo–x	algo–y	tie	majority
c50rules c50boost	4.47%	37.40%	58.14%	58.14%
c50tree c50boost	3.91%	38.33%	57.77%	57.77%
c50tree c50rules	13.02%	13.95%	73.02%	73.02%
lindiscr c50boost	3.63%	64.47%	31.91%	64.47%
lindiscr c50rules	12.09%	52.65%	35.26%	52.65%
lindiscr c50tree	10.98%	54.70%	34.33%	54.70%
ltree c50boost	13.21%	35.16%	51.63%	51.63%
ltree c50rules	27.07%	14.23%	58.70%	58.70%
ltree c50tree	25.21%	15.26%	59.53%	59.53%
ltree lindiscr	61.12%	6.51%	32.37%	61.12%
IBL c50boost	1.02%	64.65%	34.33%	64.65%
IBL c50rules	10.79%	49.77%	39.44%	49.77%
IBL c50tree	7.91%	52.00%	40.09%	52.00%
IBL lindiscr	42.33%	21.21%	36.47%	42.33%
IBL ltree	9.30%	56.65%	34.05%	56.65%
NB c50boost	2.70%	60.84%	36.47%	60.84%
NB c50rules	14.33%	51.44%	34.23%	51.44%
NB c50tree	12.09%	52.93%	34.98%	52.93%
NB lindiscr	37.30%	21.58%	41.12%	41.12%
NB ltree	5.58%	58.23%	36.19%	58.23%
NB IBL	29.12%	34.79%	36.09%	36.09%
ripper c50boost	1.58%	50.98%	47.44%	50.98%
ripper c50rules	8.09%	32.00%	59.91%	59.91%
ripper c50tree	2.70%	37.21%	60.09%	60.09%
ripper lindiscr	46.14%	17.02%	36.84%	46.14%
ripper ltree	3.44%	43.35%	53.21%	53.21%
ripper IBL	34.42%	19.44%	46.14%	46.14%
ripper NB	41.30%	16.93%	41.77%	41.77%
mean				54.14%

5.2 Results on the Meta-learning Problems

From the 8 available learning algorithms a total of $\binom{8}{2} = 28$ meta-learning problems were created. Table 2 gives the distribution of classes (algo-x, algo-y, tie) for each of these problems. The last column specifies the percentage of the majority class; this will serve as the baseline or default accuracy against which to evaluate the accuracies estimated by the learned meta-models. If their performance is deemed acceptable, these models can then be used to provide a ranking of inducers for new datasets.

We used 10-fold cross-validation to measure the accuracy of the instance-based models generated for each of the meta-learning problems. For both cases (i.e. no feature selection, feature selection) we get a significant increase over the mean default accuracy. The improvement is 27.25% without and 30.40% with feature selection (Table 3). We compare the accuracy achieved with and without feature selection, for each meta-learning problem, using a McNemar test of significance. The results are given in table 3. As we can see performance improves, by an average of 3.15%, when feature selection is performed. The difference in accuracy is statistically significant for all pairs of inducers.

Table 6 shows which characteristics were selected for some of the meta-learning problems. It is clear that the set of discriminating characteristics changes for different pairs of algorithms. Examining the selected characteristics for each pair of inducers we can only draw conclusions as to *which* factors impact their relative performance. However we cannot explain *how* these factors determine that relationship, i.e. what are the values of the characteristics for which it is better to use one inducer instead of another. To get a quantitative description of how the dataset characteristics determine inducers' superiority we plan to use a different meta-learning algorithm. Possible selections are inducers that produce a model of the classification, e.g decision trees or rule inducers.

Another way to examine the results is to explore the 'total' discriminating power of each dataset characteristic, that is how often it gets selected over *all* the meta-learning problems. Table 6 shows the relative frequency with which each feature is selected. The most often selected attribute is the noise to signal ratio, present in 25 of the 28 meta-learning problems. Although a rough approximation, since it is based on the mean attribute entropy and the mean mutual information between the attributes and the class, it is quite useful in determining the relative performance of the algorithms. The correlation histogram is another noteworthy characteristic: one of its bins shares the noise-signal ratio's extremely high selection rate, and eight of the others are above the 50% level. This seems to indicate the non negligible influence of correlated attributes on learning, due to varying degrees of sensitivity exhibited by the learners. Following is the ratio of the number of attributes to the number of instances. It is known that increasing the number of features beyond a certain point is likely to be counterproductive [6]. The number of classes is also selected very often, providing an indication that inducers react differently to variations in the number of classes. Information theoretical measures such as mean mutual information and mean attribute entropy also appear to be discriminating features considering their high selection rate. The only characteristic that seems to be completely useless is the histogram of missing values, none of its elements is ever selected. This characteristic describes the distribution of percentages of missing values of the attributes. Overall the discriminating power of each characteristic is quite different, and with the notable exception of the missing values histogram, all of them are used at least in one pair of base-inducers.

5.3 Results on the Final Suggestions

To measure the performance, in terms of the final suggestion, we first derive the true ranking of the inducers for each dataset, using the McNemar test. This give us, at least theoretically, a possible number of $2^8-1 = 255$ different combinations. In practice, in our study, only 80 of those appear. The inducer that gets the top position most often is c50boost, (26.23%). In Table 4 we give the distribution of inducer(s) that get the top ranking in more than 1% of the total number of

Table 3. Accuracies and significance levels for each of the 28 meta-learning problems.

pair	no feature selection	feature selection	significance level
c50rules c50boost	82.42%	84.65%	0.01640
c50tree c50boost	80.00%	82.79%	0.00011
c50tree c50rules	78.14%	84.47%	0.00000
lindiscr c50boost	84.84%	88.28%	0.00002
lindiscr c50rules	85.12%	88.00%	0.00030
lindiscr c50tree	85.58%	88.84%	0.00013
ltree c50boost	78.60%	80.47%	0.00721
ltree c50rules	82.14%	83.91%	0.00395
ltree c50tree	78.79%	81.86%	0.00044
ltree lindiscr	81.21%	84.65%	0.00003
IBL c50boost	85.40%	87.91%	0.00305
IBL c50rules	74.88%	80.56%	0.00000
IBL c50tree	79.53%	82.14%	0.00125
IBL lindiscr	81.49%	84.84%	0.00000
IBL ltree	74.60%	80.47%	0.00000
NB c50boost	85.02%	87.44%	0.00001
NB c50rules	84.74%	88.19%	0.00034
NB c50tree	84.47%	87.91%	0.00008
NB lindiscr	77.95%	81.58%	0.00013
NB ltree	84.37%	86.60%	0.00028
NB mlcib1	85.02%	86.42%	0.02497
ripper c50boost	85.02%	85.02%	1.00000
ripper c50rules	77.86%	83.81%	0.00000
ripper c50tree	84.09%	86.60%	0.00175
ripper lindiscr	78.70%	81.40%	0.00034
ripper ltree	75.81%	78.70%	0.00016
ripper IBL	81.30%	85.02%	0.00003
ripper NB	81.67%	84.56%	0.00007
means	81.39%	84.54%	
improvement	27.25%	30.40%	

the datasets. The next step is to perform 10-fold cross validation on the 1082 data sets that were used. In each step of the cross validation process we build all the meta-models using 90% of the datasets. The meta-models are then applied to the remaining 10% and the proposed inducer(s) is (are) compared to those which were actually in the first on the basis of McNemar tests. To estimate the accuracy we do not use a 0/1 loss function. Instead a suggestion is considered successful when the proposed inducer(s) is (are) a subset of the true set of the top inducers as determined by the McNemar tests.

In Table 5 we see the accuracy that the instance-based inducer achieves on the final suggestion of base level inducers. Accuracy without feature selection is 69.67% and increases to 76.28% with feature selection. The p-value of a binomial test of significance is 0, which means the difference is significant at any level of

Table 4. Groups of inducers that were ranked at the top for more than 1% of the datasets.

Group	Frequency	Percent
c50boost	282	26.23
ltree	140	13.02
c50rules c50tree ltree	54	5.02
c50rules c50boost c50tree	47	4.37
c50rules c50boost c50tree lindiscr ltree IBL NB ripper	47	4.37
lindiscr	41	3.81
NB	30	2.79
c50rules c50boost c50tree ltree ripper	29	2.7
c50boost ltree	29	2.7
c50rules c50boost c50tree ltree IBL ripper	27	2.51
IBL	27	2.51
c50rules c50boost c50tree ltree	25	2.33
c50rules c50boost	23	2.14
c50rules c50tree	22	2.05
c50rules	20	1.86
ltree NB	14	1.3
c50boost IBL	14	1.3
c50boost c50tree	13	1.21
lindiscr ltree	12	1.12
c50tree	11	1.02
c50rules c50boost c50tree lindiscr ltree NB ripper	11	1.02

Table 5. Accuracy of the final suggestion with and without feature selection.

no feature selection	feature selection
69.67%	76.28%

significance. To conclude, feature selection significantly improves performance with respect to the advice of the base inducer.

6 Conclusions and Future Work

Here we have continued our work on a previous system for inducer selection. The goal was twofold. First to gain a better understanding of the factors that determine the relative performance of inducers. Second to improve the performance. The meta-learning framework adopted provides great flexibility and the possibility to examine the factors that determine the relative performance of each pair of inducers.

Examining the meta-models created for each pair of inducers we saw that the factors determining the relative performance of inducers vary from pair to pair. The new subset of features for each pair of inducers not only improves the

Table 6. Characteristics selected for three of the meta-learning problems, 1 indicates selection of the corresponding characteristic, 0 elimination. The frequency column gives the frequency with which the characteristics appear in the different meta-learning models.

Attribute	IBL NB	IBL Ltree	NB C50boost	frequency %
# classes	1	1	1	82.21
# attributes	1	0	0	60.71
# instances	1	0	1	46.42
$\frac{\#\text{ attributes}}{\#\text{instances}}$	1	1	0	85.71
# unknown values	1	0	0	25.00
$\frac{\#\text{ unknown values}}{\#\text{ attributes} * \#\text{ instances}}$	1	1	1	60.71
# nominal attributes	1	1	1	71.42
max,min,mean,stdv of nominal attribute values	1101	1010	1010	54.14, 25.00, 64.28, 64.28
1..10 concentration histogram	1101110000	0101000000	1010111111	78.57, 78.57, 46.42, 42.85, 50.00, 35.71, 28.57, 21.42, 21.42, 21.42
non computable conc. histogram	0	0	1	21.42
1..10 concentration histogram with class	1001000000	0000000000	1101000000	53.57, 57.15, 3.57, 78.57, 3.57, 3.57, 3.57, 3.57, 0, 0
non computable conc. histogram with class	0	0	0	3.57
# continuous attributes	1	0	0	50.00
1..10 correlation histogram	1111111001	1111111111	1111111111	60.71, 89.28, 64.28, 50.00, 60.71, 75.00, 60.71, 53.57, 35.71, 53.57
non computable correlation histogram	0	0	1	14.28
1..10 missing values histogram	0000000000	0000000000	0000000000	0, 0, 0, 0, 0, 0, 0, 0, 0, 0
$\frac{\#\text{ continuous}}{\#\text{ attributes}}$	0	0	0	7.14
$\frac{\#\text{ nominal}}{\#\text{ attributes}}$	0	0	1	14.28
Binary Attributes	0	1	1	32.14
Fract	1	0	0	28.57
Cancor	1	1	1	64.28
Mean Skew	1	1	1	50.00
Mean Kurtosis	1	1	1	64.28
Class Entropy	0	0	1	50.00
Mean Attributes Entropy	1	1	0	78.57
Mean Mutual Information	1	1	1	75.00
Equivalent number of attributes	0	0	1	75.00
NoiseSignal Ratio	1	1	1	89.28
AttrMultiCorrel	1	1	1	67.85
SDratio	0	1	0	64.28

performance but also provides a better understanding of what is relevant and what is not.

Based on the, mainly empirical, work presented here, our next steps will be to get a better understanding of the specific conditions under which an inducer is a better choice than another one for a specific dataset. This could be achieved through the use of a more sophisticated inducer on the meta-level that actually constructs learned models. The thorough examination of the models will give a better insight on how data characteristics affect the relative performance of the algorithms and will lead to a better understanding of the weak and strong points of each inducer.

Acknowledgments. We would like to thank Johann Petrak for his Perl scripts and Joao Gama for Ltree and his implementation of linear discriminants. This work has been supported by Swiss OFES in the framework of ESPRIT IV LTR project METAL(26-357)

References

1. Agresti, A.: Categorical Data Analysis. Wiley Series in Probability and Mathematical Statistics (1990)
2. Aha, D. W.: Generalizing from Case Studies: A Case study. In: Proceedings of the 9th International Machine Learning Conference (1992) (1–10)
3. Brazdil, P., Gama, J., Henery, R.: Characterizing the applicability of classification algorithms using meta level learning. In: proceedings of the European Machine Learning Conference, ECML94, Springer Verlag (1994) 83–102
4. Brazdil, P. and Carlos, S.: Exploiting past experience in ranking classifiers. In: Proceedings of the 16th International Conference on Machine Learning, Workshop on Recent Advances in meta-Learning and Future Work (1999)
5. Cohen, W.W.: Fast Effective Rule Induction. In: Proceedings of the 12th International Conference on Machine Learning (1995)
6. Duda, R., Hart, P.: Pattern Classification and Scene Analysis. John Willey and sons (1973) 66–73
7. Gama, J., Brazdil, P.: Characterization of Classification Algorithms. In: Proceedings of the 7th Portugese Conference in AI, EPIA 95 (1995) 83–102
8. Gama, J., Brazdil, P.: Linear Tree. Intelligent Data Analysis **3** (1999) 1–22
9. Hilario, M., Kalousis, A.: Quantifying the resilience of inductive algorithms for classification. In: Proceedings of the 4th European Conference on Principles and Practice of Knowledge Discovery in Databases (2000)
10. Kalousis, A., Theoharis, T.: NOEMON: Design, implementation and performance results of an intelligent assistant for classifier selection. Intelligent Data Analysis **3(5)** (1999) 319–337
11. Kalousis, A., Hilario, M.: Knowledge discovery from incomplete data. In: Proceedings of the 2nd International Conference on Data Mining, WIT Press (2000)
12. Kalousis, A., Hilario, M.: Model Selection via Meta-learning: a Comparative Study. In: Proceedings of the 12th International IEEE Conference on Tools with AI (2000)
13. Kohavi, R., Sommerfield, D., Dougherty, J.: Data Mining Using MLC++, a Machine Learning Library in C++. Tools With AI (1996)
14. Kohavi, Ron and John, George Wrappers for Feature Subset Selection Artificial Intelligence Journal **97(1-2)** (1997) 273–324
15. Lindner, C., Studer, R.: AST: Support for Algorithm Selection with a CBR Approach. In: Proceedings of the 16th International Conference on Machine Learning, Workshop on Recent Advances in Meta-Learning and Future Work (1999)
16. Liu, H., Motoda, H.: Feature Selection for Knowledge Discovery and Data Mining. Kluwer Academic Publishers (1998)

17. Michie, D., Spiegelhalter, D.J., Taylor, C.C.: Machine Learning, Neural and Statistical Classification. Ellis Horwood Series in Artificial Intelligence (1994)
18. Nakhacizadeh, G., Schnabl, A.: Development of Multi-Criteria Metrics for Evaluation of Data Mining Algorithms. In: Proceedings of the 3rd International Conference on Knowledge Discovery and Data Mining (1997) 37–42
19. Phahringer, B., Bensusan, H., Giraud-Carrier, C.: Tell me who can learn you and I can tell you who you are: Landmarking Various Learning Algorithms. In: Proceedings of the 7th International Conference on Machine Learning Morgan Kaufman (2000) 743–750
20. Quinlan, R.: http://www.rulequest.com/see5-info.html
21. Quinlan, R.: Induction of Decision Trees. Machine Learning **1(1)** (1986) 81–106
22. Rendell, L., Seshu, R., Tcheng, D.: Layered Concept Learning and Dynamically Variable Bias Management. In: Proceedings of 10th International Joint Conference on AI (1987) 308–314
23. Salzberg, L.S.: On Comparing Classifiers: A Critique of Current Research and Methods. Data Mining and Knowledge Discovery 1(3) (1997) 317–327
24. Schaffer, C.: A Conservation Law for Generalization Performance. In: Proceedings of the 11th International Conference on Machine Learning (1994)
25. Soares, C., Brazdil, P.,: Zoomed ranking: selection of classification algorithms based on relevant performance information. In: Proceedings of the 4th European Conference on Principles Of Data Mining and Knowledge Discovery. Springer (2000) 126–135
26. Shavlik, J.W., Mooney, R.J., Towell, G.: Symbolic and Neural net learning algorithms : An experimental approach. Machine Learning **6** (1991) 111–114
27. Todorovski, L., Dzeroski, S.: Experiments in Meta-Level Learning with ILP In: Proceedings of the 3rd European Conference on Principles of Data Mining and Knowledge Discovery. Springer (1999) 98–106
28. Todorovski, L., Brazdil, P., and Soares, C.: Report on the Experiments with Feature Selection in Meta-Level Learning. In: Proceedings of the 4th European Conference on Principles on Data Mining and Knowledge Discovery, Workshop on Data Mining, Decision Support, Meta-learning and ILP (2000) 27–39
29. Weis, S.M., Kapouleas, I.,: An Empirical Comparison of Patter Recognition, Neural Nets, and Machine Learning Classification Algorithms. In: Proceedings of the 11th International Joint Conference on AI (1989) 781–787
30. Blake, C., Keogh, E., Merz, C.J.: http://www.ics.uci.edu / ~mlearn / MLRepository.html. University of California, Irvine, Dept. of Information and Computer Sciences (1998)
31. Wolpert, D,: The Lack of A Priori Distinctions Between Learning Algorithms. Neural Computation (1996) 1341–1390

Efficient Mining of Niches and Set Routines

Guozhu Dong and Kaustubh Deshpande

Dept. of CSE, Wright State University, Dayton OH 45435, USA.

Abstract. It is widely recognized that successful businesses usually fall into set routines and become limited by their past. To remain successful, they need to discover new opportunities and niches. Niches are *surprising rules that contradict the set routines*; they capture significant, representative client sectors that deserve new, more profitable treatments; they are not merely strong-rule and exception pairs. In this paper we study the efficient mining of set routines and niches. We also introduce a semantic approach to select a set of representative patterns, and present an efficient incremental algorithm to implement the approach.

Keywords: Data mining, niches, set routines, exceptions, interestingness, semantic-based selection

1 Introduction

In order to succeed, a starting business is always looking for opportunities. An established business, however, usually falls into set routines and becomes limited by its past. To remain successful, it needs to discover new opportunities and niches. Niches are *surprising rules that contradict the set routines*; they capture significant, representative client sectors that deserve new, targeted, more profitable treatments; they are not merely strong-rule and exception pairs. Niches are also useful for many other applications such as medicine, scientific discovery, and customized treatment of clients.

We illustrate the importance of niches with an example reported in KDNuggets [5]. *"Farmers Insurance found a previously unnoticed niche of sports car enthusiasts: married boomers with a couple of kids and a second family car, maybe a minivan, parked in the driveway. Farmers relaxed its underwriting rules and cut rates on certain sports cars for people who fit the profile (and presumably gained market share in this niche)."* An insurance company divides clients into different risk classes, and charges rates according to risk levels. Rules for deciding risk levels are formulated through experience, statistical analysis, or data mining; these rules may remain fixed for a long time and become set routines. A niche here can be a special segment of customers who are less risky than the company currently believes.

To mine niches, we need to capture the set routines first. For a business decision, the set routines (SRS) should correspond to a set of business rules or operational policies; in general, the SRS should correspond to a set of dominant

D. Cheung, G.J. Williams, and Q. Li (Eds.): PAKDD 2001, LNAI 2035, pp. 234–246, 2001.

trends (DTs) which are important to the task (decision) at hand. A DT can be captured by an emerging pattern (EP), namely a pattern which occurs more frequently in the undesirable instances than in the desirable instances, or vice versa. The EP should occur at a relatively high support. This corresponds to the fact that the SRS is usually formed by past experience, observations, and even data analysis, because human observations and past data mining algorithms mainly discover high support patterns.

The SRS should contain a relatively small number of DTs, such as 100 or less. More importantly, the DTs should represent different segments of the clients (instances); each DT should capture a unique segment of the instances. We will first mine a set of EPs, using ConsEPMiner [16]. Then we use a novel semantic-based approach, which minimizes overlap and maximized disjointness between DTs, to select an SRS from this set. We associate each pattern with the set of data instances containing the pattern, and we consider two patterns semantically similar if their associated data sets overlap sufficiently. Semantic-based selection ensures that different DTs capture almost disjoint segments of data, and the DTs in the SRS collectively cover as much data as possible; it also helps enhance understandability of the niches and avoid repeated computation.

The exception EPs to the SRS are mined, again using ConsEPMiner, and the semantic-based approach is used to select good representatives as niches. Experiments show that our algorithm is efficient, can find meaningful niches from real data, and can succeed in niche mining at low support levels.

To illustrate how niche mining can be done, consider an auto insurance company which has been in operation for a while. If an SRS cannot be extracted from the business manual, we simply use the company's database, using a threshold on rates to divide the clients into $Risky^{view}$ and $NonRisky^{view}$ (according to the view of the company). We extract an SRS from $Risky^{view}$ and $NonRisky^{view}$. Then we divide the clients into $Risky^{actual}$ and $NonRisky^{actual}$, using a threshold on the total number (or amount) of claims. We can mine the exceptions to the SRS using these two new data sets.

Niche mining can also be useful to a new insurance company, to help it understand how its competitors work. If the new company has access to all the data discussed above, then it can just use the procedure discussed above. Otherwise, it can still gain insights on how its competitors work, by using $Risky^{actual}$ and $NonRisky^{actual}$ for both SRS mining and niche mining.

Most past work on data mining concentrated on discovering high-support patterns or high-support relationships. In contrast, our work is concerned with finding low support exceptions to high support strong rules. Moreover, our niches are exceptions to the set routines of an organization and thus are exceptions in a global sense, and can thus provide the organization with "complete" representatives of niche opportunities. In contrast, past research on exception mining [14, 10] considered strong-rule and exception pairs in a local manner. Producing all strong-rule and exception pairs will make the results hard to understand; moreover, the mining process also becomes unnecessarily expensive, because repeated similar computation is performed for similar strong rules. Other researchers have considered mining of exceptions of other types, e.g. interesting holes in data [8]. Our work is related to interestingness of patterns [11,12,2,9].

Our semantic-based selection of representative patterns is related to [15]. That paper used the semantics-based idea of "cover" to select patterns, but it did not consider minimizing overlap and maximizing disjointness.

Our method can also be used to find semantic representatives of strong-rule and exception pairs. Structures can be added to the DTs in an SRS, e.g. by imposing an ordering; mining of niches for such extensions will be a future topic.

2 Set Routines and Niches

The way an organization operates can be understood from its past operations. We will consider one decision (e.g. is a client risky or not risky) for the organization. For each instance, the decision will be either Yes or No. We will call each of the two decisions a class, and will denote these two classes as P (positive) and N (negative). For C in $\{P, N\}$, we will use NC to mean the opposite class of C. For convenience, we will identify a class with its associated set of instances.

We assume the readers are familiar with transactions, relations, itemsets, and supports of itemsets. We will use the term instance to refer to case, transaction, vector, or tuple. We assume that numerical attributes have been discretized using some binning method such as equal-density or equal-length, and numerical values have been mapped to their containing intervals. For uniformity, we will also refer to these bins as values. Relations can now be viewed as transactions, where an item is an attribute-value pair.

For any organization, the policies regarding the decision are usually formulated under the influence of dominant trends (DTs). There are several requirements on a DT: (a) It should be a pattern (condition). (b) It should capture a significant segment of the instances. (c) It should occur much more frequently in one class than in another class.

(a) ensures that the DT can be tested on instances. (b) and (c) ensure that the DT has influenced the formulation of policies of the organization, because it differentiates between the two classes over a significant segment of instances.

For example, for insurance, statistics shows that sports car owners are usually riskier than other owners. The condition involved in this pattern is "the owner ownes sports cars". This condition captures a significant segment of clients.

An emerging pattern (EP) is a pattern meeting these requirements of a DT. EPs were introduced [3] to capture sharp differences between data classes or emerging trends in time. Suppose we are given two classes, say classes C_1 and C_2, associated with respectively data sets D_1 and D_2. Let $supp_i$(X) denote $supp_{D_i}$(X). The growth rate of an itemset X from D_1 to D_2 is defined as the ratio $\frac{supp_2(X)}{supp_1(X)}$ (letting $\frac{0}{0} = 0$ and $\frac{\neq 0}{0} = \infty$). Given a growth rate threshold $\rho > 1$, if the growth rate of X from D_1 to D_2 is $\geq \rho$, then X is called an emerging pattern from D_1 to D_2 (or from C_1 to C_2, or simply an EP of class C_2); C_1 is called the background class and C_2 the target class; we will write $X : C_2$ to signify the fact that X is an EP of class C_2. We can use EPs to make decisions.

If $X : C_2$ is an EP with a growth rate of 40, $|D_1| \approx |D_2|$, and t is an instance containing X, then the probability that t belongs to class C_2 is $97\% = \frac{40}{40+1}$.

The set routines (SRS) of an organization should consist of a set of DTs. They should satisfy these constraints[1] in order to provide an accurate model: (a) The DTs in the SRS should collectively cover as many instances as possible. (b) Different DTs in the SRS should cover disjoint subsets of instances. (c) The SRS should be relatively small, containing perhaps ≤ 100 DTs.

For example, the SRS for a car insurance company should correspond to the way the rates are determined. One DT may be {sports-car-owner : *yes*} : *Risky*. Another DT may be {points-on-license > 3} : *Risky*. A third DT may be $\{3.0 < GPA \leq 4.0\}$: *nonRisky*. There can be more DTs in the SRS.

We now formalize the meaning of cover, disjointness and overlap. For each pattern X, let $Sat(X)$ denote the set of instances (of either class P or N) satisfying (containing or covered by) X. (A more refined approach is possible, by dividing $Sat(X)$ into two classes and adjusting semantic-based selection accordingly. To focus on the spirit of the semantic-based approach, we omit the details of that refinement here.) For a set S of patterns, let $Sat(S) = \cup_{X \in S} Sat(X)$. Difference and overlap between two patterns X and Y will be measured by $|(Sat(X) - Sat(Y)) \cup (Sat(Y) - Sat(X))|$ and $|Sat(X) \cap Sat(Y)|$, resp.

Example 1. Suppose our two classes together contain the instances of Table 1 and we are given the four EPs in Table 2. Then $\text{SAT}(\{1,2\}) = \{t_1,t_2,t_3,t_6\}$, $\text{SAT}(\{2,3\}) = \{t_1,t_5\}$, and $\text{SAT}(\{\{1,2\},\{2,3\}\}) = \{t_1,t_2,t_3,t_5,t_6\}$. The overlap between $\{1,2\}$ and $\{2,3\}$ is $\{t_1\}$ and the difference between them is $\{t_2,t_3,t_5,t_6\}$.

Table 1. Transactions

Transaction id	t_1	t_2	t_3	t_4	t_5	t_6	t_7	t_8
Items	{1,2,3}	{1,2,4}	{1,2}	{2,4,5}	{2,3,4}	{1,2}	{4,5}	{1,4,5}

Table 2. EPs

EP
{1,2}
{2,4}
{2,3}
{4,5}

We now consider how to capture niches. Similar to dominant trends, niches should naturally be closely related to the decision under consideration; so we will also capture a niche by an EP. A niche should satisfy the following requirements: (a) A niche should be an exception to some DT in SRS. It should capture a subset of the instances captured by the DT and lead to a decision reversing that of the DT. (b) A niche should not be implied by other DTs of the SRS. We say that $XY : C$ is implied by an EP $Z : C$ if Z is a subset of XY.

For the auto insurance example, the EP {sports-car-owner:yes, age:[40..60], married:yes, #kids $\geq$ 2, second-family-car:yes}:NonRisky is a niche. It is an

[1] Syntactical difference is not good for capturing SRS, because syntactically disjoint patterns may be semantically similar: {small-car-owner:yes} and {age:[18..25]} are syntactically different but may cover nearly the same segment of drivers.

exception to the {sports-car-owner:yes}:Risky DT and it is not implied by any other DT in the SRS. However, the EP {age:[18..25], GPA:[3.0..4.0]}:NonRisky is not a niche: Although an exception to the DT {age:[18..25]}:Risky, it is implied by the EP {GPA:[3.0..4.0]}:NonRisky, which is in SRS.

It is sometimes possible to extract the SRS for an organization from its operational manual. If that is not possible, an SRS can be mined from operational data of the organization; we will address this problem in Section 4.

3 The Niche Mining Problem and Our Algorithm

The **niche-mining problem** is the following: Given two classes P and N, mine an SRS (if not given) and niches satisfying a minsupp threshold on DTs.

```
Niche Miner (SRS S, classes P, N, minDTsupp)
(1) if S is empty then
(2)     mine sets E1 of EPs from P to N and E2 from N to P;
(3)     select an SRS S from E1 and E2;
(4) mine exceptions to DTs in S;
(5) remove implied exceptions;
(6) select representative exceptions as niches
```

Fig. 1. Niche Miner Pseudo Code

The psuedo code of our algorithm is given in Figure 1. We first mine EPs with growth rate greater than a threshold value (e.g. 5) and remove EPs whose growth rates are relatively low (e.g. not among the top 40%). We then select DTs that are semantically distinct in terms of their SATs. Finally we mine EPs contradicting the DTs and semantically select the niches.

By selecting an SRS as a semantic representation of all possible dominant trends, we avoid the excessive computation needed over a huge number of possible DTs. This allows us to mine at lower support thresholds for the niches. Also, the resulting niches will be more understandable.

Mining EPs: We will use ConsEPMiner (Constraint Based EP Miner) [16] to mine EPs. The algorithm uses constraints, either explicitly given or inherently implied by the data/pattern type, to efficiently mine EPs, from large high dimensional data sets. It uses an improvement constraint to ensure that a representative grid of EPs (in the complete set-theoretic lattice of all EPs) is returned; the set of returned EPs is much smaller than the set of all possible EPs; more specifically, given that one EP X is chosen to be returned, then another EP Y will not be returned if Y is a superset of X but the growth rate of Y is not significantly larger (specified by the improvement constraint) than that of X, then Y will not be returned. It also uses upper estimates of supports and growth rates, obtained from the counts of candidates (regular or lookahead) already considered, to prune candidates. Pruning and dynamic reordering of items are

performed at three different stages: before counting, after the background data set is counted, after both data sets are counted. Using these ideas, ConsEPMiner overcomes the problem of combinatorial explosion of candidate itemsets.
Semantic-based Pattern Selection: In the next section we will propose efficient, semantic-based methods to select an SRS and select the representatives of exceptions as niches.
Finding niches efficiently: Given an SRS, we need to mine exception EPs that contradict the SRS, and select representative exceptions as niches. To select representative exceptions as niches, we also use the SAT-EP-Select algorithm discussed in the next section. The exception EPs are required to satisfy some appropriate growth rate threshold. Observe that the DTs usually have very high growth rates, and large reverse growth rates indicate that the exception EPs are significant reverse "trends."

Our approach to mine exception EPs is: For each DT X we reduce the data sets to contain only the transactions containing X, and then we call ConsEPMiner on the reduced data sets to find EPs that contradict X. This method is efficient since the reduced data sets are much smaller than the whole data sets. The above process of reducing data sets is called relativization; a similar technique was used in [7] for instance based classification.

A naive method to mine EPs is to call ConsEPMiner once for each class, and select niches from these common sets of EPs. This method suffers from several problems: Many of the mined EPs will be useless, since they do not contradict the EPs in SRS; moreover, to mine exceptions, we need to mine at very small threshold levels of support (such as 0.01%). At such low levels ConsEPMiner generates a huge number of EPs; it is very expensive (with respect to time) to find these EPs and to select from the set of candidate niches. One can improve this approach to make it competitive, by properly seeding ConsEPMiner with the DTs in the SRS. While this improved approach is not yet implemented, we believe that its performance might be comparable with our relativization approach. (The relativization approach may be better if the volume of data is large, as it becomes in memory for each DT after one pass for relativization.)

From the support and growth-rate of a DT X, one can determine the maximum support and growth rate for exception EPs of X. This information can help avoid useless computation: If there are no exception EPs meeting support and growth rate thresholds, then there is no need to call ConsEPMiner for X.

4 Semantic-Based Pattern Selection

When we mine a data set, we get many, perhaps millions of, EPs. The set of EPs can still number in tens of thousands, even after removing those with relatively low supports and growth rates[2]. The objective of semantic-based selection is to select a representative subset of EPs satisfying: two different EPs capture

[2] While we present the semantics-based selection algorithm for EPs, it can be easily modified to work for other types of patterns.

disjoint sets of instances whereas the selected EPs collectively cover as many instances as possible.

The exhaustive approach to selection is clearly infeasible, since the numbers of EPs and of transactions are both very large. In this paper we consider the greedy method sketched in Figure 2; we will give efficient algorithms for the key steps in Section 5. For steps 2 and 5, we use growth-rate to break ties. In step

SAT-EP-Select(data set D, EP set $\mathcal{P}$, K)
;; K is the maximum number of selected EPs
;; returns a set S of selected EPs
1) compute SAT(X) for all X in $\mathcal{P}$;
2) select an EP X from $\mathcal{P}$ with highest support;
3) let S = {X} and $\mathcal{P} = \mathcal{P} - \{X\}$;
4) while (SAT(S) can be expanded) or (|S| < K)
5) select an EP X s.t. $\frac{|Sat(X)-Sat(S)|}{|Sat(X)\cap Sat(S)|}$ is maximal for all EPs in $\mathcal{P}$;
;; if $\exists X \in \mathcal{P}$ s.t. $Sat(X) \cap Sat(S) = \emptyset$, choose X
6) let S = S ∪ {X} and $\mathcal{P} = \mathcal{P} - \{X\}$;
7) return S;

Fig. 2. Sketch of SAT-EP-Select

5, maximizing |SAT(X) - SAT(S)| to |SAT(X) ∩ SAT(S)| allows us to maximize disjointness and minimize overlap.

Example 2. We illustrate this algorithm using the transactions and EPs of Example 1. Let $K = 3$. Initially, $\mathcal{P} = \{\{1,2\},\{2,4\},\{2,3\},\{4,5\}\}$, and the *Sats* are: $Sat(\{1,2\}) = \{t_1,t_2,t_3,t_6\}$, $Sat(\{2,4\}) = \{t_2,t_4,t_5\}$, $Sat(\{2,3\}) = \{t_1,t_5\}$, and $Sat(\{4,5\}) = \{t_4,t_7,t_8\}$. We choose {1,2} as our first DT, since it has highest support; now $S = \{\{1,2\}\}$ and $\mathcal{P} = \{\{2,4\},\{2,3\},\{4,5\}\}$.

For each iteration, we need to compute, for each EP X in $\mathcal{P}$, $Sat(X) - Sat(S)$ and $Sat(X) \cap Sat(S)$ in order to find their cardinalities. For iteration 1, we get

$$|Sat(\{2,4\}) - Sat(S)| = 2,\ |Sat(\{2,4\}) \cap Sat(S)| = 1;$$
$$|Sat(\{2,3\}) - Sat(S)| = 1,\ |Sat(\{2,3\}) \cap Sat(S)| = 1;$$
$$|Sat(\{4,5\}) - Sat(S)| = 3,\ |Sat(\{4,5\}) \cap Sat(S)| = 0.$$

We choose $X = \{4,5\}$, since it is the only EP whose SAT is disjoint from SAT(S). Now, $S = \{\{1,2\}\} \cup \{\{4,5\}\} = \{\{1,2\},\{4,5\}\}$, and $\mathcal{P} = \{\{2,4\},\{2,3\}\}$.

For Iteration 2: We get $|Sat(\{2,4\}) - Sat(S)| = 1$, $|Sat(\{2,4\}) \cap Sat(S)| = 2$, $|Sat(\{2,3\}) - Sat(S)| = 1$, and $|Sat(\{2,3\}) \cap Sat(S)| = 1$. We choose $X = \{2,3\}$. Since $|S| = K$ (and $Sat(S) = \{t_1,t_2,t_3,t_4,t_5,t_6,t_7,t_8\}$ happens to be equal to the entire transaction set), we stop. So the selected EP set is $\{\{1,2\},\{2,3\},\{4,5\}\}$.

The SAT-EP-Select algorithm as given above may select many more EPs for one class than the other. One can avoid this as follows: We choose EPs by switching between the two classes; at any time, if one class C is over represented, then we select the next EP from the class NC (unless NC is exhausted already).

There are two expensive steps in this algorithm. Notice that $\mathcal{P}$ is usually a very large set, containing tens (or even hundreds) of thousands of EPs, the data set is also very large, containing tens of thousands of transactions, and the data may have very high dimension. Step 1, which computes the initial $Sats$ of all EPs in $\mathcal{P}$, can be expensive, since it must check which transactions contain which EPs. We implemented this search using a hash tree, obtaining a speed up of about 50%; the details are omitted. Step 5, which calculates $Sat(X) - Sat(S)$ and $Sat(X) \cap Sat(S)$ for each X in $\mathcal{P}$ at each iteration, can be very expensive, since SAT(X) can be as large as the number of transactions. We will discuss an incremental approach to reduce the cost next.

5 Incremental Computation of Overlap and Differences

The main idea of our technique is to avoid the repeated computations, between consecutive iterations, in computing $|Sat(X) - Sat(S)|$ and $|Sat(X) \cap Sat(S)|$ for each X in $\mathcal{P}$. To this end we store the set $Sat(X) - Sat(S)$ in a variable called SATDIF(X). SATDIF(X) is initialized to $Sat(X)$ (for $S = \{\}$).

Suppose the EP chosen for a particular iteration is Y, and the set S before Y is added in it is S_0. Let $S_Y = S_0 \cup \{Y\}$. Let SATDIF(X, S_0) denote the value of SATDIF(X) before Y is added to S, and let SATDIF(X, S_Y) denote the value of SATDIF(X) after Y is added to S.

We observe this: A transaction t in SATDIF(X, S_0) will need to be removed to get SATDIF(X, S_Y) iff t is in SATDIF(Y, S_0). This observation can be used as follows: In the incremental computation we can use the transactions in SATDIF(Y, S_0) to drive the computation of SATDIF(X, S_Y). This will improve efficiency because SATDIF(Y, S_0) is normally small, especially after a number of iterations have been executed. This idea is formalized in the algorithm below:

```
For each t in SATDIF(Y)
    For each X in P such that t is in SATDIF(X, S0)
        remove t from SATDIF(X);
```

We now illustrate how the incremental approach is more efficient using the data and EPs of Example 1. The computation with the incremental approach differs from the original naive approach, in that we replace SAT of individual EPs by their SATDIF, and that we replace SAT(S) by SATDIF(Y) for deriving SATDIF(X). More specifically, the computation in iteration 1 is identical to that for the non-incremental approach, as SATDIF(X) = SAT(X) for all X at this time. In iteration 2, this is no longer the case; for example, SAT({4,5}) is replaced by SATDIF({4,5}). The computation of SAT({2,4}) - SAT(S) is replaced by SATDIF({2,4}) - SATDIF(Y) (for Y = {4,5}); such replacement is the main reason that the incremental approach is efficient, because SATDIF(Y) $=\{t_4,t_7,t_8\}$ is much smaller than SAT(S), and because SATDIF(Y) is used to drive the modification of SATDIF of all EPs. For example, SATDIF({2,4}) is computed through $\{t_4,t_5\} - \{t_4,t_7,t_8\}$ instead of $\{t_2,t_4,t_5\} - \{t_1,t_2,t_3,t_4,t_6,t_7,t_8\}$.

We will store the set SATDIF(X) as a bit vector, and the SATDIFs of all EPs in $\mathcal{P}$ as a bit matrix, with EPs as columns and transactions as rows. Initially, position (t, e) of the matrix is 1 iff transaction t contains EP e. Table 3 below shows the initial contents of this matrix for Example 1.

To compute the cardinalities of the differences and the overlaps, we will keep two additional arrays of integers: *OldCounts* and *CurCounts*. $OldCounts(X)$ stores $|Sat(X)|$ whereas $CurCounts(X)$ stores $|Sat(X) - Sat(S)| = |\text{SATDIF}(X)|$. We can derive $|Sat(X) \cap Sat(S)|$ from $OldCounts(X) - CurCounts(X)$. The *OldCounts* array is initialized but never changed.

$CurCounts(X)$ is adjusted only when we modify SATDIF(X). Suppose Y has just been selected. For each t in SATDIF(Y), if t is in SATDIF(X), then we remove t from SATDIF(X) and decrement $CurCounts(X)$ by 1.

The algorithm, presented in terms of the SATDIF matrix, is given in Figure 3.

```
;; Suppose Y is chosen as a new DT.
For each transaction t such that the (t,Y) position of the matrix = 1
;; transaction t is contained in SatDif(Y)
    For each EP X ≠ Y
        If the (t,X) position of the matrix = 1 then
            change the value to 0 and decrement CurCounts for X;
            ;; remove t from SatDif(X) and adjust CurCounts(X)
```

Fig. 3. Sketch: Bit-Driven SAT-EP-Select

We illustrate the algorithm for $Y = \{1, 2\}$. Suppose the matrix before Y is added is given in Table 3. Because t_1 is a member of SATDIF(Y), we check if there is any other 1 in the same row; we find that there is a 1 in the column for {2,3}; we change it to 0 and decrement $CurCounts(\{2, 3\})$ by 1. Similar actions are taken for t_2, t_3, and t_6. After adding Y to S, the matrix and the *CurCounts* array become Table 4.

Table 3. SATDIFMatrix for Example 1

Trans/EPs	{1,2}	{2,4}	{2,3}	{4,5}
t_1	1	0	1	0
t_2	1	1	0	0
t_3	1	0	0	0
t_4	0	1	0	1
t_5	0	1	1	0
t_6	1	0	0	0
t_7	0	0	0	1
t_8	0	0	0	1
CurCounts	4	3	2	3
OldCounts	4	3	2	3

Table 4. SATDIFMatrix for second iteration

Trans/EPs	{1,2}	{2,4}	{2,3}	{4,5}
t_1	1	0	0	0
t_2	1	0	0	0
t_3	1	0	0	0
t_4	0	1	0	1
t_5	0	1	1	0
t_6	1	0	0	0
t_7	0	0	0	1
t_8	0	0	0	1
CurCounts	4	2	1	3
Oldcounts	4	3	2	3

Figure 4 compares the bit-driven incremental approach against the original naive approach, for waveform data with 10000 transactions, 21 attributes, and

14623 EPs. For $K = |SRS| = 100$, it took 2433 seconds for the naive approach but only 86 seconds for the incremental approach. Both approaches use hash for initial SAT computation.

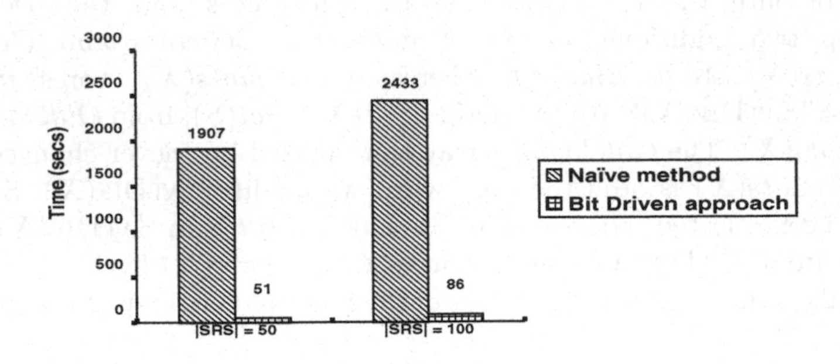

Fig. 4. Cost of SRS Selection

6 Experiments

We report results of two types of experiments: (i) results of mining SRS and niches from real data sets (mostly from the UCI Repository), (ii) results on efficiency and scalability with respect to both volume and dimensions. All experiments, including those reported in previous sections, are carried out on a single node of a multiprocessor system with 195 MHz ip27 processors and a shared main memory of 3698 MB.

SRS and Niches. Our results indicate that our algorithm can find meaningful niches from real data, at low support thresholds.

We report results on the Adult data set[3], which has 14 attributes (6 continuous and 8 nominal) and $32561 = 24720 + 7841$ instances. There are two classes defined: 1) People earning $\leq$50K (24720 instances) and 2) people earning >50K (7841 instances). The instances contain personal information (US Census), including these attributes: Age, Workclass, fnlwgt (presumably final wage tax), Education, Education-num, Marital-status, Occupation, Relationship, Race, Sex, Capital-gain, Capital-loss, Hours-per-week & Native-country; the data set was originally used to predict yearly salaries.

We found the SRS given in Table[4] 5. We first used the minsupp = 0.02 threshold, growth rate threshold of 5, and a growth rate improvement threshold of 0.05. We found around 500 EPs for each class. We then selected top 300 EPs

[3] Experiments on other data sets, including Musk (40 selected dimensions, $6598 = 5581 + 1017$ instances), Waveform (21 dimensions, $5000 = 1657 + 3343$ instances), and Arabidopsis-DNA (35 dimensions, $44484 = 2305 + 42179$ instances), are not included due to space restrictions.

[4] In the tables we will omit the attributes if such omissions will not lead to confusions.

for each class. (We found that the Adult data set has very few EPs, unlike the other data sets mentioned above.) Then we selected an SRS from these EPs.

Table 5. SRS for Adult

DTs	Supp	GR	Class
age: [-..31.6), Never-married, cg20	0.31	18.23	<= 50K
fwt30, edu-num:[13..+), relnship: Husband, White, Male	0.25	25.83	> 50K
edu-num:[10..13), Married-civ-s, occup:Exec-mngerial, White, HPW:[40.2..59.8)	0.04	17.39	> 50K
age:[31.6..46.2), edu: Prof-school	0.03	17.70	> 50K
edu-num:[7..10), Divorced, cg20, HPW:[20.6..40.2), nativ-cntry: USA	0.14	17.48	<= 50K
age:[31.6..46.2), fwt30, edu: Masters, occup: Exec-managerial, Male	0.02	18.45	> 50K
wrkclss: Private, Separated, cg20, cap-loss: [-..871.2), HPW:[20.6..40.2)	0.04	22.92	<= 50K

cg20 denotes cap-gain: [-..19999.8), fwt30 denotes fnlwgt: [-..306769)

We mined exception EPs and selected niches from them. For the first DT given above, we found 18 niches, whose supports range from 0.76% down to 0.03%. We list three of these below. (We omit the conditions in the DT.)

Table 6. Niches for First DT of Table 5

niches	Supp	GR	Class
edu: Bachelors, relnship: Not-in-family	0.0076	674.00	> 50K
edu-num:[13..+), White	0.0018	153.00	> 50K
relnship: Not-in-family, Amer-Indian-Eskimo, HPW:[20.6..40.2)	0.0003	GR:∞	> 50K

We note that, for real life data sets, niches of support at 0.01% can still be very useful – they may capture a segment of thousands of customers.

Scalability with respect to volume. To study how our algorithm performs when size of data set increases, we choose the Adult data set since it has large number of instances, which consists of 32561 instances (14 attributes). We randomly selected sub-data sets of 15K, 20K, 25K, and 30K instances. We set the maximum number of DTs to 50. The graph in Figure 5.a shows that our algorithm is scalable w.r.t. volume, and can find niches from very large data sets.

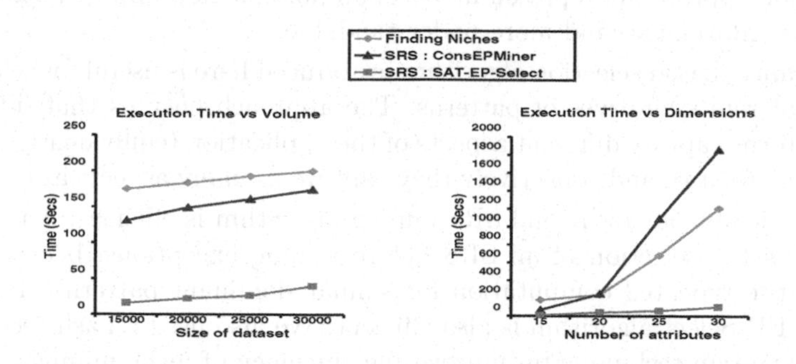

Fig. 5. Execution Time v.s. Volume and Dimensionality

The three curves show the time needed by different components of the algorithm. "SRS:ConsEPMiner" corresponds to the mining of the initial EPs for the selection of the SRS, "SRS:SAT-EP-Select" corresponds to the selection of the DTs from the initial set of EPs, and "Finding Niches" corresponds to finding the niches (including the repeated calls to the relativization, ConsEPMiner, and SAT-EP-Select procedures).

Scalability with respect to Dimensionality. To study how our algorithm performs with respect to dimensionality, we generated appropriate data sets by mutating the Musk data set as follows. The original Musk data has 166 attributes and about $6598 = 5581 + 1017$ instances. The dimensionality is high. For any given number N, we randomly choose N attributes of Musk.

However, the number of instances in Musk is small. To increase the number of instances, we add instances. We do not want to add exact copies of existing instances and we do not want to add instances which are totally different from the existing instances. Our solution is to add "mutated copies" of existing instances: We randomly choose an existing instance, and then randomly choose some attributes, and then change these attributes randomly within a range of $\pm 20\%$ of its original value. For each instance, a maximum of 25% of total number of attributes are modified. These ensure that the mutated instances are similar to existing ones but not identical.

The graph in Figure 5.b shows performance of our algorithm w.r.t. the number of dimensions: it is fast and can efficiently deal with high dimensions.

7 Concluding Remarks

In this paper we proposed a way to capture set routines (SRS) and niches, and introduced algorithms to efficiently mine SRS and niches. SRS allows one to understand in a global sense how an organization operates with respect to a decision. By mining niches together with an SRS, we ensure that the niches can indeed provide appropriate representatives of all possible new opportunities, and that they are informative and more understandable.

The semantic-based selection algorithm introduced here is useful for selecting a good set of representatives of patterns. The approach ensures that different selected patterns capture different aspects of the application (equivalently, different segments of data), and collectively they capture as many aspects as possible.

Algorithmically, our niche and SRS mining algorithm is efficient. An important reason is the selection of an SRS before mining exceptions, because this helps avoid the repeated computation for similar dominant patterns. The bit-driven SAT-EP-Select algorithm is also efficient. We also used a hash technique and a relativization technique to improve the efficiency of niche mining.

For future research, the following problems can be considered: How to push the semantic-based selection into a tree-based pattern mining algorithm? How to use niches to improve prediction accuracy in the classification process? It is also interesting to generalize our SAT-EP-Select algorithm, by considering it as a clustering problem for extremely high dimensions.

References

1. R. J. Bayardo. Efficiently mining long patterns from databases. In *Proceedings of ACM-SIGMOD Int'l Conference on Management of Data*, pages 85–93, 1998.
2. G. Dong and J. Li. Interestingness of discovered association rules in terms of neighborhood-based unexpectedness. In PAKDD, 1998.
3. G. Dong and J. Li. Efficient mining of emerging patterns: Discovering trends and differences. In *Proc. of ACM SIGKDD* 1999.
4. G. Dong, X. Zhang, L. Wong, and J. Li. CAEP: Classification by aggregating emerging patterns. In *Proc. Int'l Conference on Discovery Science*, 1999.
5. KDNuggets. Data mining digs in, July 1999, American demographics. Also KD-Nuggets 99:17, http://www.kdnuggets.com/news/99/n17/i7.html, edited by Gregory Piatetsky-Shapiro.
6. J. Li, G. Dong, and K. Ramamohanarao. Instance-based classification by emerging patterns. In *PKDD 2000.*
7. J. Li, G. Dong, and K. Ramamohanarao. Making use of the most expressive jumping emerging patterns for classification. In *PAKDD 2000.*
8. B. Liu, L.-P. Ku, & W. Hsu. Discovering interesting holes in data. In *IJCAI 1997.*
9. Bing Liu, Wynne Hsu, Lai-Fun Mun, and Hing-Yan Lee. Finding interesting patterns using user expectations. *TKDE*, 11(6):817–832, 1999.
10. H. Liu, H. Lu, L. Feng, and F. Hussain. Efficient search of reliable exceptions. In *PAKDD*, 1999.
11. G. Piatesky-Shapiro and C. J. Matheus. The interestingness of deviations. In *Proc. AAAI KDD Workshop*, 1994.
12. S. Ramaswamy, S. Mahajan, and A. Silberschatz. On the discovery of interesting patterns in association rules. In *VLDB*, 1998.
13. Ron Rymon. Search through systematic set enumeration. In *Proc. the Third Int'l Conference on Principles of Knowledge Representation and Reasoning*, 1992.
14. E. Suzuki and Y. Kodratoff. Discovery of surprising exception rules based on intensity of implication. In *PKDD*, pages 10–18, 1998.
15. H. Toivonen, M. Klemettinen, P. Ronkainen, K. Hätönen, and H. Mannila. Pruning and grouping discovered association rules. In *MLnet Workshop on Statistics, Machine Learning, and Discovery in Databases*, 1995.
16. X. Zhang, G. Dong, and K. Ramamohanarao. Exploring constraints to efficiently mine emerging patterns from large high-dimensional datasets. In *KDD*, 2000.

Evaluation of Interestingness Measures for Ranking Discovered Knowledge

Robert J. Hilderman[1] and Howard J. Hamilton[2]

[1] Saskatchewan Population Health and Evaluation Research Unit
[2] Department of Computer Science
University of Regina
Regina, Saskatchewan, Canada S4S 0A2
{Robert.Hilderman,Howard.Hamilton}@uregina.ca

Abstract. When mining a large database, the number of patterns discovered can easily exceed the capabilities of a human user to identify interesting results. To address this problem, various techniques have been suggested to reduce and/or order the patterns prior to presenting them to the user. In this paper, our focus is on ranking summaries generated from a single dataset, where attributes can be generalized in many different ways and to many levels of granularity according to taxonomic hierarchies. We theoretically and empirically evaluate thirteen diversity measures used as heuristic measures of interestingness for ranking summaries generated from databases. The thirteen diversity measures have previously been utilized in various disciplines, such as information theory, statistics, ecology, and economics. We describe five principles that any measure must satisfy to be considered useful for ranking summaries. Theoretical results show that only four of the thirteen diversity measures satisfy all of the principles. We then analyze the distribution of the index values generated by each of the thirteen diversity measures. Empirical results, obtained using synthetic data, show that the distribution of index values generated tend to be highly skewed about the mean, median, and middle index values. The objective of this work is to gain some insight into the behaviour that can be expected from each of the measures in practice.

1 Introduction

When mining a large database, the number of patterns discovered can easily exceed the capabilities of a human user to identify interesting results. To address this problem, various techniques have been suggested to reduce and/or order the patterns prior to presenting them to the user. For example, in [3], it is shown that the most interesting rules may reside along a support/confidence border. A technique is described in [20] that discovers interesting rules via an interactive process that seeks to classify rules that are not interesting. In [8], a measure is described that determines the interestingness (called surprise there) of discovered knowledge via the explicit detection of Simpson's Paradox. An approach is described in [7] that utilizes a distance metric to evaluate the importance of a rule by considering its unexpectedness in terms of other rules in its neighborhood.

D. Cheung, G.J. Williams, and Q. Li (Eds.): PAKDD 2001, LNAI 2035, pp. 247–259, 2001.

Our focus is on the use of diversity measures for ranking summaries generated from a single dataset, where attributes can be generalized in many different ways and to many levels of granularity according to taxonomic hierarchies. We introduced this use of diversity measures in [10] and [11]. An empirical analysis found that highly ranked, concise summaries provided a reasonable starting point for further analysis of discovered knowledge. It was also shown that for selected sample datasets, the order in which some of the measures rank summaries is highly correlated, but the rank ordering can vary substantially when different measures are used. In [12], the notion of a summary was extended to include other forms of knowledge representation, and we showed that these other forms are also amenable to ranking using diversity measures. And significant progress has been made into more theoretical issues regarding formal principles for diversity measures used as measures of interestingness in data mining applications [14].

In this paper, we evaluate thirteen diversity measures as heuristic measures of interestingness for ranking summaries in data mining applications. We describe five principles that any measure must satisfy to be considered useful for ranking summaries. Our theoretical results show that only four of the thirteen diversity measures satisfy all of the principles. We then analyze the distribution of the index values generated by each of the thirteen diversity measures. Empirical results, obtained using synthetic data, show that the distribution of index values generated tend to be highly skewed about the mean, median, and middle index values. The objective of this work is to gain some insight into the behaviour that can be expected from each of the measures in practice so that when choosing a candidate interestingness measure, we can determine which of the five principles are satisfied, and then knowing the behavioural characteristics of each measure, judge the suitability of the candidate interestingness measure for the intended application.

The remainder of the paper is organized as follows. In Section 2, we describe several forms of knowledge representation, which we collectively refer to as summaries, and motivate the need for ranking discovered knowledge. In Section 3, we provide a brief overview of thirteen diversity measures introduced and evaluated as heuristic measures of interestingness in previous work. In Section 4, we describe five principles that useful diversity measures must satisfy, and identify those diversity measures satisfying the five principles. In Section 5, we present experimental results describing the distribution of index values generated by each of the thirteen measures. We conclude in Section 6 with a summary of our work and suggestions for future research.

2 Background and Motivation

Let a *summary* S be a relation defined on the columns $\{(A_1, D_1), (A_2, D_2), \ldots, (A_n, D_n)\}$, where each (A_i, D_i) is an attribute-domain pair. Also, let $\{(A_1, v_{i1}), (A_2, v_{i2}), \ldots, (A_n, v_{in})\}$, $i = 1, 2, \ldots, m$, be a set of m unique tuples, where each (A_j, v_{ij}) is an attribute-value pair and each v_{ij} is a value from the domain D_j associated with attribute A_j. Let attribute A_k be a derived attribute whose values v_{ik}, from the domain D_k, for each attribute-value pair (A_k, v_{ik}) is an aggregation of values from the the unconditioned data present in the original database. For example, a sample summary is shown in Table 1. Table 1 is a generalized relation in which retail sales transactions have been aggregated to show the derived attributes *Quantity*, *Amount*, and *Count* (i.e., number of transactions) by *Region*.

Table 1. A generalized relation

Region	*Quantity*	*Amount*	*Count*
North	12	\$150.00	7
South	5	\$325.00	2
West	8	\$200.00	4
East	11	\$275.00	3

The summary definition given above can also be naturally extended to include summaries that are multi-dimensional. For example, another sample summary, is shown in Figure 1. Figure 1 shows a data cube in which retail sales transactions have been aggregated in three dimensions, where the *Item* attribute is on the vertical dimension, *Transact.Loc* is on the horizontal, and *Cust.Loc* is on the diagonal. *Transact.Loc* is the city where the sales transaction was processed, and *Cust.Loc* is the city where the sales transaction was initiated. Here we show each cell containing two values (due to space limitations); the top value is the quantity of items aggregated from sales transactions (i.e., *Quantity*), and the bottom value is the number of transactions aggregated (i.e., *Count*).

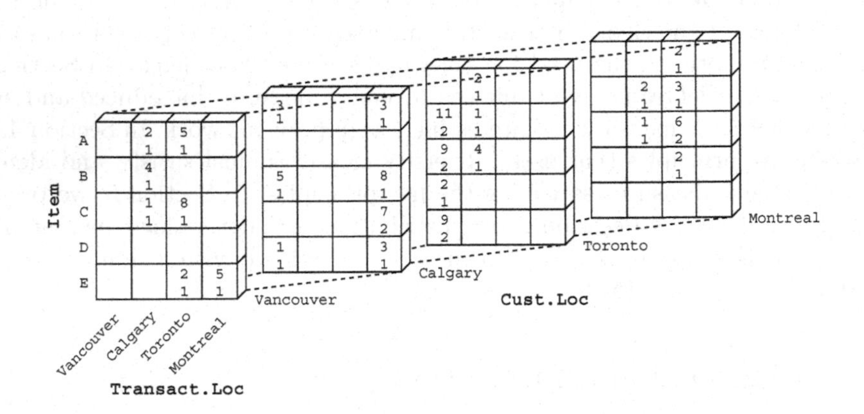

Fig. 1. A data cube

Of course, numerous methods could be used to guide the generation of summaries, such as concept hierarchies [5], domain generalization graphs [15], Galois lattices [9], conceptual graphs [4], and formal concept analysis [22]. Also, summaries could more generally include many other forms of knowledge representation, such as database views, association rules, itemsets, and web search results.

However, when given hundreds, or even thousands of summaries (possibly multi-dimensional), it is simply not feasible to determine the most interesting summaries or dimensions using a manual technique. What is needed are effective measures of interestingness to assist in the interpretation and evaluation of the discovered knowledge. The development of such measures is currently an active research area in KDD. Such measures are broadly classified as either objective or subjective. *Objective measures* are based upon the structure of discovered patterns, such as the frequency with which combinations of items appear in sales transactions [1]. *Subjective measures* are based upon user beliefs or biases regarding relationships in the data, such as an approach utilizing Bayes Rule to revise prior beliefs [18]. Here we focus on objective measures of interestingness.

3 Objective Interestingness Measures

The tuples in a summary or dimension generated from a database are unique, and therefore, can be considered to be a population with a structure that can be described by some frequency or probability distribution. Here, we review thirteen diversity measures, described in detail in [10], and shown in Figure 2, that evaluate the frequency or probability distribution of the values in a derived attribute to assign a single real-valued index that represents its interestingness relative to other summaries or dimensions generated from the same database.

In Figure 2, let m be the total number of tuples in a summary. Let n_i be the value contained in the derived attribute for tuple t_i. Let $N = \sum_{i=1}^{m} n_i$ be the total of the derived attribute. Let p be the actual probability distribution of the tuples based upon the values n_i. Let $p_i = n_i/N$ be the actual probability for tuple t_i. Let q be a uniform probability distribution of the tuples. Let $\bar{u} = N/m$ be the value for tuple t_i, $i = 1, 2, \ldots, m$ according to the uniform distribution q. Let $\bar{q} = 1/m$ be the probability for tuple t_i, for all $i = 1, 2, \ldots, m$ according to the uniform distribution q. Let r be the probability distribution obtained by combining the values n_i and $\bar{u}$. Let $r_i = (n_i + \bar{u})/2N$, be the probability for tuples t_i, for all $i = 1, 2, \ldots, m$ according to the distribution r.

The measures shown in Figure 2 are well-known measures of dispersion, dominance, inequality, and concentration that have previously been successfully applied in several areas of the social, ecological, information, and computer sciences. Although the terminology varies depending upon the application, the concept of diversity has been considered a useful one for analyzing many phenomena. For example, in ecology, various measures of diversity have been proposed and studied to aid in understanding the variability of populations of organisms within different types of habitat [17]. Diversity measures have also been used by economists and social scientists to study the distribution of income between different socioeconomic groups and geographical regions [2]. In information theory, diversity

$$I_{Variance} = \frac{\sum_{i=1}^{m} (p_i - \bar{q})^2}{m-1}$$

$$I_{Simpson} = \sum_{i=1}^{m} p_i^2$$

$$I_{Shannon} = -\sum_{i=1}^{m} p_i \log_2 p_i$$

$$I_{McIntosh} = \frac{N - \sqrt{\sum_{i=1}^{m} n_i^2}}{N - \sqrt{N}}$$

$$I_{Lorenz} = \bar{q} \sum_{i=1}^{m} (m - i + 1) p_i$$

$$I_{Gini} = 0.5 \left(\sum_{i=1}^{m} \sum_{j=1}^{m} |p_i \bar{q} - p_j \bar{q}| \right)$$

$$I_{Berger} = \max(p_i)$$

$$I_{Schutz} = \frac{\sum_{i=1}^{m} |p_i - \bar{q}|}{2m\bar{q}}$$

$$I_{Bray} = \frac{\sum_{i=1}^{m} \min(n_i, \bar{u})}{N}$$

$$I_{Whittaker} = 1 - \left(0.5 \sum_{i=1}^{m} |p_i - \bar{q}| \right)$$

$$I_{MacArthur} = \left(-\sum_{i=1}^{m} r_i \log_2 r_i \right) - \left(0.5 \left(-\sum_{i=1}^{m} p_i \log_2 p_i \right) + \log_2 m \right)$$

$$I_{Theil} = \frac{\sum_{i=1}^{m} |p_i \log_2 p_i - \bar{q} \log_2 \bar{q}|}{m\bar{q}}$$

$$I_{Atkinson} = 1 - \left(\prod_{i=1}^{m} \frac{p_i}{\bar{q}} \right)^{\bar{q}}$$

Fig. 2. Thirteen diversity measures

measures are used to measure the information content in messages [21]. Diversity measures have been used to describe the linguistic differences between the inhabitants of neighboring geographic regions [16]. More general treatments attempt to define the concept of diversity and develop a related theory of diversity measurement [19,23].

4 Theoretical Results

We now describe principles of interestingness against which the utility of candidate interestingness measures can be assessed. We do this through the mathematical formulation of five principles that must be satisfied by any acceptable diversity measure for ranking the interestingness of discovered knowledge using our, or a similar, technique. Proofs are omitted due to space considerations, so refer to [13] and [14] for complete details. We study functions f of m variables, $f(n_1, \ldots, n_m)$, where f denotes a general measure of diversity, m and each n_i (n_i assumed to be non-zero) are as defined in the previous section, and $(n_1, \ldots, n_m)$ is a vector corresponding to the values in a derived numeric measure attribute (e.g., the *Count* values from the examples in Section 2)for some arbitrary summary whose values are arranged in descending order such that $n_1 \geq \ldots \geq n_m$

(except for discussions regarding I_{Lorenz}, which requires that the values be arranged in ascending order). The principles presented here are for ranking the interestingness of summaries generated from a single dataset, so we assume that N is fixed. We justify the non-zero assumption for the n_i's, as follows. If the value of the *Count* attribute for a particular tuple is zero, there are two possible reasons. Either the combination of domain values being counted in the tuple can occur in practice, but no occurrences have been encountered during the mining process, or else the combination of domain values being summarized cannot occur in practice, and no occurrences will ever be encountered (i.e., such an entity does not exist). So, to preserve and simplify the general applicability of our technique, we make no assumptions regarding the possibility of occurrence of particular combinations of domain values. We now begin by specifying two fundamental principles.

Minimum Value Principle (P1). Given a vector $(n_1, \ldots, n_m)$, where $n_i = n_j$, $i \neq j$, for all i, j, $f(n_1, \ldots, n_m)$ attains its minimum value.

P1 specifies that the minimum interestingness should be attained when the tuple counts are all equal (i.e., uniformly distributed). For example, given the vectors $(2, 2)$, $(50, 50, 50)$, and $(1000, 1000, 1000, 1000)$, we require that the index value generated by f be the minimum possible for the respective values of m and N.

Maximum Value Principle (P2). Given a vector $(n_1, \ldots, n_m)$, where $n_1 = N - m + 1$, $n_i = 1$, $i = 2, \ldots, m$, and $N > m$, $f(n_1, \ldots, n_m)$ attains its maximum value.

P2 specifies that the maximum interestingness should be attained when the tuple counts are distributed as unevenly as possible. For example, given the vectors $(3, 1)$, $(148, 1, 1)$, and $(3997, 1, 1, 1)$, where $m = 2, 3, \text{and } 5$, respectively, and $N = 4, 150, \text{and } 4000$, respectively, we require that the index value generated by f be the maximum possible for the respective values of m and N.

The behaviour of a measure relative to satisfying both *P1* and *P2* is significant because it reveals an important characteristic about its fundamental nature as a measure of diversity. A measure of diversity can generally be considered either a *measure of concentration* or a *measure of dispersion*. A measure of concentration can be viewed as the opposite of a measure of dispersion, and we can convert one to the other via simple transformations. For example, if g corresponds to a measure of dispersion, then we can convert it to a measure of concentration f, where $f = \max(g) - g$. Here we only consider measures of concentration. A measure was considered to be a measure of concentration if it satisfied P1 and P2 without transformation. A measure was considered to be a measure of dispersion if it satisfied P1 and P2 following transformation. All measures of dispersion were transformed into measures of concentration prior to our analysis.

Skewness Principle (P3). Given a vector $(n_1, \ldots, n_m)$, where $n_1 = N - m + 1$, $n_i = 1$, $i = 2, \ldots, m$, and $N > m$, and a vector $(n_1 - c, n_2, \ldots, n_m, n_{m+1}, \ldots, n_{m+c})$, where $n_1 - c > 1$ and $n_i = 1$, $i = 2, \ldots, m + c$, $f(n_1, \ldots, n_m) > f(n_1 - c, n_2, \ldots, n_m, n_{m+1}, \ldots, n_{m+c})$.

P3 specifies that a summary containing m tuples, whose counts are distributed as unevenly as possible, will be more interesting than a summary containing $m + c$ tuples, whose counts are also distributed as unevenly as possible. For example, given the vectors $(999, 1)$ and $(997, 1, 1, 1)$ (i.e., $c = 2$), we require that $f(999, 1) > f(997, 1, 1, 1)$.

Permutation Invariance Principle (P4). Given a vector $(n_1, \ldots, n_m)$ and any permutation $(i_1, \ldots, i_m)$ of $(1, \ldots, m)$, $f(n_1, \ldots, n_m) = f(n_{i_1}, \ldots, n_{i_m})$.

P4 specifies that every permutation of a given distribution of tuple counts should be equally interesting. That is, interestingness is not a labeled property, it is only determined by the distribution of the counts. For example, given the vector $(2, 4, 6)$, we require that $f(2, 4, 6) = f(4, 2, 6) = f(4, 6, 2) = f(2, 6, 4) = f(6, 2, 4) = f(6, 4, 2)$.

Transfer Principle (P5). Given a vector $(n_1, \ldots, n_m)$ and $0 < c < n_j$, $f(n_1, \ldots, n_i + c, \ldots, n_j - c, \ldots, n_m) > f(n_1, \ldots, n_i, \ldots, n_j, \ldots, n_m)$.

P5, adapted from [6], specifies that when a strictly positive transfer is made from the count of one tuple to another tuple whose count is greater, then interestingness increases. For example, given the vectors $(10, 7, 5, 4)$ and $(10, 9, 5, 2)$, we require that $f(10, 9, 5, 2) > f(10, 7, 5, 4)$.

Those measures satisfying the above principles of interestingness are shown in Table 2. In Table 2, the *P1* to *P5* columns describe the five principles, and a measure that satisfies a principle is indicated by the *bullet* symbol (i.e., •).

Table 2. Measures satisfying the five principles

Measure	*P1*	*P2*	*P3*	*P4*	*P5*
$I_{Variance}$	•	•	•	•	•
$I_{Simpson}$	•	•	•	•	•
$I_{Shannon}$	•	•	•	•	•
$I_{McIntosh}$	•	•	•	•	•
I_{Lorenz}	•	•			•
I_{Gini}	•	•		•	•
I_{Berger}	•	•	•	•	
I_{Schutz}	•	•		•	
I_{Bray}	•	•		•	
$I_{Whittaker}$	•	•		•	
$I_{MacArthur}$	•	•		•	•
I_{Theil}	•			•	
$I_{Atkinson}$	•	•		•	•

5 Experimental Results

We now analyze the distribution of the index values generated by each of the thirteen measures. Input data consists of two populations of vectors shown in Table 3, where index values for 16,928 vectors (i.e., all possible ordered arrangements of a population of 50 objects among 10 classes) and 2,611 vectors (i.e., all possible ordered arrangements of a population of 50 objects among 5 classes) were generated. The choice of vectors to evaluate here was made somewhat arbitrarily, but it does provide a large, controlled population of index values in which a gradual change in evenness occurs from the most highly skewed distribution in the first vector, to the uniform distribution in the last vector.

Table 3. Ordered arrangements of two populations

50 objects / 10 classes	*50 objects / 5 classes*
(41, 1, 1, 1, 1, 1, 1, 1, 1, 1)	(46, 1, 1, 1, 1)
(40, 2, 1, 1, 1, 1, 1, 1, 1, 1)	(45, 2, 1, 1, 1)
(39, 3, 1, 1, 1, 1, 1, 1, 1, 1)	(44, 3, 1, 1, 1)
⋮	⋮
(6, 6, 5, 5, 5, 5, 5, 5, 4, 4)	(11, 11, 10, 10, 8)
(6, 5, 5, 5, 5, 5, 5, 5, 5, 4)	(11, 10, 10, 10, 9)
(5, 5, 5, 5, 5, 5, 5, 5, 5, 5)	(10, 10, 10, 10, 10)

Histograms of the absolute frequencies of the index values for the vectors in Table 3 were generated for each measure. Again, due to space limitations, we cannot show all of these histograms. However, sample histograms of the index values generated for the population of 50 objects among 10 classes by $I_{Variance}$ and I_{Schutz} are shown in Figures 3 and 4, respectively. In Figures 3 and 4, the horizontal and vertical axes describe intervals for the index values generated and the number of index values that fall in each interval, respectively. For example, the histogram for $I_{Variance}$ shows that 68 index values were generated on the interval $(0.000, 0.0009]$, 1,106 on $(0.0009, 0.003]$, 2,464 on $(0.003, 0.005]$, 3,006 on $(0.005, 0.007]$, 2,581 on $(0.007, 0.008]$, 2,055 on $(0.008, 0.010]$, 1,549 on $(0.010, 0.012]$, and 4,099 on the remaining intervals in $(0.012, 0.065]$. A curve describing the standard normal distribution (SND) of the index values is superimposed over the observed frequencies.

To provide a summary description of each histogram, we can use the skewness and kurtosis for the distribution of index values. *Skewness* is a measure of the symmetry of a distribution. It has a value of zero when the distribution is a symmetrical curve (i.e., as in a SND). If the skewness is different from zero, then the distribution is asymmetrical. A positive (negative) value indicates the index values are clustered more to the left (right) of the mean, with most of the extreme index values to the right (left) of the mean. In general, for positive (negative) skewness, we have mode $\leq$ median $\leq$ mean (mean $\leq$ median $\leq$ mode). *Kurtosis* is a measure of the relative peakedness of a distribution and indicates the extent to which outliers cause the distribution to differ from the SND. When a distribution follows the SND, it has value of zero. When the value is greater than (less than) zero, the distribution has a sharper (flatter) peak than the SND and is more (less) prone to containing outliers.

The skewness and kurtosis for all measures are shown in Table 4. In Table 4, mnemonics are provided as an aid to interpreting the curves described by the

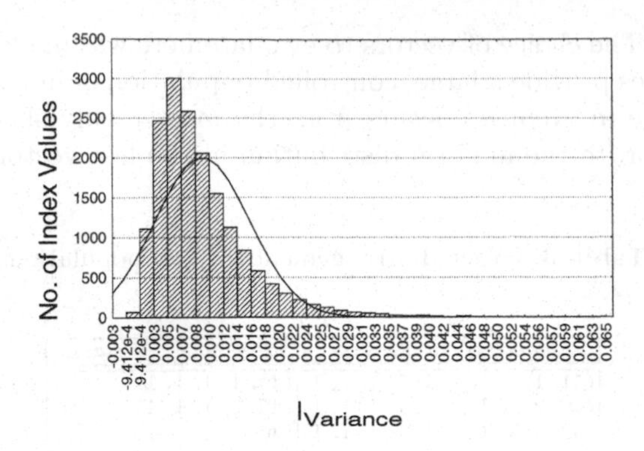

Fig. 3. Histogram for $I_{Variance}$

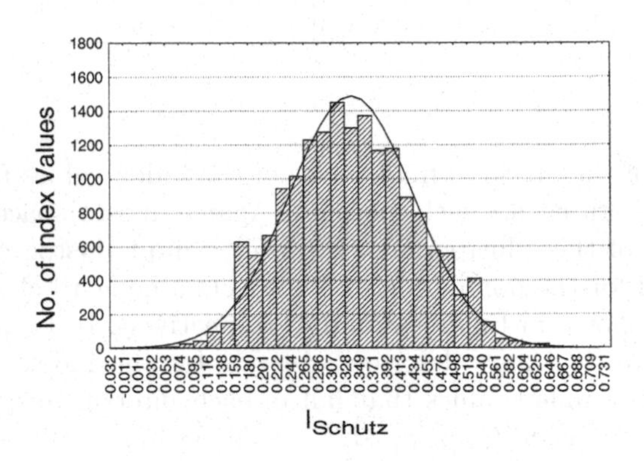

Fig. 4. Histogram for I_{Schutz}

values. The skewness mnemonics describe the symmetry of the frequency distribution in relation to the mean (i.e., AL = asymmetrical left, AR = asymmetrical right, NS = near symmetrical, and S = symmetrical) and the kurtosis mnemonics describe the relative peakedness of the frequency distribution in relation to the SND (i.e., SP = sharp peaked, NSN = near standard normal, MP = more peaked, and LP = less peaked). For example, the histogram for $I_{Variance}$, shown in Figure 3, has a skewness and kurtosis of approximately 1.8 and 5.6, respectively. This means that the distribution of index values is asymmetrical to the left of the mean (i.e., AL) and more sharply peaked than the SND (i.e., SP). Similarly, in the histogram for I_{Schutz}, shown in Figure 5, the distribution of index values is near symmetrical (i.e., NS) and less peaked than the SND (i.e., LP). The other measures in Table 4 can also be interpreted similarly.

We now determine the number of index values generated by each measure that are less than and greater than the middle index value (i.e., $(minimum + maximum)/2$), and less than and greater than the median (i.e., the value for which 50% of the generated index values lie below and 50% lie above). Our belief

Table 4. Skewness and kurtosis of the index values for the two populations

Measure	50 objects / 10 classes				50 objects / 5 classes			
	Skewness		Kurtosis		Skewness		Kurtosis	
$I_{Variance}$	1.84421	AL	5.571732	SP	1.55959	AL	3.273237	SP
$I_{Simpson}$	1.84421	AL	5.571732	SP	1.55959	AL	3.273237	SP
$I_{Shannon}$	-0.95761	AR	1.357844	MP	-1.03452	AR	1.391038	MP
$I_{McIntosh}$	-1.24351	AR	2.317341	SP	-1.13072	AR	1.496420	SP
I_{Lorenz}	0.14435	S	-0.232495	NSN	0.02128	S	-0.317871	NSN
I_{Gini}	-0.14435	S	-0.232495	NSN	-0.02128	S	-0.317871	NSN
I_{Berger}	0.97607	AL	1.139526	SP	0.75039	AL	0.264196	SP
I_{Schutz}	0.13192	NS	-0.130277	LP	0.27521	NS	-0.076436	LP
I_{Bray}	-0.13192	NS	-0.130277	LP	-0.27521	NS	-0.076436	LP
$I_{Whittaker}$	-0.13192	NS	-0.130277	LP	-0.27521	NS	-0.076436	LP
$I_{MacArthur}$	0.68369	AL	0.485805	MP	0.86586	AL	0.883313	MP
I_{Theil}	-0.05563	S	-0.236451	NSN	0.68371	AL	1.112360	MP
$I_{Atkinson}$	0.16650	NS	-0.422023	LP	0.30949	AL	-0.476633	LP

is that a useful measure of interestingness should generate index values that are reasonably distributed throughout the range of possible values (such as in a SND). Again, we analyze the index values generated from the two populations shown in Table 3, with the results shown in Tables 5 and 6. In Tables 5 and 6, the *Minimum* and *Maximum* columns describe the minimum and maximum index values generated by each measure, respectively, the *Middle* column describes the middle index value, the < *Middle* and > *Middle* columns describe the number of index values less than and greater than the middle index value, respectively, and the *Median* column describes the median index value. For example, for the $I_{Variance}$ measure, the minimum and maximum index values are 0.0 and 0.064, respectively, the middle index value is 0.032, 16,761 (167) index values lie below (above) the middle index value, and the median index value is 0.00791. The distribution of index values in Tables 5 and 6 is highly skewed about the middle and median values for most of the measures. Isolated exceptions include I_{Bray} and $I_{Whittaker}$ in Table 5, and I_{Lorenz} and I_{Gini} in Table 6.

Table 5. Distribution of index values for 50 objects among 10 classes

Measure	Minimum	Maximum	Middle	< Middle	> Middle	Median
$I_{Variance}$	0.0	0.064	0.032	16761	167	0.007911
$I_{Simpson}$	0.1	0.676	0.388	16761	167	0.1712
$I_{Shannon}$	1.250664	3.321928	2.286295	613	16315	2.860161
$I_{McIntosh}$	0.207096	0.7964	0.50175	509	16419	0.682799
I_{Lorenz}	0.214	0.55	0.37	12353	4575	0.338
I_{Gini}	0.107	0.275	0.185	4786	12142	0.169
I_{Berger}	0.14	0.82	0.46	15836	1092	0.28
I_{Schutz}	0.0	0.72	0.36	10751	6177	0.34
I_{Bray}	0.28	1.0	0.64	7549	9379	0.66
$I_{Whittaker}$	0.28	1.0	0.64	7549	9379	0.66
$I_{MacArthur}$	0.0	0.420842	0.21042	15683	1245	0.114606
I_{Theil}	0.0	2.141432	1.07072	5550	11378	1.21593
$I_{Atkinson}$	0.0	0.71	0.35503	11432	5496	0.296977

Table 6. Distribution of index values for 50 objects among 5 classes

Measure	*Minimum*	*Maximum*	*Middle*	< *Middle*	> *Middle*	*Median*
$I_{Variance}$	0.0	0.162	0.081	2507	104	0.0258
$I_{Simpson}$	0.2	0.848	0.524	2507	104	0.3032
$I_{Shannon}$	0.562179	2.321928	1.44205	164	2447	1.940238
$I_{McIntosh}$	0.092165	0.643839	0.36800	200	2411	0.523381
I_{Lorenz}	0.24	0.6	0.42	1496	1115	0.412
I_{Gini}	0.12	0.300	0.21	1183	1428	0.0.206
I_{Berger}	0.2	0.92	0.56	2180	431	0.42
I_{Schutz}	0.0	0.72	0.36	1850	761	0.3
I_{Bray}	0.28	1.0	0.64	939	1672	0.7
$I_{Whittaker}$	0.28	1.0	0.64	939	1672	0.7
$I_{MacArthur}$	0.0	0.427524	0.213765	2425	186	0.099571
I_{Theil}	0.0	1.759749	0.879875	2357	254	0.566115
$I_{Atkinson}$	0.0	0.784944	0.39247	1964	647	0.283374

6 Conclusion and Future Research

The use of diversity measures for ranking the interestingness of summaries generated from databases is a new application area. Here we theoretically and experimentally analyzed thirteen diversity measures. Five principles of interestingness for useful diversity measures were described. Theoretical results showed that only four of the thirteen diversity measures satisfied all five principles. Experimental results showed that the distribution of index values, in relation to the mean, is least skewed for I_{Lorenz}, I_{Gini}, I_{Schutz}, I_{Bray}, and $I_{Whittaker}$, but these measures are poorly behaved, containing a sharp peak, or multiple sharp peaks, in the frequency distribution of the index values. The remaining eight measures were skewed asymmetrically in relation to the mean, and more or less peaked than the SND. The experimental results also show that the distribution of the index values is highly skewed, in relation to the middle and median values, for most of the measures.

Future research will focus on extending the theory of interestingness for diversity measures used to rank summaries. New principles will be developed for ranking the interestingness of summaries generated from different sources (i.e., related, but physically, logically, or temporally independent databases).

References

1. R. Agrawal and R. Srikant. Fast algorithms for mining association rules. In *Proceedings of the 20th International Conference on Very Large Databases (VLDB'94)*, pages 487–499, Santiago, Chile, September 1994.
2. A.B. Atkinson. On the measurement of inequality. *Journal of Economic Theory*, 2:244–263, 1970.
3. R.J. Bayardo and R. Agrawal. Mining the most interesting rules. In *Proceedings of the Fifth International Conference on Knowledge Discovery and Data Mining (KDD'99)*, pages 145–154, San Diego, California, August 1999.

4. I. Bournaud and J.-G. Ganascia. Accounting for domain knowledge in the construction of a generalization space. In *Proceedings of the Third International Conference on Conceptual Structures*, pages 446–459. Springer-Verlag, August 1997.
5. C.L. Carter and H.J. Hamilton. Efficient attribute-oriented algorithms for knowledge discovery from large databases. *IEEE Transactions on Knowledge and Data Engineering*, 10(2):193–208, March/April 1998.
6. H. Dalton. The measurement of the inequality of incomes. *Economic Journal*, 30:348–361, 1920.
7. G. Dong and J. Li. Interestingness of discovered association rules in terms of neighborhood-based unexpectedness. In X. Wu, R. Kotagiri, and K. Korb, editors, *Proceedings of the Second Pacific-Asia Conference on Knowledge Discovery and Data Mining (PAKDD'98)*, pages 72–86, Melbourne, Australia, April 1998.
8. A.A. Freitas. On objective measures of rule surprisingness. In J. Zytkow and M. Quafafou, editors, *Proceedings of the Second European Conference on the Principles of Data Mining and Knowledge Discovery (PKDD'98)*, pages 1–9, Nantes, France, September 1998.
9. R. Godin, R. Missaoui, and H. Alaoui. Incremental concept formation algorithms based on galois (concept) lattices. *Computational Intelligence*, 11(2):246–267, 1995.
10. R.J. Hilderman and H.J. Hamilton. Heuristic measures of interestingness. In J. Zytkow and J. Rauch, editors, *Proceedings of the Third European Conference on the Principles of Data Mining and Knowledge Discovery (PKDD'99)*, pages 232–241, Prague, Czech Republic, September 1999.
11. R.J. Hilderman and H.J. Hamilton. Heuristics for ranking the interestingness of discovered knowledge. In N. Zhong and L. Zhou, editors, *Proceedings of the Third Pacific-Asia Conference on Knowledge Discovery and Data Mining (PAKDD'99)*, pages 204–209, Beijing, China, April 1999.
12. R.J. Hilderman and H.J. Hamilton. Applying objective interestingness measures in data mining systems. In *Proceedings of the 4th European Symposium on Principles of Data Mining and Knowledge Discovery (PKDD'00)*, pages 432–439, Lyon, France, September 2000.
13. R.J. Hilderman and H.J. Hamilton. Principles for mining summaries: Theorems and proofs. Technical Report CS 00-01, Department of Computer Science, University of Regina, February 2000. Online at http://www.cs.uregina.ca/research/Techreport/0001.ps.
14. R.J. Hilderman and H.J. Hamilton. Principles for mining summaries using objective measures of interestingness. In *Proceedings of the Twelfth IEEE International Conference on Tools with Artificial Intelligence (ICTAI'00)*, pages 72–81, Vancouver, Canada, November 2000.
15. R.J. Hilderman, H.J. Hamilton, and N. Cercone. Data mining in large databases using domain generalization graphs. *Journal of Intelligent Information Systems*, 13(3):195–234, November 1999.
16. S. Lieberson. An extension of Greenberg's linguistic diversity measures. *Language*, 40:526–531, 1964.
17. A.E. Magurran. *Ecological diversity and its measurement.* Princeton University Press, 1988.
18. B. Padmanabhan and A. Tuzhilin. A belief-driven method for discovering unexpected patterns. In *Proceedings of the Fourth International Conference on Knowledge Discovery and Data Mining (KDD'98)*, pages 94–100, New York, New York, August 1998.
19. G.P. Patil and C. Taillie. Diversity as a concept and its measurement. *Journal of the American Statistical Association*, 77(379):548–567, 1982.

20. S. Sahar. Interestingness via what is not interesting. In *Proceedings of the Fifth International Conference on Knowledge Discovery and Data Mining (KDD'99)*, pages 332–336, San Diego, California, August 1999.
21. C.E. Shannon and W. Weaver. *The mathematical theory of communication*. University of Illinois Press, 1949.
22. G. Stumme, R. Wille, and U. Wille. Conceptual knowledge discovery in databases using formal concept analysis methods. In J. Zytkow and M. Quafafou, editors, *Proceedings of the Second European Conference on the Principles of Data Mining and Knowledge Discovery (PKDD'98)*, pages 450–458, Nantes, France, September 1998.
23. M.L. Weitzman. On diversity. *The Quarterly Journal of Economics*, pages 363–405, May 1992.

Peculiarity Oriented Mining and Its Application for Knowledge Discovery in Amino-Acid Data

Ning Zhong[1], Muneaki Ohshima[1], and Setsuo Ohsuga[2]

[1] Dept. of Information Eng., Maebashi Institute of Technology
[2] Dept. of Information and Computer Science, Waseda University

Abstract. The paper proposes a way of peculiarity oriented mining and its application for knowledge discovery in the amino-acid data set. We introduce the *peculiarity rules* as a new type of association rules, which can be discovered from a relatively small number of *peculiar* data by searching the relevance among the peculiar data. We argue that the peculiarity rules represent a typically unexpected, interesting regularity hidden in the amino-acid data set.

1 Introduction

Peculiarity is a kind of interestingness. Peculiarity relationships/rules (with common sense) may be hidden in a relatively small number of data. Generally speaking, hypotheses (knowledge) generated from databases can be divided into the following three types:

- Incorrect hypotheses.
- Useless hypotheses.
- *New, surprising, interesting hypotheses.*

The purpose of data mining is to discover new, surprising, interesting knowledge hidden in databases. Hence, the evaluation of interestingness (including peculiarity, surprisingness, unexpectedness, usefulness, novelty) should be done in pre-processing and/or post-processing of the knowledge discovery process [8,18].

In the paper, we discuss a way of mining *peculiarity rules* from the amino-acid data set. Section 2 introduces the *peculiarity rules* as a new type of association rules, which can be discovered from a relatively small number of the *peculiar* data by searching the relevance among the peculiar data. Sections 3 describes a method of finding the peculiar data/rules. Then in Section 4, we discuss a result of mining from the amino-acid data set. We shows that the peculiarity rules represent a typically unexpected, interesting regularity hidden in the amino-acid data set. Finally, Section 5 gives conclusions and outlines further research directions.

D. Cheung, G.J. Williams, and Q. Li (Eds.): PAKDD 2001, LNAI 2035, pp. 260–269, 2001.

2 Association Rules vs. Peculiarity Rules

Association rules are an important class of regularity hidden in transaction databases [1,2]. The intuitive meaning of such a rule is that transactions of the database which contain X tend to contain Y. So far, three categories of the association rules, the *general rule*, the *exception rule*, and the *peculiarity rule* have been investigated [10,18].

A *general rule* is a description of a regularity for numerous objects and represents the well-known fact with common sense, while an *exception rule* is for a relatively small number of objects and represents exceptions to the well-known fact. Usually, the exception rule should be associated with a general rule as a set of rule pairs. For example, the rule "using a seat belt is risky for a child" which represents exceptions to the general rule with common sense "using a seat belt is safe".

Zhong et al proposed *peculiarity rules* as a new type of association rules [18]. A *peculiarity rule* is discovered from the *peculiar* data by searching the relevance among the *peculiar* data. Roughly speaking, a data is *peculiar* if it represents a peculiar case described by a relatively small number of objects and is very different from other objects in a data set. Although it looks like the exception rule from the viewpoint of describing a relatively small number of objects, the peculiarity rule represents the well-known fact with common sense, which is a feature of the general rule [2].

We argue that the *peculiarity rules* are a typical regularity hidden in a lot of scientific, statistical, and transaction databases. Sometimes, the general rules that represent the well-known fact with common sense cannot be found from numerous scientific, statistical or transaction data, or although they can be found, the rules may be uninteresting ones to the user since data are rarely specially collected/stored in a database for the purpose of mining knowledge in most organizations. Hence, the evaluation of interestingness (including surprisingness, unexpectedness, peculiarity, usefulness, novelty) should be done before and/or after knowledge discovery [4,8,9]. In particular, unexpected (common sense) relationships/rules may be hidden in a relatively small number of data. Thus, we may focus on some interesting data (the peculiar data), and then we find more novel and interesting rules (peculiarity rules) from the data. For example, the following rules are the peculiarity ones that can be discovered from a relation called *Japan-Geography* (see Table 1) in a *Japan-Survey* database:

$$rule_1 : \mathit{ArableLand(large)}\ \&\ \mathit{Forest(large)} \rightarrow \mathit{PopulationDensity}(low).$$
$$rule_2 : \mathit{ArableLand(small)}\ \&\ \mathit{Forest(small)} \rightarrow \mathit{PopulationDensity}(high).$$

In order to discover the rules, we first need to search the peculiar data in the relation *Japanese-Geography*. From Table 1, we can see that the values of the attributes *ArableLand* and *Forest* for Hokkaido (i.e. 1209 Kha and 5355 Kha) and for Tokyo and Osaka (i.e. 12 Kha, 18 Kha, and 80 Kha, 59 Kha) are very different from other values in the attributes. Hence, the values are regarded as the peculiar data. Furthermore, $rule_1$ and $rule_2$ are generated by searching the relevance

among the peculiar data. Note that we use the qualitative representation for the quantitative values in the above rules. The transformation of quantitative to qualitative values can be done by using the following background knowledge on information granularity:
Basic granules:

bg_1 = {*high, low*}; bg_2 = {*large, small*};
bg_3 = {*many, little*}; bg_4 = {*far, close*};
bg_5 = {*long, short*};
....

Specific granules:

biggest-cities = {*Tokyo, Osaka*};
kanto-area = {*Tokyo, Tiba, Saitama, ...*};
kansei-area = {*Osaka, Kyoto, Nara, ...*};
....

That is, *ArableLand* = *1209*, *Forest* = *5355* and *PopulationDensity* = *67.8* for Hokkaido are replaced by the granules, "large" and "low", respectively. Furthermore, Tokyo and Osaka are regarded as a neighborhood (i.e. the biggest cities in Japan). Hence, $rule_2$ is generated by using the peculiar data for both Tokyo and Osaka as well as their granules (i.e. "small" for *ArableLand* and *Forest,* and "high" for *PopulationDensity*).

Table 1. Japan-Geography

Region	Area	Population	PopulationDensity	PeasantFamilyN	ArableLand	Forest	...
Hokkaido	82410.58	5656	**67.8**	93	**1209**	**5355**	...
Aomori	9605.45	1506	156.8	87	169	623	...
...	...	...	...	...	...	...	...
Tiba	5155.64	5673	1100.3	116	148	168	...
Tokyo	2183.42	11610	**5317.2**	21	**12**	**80**	...
...	...	...	...	...	...	...	...
Osaka	1886.49	8549	**4531.6**	39	**18**	**59**	...
	...	...	...	...	...	...	...

3 Peculiarity Oriented Mining

This section describes a way of mining peculiarity rules.

3.1 Finding the Peculiar Data

There are many ways of finding the peculiar data. In this section, we describe an attribute-oriented method.

Let $X = \{x_1, x_2, \ldots, x_n\}$ be a data set related to an attribute in a relation, and n is the number of different values in an attribute. The peculiarity of x_i can be evaluated by the *Peculiarity Factor*, $PF(x_i)$,

$$PF(x_i) = \sum_{j=1}^{n} \sqrt{N(x_i, x_j)}. \quad (1)$$

It evaluates whether x_i occurs relatively small number and is very different from other data x_j by calculating the sum of the square root of the conceptual distance between x_i and x_j. The reason why the square root is used in Eq. (1) is that we prefer to evaluate more near distances for relatively large number of data so that the peculiar data can be found from relatively small number of data.

Major merits of the method are

- It can handle both the continuous and symbolic attributes based on a unified semantic interpretation;
- Background knowledge represented by binary neighborhoods can be used to evaluate the peculiarity if such background knowledge is provided by a user.

If X is a data set of a continuous attribute and no background knowledge is available, in Eq. (1),

$$N(x_i, x_j) = |x_i - x_j|. \quad (2)$$

Table 2 shows an example for the calculation. On the other hand, if X is a data set of a symbolic attribute and/or the background knowledge for representing the conceptual distances between x_i and x_j is provided by a user, the peculiarity factor is calculated by the conceptual distances, $N(x_i, x_j)$. Table 3(a) shows an example in which the binary neighborhoods shown in Table 3(b) are used as the background knowledge for representing the conceptual distances of different type of restaurants [6,14]. However, all the conceptual distances are 1, as default, if background knowledge is not available.

Table 2. An example of peculiarity factors for a continuous attribute

Region	ArableLand
Hokkaido	**1209**
Tokyo	12
Osaka	18
Yamaguchi	162
Okinawa	147

⟹

PF
134.1
60.9
60.3
60.5
59.4

After the evaluation for the peculiarity, the peculiar data are extracted by using a threshold value,

$$threshold = mean\ of\ PF(x_i) + \alpha \times standard\ deviation\ of\ PF(x_i), \quad (3)$$

where α can be adjusted by a user, and $\alpha = 1$ as default. That is, if $PF(x_i)$ is over the threshold value, x_i is a peculiar data.

Based on the preparation stated above, the process of finding the peculiar data can be outlined as follows:

Table 3. An example of peculiarity factors for a symbolic attribute (Restaurants) and their conceptual distances

(a) Peculiarity factors

Restaurant	Type		PF
Wendy	American		6.6
Le Chef	French		7.2
Great Wall	Chinese	⟹	4.7
Kiku	Japanese		5.5
South Sea	Chinese		4.7

(b) Conceptual distances

Type	Type	N
Chinese	Japanese	1
Chinese	American	3
Chinese	French	4
American	French	2
American	Japanese	3
French	Japanese	3

Step 1. Calculate the peculiarity factor $PF(x_i)$ in Eq. (1) for all values in a data set (i.e. an attribute).
Step 2. Calculate the threshold value in Eq. (3) based on the peculiarity factor obtained in *Step* 1.
Step 3. Select the data that is over the threshold value as the peculiar data.
Step 4. If the current peculiarity level is enough, then goto *Step* 6.
Step 5. Remove the peculiar data from the data set and thus, we get a new data set. Then go back to *Step* 1.
Step 6. Change the granularity of the peculiar data by using background knowledge on information granularity if the background knowledge is available.

Furthermore, the process can be done in a parallel-distributed mode for multiple attributes, relations and databases since this is an attribute-oriented finding method.

3.2 Relevance among the Peculiar Data

A peculiarity rule is discovered from the peculiar data by searching the relevance among the peculiar data. Let $X(x)$ and $Y(y)$ be the peculiar data found in two attributes X and Y respectively. We deal with the following two cases:

– If both $X(x)$ and $Y(y)$ are symbolic data, the relevance between $X(x)$ and $Y(y)$ is evaluated in the following equation:

$$R_1 = P_1(X(x)|Y(y))P_2(Y(y)|X(x)). \tag{4}$$

That is, the larger the product of the probabilities of P_1 and P_2, the stronger the relevance between $X(x)$ and $Y(y)$.

- If both $X(x)$ and $Y(y)$ are continuous attributes, the relevance between $X(x)$ and $Y(y)$ is evaluated by using the method developed in our KOSI system [17]

Furthermore, Eq. (4) is suitable for handling more than two peculiar data found in more than two attributes if $X(x)$ (or $Y(y)$) is a granule of the peculiar data.

4 Application in Amino-Acid Data Mining

Some of databases such as Japan-survey, web-log, weather, supermarket, and amino-acid data have been tested or have been testing for our approach. This section discusses a result of mining from the amino-acid data set [12].

The amino-acid data set can be divided into two groups: amino-acid matrix (including VH and VL amino-acid matrixes) and experimental data (including combining coefficients and coefficients related thermodynamics). The main features of the data set can be summarized as follows:

- The number of the attributes is quite many. That is, the number of the attributes with respect to the amino-acid matrix is 230, the number of the attributes with respect to experimental data is 7.
- The number of the instances is relatively small and only small number of data in the amino-acid matrix changes.

The objective of data mining is to find the association between the amino-acid matrix and experimental data. That is, how experimental data change when amino-acid data are changed.

At first, we find the peculiar data in all attributes respectively by using the method stated in Section 3. As a result, the data denoted in a bold type style in Tables 4, 5 and 6 are the peculiar data. Note that in Tables 4, 5 and 6, the last tuple T is threshold calculated in Eq. (3) and α is 1.

From the tuple 23 (i.e. No 23) in Tables 4, 5 and 6, we can see that the value 42 in the attribute *Ka* (combining coefficients) is a peculiar data and the maximum one in *Ka*, and no any change in the amino-acid matrix. Therefore, we focus on the attribute *Ka* and search the minimum value in *Ka*. In other words, we want to find *how coefficients related thermodynamics and the amino-acid matrix change when combining coefficients have big change.* We found that

- The value 0.04 in the attribute *Ka* (the tuple 26, i.e. No 26) is the minimum one;
- In the same tuple (the tuple 26, i.e. No 26), the values related thermodynamics: -32.6 in *DG,* -53.4 in *DH,* -0.92 in *DCp* are peculiar data;
- The value *a* in 32 of VL amino-acid matrix is also a peculiar one.

Furthermore, we found that there is a functional relationship between *Ka* and *DG* [17]. Therefore, we just use one attribute, *Ka* or *DG*, when generating a peculiarity rule.

In summary, the discovered rules are

If the value in 32 of VL amino-acid matrix is changed to **a**,
Then the value of Ka *is the minimum one and the values of* DH *and* DCp *are peculiar ones.*

or

If the value of Ka *is the minimum one and the values of* DH *and* DCp *are peculiar ones,*
Then the value in 32 of VL amino-acid matrix is changed to **a**.

The result has been evaluated by an expert [12]. According to his opinion, the discovered rules are reasonable and interesting.

We argue that the peculiarity rules represent a typically unexpected, interesting regularity hidden in the amino-acid data set. The rules are peculiar ones rather than exceptions because of semantic common sense.

Table 4. VH amino-acid matrix with peculiar data and their PF values

No	31	PF(x)	32	PF(x)	33	PF(x)	50	PF(x)	53	PF(x)	56	PF(x)	58	PF(x)	98	PF(x)	99	PF(x)
1	**a**	**34**	d	4	y	4	y	3	y	4	s	1	y	3	w	1	d	2
2	s	1	**a**	**33**	y	4	y	3	y	4	s	1	y	3	w	1	d	2
3	s	1	**e**	**34**	y	4	y	3	y	4	s	1	y	3	w	1	d	2
4	s	1	**n**	**34**	y	4	y	3	y	4	s	1	y	3	w	1	d	2
5	s	1	d	4	**a**	**34**	y	3	y	4	s	1	y	3	w	1	d	2
6	s	1	d	4	**l**	**34**	y	3	y	4	s	1	y	3	w	1	d	2
7	s	1	d	4	**f**	**34**	y	3	y	4	s	1	y	3	w	1	d	2
8	s	1	d	4	**w**	**34**	y	3	y	4	s	1	y	3	w	1	d	2
9	s	1	d	4	y	4	**a**	**34**	y	4	s	1	y	3	w	1	d	2
10	s	1	d	4	y	4	**l**	**34**	y	4	s	1	y	3	w	1	d	2
11	s	1	d	4	y	4	**f**	**34**	y	4	s	1	y	3	w	1	d	2
12	s	1	d	4	y	4	y	3	**a**	**34**	s	1	y	3	w	1	d	2
13	s	1	d	4	y	4	y	3	**l**	**34**	s	1	y	3	w	1	d	2
14	s	1	d	4	y	4	y	3	**p**	**34**	s	1	y	3	w	1	d	2
15	s	1	d	4	y	4	y	3	**w**	**34**	s	1	y	3	w	1	d	2
16	s	1	d	4	y	4	y	3	y	4	**a**	**34**	y	3	w	1	d	2
17	s	1	d	4	y	4	y	3	y	4	s	1	**a**	**34**	w	1	d	2
18	s	1	d	4	y	4	y	3	y	4	s	1	**l**	**34**	w	1	d	2
19	s	1	d	4	y	4	y	3	y	4	s	1	**f**	**34**	w	1	d	2
20	s	1	d	4	y	4	y	3	y	4	s	1	y	3	**a**	**34**	d	2
21	s	1	d	4	y	4	y	3	y	4	s	1	y	3	w	1	**a**	**33**
22	s	1	**a**	**33**	y	4	y	3	y	4	s	1	y	3	w	1	**a**	**33**
23	s	1	d	4	y	4	y	3	y	4	s	1	y	3	w	1	d	2
24	s	1	d	4	y	4	y	3	y	4	s	1	y	3	w	1	d	2
25	s	1	d	4	y	4	y	3	y	4	s	1	y	3	w	1	d	2
26	s	1	d	4	y	4	y	3	y	4	s	1	y	3	w	1	d	2
27	s	1	d	4	y	4	y	3	y	4	s	1	y	3	w	1	d	2
28	s	1	d	4	y	4	y	3	y	4	s	1	y	3	w	1	d	2
29	s	1	d	4	y	4	y	3	y	4	s	1	y	3	w	1	d	2
30	s	1	d	4	y	4	y	3	y	4	s	1	y	3	w	1	d	2
31	s	1	d	4	y	4	y	3	y	4	s	1	y	3	w	1	d	2
32	s	1	d	4	y	4	y	3	y	4	s	1	y	3	w	1	d	2
33	s	1	d	4	y	4	y	3	y	4	s	1	y	3	w	1	d	2
34	s	1	d	4	y	4	y	3	y	4	s	1	y	3	w	1	d	2
35	s	1	d	4	y	4	y	3	y	4	s	1	y	3	w	1	d	2
T		7.44		16.75		16.97		14.34		16.97		7.44		14.34		7.44		10.97

Table 5. VL amino-acid matrix with peculiar data and their PF values

No	31	PF(x)	32	PF(x)	50	PF(x)	53	PF(x)	91	PF(x)	92	PF(x)	96	PF(x)
1	n	2	n	2	y	2	q	2	s	1	n	2	y	1
2	n	2	n	2	y	2	q	2	s	1	n	2	y	1
3	n	2	n	2	y	2	q	2	s	1	n	2	y	1
4	n	2	n	2	y	2	q	2	s	1	n	2	y	1
5	n	2	n	2	y	2	q	2	s	1	n	2	y	1
6	n	2	n	2	y	2	q	2	s	1	n	2	y	1
7	n	2	n	2	y	2	q	2	s	1	n	2	y	1
8	n	2	n	2	y	2	q	2	s	1	n	2	y	1
9	n	2	n	2	y	2	q	2	s	1	n	2	y	1
10	n	2	n	2	y	2	q	2	s	1	n	2	y	1
11	n	2	n	2	y	2	q	2	s	1	n	2	y	1
12	n	2	n	2	y	2	q	2	s	1	n	2	y	1
13	n	2	n	2	y	2	q	2	s	1	n	2	y	1
14	n	2	n	2	y	2	q	2	s	1	n	2	y	1
15	n	2	n	2	y	2	q	2	s	1	n	2	y	1
16	n	2	n	2	y	2	q	2	s	1	n	2	y	1
17	n	2	n	2	y	2	q	2	s	1	n	2	y	1
18	n	2	n	2	y	2	q	2	s	1	n	2	y	1
19	n	2	n	2	y	2	q	2	s	1	n	2	y	1
20	n	2	n	2	y	2	q	2	s	1	n	2	y	1
21	n	2	n	2	y	2	q	2	s	1	n	2	y	1
22	n	2	n	2	y	2	q	2	s	1	n	2	y	1
23	n	2	n	2	y	2	q	2	s	1	n	2	y	1
24	**a**	**34**	n	2	y	2	q	2	s	1	n	2	y	1
25	**d**	**34**	n	2	y	2	q	2	s	1	n	2	y	1
26	n	2	**a**	**34**	y	2	q	2	s	1	n	2	y	1
27	n	2	**d**	**34**	y	2	q	2	s	1	n	2	y	1
28	n	2	n	2	**a**	**34**	q	2	s	1	n	2	y	1
29	n	2	n	2	**f**	**34**	q	2	s	1	n	2	y	1
30	n	2	n	2	y	2	**a**	**34**	s	1	n	2	y	1
31	n	2	n	2	y	2	**e**	**34**	s	1	n	2	y	1
32	n	2	n	2	y	2	q	2	**a**	**34**	n	2	y	1
33	n	2	n	2	y	2	q	2	s	1	**d**	**34**	y	1
34	n	2	n	2	y	2	q	2	s	1	**a**	**34**	y	1
35	n	2	n	2	y	2	q	2	s	1	n	2	**f**	**34**
T		11.26		11.26		11.26		11.26		7.44		11.26		7.44

Table 6. Experimental data and their PF values

No	Ka x107/M-1	PF(x)	DG/Jmol-1	PF(x)	DH/Jmol-1	PF(x)	TDS/Jmol-1	PF(x)	DCp/Jmol-1K-1	PF(x)
1	9.6	74.38	-46.3	44.2	-97.9	119.79	-51.6	113.6	-2.25	19.67
2	10	74.85	-46.4	43.95	-112.9	145.52	-66.5	142.47	-2.15	19.71
3	16.9	92.03	-47.7	49.28	-108.7	136.72	-61	129.63	-2.26	20.02
4	22	108.06	-48.5	54.07	-115.8	154.33	-67.3	144.9	-2.25	19.67
5	ND	0	ND	0	ND	0	ND	0	ND	0
6	1.5	83.08	-41.4	63.93	-67.3	138.89	-25.9	123.67	-1.8	18.15
7	7.1	75.62	-45.6	46.3	-73.2	131.2	-27.6	121.36	-1.81	18.19
8	2.3	80.84	-42.6	58.04	-65.6	141.15	-23	128.26	-1.78	18.19
9	ND	0	ND	0	ND	0	ND	0	ND	0
10	0.37	86.69	-38	80.07	-53.9	163.18	-15.9	145.42	-1.38	18.4
11	2.9	79.65	-43.1	55.11	-59.8	150.34	-16.7	143.06	-1.27	20.37
12	0.19	87.62	-36.4	87.81	-60.2	149.47	-23.8	126.77	ND	0
13	11	76.25	-46.4	43.95	-81.9	123	-35.4	114.99	-1.43	18.35
14	15	85.34	-47.2	45.93	-84.4	119.01	-37.2	113.34	-0.98	23.6
15	13	80.01	-46.8	44.58	-98.6	120.69	-51.8	113.9	-1.38	18.4
16	12	77.05	-47	45.42	-90.5	118.37	-42.5	112.02	-1.4	18.07
17	2.6	80.08	-43.1	55.11	-72.3	132.28	-29.2	119.57	-1	23.32
18	3.4	79.3	-43.5	54.23	-62.3	146.21	-18.8	137.83	-0.92	24.47
19	23	111.74	-48.5	54.07	-85.3	119.49	-36.8	113.47	-1.58	18.19
20	ND	0	ND	0	ND	0	ND	0	ND	0
21	21	105.05	-48.2	52.46	-113.7	147.16	-65.5	140.19	-1.9	18.69
22	4	79.16	-44.1	52.63	-114.2	148.64	-70.1	154.37	-2.1	19.55
23	**42**	**169.11**	-50.2	67.3	-91.5	117.98	-41.3	112.39	-1.4	18.07
24	34	147.06	-49.5	62.16	-106.3	130.92	-56.8	122.24	-2.42	23.28
25	8.8	74.45	-46.1	44.36	-105.8	129.88	-59.7	126.2	-2.31	21.13
26	**0.04**	88.67	**-32.6**	**105.24**	**-53.4**	**164.42**	-20.8	133.17	**-0.92**	**24.47**
27	0.97	84.53	-40.5	68.11	-53.6	163.73	-13.1	154.83	-1.59	18.17
28	0.64	85.6	-39.4	73.35	-84.4	119.01	-45	111.82	-1.02	23.15
29	9.2	74.25	-46.1	44.36	-76	129.01	-30.1	119.02	-1.64	18.26
30	7.8	75.17	-45.7	45.96	-96.8	119.11	-51.1	113.54	-2.25	19.67
31	6.9	75.93	-45.5	46.87	-105	128.96	-59.5	125.84	-2.39	22.67
32	15.6	87.28	-47.5	48	-101.6	125.09	-54.1	118.04	-2.24	19.81
33	14	82.61	-47.2	45.93	-93.6	117.85	-46.4	111.9	-1.4	18.07
34	ND	0	ND	0	ND	0	ND	0	ND	0
35	12	77.05	-46.8	44.58	-94.6	118.08	-47.8	112.52	-1.97	19.04
T		112.48		71.65		164.16		153.35		24.44

5 Conclusion

We presented a way of peculiarity oriented mining and its application for knowledge discovery in the amino-acid data set. The peculiarity rules are defined as a new type of association rules that is a kind of regularity hidden in a relatively small number of peculiar data. We showed that the peculiarity rules represent a typically unexpected, interesting regularity hidden in the amino-acid data set.

Since this project is very new, we just had a preliminary result in knowledge discovery from the amino-acid data set. Our future work includes using more domain knowledge in the knowledge discovery process, mining in multiple information sources, and developing an agent-based mining system.

Acknowledgements. The authors would like to thank Prof. S. Tsumoto and Prof. K. Tsumoto for providing the amino-acid data set and background knowledge, and evaluating the experimental results.

References

1. Agrawal R. et al. "Database Mining: A Performance Perspective", *IEEE Trans. Knowl. Data Eng.*, 5(6) (1993) 914-925.
2. Agrawal R. et al. "Fast Discovery of Association Rules", *Advances in Knowledge Discovery and Data Mining*, AAAI Press (1996) 307-328.
3. Fayyad, U.M., Piatetsky-Shapiro, G et al (eds.) *Advances in Knowledge Discovery and Data Mining*, AAAI Press (1996).
4. Freitas, A.A. "On Objective Measures of Rule Surprisingness" J. Zytkow and M. Quafafou (eds.) *Principles of Data Mining and Knowledge Discovery*, LNAI 1510, Springer (1998) 1-9.
5. Johnson, R.A. and Wichern, D.W. *Applied Multivariate Statistical Analysis*, Prentice Hall (1998).
6. Lin, T.Y. "Granular Computing on Binary Relations 1: Data Mining and Neighborhood Systems", L. Polkowski and A. Skowron (eds.) *Rough Sets in Knowledge Discovery 1*, In Studies in Fuzziness and Soft Computing series, Vol. 18, Physica-Verlag (1998) 107-121.
7. Lin, T.Y., Zhong, N., Dong, J., and Ohsuga, S. "Frameworks for Mining Binary Relations in Data", L. Polkowski and A. Skowron (eds.) *Rough Sets and Current Trends in Computing*, LNAI 1424, Springer (1998) 387-393.
8. Liu, B., Hsu W., and Chen, S. "Using General Impressions to Analyze Discovered Classification Rules", *Proc. Third International Conference on Knowledge Discovery and Data Mining (KDD-97)*, AAAI Press (1997) 31-36.
9. Silberschatz, A. and Tuzhilin, A. "What Makes Patterns Interesting in Knowledge Discovery Systems", *IEEE Trans. Knowl. Data Eng.*, 8(6) (1996) 970-974.
10. Suzuki E.. "Autonomous Discovery of Reliable Exception Rules", *Proc Third Inter. Conf. on Knowledge Discovery and Data Mining (KDD-97)*, AAAI Press (1997) 259-262.
11. Thrun, S. et al. "Automated Learning and Discovery", AI Magazine (Fall 1999) 78-82.
12. Tsumoto, K. and Kumagai, I. "Thermodynamic and Kinetic Analyses of The Antigen-Antibody Interaction Using Mutants", *Research Report of JSAI SIG-KBS-A002* (2000) 83-88.

13. Wrobel, S. "An Algorithm for Multi-relational Discovery of Subgroups", J. Komorowski and J. Zytkow (eds.) *Principles of Data Mining and Knowledge Discovery*. LNAI 1263, Springer (1997) 367-375.
14. Yao, Y.Y. "Granular Computing using Neighborhood Systems", Roy, R., Furuhashi, T., and Chawdhry, P.K. (eds.) *Advances in Soft Computing: Engineering Design and Manufacturing*, Springer (1999) 539-553.
15. Yao, Y.Y. and Zhong, N. "Potential Applications of Granular Computing in Knowledge Discovery and Data Mining", *Proc. The 5th International Conference on Information Systems Analysis and Synthesis (IASA'99)*, edited in the invited session on Intelligent Data Mining and Knowledge Discovery (1999) 573-580.
16. Zadeh, L. A. "Toward a Theory of Fuzzy Information Granulation and Its Centrality in Human Reasoning and Fuzzy Logic", *Fuzzy Sets and Systems*, Elsevier, 90 (1997) 111-127.
17. Zhong, N. and Ohsuga, S. "KOSI - An Integrated System for Discovering Functional Relations from Databases", *Journal of Intelligent Information Systems*, Vol.5, No.1, Kluwer (1995) 25-50.
18. Zhong, N., Yao, Y.Y., and Ohsuga, S. "Peculiarity Oriented Multi-Database Mining", J. Zytkow and Jan Rauch (eds.) *Principles of Data Mining and Knowledge Discovery*. LNAI 1704, Springer (1999) 136-146.
19. Zhong, N., Skowron, A., and Ohsuga, S. (eds.) *New Directions in Rough Sets, Data Mining, and Granular-Soft Computing*, LNAI 1711, Springer (1999).
20. Zhong, N. "MULTI-DATABASE MINING: A Granular Computing Approach", *Proc. 5th Joint Conference on Information Sciences (JCIS'00)* in special session on Granular Computing and Data Mining (GrC-DM) (2000) 198-201.

Mining Sequence Patterns from Wind Tunnel Experimental Data for Flight Control

Zhenyu Liu[1], Wesley W. Chu[1], Adam Huang[2], Chris Folk[2], and Chih-Ming Ho[2]

University of California, Los Angeles
Los Angeles, CA90095, USA
{vicliu, wwc}@cs.ucla.edu, {pohao, chrisf, chihming}@ucla.edu

Abstract. This paper presents a sequence pattern mining technique to mine data generated from a wind tunnel experiment. The goal is to discover the nonlinear input-output relationship for a delta wing aircraft. In contrast to categorical datasets, the output variable(s) in this dataset is continuous and takes distinct values, which is common in physical experiments. Directly applying existing decision tree or rule induction mining methods fails to discover sufficient knowledge. Therefore, we propose to extend current techniques by constructing sequence patterns that represent the output variations in certain ranges of selective inputs. Similar sequence patterns are clustered based on a weighted variance measure. Rules can then be derived from similar sequences to predict the output. This technique has been applied to the experimental data and generates rules useful for flight control.

1 Introduction

Existing data mining methods such as decision tree induction[11], rule derivation [1] or Bayesian learning [3], have largely focused on datasets with nonnumeric or categorical variables. Therefore, these methods are suitable for such applications as product forecasting or cross-selling where categorical variables prevail. However, data generated from scientific experiments are different from conventional datasets in the following aspects:

- Numerical variables involved are continuous and may take distinct real numbers within valid ranges.
- Strong casual relationship exists among these numerical variables. The outcome of one output variable is often correlated with all the input variables.

Therefore, new approaches are required to discover knowledge from these experimental data.

In this paper, we are focused on a dataset generated by the MEMS UAV (Uninhabited Aerial Vehicle) project in the Mechanical and Aerospace Engineering Department at UCLA. The data is collected from a wind tunnel in which a delta wing

---

[1] Computer Science Department
[2] Mechanical and Aerospace Engineering Department

D. Cheung, G.J. Williams, and Q. Li (Eds.): PAKDD 2001, LNAI 2035, pp. 270-281, 2001.

aircraft model is mounted [7]. Each tuple correlates one particular input configuration of the aircraft model with the corresponding force loading outputs. The goal of mining this dataset is to derive the highly nonlinear input-output relationship for the aircraft. Such knowledge will be useful for flight control. The preliminary dataset contains 192 tuples summarized from the wind tunnel experiments and provides insights into aircraft maneuvering via MEMS devices.

Traditional mining methods generate knowledge to predict output variables base on a subset of input variables. This approach is acceptable in many business-related applications where a portion of the inputs is sufficient to predict the output. In physical system control such as the delta wing aircraft with MEMS actuators, however, output variables (e.g. force and moments) are highly dependent on all input variables (angle of attack, stream velocity, actuation position, see Figure 1 and Figure 2). Under such environments, existing algorithms are unable to derive input-output relationship that covers all the cases.

To remedy this problem, we transform the original dataset by merging the output with several inputs into a composite output variable called sequence. More precisely, a sequence is defined as the output variation in a certain range of selected inputs. The transformed results enable us to cluster similar sequences via a bottom-up algorithm. Existing methods, e.g. rule induction, can then be applied to these sequence clusters. Using such an approach, we are able to derive fairly complete input-output relationship for the wind tunnel experimental data.

Scientific discovery research has been existing for more a decade. Its goal is to find knowledge that is novel, interesting, plausible, and understandable [14]. From this general perspective, scientific discovery shares common characteristics with that of knowledge discovery (data mining) in business applications. This work is strongly influenced by the scientific discovery viewpoint and yet leveraged on the existing data mining techniques in discovering interesting patterns from a scientific dataset. The resulting rules are special cases of the qualitative and quantitative laws in the general scientific discovery framework [8].

The rest of the paper is organized as follows. Section 2 gives a brief background on the aircraft control principles and shows the deficiencies of directly applying traditional methods. In Section 3, we propose the sequence clustering technique and apply it to the wind tunnel experimental dataset. Section 4 concludes the paper and provides future research directions.

2 Control of a Delta Wing Aircraft

MEMS UAV (Uninhabited Aerial Vehicle), an ongoing project at UCLA, has demonstrated the possibility of using MEMS micron-scale actuation devices to control macro-scale machines, e.g., an aircraft. Such a design has numerous advantages in reducing weight, overall power consumption and radar cross-section. The project uses the "vortex" control method to provide forces and moments for controlling the aircraft. Typically, a delta wing aircraft will produce pairs of vortices above the wings (Figure 1). These vortices are sources of low-pressure flows that provide "suction", which produces a portion of lift for the aircraft. Airflow blows toward the delta wing, first hitting the lower surface and then moving up toward the upper surface, eventually detaching near the leading edge and creating the vortices

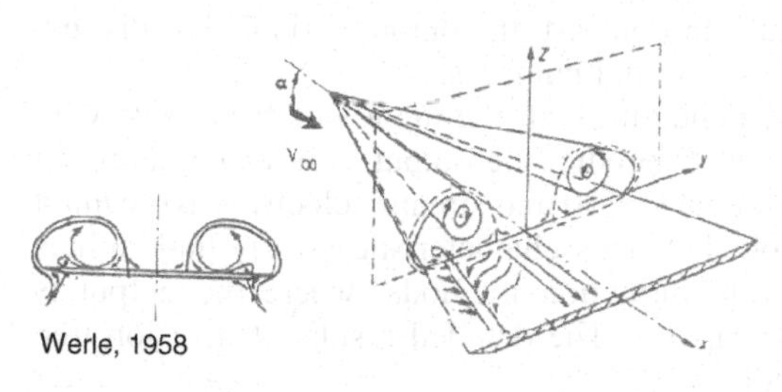

Fig. 1. Pairs of vortices above the delta wing

pair. Numerous researches have shown that the genesis location of these vortices, i.e. the detaching positions, is very important to the characteristics of the resulting large primary vortices [9, 10]. By placing MEMS actuators near this location, the symmetry of these high "suction" vortices is broken. As a result, aerodynamic loadings on the aircraft can be controlled.

2.1 Problem Description

The key for aircraft control is accurately predicting the aircraft's force loadings based on certain environment settings and an actuation position. Load measurements for the delta wing, with a six-component force balance, are divided into two categories: forces and moments. Each category has three variables, corresponding to the three dimensions. Environment settings include the wind tunnel stream velocity and the aircraft's angle of attack. Figure 2 visually interprets these terms. As shown in the figure, the delta wing is equipped with rounded leading edges. Note that the actuation position is one point at each cross section, forming a straight line along the whole leading edge. This position is represented by an angle value, ranging from 0° to 180°.

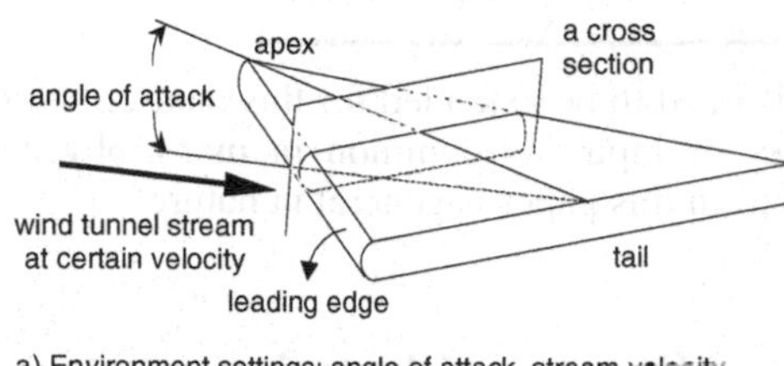

a) Environment settings: angle of attack, stream velocity.

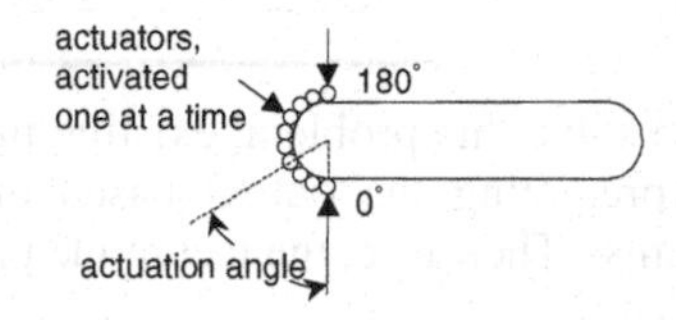

b) Cross section view of the delta wing with actuators equipped on the leading edge, from 0° to 180°. The actuation angle affects the force and moment output of the delta wing aircraft.

Fig. 2. Input variables for a delta wing aircraft

In the rest of the paper, the variables about the environment settings and the actuation angle are referred to as 'input variables', whereas the variables about the force balance outputs as 'output variables'.

Wind tunnel experimental results have shown drastic variances of the force balance outputs with different environment and actuation settings. Table 1 shows part of the dataset that relates the rolling moment output (one component in the force balance outputs) with corresponding input values.

2.2 Data Characteristics

Due to current experiment design, the input variables (i.e. angle of attack, stream velocity and actuation angle) only take a small number of distinct values. Therefore, these inputs can be treated as categorical. In contrast, the output variable has distinct values for all tuples and ranges from the set of real numbers.

Furthermore, the output variable is dependent on all the input variables. Knowing even two of the three inputs is insufficient to predict the output. For example, using two variable combinations like "angle of attack = 20 and stream velocity = 10" cannot predict the rolling moment output (Table 1). This characteristic greatly undermines the effectiveness of decision tree or rule induction methods, where the output is predicted only based on a subset of the inputs. The detailed results of decision tree and rule mining are shown in the appendix.

Table 1. Wind tunnel experiment results of the delta wing aircraft. Angle of attack, stream velocity and actuation angle are the input variables and rolling moment output is the output variable

Angle of attack (°)	Stream velocity (m/s)	Actuation angle (°)	Rolling moment output
		...	
20	10	40	-0.00485
20	10	60	0.00092
20	10	80	0.00026
20	10	100	-0.00621
20	10	120	0.00011
20	10	140	-0.00626
20	15	40	-0.01179
20	15	...	...
20	15	140	-0.00361
		...	

To solve this problem, existing methods need to be extended for this dataset. Note that predicting the output based on a set of inputs is common in many physical systems. Therefore, the technique presented in this paper is general in nature.

3 Discovering Rules on the Basis of Sequence Clustering

The basic idea of our technique is as follows. The output value may not be determined based on a subset of the inputs. However, the output variation in certain ranges of selected inputs may follow certain sequence patterns. We shall first extract such sequence patterns from the raw data. A sequence clustering hierarchy can be built in a bottom-up fashion based on inter-cluster errors (*ice*). Such a hierarchy provides cluster candidates. A weighted variance (*wvar*) measure is used to describe the sequences closeness within each candidate. The clustering is finalized by selecting candidate clusters whose *wvar*s are below a user-specified threshold. Sequences in such clusters are considered similar and approximated by the corresponding sequence mean. Rules can be then derived on each cluster to represent the input-output relationship.

3.1 Definition of a Sequence

Consider a dataset *D* with input variables X_1, ..., X_n, an output variable *Y* and a predicate *p* defined on the inputs. A *sequence of Y w.r.t X_i ($1 \le i \le n$) characterized by p* is a set of 2-tuples: $\{<y_1, x_{i_1}>, <y_2, x_{i_2}>, \ldots, <y_m, x_{i_m}>\}$ calculated by $\Pi_{Y,X_i}(\sigma_p(D))^1$, where $y_1, \ldots, y_m, x_{i_1}, \ldots, x_{i_m}$ are specific values of *Y* and X_i, respectively. Without losing generality, we can assume $x_{i_1} \le x_{i_2} \le \ldots \le x_{i_m}$. For example, Table 1 contains a sequence of the rolling moment w.r.t the actuation angle: {<-0.00485, 40>,<0.00092, 60>, <0.00026, 80>, <-0.00621, 100>, <0.00011, 120>, <-0.00626, 140>} characterized by "angle of attack = 20 AND stream velocity = 10".

Note that this definition is slightly different from those in existing research, e.g., [12, 13], where the time variable is implicitly used as X_i in our definition.

3.2 Clustering Hierarchy Generation

Since sequences are objects with no total order, we use a bottom-up clustering strategy, MDC [15], to build a hierarchical cluster over the sequence set. Given *s* sequences, *s* initial clusters are built each containing one sequence. The algorithm merges two closest clusters at each step and finishes constructing a binary-tree after the s-1_{th} iteration.

Let us now apply this clustering strategy to the experimental data:

1. Extract sequences of the rolling moment output w.r.t the actuation angle (Table 1). Such sequences are characterized by predicates in the form: "angle of attack = α AND stream velocity = *v*", where α and *v* range from {5°, 10°, 15°, 20°, 25°, 30°, 35°} and {10 m/s, 15 m/s, 20 m/s}, respectively.
2. In order to discover more frequently occurred patterns from these sequences, we normalize on the output variable so that for a particular angle of attack and stream velocity the difference between the maximum and the minimum output is 1.
3. Euclidean distance is used as the distance measure between two sequences S_i and S_j

$$|S_i - S_j| = [\sum_{k=1}^{m} (s_{i_k} - s_{j_k})^2]^{1/2} \quad (1)$$

Here m is the length of each sequence, while s_{i_k}, s_{j_k} ($1 \le k \le m$) are the k_{th} output values in sequence S_i and S_j, respectively.
4. An inter-cluster error (*ice*) measure [15] is used to calculate the distance between two sequence clusters C_1 and C_2 ($|C|$ denotes the size of *C*):

$$ice(C_1, C_2) = \frac{1}{|C_1|}\frac{1}{|C_2|}\sum_{i=1}^{|C_1|}\sum_{j=1}^{|C_2|} |S_i - S_j|,\ S_i \in C_1,\ S_j \in C_2 \quad (2)$$

The resulting clustering hierarchy is shown in Figure 3. Each leaf is a sequence characterized by the corresponding label. For example, the leaf "angle of attack = 20° AND stream velocity = 10" represents the sequence {<-0.00485, 40>, <0.00092, 60>, <0.00026, 80>, <-0.00621, 100>, <0.00011, 120>, <-0.00626, 140>}.

3.3 Cluster Selection Based on Weighted Variance

Each branch node in the generated hierarchy represents a candidate cluster. We shall introduce the notion of weighted variance (*wvar*) to measure the closeness within a candidate cluster. Candidates with *wvar*s lower than a specified threshold will be chosen as final clusters. For a cluster $C = \{S_1, S_2, \ldots, S_l\}$, *wvar(C)* should be proportional to the cluster's standard deviation:

$$\sigma_S = \left[\frac{1}{l} \sum_{i=1}^{l} | S_i - E_S |^2 \right]^{1/2}, \text{ where } E_S \text{ is the mean of } S_1, \ldots, S_l. \quad (3)$$

For two clusters with the same standard deviation but different amplitude ranges, we introduce the amplitude measure *amp*(*C*) to provide weighted preference based on a cluster *C*'s amplitude range:

$$amp(C) = \max_{1 \le i \le l}(\max_{1 \le k \le m} s_{i_k}) - \min_{1 \le i \le l}(\min_{1 \le k \le m} s_{i_k}), \quad (4)$$

where m is the length of each sequence and s_{i_k} $(1 \le i \le l,\ 1 \le k \le m)$ is the k_{th} output value in sequence S_i.

Therefore, we define the weighted variance for a cluster C as:

$$wvar(C) = \frac{\sigma_S}{amp(C)}, \text{ if } amp(C) \ne 0; \text{ otherwise } 0 \quad (5)$$

Each branch node in Figure 3 represents a candidate cluster and is labeled by that cluster's *wvar*. All the sequences in a branch with *wvar* below certain user-specified threshold are considered similar. A smaller *wvar* threshold yields smaller cluster sizes and more accurate approximation by the sequence mean. By setting such a threshold as "*wvar* < 0.32", the final clusters are chosen as Figure 4.

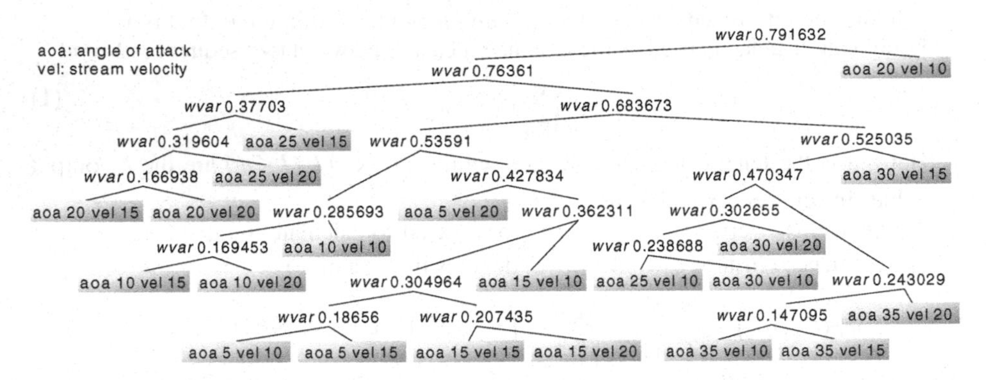

Fig. 3. Clustering of sequences extracted from the experimental flight data

Based on the results in Figure 4, traditional mining methods such as rule induction can be applied to generate more complete knowledge.

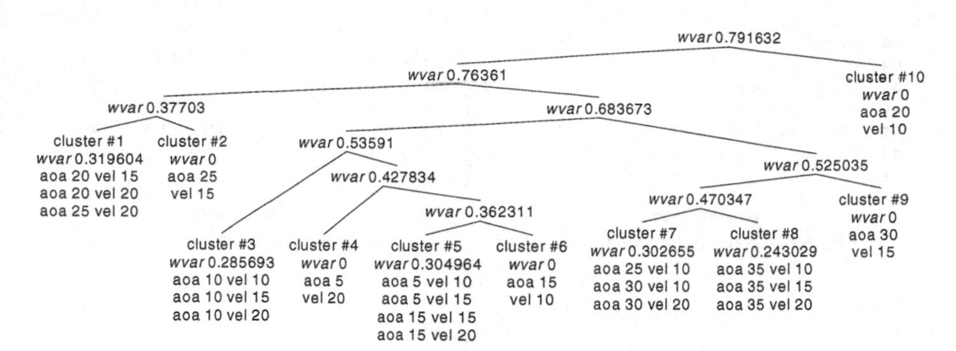

Fig. 4. The pruned clustering hierarchy of Figure 3 with *cae* < 0.32

3.4 Rule Derivation from Similar Sequences

Based on the clustering result, forward inference rules (also referred to as classification or discriminant rules [5, 6]) can be derived in the following form:

$$\text{IF } p \text{ THEN } mean(\text{cluster } \#i),\ wvar(\text{cluster } \#i),\ \text{confidence: P[cluster } \#i \mid p] \quad \textbf{(6)}$$

Here p is a predicate defined on the input variables, *mean*(cluster #*i*) is the sequence mean calculated on cluster #*i*, *wvar*(cluster #*i*) is the cluster's weighted variance and P[cluster #*i* | *p*] is the conditional probability of cluster #*i* given *p*.

To derive such forward inference rules, an algorithm should search over all possible input variable predicates and select those predicates that yield rule supports and confidences above certain thresholds. Pruning strategies are used in this process to reduce the search space. For efficient algorithms on forward inference rule generation, see [4, 5].

For example, to generate rules on cluster #5 (Figure 4), we set the minimum support as "2" and minimum confidence as "60%". The forward inference rules generated are:

IF angle of attack=5° THEN *mean*(cluster #5), *wvar* 0.304964, confidence 66.7%.

IF angle of attack=15° THEN *mean*(cluster #5), *wvar* 0.304964, confidence 66.7%

Figure 5(a) displays the rolling moment output values of the four sequences in cluster #5. Figure 5(b) shows the corresponding sequence mean.

Similarly, the following rule can be generated from cluster #8:

IF angle of attack=35° THEN *mean*(cluster #8), *wvar* 0.243029, confidence 100%.

The sequences and mean of cluster #8 are shown in Figure 6(a) and Figure 6(b), respectively.

3.5 Application of Derived Rules

Since the clustering result summarizes the raw data, we can derive rules on all the clusters. The resulting rule set gives much better coverage over the entire case space, and therefore is more useful for flight control.

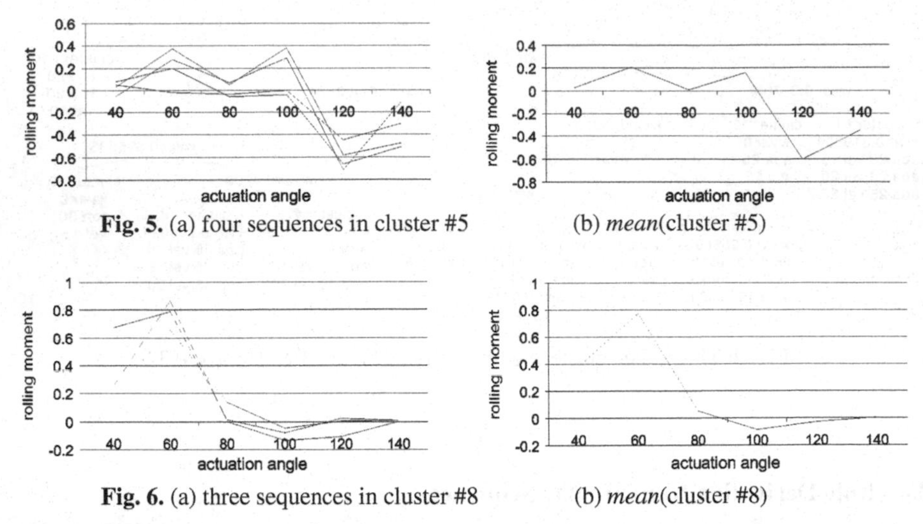

Fig. 5. (a) four sequences in cluster #5 (b) *mean*(cluster #5)

Fig. 6. (a) three sequences in cluster #8 (b) *mean*(cluster #8)

Currently, forward inference rules can be used to predict force balance outputs based on input settings. An average sequence is first predicted using angle of attack and/or stream velocity. For example, given an angle of attack at 35°, the sequence in Figure 6(b) is selected. Using this sequence, the rolling moment output can be determined at each actuation angle. Note that the sequences have been normalized before clustering (see Sect. 3.2). The output value should be multiplied by the corresponding normalization factor.

We are planning a series of wind tunnel experiments to learn the dynamic characteristics of the delta wing aircraft. Not only the force balance outputs but also the aircraft's moving dynamics (e.g. speed and acceleration) will be recorded. The augmented dataset will allow us to derive rules providing more insight into flight dynamics. For example, rules can be generated predicting the variation of the rolling speed with respect to the actuation setting and/or the roll angle. Such rules can guide us to select the proper actuation schema to achieve a desirable control effect.

4 Conclusion and Future Work

Traditional mining methods fail to derive sufficient input-output relationship for predicting physical system behavior. In this paper, we propose a novel knowledge discovery technique based on sequence patterns. A sequence is defined as the output variation in certain ranges of selected inputs. A sequence clustering hierarchy can be built in a bottom-up fashion, using inter-cluster error (*ice*) as the distance measure. Based on the hierarchy, similar sequences are grouped in to clusters. The sizes of these clusters are controlled by the weighted variance (*wvar*) measure. Each cluster is represented by the sequence mean of that cluster. Forward inference rules are then derived on each cluster to represent the input-output relationship. We have applied this technique to the wind tunnel experimental data and derive useful knowledge for MEMS-based aircraft control.

From the experiment design aspect, we plan to expand the current wind tunnel experiments to include dynamic behaviors. Such experimental data will allow us to mine input-output relationship under dynamic environments. From the algorithm development aspect, the current sequence definition needs to be extended to include multiple input and output variables, which will widen the scope of frequent patterns. Further, we need to extend the proposed sequence clustering and rule derivation technique to future augmented datasets and reduce the computation complexity.

Reference:

1. R. Agrawal, H. Mannila, R. Srikant, H. Toivonen, and A.I. Verkamo. Fast discovery of association rules. In U. M. Fayyad, G. Piatetsky-Shapiro, P. Smyth and R. Uthurusamy, editors, *Advances in Knowledge Discovery and Data Mining*, pages 399-421. AAAI/MIT Press, 1996.
2. W.W. Chu and K. Chiang. Abstraction of high level concepts from numerical values in databases. In *Proc. of the AAAI Workshop on Knowledge Discovery in Databases*, Seattle, MA, July 1994.
3. C. Elkan. Boosting and naive bayesian learning. Technical report no cs97-557. Dept. of Computer Science and Engineering, UCSD, September 1997.
4. G. Giuffrida, L. G. Cooper, and W. W. Chu. A scalable bottom-up data mining algorithm for relational databases. In *10th Int'l Conf. on Scientific and Statistical Database Management (SSDBM '98)*, Capri, Italy, July 1998.
5. G. Giuffrida, W. W. Chu, and D. M. Hanssens. NOAH: An algorithm for mining classification rules from datasets with large attribute space. In *Proc. of 12th Int'l Conf. on Extending Database (EDBT)*, Konsta, Germany, March 2000.
6. J. Han and Y. Fu. Exploration of the power of attribute-oriented induction in data mining. In U. M. Fayyad, G. Piatetsky-Shapiro, P. Smyth and R. Uthurusamy, editors, *Advances in Knowledge Discovery and Data Mining*, pages 399-421. AAAI/MIT Press, 1996.
7. A. Huang, C. Folk, C. Silva, B. Christensen, Y. Chen, C.M. Ho, F. Jiang, C. Grosjean, Y.C. Tai, G.B. Lee, M. Chen, S. Newbern. Applications of MEMS devices to delta wing aircraft: from concept development to transonic flight test. In *39th AIAA Aerospace Sciences Meeting and Exhibit*, Reno, U.S.A., January, 2001.
8. P. Langley. The computer-aided discovery of scientific knowledge. In *Proc. of the 1st Int'l Conf. on Discovery Science (DS'98)*, Fukuoka, Japan, December 1998.
9. G.B. Lee, S. Chiang, Y.C. Tai, T. Tsao, and C.M. Ho, Robust vortex control of a delta wing using distributed MEMS actuators, accepted, *Journal of Aircraft* (2000).
10. J.F. Marchman III, "Effect of heating on leading edge vortices in subsonic flow," *Journal of Aircraft*, Vol. 12, No. 2, pp. 121-123 (1975).
11. M. Mehta, R. Agrawal, and J. Rissanen. SLIQ: A fast scalable classifier for data mining. In *Proc. of the 5th Int'l Conf. on Extending Database Technology (EDBT)*, Avignon, France, March 1996.
12. S. Park, W.W. Chu, J. Yoon, and C. Hsu. Efficient searches for similar subsequences of different lengths in sequence databases. In *Proc. of the 16th IEEE Int'l Conf. on Data Engineering (ICDE)*, San Diego, CA, February, 2000.
13. R. Srikant and R. Agrawal. Mining sequential patterns: generalizations and performance improvements. In *Proc. of the 5th Int'l Conf. on Extending Database Technology (EDBT)*, Avignon, France, March 1996.
14. R.E. Valdés-Pérez. Knowledge discovery tools for science applications. In *Communications of the ACM*, 42(11):37-41, November 1999.
15. G. Zhang and W.W. Chu. MDC: a mixed-type data clustering algorithm. Technical report, Computer Science Department, UCLA.

Acknowledgement. The authors would like to thank Giovanni Giuffrida and Qinghua Zou for many fruitful discussions and valuable comments on this paper. The wind tunnel experiments discussed in this paper are supported by DARPA MTO.

Appendix

Conventional data mining research mostly focused to datasets with only categorical variables. To apply existing methods on this particular dataset, we need to first discretize the output variable that takes continuous values. Such discretization methods are discussed in [6]. The basic idea is to leverage on a concept hierarchy generated manually by experts or automatically from the data distribution. The Distance Sensitive Clustering (DISC) method is used [2] to build a Type Abstraction Hierarchy (TAH) for the output variable. Each node in such a hierarchy corresponds to a value range. The whole data range can be partitioned using the ranges of nodes at a certain level. For example, the rolling moment output data in Table 1 can be partitioned into six clusters: [-0.01179, -0.00380], [-0.00380, -0.00138], [-0.00138, -0.00024], [-0.00024, 0.00028], [0.00028, 0.00124], [0.00124, 0.00537].

In the following sections, we apply two common methods to classify the rolling moment output: decision tree and association rule induction.

Decision Tree

The dataset was first run on a decision tree generation algorithm provided by IBM Intelligent Miner [11]. The class variable is the discretized rolling moment output. The input variables include angle of attack, stream velocity and actuation angle. The tool generated a four-level decision tree after pruning. To test the effectiveness of the result, the tree was directly applied back to predict the training dataset. Table 2 shows the prediction confusion matrix. The number in the (i_{th}, j_{th}) table entry represents the percentage of tuples that belongs to the i_{th} class yet predicted as the j_{th} class.

The high error rate attributes to the high dependency of the output variable on all the input variables. A single input variable has low predictive power on the output when taken alone. Therefore, univariate splitting, the basic philosophy behind decision tree induction, makes the method unsuitable for this kind of dataset.

Table 2. Confusion matrix based on the pruned decision tree. Error rate: 46.35%

Predicted Classes / Actual Classes	[-0.01179, -0.00380]	[-0.00380, -0.00138]	[-0.00138, -0.00024]	[-0.00024, 0.00028]	[0.00028, 0.00124]	[0.00124, 0.00537]	Total
[-0.01179, -0.00380]	6.25%	0%	0%	1.56%	0%	1.04%	8.85%
[-0.00380, -0.00138]	1.56%	0%	0%	6.77%	0%	1.56%	9.90%
[-0.00138, -0.00024]	2.60%	0%	0%	10.42%	0%	2.60%	15.62%
[-0.00024, 0.00028]	3.02%	0%	0%	38.54%	0%	0.52%	42.18%
[0.00028, 0.00124]	4.17%	0%	0%	6.25%	0%	1.04%	11.46%
[0.00124, 0.00537]	1.04%	0%	0%	2.08%	0%	8.85%	11.98%
Total	18.75%	0%	0%	65.62%	0%	15.62%	100%

Association Rules

Our second effort was to run the dataset on association rule derivation methods. Since we are concerned about using input variables to predict the output variable, we concentrate only on rules that have input variables in their left-hand-sides and the output variable as their right-hand-sides. The Apriori algorithm [1] has been tested on the dataset after discretization. The minimum support and confidence are set to 3% and 70%, respectively. All the resulting rules that satisfy the above restriction are listed in table 3.

Table 3. Rules generated by Apriori associate the output variable with the input variables

#	Support (%)	Confidence (%)	Rule body	Rule head i.e. the rolling moment
1	11.9792	95.8300	angle of attack = 0	[-0.00024, 0.00028]
2	4.1667	100.0000	angle of attack = 0 AND stream velocity = 10	[-0.00024, 0.00028]
3	3.6458	87.5000	angle of attack = 0 AND stream velocity = 15	[-0.00024, 0.00028]
4	4.1667	100.0000	angle of attack = 0 AND stream velocity = 20	[-0.00024, 0.00028]
5	9.8958	79.1700	angle of attack = 5	[-0.00024, 0.00028]
6	3.1250	75.0000	angle of attack = 5 AND stream velocity = 10	[-0.00024, 0.00028]
7	3.1250	75.0000	angle of attack = 5 AND stream velocity = 15	[-0.00024, 0.00028]
8	3.6458	87.5000	angle of attack = 5 AND stream velocity = 20	[-0.00024, 0.00028]
9	3.1250	75.0000	stream velocity = 10.00 AND actuation angle = 0	[-0.00024, 0.00028]

The knowledge provided by those rules suffer from the following shortcomings:

1. **Low coverage.** The nine rules in Table 3 cover only 28.125% of the original dataset, whereas 71.875% of the cases encountered cannot be predicted. Due to the low coverage over the entire case space, this rule set cannot provide sufficient information about the input-output relationship. Thus, it is insufficient for flight control.

2. **Unable to handle control-sensitive regions.** When the angle of attack is above 15°, the output variable is more sensitive to the inputs. That is, the output in this region has larger magnitudes and greater variances. However, the rules (Table 3) derived by Apriori are mostly in the insensitive region (i.e. angle of attack below 15°) since data in this region is less variant and tends to give higher rule supports and confidences.

The reason that association rules fail to capture the sensitive region is due to the basic rule form: " IF $X_1 = x_1$ AND … AND $X_t = x_t$ THEN $Y = y$". Here X_1, … , X_t are input variables and Y is the output. For a dataset with n input variables, t is usually less than n. Otherwise a rule simply reiterates a tuple in the dataset. However, a t less than n means omitting certain input variables. In the sensitive region, omitting any input variable in the left-hand-side may be disastrous since the right-hand-side cannot be concentrated in one category. This is best illustrated by the real data shown below.

Thus, using conventional rule induction results in the following dilemma: rules generated either reiterate the original tuples, or have undesirably low confidences.

Table 4. Droping any one variable in "angle of attack = 20° AND stream velocity = 15 AND actuation angle = 60°" generates three 2-variable combinations. Each of these 2-variable combinations corresponds to a sub table listed below. From these tables, we note that no rules containing only two variables in the left-hand-side, e.g."IF angle of attack = 20° AND stream velocity = 15 THEN Rolling moment = ... ", will have a big confidence measure

Angle of attack (°)	Stream velocity (m/s)	Actuation angle (°)	Rolling moment
20	15	40	[-0.01179, -0.00380]
20	15	60	[-0.00380, -0.00138]
20	15	80	[-0.01179, -0.00380]
20	15	100	[0.00124, 0.00537]
20	15	120	[0.00028, 0.00124]
20	15	140	[-0.00380, -0.00138]

a) angle of attack = 20 AND stream velocity = 15

Angle of attack (°)	Stream velocity (m/s)	Actuation angle (°)	Rolling moment
0	15	60	[0.00028, 0.00124]
5	15	60	[-0.00024, 0.00028]
10	15	60	[-0.00024, 0.00028]
15	15	60	[0.00124, 0.00537]
20	15	60	[-0.00380, 0.00138]
25	15	60	[0.00028, 0.00124]
30	15	60	[0.00124, 0.00537]
35	15	60	[0.00124, 0.00537]

b) stream velocity = 15 AND actuation angle = 60

Angle of attack (°)	Stream velocity (m/s)	Actuation angle (°)	Rolling moment
20	10	60	[0.00028, 0.00124]
20	15	60	[-0.00380, -0.00138]
20	20	60	[-0.01179, -0.00380]

c) angle of attack = 20 AND actuation angle = 60

Scalable Hierarchical Clustering Method for Sequences of Categorical Values*

Tadeusz Morzy, Marek Wojciechowski, and Maciej Zakrzewicz

Poznan University of Technology
Institute of Computing Science
ul. Piotrowo 3a, 60-965 Poznan, Poland
Tadeusz.Morzy@put.poznan.pl
Marek.Wojciechowski@cs.put.poznan.pl
Maciej.Zakrzewicz@cs.put.poznan.pl

Abstract. Data clustering methods have many applications in the area of data mining. Traditional clustering algorithms deal with quantitative or categorical *data points*. However, there exist many important databases that store *categorical data sequences*, where significant knowledge is hidden behind sequential dependencies between the data. In this paper we introduce a problem of clustering categorical data sequences and present an efficient scalable algorithm to solve the problem. Our algorithm implements the general idea of agglomerative hierarchical clustering and uses frequently occurring subsequences as features describing data sequences. The algorithm not only discovers a set of high quality clusters containing similar data sequences but also provides descriptions of the discovered clusters.

1 Introduction

Clustering is one of the most popular unsupervised data analysis methods that aims at identifying groups of similar objects based on the values of their attributes [14][15]. Many clustering techniques have been proposed in the area of machine learning [7][12][14] and statistics [15]. Those techniques can be classified as *partitional* and *hierarchical*. Partitional clustering obtains a partition of data objects into a given number of clusters optimizing some clustering criterion. Hierarchical clustering is a set of partitions forming a cluster hierarchy. An *agglomerative* hierarchical clustering starts with clusters containing single objects and then merges them until all objects are in the same cluster. In each iteration two most similar clusters are merged. *Divisive* hierarchical clustering starts with one cluster and iteratively divides it into smaller pieces.

Emerging data mining applications place additional requirements on clustering techniques, namely: scalability with database sizes, effective treatment of high dimensionality and interpretability of results [1]. Recently, the problem of data clustering has been redefined in the data mining area. The concept of *cluster mining*

* This work was partially supported by the grant no. KBN 43-1309 from the State Committee for Scientific Research (KBN), Poland.

D. Cheung, G.J. Williams, and Q. Li (Eds.): PAKDD 2001, LNAI 2035, pp. 282-293, 2001.

[18] is used to represent a method which analyzes very large data sets to efficiently identify a small set of high-quality (statistically strong) clusters of data items. Cluster mining does not aim at partitioning of all the data items – instead, less frequent noise and outliers are ignored. In other words, cluster mining finds only the highest density areas hidden in the data space.

A number of efficient and scalable clustering algorithms for clustering *data points* represented by multidimensional quantitative values [1][6][10][22], as well as by sets of categorical values (product names, URLs, etc.) [8][9][11][13][19][21] have been proposed so far in the data mining area. However, we notice that there exist many important databases that store *categorical data sequences*, where significant knowledge is hidden behind sequential dependencies between the data (credit card usage history, operating system logs, database redo logs, web access paths, etc.). The existing clustering algorithms either cannot be easily transformed to deal with categorical data sequences or cannot take into account the sequential dependencies.

Categorical data sequence clustering can have many applications in the area of behavioral segmentation – e.g. *web users segmentation* [18]. The problem of web users segmentation is to use web access log files to partition a set of users into clusters such that the users within a cluster are more similar to each other than users from different clusters. The discovered clusters can then help in on-the-fly transformation of the web site content. In particular, web pages can be automatically linked by additional hyperlinks. The idea is to try to match an active user's access pattern with one or more of the clusters discovered from the web log files. Pages in the matched clusters that have not been explored by the user may serve as navigational hints for the user to follow.

In this paper we define the problem of clustering categorical data sequences and we propose an efficient algorithm to solve the problem. The algorithm employs the idea of *agglomerative hierarchical clustering*, which consists in merging pairs of similar clusters to form new larger clusters. We have taken the following assumptions: 1. the simplest cluster (elementary cluster) is a set of data sequences containing a common subsequence, 2. a significant cluster is a cluster that contains a large number of data sequences, 3. two clusters can be merged if a large number of their corresponding sequences overlap. Our algorithm starts with a set of significant elementary clusters and merges them iteratively until a user defined stop condition is satisfied.

Let us illustrate the approach with the following example (Fig. 1). We are given a web log file, which records paths used by users for navigation (e.g. the user s1 has visited the URLs: A, B, C, Z, and then D). Assume we are interested in discovering groups of users (sequences), whose behavior is similar to each other, i.e. who visit identical pages in the identical order. First, we create the elementary clusters c1, c2, c3, c4 that contain overlapping sequences. Then we notice that we can merge the clusters c1, c2, and c3 since the sequences they contain overlap between the clusters. Finally, the algorithm ends with two clusters, which represent web users of similar behavior.

The paper is organized as follows. Section 2 discusses related work. In Section 3, the basic definitions and the formulation of the problem are given. Section 4 contains the problem decomposition and the description of the algorithm for pattern-oriented clustering. Experimental results concerning the proposed clustering method are presented in Section 5. We conclude with a summary in Section 6.

s1: A->B->C->Z->D	s4: L->M->N->O
s2: P->Q->R	s5: E->F->G->A->B
s3: C->D->E->F	s6: Q->X->R ->S

c1:	c2:	c3:	c4:
s1: A->B->**C**->Z->**D**	s3: C->D->**E->F**	s1: **A->B**->C->Z->D	s2: P->**Q->R**
s3: **C->D**->E->F	s5: **E->F**->G->A->B	s5: E->F->G->**A->B**	s6: **Q**->X->**R** ->S

c1:	c4:
s1: **A->B**->**C**->Z->**D**	s2: P->**Q->R**
s3: **C->D->E->F**	s6: **Q**->X->**R** ->S
s5: **E->F**->G->**A->B**	

Fig. 1. Behavioral segmentation example

2 Related Work

Many clustering algorithms have been proposed in the area of machine learning [7][12][14] and statistics [15]. Those traditional algorithms group the data based on some measure of similarity or distance between data points. They are suitable for clustering data sets that can be easily transformed into sets of points in n-dimensional space, which makes them inappropriate for categorical data.

Recently, several clustering algorithms for categorical data have been proposed. In [13] a method for hypergraph-based clustering of transaction data in a high dimensional space has been presented. The method used frequent itemsets to cluster items. Discovered clusters of items were then used to cluster customer transactions. [9] described a novel approach to clustering collections of sets, and its application to the analysis and mining of categorical data. The proposed algorithm facilitated a type of similarity measure arising from the co-occurrence of values in the data set. In [8] an algorithm named CACTUS was presented together with the definition of a cluster for categorical data. In contrast with the previous approaches to clustering categorical data, CACTUS gives formal descriptions of discovered clusters.

In [21] the authors replace pairwise similarity measures, which they believe are inappropriate for categorical data, with a clustering criterion based on the notion of large items. An efficient clustering algorithm based on the new clustering criterion is also proposed.

The problem of clustering sequences of complex objects was addressed in [16]. The clustering method presented there used class hierarchies discovered for objects forming sequences in the process of clustering sequences seen as complex objects. The approach assumed applying some traditional clustering algorithm to discover classes of sub-objects, which makes it suitable for sequences of objects described by numerical values, e.g. trajectories of moving objects.

The most similar approach to ours is probably the approach to document clustering proposed in [5]. The most significant difference between their similarity measure and ours is that we look for the occurrence of variable-length subsequences and concentrate only on frequent ones.

Our clustering method can be seen as a scalable version of a traditional agglomerative clustering algorithm. Scaling other traditional clustering methods to large databases was addressed in [4], where the scalable version of the K-means algorithm was proposed.

Most of the research on sequences of categorical values concentrated on the discovery of frequently occurring patterns. The problem was introduced in [2] and then generalized in [20]. The class of patterns considered there, called sequential patterns, had a form of sequences of sets of items. The statistical significance of a pattern (called support) was measured as a percentage of data sequences containing the pattern.

In [17] an interesting approach to sequence classification was presented. In the approach, sequential patterns were used as features describing objects and standard classification algorithms were applied. To reduce the number of features used in the classification process, only distinctive (correlated with one class) patterns were taken into account.

3 Pattern-Oriented Agglomerative Hierarchical Clustering

Traditional agglomerative hierarchical clustering algorithms start by placing each object in its own cluster and then iteratively merge these atomic clusters until all objects are in a single cluster. Time complexity of typical implementations vary with the cube of the number of objects being clustered. Due to the poor scalability, traditional hierarchical clustering cannot be applied to large collections of data, such as databases of customer purchase histories or web server access logs. Another very important issue that has to be addressed when clustering large data sets is automatic generation of conceptual descriptions of discovered clusters. Such descriptions should summarize clusters' contents and have to be comprehensible to humans.

To improve performance of hierarchical clustering for large sets of sequential data, we do not handle sequences individually but operate on groups of sequences sharing a common subsequence. We concentrate only on the most frequently occurring subsequences, called frequent patterns (sequential patterns). We start with initial clusters associated with frequent patterns discovered in the database. Each of the initial clusters (clusters forming the leaf nodes of the hierarchy built by the clustering process) consists of sequences containing the pattern associated with the cluster. Clusters being results of merging of smaller clusters are described by sets of patterns and consist of sequences that contain at least one pattern from the describing set.

Definition 3.1. Let $L = \{l_1, l_2, ..., l_m\}$ be a set of literals called items. A *sequence* $S = <X_1 X_2 ... X_n>$ is an ordered list of sets of items such that each set of items $X_i \subseteq L$. Let the database D be a set of sequences.

Definition 3.2. We say that the sequence $S_1 = <Y_1 Y_2 ... Y_m>$ *supports* the sequence $S_2 = <X_1 X_2 ... X_n>$ if there exist integers $i_1 < i_2 < ... < i_n$ such that $X_1 \subseteq Y_{i1}, X_2 \subseteq Y_{i2}, ..., X_n \subseteq Y_{in}$. We also say that the sequence S_2 is a *subsequence* of the sequence S_1 (denoted by $S_2 \subset S_1$).

Definition 3.3. A *frequent pattern* is a sequence that is supported by more than a user-defined minimum number of sequences in D. Let P be a set of all frequent patterns in D.

Definition 3.4. A *cluster* is an ordered pair $<Q,S>$, where $Q \subseteq P$ and $S \subseteq D$, and S is a set of all database sequences supporting at least one pattern from Q. We call Q a *cluster description*, and S a *cluster content*. We use a dot notation to refer to a description or a content of a given cluster ($c.Q$ represents the description of a cluster c while $c.S$ represents its content).

Definition 3.5. A cluster c is called an *elementary cluster* if and only if $| c.Q | = 1$.

Definition 3.6. A *union* c_{ab} of the two clusters c_a and c_b is defined as follows:
$c_{ab} = union(c_a, c_b) = < c_a.Q \cup c_b.Q \, , c_a.S \cup c_b.S >$.

Elementary clusters form leaves in the cluster hierarchy. The clustering algorithm starts with the set of all elementary clusters. Due to the above formulations, input sequences that do not support any frequent pattern will not be assigned to any elementary cluster and will not be included in the resulting clustering. Such sequences are treated as outliers. The above definitions also imply that in our approach clusters from different branches of the cluster hierarchy may overlap. We do not consider it to be a serious disadvantage since sometimes it is very difficult to assign a given object to exactly one cluster, especially when objects are described by categorical values. In fact, if two clusters overlap significantly, then it means that patterns describing one of the clusters occur frequently in sequences contained in the other cluster and vice versa. This means that such clusters are good candidates to be merged to form a new larger cluster. The measures of similarity between clusters we propose in the paper are based on this observation. The cluster similarity measures we consider in this paper are based on the co-occurrence of the frequent patterns.

Definition 3.7. The *co-occurrence* of two frequent patterns p_1 and p_2 is a Jaccard coefficient [14] applied to the sets of input sequences supporting the patterns:

$$co(p_1, p_2) = \frac{|\{ s_i \in D : s_i \supset p_1 \wedge s_i \supset p_2 \}|}{|\{ s_i \in D : s_i \supset p_1 \vee s_i \supset p_2 \}|} \quad (1)$$

The similarity of two elementary clusters is simply the co-occurrence of patterns from their descriptions. The first of our inter-cluster similarity measures for arbitrary clusters is the extension of the above pattern co-occurrence measure i.e. similarity between two clusters c_a and c_b is a Jaccard coefficient applied to cluster contents:

$$f_J(c_a, c_b) = \frac{|c_a.S \cap c_b.S|}{|c_a.S \cup c_b.S|} \quad (2)$$

The second inter-cluster similarity measure we consider in the paper is defined as the average co-occurrence of pairs of patterns between two clusters' descriptions (group-average similarity):

$$f_{GA}(c_a, c_b) = avg(co(p_i, p_j)), where: p_i \in c_a.Q \wedge p_j \in c_b.Q) \quad \textbf{(3)}$$

Traditional hierarchical clustering builds a complete cluster hierarchy in a form of a tree. We add two stop conditions that can be specified by a user: the required number of clusters and the inter-cluster similarity threshold for two clusters to be merged. The first stop condition is suitable when a user wants to obtain the partitioning of the data set into the desired number of parts, the second is provided for cluster mining, which identifies high quality clusters.

Problem Statement. Given a database $D = \{s_1, s_2, ..., s_k\}$ of data sequences, and a set $P = \{p_1, p_2, ..., p_m\}$ of frequent patterns in D, the problem is to build a cluster hierarchy starting from elementary clusters as leaves, iteratively merging the most similar clusters until the required number of clusters is reached or there are no pairs of clusters exceeding the specified similarity threshold.

4 Algorithms

In this section, we describe a new clustering algorithm POPC for clustering large volumes of sequential data (POPC stands for Pattern-Oriented Partial Clustering). The algorithm implements the general idea of agglomerative hierarchical clustering. As we mentioned before, instead of starting with a set of clusters containing one data sequence each, our algorithm uses previously discovered frequent patterns and starts with clusters containing data sequences supporting the same frequent pattern. We assume that a set of frequent patterns has already been discovered and we do not include the pattern discovery phase in our algorithm. The influence of the pattern discovery process on the overall performance of our clustering method is described in the next section.

The POPC algorithm is database-oriented. It assumes that the input data sequences and the contents of clusters to be discovered are stored on a hard disk, possibly managed by a standard DBMS. Only the structures whose size depends on the number of patterns used for clustering and not on the number of input data sequences are stored in the main memory. These structures are similarity and co-occurrence matrices and cluster descriptions.

We introduce two variants of our algorithm based on two different cluster similarity measures: POPC-J using the Jaccard coefficient of the clusters' contents, and POPC-GA using the group average of co-occurrences of patterns describing clusters. First we present the generic POPC algorithm and then we describe elements specific to particular variants of the algorithm.

4.1 Generic POPC Algorithm

The algorithm for partial clustering based on frequently occurring patterns is decomposed into two following phases:

- *Initialization Phase*, which creates the initial set of elementary clusters and computes the co-occurrence matrix between patterns which serves as the similarity matrix for the initial set of clusters,
- *Merge Phase*, which iteratively merges the most similar clusters.

4.1.1 Initialization Phase

In this phase, the initial set of clusters C_1 is created by mapping each frequent pattern into a cluster. During the sequential scan of the source database, the contents of initial clusters are build. At the same time the co-occurrence matrix for the frequent patterns is computed. This is the only scan of the source database required by our algorithm. (It is not the only place where access to disk is necessary because we assume that cluster contents are stored on disk too.) Figure 2 presents the Initialization Phase of the clustering algorithm.

```
C1 = {ci: ci.Q={pi}, ci.S=∅};
UNION[][] = {0}; INTERSECT[][] = {0};
for each sj∈D do
begin
   for each pi∈P do
      if sj supports pi then
         ci.S = ci.S ∪ { sj };
      end if;
   for each pi, pk ∈P do
      if sj supports pi or sj supports pk then
         UNION[i][k]++; UNION[k][i]++;
         if sj supports pi and sj supports pk then
            INTERSECT[i][k]++; INTERSECT[k][i]++;
         end if;
      end if;
end;
for each pi, pk ∈P do
   CO[i][k] = INTERSECT[i][k] / UNION[i][k];
M1 = CO;
```

Fig. 2. Initialization phase

To compute the pattern co-occurrence matrix *CO*, for each pair of patterns we maintain two counters to count the number of sequences supporting at least one of the patterns and both of the patterns respectively. Those counters are represented by temporary matrices *UNION* and *INTERSECT*, and are used to evaluate the coefficients in the matrix *CO* after the database scan is completed. The similarity matrix M_1 for the initial set of clusters C_1 is equal to the pattern co-occurrence matrix.

4.1.2 Merge Phase

Figure 3 presents the Merge Phase of the clustering algorithm. This phase of the algorithm iteratively merges together pairs of clusters according to their similarity values. In each iteration *k*, the two most similar clusters $c_a, c_b \in C_k$ are determined, and replaced by a new cluster $c_{ab} = union(c_a, c_b)$. The actual merging is done by the function called *cluster*, described in detail in Section 4.1.4. When the new cluster is created, the matrix containing similarity values has to be re-evaluated. This operation is performed by means of the function called *simeval*, described in Section 4.1.3.

```
k = 1;
while |Ck| > n and exist ca,cb ∈ Ck
such that f(ca,cb) > min_sim do begin
    Ck+1 = cluster(Ck, Mk);
    Mk+1 = simeval(Ck+1, Mk);
    k++;
end;
Answer = Ck;
```

Fig. 3. Merge phase

The Merge Phase stops when the number of clusters reaches *n* (the required number of clusters) or when there is no such pair of clusters $c_a, c_b \in C_k$ whose similarity is greater than *min_sim* (the similarity threshold).

4.1.3 Similarity Matrix Evaluation: *Simeval*

Similarity matrix M_l stores the values of the inter-cluster similarity function for all possible pairs of clusters in the *l*-th algorithm iteration. The cell $M_l[x][y]$ represents the similarity value for the clusters c_x and c_y from the cluster set C_l. The function *simeval* computes the values of the similarity matrix M_{l+1}, using both the similarity matrix M_l and the current set of clusters. Notice that in each iteration, the similarity matrix need not be completely re-computed. Only the similarity values concerning the newly created cluster have to be evaluated. Due to diagonal symmetry of the similarity matrix, for *k* clusters, only $(k-1)$ similarity function values need to be computed in each iteration.

In each iteration, the size of the matrix decreases since two rows and two columns corresponding to the clusters merged to form a new one are removed and only one column and one row are added for a newly created cluster.

4.1.4 Cluster Merging: *Cluster*

In each iteration, the number of processed clusters decreases by one. The similarity-based merging is done by the function called *cluster*. The function *cluster* scans the similarity matrix and finds pairs of clusters, such that their similarity is maximal. If there are many pairs of clusters that reach the maximal similarity values, then the function *cluster* selects the one that was found as first. The function *cluster* takes a set of clusters C_k as one of its parameters and returns a set of clusters C_{k+1} such that $C_{k+1} = (C_k \setminus \{c_a, c_b\}) \cup \{c_{ab}\}$, where $c_a, c_b \in C_k$ are clusters chosen for merging and $c_{ab} = union(c_a, c_b)$.

4.2 Algorithm Variants POPC-J and POPC-GA

The POPC-J version of the algorithm optimizes the usage of main memory. The pattern co-occurrence matrix is used only in the Initialization Phase as the initial cluster similarity matrix. It is not needed in the Merge Phase of the algorithm because the similarity function based on Jaccard coefficient does not refer to the co-occurrences of patterns describing clusters. Thus the co-occurrence matrix *CO* becomes the initial cluster similarity matrix M_1 and no copying is done, which reduces the amount of main memory used by the algorithm.

In case of the POPC-GA version of the algorithm the initial cluster similarity matrix M_1 is created as a copy of the pattern co-occurrence matrix *CO*, since the latter is used in each iteration of the Merge Phase to compute the similarity between the newly created cluster and the rest of clusters. The advantage of this version of the POPC algorithm is that it does not use clusters' contents to evaluate similarity between clusters in the Merge Phase (only clusters' descriptions and the pattern co-occurrence matrix are used). Due to this observation, the POPC-GA does not build clusters' contents while the Merge Phase progresses. This is a serious optimization as compared to the POPC-J algorithm which has to retrieve clusters' contents from disk to re-evaluate the similarity matrix. Contents of discovered clusters can be built in one step after the Merge Phase completes according to the following SQL query (using the clusters' descriptions maintained throughout the algorithm and sets of input sequences supporting given patterns, built in the Initialization Phase):

```
select distinct d.cluster_id, p.sequence_id
from CLUSTER_DESCRIPTIONS d , PATTERNS p
where d.pattern_id = p.pattern_id.
```

Each row in the CLUSTER_DESCRIPTIONS table contains information about the mapping of one pattern to the description of one cluster, while each row in the PATTERNS table contains information that a given data sequence supports a given pattern.

5 Experimental Results

To assess the performance and results of the clustering algorithm, we performed several experiments on a PC machine with Intel Celeron 266MHz processor and 96 MB of RAM. The data were stored in an Oracle8i database on the same PC machine. Experimental data sets were created by synthetic data generator *GEN* from Quest project [3].

First of all, we compared the sets of clusters generated by the two versions of the POPC algorithm. The difference was measured as a percentage of all pairs of sequences from all clusters discovered by one version of the algorithm that were not put into one cluster by the other version of the algorithm. This measure is asymmetric but we believe that it captures the difference between two results of clustering. We performed several tests on a small data set consisting of 200 input sequences, using about 100 frequent patterns. As a stop condition for both versions of the algorithm we chose the desired number of clusters ranging from 5 to 45. An average difference between clustering results generated by the two methods was less than 10%.

In the next experiment we compared the performance of the two POPC variants and tested their scalability with respect to the size of the database (expressed as a number of input sequences) and the number of frequent patterns used for clustering.

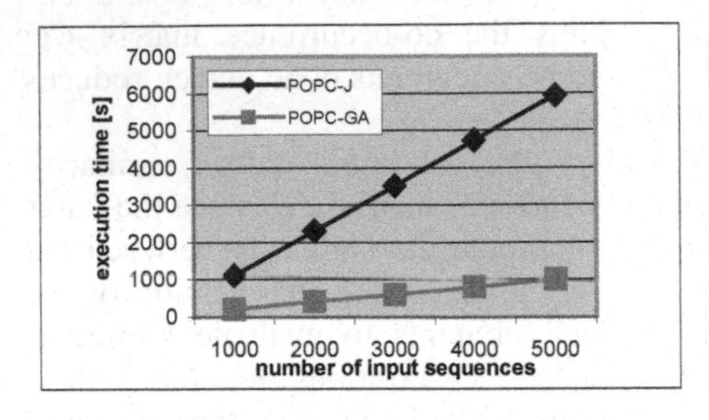

Fig. 4. Execution time for different data-base sizes

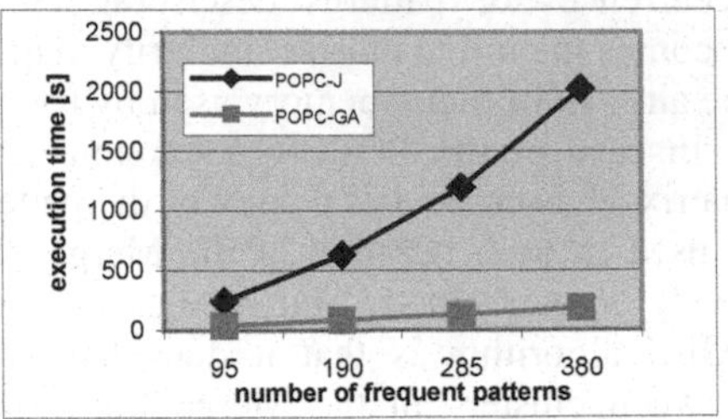

Fig. 5. Execution time for different number of frequent patterns used for clustering

Figure 4 shows the performance of the clustering algorithm for different database sizes expressed as the number of sequences in the database. In the experiment, for all the database sizes the data distribution was the same, which resulted in the same set of patterns used for clustering. Both versions of the algorithm scale linearly with the number of source sequences, which makes them suitable for large databases. The key factor is that the number of frequent patterns (equal to the number of initial clusters in our approach) does not depend on the database size but on the data distribution only. The execution time depends linearly on the number of input sequences, because the number of sequences supporting a given frequent pattern (for the same support threshold) grows linearly as the number of sequences in the database increases.

Figure 5 illustrates the influence of the number of frequent patterns used for clustering on the execution time of our algorithm. The time requirements of the algorithm vary with the cube of the number of patterns (the maximal possible number of iterations in the Merge Phase is equal to the number of patterns decreased by 1, in each iteration the cluster similarity matrix has to be scanned, the initial size of the matrix is equal to the square of the number of patterns). We performed experiments on a small database consisting of 200 sequences, using from 95 to 380 patterns. The experiments show that in practice the algorithm scales well with the number of patterns. This is true especially for the POPC-GA version of the algorithm, for which the cost of the Initialization Phase dominates the efficient Merge Phase.

Experiments show that both methods are scalable, but POPC-GA significantly outperforms POPC-J thanks to the fact that it does not have to retrieve clusters' contents from the database in the Merge Phase of the algorithm.

The execution times presented in the charts do not include the time needed to discover the set of frequent patterns. The cost of this pre-processing step depends strongly on the data distribution and the support threshold for patterns to be called frequent. Nevertheless, the time required to discover frequent patterns depends linearly on the database size, which preserves the overall linear scalability of the clustering method with respect to the database size. In our experiments we used the GSP algorithm [20] for pattern discovery. The time required for this step varied from

5 to 17 seconds for database sizes from 1000 to 5000 input sequences and the support threshold of 6%. This means that the time needed for pattern discovery does not contribute significantly to the overall processing time.

6 Concluding Remarks

We considered the problem of hierarchical clustering of large volumes of sequences of categorical values. We introduced two variants of the algorithm using different similarity functions to evaluate the inter-cluster similarity, which is a crucial element of the agglomerative hierarchical clustering scheme. Both of the proposed similarity measures were based on the co-occurrence of frequent patterns.

Both versions of the algorithm scale linearly with respect to the size of the source database, which is very important for large data sets. Both methods generate similar sets of clusters but the POPC-GA variant is much more efficient than POPC-J.

An important feature of the algorithm is that it does not only discover the clusters but also delivers the description of each cluster in form of patterns that are "popular" within the set of sequences forming the cluster.

In our approach clusters at any level of a cluster hierarchy can overlap. However, our method can easily be modified to generate disjoint clusters by using such techniques as placing each sequence into a cluster from whose description the sequence supports the highest number or percentage of patterns.

References

1. Agrawal R., Gehrke J., Gunopulos D., Raghavan P.: Automatic Subspace Clustering of High Dimensional Data for Data Mining Applications. Proceedings of the 1998 ACM SIGMOD International Conference on Management of Data (1998)
2. Agrawal R., Srikant R.: Mining Sequential Patterns. Proceedings of the 11th International Conference on Data Engineering (1995)
3. Agrawal, R.; Mehta, M.; Shafer, J.; Srikant, R.; Arning, A.; Bollinger, T.: The Quest Data Mining System. Proceedings of the 2nd International Conference on Knowledge Discovery and Data Mining (1996)
4. Bradley P.S., Fayyad U.M., Reina C.: Scaling Clustering Algorithms to Large Databases. Proceedings of the 4th International Conference on Knowledge Discovery and Data Mining (1998)
5. Broder A., Glassman S., Manasse M., Zweig G.: Syntactic clustering of the Web. Computer Networks and ISDN Systems 29, Proceedings of the 6th International WWW Conference (1997)
6. Ester M., Kriegel H-P., Sander J., Xu X.: A Density-Based Algorithm for Discovering Clusters in Large Spatial Databases with Noise. Proceedings of the 2nd International Conference on Knowledge Discovery and Data Mining (1996)
7. Fisher D.H.: Knowledge acquisition via incremental conceptual clustering. Machine Learning 2 (1987)
8. Ganti V., Gehrke J., Ramakrishnan R.: CACTUS-Clustering Categorical Data Using Summaries. Proceedings of the 5th ACM SIGKDD International Conference on Knowledge Discovery and Data Mining (1999)

9. Gibson D., Kleinberg J.M., Raghavan P.: Clustering Categorical Data: An Approach Based on Dynamical Systems. Proceedings of the 24th International Conference on Very Large Data Bases (1998)
10. Guha S., Rastogi R., Shim K.: CURE: An Efficient Clustering Algorithm for Large Databases. Proceedings of the 1998 ACM SIGMOD International Conference on Management of Data (1998)
11. Guha S., Rastogi R., Shim K.: ROCK: A Robust Clustering Algorithm for Categorical Attributes. Proceedings of the 15th International Conference on Data Engineering (1999)
12. Hartigan J.A.: Clustering Algorithms. John Wiley & Sons, New York (1975)
13. Han E., Karypis G., Kumar V., Mobasher B.: Clustering based on association rules hypergraphs. Proceedings of the Workshop on Research Issues on Data Mining and Knowledge Discovery (1997)
14. Jain A.K., Dubes R.C.: Algorithms for Clustering Data. Prentice Hall (1988)
15. Kaufman L., Rousseeuw P.: *Finding Groups in Data.* John Wiley & Sons, New York (1989)
16. Ketterlin A.: Clustering Sequences of Complex Objects. Proceedings of the 3rd International Conference on Knowledge Discovery and Data Mining (1997)
17. Lesh N., Zaki M.J., Ogihara M.: Mining Features for Sequence Classification. Proceedings of the 5th ACM SIGKDD International Conference on Knowledge Discovery and Data Mining (1999)
18. Perkowitz M., Etzioni O.: Towards Adaptive Web Sites: Conceptual Framework and Case Study. Computer Networks 31, Proceedings of the 8th International WWW Conference (1999)
19. Ramkumar G. D., Swami A.: Clustering Data Without Distance Functions. Bulletin of the IEEE Computer Society Technical Committee on Data Engineering, Vol.21 No. 1 (1998)
20. Srikant R., Agrawal R.: Mining Sequential Patterns: Generalizations and Performance Improvements. Proceedings of the 5th International Conference on Extending Database Technology (1996)
21. Wang K., Xu C., Liu B.: Clustering Transactions Using Large Items. Proceedings of the 1999 ACM CIKM International Conference on Information and Knowledge Management (1999)
22. Zhang T., Ramakrishnan R., Livny M.: Birch: An efficient data clustering method for very large databases. Proceedings of the 1996 ACM SIGMOD International Conference on Management of Data (1996)

FFS – An I/O-Efficient Algorithm for Mining Frequent Sequences

Minghua Zhang, Ben Kao, Chi-Lap Yip, and David Cheung

Department of Computer Science and Information Systems
The University of Hong Kong
{mhzhang, kao, clyip, dcheung}@csis.hku.hk

Abstract. This paper studies the problem of mining frequent sequences in transactional databases. In [1], Agrawal and Srikant proposed the AprioriAll algorithm for extracting frequently occurring sequences. AprioriAll is an iterative algorithm. It scans the database a number of times depending on the length of the longest frequent sequences in the database. The I/O cost is thus substantial if the database contains very long frequent sequences. In this paper, we propose a new I/O-efficient algorithm FFS. Experiment results show that FFS saves I/O cost significantly compared with AprioriAll. The I/O saving is obtained at a cost of a mild overhead in CPU cost.

Keywords: data mining, sequence, AprioriAll, FFS

1 Introduction

Data mining has recently attracted considerable attention from database practitioners and researchers because of its applicability in many areas such as decision support, market strategy and financial forecasts. Combining techniques from the fields of machine learning, statistics and databases, data mining enables us to find out useful and invaluable information from huge databases. One of the many data mining problems is the extraction of frequent sequences from transactional databases. The goal is to discover frequent sequences of events. For example, an on-line bookstore may find that most customers who have purchased the book "The Gunslinger" are likely to come back again in the future to buy "The Gunslinger II" in another transaction. Knowledge of this sort enables the store manager to conduct promotional activities and to come up with good marketing strategies.

The problem of mining frequent sequences was first introduced by Agrawal and Srikant [1]. In their model, a database is a collection of transactions. Each transaction is a set of items (or an itemset) and is associated with a customer ID and a time ID. If one groups the transactions by their customer IDs, and then sorts the transactions of each group by their time IDs in increasing value, the database is transformed into a number of customer sequences. Each customer sequence shows the order of transactions a customer has conducted. Roughly

D. Cheung, G.J. Williams, and Q. Li (Eds.): PAKDD 2001, LNAI 2035, pp. 294–305, 2001.

speaking, the problem of mining frequent sequences is to discover "subsequences" (of itemsets) that occur frequently enough among all the customer sequences.

In [1], an algorithm, `AprioriAll`, was proposed to solve the problem of mining frequent sequences. `AprioriAll` is a multi-phase iterative algorithm. It scans the database a number of times. Very similar to the structure of the Apriori algorithm [9] for mining association rules, `AprioriAll` starts by finding all frequent 1-sequences[1] from the database. A set of candidate 2-sequences are then generated. The frequencies of the candidate sequences are then counted by scanning the database once. Those frequent 2-sequences are then used to generate candidate 3-sequences, and so on. In general, `AprioriAll` uses a function called `apriori-generate` to generate candidate $(k+1)$-sequences given the set of all frequent k-sequences. The algorithm terminates when no more frequent sequences are discovered during a database scan.

It is not difficult to see that the number of database scans required by `AprioriAll` is determined by the length of the longest frequent sequences in the database. If the database is huge and if it contains very long frequent sequences, the I/O cost of `AprioriAll` is high.

The goal of this paper is to analyze and to improve the I/O requirement of the `AprioriAll` algorithm. We propose a new candidate generation function `FGen`. Unlike `apriori-generate`, which only generates $(k+1)$-sequences given the set of frequent k-sequences, `FGen` generates candidate sequences of various lengths when provided with a set of frequent sequences of various lengths. Our strategy for an I/O-efficient algorithm (called `FFS`) goes as follows. First, we apply `AprioriAll` on a small sample of the database to obtain an *estimate* of the set of frequent sequences ($\hat{L}$). Next, we scan the database to (i) discover the set of all frequent 1-sequences, and (ii) verify which sequences in the estimate $\hat{L}$ are frequent. $\hat{L}$ is then updated to contain the resulting frequent sequences (length 1 and above). After that, `FGen` is applied to $\hat{L}$ to obtain a candidate sequence set. We then scan the database to determine which candidate sequences are frequent. The result is used to update $\hat{L}$. We repeat the above procedure of candidate-generation-verification until no new frequent sequences are discovered.

We remark that the initial estimate of the set of frequent sequences ($\hat{L}$) can be obtained in many different ways. For example, if the database is periodically updated, and frequent sequences are mined regularly, the result of a previous mining exercise can well be used as $\hat{L}$ of the next mining exercise. In such a case of incremental update, `FFS` does not even require the sampling phase.

In later sections, we will prove that `FFS` is correct and that the set of candidates generated by `FGen` is a subset of those generated by `apriori-generate`. We will show that, in many cases, the I/O cost of `FFS` is significantly less than that of `AprioriAll`. We will show how the performance gain `FFS` achieves depends on the accuracy of the initial estimate, $\hat{L}$. As an extreme case, `FFS` requires only one or two database scans if $\hat{L}$ covers all frequent sequences of the database. This number is independent of the length of the longest frequent sequences. For a database containing long frequent sequences, the I/O saving is significant.

[1] A k-sequence is a sequence of k and only k itemsets.

The rest of this paper is organized as follows. In Section 2, we review some related works. In Section 3 we give a formal definition of the problem of mining frequent sequences. In Section 4, we briefly review the `AprioriAll` algorithm. Section 5 presents the `FFS` algorithm and the `FGen` function. Experiment results comparing the performance of `FFS` and `AprioriAll` are shown in Section 6. Finally, we conclude the paper in Section 7.

2 Related Work

Agrawal and Srikant [1] first studied the problem of mining frequent sequences. Among three algorithms they proposed, `AprioriAll` has the best overall performance. As we have discussed, the I/O cost of `AprioriAll` depends on the length of the longest frequent sequences. `AprioriAll` would not be very efficient if the database contains long frequent sequences. We will give a more detailed discussion of `AprioriAll` in Section 4.

In [10], a faster version of `AprioriAll` called `GSP` was proposed. `GSP` shares a similar structure of `AprioriAll` in that it works iteratively. A performance study shows that `GSP` is much faster than `AprioriAll`. `GSP` can also be used to solve other generalized versions of the frequent-sequence mining problem. For example, a user can specify a sliding time window. Items that occur in transactions that are within the sliding time window could be considered as to occur in the same transaction. Also, the problem of mining multi-level frequent sequences is addressed. Although not shown in this paper, our approach of improving the I/O efficiency of `AprioriAll` can also be applied to `GSP`. Due to space limitation, that modification to `GSP` is not explicitly discussed in this paper.

A very interesting I/O-efficient algorithm, SPADE, was proposed by Zaki [11]. SPADE works on a "vertical" representation of the database, and it only needs three database scans to discover frequent sequences. While SPADE is an efficeint algorithm, it requires the availability of the "vertical" database. ISM [8], an algorithm for incremental sequence mining, is based on SPADE; it also provides some kind of interactivity.

In [4], Garofalakis et al. proposed the use of regular expressions as a tool for end-users to specify the kinds of frequent sequences that a system should return. Algorithms are proposed to mine frequent sequences with regular expression constraints.

In [3], Chen et al. studied the problem of mining path traversal patterns for the World Wide Web. The goal is to discover the frequently occurring patterns of Web page visits. We can consider the path-traversal-pattern mining problem a special case of the frequent-sequence mining problem with which each "transaction" contains a lone "item" (a page visit).

3 Problem Definition

In this section, we give a formal problem statement of mining frequent sequences. We also define some notations to simplify our discussion.

Let $I = \{i_1, i_2, \ldots, i_m\}$ be a set of literals called items. An itemset X is a set of items (hence, $X \subseteq I$). A sequence $s = \langle t_1, t_2, \ldots, t_n \rangle$ is an ordered set of itemsets. The length of s (represented by $|s|$) is defined as the number of itemsets contained in s. A sequence of length k is called a k-sequence.

Consider two sequences $s_1 = \langle a_1, a_2, \ldots, a_m \rangle$ and $s_2 = \langle b_1, b_2, \ldots, b_n \rangle$. We say that s_1 contains s_2 if there exist integers $j_1, j_2, \ldots, j_n$, such that $1 \leq j_1 < j_2 < \ldots < j_n \leq m$ and $b_1 \subseteq a_{j_1}, b_2 \subseteq a_{j_2}, \ldots, b_n \subseteq a_{j_n}$. We represent this relationship by $s_2 \sqsubseteq s_1$.

As an example, the sequence $s_2 = \langle \{a\}, \{b, c\}, \{d\} \rangle$ is contained in $s_1 = \langle \{e\}, \{a, d\}, \{g\}, \{b, c, f\}, \{d\} \rangle$ because $\{a\} \subseteq \{a, d\}$, $\{b, c\} \subseteq \{b, c, f\}$, and $\{d\} \subseteq \{d\}$. Hence, $s_2 \sqsubseteq s_1$. On the other hand, $s_2 \not\sqsubseteq s_3 = \langle \{a, d\}, \{d\}, \{b, c, f\} \rangle$ because $\{b, c\}$ occurs before $\{d\}$ in s_2, which is not the case in s_3.

Given a sequence set V and a sequence s, if there exists a sequence $s' \in V$ such that $s \sqsubseteq s'$, we write $s \vdash V$. In words, we say s is contained in some sequence of V.

Given a sequence set V, a sequence $s \in V$ is *maximal* if s is not contained in any other sequences in V except s itself.

A database $\mathcal{D}$ consists of a number of sequences. The support of a sequence s is defined as the fraction of all sequences in $\mathcal{D}$ that contain s. We use $sup(s)$ to denote the support of s. If the support of s is no less than a user specified support threshold ρ_s, s is a frequent sequence. The problem of mining frequent sequences is to find all *maximal* frequent sequences given a sequence database $\mathcal{D}$.

We use the symbol L_i to denote the set of all length-i frequent sequences. Also, we use L to denote the set of all frequent sequences. That is $L = \bigcup_{i=1}^{\infty} L_i$.

4 `AprioriAll`

In this section we review the `AprioriAll` algorithm. `AprioriAll` solves the problem of mining frequent sequences in the following five phases.

1. Sort Phase. This phase transforms a transaction database to a sequence database by sorting the database with customer-ID as the major key and transaction time as a minor key.
2. Litemset (Large Itemset) Phase. In this phase, the set of frequent itemsets *LIT* is found using the `Apriori` algorithm. The support of an itemset X is defined as the fraction of sequences in the database which contain a transaction T such that $X \subseteq T$.[2] By definition, the set of frequent 1-sequences is simply $\{< t > | t \in LIT\}$. Hence, all frequent 1-sequences are found in the Litemset Phase.
3. Transformation Phase. In this phase, every frequent itemset found in the Litemset Phase is mapped to a unique integer. Each transaction in the database is then replaced by the set of all frequent itemsets (identified by

[2] Note that this definition of support is sligtly different from that of the traditional association rule mining model.

their unique integers) that are contained in that transaction. If a transaction does not contain any frequent itemset, the transaction is simply thrown away.

4. Sequence Phase. All frequent sequences are found in this phase. Similar to `Apriori`, `AprioriAll` is an iterative algorithm. It uses a function `apriori-generate` to generate candidate sequences, given a set of frequent sequences. Candidates generated in an iteration are of the same length. In general, during the i-th iteration, `apriori-generate` is applied to the set L_i to generate candidate sequences of length $i+1$. The database is scanned to count the supports of the candidates. Those frequent ones are put into the set L_{i+1}. The iteration terminates when no new frequent sequences are found. To generate candidate sequences in the i-th iteration, `apriori-generate` considers every pair of frequent sequences s_1 and s_2 from L_i, such that:

$$s_1 = \langle X_1, X_2, ..., X_{i-1}, A \rangle, \quad s_2 = \langle X_1, X_2, ..., X_{i-1}, B \rangle,$$

where $X_1, ..., X_{i-1}, A, B$ are all itemsets. That is, the first $i-1$ itemsets in s_1 and s_2 are exactly the same. Two new sequences:

$$m_1 = \langle X_1, X_2, ..., X_{i-1}, A, B \rangle \text{ and } m_2 = \langle X_1, X_2, ..., X_{i-1}, B, A \rangle$$

are generated. `AprioriAll` then checks to see if all length-i subsequences[3] of m_1 are in L_i. If so, m_1 is put into the candidate set. The sequence m_2 is similarly checked. Note that if the length of the longest frequent sequence is n, `AprioriAll` would scan the database at least $n-1$ times in this phase.

5. Maximal Phase. Frequent sequences which are not maximal (i.e., they are contained by some other frequent sequences) are deleted in this phase.

5 `FFS` and `FGen`

Among the five phases of `AprioriAll`, the Litemset Phase (for finding frequent itemsets) and the Sequence Phase (for finding frequent sequences) are the most I/O intensive. To reduce the I/O cost, one needs to find efficient algorithms for these two phases. Algorithms like DIC [2], Pincer-Search [6], and FlipFlop [7] are example I/O efficient algorithms for finding frequent itemsets. They can be used to improve the efficiency of the Litemset Phase. In this paper, we focus on improving the Sequence Phase. This section introduces our algorithm `FFS` and its candidate generating function `FGen`.

To reduce the I/O cost of the Sequence Phase, `FFS` first finds a *suggested frequent sequence set*, or an estimate. We denote this set by $\hat{L}$. If the database is regularly updated and that frequent sequences are mined periodically, then the result obtained from a previous mining exercise can be used as the estimate. If such an estimate is not readily available, we could mine a small sample (let's say 10%) of the database to obtain $\hat{L}$.

[3] A subsequence of a given sequence s is a sequence obtained by deleting one or more itemsets from s.

As has been explained in the previous section, all length-1 frequent sequences (i.e., the set L_1) are already found during the Litemset Phase. FFS then concatenates every possible pair of sequences from L_1 to form a set of length-2 candidates. (Essentially, we are applying apriori-generate to L_1 just like AprioriAll does.) The database is then scanned to verify which length-2 candidate sequences are frequent as well as which sequences in $\hat{L}$ are frequent. The set $\hat{L}$ is then updated to contain only those frequent sequences found. After that, FFS iterates the following two steps: (1) apply FGen on $\hat{L}$ to obtain a candidate set C; (2) scan the database to find out which sequences in C are frequent. Those frequent candidate sequences are added to $\hat{L}$. This *successive refinement* procedure stops when FFS cannot find any new frequent sequences in an iteration.

Figure 1 shows the algorithm FFS. The algorithm takes as inputs a transformed database (*ConvertedD*) obtained from the Transformation Phase, the set of length-1 frequent sequences (L_1) obtained from the Litemset Phase, and a suggested set of frequent sequences (S_{est}) obtained perhaps by applying AprioriAll on a database sample. FFS maintains a set *MFSS* which contains all and only those *maximal* sequences of the set of frequent sequences known at any instant. Since all subsequences of a frequent sequence are frequent, *MFSS* is sufficient to represent the set of all frequent sequences known.

```
1  Algorithm FFS(ConvertedD, ρ_s, L_1, S_est)
2     MFSS := L_1
3     CandidateSet := {⟨t_1, t_2⟩|⟨t_1⟩, ⟨t_2⟩ ∈ L_1} ∪ {s|s ⊢ S_est, |s| > 2}
4     Scan ConvertedD to get support of every sequence in CandidateSet
5     NewFrequentSequences := {s|s ∈ CandidateSet, sup(s) >= ρ_s}
6     AlreadyCounted := {s|s ⊢ S_est, |s| > 2}
7     Iteration := 3
8     while(NewFrequentSequences ≠ ∅)
9        //Max(S) returns the set of all maximal sequences is S
10       MFSS := Max(MFSS ∪NewFrequentSequences)
11       CandidateSet := FGen(MFSS, Iteration, AlreadyCounted)
12       Scan ConvertedD to get support of every sequence in CandidateSet
13       NewFrequentSequences := {s|s ∈ CandidateSet, sup(s) >= ρ_s}
14       Iteration := Iteration+1
15    Return MFSS
```

Fig. 1. Algorithm FFS

The most important component of FFS is the FGen function (see Figure 2). The function takes three parameters, namely, *MFSS* — the set of all maximal frequent sequences known so far; *Iteration* — a loop counter that FFS maintains; and *AlreadyCounted* — a set of sequences whose supports have already been counted.

FGen generates candidate sequences given a set of frequent sequences by "joining" *MFSS* with itself (lines 3–6). A candidate sequence m is removed (from the candidate set) if any one of the following conditions is true (lines 7–11):

- m is contained in some sequence already known to be frequent. (Since m must then be frequent, we do not need to count its support.)
  m's support has already been counted.
- some of m's subsequences are not known to be frequent.

```
1  Function FGen(MFSS, Iteration, AlreadyCounted)
2     CandidateSet := ∅
3     for each pair of s1, s2 ∈ MFSS such that |s1| > Iteration-2, |s2| > Iteration-2
      and that s1, s2 share at least one common subsequence of length ≥ Iteration-2
4        for each common subsequence s of s1, s2 such that |s| ≥ Iteration-2
5           NewCandidate := { ⟨t1, s, t2⟩ | ⟨t1, s⟩ ⊑ s1, ⟨s, t2⟩ ⊑ s2
                                            or ⟨t1, s⟩ ⊑ s2, ⟨s, t2⟩ ⊑ s1, }
6           CandidateSet := CandidateSet ∪ NewCandidate
7     for each sequence s ∈ CandidateSet
8        if (s ⊢ MFSS) delete s from CandidateSet
9        if s ∈ AlreadyCounted delete s from CandidateSet
10       for any subsequence s' of s with length |s| − 1
11          if (s' ⊬ MFSS) delete s from CandidateSet
12    AlreadyCounted := AlreadyCounted ∪CandidateSet
13    for each sequence s ∈ AlreadyCounted
14       if (|s| = Iteration) delete s from AlreadyCounted
15    Return CandidateSet
```

Fig. 2. Function FGen

5.1 Theorems

In this subsection, we summarize a few properties of FFS and FGen in the following theorems. Due to space limitation, we refer the readers to [12] for the proofs. For convenience, we use the symbol L to represent the set of all frequent sequences found by AprioriAll, $C_{AprioriAll}$ to represent the set of all candidate sequences generated (and whose supports are counted) by apriori-generate, and C_{FGen} to represent the set of all candidate sequences generated (and whose supports are counted) by FGen. Also, if X is a sequence set, we use $Max(X)$ to represent the set of all maximal sequences in X.

Theorem 1 *When FFS terminates, MFSS = Max(L).*

Since $Max(L)$ is the set of maximal frequent sequences found by AprioriAll and *MFSS* is the set of maximal frequent sequences found by FFS, Theorem 1 says that AprioriAll and FFS discover the same set of maximal frequent sequences. Therefore, if AprioriAll finds out all maximal frequent sequences in the database, so does FFS. Hence, FFS is correct.

Theorem 2 $C_{FGen} \subseteq C_{AprioriAll}$.

Theorem 2 says that the set of candidates generated by FGen is a subset of that generated by AprioriAll. Thus FGen does not generate any unnecessary candidates and waste resources for counting their supports.

6 Performance

We performed a number of experiments comparing the performance of `AprioriAll` and `FFS`. Our goals are to study how much I/O cost `FFS` could save, and how effective sampling is in discovering an initial estimate of the set of frequent sequences required by `FFS`. In this section, we present some representative results from our experiments.

We used synthetic data as the test databases. The generator is obtained from the IBM Quest data mining project. Readers are referred to [5] for the details of the data generator. The values of the parameters we used in the data generation are listed in Table 1.

Table 1. Parameters and values for data generation

Parameter	Description	value
$\|D\|$	Number of customers (size of Database)	100,000
$\|C\|$	Average number of transactions per Customer	10
$\|T\|$	Average number of items per transaction	5
$\|S\|$	Average length of maximal potentially large Sequences	4
$\|I\|$	Average size of Itemsets in maximal potentially large Sequences	1.25
N_S	Number of maximal potentially large Sequences	5,000
N_I	Number of maximal potentially large Itemsets	25,000
N	Number of items	10,000

6.1 Coverages and I/O Savings

Recall that `FFS` requires a suggested frequent sequence set, S_{est}. In this subsection, we study how the "coverage" of S_{est} affects the performance of `FFS`. By coverage, we mean the fraction of the *real* frequent sequences that are contained in S_{est}. It is defined as

$$coverage = \frac{|\{s | s \vdash S_{est}\} \cap (\cup_{i=3}^{\infty} L_i)|}{|\cup_{i=3}^{\infty} L_i|}$$

where L_i represents the set of all frequent sequences of length i. For our definition of coverage, we only consider those frequent sequences that are of length 3 or longer. This is because the set L_1 is already discovered in the Litemset Phase and that all length-2 candidate sequences *will* be checked by `FFS` during its first scan of the database (see Figure 1, line 3). Therefore, whether S_{est} contains frequent 1-sequences or frequent 2-sequences are immaterial, and the number of I/O passes required by `FFS` is not affected.

In our first experiment, we generated a database using the parameter values listed in Table 1. We then applied `AprioriAll` on the database to obtain L, the set of all frequent sequences. We then randomly selected a subset of sequences from L, forming an estimated set S_{est}. `FFS` was then applied on the database using S_{est}. The number of I/O passes used by `AprioriAll` and `FFS` in the Sequence Phase were then compared. We repeated the experiment using different support thresholds, ρ_s. The results of the experiments (for ρ_s = 0.75%, 0.5%, and 0.4%) are shown in Figure 3.

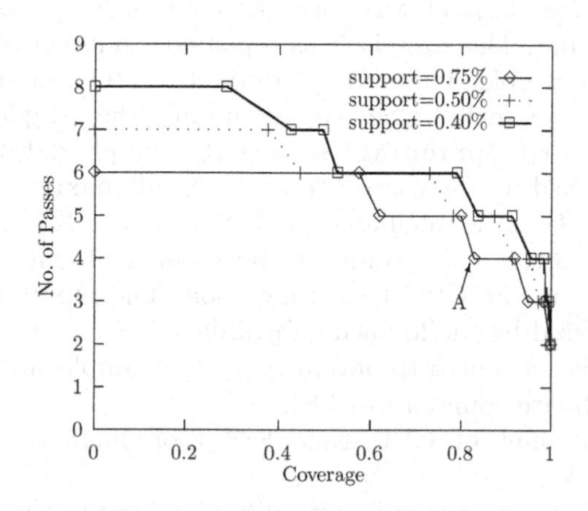

Fig. 3. Number of I/O passes vs. coverage under different support thresholds

In Figure 3, we use three kinds of points ('◇', '+', '□') to represent three support thresholds 0.75%, 0.50%, 0.40% respectively. For example, point A is represented by ◇ and its coordinate is (0.83,4). This means that when the support threshold was 0.75% and the coverage of S_{est} was 0.83, `FFS` took 4 I/O passes. We see that when coverage increases, the number of I/O passes required by `FFS` decreases.

We observe that the curves follow a general trend:

- When coverage = 0 (e.g., when S_{est} is empty), `FFS` degenerates to `AprioriAll`. They thus require the same number of I/O passes.
- When coverage is small, S_{est} would contain very few long frequent sequences. This is because if S_{est} covers a long frequent sequence s, it also covers every subsequence of s. These subsequences are frequent and if s is long they are numerous. The coverage of S_{est} would thus be high. Since few long frequent sequences are covered by S_{est}, quite a number of I/O passes are required to discover them. Hence, with a small coverage, `FFS` does not reduce the I/O cost at all.
- When coverage is moderate, `FFS` becomes more effective. The amount of I/O saving increases with the coverage.
- When coverage is 100% (i.e., S_{est} covers all frequent sequences in the database), `FFS` requires only two passes over the database: one pass to verify that the sequences in S_{est} are all frequent, another pass to verify that no more frequent sequences can be found.

6.2 Sampling

One way to obtain the set S_{est} is to apply AprioriAll on a database sample. In the next set of experiments, we study this sampling approach. We first applied AprioriAll on our synthetic dataset to obtain the number of I/O passes it required in the Sequence Phase. Then a random sample of the database was drawn on which AprioriAll was applied to obtain S_{est}. We then executed FFS using the S_{est} found. This exercise was repeated a number of times, each with a different random sample. The average amount of I/O cost required by FFS was noted. This I/O cost includes the cost of mining the sample. For example, if a 1/8 sample was used, AprioriAll scaned the sample database 6 times to get S_{est}, and FFS scaned the database 3 times to get all maximal frequent sequences, then the total I/O cost is calculated as: $1/8 \times 6 + 3 = 3.75$ passes.

Besides I/O cost, we also compare the number of candidates the two algorithms counted and the CPU time they took. The number of candidates FFS counted is measured by the following formula:

(# of candidate counted to obtain S_{est}) $\times$ (sample size)
+(# of candidate counted in FFS).

Similarly, the amount of CPU time FFS took includes the CPU time for mining the sample.

The result of the experiment (with different sample sizes) is shown in Table 2. Note that the average number of candidates counted and the average CPU cost of FFS are shown in relative quantities (with those of AprioriAll set to 1).

Table 2. Performance of FFS vs sample size ($\rho_s = 0.75\%$)

sample size	1/128	1/64	1/32	1/16	1/8	1/4	1/2	0(AprioriAll)
avg. coverage	0.673	0.703	0.763	0.806	0.855	0.904	0.946	N/A
avg. I/O cost	5.095	4.859	4.541	4.448	4.357	4.691	5.938	6
avg. # of cand.	1.087	1.069	1.078	1.083	1.133	1.254	1.502	1
avg. CPU cost	1.152	1.136	1.138	1.157	1.212	1.355	1.599	1

From Table 2, we see that FFS needed fewer I/O passes than AprioriAll (6 passes). As the sample size increases, the coverage of S_{est} becomes higher, and fewer I/O passes are needed for FFS to discover all the frequent sequences given S_{est}. This accounts for the drop of I/O cost from 5.095 passes to 4.357 passes as the sample size increases from 1/128 to 1/8. As the sample size increases further, however, the I/O cost of mining the sample becomes substantial. The benefit obtained by having a better-coverage S_{est} is outweighted by the penalty of mining the sample. Hence, the overall I/O cost increases as the sample size increases from 1/8 to 1/2. Also, as the sample size increases, more work is spent on candidate counting, and the CPU cost increases. For a 1/16 sample, for example, FFS reduces about $\frac{6-4.448}{6} = 26\%$ of the I/O cost at the expense of a 16% increment in CPU cost.

We can further improve the performance of FFS by using a slightly smaller support threshold (ρ_{s_sample}) to mine the sample. The idea is that by using a small ρ_{s_sample}, more sequences will be included in S_{est}. This will potentially improve the coverage of S_{est}, and hence a larger reduction in I/O cost is achieved. The disadvantage of this trick, however, is a larger CPU cost, since more support counting would have to be done to verify which sequences in S_{est} are frequent or not.

Table 3 shows the performance of FFS using different values of ρ_{s_sample}. In this set of experiments, the sample size is fixed at 1/16 and ρ_s is 0.75%.

Table 3. Performance of FFS with different values of ρ_{s_sample} (ρ_s = 0.75%, sample size = 1/16)

ρ_{s_sample}	0.750%	0.675%	0.600%	0.525%	0.450%	0.375%	AprioriAll
avg. coverage	0.806	0.939	0.986	0.998	1.000	1.000	N/A
avg. I/O cost	4.448	3.731	3.215	2.544	1.913	1.450	6
avg. # of cand.	1.083	1.127	1.183	1.255	1.304	1.322	1
avg. CPU cost	1.157	1.149	1.150	1.162	1.175	1.173	1

From Table 3, we see that when ρ_{s_sample} decreases, the I/O cost of FFS decreases dramatically, while the CPU cost increases only very mildly. For example, when ρ_{s_sample} = 0.450%, the average coverage reaches 100%, and the average I/O cost is 1.913. Compared with ρ_{s_sample} = 0.750% (i.e. $\rho_{s\_sample} = \rho_s$), we saved an additional $(4.448 - 1.913)/6 = 42\%$ I/O cost at the expense of an additional $(1.175 - 1.157)/1 = 1.8\%$ increment in CPU cost.

Notice that 1.913 passes of I/O already included the I/O cost of mining the 1/16 sample. We looked into our experimental result and we found that in some instances, FFS only took 1 pass to complete the finding of frequent sequences given an S_{est}. With a small ρ_{s_sample}, S_{est} includes many infrequent sequences. During the first iteration of FFS (after sampling), the support of these infrequent sequences are also counted. During the second iteration of FFS, candidate sequences are generated. However, in those instances, all the candidate sequences were in the set S_{est}. Since their supports were already counted in the first iteration, no database scan is needed. FFS thus terminated with only 1 pass over the database (plus the I/O needed to mine the sample). This is why the average I/O cost for ρ_{s_sample}= 0.45% is less than 2.

7 Conclusion

In this paper, we proposed a new I/O efficient algorithm FFS to solve the problem of mining frequent sequences. A new candidate generation method FGen was proposed which can generate candidate sequences of multiple lengths given a set of suggested frequent sequences. We performed experiments to compare the performance of FFS and AprioriAll. We showed that FFS saves I/O passes significantly, especially when an estimate (S_{est}) of the set of frequent sequences with a good coverage is available. We showed how mining a small sample of the database leads to a good S_{est}. By using a smaller support threshold (ρ_{s_sample}) in mining the sample, we showed that FFS outperforms AprioriAll by a wide margin. The I/O saving is obtained, however, at a mild CPU cost.

References

1. Rakesh Agrawal and Ramakrishnan Srikant. Mining sequential patterns. In *Proc. of the 11th Int'l Conference on Data Engineering*, Taipei, Taiwan, March 1995.
2. Sergey Brin, Rajeev Motwani, Jeffrey D. Ullman, and Shalom Tsur. Dynamic itemset counting and implication rules for market basket data. In *Proceedings of the ACM SIGMOD Conference on Management of Data*, 1997.
3. Ming-Syan Chen, Jong Soo Park, and Philip S. Yu. Efficient data mining for path traversal patterns. *IEEE Transactions on Knowledge and Data Engineering*, 10(2), March/April 1998.
4. Minos N. Garofalakis, Rajeev Rastogi, and Kyuseok Shim. SPIRIT: Sequential pattern mining with regular expression constraints. In *Proceedings of the 25th International Conference on Very Large Data Bases*, Edinburgh, Scotland, UK, September 1999.
5. http://www.almaden.ibm.com/cs/quest/.
6. Dao-I Lin and Zvi M. Kedem. Pincer-search: A new algorithm for discovering the maximum frequent set. In *Proceedings of the 6th Conference on Extending Database Technology (EDBT)*, Valencia, Spain, March 1998.
7. K.K. Loo, C.L. Yip, Ben Kao, and David Cheung. Exploiting the duality of maximal frequent itemsets and minimal infrequent itemsets for I/O efficient association rule mining. In *Proc. of the 11th International Conference on Database and Expert Systems Conference*, London, Sept. 2000.
8. S. Parthasarathy, M. J. Zaki, M. Ogihara, and S. Dwarkadas. Incremental and interactive sequence mining. In *Proceedings of the 1999 ACM 8th International Conference on Information and Knowledge Management (CIKM'99)*, Kansas City, MO USA, November 1999.
9. T. Imielinski R. Agrawal and A. Swami. Mining association rules between sets of items in large databases. In *Proc. ACM SIGMOD International Conference on Management of Data*, page 207, Washington, D.C., May 1993.
10. Ramakrishnan Srikant and Rakesh Agrawal. Mining sequential patterns: Generalizations and performance improvements. In *Proc. of the 5th Conference on Extending Database Technology (EDBT)*, Avignion, France, March 1996.
11. Mohammed J. Zaki. Efficient enumeration of frequent sequences. In *Proceedings of the 1998 ACM 7th International Conference on Information and Knowledge Management(CIKM'98)*, Washington, United States, November 1998.
12. Minghua Zhang, Ben Kao, C.L. Yip, and David Cheung. FFS – An I/O efficient algorithm for mining frequent sequences. Technical Report CS Technical Report TR-2000-6, University of Hong Kong, 2000.

Sequential Index Structure for Content-Based Retrieval

Maciej Zakrzewicz

Poznan University of Technology, Institute of Computing Science
Piotrowo 3a, 60-965 Poznan , Poland
mzakrz@cs.put.poznan.pl

Abstract. Data mining applied to databases of data sequences generates a number of sequential patterns, which often require additional processing. The post-processing usually consists in searching the source databases for data sequences which contain a given sequential pattern or a part of it. This type of content-based querying is not well supported by RDBMSs, since the traditional optimization techniques are focused on exact-match querying. In this paper, we introduce a new bitmap-oriented index structure, which efficiently optimizes content-based queries on dense databases of data sequences. Our experiments show a significant improvement over traditional database accessing methods.

1 Introduction

Mining of sequential patterns consists in identifying trends in databases of data sequences [1,6]. A *sequential pattern* represents a frequently occurring subsequence. An example of a sequential pattern that holds in a video rental database is that customers typically rent "Star Wars", then "Empire Strikes Back", and then "Return of the Jedi". Note that 1. these rentals need not be consecutive , and 2. during a single visit, a customer may rent a set of videos, instead of a single one. Post-processing of discovered sequential patterns usually consists in searching the source databases for data sequences, containing a given sequential pattern. For example, when we discover an interesting sequential pattern in the video rental database, we would probably like to find all customers, who satisfy (contain) the pattern. We will refer to these types of searching as to *content-based sequence retrieval.*

SID	TS	L
1	1	A
1	1	B
1	2	C
1	3	D
2	1	A
2	2	E
2	2	C
2	3	F

```
SELECT SID
FROM   R R1, R R2, R R3
WHERE  R1.SID=R2.SID
  AND  R2.SID=R3.SID
  AND  R1.TS<R2.TS
  AND  R2.TS<R3.TS
  AND  R1.L='A'
  AND  R2.L='E'
  AND  R3.L='F';
```

Fig. 1. The relation of data sequences and the content-based sequence retrieval query

In most cases, data sequences (and sequential patterns) are stored in relational databases. Consider the following example of using the relational approach to content-based sequence retrieval. The relation *R(SID,TS,L)* stores data sequences. Each tuple contains the sequence identifier (*SID*), the timestamp (*TS*), and the item (*L*). Our relation *R* describes two data sequences: {A,B}→{C}→{D}, {A}→{E,C}→{F}. Let the

D. Cheung, G.J. Williams, and Q. Li (Eds.): PAKDD 2001, LNAI 2035, pp. 306-311, 2001.

searched data subsequence be: {A}→{E}→{F}. Figure 1 gives the relation *R* and the *SQL* query, which implements the content-based sequence retrieval problem.

Since mined databases tend to be very large, there is a problem of optimizing the database access while performing content-based sequence retrieval, e.g. by means of the above *SQL* query. Database research has developed many *indexing techniques*, like B+ trees [3], bitmapped indexes [8], k-d trees [2], R trees [5], to optimize queries based on exact matches of single tuples. However, these techniques do not significantly improve content-based sequence retrieval queries, which deal with partial matches of multi-tuple sequences. There are also proposals for set-based indexing [7][4], which is used to improve subset searching. However, these methods work for retrieval of unordered sets of items only.

In order to realize the shortcomings of the existing indexing methods, let us consider applying B+ tree and set-based indexes to execute the query from Figure 1. Using a B+ tree index, tuples containing all items of each data sequence are joined first (*SID* attribute), and then verified whether they contain given items in the given order. This can be fairly ineffective since a data sequence may span across many disk block, what results in multiple scanning of each block of the relation. Using a set-based index, the sequence identifiers (*SID* attribute) of all sequences, which contain the searched items in any order, are found, and then the sequences are read from the relation (perhaps with help of a B+ tree) to verify the ordering of their items. This approach gives much better results, as compared to a B+ tree index, but a significant overhead comes from reading and verifying the sequences having incorrect ordering.

In this paper we consider content-based retrieval of data sequences from *dense databases*, characterized by relatively small number of items, which occur frequently in various order (e.g. web logs), and therefore a set-based index is not efficient. We introduce a new bitmap-oriented indexing method, which optimizes the problem. The basic idea behind our method, as compared to set-based indexes, is that the index structure includes not only the items of a sequence, but also the ordering of the items.

Basic definitions. Let $L=\{l_1,l_2,...,l_m\}$ be a set of literals called items. *Data sequence* $S=<X_1 X_2... X_n>$ is an ordered list of sets of items such that each set of items $X_i \subseteq L$. X_i is called a *sequence element*. All items in a sequence element are unordered. We say that a data sequence $<X_1 X_2 ... X_n>$ *is contained* in another data sequence $<Y_1 Y_2 ... Y_n>$ if there exist integers $i_1 < i_2 < ... < i_n$ such that $X_1 \subseteq Y_{i1}$, $X_2 \subseteq Y_{i2}$, ..., $X_n \subseteq Y_{in}$. Let *D* be a database of variable length data sequences. Let *S* be a data sequence. The problem of content-based sequence retrieval consists in finding in *D* all data sequences, which contain the data sequence *S*.

2 Preliminaries

Data sequences may contain categorical items of various data types. For sake of convenience, we convert the items to integers by means of an *item mapping function*. An *item mapping function* $fi(x)$, where x is a literal, is a function which transforms a literal into an integer value. For example, for a set of literals $L = \{A, B, C, D, E, F\}$, an item mapping function can take the values: $fi(A)=1$, $fi(B)=2$, $fi(C)=3$, $fi(D)=4$, $fi(E)=5$, $fi(F)=6$.

Similarly, we use an *order mapping function* to express data sequence ordering relations by means of integer values. Thus, we will be able to represent data sequence items as well as data sequence ordering uniformly. An *order mapping function* $fo(x,y)$, where x and y are literals and $fo(x,y) \neq fo(y,x)$, is a function which transforms a data sequence $<\{x\}\{y\}>$ into an integer value. For the set of literals used in the previous example, an order mapping function can be: $fo(x,y) = 6*fi(x) + fi(y)$, e.g. $fo(C,F) = 24$.

Using the above definitions, we will be able to transform data sequences into item sets, which are easier to manage, search and index. An item set representing a data sequence is called an *equivalent set*. An *equivalent set* E for a data sequence $S = <X_1\ X_2 \ldots X_n>$ is defined as:

$$E = \left(\bigcup_{x \in X_1 \cup X_2 \cup \ldots \cup X_n} \{fi(x)\} \right) \cup \left(\bigcup_{\substack{x,y \in X_1 \cup X_2 \cup \ldots \cup X_n: \\ x \text{ precedes } y}} \{fo(x,y)\} \right) \tag{1}$$

where: $fi()$ is an item mapping function and $fo()$ is an order mapping function. For example, for the data sequence $S = <\{A,B\}\{C\}\{D\}>$ and the presented item mapping function and order mapping function, the equivalent set E is evaluated as follows:

$$E = \left(\bigcup_{x \in \{A,B,C,D\}} \{fi(x)\} \right) \cup \left(\bigcup_{x,y \text{ in } \{<\{A\}\{C\}>,<\{B\}\{C\}>,<\{A\}\{D\}>,<\{B\}\{D\}>,<\{C\}\{D\}>\}} \{fo(x,y)\} \right) =$$

$$= \{fi(A)\} \cup \{fi(B)\} \cup \{fi(C)\} \cup \{fi(D)\} \cup \{fo(A,C)\} \cup \{fo(B,C)\} \cup \cup \{fo(A,D)\} \cup \{fo(B,D)\} \cup \{fo(C,D)\} = \{1,2,3,4,9,15,10,16,22\} \tag{2}$$

Notice that for any two data sequences S_1 and S_2, we have: S_2 contains S_1 if $E_1 \subseteq E_2$, where E_1 is the equivalent set for S_1, and E_2 is the equivalent set for S_2. This property is not reversible.

Since the size of an equivalent set quickly increases while increasing the number of the original sequence elements, we split data sequences into *partitions*, which are small enough to process and encode. We say that a data sequence $S = <X_1\ X_2 \ldots X_n>$ is *partitioned* into data sequences $S_1 = <X_1 \ldots X_{a1}>$, $S_2 = <X_{a1+1} \ldots X_{a2}>$,..., $S_k = <X_{aj+1} \ldots X_n>$ *with level* β if for each data sequence S_i the size of its equivalent set $|E_i| < \beta$ and for all $x,y \in X_1 \cup X_2 \cup \ldots \cup X_n$, where x precedes y, we have: either $<\{x\}\{y\}>$ is contained in S_i or $\{x\}$ is contained in S_i, and $\{y\}$ is contained in S_j, where $i<j$ (β should be greater than maximal item set size). For example, partitioning the data sequence $S=<\{A,B\}\{C\}\{D\}\{A,F\}\{B\}\{E\}>$ with level 10 results in two data sequences: $S_1=<\{A,B\}\{C\}\{D\}>$ and $S_2=<\{A,F\}\{B\}\{E\}>$, since the sizes of the equivalent sets are respectively: $|E_1| = 9$ ($E_1 = \{1,2,3,4,9,15,10,16,22\}$), and $|E_2| = 9$ ($E_2 = \{1,6,2,5,8,38,11,41,17\}$). Notice that for a data sequence S partitioned into $S_1, S_2, \ldots, S_k$, and a data sequence Q, we have: S contains Q if there exists a partitioning of Q into $Q_1, Q_2, \ldots, Q_m$, such that Q_1 is contained in S_{i1}, Q_2 is contained in S_{i2}, ..., Q_m is contained in S_{im}, and $i_1 < i_2 < \ldots < i_m$.

Our index structure will consist of equivalent sets stored for all data sequences, optionally partitioned to reduce the complexity. To reduce storage requirements, equivalent sets will be stored in database in the form of *bitmap signatures*. The *bitmap signature* of a set X is an N-bit binary number created, by means of bit-wise OR operation, from the hash keys of all data items contained in X. The *hash key* of the item $x \in X$ is an N-bit binary number defined as follows: `hash_key(X) = 2`$^{(X \bmod n)}$. The bitmap signature of the set X is the bit-wise OR of all items' hash keys. Notice that for any two sets X and Y, if $X \subseteq Y$ then: `bit_sign(X) AND bit_sign(Y) = bit_sign(X)`, where AND is a bit-wise AND operator. This property is not reversible.

3 Sequential Index Construction Algorithm

The sequential index construction algorithm iteratively processes all data sequences in the database. First, the data sequences are partitioned with the given level β. Then, for each partition of each data sequence, the equivalent set is evaluated. In the next step, for each equivalent set, its N-bit bitmap signature is generated and stored in the database. The formal description of the algorithm is given below.

```
for each data sequence S ∈ D do begin
 partition S into partitions S1, S2, ..., S3 with level β;
  for each partition Si do begin
    evaluate equivalent set Ei for Si;
     bitmapi = bit_sign(Ei); store bitmapi in the database;
  end;
end;
```

Example. Assume that β=10, N=16, and the database D contains three data sequences: S_1 = <{A,B}{C} {D}{A,F}{B}{E}>, S_2=<{A}{C,E}{F}{B}{E}{A,D}>, S_3 = <{B,C,D},{A}>. First, we partition the data sequences with β=10. Notice that S_3 is, in fact, not partitioned since its equivalent set is small enough. The symbol $S_{i,j}$ denotes j-th partition of the i-th data sequence: $S_{1,1}$= <{A,B}{C}{D}>, $S_{1,2}$= <{A,F}{B}{E}>, $S_{2,1}$= <{A}{C,E}{F}>, $S_{2,2}$= <{B}{E}{A,D}>, $S_{3,1}$= <{B,C,D}{A}>. Then we evaluate the equivalent sets for the partitioned data sequences. We use the example item mapping function and order mapping function. The symbol $E_{i,j}$ denotes the equivalent set for $S_{i,j}$.

$E_{1,1}$ = {1, 2, 3, 4, 9, 15, 10, 16, 22}
$E_{1,2}$ = {1, 6, 2, 5, 8, 38, 11, 41, 17}
$E_{2,1}$ = {1, 3, 5, 6, 11, 9, 36, 24}
$E_{2,2}$ = {2, 5, 1, 4, 17, 13, 16, 31, 36}
$E_{3,1}$ = {2, 3, 4, 1, 13, 19, 25}

In the next step, we generate 16-bit bitmap signatures for all equivalent sets.

```
bit_sign(E1,1) = 1000011001011111
bit_sign(E1,2) = 0000101101100110
bit_sign(E2,1) = 0000101101111010
bit_sign(E2,2) = 1010000000110111
bit_sign(E3,1) = 0010001000011110
```

Finally, the sequential index is stored in the database in the following form: {sid=1: bit_sign={1000011001011111, 0000101101100110}, sid=2: bit_sign={0000101101111010, 1010000000110111}, sid=3: bit_sign={0010001000011110}}

4 Using Sequential Index for Content-Based Retrieval

During content-based sequence retrieval, the bitmap signatures for all data sequences are scanned. For each data sequence, the test of a searched subsequence mapping is performed. If the searched subsequence can be successfully mapped to the data sequence partitions, then the data sequence is read from the database. Due to the ambiguity of bitmap signature representation, additional verification of the retrieved data sequence is required. The verification can be performed using the traditional B+ tree method, since it consists in reading the data sequence from the database and checking whether it contains the searched subsequence. The formal description of the algorithm is given below. We use a simplified notation of $Q[i\_start..i\_end]$ to denote a partition $<X_{i\_start}\ X_{i\_start+1} \ldots X_{i\_end}>$ of a sequence $Q = <X_1\ X_2 \ldots X_n>$, where $1 \le i\_start \le i\_end \le n$. The symbol & denotes bit-wise AND operation.

```
for each sequence identifier sid do begin
 j = 1;  i_end = 1;
 repeat
  i_start = i_end;
  evaluate equiv. set E_Q for Q[i_start..i_end]; mask=bit_sign(E_Q);
  while mask & bit_sign(E_sid,i)<>mask and j<=#partitions for sid do j++;
  if j<= number of partitions for sid then repeat
   i_end++; generate equivalence set E_Q for Q[i_start..i_end];
   mask = bit_sign(E_Q);
  until mask & bit_sign(E_sid,i) <> mask or i_end = size of Q;
 until i_start = i_end or j > number of partitions for sid;
 if j <= number of partitions then return(sid);
end;
```

Example. Assume that we look for all data sequences, which contain the subsequence <{*F*}{*B*}{*D*}>. We begin with *sid*=1. We find that <{*F*}> matches the first partition. So, we check whether <{*F*},{*B*}> also matches this partition. Accidentally it does, but when we try <{*F*},{*B*},{*D*}>, we find that it does not match the first partition. Then we move to the second partition to check whether <{*D*}> matches the partition. This test fails and since we have no more partitions, we reject *sid*=1 (this data sequence does not contain the given subsequence). In the next step, we check *sid*=2. We find that <{*F*}> matches the first partition. So, we check whether <{*F*},{*B*}> also matches this partition. It does not, so we move to the second partition and find that <{*B*}> matches the partition. Then we must check whether <{*B*},{*D*}> also matches the partition. This time the check is positive and since we have matched the whole subsequence, we return *sid*=2 as a part of the result. The data sequence will be verified later. Finally, we check *sid*=3. We find that <{*F*}> does not match the first partition. Since we have no more partitions, we reject *sid*=3 (this data sequence does not contain the given subsequence). So far, the result of our index scanning is the data sequence identified by *sid*=2. We still need to read and verify, whether the sequence really contains the searched subset. Here it does - the result is returned to a user.

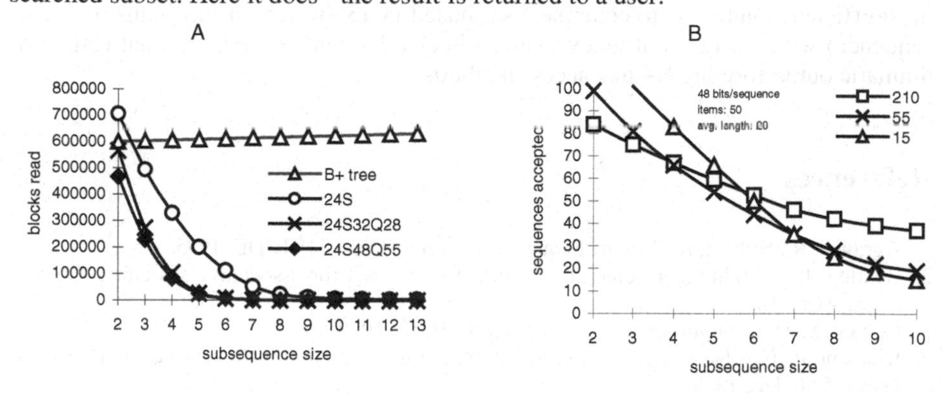

Fig. 2. Experimental results

5 Experimental Results

We performed experiments on Oracle8 (randomly generated data sets - uniform item distribution). Figure 2A shows the number of disk blocks read to retrieve data sequences containing subsequences of various lengths. The data set contained 50000 data sequences, having 20 items of 50 in average. The compared database accessing methods were: B+ tree index (B+ tree), 24-bit set-based bitmap index (24S), 32-bit sequential index ($\beta = 28$) built on top of 24-bit set-based bitmap index (24S32Q28), and 48-bit sequential index ($\beta = 55$) built on top of 24-bit set-based bitmap index (24S48Q55). Our index achieved a significant improvement for the searched subsequences of length greater than 4.

We also analyzed the influence of the partitioning level on the index performance. Figure 2B illustrates the filtering factor for three sequential indexes built on bitmap signatures of total size of 48 bits, but with different partitioning. We noticed that partitioning data sequences into a large number of partitions results in performance increase for long subsequences, but worsens the performance for short subsequences. Using a small number of data sequence partitions results in more "stable" performance, but worse for long subsequences.

6 Final Conclusions

Content-based sequence retrieval is specific in the sense that it requires complicated *SQL* queries and database access methods. In this paper we have introduced the new indexing method, called sequential indexing, which can replace a B+ tree index and set-based indices for dense databases. During experiments, we have found that the most efficient solution is to combine a set-based index (which checks items of a data sequence) with a sequential index (which checks the items ordering), what results in dramatic outperforming B+ tree access methods.

References

1. Agrawal, R., Srikant, R., Mining Sequential Patterns, Proc. 11th ICDE, 1995
2. Bentley, J.L., Multidimensional binary search trees used for associative searching, Comm. of the ACM 18
3. Comer D., The Ubiquitous B-tree, Comput. Surv. 11, 1979
4. Diamantini, C., Panti, M., A Conceptual Indexing Method for Content-Based Retrieval, Proc. 15th ICDE, 1999
5. Guttman, A., R-trees: A dynamic index structure for spatial searching, Proc. ACM SIGMOD Conf., 1984
6. Mannila H., Toivonen H., Verkamo A.I., Discovering frequent episodes in sequences, Proc. 1st KDD, 1995
7. Morzy, T., Zakrzewicz, M., Group Bitmap Index: A Structure for Association Rules Retrieval, Proc. 4th KDD, 1998
8. O'Neil, P, Model 204 Architecture and Performance, Springer-Verlag Lecture Notes in Computer Science 359, 2nd HTPS, 1987

The S^2-*Tree*: An Index Structure for Subsequence Matching of Spatial Objects

Haixun Wang and Chang-Shing Perng

IBM T. J. Watson Research Center
Yorktown Heights, NY 10598
{haixun,perng}@us.ibm.com

Abstract. We present the S^2-*Tree*, an indexing method for subsequence matching of spatial objects. The S^2-*Tree* locates subsequences within a collection of *spatial sequences*, i.e., sequences made up of spatial objects, such that the subsequences match a given query pattern within a specified tolerance. Our method is based on (i) the string-searching techniques that locate substrings within a string of symbols drawn from a discrete alphabet (e.g., ASCII characters) and (ii) the spatial access methods that index (unsequenced) spatial objects. Particularly, the S^2-*Tree* can be applied to solve problems such as subsequence matching of time-series data, where features of subsequences are often extracted and mapped into spatial objects. Moreover, it supports queries such as "what is the longest common pattern of the two time series?", which previous subsequence matching algorithms find difficult to solve efficiently.

1 Introduction

The sequence of objects can endow it with some special significance that an unsequenced grouping of the same objects could never convey. In this paper, we focus on the design of fast searching methods that will search a database of sequences of text, spatial, or multimedia objects to locate those that match a query subsequence of objects. Such sequences can be 1-dimensional time series, digitized voice or music, video clips, trail of mobile objects:

- Time series databases. The efficient matching [3] of time series data often relies on some distance-preserving transform, such as the Discrete Fourier transform (DFT), which extracts the first f DFT coefficients and map them into points in the f-dimensional feature space.
- Content-based image querying. A similarity retrieval algorithm for image databases[11] extracts image regions and uses Harr wavelet[12] to compute their signatures by mapping them to some multidimensional space.
- Content-based analysis, indexing, and retrieval of audio or video sequences. For instance, *VideoTrails*' approach to analyzing a video clip involves first generating a trail of points in a multidimensional space where each point is derived from physical features of a single frame in the video clip[13].
- Spatio-temporal databases, which deal with geometries changing over time[16].

D. Cheung, G.J. Williams, and Q. Li (Eds.): PAKDD 2001, LNAI 2035, pp. 312–323, 2001.

The above databases and the queries processed against them have the following characteristics in common:

- Entities in the databases are either spatial objects, or can be converted into spatial objects through feature extraction. Examples of feature vectors are color histograms[5], Fourier vectors, text descriptors[15], etc.
- There exists an order among the entities. Objects in time series databases, audio, and video sequences are ordered by time. In content-based image querying, an image is often decomposed into a set of sub images which can be partially ordered by their relative positions in the original image. In this paper, we assume there is a total order among the entities.

Taking advantage of the above characteristics, we find that the task of subsequence matching against sequences of text, spatial, or multimedia objects is essentially the following problem: Given a query spatial sequence, search a set of spatial sequences to locate subsequences that are similar to the query sequence.

However, traditional database indexing techniques are inadequate for this purpose. There is currently much excellent work in indexing multidimensional data, including geometric hashing[14], grid-based index structures[9], and the R-tree family[7,6] index structures. These spatial access methods are designed to index *unsequenced* objects. The order among the entities is not taken into consideration when the index structures are created and hence no effective retrieval method in terms of subsequence matching of spatial objects is supported.

Our work extends the substring matching technique to spatial sequences. We propose a new index structure, the S^2-*Tree*, that can be applied to search databases of different contents when the features of the data are extracted into sequences of spatial objects. It also supports new SQL predicates, for example, *sound like* and *look like*, which are similar to the standard *like* predicate for substring matching, for queries in multimedia databases. In this paper, we focus on time series, where temporal patterns are usually mapped to feature vectors in some high-dimensional space[3].

The organization of the rest of the paper is as follows. We first review some background material, including spatial access methods and the suffix tree. In Section 3 we propose S^2-*Tree* for fast subsequence matching of spatial objects. In Section 4, we use S^2-*Tree* for subsequence matching in time series databases. Section 5 contains experiments that show the effectiveness of our algorithms.

2 Background

The R-tree[7] can be viewed as an extension of the B-tree to multi-dimensions. The R^*-tree[6] improves the R-tree by introducing a policy called *forced reinsert*. It also refines the node splitting policy by taking overlapping area and region perimeter into consideration. Another modification of the R-tree, called X-tree[4], is well suited for indexing high-dimensional data. The main idea of the X-tree is to avoid overlap of bounding boxes in the directory by using a new organization of the directory which is optimized for high dimensional space.

Instead of allowing splits that introduce high overlaps, X-tree postpones node splitting by introducing supernodes, i.e., nodes larger than the usual block size.

A *suffix tree*[1] embodies a compact index to all the distinct, non-empty substrings of a given string. The overall space requirement of the suffix tree is linear in the length of the string it represents. Various approaches of building the substring index in linear time have been developed. McCreight's algorithm builds a suffix tree in linear time and is space efficient[1]. Ukkonen [2] developed a linear-time, on-line suffix tree construction algorithm.

3 The S^2-*Tree*

The S^2-*Tree*[1] is motivated by the fact that searching substrings in a suffix tree takes an average $O(a + \log n)$ disk accesses, where a is the size of the answer set. It would be desirable if the same technique can be used to solve the problem of subsequence matching of spatial objects and reduce its complexity.

The major differences between spatial sequences and text strings are: (i) The alphabet of text strings usually consists of only a few discrete symbols (e.g. ASCII set); spatial sequences do not have a pre-defined "alphabet"; (ii) There is no relationship among symbols in a text string. While relationships could exist between two spatial objects, for example, *contains* and *overlaps*.

The S^2-*Tree* bridges the gap between spatial sequences and text strings by creating an alphabet that encodes spatial objects as well as the (containment) relationship. The S^2-*Tree* is a combination of two trees (i) The X-tree, which provides a clustering method, according to which objects are converted into binary encodings that embody the containment relationship. (ii) The suffix tree, which implements subsequence matching on the binary sequences.

For the rest of the paper, we shall use the following notational conventions. Unless otherwise specified, we use uppercase letters, Q, R, S, to denote spatial sequences and we use lowercase letters, a, b, c, to denote *minimum bounding rectangles* (MBRs) of spatial objects.

$\|S\|$	the length of spatial sequence S.
$S[i]$	the ith entry of spatial sequence S.
$S[i, j]$	a subsequence that includes entries in position i through j.
$a \subseteq b$	MBR b contains MBR a.
a'	a binary encoding of MBR a
S'	a binary encoding of spatial sequence S.

Given two spatial sequences P and Q, we say P *matches* Q if P and Q are of the same length and each spatial object of P is contained by the corresponding spatial object of Q, i.e., $P[i] \subseteq Q[i]$, for all $1 \leq i \leq |P|$. Now, the problem of subsequence matching of spatial objects can be defined as follows:

- We have N spatial sequences $S_1, \cdots, S_n$, each of potentially different length.
- We have a query subsequence $\mathcal{Q}$ of length $|\mathcal{Q}|$.

[1] S^2-*Tree* stands for Spatial Suffix Tree

- We want to find all the sequences S_i, along with the correct offset k, such that the subsequence $l = S_i[k, k+|\mathcal{Q}|-1]$ is enveloped by the query sequence, i.e., each spatial object in l is inside its corresponding spatial object in $\mathcal{Q}$.

When a query subsequence $\mathcal{Q}$ is given, a *matching tolerance* is specified *implicitly* at the same time. A bigger MBR represents a higher tolerance. The user has the freedom to enlarge/reduce the size of *each* MBR in the query subsequence, i.e., the tolerance can be customized for different portions of the query subsequence. For example, to specify "don't care", the user can simply make some MBRs in the query subsequence as large as the universe so that each spatial object in the database matches the MBRs at those positions.

3.1 Three Steps to Constructing the S^2-*Tree*

Creating an index structure. Given a set of spatial sequences, we add all the spatial objects in those sequences into a multidimensional index structure. The following are the major concerns when we choose our spatial access method:

- *Dimensionality.* One goal of the S^2-*Tree* is to index features extracted from databases of different contents. These features can have 3-20 dimensions. The index structure should be able to handle dimensions in this range.
- *Overlap.* Overlap is the percentage of the volumn covered by more than one directory MBR. The S^2-*Tree* maps an object into binary strings according to the hierarchy of the index structure. Minimizing the overlap is equivalent to minimizing the number of different mappings for each spatial object.
- *Space Utilization.* The size of the index structure is another concern. Since the length of the encoded binary string depends on both the width and depth of the tree structure, maximizing storage utilization is equivalent to minimizing the total number and length of the binary strings.

After comparing the R-tree family access methods, we chose the X-tree for our index structure. The notion of supernode introduced by the X-tree creates a balance between overlap and space utilization and is more suitable for our purpose.

Encoding the X-tree. The root node of the X-tree is labeled with ϵ, the empty string. An edge connecting a node with its kth child is labeled k (in binary, $k \geq 0$). Nodes other than the root are labeled with the concatenation of the labels on the edges connecting the root to that node. Thus, we have generated an alphabet $\mathcal{A} = \{$*labels on all the nodes*$\}$. The following property holds for the prefix relationship among the symbols in the alphabet:

Theorem 1. *If $\alpha, \beta \in \mathcal{A}$ and α is a prefix of β, then the MBR of the node labeled with α contains the MBR of the node labeled with β.*

Proof: α is a prefix of β, according to the encoding method, the node labeled with α must be an ancestor of the node labeled with β. The property holds because in the X-tree, the MBR of a child node is contained by the MBR of its parent node.

For instance, we have a spatial sequence $S = abcdefghijklmnop$. Each symbol in the sequence is a point in a 2-dimensional space. Figure 1(a) shows these points and Figure 1(b) is the corresponding X-tree built on these points (minimal and maximal number of entries per node are 2 and 3, max overlap 20%).

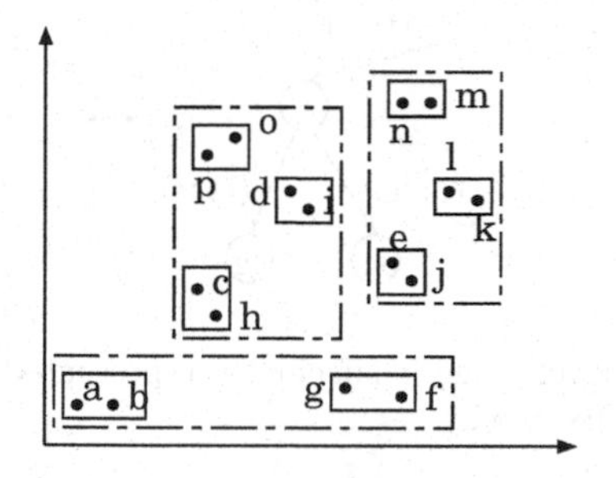

(a) Some 2-dimensional points organized into an X-tree.

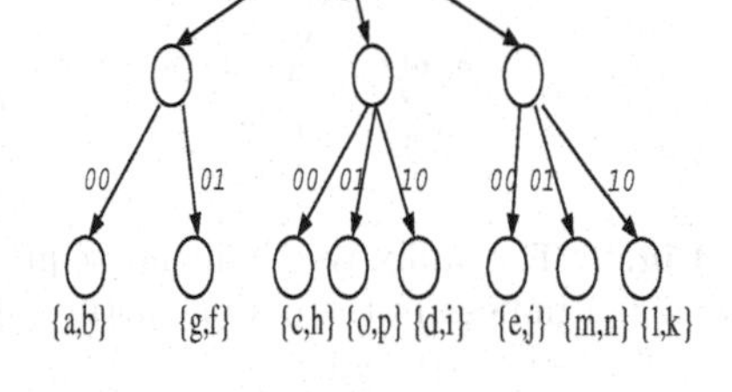

(b) A labeled X-tree. The leaves contain pointers to the spatial objects.

Fig. 1. Using X-tree to cluster spatial objects.

Since the maximum branching factor of the X-tree in Figure 1(b) is 3, we need no more than two bits to label an edge. Each node is coded as the concatenation of the labels on the path from the root to that node. Thus the leftmost leaf node to which points a and b belong is 0000, and the rightmost leaf node 1010. The alphabet $\mathcal{A}$, composed of all the codes of the nodes, is as follows:

$$\mathcal{A} = \{\epsilon, 00, 01, 10, 0000, 0001, 0100, 0101, 0110, 1000, 1001, 1010\}$$

Creating the suffix tree. Representing each spatial object in the original sequence with the code of the node it belongs to, we transforms S into S' using alphabet $\mathcal{A}$ (the '$' sign at the end of S' marks the end of the sequence):

$$S' = 0000 \cdot 0000 \cdot 0100 \cdot 0110 \cdot 1000 \cdot 0001 \cdot 0001 \cdot 0100 \cdot 0110 \cdot 1000 \cdot 1010 \cdot 1010 \cdot 1001 \cdot 1001 \cdot 0101 \cdot 0101 \cdot \$$$

or, if we write the binary code in decimal numbers:

$$S' = 0 \cdot 0 \cdot 4 \cdot 6 \cdot 8 \cdot 1 \cdot 1 \cdot 4 \cdot 6 \cdot 8 \cdot 10 \cdot 10 \cdot 9 \cdot 9 \cdot 5 \cdot 5 \cdot \$$$

We construct a suffix tree in linear time for sequence S' using McCreight's algorithm[1]. A partial suffix tree is shown in Figure 2. The pairs on the edges are the indices in sequence S'. For instance, $(3, 5)$ represents subsequence $S'[3, 5]$, i.e., subsequence $4 \cdot 6 \cdot 8$ in decimal, or $0100 \cdot 0110 \cdot 1000$ in binary. The leaves are labeled with the start positions in S' of the suffixes that they represent. For example, subsequence $S'[3, 5] = 4 \cdot 6 \cdot 8$ can be found at offset 3 and 8 of S'.

To construct a suffix tree for a set of spatial sequences, we glue them together into a long sequence by a special symbol '$', and construct the suffix tree for the concatenated sequence.

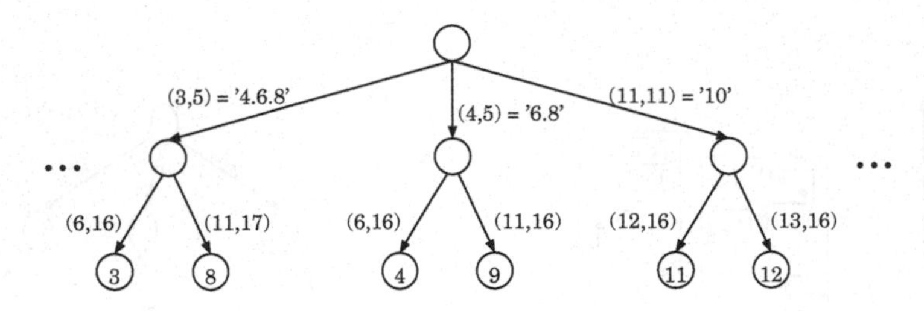

Fig. 2. The suffix tree built for sequence S'. Each edge is succinctly represented by a pair of indices in the data structure of the suffix tree.

3.2 Encoding the Query Sequence

Given a query sequence $\mathcal{Q}$, we need to encode it into binary strings over alphabet $\mathcal{A}$ before we can do the search. A spatial object, or its MBR, may correspond to several symbols in the alphabet. We use procedure `EncodeSpatialObject(mbr, root, set)`, where `root` is the root of the X-tree, to encode the `mbr` into a set of symbols. Figure 3 shows the algorithm. It performs a depth-first search of the X-tree, looking for the uppermost internal nodes contained entirely inside `mbr`, or leaf nodes that intersect `mbr`. For instance, a spatial object which contains the MBR of the root node will be encoded into a single symbol, ϵ.

```
Procedure EncodeSpatialObject(mbr, Node, symbol_set)
Input:            mbr is the MBR of a spatial object, Node is a node of the X-tree
Input & Output:   symbol_set is a set of binary characters in alphabet A

01 if mbr contains the MBR of Node or Node is a leaf node then
02      add the label of Node to symbol_set
03 else
04      for each child node c of Node do
05           if the MBR of node c intersects mbr then
06                EncodeSpatialObject(mbr, c, symbol_set)
07           end if
08      end for
09 end if
```

Fig. 3. EncodeSpatialObject() encodes a spatial object into a set of symbols.

`EncodeSpatialObject()` maps each spatial object in the query sequence $\mathcal{Q}$ into a set of symbols, and we get a list of symbol sets, L, for the entire sequence $\mathcal{Q}$. Encoding fails if any member of list L is an empty set. The user then will have to raise the tolerance, i.e., to enlarge the MBR at the corresponding position in the query sequence, in order to find any match in the database.

3.3 Subsequences Matching in the S^2-*Tree*

The search algorithm (Figure 5) works as follows: Given L (the list of symbol-sets corresponding to a query sequence), the algorithm performs a depth-first search of the suffix tree. At a certain node, if all the symbol sets have been matched ($N = 0$), then the labels of the leaf nodes that are descendants of that node will be the offsets of the query sequence $\mathcal{Q}$ found in S'. Otherwise, it calls `Search()` recursively on the sub-nodes whose labels match a prefix of L (line 07).

Procedure symbolMatch(seq, L)
Input: `seq` is a symbol sequence; `L` is a sequence of sets, and $|L| = |seq|$
Output: SUCCESS if matches

```
01 for i=1 to |seq| do
02        if none of the symbols in set L[i] is a prefix of the symbol seq[i]
03              return FAIL
04        end if
05 end for
06 return SUCCESS
```

Fig. 4. symbolMatch() decides whether a label can be matched by a symbol-set.

In contrast to substring matching in traditional suffix trees, our searching algorithm will traverse *multiple* sub-branches of a node when more than one subsequence on the edges matches a prefix of L. We use `symbolMatch()` in Figure 4 to determine the compatibility between a list of symbols and a list of symbol-sets generated by `EncodeSpatialObject()`. Instead of exact matching, the partial order in the alphabet is used to decide whether the MBR of a symbol is contained in another MBR, thus allowing matching with a flexible tolerance.

It is easy to prove that the above searching algorithm is *correct*, that is, it never misses qualifying subsequences. This is simply because neither the X-tree nor the suffix tree allows false dismissals. However, the result returned in the `offset_set` by the `Search()` procedure may contain some "false alarms", and they are discarded in the post-processing step (Section 3.4).

3.4 Minimizing False Alarms

The `Search()` procedure returns a superset of the qualifying subsequences. To filter out false alarms, we check each offset returned by `Search()` to see if we have a valid match.

This post-processing step is time-consuming when we have a high percentage of false alarms. False alarms are introduced during the encoding of the query sequence. Suppose α is one of the symbols which encodes a spatial object s in the query sequence, then α corresponds to a node N in the X-tree. If

- N *is a leaf node.* The MBR of N contains the MBR of s. Suppose N consists of k spatial objects, we may introduce as many as k false alarms by encoding s into α since it is possible that none of them is actually contained by s.

```
Procedure Search(Node, L, offset_set)
Input:   Node is a node in the suffix tree; L is a sequence of symbol sets
Output:  offset_set is the set of all the start positions in S of the subsequence
         match

01 N ⇐ |L|
02 if N = 0 then
03       add all the labels on the leaves that are descendants of Node to offset_set
04 end if
05 for each child node c of Node do
06       S'(i,j) ⇐ the pair index on the edge linking Node and its child c
07       if symbolMatch(S'[i,j], L[1, j-i+1]) = SUCCESS then
08             Search(c, L[j-i+2, N], offset_set)
09       end if
10 end for
```

Fig. 5. Search() performs subsequence matching on a suffix tree.

- *N is not a leaf node.* No false alarm is introduced by this encoding. The MBR of the spatial object contains the MBR of node N, which means it contains all the MBRs of the spatial objects in the leaf nodes under node N.

Thus, the number of false alarms is affected by the number of the objects contained in the leaf nodes of the X-tree. However, reducing the size of leaf nodes will bring the following disadvantages to subsequence matching in the suffix tree:

- Smaller block size means we need more nodes to hold all the spatial objects. This translates to a larger alphabet, and a larger suffix tree.
- Usually, the size of the MBR of the leaf node will decrease if it holds fewer objects. Thus, a spatial object will possibly be encoded into more symbols, and the symbolMatch() procedure will find more matchings, which means we need to traverse more sub-branches in the suffix tree.

Hence, we need a balance between the number of entries in the leaf nodes and the potential size of the symbol set. To reduce the size of the symbol set, we prune the X-tree bottom-up. The pruning process picks on level $h-1$ a node that has the smallest MBR among all the nodes on level $h-1$, where h is the height of the X-tree, and removes all its child nodes. Now this node becomes a new leaf node, which represents all the entries in its former child nodes. Thus, we reduce the size of the symbol set by increasing the number of the entries in the leaf node. We repeat this process until the size of the alphabet falls below a threshold value. We will study the relationship between the number of entries in the leaf node and the number of false alarms further in Section 5.

3.5 Repeated Subsequences

The detection of repeated patterns in sequences is an important activity which crops up in a variety of different situations. However, most similarity-based time series matching algorithms find it difficult to answer queries such as "what is the longest common pattern of the two time series?"

It is much easier to use the S^2-*Tree* for this purpose. To find the *longest common subsequences*, it suffices to keep track of the maximum length of sequences represented by the internal-nodes during the construction of the suffix tree. To find the *most frequently repeated subsequences* of length k, it suffices to perform a walk of the suffix tree and compare the number of descendents of all the nodes which represent strings of length k.

4 Subsequence Matching in Time-Series Databases

Time series databases naturally arise in business as well as scientific decision-support applications. Most current time series subsequence matching algorithms can be seen as consisting of two phases:

1. Converting time series to points in feature space using DFT or other feature extraction methods.
2. Using spatial access methods (e.g., R^*-tree) to store and retrieve the features.

DFT is used in [3] to map a time series to the frequency domain. It uses a moving window of size w, and features of the subsequence inside the moving window are extracted. Thus, a data sequence S is mapped to a trail consisting of $|S| - w + 1$ points in feature space. The trail is then divided into sub-trails, and the MBRs of the sub-trails are managed by the R^*-tree.

We noticed several limitations of this pioneering work:

1. The effectiveness of the *ST-index* is affected by the length of the query pattern. Since the *ST-index* 'knows' only about subsequences of length w, query patterns of length longer than w will be broken into sub-queries of length w. It then searches for subsequences that match at least one the sub-queries. This approach will clearly enlarge the searching space.
2. The *ST-index* uses a fixed tolerance, ϵ, for the entire query pattern. Users might want to have different ϵ, (e.g., "don't care"), for different parts.
3. It is very difficult for the *ST-index* to detect "the most frequently repeated sequences of length k?" or "the longest common pattern?".
4. The problems of amplitude scaling and offset translation are not addressed.

To overcome these limitations, we use the S^2-*Tree* in phase 2 for subsequence matching of time series data. The S^2-*Tree* naturally overcomes the first three limitations mentioned above.

However, in order to solve the problem of amplitude scaling and offset translation, we need an improved feature extraction method in phase 1. In [10], we proposed a new feature extraction algorithm called Landmarks, which extracts features that are invariant under certain transformations.

5 Experimental Evaluations

We implemented the S^2-*Tree* and ran our experiments on stock price spreadsheets from Yahoo!. Our environment is a SPARC 20 machine running Solaris 2.7 with 128M memory.

5.1 False Alarms and the Size of the Alphabet

Figure 6(a) shows that the false alarms drop dramatically when the size of the alphabet increases. (The database contains 10,000 spatial objects.) However, when the size of the alphabet continues to increase, the percentage of false alarms rebounds. The reason of this phenomenon is explained in Section 3.4.

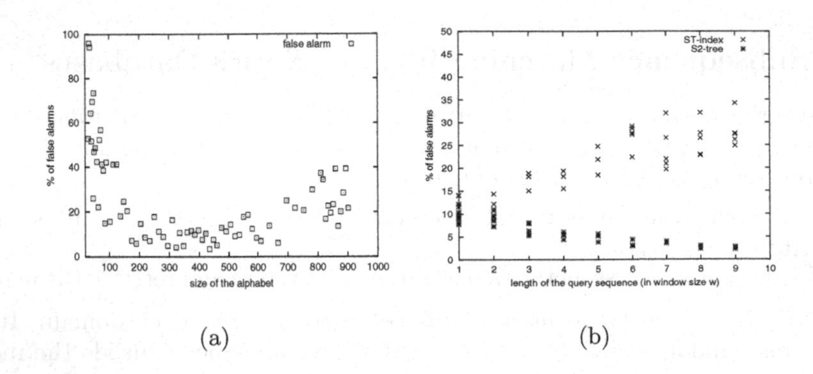

Fig. 6. (a) Relationship between the size of the encoding alphabet and the percentage of false alarms. (b) Percentage of false alarms varying the length of the query sequence.

5.2 Performance Comparison

The use of the S^2-*Tree* index structure in phase 2 is independent of the feature extraction methods applied on the time series data in phase 1. The ST-index uses the first 3 DFT coefficients to map stock prices into the feature space.

Figure 6(b) shows the impact of the length of the query sequence on the number of false alarms. These experiments were carried out on points extracted by the DFT with window size $w = 64$. For each point in the query sequence, the S^2-*Tree* used a universal tolerance ϵ, which was also used by the ST-index method in comparison. Since the ST-index method index only patterns of length w – the size of the moving window – false alarms increase when the length of the query sequence becomes longer. (For query sequence longer than w, the ST-index uses the *'MultiPiece'* algorithm, which searches space volume considerably smaller than the *'PrefixSearch'* algorithm, but the percentage of false alarm still increases.) To search for a longer query sequence using the S^2-*Tree*, however, the percentage of false alarms decreases. This is because when we go deeper in the suffix tree an internal node will have a smaller number of leaf nodes under it.

The S^2-*Tree* outperforms ST-index for subsequence matching of time-series data. Figure 7(a) shows the relative response time of the ST-index method (T_s) vs. the S^2-*Tree* method (T_{ss}), both using DFT to extract features from the stock prices. The advantage of the S^2-*Tree* is also demonstrated by using the Landmarks method[10] for feature extraction in Figure 7(b), where T_s is the response time of the ST-index method, and T_{ms} that of the S^2-*Tree* method using

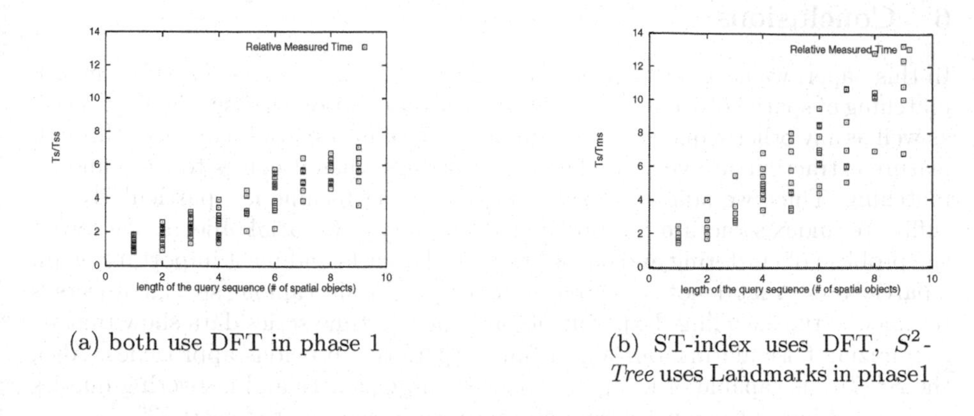

(a) both use DFT in phase 1

(b) ST-index uses DFT, S^2-*Tree* uses Landmarks in phase1

Fig. 7. ST-index vs. S^2-*Tree*, using different methods in phase 1.

the Landmarks for feature extraction. The advantage of the S^2-*Tree* method is obvious when the query sequence is mapped to more than one spatial object.

5.3 Longest Common Subsequences of Two or More Sequences

We use the S^2-*Tree* to search for longest common subsequences of two or more sequences. Figure 8 shows two stock price curves that have the "double bottoms" characteristics. The Landmarks features of the curves are extracted into a 2-dimensional space: *pv* is the percentage of change between the previous landmark and the current one; *vr* is the ratio of changes between the previous period and the current one (for detail, see [10]). The S^2-*Tree* successfully retrieved the "double bottoms" as the longest common subsequences of the two curves.

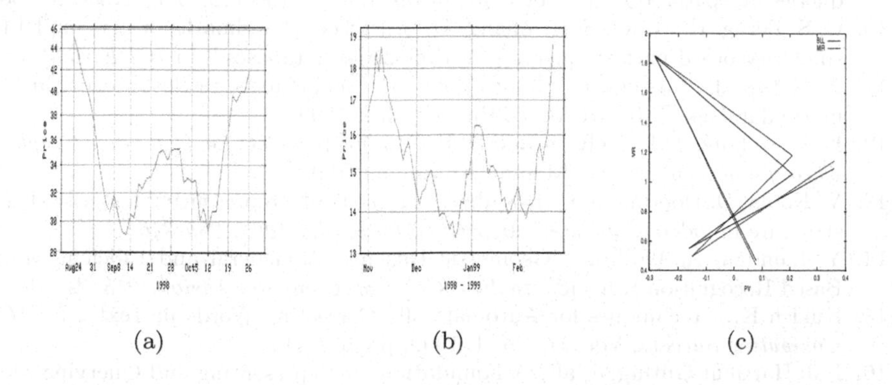

(a) (b) (c)

Fig. 8. (a) BLL 8/98-10/98 (b) MIR 11/98-2/99 (c) Landmarks features.

6 Conclusions

In this paper, we have developed an index method, the S^2-*Tree*, for subsequence matching of spatial objects. The insight is the observation that spatial sequences, as well as any other sequences that can be mapped into spatial sequences through feature extraction, are very similar to text strings when it comes to subsequence matching. Thus, we adapt the substring matching techniques, particularly the suffix tree index structure, to subsequence matching of spatial objects. We solved the problem of clustering and encoding spatial objects and most important of all, a partial order that denotes the containment relationship among spatial objects is retained in the encoding. Experiments on indexing time-series data show that by minimizing false alarms, our algorithm outperforms previous approaches. Also, the S^2-*Tree* is capable of locating repeated subsequences and answering queries such as "What is the longest common pattern in the two time-series?".

References

1. McCreight, E.M. (1976) "A space-economical suffix tree construction algorithm," In *Journal of the ACM,* Vol. 23, No.2, pp. 262-72, April 1976.
2. Ukkonen, E. (1992) "Constructing suffix-trees on-line in linear time," In *Algorithms, Software, Architecture: Information Processing 92,* Vol. 1, pp. 484-92.
3. Christos Faloutsos, M. Ranganathan, Yannis Manolopoulos: "Fast subsequence matching in time-series databases," In *SIGMOD Conference 1994:* pp. 419-429.
4. S. Berchtold, D. A. Keim, H. Kriegel: "The X-tree: an index structure for high-dimensional data." In *VLDB 1996*, Mumbai(Bombay), India.
5. Shawney H., Hafner J.: "Efficient color histogram indexing", In *Proc. Int. Conf. on Image Processing*, 1994.
6. Beckmann N., Kriegel H.-P., Schneider R., Seeger B.: "The R*-tree: an efficient and robust access method for points and rectangles", In *SIGMOD 90.*
7. Guttman A.: "R-trees: a dynamic index structure for spatial searching", In *Proc. ACM SIGMOD Int. Conf. on Management of Data*, Boston, MA, 1984, pp. 47-57.
8. V. Gaede, O. Günther: "Multidimensional access methods", In *ACM Computing Surveys*, Volumn 30, Number 2, June 1998.
9. K. Hinriches, J. Nievergelt: "The grid file: a data structure to support proximity queries on spatial objects." In *Proc. of the WG'83*, 100-113, Linz, Austria, 1983.
10. C. S. Perng, H. Wang, S. R. Zhang, D. S. Parker: "Landmarks: a new model for similarity-based pattern querying in time series databases." In *ICDE 2000.*
11. A. Natsev, R. Rastogi, K. Shim: "WALRUS: a similarity retrieval algorithm for image databases." In *SIGMOD*, Philadelphia, 1999.
12. E. J. Stollnitz, T. D. DeRose, and D. H. Salesin: *Wavelets for Comptuer Graphics: Theory and Applications.* Morgan Kaufmann, 1996.
13. V. Kobla, D. Doermann, C. Faloutsos: "VideoTrails: representing and visualizing structure in video sequences." In *ACM Multimedia 1997: 335-346.*
14. Y. Lamdan, H. Wolfson, "Geometric Hashing: A General and Efficient Model Based Recognition Scheme." In *Intl. Conf. on Computer Vision, 218–249*, 1988.
15. Kukich K.: "Techniques for Automatically Correcting Words in Text", In *ACM Computing Surveys*, Vol. 24, No. 4, 1992, pp. 377-440.
16. Ralf Hartmut Güting, et al: "A Foundation for Representing and Querying Moving Objects", To appear in *ACM Transactions on Database Systems.*

Temporal Data Mining Using Hidden Markov-Local Polynomial Models

Weiqiang Lin[1], Mehmet A. Orgun[1], and Graham J. Williams[2]

[1] Department of Computing, Macquarie University Sydney, NSW 2109, Australia,
{wlin,mehmet}@ics.mq.edu.au

[2] CSIRO Mathematical and Information Sciences, GPO Box 664, Canberra ACT 2601,
Australia, Graham.Williams@cmis.csiro.au

Abstract. This study proposes a data mining framework to discover qualitative and quantitative patterns in discrete-valued time series (DTS). In our method, there are three levels for mining similarity and periodicity patterns. At the first level, a structural-based search based on distance measure models is employed to find pattern structures; the second level performs a value-based search on the discovered patterns using local polynomial analysis; and then the third level based on hidden Markov-local polynomial models (HMLPMs), finds global patterns from a DTS set. We demonstrate our method on the analysis of "Exchange Rates Patterns" between the U.S. dollar and the United Kingdom Pound.

Keywords: temporal data mining, discrete-valued time series, similarity patterns, periodicity analysis, local polynomial modelling, hidden Markov models.

1 Introduction

Temporal data mining is concerned with discovering qualitative and quantitative temporal patterns in a temporal database or in a discrete-valued time series (DTS) dataset. DTS commonly occur in temporal databases (e.g., the weekly salary of an employee, or a daily rainfall at a particular location). We identify two kinds of major problems that have been studied in temporal data mining:

1. The similarity problem: finding fully or partially similar patterns in a DTS, and
2. The periodicity problem: finding fully or partially periodic patterns in a DTS.

Although there are various results to date on discovering periodic patterns and similarity patterns DTS datasets (e.g. [4]), a general theory and general method of data analysis of discovering patterns for DTS data analysis is not well known.

Our proposed framework is based on a new model for discovering patterns by using hidden Markov models and local polynomial modelling. The first step of the framework consists of a distance measure function for discovering structural patterns (shapes). In this step, the rough shapes of patterns are only decided from the DTS and a distance measure is employed to compute the nearest neighbors (NN) to, or the closest candidates of, given patterns among the similar ones selected. In the second step, the degree of similarity and periodicity between the extracted patterns is measured based on local polynomial

D. Cheung, G.J. Williams, and Q. Li (Eds.): PAKDD 2001, LNAI 2035, pp. 324–335, 2001.

models. The third step of the framework consists of a hidden Markov-local polynomial model for discovering all levels patterns based on results from the first two steps.

The paper is organised as follows. Section 2 presents the definitions and basic methods of hidden Markov models and local polynomial modelling. Section 3 presents our new method of hidden Markov-local polynomial models (HMLPM). Section 4 applies new models to "Daily Foreign Exchange Rates" data and section 5 discusses related work. The final section concludes the paper with a short summary.

2 Definitions and Basic Methods

We first give a definition of what we mean by DTS and some other notations will be introduced later. The basic models will be given here and studied in detail in the rest of the paper.

Definition 1 *Suppose that* $\{\Omega, \Gamma, \Sigma\}$ *is a probability space and* T *is a discrete-valued time index set. If for any* $t \in T$ *there exists a random variable* $\xi_t(\omega)$ *defined on* $\{\Omega, \Gamma, \Sigma\}$ *then the family of random variables* $\{\xi_t(\omega), t \in T\}$ *is called* ***a discrete-valued time series (DTS).***

2.1 Definitions and Properties

We consider the bivariate data $(X_1, Y_1), \ldots, (X_n, Y_n)$ which form an independent and identically distributed sample from a population (X, Y). For given pairs of data (X_i, Y_i), for $i = 1, 2, \ldots, N$, we can regard the data as generated from the model

$$\mathbf{Y} = m(\mathbf{X}) + \sigma(\mathbf{X})\varepsilon$$

where $E(\varepsilon) = 0$, $Var(\varepsilon) = 1$, and X and ε are independent.

We assume that for every successive pair of two time points in DTS, $t_{i+1} - t_i = f(t)$ is a function (in most cases, $f(t)$ = constant). For every successive three observations: X_j, X_{j+1} and X_{j+2}, the triple value of (Y_j, Y_{j+1}, Y_{j+2}) has only nine distinct states (called local features) depending on changes in value.

Let state: S_s be the same state as the prior one, S_u the go-up state compared with the prior one and S_d the go-down state compared with the prior one, then we have state-space $\mathcal{S}$ = {s1, s2, s3, s4, s5, s6, s7, s8, s9} = {(Y_j, S_u, S_u), (Y_j, S_u, S_s), (Y_j, S_u, S_d), (Y_j, S_s, S_u), (Y_j, S_s, S_s), (Y_j, S_s, S_d), (Y_j, S_d, S_u), (Y_j, S_d, S_s), (Y_j, S_d, S_d) }.

A sequence is called a *full periodic sequence* if every point in time contributes (precisely or approximately) to the cyclic behavior of the overall time series (that is, there are cyclic patterns with the same or different periods of repetition).

A sequence is called a *partial periodic sequence* if the behavior of the sequence is periodic at some but not all points in the time series.

Definition 2 *Let* $h = \{h_1, h_2, \ldots\}$ *be a sequence. If for every* $h_j \in h$, $h_j \in \mathcal{S}$, *then the sequence* h *is called a Structural Base sequence and a subsequence of* h *is called a sub-Structural Base sequence. If any subsequence* h_{sub} *of* h *is a periodic sequence, then* h_{sub} *is called a sub-structural periodic sequence,* h *also is a structural periodic sequence (existence periodic pattern(s)).*

Definition 3 *Let $y = \{y_1, y_2, \ldots\}$ be a real valued sequence. Then y is called a value-point process. For y_j with $0 \leq y_j < 1$ (mod 1) for all j, we say that y is uniformly distributed if every subinterval of [0, 1] gets its fair share of the terms of the sequence in the long run.*

Definition 4 *Let $y = \{y_1, y_2, \ldots\}$ be a sequence of real numbers with $I - \delta < y_k < I + \delta$, for all k, where I is a constant and δ is an allowable variable parameter. We say that y has an approximate constant sequence distribution of $y = \{I, I, \ldots\}$. In general, if $h(t) - \delta < y_k < h(t) + \delta$ for all k, we say that y has an approximate distribution function h(t).*

2.2 Hidden Markov Models (HMMs)

In a hidden Markov model (HMM) an underlying and unobserved sequence of states follows a Markov chain with a finite state space and the probability distribution of the observation at any time is determined only by the current state of that Markov chain. In this subsection we briefly introduce the hidden Markov time series models which is limited to standard results taken from the literature. We have in particular used those of Baldi and Brunak [13].

Let $\{S_t : t \in \mathsf{N}\}$ be an irreducible homogeneous Markov chain on the state space $\{1, 2, \ldots, m\}$, with transition probability matrix Δ. That is, $\Delta = (\eta_{ij})$, where for all states i and j, and times t:

$$\eta_{ij} = \mathsf{P}(\, S_t = j \mid S_{t-1} = i \,)$$

For $\{S_t\}$, there exists a unique, strictly positive, stationary distribution $\gamma = (\gamma_1, \ldots, \gamma_m)$, where we suppose $\{S_t\}$ is stationary, so that γ is, for all t, the distribution of S_t

Suppose there exists a nonnegative random process $\{\xi_t;\ t \in \mathsf{N}\}$ such that, conditional on $S^{(T)} = \{S_t : t = 1, \ldots, T\}$, the random variables $\{\xi_t : t = 1, \ldots, T\}$ are mutually independent and, if $S_t = i$, ξ_t takes the value v with probability π^t_{vi}. That is, for $t = 1, \ldots, T$, the distribution of ξ_t conditional on $S^{(T)}$ is given by

$$\mathsf{P}(\xi_t = v | S_t = i) = \pi^t_{vi}$$

where the probabilities π^t_{vi} as the "state-dependent probabilities". If the probabilities π^t_{vi} do not depend on t, the subscript t will be omitted.

2.3 Local Polynomial Models (LPMs)

The key idea of local modelling is explained in the context of least squares regression models. We use standard results from the local polynomial analysis theory which can be found from the literature on linear polynomial analysis (e.g, [8]). Recall the data model function given earlier: $\mathbf{Y} = m(\mathbf{X}) + \sigma(\mathbf{X})\varepsilon$ where $E(\varepsilon) = 0$, $Var(\varepsilon) = 1$, and X and ε are independent [1]. We approximate the unknown regression function $m(x)$ locally by a polynomial of order p in a neighbourhood of x_0,

[1] We always denote the conditional variance of Y given $X = x_0$ by $\sigma^2(x_0)$ and the density of X by $f(\cdot)$

$$m(x) \approx m(x_0) + m'(x_0)(x - x_0) + \ldots, + \frac{m^{(p)}(x_0)}{p!}(x - x_0)^p.$$

This polynomial is fitted locally by a weighted least squares regression problem:

$$\text{minimize}\{\sum_{i=1}^{n}\{Y_i - \sum_{j=0}^{p}\beta_j(X_i - x_0)^j\}^2 K_\delta(X_i - x_0)\},$$

where δ is the same δ as in definition 4, and $K_\delta(\cdot)$ with K a kernel function assigning weights to each datum point [2].

3 Hidden Markov-Local Polynomial Models (HMLPMs)

A real-world temporal dataset may contain different kinds of patterns such as complete and partial similarity patterns and periodicity patterns, and complete or partial different order patterns. There are many different techniques for efficient sequence or subsequence matching to find patterns in discrete-valued time series database (DTSB) (e.g, [1]). A limitation of those techniques is also that they do not provide a coherent language for expressing prior knowledge and handling uncertainty in the matching process. Also the existence of different patterns does not guarantee the existence of an explicit model.

In this section we introduce our new data mining model for pattern analysis in a DTS by a combination of the hidden Markov models (HMMs) and local polynomial models (LPMs), called hidden Markov-local polynomial models (HMLPMs). HMMs have been successfully used in many applications, such as in isolated word recognition (see [7]), but they have two major limitations. One is HMMs often have a large number of unstructured parameters, and the other is they cannot express dependencies between hidden states. In order to overcome the limitations of HMMs we apply local polynomial modelling techniques to relax the restrictive form of a HMM. We combine HMMs and LPMs to form hybrid models that contain the expressive power of artificial LPMs with the sequential time series aspect of HMMs.

For building up our new data mining model we divide the data sequence or data vector sequence into two groups: (1) the structural-base data group and (2) the pure value-based data group. In group one we only consider the data sequence as a 9-state structural sequence by applying a distance measure function for performing structural pattern search. In group two, we use local polynomial techniques on the pure value-based sequence data for discovering pure value-based patterns. Then we combine those two groups by using hidden Markov models to obtain the final results.

3.1 Modelling DTS

Without loss of generality we assume that for each successive pair of time points in a DTS, we have t_{i+1} - t_i = c (a unit constant). According to our method the structural base sequence and value-point process data model become:

$$\mathbf{U} = m(\mathbf{V}) + \sigma(\mathbf{V})\varepsilon$$

where $\mathbf{U}$ is the number of y_j of a given sample sequence.

[2] In section 4, we choose Epanechnikov kernel function: $K(z) = \frac{3}{4}(1 - z^2)$ for our experiments in pure-value pattern searching.

Firstly we may view the structural base as a set of vector sequence $\{\mathbf{V_1}, \cdots, \mathbf{V_m}\}$, where each $\mathbf{V_i} = (s1, s2, s3, s4, s5, s6, s7, s8, s9)^T$ denotes the 9-dimensional observation on an object that is to be assigned to a prespecified group.

Then we may also view the value-point process model as a local polynomial model:

$$y(x) = \beta_0 + \beta_1(x - x_0) + \ldots, +\beta_p(x - x_0)^p + \varepsilon.$$

It is more convenient to work with matrix notation for the solution to the above least squares problem in section 2.3. Let

$$\mathbf{X} = \begin{pmatrix} 1 \ (X_1 - x_0) \cdots (X_n - x_0)^p \\ \vdots \quad \vdots \quad\quad \vdots \\ 1 \ (X_1 - x_0) \cdots (X_n - x_0)^p \end{pmatrix},$$

and put $\mathbf{Y} = (Y_1, \cdots, Y_n)^T$ and $\hat{\beta} = (\hat{\beta}_0, \cdots, \hat{\beta}_p)^T$.

Further, let $\mathbf{W}$ be the $n \times n$ diagonal matrix of the weights:

$$\mathbf{W} = diag\{K_\delta(X_i - x_0)\}.$$

The solution vector is provided by weighted least squares theory and is given by

$$\hat{\beta} = (\mathbf{X}^T\mathbf{W}\mathbf{X})^{-1}\mathbf{X}^T\mathbf{W}\mathbf{Y}.$$

Then the problem of value-point pattern discovery can be formulated as the local polynomial analysis of discrete-valued time series.

3.2 Structural Pattern Discovery

We now introduce an approach to discovering patterns in structural base sequences which uses a distance measure function with its density estimator.

From the point of view of our method in structural sequence data analysis, we use squared distance functions which are provided by a class of positive semidefinite quadratic forms. Specifically, if $\mathbf{u} = (u_1, u_2, \cdots, u_9)$ denotes the 9-dimensional observation of each different distance of patterns in a state on an object that is to be assigned to one of the g prespecified groups, then, for measuring the squared distance between u and the centroid of the ith group, we can consider the function [3]

$$D^2(i) = (\mathbf{u} - \bar{\mathbf{y}})^{'}\mathbf{M}(\mathbf{u} - \bar{\mathbf{y}})$$

where $\mathbf{M}$ is a positive semidefinite matrix to ensure the $D^2(i) \geq 0$.

3.3 Point-Value Pattern Discovery

Here we introduce an enhancement to the local polynomial modelling approach through functional data analysis. On the value-point pattern discovery, given the bivariate data

$(X_1, Y_1), \cdots, (X_n, Y_n)$, one can replace the weighted least squares regression function in section 2.3 by

$$\sum_{i=1}^{n} \ell\{Y_i - \sum_{j=0}^{p} \beta_j (X_i - x_0)^j\} K_h(X_i - x_0)$$

where $\ell(\cdot)$ is a loss function. For the purpose of predicting future values we use a special case of the above function with $\ell_\alpha(t) = |t| + (2\alpha - 1)t$ [3].

3.4 Using HMLPMs for Pattern Discovery

For using HMLPMs in pattern discovery we combine the above two kinds of pattern discovery. In structural pattern searching let the structural sequence $\{V_t : t \in \mathsf{N}\}$ be an irreducible homogeneous Markov chain on the state space $\{s1, s2, \ldots, s9\}$, with the transition probability matrix Δ (see section 2.1 for details).

In value-point pattern searching suppose the pure valued data sequence is a non-negative random process $\{C_t;\ t \in \mathsf{N}\}$ such that, conditional on $V^{(T)} = \{V_t : t = 1, \ldots, T\}$, the random variables $\{C_t : t = 1, \ldots, T\}$ are mutually independent and, if $S_t = i$, C_t takes the value v with probability π_{vi}^t. That is, for $t = 1, \ldots, T$, the distribution of C_t conditional on $V^{(T)}$ is given by

$$\mathsf{P}(C_t = v | V_t = i) = \pi_{vi}^t$$

Suppose that if $V_t = i$, ξ_t has a local polynomial distribution with parameters $n_{p,t}$ (a known positive integer) and p_i. That is, the conditional local polynomial distribution of ξ_t has parameters $n_{p,t}$ and $m(t)$, where

$$m(t) = \sum_{i=1}^{m} p_i W_i(t),$$

and $W_i(t)$ is, as before, the indicator of the event $\{V_t = i\}$. Then we have "state-dependent probabilities" for each nine states ($v = 0, 1, \ldots, n_{p,t}$)

The models $\{\xi_t\}$ are defined as hidden Markov-local polynomial models. In this case there are m^2 parameters: m parameters λ_i or p_i, and $m^2 - m$ transition probabilities η_{ij}, e.g. the off-diagonal elements of Δ, to specify the "hidden Markov chain" $\{S_t\}$.

4 Experimental Results

This section presents selected experimental results. There are three steps of experiments for the investigation of "Daily Foreign Exchange Rates"[4] analysis of "Exchange Rates Patterns" between the U. S. dollar and the U. K. pound. The data consist of daily exchange rate for each business day between 2 January 1971 and 21 June 1999. The time series is plotted in figure 1.

[3] This is often called *quantile regression*.

[4] The Federal Reserve Bank of New York for trade weighted value of the dollar = index of weighted average exchange value of U.S. dollar against the United Kingdom Pound: `http://www.frbchi.org/econinfo/finance/finance.html`.

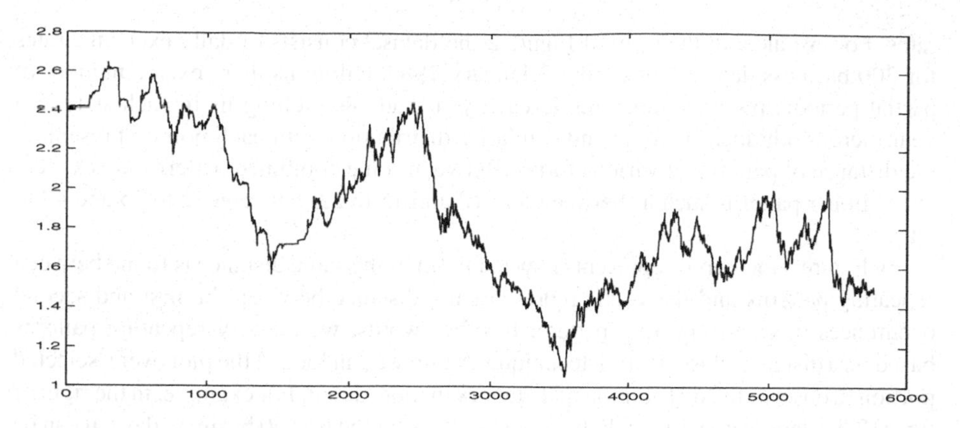

Fig. 1. 5764 working days exchange rates between the U. S. dollar and the U. K. pound, since 1971.

4.1 On Structural Pattern Searching

We investigate the sample of the structural base to test naturalness of the similarity and periodicity on the Structural Base distribution. The size of this discrete-valued time series is about 5764 points. We consider 9 states in the state-space of structural distribution: $\mathcal{S} = \{s1, s2, s3, s4, s5, s6, s7, s8, s9\}$.

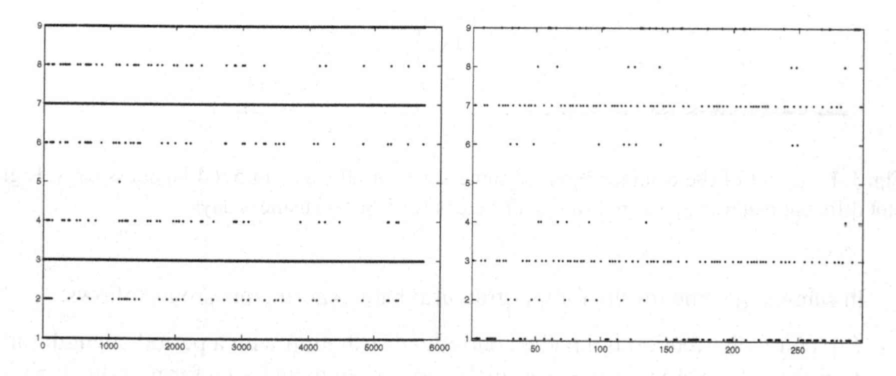

Fig. 2. Left: plot of the distance between same state for all 9 states in 5764 business days. Right: plot of the distance between same state for all 9 states in first 300 business days.

In Figure 2, each point represents the occurence of one of the nine transition states, retaining the original order of the states. There exist two approximation uniformly distributed on state 3 and state 7 if the observations are big enough. Figure 2 also explains two facts: (1) there exists a hidden periodic distribution which corresponds to patterns on the same line with different distances, and (2) there exist partial periodic patterns on and between the same lines. To explain this further, we can look at the plot of distances between the patterns at a finer granularity over a selected portion of the daily exchange rates. For instance, in the right of Figure 2 the dataset consists of daily exchange rates for 300 business days starting from 3 January 1983, telling us there exist a number of partial periodic patterns appearing in each year and, also telling us in each state in a year there is a hidden periodic and similarity distribution with each point representing the distance of patterns of various forms. Between some combined pattern classes there exist similar patterns such as between 5 to 10 and 15 to 22; between 32 to 35 and 42 to 44.

In Figure 3 the *x*-axis represents how many times the same distance is found between repeating patterns and the *y*-axis represents the distance between the first and second occurences of each repeating pattern. In other words, we classify repeating patterns based on a distance classification technique. Again we can look at the plot over a selected portion to observe the distribution of distances in more detail. For example, in the right of figure 3 the dataset consists of daily exchange rates for the first 50 business days. It can be observed that the distribution of distances is a cubic curve distribution: $y = \frac{1}{ax^2+bx+c}$, where Δ = ax^2 + bx + c and Δ, b < 0, a > 0.

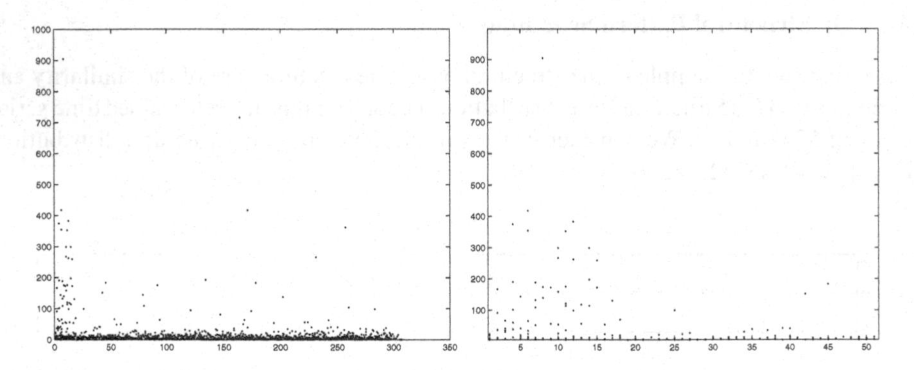

Fig. 3. Left: plot of the distance between same state for all states in 5764 business days. Right: plot different pattern appear in different distances for first 50 business days.

In summary, some results for the structural base experiments are as follows:

- Structural distribution is a hidden periodic distribution with a periodic length function $f(t)$ (there are techniques available to approximate to the form of this function such as higher-order polynomial functions).
- There exist some partial periodic patterns based on a distance shifting.
- For all kinds of distance functions there exist a cubic curve: $y = \frac{1}{ax^2+bx+c}$, where Δ = ax^2 + bx + c and Δ, b < 0, a > 0.
- there exists an approximate uniform distribution in state 3 and state 7.

4.2 On Value-Point Pattern Searching

We now illustrate our new method to construct predictive intervals on the value-point sequence for searching periodic and similarity patterns. The linear regression of value-point of X_t against X_{t-1} explains about 99% of the variability of the data sequence, but it does not help us much in analysis and predicting future exchange rates. In the light of our structural base experiments, we have found that the series $Y_t = X_t - X_{t-2}$ has non-trivial autocorrelation. The correlation between Y_t and Y_{t-1} is 0.5268. Then the observations can be modelled as a polynomial regression function, say

$$Y_t = X_t - X_{t-2} + \sigma(X_t)\varepsilon_t, \qquad t = 1, 2, \ldots, N$$

and then the following new series

$$y(t) = Y(t) + Y(t-1) + \varepsilon_{t'} \qquad t = 1, 2, \ldots, N$$

may be obtained. We also consider the $\varepsilon(t)$ as an auto-regression $AR(2)$ model

$$\varepsilon_{t'} = a\varepsilon_{t'-1} + b\varepsilon_{t'-2} + e_{t'}$$

where a, b are constants dependent on sample dataset, and $e_{t'}$ with a small variance constant which can be used to improve the predictive equation. Our analysis is focused on the series Y_t which is presented in the left of Figure 4. It is scatter plot of lag 2 differences: Y_t against Y_{t-1}.

We obtain the exchange rates model according to nonparametric quantile regression theory:

$$Y_t = 0.488Y_{t-1} + \varepsilon_t$$

From the distribution of ε_t, the $\varepsilon(t)$ can be modelled as an $AR(2)$

$$\varepsilon_t = 0.261\varepsilon_{t-1} - 0.386\varepsilon_{t-2} + e_t$$

with a small $\mathsf{Var}(e_t)$(about 0.00093) to improve the predictive equation.

For prediction of future exchange rates for the next 210 business days, we use the simple equation $Y_t = 0.488Y_{t-1}$ with an average error of 0.00135. In the right of Figure 4 the actually observed series and predicted series are shown.

Some results for the value-point of experiments are as follows:

- There does not exist any full periodic pattern, but there exist some partial periodic patterns based on a distance shifting.
- There exist some similarity patterns with a small distance shifting.

4.3 Using HMLPMs for Pattern Searching

Let $\{S_t : S_t \in \mathcal{S}, t \in \mathsf{N}\}$ be an irreducible homogeneous Markov chain on the state space $\{s1, s2, s3, s4, s5, s6, s7, s8, s9\}$, with transition probability matrix (TPM) (or, stochastic matrix) Δ:

$$\Delta = \begin{vmatrix} 0.5186 & 0.0208 & 0.4606 & 0 & 0 & 0 & 0 & 0 & 0 \\ 0 & 0 & 0 & 0.5161 & 0.0323 & 0.4516 & 0 & 0 & 0 \\ 0 & 0 & 0 & 0 & 0 & 0 & 0.5146 & 0.0362 & 0.4492 \\ 0.4118 & 0.0588 & 0.5294 & 0 & 0 & 0 & 0 & 0 & 0 \\ 0 & 0 & 0 & 1 & 0 & 0 & 0 & 0 & 0 \\ 0 & 0 & 0 & 0 & 0 & 0 & 0.5 & 0 & 0.5 \\ 0.5355 & 0.0260 & 0.4385 & 0 & 0 & 0 & 0 & 0 & 0 \\ 0 & 0 & 0 & 0.4359 & 0 & 0.5641 & 0 & 0 & 0 \\ 0 & 0 & 0 & 0 & 0 & 0 & 0.4962 & 0.0342 & 0.4696 \end{vmatrix}$$

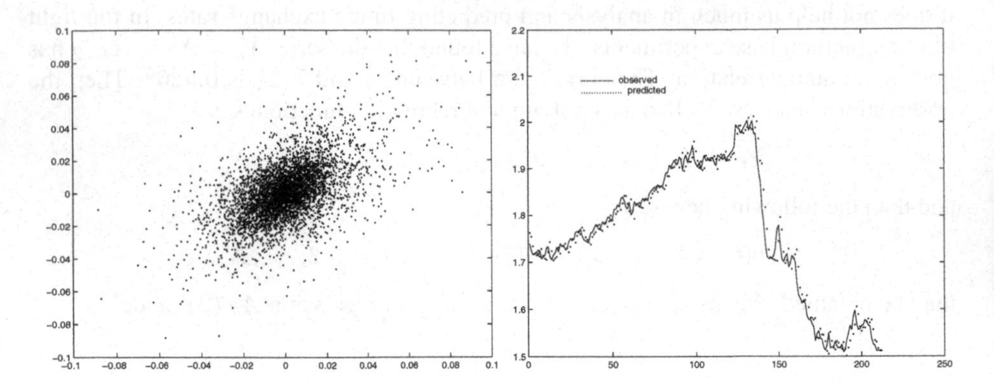

Fig. 4. Left: Scatter plot of lag 2 differences: Y_t against Y_{t-1}. Right:Plot of future exchange rates only for 210 business days by using the simple equation $Y_t = 0.488\ Y_{t-1}$.

We are interested in the future of distribution of TPM, $f(t) = \Delta^t$. According to the Markov property, the TPM: $\lim_{t\to\infty} \Delta^t = 0$. This means that the TPM is non-recurrent of a state si to a state sj. In other words, we cannot use present exchange rate to predict future exchange rate of some period after, but we are only able to predict near future exchange rate.

Suppose that our prediction of future exchange rate of value-point sequence is a nonnegative random process $\{C_t;\ t \in \mathsf{N}\}$, and satisfy $Z_t = \alpha Z_{t-1} + \theta_t$.

Suppose the distribution of sequence of transition probability matrix (TPM) under time order $\Delta_1, \Delta_2, \cdots, \Delta_t; t \in \mathsf{N}$ corresponding to the prediction value-point $Z_t = Y_t - Y_{t-2}$.

We have main combined-results on exchange rates as follows:

- We are only able to predict a short future period by using all present information.
- There does not exist any full periodic pattern but there exist some partial periodic patterns.
- There exist some similarity patterns with a small distance shifting.

5 Related Work

According to pattern theory objectives in pattern searching can be classified into three categories:

- Create a representation in terms of algebraic systems with probabilistic superstructures intended for the representation and understanding of patterns in nature and science.
- Analyse the regular structures from the perspectives of mathematical theory.
- Apply regular structures to particular applications and implement the structures by algorithms and code.

In recent years various studies have considered temporal datasets for searching different kinds of and/or different levels of patterns. These studies have only covered one or two of the above categories. For example, many researchers use statistical techniques such as Metric-distance based techniques, Model-based techniques, or a combination of techniques (e.g, [14]) to search for different pattern problems such as in periodic patterns searching (e.g., [9]) and similarity pattern searching (e.g., [6]).

Some studies have covered the above three categories for searching patterns in data mining. For instance [2] presents a "shape definition language", called $\mathcal{SDL}$, for retrieving objects based on shapes contained in the histories associated with these objects. Also [12] present a logic algorithm for finding and representing hidden patterns. In [5], authors described adaptive methods which are based on similar methods for finding rules and discovering local patterns.

Our work is different from these works. First, we use a statistical language to perform all the search work. Second, we divide the data sequence or, data vector sequence, into two groups: one is the structural base group and the other is the pure value based group. In group one our techniques are similar to Agrawal's work but we only consider three state changes (i.e., up (value increases), down (value decreases) and same (no change)) whereas Agarwal considers eight state changes (i.e., up (slightly increasing value), Up (highly increasing value), down (slightly increasing value) and so on). In this group, we also use distance measuring functions on structural based sequences which is similar to [12]. In group two we apply statistical techniques such as local polynomial modelling to deal with pure data which is similar to [5]. Finally, our work combines significant information of two groups to get global information which is behind the dataset.

6 Concluding Remarks

This paper has presented a new approach combining hidden Markov models and local polynomial analysis to form new models of application of data mining. The rough decision for pattern discovery comes from the structural level that is a collection of certain predefined similarity patterns. The clusters of similarity patterns are computed in this level by the choice of certain distance measures. The point-value patterns are decided in the second level and the similarity and periodicity of a DTS are extracted. In the final level we combine structural and value-point pattern searching into the HMLPM model to obtain a global pattern picture and understand the patterns in a dataset better. Another approach to find similar and periodic patterns has been reported else where [10,11]. With these the model used is based on hidden periodicity analysis and plocal polynormial analysis. However, we have found that using different models at different levels produces better results.

The "Daily Foreign Exchanges Rates" data was used to find the similar patterns and periodicities. The existence of similarity and partially periodic patterns are observed even though there is no clear full periodicity in this analysis.

The method guarantees finding different patterns if they exist with structural and valued probability distribution of a real-dataset. The results of preliminary experiments are promising and we are currently applying the method to large realistic data sets such as two kinds of diabetes dataset.

References

1. Rakesh Agrawal, King-Ip Lin, Harpreet S. Sawhney, and Kyuseok Shim. Fast similarity search in the presence of noise, scaling, and translation in time-series databases. In *Proceedings of the 21st VLDB Conference*, Zürich, Switzerland, 1995.
2. Rakesh Agrawal, Giuseppe Psaila, Edward L.Wimmers, and Mohamed Zait. Querying shapes of histories. In *Proceedings of the 21st VLDB Conference*, 1995.
3. T. W. Anderson. *An introduction to Multivariate Statistical Analysis*. Wiley, New York, 1984.
4. C. Bettini. Mining temportal relationships with multiple granularities in time sequences. *IEEE Transactions on Data & Knowledge Engineering*, 1998.
5. H. Mannila G.Renganathan G. Das, K. Lin and P. Smyth. Rule discovery from time series. In *Proceedings of the international conference on KDD and Data Mining(KDD-98)*, 1998.
6. D.Gunopulos G.Das and H. Mannila. Finding similar time seies. In *Principles of Knowledge Discovery and Data Mining '97*, 1997.
7. P. Guttort. *Stochastic Modeling of Scientific Data*. Chapman & Hall, London, 1995.
8. J.Fan and I.Gijbels, editors. *Local polynomial Modelling and Its Applications*. Chapman and hall, 1996.
9. Cen Li and Gautam Biswas. Temporal pattern generation using hidden markov model based unsuperised classifcation. In *Proc. of IDA-99*, pages 245–256, 1999.
10. Wei Q. Lin and Mehmet A.Orgun. Temporal data mining using hidden periodicity analysis. In *Proceedings of ISMIS2000*, University of North Carolina, USA, 2000.
11. Wei Q. Lin, Mehmet A.Orgun, and Graham Williams. Temporal data mining using multilevel-local polynomial models. In *Proceedings of IDEAL2000*, The Chinese University of Hongkong, Hong Kong, 2000.
12. S.Jajodia andS.Sripada O.Etzion, editor. *Temporal databases: Research and Practice*. Springer-Verlag,LNCS1399, 1998.
13. P.Baldi and S. Brunak. *Bioinformatics & The Machine Learning Approach*. The MIT Press, 1999.
14. Z.Huang. Clustering large data set with mixed numeric and categorical values. In *1st Pacific-Asia Conference on Knowledge Discovery and Data Mining*, 1997.

Patterns Discovery Based on Time-Series Decomposition

Jeffrey Xu Yu[1], Michael K. Ng[2], and Joshua Zhexue Huang[3]

[1] Department of System Engineering & Engineering Management
The Chinese University of Hong Kong
Shatin, N.T., Hong Kong
yu@se.cuhk.edu.hk
[2] Department of Mathematics, The University of Hong Kong
Pokfulam Road, Hong Kong
mng@maths.hku.hk
[3] E-Business Technology Institute, The University of Hong Kong
Pokfulam Road, Hong Kong
jhuang@eti.hku.hk

Abstract. Complete or partial periodicity search in time-series databases is an interesting data mining problem. Most previous studies on finding periodic or partial periodic patterns focused on data structures and computing issues. Analysis of long-term or short-term trends over different time windows is a great interest. This paper presents a new approach to discovery of periodic patterns from time-series with trends based on time-series decomposition. First, we decompose time series into three components, *seasonal, trend and noise.* Second, with an existing partial periodicity search algorithm, we search either partial periodic patterns from trends without seasonal component or partial periodic patterns for seasonal components. Different patterns from any combination of the three decomposed time-series can be found using this approach. Examples show that our approach is more flexible and suitable to mine periodic patterns from time-series with trends than the previous reported methods.

1 Introduction

Time-series data is frequently encountered in real world applications. Stock price, economic growth and weather forecast are typical examples. A time-series represents a set of consecutive observations or measurements taken at certain time intervals. The observations can be real numerical values, for instance, a stock price, or categories, for instance, different medicines taken by a patient over a treatment period. The continuous values of a time-series can be converted to categories if needed.

Discovery of interesting patterns from a large number of time-series is required in many applications. In disease management, for example, the patterns of the best treatment on diabetes in different age groups and genders can be

D. Cheung, G.J. Williams, and Q. Li (Eds.): PAKDD 2001, LNAI 2035, pp. 336–347, 2001.

discovered from the health insurance time-series data [8]. Such information is extremely useful to the government in developing sound health policies. In astronomical science, the microlensing event patterns are hidden in millions of time series measured from 20 million stars over many years [12]. The discovery of such pattern provides a strong evidence on the existence of the "dark matter" in the universe.

One of the tasks for time-series data mining is search for periodic patterns [7,6]. The problem is interesting because many patterns in time-series data are periodic or partially periodic. Han et al recently described some methods for periodic pattern discovery from time-series with observations as categories [7,6]. When their methods are applied to time-series with observations as continuous values, a conversion of values to categories is needed. Such conversion can be easily done on the time-series without trends and their methods are directly applicable to the converted time series. However, problems occur when the time-series have trends because some hiding cyclic patterns may be converted to different categories due to the trend effect. Direct application of Han et al's methods to these time-series will miss a lot of partially periodic patterns. In fact, many time-series in the real world have trends. For example, the stock value of an equity fund is increasing in the long run although its price goes up and down over time. Our motivation in this work is to enhance Han et al's methods [7,6] and make them applicable to the time-series with trends.

In this paper, we present a method to discover periodic patterns from time-series with trends. We first use a time-series decomposition technique to decompose time-series with trends into three components, the seasonal component, the trend and the noise. Then, we apply the algorithms in [7,6] to the seasonal time-series to discover the periodic patterns. After that, we divide the trend component into different categories of regions, such as "increasing", "flat" and "decreasing". Finally, we combine the periodic patterns with the trend categories to form conditional patterns. For example, pattern C(p,q,s) occurs when the trend is "decreasing". Here p is the period of the pattern, q is the offset indicating the first time stamp at which the pattern occurs and s is the number of cycles of the pattern.

Our contribution in this paper is to introduce the time-series decomposition technique, in particular, the STL method, into the process of discovery of periodic patterns from time-series with trends. Our approach has two major advantages: (1) we can discover periodic patterns from time-series with trends, while the previous methods will miss some partially periodic patterns in the areas where the trend increases or decreases significantly, and (2) we discover periodic patterns with trend conditions which can reveal more useful information about the patterns. For example, pattern A occurs when the trend increases. However, such information is not given in the previous approach. For time-series without trends, our approach is similar to those in [7,6] except that we consider noise. In this sense, our patterns are noise-resistant.

1.1 Related Work

Most of the reported work concentrated on symbolic patterns. In [1], an Apriori-like technique was developed for mining sequential patterns. In [11], the authors studied frequent episodes in sequences, where episodes are essentially acyclic graphs of events whose edges specify the temporal before-and-after relationship but without timing-interval restrictions. In [2], a generalization of inter-transaction rules was studied where the left hand and right-hand sides of a rule are episodes with time-interval restrictions. In [4], multi-dimensional inter-transaction rule mining was studied. The mining of cyclic association rules considered the mining of some patterns of a range of possible periods [13]. Because cyclic association rules are partial periodic patterns with perfect periodicity in the sense that each pattern re-occurs in every cycle with 100% confidence, partial periodic patterns were studied in [7,6]. Recently, in [9], the authors studied how to identify representative trends in massive time-series data sets using sketches. Informally, an interval of observations in as time-series is defined as a representative trend if its distance from other intervals satisfy certain properties, for suitably defined distance functions between time-series intervals. The problems with [13,7,6,9] are that they mine patterns based on what appears in the time-series data as they are. They do not mine patterns based on the global cycle-trends, seasonal and irregular patterns, as a whole and individually.

1.2 Paper Organization

The definitions of periodic patterns in time-series data are given in Section 2. In Section 3, we present the time series decomposition techniques and describe the STL approach for time series decomposition in details. In Section 4, we discuss the periodic pattern discovery methods from decomposed time-series and show some real examples. Our concluding remarks on this work are given in Section 5.

2 Periodic Patterns in Time-Series Database

A time-series is a set of observations x_t, each one being recorded at a specific time stamp t. A discrete-time time-series is one in which the set of times at which observations are made is a discrete set, as is the case for example when observations are made at fixed time intervals. In this paper, we consider a time-series of real numbers:

$$x_1, x_2, \cdots, x_n, \quad x_i \in \mathbb{R}.$$

For any time-series, the time stamps at t_1 and t_2 are called similar if their time-related values at these two time stamps are the same, i.e., $x_{t_1} = x_{t_2}$. We follow the similar definitions given in [13,7,6] and define a periodic pattern as follows.

Definition 1. *For any given time-series $\{x_t\}_{t=1}^{n}$, if there exist positive integers p and q $(0 \le p, q < n)$ and s $(0 \le s \le n/p)$ such that all the time stamps $pr + q$ are similar for $0 \le r \le s-1$, we call*

$$\{x_q, x_{p+q}, x_{2p+q}, \cdots, x_{(s-1)p+q}\}$$

a cycle, denoted by $C = (p, q, s)$. Here p is the length or period of the cycle, q is the offset indicating the first time stamp at which the cycle occurs and s is the number of time stamps which are similar in the cycle.

If there are m cycles with the same period p, then these m cycles form a partial periodic pattern with period p. However, if the number of cycles equals to the pattern length, then a complete periodic pattern is formed.

Definition 2. *For any given time-series $\{x_t\}_{t=1}^{n}$, if there exist m cycles*

$$C_1 = (p, q_1, s_1), C_2 = (p, q_2, s_2), \cdots, C_m = (p, q_m, s_m)$$

such that the number of cycles m in a pattern equals to the pattern length $\max_{1\leq i\leq m} q_i - \min_{1\leq i\leq m} q_i + 1$, then we refer to such a pattern $C_1, C_2, \cdots, C_m$ as a complete periodic pattern, denoted by

$$\mathcal{C}(p, \min_{1\leq i\leq m}\{q_i\}, \min_{1\leq i\leq m}\{s_i\}).$$

Definition 3. *A time-series contains a cycle $C = (p, q, s)$ with a confidence α if there are $\alpha \cdot s$ time stamps which are similar. Similarly, a time-series contains a partial periodic pattern with a confidence β if there are $\beta \cdot m$ cycles with the period p.*

In [7,6], J. Han et al. developed an efficient method for mining periodicity in time-series database by using Apriori mining technique. They also showed that data cube structure provides an efficient and effective structure. However, their algorithm is not effective when we apply to a time-series contains a "trend component". For instance, consider the data sequence in Figure 1 which shows 468 monthly observations on Atmospheric concentrations of CO_2 from 1959 to 1997. The top panel is the original time-series. If we apply their algorithm to the time series in the top panel, we cannot find any periodic patterns. The main reason is that there is a hidden trend in the data sequence. Based on these observations, we propose to identify a "trend component" in a time-series before we apply data mining algorithm to find periodic patterns.

3 Time-Series Decomposition

In this section, we provide some notations and background information for time-series decomposition [10]. Time-series decomposition assumes that the data are made up as patterns and errors. Typically, there are three patterns, namely, cyclical pattern, trend pattern and seasonal pattern.

- A cyclical pattern exists when the data exhibit rises and falls that are not of a fixed period.
- A trend pattern exists when there is a long-term increase or decrease in the data.
- A seasonal pattern exists when a series is influenced by seasonal factors.

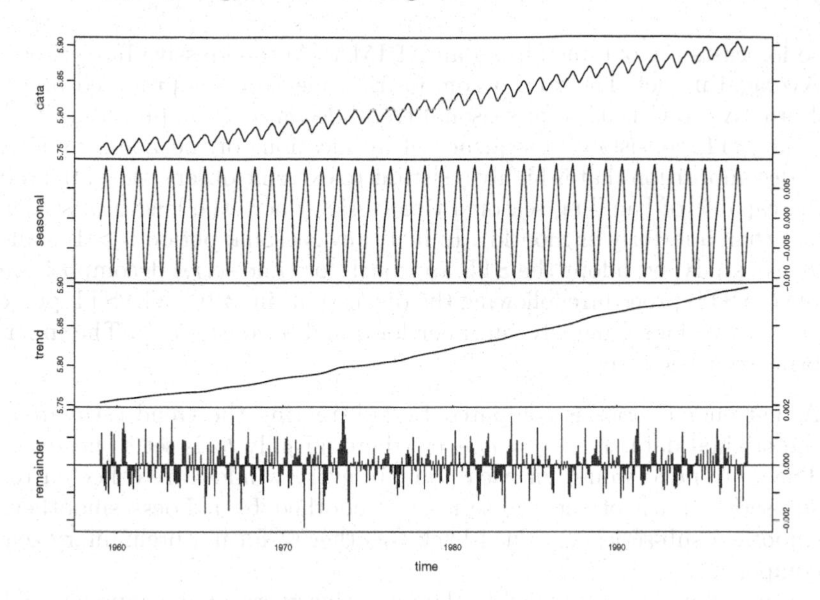

Fig. 1. Atmospheric concentrations of $CO2$: A time-series of 468 observations; monthly from 1959 to 1997.

An error is the difference between the combined effect of the above trend-cycle-seasonality and the actual data. It is often called the irregular or the remainder component. Because the distinction between trend and cycle is somewhat artificial, most decomposition procedures treat the trend and cycle as a single component called trend-cycle. An example of time-series decomposition is shown in Figure 1.

The general mathematical representation of a time-series decomposition approach is given as $x_t = f(y_t, h_t, e_t)$ where x_t is the time series value (actual data) at period t, y_t is the seasonal component (or index) at period t, h_t is the trend-cycle component at period t, and e_t is the irregular component at period t. Two common approaches are additive decomposition and multiplicative decomposition.

- additive decomposition: $x_t = y_t + h_t + e_t$.
- multiplicative decomposition: $x_t = y_t \times h_t \times e_t$.

In fact, logarithms turn a multiplicative relationship into an additive relationship. Therefore, it is possible to fit a multiplicative relationship by fitting an additive relationship to the logarithms of the data.

The classical time-series decomposition was developed in 1920s, there are many variants being developed [10]. The Census II method has been developed by the U.S. Bureau of the Census. One of the most widely used variants is X-12-ARIMA [14]. X-12-ARIMA uses shorter weighted moving averages to provide estimates for the observations at the beginning and end of the series, and provides the facility to extend the original series with forecasts to ensure that more of the observations are adjusted using the full weighted moving averages.

These forecasts are obtained using an ARIMA (Autoregressive Integrated Moving Average) model. The STL decomposition method was proposed in 1990 as an alternative to Census II for seasonal-trend decomposition procedure based on Loess [3]. STL consists of a sequence of applications of the Loess smoother to give a decomposition that is highly resistant to extreme observations. In addition, STL is capable of handling seasonal time-series where the length of seasonality is other than quarterly or monthly. In fact, any seasonal period $n > 1$ is allowed.

In this study, we adopt the STL approach for time-series decomposition. We outline the STL procedure following the discussions in [3,10]. The STL procedure consists of two loops, namely, an inner loop and an outer loop. The inner loop performs six basic steps.

1. A de-trended series is computed by subtracting the trend estimated from the original data. $x_t - h_t = y_t + e_t$ where initially h_t is set to be zero.
2. The de-trended values for each point in every window are collected to form sub-series. Each of the sub-series is smoothed by a Loess smoother. The smoothed sub-series are glued back together to form a preliminary seasonal component.
3. A moving average is applied to the preliminary seasonal component. The result is in turn smoothed by a Loess smoother again. The purpose of this step is to identify any trend-cycle that may have contaminated the preliminary seasonal component in the previous step.
4. The seasonal component is estimated as the difference between the preliminary seasonal component of the second step and the seasonal component in of the third step.
5. A seasonally adjusted series is computed by subtracting the result of the fourth step from the original data ($x_t - y_t = h_t + e_t$).
6. The seasonally adjusted series is smoothed by Loess to give the trend component h_t.

The outer loop begins with one of two iterations of the inner loop. The resulting estimates of trend-cycle and seasonal components are then used to calculate the irregular component: $e_t = x_t - h_t - y_t$. Large values of e_t indicate an extreme observation. These are identified and a weight is calculated. That concludes the outer loop. Further iterations of the inner loop use the weights in the second step of and the sixth step of the inner loop to downweight the effect of extreme values. Further iterations of the inner loop begin with the trend component from the previous iteration.

4 Patterns Discovery from Decomposed Time-Series

In this section, we make use the decomposition of time-series to mine periodic patterns. The outline of our approach is given below.

- **Step-1**: Decompose the time-series $\{x_t\}_{t=1}^n$ into three parts: $x_t = y_t + h_t + e_t$ using the STL procedure as discussed in the previous section.
- **Step-2**: Mine the periodic patterns for the time-series $\{y_t\}_{t=1}^n$, using an existing approach [7,6].

– **Step-3:** Mine the rules from the combination of the trend from h_t, the periodic patterns from y_t and the error from e_t.

Table 1. A time sequence example.

t	1	2	3	4	5	6
x_t	1.0000	2.7071	4.0000	4.7071	5.0000	5.2929
t	7	8	9	10	11	12
x_t	6.0000	7.2929	9.0000	10.707	12.000	12.7071

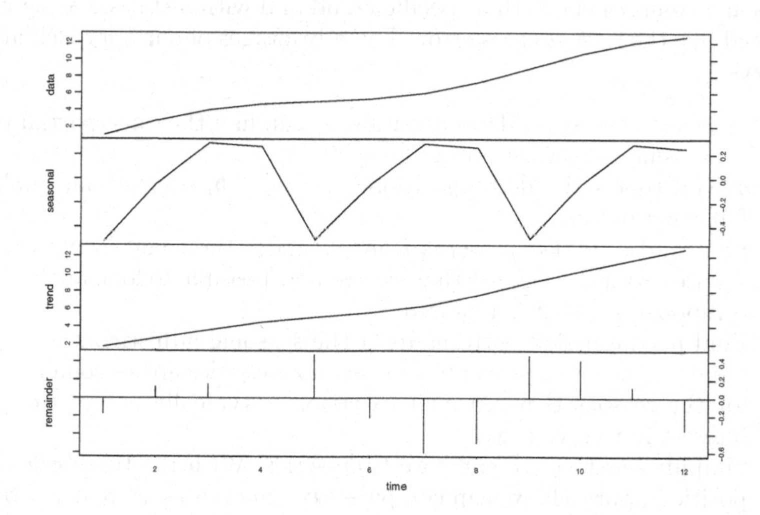

Fig. 2. Decomposition for Table1.

Consider a simple time-series, $\{x_t\}_{t=1}^{12}$, in Table 1, where each x_t is generated as $(t-1)+\sin(\pi(t-1)/4)$. Suppose that we quantify them as $\langle 1, 3, 4, 5, 5, 6, 7, 9, 10, 12, 13\rangle$. Obviously, there are no any partial cyclic patterns from this sequence due to the trend. After STL decomposition, Figure 2 shows the three time-series, namely, $\{y_t\}_{t=1}^{12}$ (seasonal), $\{h_t\}_{t=1}^{12}$ (trend), $\{e_t\}_{t=1}^{12}$ (error), where $x_t = y_t + h_t + e_t$. Suppose that the quantified sequence of y_t is $\langle 0, 5, 8, 8, 0, 5, 8, 8, 0, 5, 8, 8\rangle$. Applied the partial periodic mining algorithm [7] to y_t, we can identity periodic patterns, for example, a complete periodic pattern $\mathcal{C}(4, 1, 3)$. However, the interpretation of periodic patterns from the seasonal time-series, y_t is different. As can be seen from Figure 2, the values of seasonal component is in the range -5.0 and 3.0. As mentioned by Hyndman [10], for additive decomposition, the seasonal component is added to the trend, and can be positive or negative. The seasonal component represents the amount to be added on average each season. If this is negative, it means the season is lower than average. If it is positive, it means the season is higher than average. Therefore, the patterns we find so far are based on it under the trend. In addition, we can classify a trend as "slowly increasing", "increasing", "increasing quickly", "no change", etc. It is worth noting that there are possible periodic patterns in a trend where periodic patterns are for a long term. A periodic pattern $\mathcal{C}(p,q,s)$ appears in y_t with the trend h_t needs to be justified with statistical significance. For instance, the least squares errors for the above time-series y_t is $\sum_{t=t_1}^{t_{12}} e_t^2 = 1.47$. In a similar fashion, for the atmospheric concentrations of CO2 shown in the top panel of Figure 1, there are no particular periodic patterns existing in the original time-series. However, after STL decomposition, we find a period pattern over years. The least squares error is 21.27.

In general, the following statement can be made on periodic pattern mining on the seasonal component. The periodic patterns are found on a time-series, y_t (seasonal component), with a specific trend and with a statistical significance associated like the least squares error. The advantages of our approach are given as follows.

- Without any time-series decomposition, it can find the same partial periodic patterns using the algorithm in [7].
- After STL time-series decomposition, $x_t = y_t + h_t + e_t$, it can further find the following patterns.
  - Find partial period patterns from y_t under the trend h_t with the least squares error. It suggests that we are now possible to look at the seasonal components if that is required.
  - Find partial period patterns from the seasonal adjustment $x_t - y_t$. It is required because many published economic series are seasonally adjusted for the reasons that seasonal variation is typically not of the primary interest in that context.
  - Simplify our tasks to compare time-series. With the time-series decomposition approach, we can compare two time-series, x'_t and $x"_t$ based on their original time-series, the trend component, the seasonal component, and any meaningful combinations.

However, care must be taken in using the periodic patterns discovery technique based on time-series decomposition. Figure 3, 4, 5 and 6 show four different stocks taken in different length of period from the 1997 CRSP US stock databases, their trend components, seasonal components and errors.

In the top panel of Figure 3, we can see some possible periodic patterns in the original time-series. After decomposition, the periodic patterns in the seasonal component are more visible. Both the original time-series and the seasonal component have some partial periodic components. But they are different. First, the fluctuation is different. Second, the patterns from the seasonal component are based on the seasonality, where the patterns from the original time-series included both factors of long-term trend and seasonality. Third, the patterns from the seasonal component is found under the trend with a statistical significance like the least squares error. After removing the seasonal component, the trend

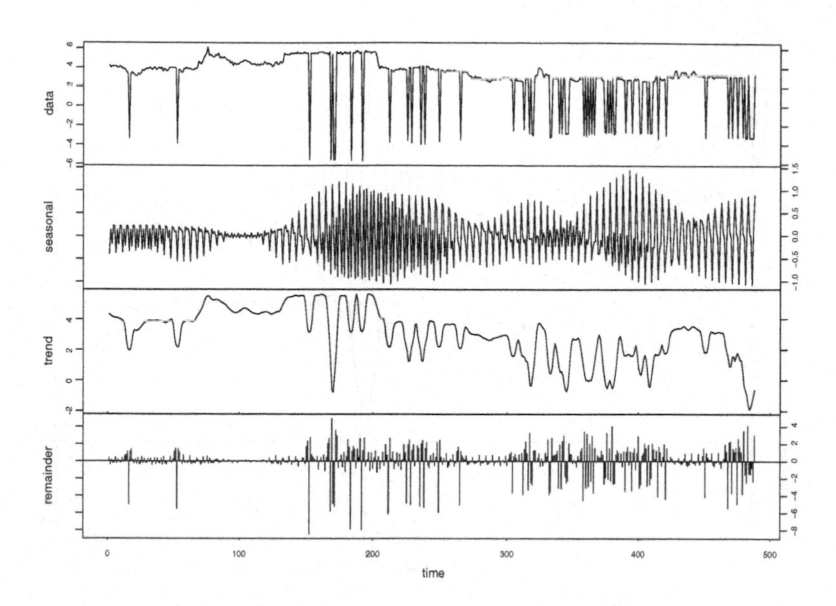

Fig. 3. Stock price for Stock-A. The time dimension is the i-th working day from Nov. 1, 1988 to Oct. 5, 1990. (5 working days a week), taken from [5].

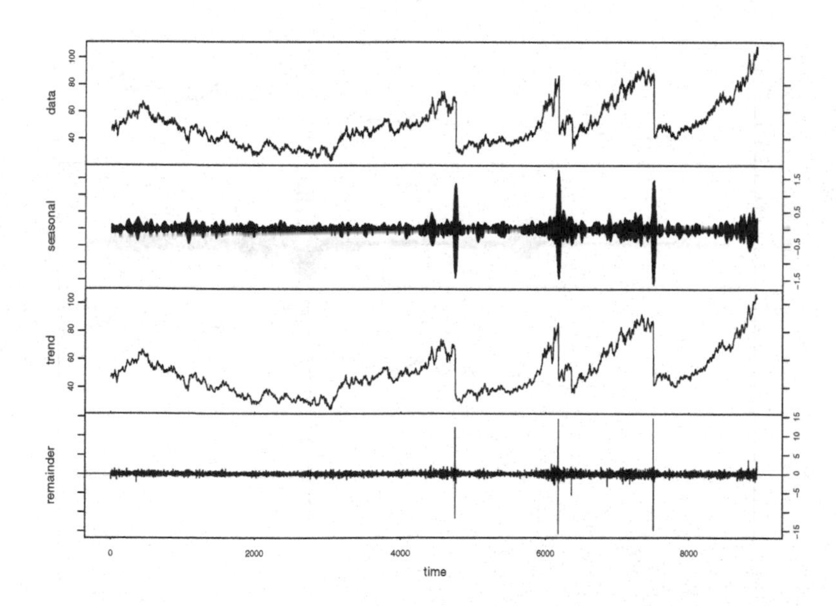

Fig. 4. Stock price for Stock-B. The time dimension is the i-th working day from Jul. 2, 1962 to Dec. 31, 1997. (5 working days a week), taken from [5].

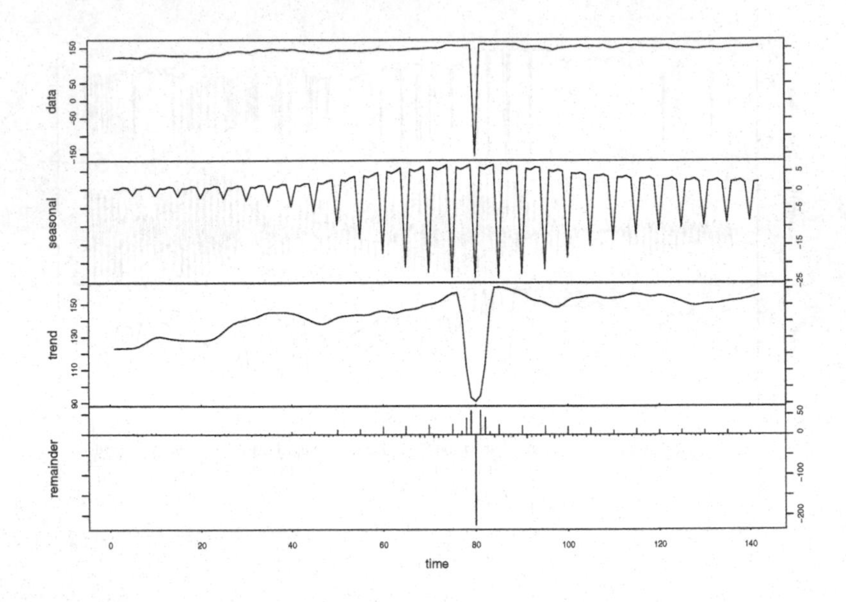

Fig. 5. Stock price for Stock-C. The time dimension is the i-th working day from Jun. 11, 1997 to Dec. 31, 1997. (5 working days a week), taken from [5].

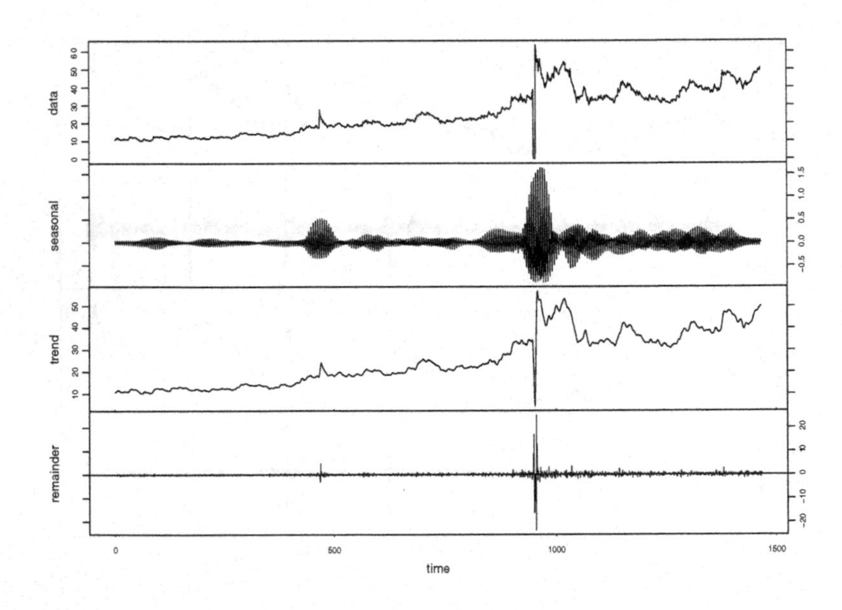

Fig. 6. Stock price for Stock-D. The time dimension is the i-th working day from Jul. 2, 1962 to Apr. 30, 1968. (5 working days a week), taken from [5].

is much smooth. Some periodic patterns can be found in the trend component. The trend component helps finding other stocks with the similar trend, and help comparing their seasonal behaviors. As can be seen from this example, we are possible to search patterns in the original time-series, its trend component and its seasonal component.

In Figure 4, we observe some periodic patterns in the original time-series. However, they are very sensitive to the trend. We find that after the decomposition, the "trend" time-series contains "up" and "down" movements. If the period that a user chooses matches the trend, it is possible to find more periodic patterns. But, in fact, the period is not easy to determine. People usually use week, month, quarter or year as period to search periodic patterns, which might be difficult to find patterns in this case. Searching periodic patterns in the "seasonal" time-series does not completely remove the sensitivity of the period. But the issue become less important. In addition, some existing statistic approaches can be used to determine the period for time-series decomposition.

In Figure 5, we cannot conclude any periodic patterns in original time series and the "seasonal" time series obtained by decomposition. The main reason is that there is sudden change of level in the original time series. This suggests us to partition the original time series into two parts and study each of them. This approach has also been proposed in non-parametric regression in statistics.

In Figure 6, we can view this case as a combination of the two examples in Figures 4 and 5. Based on the above observations, we find that Han's partial periodic algorithm cannot handle all real applications very well. However, our approach using decomposition of time series and partitioning of time series provides users with more flexibility to discover useful periodic patterns.

5 Concluding Remarks

In this paper, we present an approach for periodic pattern searching based on time-series decomposition, STL, which is capable of handling any length of seasonality. The computing issue is not the main issue in this paper. In this study, we focus on the flexibility of periodic pattern search, in particular when there are trends. It is because in many real applications, there are some typical trends. For example, the airline passengers increase for many years, the web-access rapidly increases. It is difficult to find periodic patterns from those time-series data. With decomposition, we are possible to search periodic patterns in the seasonal component and the trend component as well as in the original time-series. We conducted a preliminary study using US CRSP STOCK database [5]. Our approach has flexibility to find periodic patterns in different contexts.

References

1. R. Agrawal and R. Srikant. Mining sequential patterns. In *Proc. Intl. Conf. Data Engineering*, Taipei, Taiwan, 1995.
2. C. Bettini, X. Wang, and S. Jajodia. Mining temporal relationships with multiple granularities in time sequences. *Data Engineering*, 21(1):32–38, March 1998.

3. R.B. Cleveland, W.S. Cleveland, J.E. McRae, and I. Terpenning. STL: A seasonal-trend decomposition procedure based on loess (with discussion). *Journal of Official Statistis*, pages 3–73, 1900.
4. Ling Feng, Hongjun Lu, Jeffrey Xu Yu, and Jiawei Han. Mining inter-transaction associations with templates. In *Proceedings of CIKM'99*, 1999.
5. University of Chicago Graduate School of Business. 1997 crsp us stock databases. *http://www.crsp.com*, 1997.
6. Jiawei Han, Guozhu Dong, and Yiwen Yin. Efficient mining of partial periodic patterns in time series database. In *Proceedings of ICDE'99*, 1999.
7. Jiawei Han, Wan Gong, and Yiwen Yin. Mining segment-wise periodic patterns in time-related databases. In *Proceedings of KDD'98*, 1998.
8. Hongxing He, Hari Koesmarno, Thach Van, and Zhexue Huang. Data mining in disease management: A diabetes case study. In *Proceedings of PRICAI'00*, 2000.
9. Piotr Indyk, Nick Koudas, and S. Muthukrishnan. Identifying representative trends in massive time series data sets using sketches. In *Proceedings of the 26th VLDB*, 2000.
10. Spyros Makridakis, Steven C. Wheelwright, and Rob J. Hyndman. *Forecasting Methods and Applications (Third Edition).* John Wiley & Sons. Inc., 1998.
11. H. Mannila and H. Toivonen. Discovering generalized episodes using minimal occurrences. In *Proc. 2nd Intl. Conf. Knowledge Discovery and Data Mining*, pages 146–151, 1996.
12. M.K. Ng and Z. Huang. Data-mining massive time series astronomical data: challenges, problems and solutions. *Information and Software Technology*, 41:545–556, 1999.
13. Banu Ozden, Sridhar Ramaswamy, and Avi Silberschatz. Cyclic association rules. In *Proceedings of ICDE'98*, 1998.
14. Time Series Staff. X-12-ARIMA reference manual (version 0.2.7). *U.S. Census Bureau*, 2000.

Criteria on Proximity Graphs for Boundary Extraction and Spatial Clustering

Vladimir Estivill-Castro[1], Ickjai Lee[1], and Alan T. Murray[2]

[1] Department of Computer Science & Software Engineering
The University of Newcastle, Callaghan, NSW 2308, Australia
{vlad, ijlee}@cs.newcastle.edu.au
[2] Department of Geography
The Ohio State University, Columbus, OH 43210-1361, USA
murray.308@osu.edu

Abstract. Proximity and density information modeling of 2D point-data by Delaunay Diagrams has delivered a powerful exploratory and argument-free clustering algorithm [6] for geographical data mining [13]. The algorithm obtains cluster boundaries using a Short-Long criterion and detects non-convex clusters, high and low density clusters, clusters inside clusters and many other robust results. Moreover, its computation is linear in the size of the graph used. This paper demonstrates that the criterion remains effective for exploratory analysis and spatial data mining where other proximity graphs are used. It also establishes a hierarchy of the modeling power of several proximity graphs and presents how the argument free characteristic of the original algorithm can be traded for argument tuning. This enables higher than 2 dimensions by using linear size proximity graphs like k-nearest neighbors.

1 Introduction

In Geographical data mining [13], *spatial clustering* consists of partitioning a set $P = \{p_1, p_2, \ldots, p_n\}$ of geo-referenced point-data in a two-dimensional study region R, into homogeneous sub-sets due to spatial proximity. Spatial proximity indicates relative closeness rather than absolute closeness. In particular, one peculiarity of spatial data typically manipulated in a Geographical Information System, is spatial heterogeneity. For example, spatial heterogeneity indicates the intrinsic uniqueness of each location on 2D-space. But, statistical measurements are dependent upon absolute location in R (as opposed to relative location). Hence, the same statistical property has different interpretation over R. For instance, Fig. 1 shows two points p_1 and p_6 equidistant from p_5. Clustering based merely on distance would place p_1 and p_6 in the same cluster as p_5. But, the length of edge $e_{15} = (p_1, p_5)$ is relatively long with respect to the length of edge $e_{56} = (p_5, p_6)$. Site p_5 is likely to belong to the same cluster as p_6. So, although the lengths of the edges e_{15} and e_{56} are equal, they have different relative interpretations (relative to other points in R). In spatial settings (and in spatial clustering) relative proximity is more important than absolute proximity.

D. Cheung, G.J. Williams, and Q. Li (Eds.): PAKDD 2001, LNAI 2035, pp. 348–357, 2001.

Spatial clusters are spatial phenomena. The mixture of global and local effects defines clusters and contributes to their characteristics such as location, scatter, number of clusters, size and distribution. These properties are unknown prior to clustering and prior to Knowledge Discovery. Thus, spatial clustering must reveal them from the data rather than user pre-specified arguments or assumptions. This is the key difference of ESDA (Exploratory Spatial Data Analysis) from model-driven confirmatory analysis. ESDA explores arguments to be set according to the spatial heterogeneity in the data.

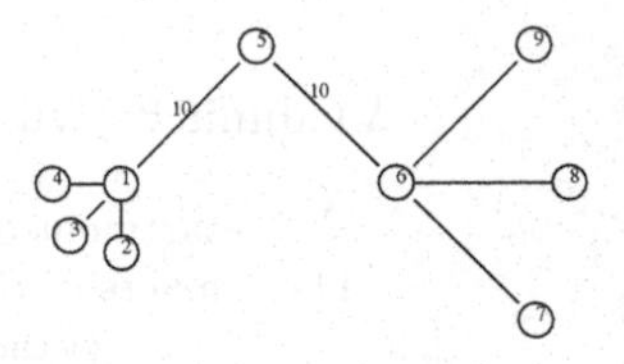

Fig. 1. An example of spatial heterogeneity.

The problem of clustering becomes the matter of identifying neighbors (building a proximity graph) and quantifying their relative closeness and remoteness (establishing a criterion function). This is a two-step process: preprocessing and grouping. Preprocessing structures raw input data and produces a clustering schema. The schema will guide grouping in the next step. The structuring includes (explicitly or implicitly) an underlying proximity graph [2,3,5,6], values for arguments [11] and mathematical summaries [14,15]. During the grouping process, points are combined by a certain criterion function based on the schema generated in the previous stage. Clustering criteria reflect an implicit or explicit inductive bias to suggest the information that is to be inferred and vary from method to method. These inference criteria allow to estimate the cluster membership. All other inference criteria to the one used here seem to focus on the global peaks. Some criteria [2,3,4,8,11] use a global argument. A global value ignores spatial heterogeneity. More seriously, most are neither data-driven nor using a mixture of both global and local effects. These peak criteria find the largest peaks, but have difficulty finding relative smaller peaks. Geographically, these are sparse clusters that are also of interest besides global high-density clusters.

Recently, Estivill-Castro and Lee [6] proposed an effective and efficient clustering criterion, (referred here as the Short-Long criterion) overcoming the shortcomings of peak-inference clustering and satisfying the special needs of geo-information. The main idea of boundary-based clustering is detection of the sharp changes in point density that form cluster boundaries. If such changes are significant in the global sense of view, then they are reported as cluster boundaries. Thresholds for significance test are learned from data, but vary over R, thus localized spatial concentrations are correctly reported without any prior knowledge. The original algorithm uses the Short-Long criterion with the Delaunay Diagram as an underlying proximity graph. Here we extend the criterion to work well with other proximity graphs. In particular, the Delaunay Diagram is linear in size for 2D but quadratic in size for 3D. However, other proximity graphs, (like k-nearest neighbor and k-cone spanner graphs) are linear in size for all dimensions. Thus, the main goal of this paper is to compare and contrast the clustering results with different proximity graphs when they are applied to the Short-Long criterion. Proximity graphs are ranked for their value as exploration tools for very large spatial data sets, where the user may trade the argument free use of the Delaunay Diagram, for an argument driven exploration of the data.

Next, we introduce the working principle of the Short-Long criterion in Section 2. Section 3 summarizes eight popular spatial proximity graphs. We compare and contrast their results. Section 4 summarizes the results of experimental evaluations. Finally, the last section draws conclusions.

2 Short-Long Criterion

The proximity graph named Delaunay Diagram inspired the Short-Long criterion. In proximity graphs, vertices represent data points and edges connect pairs of points to model spatial proximity and adjacency. They encode explicitly a discrete relation IS_NEIGHBOR $\subset P \times P$. By assigning lengths to the edges of the relation, not only we encode proximity, but also density.

Cluster boundaries occur where there is high discrepancy (great variability) among the lengths of incident edges on a vertex p in a proximity graph. This is because both sides of cluster boundaries have different densities (sparse and dense). Thus, border points in proximity graphs may have two different types of incident edges: short edges and long edges. The former connect points within a cluster (intra-cluster links) and the latter straddle between clusters (inter-cluster links) or between a cluster and noise/outliers. Exceptionally long incident edges to a border point are indicative of inter-cluster links. These relatively close and remote neighbors to border points contribute to characterization of border points. Interestingly, border points share this characteristic with points on bridges. Bridge points also have short edges linking to other neighboring bridge points or border points and have long edges connecting to noise. The

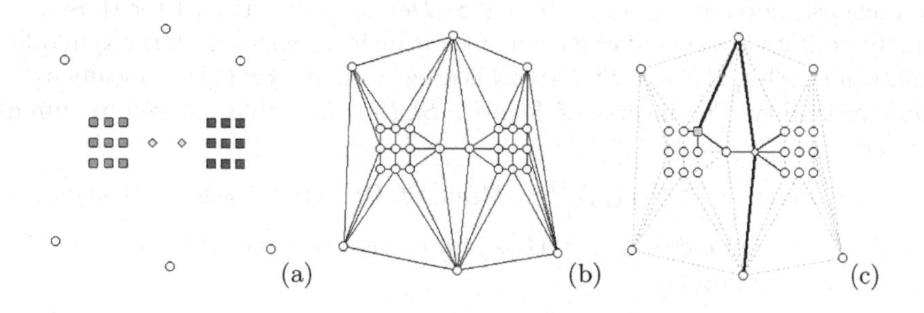

Fig. 2. An example of border and bridge points: (a) Data set ($n = 26$). (b) Delaunay Diagram. (c) Neighbourhoods of a border point and a bridge point.

data shown in Fig. 2(a) seems to have two clusters linked by two bridge points. Fig. 2(c) highlights a border point (green rectangle) and a bridge point (yellow rhombus) with incident edges in solid lines. Both have exceptionally long edges in thick solid lines and exceptionally short edges in thin solid lines. In order to differentiate these bridge points from border points, the Short-Long criterion temporarily considers exceptionally short edges as inter-cluster links. Later, it performs connected component analysis to recuperate all short edges that are intra-cluster links (typically incident to border points) and to permanently remove bridges.

More formally, for a point p_i, let $Local\_Mean(p_i)$ be the mean length of edges incident to p_i and let $Local\_St\_Dev(p_i)$ be the standard deviation in the length of these incident edges. An edge e_j incident to point p_i could be classified as an inter-cluster edge if

$$|e_j| > Local\_Mean(p_i) + l \times Local\_St\_Dev(p_i). \tag{1}$$

That is, edges are considered long and inconsistent if their lengths are away from the local mean. However, $Local\_Mean(p_i)$ and $Local\_St\_Dev(p_i)$ represent only local properties. We need to incorporate global trend to Equation (1) to correctly identify globally localized spatial clusters. If the factor l in Equation (1) scales the threshold globally, it will apply absolute information to all p_i but not relative information. This is because $Local\_St\_Dev(p_i)$ is expressed in the same units as $Local\_Mean(p_i)$. That is, we would have a uniform (absolute) acceptance interval for intra-cluster edges, for all p_i. However, a smaller value of $Local\_Mean(p_i)$ is indicative of p_i as internal to a cluster, since smaller $Local\_Mean(p_i)$ implies relative closeness around p_i to its neighbors. Thus, we require the value of l to be greater than 1 in order to widen the acceptance interval and thus preserve a certain level of heterogeneity around p_i for smaller $Local\_Mean(p_i)$. Conversely, we would require the value of l to be less than 1 to restrict the acceptance interval for larger values of $Local\_Mean(p_i)$. Thus, more edges incident to border points or noise/outliers are removed. Let $Relative\_St\_Dev(p_i)$ denote the ratio of $Local\_St\_Dev(p_i)$ and $Mean\_St\_Dev(P)$, where $Mean\_St\_Dev(P)$ is the average of the $Local\_St\_Dev(p_i)$. Then, $Relative\_St\_Dev(p_i)$ provides a ratio of local deviation against global deviation. $Relative\_St\_Dev(p_i)$ is less than 1 for those p_i that locally exhibits less spread of length of their incident edges than the generality of points in P. Also, $Relative\_St\_Dev(p_i)$ is greater than 1 for p_i that locally exhibits more variability. The inverse of $Relative\_St\_Dev(p_i)$ fulfills the role required for the factor l.

$$l = Relative\_St\_Dev(p_i)^{-1} = Mean\_St\_Dev(P) \ / \ Local\_St\_Dev(p_i). \tag{2}$$

Finally, the acceptance interval $AI(p_i)$ for each p_i is obtained by replacing Equation (2) in Equation (1).

$$\begin{aligned} &Local\_Mean(p_i) - Mean\_St\_Dev(P) \leq \\ &AI(p_i) \leq Local\_Mean(p_i) + Mean\_St\_Dev(P). \end{aligned} \tag{3}$$

Note that, the acceptance interval $AI(p_i)$ utilizes both local trend information and global trend information, where $Local\_Mean(p_i)$ represents inverse local strength and $Mean\_St\_Dev(P)$ denotes the global degree of variation. Further, $AI(p_i)$ is not static, but rather dynamic over R.

The Short-Long criterion is applied to a proximity graph for the grouping process. Readers may refer to the original paper [6] for details of the process.

3 Proximity Graphs Model Spatial Proximity

We now list 8 common families of proximity graphs. They derive from several modeling considerations, including modeling proximity and topology. For example, one possible way of capturing topological relations such as ADJACENT_TO amongst point data is to perform point-to-area transformations. A widely adopted point-to-area transformation is to assign every location in R to the nearest p_i in P. This creates regions in R. The resulting tessellation is the well-known Voronoi Diagram denoted by $VD(P)$. As a consequence, two points are neighbors if and only if their Voronoi regions share a common Voronoi boundary. The explicit representation of this IS_NEIGHBOR relation is another tessellation (Delaunay Triangulation denoted by $DT(P)$). Many researchers [2,4,8] use $DT(P)$ as a proximity graph for clustering. Recently, Estivill-Castro and Lee [5] used the Delaunay Diagram, denoted by $DD(P)$, since this removes ambiguous diagonals when more than three points are co-circular.

Other proximity graphs are based on similar local closeness criteria. The Gabriel Graph of P, denoted by $GG(P)$, has an edge $e = (p_i, p_j)$ if and only if all other points in P - $\{p_i, p_j\}$ lie outside the circle having e as diameter. The Relative Neighborhood Graph of P, denoted by $RNG(P)$, is based on the notion of "relative close" neighbors. Two points $p_i, p_j \in P$ define an edge if they are relatively close enough as they are to any other point [12]. Matula and Sokal [10] believe that Gabriel Graphs provide sufficient but not excessive interconnections.

The Minimum Spanning Tree of P, denoted by $MST(P)$, has long been used for single-linkage clustering (in fact, $MST(P)$ is not unique, there may be several Minimum Spanning Trees for the same data). Since $MST(P)$ is a sub-graph of $RNG(P)$ which in itself is a sub-graph of $GG(P)$. And $GG(P)$ is a sub-graph of $DT(P)$ [12], each encodes less proximity information along this family. The simplicity of $MST(P)$ makes its local statistical properties vulnerable to small changes. More seriously, $MST(P)$ does not reflect local optimization information like $DT(P)$, $GG(P)$ or $RNG(P)$, but rather global optimization. Thus, it ignores local variations to a larger degree.

The Greedy Triangulation of P, denoted by $GT(P)$, is another type of spatial tessellation implicitly encoding proximity. $GT(P)$ is obtained by repeatedly inserting the shortest edge that does not intersect any previously inserted edge. A *greedy* edge e represents spatial proximity in the sense that e is the strongest interaction that does not interfere previously chosen stronger interactions.

Another popular model for spatial proximity is using a metric and the corresponding distance concept (this is the implicit philosophy within methods like DBSCAN [3]). Namely, points are considered as neighbors if and only if they lie within a certain distance d. The corresponding proximity graph is denoted by d-$DG(P)$. However, this has a number of critical problems, the most crucial is that globally setting the value of the argument d ignores spatial heterogeneity.

An alternative is to assign the same number of neighbors to each point, namely k-nearest neighbors, denoted by k-$NN(P)$. This has been used in spatial data mining in CHAMELEON [9] and could be seen as a variant of d-distance neighboring where is not globally fixed, but varies over R. All points in P are assigned a globally fixed number k of neighbors.

Table 1. Characteristics of proximity graphs.

	DD	GG	RNG	MST	GT	d-DG	k-NN	k-CSG
Symmetry	√	√	√	√	√	√	×	×
Planar tessellation	√	×	×	×	√	×	×	×
Planar graph	√	√	√	√	√	×	×	√
Spatial heterogeneity	√	√	√	√	√	×	×	×
Neighbor visibility	√	√	√	√	√	×	×	√
Scale-independence	√	√	√	√	√	×	√	√
Argument-free	√	√	√	√	√	×	×	×
Variable # of neighbors	√	√	√	√	√	√	×	√
Local optimization	√	√	√	×	×	√	√	√

One variant is the family of k-cone spanner graphs of P, denoted by k-$CSG(P)$. This captures the nearest neighbors in pre-specified k directional cones. Thus, k-nearest neighbors, each the nearest in each of k directions.

The graphs k-$NN(P)$ and k-$CSG(P)$ are argument-dependent (need a value for k) and directed (some edges are not symmetric). Table 1 summarizes the characteristics of proximity graphs. In two dimensions, they all have linear size, however, as we move up to 3D this is not the case. The graphs k-$NN(P)$ and k-$CSG(P)$ are attractive as we move to higher dimensions since they remain linear in size (when k is so small with respect to n that it can be regarded as a constant). Moreover, the field of Computational Geometry has developed $O(n \log n)$ time algorithms to compute these proximity graphs. Thus, the extension of the Short-Long criteria proposed here is scalable to large data sets.

4 Performance Evaluation

When applied to Delaunay Diagrams, the Short-Long criterion robustly obtains clusters of arbitrary shapes, clusters of different densities, clusters of variable sizes, clusters in the presence of noise, sparse clusters adjacent to high-density clusters, closely located high-density clusters, clusters linked by multiple bridges and clusters in the presence of obstacles. However, it would be rather limiting if one is forced to use Delaunay Diagrams. We report on experiments [7] to contrast the clustering results of the combination of the Short-Long criterion discussed in Section 2 and the proximity graphs discussed in Section 3. The cases represent benchmark 2D data where mere-density clustering fails [6].

Close clusters of several levels of density. An initial example is sparse clusters surrounding closely high-density clusters. In this type of dataset, $DD(P)$ produces the best result. Boundaries are well detected. Clusters surrounding higher density cluster are detected and kept connected, but disconnected from the internal high-density clusters. The results of $GT(P)$ approximate those of $DD(P)$, but points on less dense clusters may be left isolated.

In this type of data, using $GG(P)$ successfully identifies high-density clusters, but fails to reliably detect the sparser cluster in that recuperation is incomplete (more points left isolated). Using $RNG(P)$ deteriorates the results more. Now, not only are sparser cluster undetected, but the high-density clusters are divided into meaningless sub-groups.

Not surprisingly, $MST(P)$ reports poor results. The main reason for this is that $MST(P)$ has lost proximity information in its $n-1$ interconnections. Note that $DD(P)$ has a small constant factor more edges ($3n-6$ at most) and encodes far much more complete spatial proximity. The simple proximity information makes $MST(P)$ extremely vulnerable to small variations, and it is not surprising that $MST(P)$ is known to be fast but also known to be a fragile clustering method. Intermediate quality is obtained with $GG(P)$ and $RNG(P)$. These proximity graphs tend to exclude relatively heterogeneous interconnections from $DT(P)$ such as edges between the sparse cluster and the high-density clusters. This reduction eventually decreases *Mean_St_Dev*(P) and thus narrows $AI(p_i)$ for each p_i (refer to Equation (3)). As a consequence, relative heterogeneity around large *Local_Mean*(p_i) (within the sparse cluster) is not preserved, which causes the sparse cluster to be fragmented. In general, the sub-graphs of $DT(P)$ miss some proximity information, and complicate detection of sparser clusters.

The family of graphs k-$NN(P)$ provides good results when the best value for k is used. It successfully detects the global hot spots, but still has difficulty to identify the surrounding sparser concentration. This is to be expected, but still a valuable result in higher dimensions when Delaunay Diagrams are quadratic. This is because the proximity information in k-$NN(P)$ is biased to one direction for points in the border of a cluster. That is, edges connect to the dense side of the border where the cluster is. Non-connectivity between the sparse cluster and the dense clusters lowers *Mean_St_Dev*(P), which eventually narrows $AI(p_i)$ for each p_i. Thus, the Short-Long criterion considers local variations around the sparse cluster as too heterogeneous.

After several argument tuning steps (locating the value k), the best result of k-$CSG(P)$ are good (as k-$NN(P)$). Interestingly, some points belonging to sparser cluster are connected to high-density clusters.

Argument tuning (locating a value for d) is necessary for the best result of d-$DG(P)$. This extracts dense clusters but sparser clusters are not identified, when intra-cluster distance within sparser clusters is similar to inter-cluster distance between high-density clusters and sparser clusters. Thus, if the value for the argument d is large enough, the sparse cluster and the high-density clusters have many intra-connections. This lowers *Mean_St_Dev*(P) significantly. Consequently, d-$DG(P)$ reports all the points as a single cluster. Inversely, if the value of d is small enough just to avoid the total merge, less dense surrounding points are not identified as related into a cluster.

Narrow links of high-density clusters. This situation may be seen as clusters linked by narrow multiple bridges. Unlikely to the first example, proximity graphs are robust to multiple bridges except for $RNG(P)$ and $MST(P)$. This is a great improvement over typical mere-density clustering. The graph k-$CSG(P)$ does not work as well for this case, but still separates some bridges.

Narrow gap of high-density clusters. In many real world settings, clusters of different densities are closely located. For example, two highly populated cities lying on opposite side of a river, or densely located troops around national boundaries. Datasets emulating these scenarios were tested. The two spatial tessellations, $DD(P)$ and $GT(P)$ hold enough interconnections and the clustering results pinpoint closely located dense clusters while still identifying other relatively sparse clusters.

Again, the argument k has to be tuned by the user for k-$NN(P)$ and k-$CSG(P)$. But, the results are very satisfactory and parameter-specific neighboring k-$NN(P)$ and k-$CSG(P)$ find the clusters. These graphs have produced good results, but at the cost of user direction and exploration of the argument values.

For the same reason explained in the first example before, large values of d detect the sparser clusters but merge closely located dense clusters into one. Also, a smaller value of d is ineffective because the result leaves points within sparser cluster isolated. The proper sub-graphs of $DT(P)$ offer less quality results as they are more distant from $DT(P)$. We can see that the proper sub-graphs of $DT(P)$ may work for the resolution of multiple bridges, but fail to detect clusters with different densities due to loss of proximity and density information.

Up to this point it seems that $GT(P)$ is as good as $DD(P)$ for the Short-Long criterion. These two graphs are approximations to the Minimum Weight Triangulation. Thus, they may be similar in some cases, and when that happens, resulting clustering is naturally very similar. However, our experiments showed cases where $GT(P)$ is radically different to $DT(P)$. In such cases, the same problems of single-linkage clustering appear, creating greedily long paths of artificial bridges using short edges. Thus, $GT(P)$ is not better than $DT(P)$.

Experiments conducted in this section indicate that the Short-Long criterion works well with properly modeled proximity graphs holding enough information but not excessive. Our experiments also show that the criterion may work well with the argument-specific graphs k-NN and k-CSG with careful tuning of the argument value. Since typical mere-density clustering fails to detect these examples [6], experiments demonstrate the robustness of the Short-Long criterion and its applicability to other proximity graphs.

Sensitivity to Arguments k and d. We proceeded to analyze sensitivity with respect to the values of argument k and d. A change of these values results in rather different proximity graphs and thus clustering results. In the graph k-NN, for the smaller values of k, high-density clusters are broken up into less meaningful sub-groups. This is due to the fact that smaller values of k only capture a few very close neighbors for each point, thus neighbors for each point are relatively homogeneous in terms of their lengths of edges. This homogeneity lowers the global indicator of heterogeneity and narrows the acceptance interval for each point, which eventually causes homogeneous local variations to be fragmented. In order to avoid this fragmentation, we may tune the argument k to higher values. In this case, too many interconnections merge the dense clusters and the sparse cluster into one large cluster. In addition, the higher values of k connect bridge points together. Exploring different values of k allows users to investigate the cohesion of clusters relative to others as well as the emergence of bridges, geographical features. Bridges are to lines what clusters are to areas.

Similar merge and split happen to k-$CSG(P)$ when we tune the value of k. As we increase the value of k, more points merge. Large values of k detect sparse clusters but merge the closely located high-density clusters. Again, this type of analysis can assess the relative cohesion between clusters relative to other clusters. It can identify where a cluster is weaker, about to split. The analysis can indicate points that are just hanging to a cluster and are likely to have/receive less influence on fellow cluster members.

However, the results with exploring along d with d-$DG(P)$ are not as encouraging. When a value d allows to detect densely populated clusters, it then fails to detect the sparser clusters. If we increase the value of d until it is long enough for points within sparser cluster to reach their neighbors, then all points belong to the same cluster. The increase of d merges not only bridges, but closely located high-density clusters.

The sensitivity analysis illustrates the trade-offs of tuning arguments. Several trial and error steps to find best-fit arguments require exploration time, and user bias may be introduced. Argument-specific proximity graphs demand user exploration of argument values for clustering massive data sets. They offer in exchange information on where the clusters are about to break, information on bridges and other elements connecting or about to merge clusters.

In fact, we have seen that d-$DG(P)$ is one of the poorest performers (also poor are $MST(P)$ and $RNG(P)$). In some cases, not any value of d results to be appropriate nor satisfactory. This is disappointing for algorithms like DBSCAN based on this type of proximity representation. They not only require the use of queries into data structures like R-Trees to obtain the proximity information, but also the use of visualization tools like OPTICS [1] to aid the user in tuning the arguments for d. They may in fact miss many clusters.

5 Final Remarks

Edges in proximity graphs represent interactions between neighboring points, thus they encode spatial proximity. Assigning a weight proportional to the length of the edges provides spatial density. In practice, the more edges a proximity graph has, the more proximity information the graph possesses. However, some of these edges may become irrelevant. Theoretically, the set P (of size n) encodes all the proximity information, while the explicit complete graph, although a proximity graph, it is certainly redundant since it has quadratic size. We have shown that the Short-Long criterion is applicable to a large family of proximity graphs to obtain satisfactory clustering. That is, the Short-long criterion does not require maximal information. It just requires enough interconnections to detect spatially aggregated concentrations. This is very important for Data Mining applications, since the intent is to apply the Short-Long criteria to proximity graphs that take sub-quadratic time to compute and thus use sub-quadratic space. Such is the case of Delaunay Diagrams and its sub-graphs presented here which require $\Theta(n \log n)$ time and space in two dimensions. But, as we progress to three dimensions this is no longer the case, and graphs like k-NN and k-CSG are sub-quadratic in time and space for any dimension. We have shown that the Short-Long criteria will be effective for these proximity graphs, at the expense of more user participation in setting argument values. This exploration can detect cohesion of clusters relatively to other clusters, potential weakness on a cluster or its likely split as well as items about to depart from the cluster.

References

1. M. Ankerst, M. M. Breunig, H. P. Kriegel, and J. Sander. OPTICS: Ordering Points to Identify the Clustering Structure. In *Proc. ACM-SIGMOD-99*, pages 49–60, 1999.
2. C. Eldershaw and M. Hegland. Cluster Analysis using Triangulation. In *CTAC97*, pages 201–208. World Scientific, Singapore, 1997.
3. M. Ester, H. P. Kriegel, J. Sander, and X. Xu. A Density-Based Algorithm for Discovering Clusters in Large Spatial Databases with Noise. In *Proc. 2nd Int. Conf. KDDM*, pages 226–231, 1996.
4. V. Estivill-Castro and M. E. Houle. Robust Clustering of Large Geo-referenced Data Sets. In *Proc. 3rd PAKDD*, pages 327–337, 1999.
5. V. Estivill-Castro and I. Lee. AMOEBA: Hierarchical Clustering Based on Spatial Proximity Using Delaunay Diagram. In *Proc. 9th Int. SDH*, pages 7a.26–7a.41, 2000.
6. V. Estivill-Castro and I. Lee. AUTOCLUST: Automatic Clustering via Boundary Extraction for Mining Massive Point-Data Sets. In *Proceedings of GeoComputation 2000*, 2000.
7. V. Estivill-Castro, I. Lee, and A. T. Murray. Spatial Clustering Analysis with Proximity Graphs Based on Cluster Boundary Characteristics. Technical Report 2000-07, http://www.cs.newcastle.edu.au, Department of CS & SE, University of Newcastle, 2000.
8. I. Kang, T. Kim, and K. Li. A Spatial Data Mining Method by Delaunay Triangulation. In *Proc. 5th Int. ACM-Workshop on Advances in GIS*, pages 35–39, 1997.
9. G. Karypis, E. Han, and V. Kumar. CHAMELEON: A Hierarchical Clustering Algorithm Using Dynamic Modeling. *IEEE Computer*, 32(8):68–75, 1999.
10. D. W. Matula and R. R. Sokal. Properties of Gabriel Graphs Relevant to Geographic Variation Research and the Clustering of Points in the Plane. *Geographical Analysis*, 12:205–222, 1980.
11. R. T. Ng and J. Han. Efficient and Effective Clustering Method for Spatial Data Mining. In *20th VLDB*, pages 144–155, 1994.
12. A. Okabe, B. N. Boots, and K. Sugihara. *Spatial Tessellations: Concepts and Applications of Voronoi Diagrams.* John Wiley & Sons, West Sussex, 1992.
13. S. Openshaw. Geographical data mining: key design issues. In *Proceedings of GeoComputation 99*, 1999.
14. W. Wang, J. Yang, and R. Muntz. STING+: An Approach to Active Spatial Data Mining. In *Proc. ICDE*, pages 116–125, 1999.
15. T. Zhang, R. Ramakrishnan, and M. Livny. BIRCH: An Efficient Data Clustering Method for Very Large Databases. In *Proc. ACM SIGMOD*, pages 103–114, 1996.

Micro Similarity Queries in Time Series Database

Xiao-ming Jin, Yuchang Lu, and Chunyi Shi

The State Key Laboratory of Intelligent Technology and System
Computer Science and Technology Dept., TsingHua University
BeiJing China, 100084
xmjin00@mails.tsinghua.edu.cn
lyc@tsinghua.edu.cn
scy@est4.cs.tsinghua.edu.cn

Abstract. Currently there is no model available that would facilitate the task of finding similar time series based on partial information that interest users. We studied a novel query problem class that we termed micro similarity queries (MSQ) in this paper. We present the formal definition of MSQ. A method is investigated for the purpose of efficient processing of MSQ. We evaluated the behavior of MSQ problem and our query algorithm with both synthetic data and real data. The results show that the knowledge revealed by MSQ corresponds with the subjective feeling of similarity based on singular interest.

Keyword: Micro similarity query, Micro nearest neighbor queries, Micro range query, Time series analysis, Data mining algorithm.[1]

1. Introduction

Time series constitute a large part of data stored in many information systems. Algorithms that solve the problem of finding similar series only based on the partial interesting information are crucial to many data mining applications.

There have been several efforts to develop similarity query model for time series data. [4] uses real-valued functions for representing and querying. In [5], non-overlapping ordered similar subsequence is used. Fourier transform [1] and singular value decomposition [6] are used to provide better performance.

These similarity query models mainly focus on the overall and the most remarkable series behaviors. All information, or partial information that is chosen just by ignoring parts of minor energy, is used during the query process. An example is "find the stocks with similar price movement". Using this kind of query models, the local or minor series behaviors are seen as useless information and sometimes are ignored for the sake of efficiency. However, in many applications, this kind of unconsidered behaviors is not what we can ignore but what interested us. For example, in data mining problems in stock market, we are not only care about the overall price movements, but also local or minor ones or the combination of them. A

[1] The research has been supported in part of Chinese national fund of natural science (no. 79990580), Chinese national key fundamental research program (no, G1998030414) and 985 Program of Tsinghua University.

D. Cheung, G.J. Williams, and Q. Li (Eds.): PAKDD 2001, LNAI 2035, pp. 358-363, 2001.

possible problem is "identify stocks whose price usually fluctuate at same time and with same amplitude".

We believe that the queries based on these series behavior can be widely used in many data mining applications. To the best of our knowledge, this has not been well considered. And the problem has not been formally defined.

Cluster identifiers of sub-series [3] and subsequence matching [2] can also be used in similar queries by local features. However, the interesting behaviors are always influenced, or even flooded by the overall and most remarkable series movement. The patterns these methods reveal may contain useless information. This makes it difficult or even impossible to discover really useful knowledge.

For this reason, we propose a novel query problem class that we termed micro similarity queries (MSQ) and a method for it. We evaluated the behavior of MSQ problem and our query algorithm with both synthetic data and real data, which showed the query results closely correspond with the subjective feeling of similarity between the time series based on singular interest.

This paper is organized as follows: Section 2 formally defines the MSQ problem. Section 3 describes our solution to this problem. Section 4 presents our experimental results. Finally, Section 5 offers some concluding remarks.

2. Formal Definition of the Problem

A time series O(n) is a sequence of real numbers, each number representing a value at a time point. In this paper, the time series to be compared are with the same length.

The formal definition is as follow:

Definition 2.1: Choose a rule F to decompose O(n), $(O(n)) \overset{F}{\longleftrightarrow} (O_k(n))$ and $O(n) = \sum_k O_k(n)$, $(O_k(n))$ is said to be the k-th F-based decomposition of O(n).

Definition 2.2: Let $O_m(n)$ be the m-th sub series of O(n) and $(O_{mk}(n))$ be the F based decomposition of $O_m(n)$. Choose a one-to-one series-value mapping rules T, map each $O_{mk}(n)$ to a value $O_{mk} \overset{T}{\longrightarrow} S_k(m)$. Then all $S_k(m)$ combine a sequence S_k with the original order. The sequence S_k is said to be the k-th micro representing sequence.

Definition 2.3: Let sequence S_k be the k-th micro representing sequence. $K=(K_1,K_2,\ldots,K_W)$, where K_n is the order of n-th interesting micro representing sequence. The degree of user's interest to the k-th micro representing sequence is A_k. The sequence

$$S(m) = A_{K_{MOD(m,W)}} S_{K_{MOD(m,W)}}(K_{\lceil m/W \rceil})$$

is said to be the full micro representing sequence of time series O(n) (FMR(O)). Here MOD(m,K) represents the remainder of m/K. $K=(K_1,\ldots,K_W)$ and $A=(A_1,\ldots,A_r)$ represent user's interest.

Definition 2.4: Let X, Y be two time series. The micro distance MD (X,Y) is

MD(X,Y)=D (FMR(X),FMR(Y)).

Here D(X,Y) is a distance measurement model currently in use. For example, we can use Euclidean distance: $D(X,Y)=L_2(X,Y)$.

Definition 2.5: Given time series set SS and query time series Q, the micro nearest neighbor queries (MNNQ) problem is

MNN(Q)={ X•SS|∀Y•SS: MD(X,Q)≤ MD (Y, Q)}

Definition 2.6: Given a threshold ε, time series set SS and query time series Q, the micro range queries (MRQ) problem is

MR(Q)={X•SS|MD(X,Q)•ε }

3. Method for MSQ Problem

3.1 Decomposition and Representation

We use a discrete transform for time series decomposing and representing and use L_2 distance to measure the distance of FMR sequence. Here we would like a transform (1) that is easy to compute and (2) that can represent interesting pattern easily and efficiently. We have chosen the Discrete Cosine Transform (DCT) in our experiments because it is wildly used in many areas and it does a good job of meet our requirements.

Let O(n) be a time series with length N. We first divide input time series by sliding a window of width r. Then the j-th window of O is a contiguous sub series O_j=(O(jr-r+1),…,O(jr)), q≥j≥1 where q=N/r.

Let H be the transform matrix. Then the r point discrete transform of sub series O_j is defined to be series $T_j(k)$, given by $T_j= H\cdot O_j$

Based on the results of discrete transform, the decomposing version of O(n) in our method is: $O_k(n) = T_{\lceil n/r \rceil}(k) \cdot IB_k(n - \lfloor n/r \rfloor \cdot r)$ (K>k≥0) where IB_m is $[H^{-1}_{0,m} \quad H^{-1}_{1,m} \quad \dots \quad H^{-1}_{r-1,m}]^T$. The k-th micro representing sequence is: $S_k(m) = T_m(k)$ .FMR(O) can be calculated as: $S(m) = A_{K_{MOD(m,W)}} T_{\lceil m/W \rceil}(K_{MOD(m,W)})$

We can calculate the Euclidean distance between FMR (X)=XS and FMR (Y)=YS to determine MD(X,Y). This makes it easy to apply the current index structures and query methods to MSQ problem. Let DT(Xj)=XTj•DT(Yj)=YTj. It's easy to proof:

$$MD(X,Y) = L_2(XS,YS) = \sqrt{\sum_{j=1}^{q}\sum_{n=1}^{W} (XTj(K_n) - YTj(K_n))^2 A_{K_n}^{\ 2}}$$

Our method can be irrelevant with baseline if $XT_j(0)$ and $YT_j(0)$ are not used in the process of calculating MD. The following MD is irrelevant with scaling factors.

$$MD(X,Y) = \sqrt{\sum_{j=1}^{q}\sum_{n=1}^{W} \left(\frac{XT_j(K_n)}{XT_j(0)} - \frac{YT_j(K_n)}{YT_j(0)}\right)^2 A_{K_n}^{\ 2}}$$

3.2 Indexing and Query Processing

Our overall querying strategy is as follows.

Step 1 - According to their applications, users need express their interest by K and A.

Step 2 - feature retrieving: We obtain the transform coefficients needed by applying transform equation on each sub series. Then, we use the selected transform coefficients to construct FMR of each time series.

Step 3 - Index construction: Insert all FMR into a multidimensional index for supporting efficient querying. The MNNQ and MRQ problems can be simplified to nearest neighbor query and range query. Therefore, our approach can be applied with any index structure that was designed for nearest neighbor query or range query such as R*-tree [7] and SS-tree [8].

Step 4 - After the index has been built, we can carry out MRQ or MNNQ. Any algorithms that fit for range query and nearest neighbor query can be used here.

4. Experimental Results

In our experiments, DCT was used, K=(2), A=(1). This means we want to find time series that always show similar up/down behaviors to the query time series in each window.

For experiments with synthetic data, we use a generated random walk time series Q(n) with 180 data points as the query time series. Then we query it within data set consisting 5 time series that were generated using the following function:

$$S_m(n) = Q(n) + A_m \sin(2\pi n / T_m) + \varepsilon$$

Here ε is Gaussian noise. An example data set is depicted in fig. 1, the coefficients used are showed in table 1 together with the distance D and micro distance MD. In this experiment, the length of sub series was set to 8.

From the visual analysis, as the frequency $1/T_m$ increases, the micro distance increases while the global distance decrease, i.e. for the time series in which change of lower frequency sinusoid was applied, MD become smaller while the distance D become larger. Our algorithm presents the same results.

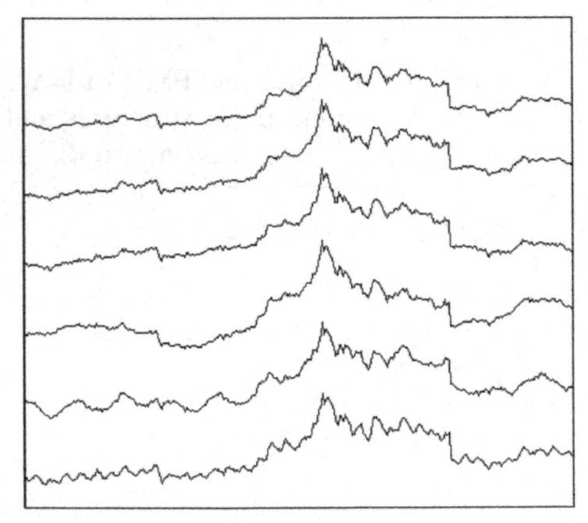

Fig. 1. Synthetic Date Set

Table 1. Experimental Results for Synthetic Data

m	A_m	T_m	MD (Q,S_m) (10^{-5})	D (Q,S_m) (10^{-2})
1	4	160	0.87	20.09
2	4	80	0.96	17.58
3	4	20	1.04	17.4
4	2	4	2.04	8.78
5	1	1.3	4.02	5.26

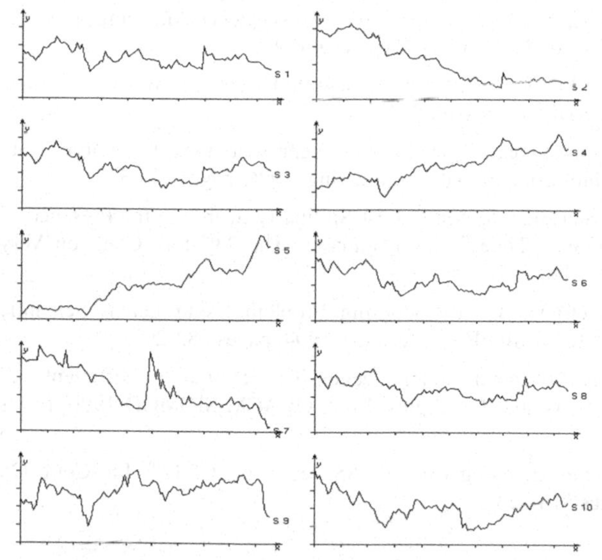

Fig. 2. Query Results for Real Data

Table 2. Experimental Results for Real Data

m	MD (S_1,S_m) (10^{-5})	D (S_1,S_m) (10^{-2})
2	0.19	0.81
3	0.82	0.24
4	1.42	0.73
5	1.95	0.23
6	2.05	0.35
7	2.76	0.27
8	3.39	0.69
9	3.85	0.66
10	5.53	1.47

Real data we used in our experiments was extracted from different equities of Shenzhen stock market from 10/1/1997 to 8/1/2000. They have been collected daily over the time period. Totally 100 time series were used. All FMRs were extracted by a sliding window of size = 5 which accorded to the number of exchanging days in one week and were inserted into an R*-Tree.

One of our results is shown in fig. 2. S1 is the query time series, others is the most nearest 9 series sorted by the distance from S1. The micro distance is listed in table 2. The results reveal a new kind of associations between the companies, which can't be discovered by other query model focus on overall time series behavior.

5. Conclusion

In this paper, we studied a novel query problem to find the association in time series database, which can't be discovered by other query model that focus on overall and the most remarkable time series behavior. We believe this kind of queries can be widely used in many data mining application, and will discover novel and interesting knowledge.

Application knowledge should be used during the procedure of definition of interesting pattern, coefficients choosing, and analysis of the results. Whether the algorithm will find the time series we want is mainly depend on the well involvement of application knowledge.

References

1. Rakesh A, Christos F, Efficient Similarity Search in Sequence Databases, FODO 1993, pages 69-84.

2. C. Faloutsos, M. Ranganathan, Y. Manolopoulos, Fast Subsequence Matching in Time-Series Database, Proc. of ACM SIGMOD 1994, Page 419-429.

3. Gautam D, King-IP L, Heikki M, Gopal R, Padhraic S. Rule Discovery from Time Series. Proc. of the 4th Intl. Conf. on KDD, 1998, pages 16-22.

4. Hagit S, Stanley B. Z. Approximate Queries and Representations for Large Data Sequences. Proc. of the 12th Intl. Conf. on data engineering, 1996, pages 536-545.

5. R. Agrawal, K.I. Lin, H. S. Sawhney, K. Shim, Fast Similarity Search in the Presence of Noise, Scaling and Translation in Time-Series Databases, The 23rd Intl. Conf. on Very Large Data Bases, 1995, pages 490-501.

6. D. Hull. Improving Text Retrieval for the Routing Problem Using Latent Semantic Indexing. In Proc. Of the 17th ACM-SIGIR Conference, 1994, pages 282-291.

7. N. Beckmann, H. Kriegel, R. Schneider, and B.Seeger. The R* tree: An Efficient and Robust Access Method for Points and Rectabgles. Proc. Of ACM SIGMOD 1990, pages 322-331

8. D. White and R. Jain. Similarity Indexing with the SS-tree. Proc. Of 12th Int. Conf. On Data Engineering, 1996, pages 516-523.

Mining Optimal Class Association Rule Set

Jiuyong Li, Hong Shen, and Rodney Topor

School of Computing and Information Technology
Griffith University
Nathan Qld 4111 Australia
{jiuyong,hong,rwt}@cit.gu.edu.au

Abstract. We define an optimal class association rule set to be the minimum rule set with the same prediction power of the complete class association rule set. Using this rule set instead of the complete class association rule set we can avoid redundant computation that would otherwise be required for mining predictive association rules and hence improve the efficiency of the mining process significantly. We present an efficient algorithm for mining the optimal class association rule set using an upward closure property of pruning weak rules before they are actually generated. We have implemented the algorithm and our experimental results show that our algorithm generates the optimal class association rule set, whose size is smaller than $\frac{1}{17}$ of the complete class association rule set on average, in significantly less time than generating the complete class association rule set. Our proposed criterion has been shown very effective for pruning weak rules in dense databases.

1 Introduction

1.1 Mining Predictive Association Rules

The goal of association rule mining is to find all rules satisfying some basic requirement, such as the minimum support and the minimum confidence. It was initially proposed to solve market basket problem in transaction databases, and has then been extended to solve many other problems such as classification problem. A set of association rules for the purpose of classification is called *predictive association rule set*. Usually, predictive association rules are based on attribute value (relational) databases, where the consequences of rules are pre-specified categories. Clearly, an attribute value database can be mapped to a transaction database when an attribute and attribute value pair is considered as an item. After having mapped an attribute value database into a transaction database, a class association rule set is a subset of association rules with the specified targets (classes) as their consequences. Generally, mining predictive association rules undergoes the following two steps.

1. Find all class association rules from a database, and then
2. Prune and organize the found class association rules and return a sequence of predictive association rules.

D. Cheung, G.J. Williams, and Q. Li (Eds.): PAKDD 2001, LNAI 2035, pp. 364–375, 2001.

In this paper, we focus on the first step. There are two problems in finding all class association rules.

- It may be hard to find the all class association rule set in dense databases due to the huge number of class association rules. For example, many databases support more than 80,000 class association rules as in [12].
- Too many class association rules will reduce the overall efficiency of mining predictive association rule set. This is because the set of found class association rules is the input of the second step processing whose efficiency is mainly determined by the number of input rules.

To avoid the above problems, it is therefore necessary to find a small subset but with the same prediction accuracy of the complete class association rule set, so that this subset can replace the complete class association rule set. Our proposed *o*ptimal class association rule set is the smallest subset with the same prediction power, which will be formally defined in Section 2, of the complete class association rule set. We present an efficient algorithm to generate the optimal class association rule set that takes the advantage of an upward closure property to prune those complex rules that have lower accuracy than their simple form rules have before they are actually generated in dense databases. Our algorithm avoids redundant computation of mining the complete class association rule set from dense databases and improves efficiency of the mining process significantly.

1.2 Related Work

Mining association rules [1] is a central task of data mining and has shown applications in various areas [7,3,12]. Currently most algorithms for mining association rules are based on Apriori [2], and used the so-calle "downward closure" property which states that all subsets of a frequent itemset must be frequent. Example of these algorithms can be found in [10,14,17]. A symmetric expression of downward closure property is *upward closure* property — all supersets of an infrequent itemset must be infrequent. We will use this property throughout the paper.

Finding classification rules has been an important research focus in the machine learning community [18,8]. Mining classification rules can be viewed as a special form of mining association rules, since a set of association rules with prespecified targets can be used for classification. Techniques for mining association rules have already been applied to mining classification rules [3,12]. Particularly, results in [12] are very encouraging, since it can build more accurate classifiers than those from C4.5 [18]. However, the algorithm in [12] is not very efficient since it uses Apriori-like algorithm to generate the class association rules, which may be very large when the minimum support is small. In this paper will show that we can use a much smaller class association rule set to replace this set while not losing accuracy (prediction power).

Generally speaking, class association rule set is a type of target-constraint association rules. Constraint rule sets [5] and optimal rule sets [4] belong to this type. Problems with these rule sets are that they either exclude some useful predictive association rules, or contain many redundant rules that are of no use for prediction. Moreover, algorithms for mining these rule sets handle only one target at one time (building one enumeration tree), so they cannot be efficiently used for mining class association rules that are on multiple targets, especially when the number of targets is large. Our optimal class association rule set differs from these rule sets at that it is minimal in size and keeps high prediction accuracy. We propose an algorithm that finds this rule set with respect to all targets at once.

In this paper we only address the first step of mining predictive association rules. Related work on pruning and organizing the found class association rules can be referred to [9,13,16].

1.3 Contributions

Contributions in this paper are the following.

1. We propose the concept of optimal class association rule set for predictive association rule mining. It is the minimum subset of complete class association rule set with the same prediction power as the complete class rule set, and can be used as a substitute of the complete class association rule set.
2. We present an efficient algorithm for mining the optimal class association rule set. This algorithm is different from Apriori at that 1) it uses an additional upward closure property for forward pruning weak rules (pruning before they are generated), and 2) it integrates frequent sets mining and rule finding together.
   Unlike the existing constraint and optimal rule mining algorithms, our algorithm finds strong (optimal) rules with all possible targets at one time.

2 Optimal Class Association Rule Set

Given attribute-value database D with n attribute domains. A record of D is a n-tuple. For the convenience of description, we consider a record as a set of attribute and value pairs, denoted by T. A *pattern* is a subset of a record. We say a pattern is a *k-pattern* if it contains k attribute and value pairs. An *implication* in database D is $A \Rightarrow c$, where A is a pattern, called *antecedent*, and c is an attribute value, named *consequence*. Exactly, the consequence is an attribute and value pair, but in class association rule mining, the target attribute is usually specified, so we can use its value directly without confusing. The *support* of pattern A is defined to be the ratio of the number of records containing A to the number of all records in D, denoted by $sup(A)$. The support of implication $A \Rightarrow c$ is defined to be the ratio of the number of records containing both A and c to the number of all records in D, denoted by $sup(A \Rightarrow c)$. The *confidence* of the implication $A \Rightarrow c$ is defined to be the ratio of $sup(A \Rightarrow c)$ to $sup(A)$, represented by $conf(A \Rightarrow c)$.

A *class association rule* is defined to be an implication with a pre-specified target (a value of target attribute) as its consequence and its support and confidence are above given thresholds from a database respectively. Given a target attribute, minimum support σ and minimum confidence ψ, a *complete class association rule set* is a set of all class association rules, denoted by $R_c(\sigma, \psi)$.

Our goal in this section is to find the minimum subset of the complete class association rule set that has the same prediction power as the complete class association rule set.

To begin with, let us have a look at how a rule makes prediction. Given a rule r, we use $cond(r)$ to represent its antecedent (conditions), and $cons(r)$ to denote its consequence. Given a record T in a database D, we say rule r can make prediction on T if $cond(r) \subset t$, denoted by $r(T) \rightarrow cons(r)$. If $cons(r)$ is the category (target attribute value) of record T, then this is a correct prediction. Otherwise, a wrong prediction.

Then we consider the accuracy of a prediction. We begin by defining the accuracy of a rule. Confidence is not the accuracy of a rule, or more precisely, not the prediction accuracy of a rule, but the sample accuracy, since it is obtained from the sampling (training) data. Suppose that all instances in a database are independent of one another. Statistical theory supports the following assertion [15]: $acc_t(r) = acc_s \pm z_N \sqrt{\frac{acc_s(1-acc_s)}{n}}$, where acc_t is the true (prediction) accuracy, acc_s is the accuracy over sampling data, n is the number of sample data ($n \geq 30$), and z_N is a constant relating to confidence interval. For example, $z_N = 1.96$ if confidence interval is 95%. We use pessimistic estimation as the prediction accuracy of a rule. That is $acc(r) = conf(r) - z_N \sqrt{\frac{conf(r)(1-conf(r))}{|cov(r)|}}$, where $cov(r)$ is the covered set of rule r that is defined in the next section. If $n < 30$, then we use Laplace accuracy instead [8], that is $acc(r) = \frac{sup(r)*|D|+1}{|cov(r)|+p}$, where p is the number of target attribute values (classes).

After we have obtained the prediction accuracy of a rule, we can estimate the accuracy of a prediction as follows: the accuracy of a prediction equals to the prediction accuracy of the rule making such prediction, denoted by $acc(r(T) \rightarrow c)$.

In the following part, we will discuss a prediction made by a rule set, and how to compare the prediction power of two rule sets.

Given a rule set R and an input T, there may be more than one rule in R that can make prediction, such as, $r_1(T) \rightarrow c_1, r_2(T) \rightarrow c_2, \ldots$. We say that the prediction made by R is the same as the prediction made by r if r is the rule with the highest prediction accuracy of all r_i where $cond(r_i) \subset t$. The accuracy of such prediction equals to the accuracy of rule r. In case if there are more than one rule with the same highest prediction accuracy, we choose the one with the highest support among them. When the predicting rules have the same accuracy and support, then we choose the one with the shortest antecedent. If there is no prediction made by R, then we say the rule set gives arbitrary prediction with accuracy zero.

To compare prediction power of two rule sets, we define

Definition 1. *Prediction power*
Given rule sets R_1 and R_2 from database D, we say that R_2 has at least the same power as R_1 iff, for all possible input, both R_1 and R_2 give the same prediction and prediction accuracy of R_2 is at least the same as that of R_1.

It is clear that not all rule sets are comparable in their prediction power. Suppose that rule set R_2 has more power than rule set R_1. Then for all input T, if there is rule $r_1 \in R_1$ giving prediction c with accuracy κ_1, then there must be another rule $r_2 \in R_2$ so that $r_2(T) \to c$ with accuracy $\kappa_2 \geq \kappa_1$.

We represent that rule set R_2 has at least the same power as rule set R_1 by $R_2 \geq R_1$. It is clear that R_2 has the same power as R_1 iff $R_2 \geq R_1$ and $R_1 \geq R_2$.

Now, we can define our optimal class association rule set.

Given two rules r_1 and r_2, we say that r_2 is *stronger* than r_1 iff $r_2 \subset r_1 \wedge acc(r_2) > acc(r_1)$, denoted by $r_2 > r_1$. Specifically, we mean $cond(r_2) \subset cond(r_1)$ and $cons(r_2) = cons(r_1)$ when we say $r_2 \subset r_1$. Given a rule set R, we say a rule in R is *strong* if there is no other rule in R that is stronger than it. Otherwise, the rule is *weak*. Thus, we have the definition for optimal class association rule set.

Definition 2. *Optimal class association rule set*
Rule set R_o is optimal for class association over database D iff (1) $\forall r \in R_o, \nexists r' \in R_o$ such that $r < r'$ and (2) $\forall r' \in R_c - R_o, \exists r \in R_o$ such that $r > r'$.

It is not hard to prove that the optimal class association rule set is unique at given minimum support and minimum confidence from a database. Let $R_o(\sigma, \psi)$ stand for the optimal class association rule set on database D at given minimum support σ and minimum confidence ψ. Then $R_o(\sigma, \psi)$ contains all strong rules from the complete class association rule set $R_c(\sigma, \psi)$.

Finally, we consider the prediction power of the optimal class association rule set we are concerned with.

Theorem 1. *The optimal class association rule set is the minimum subset of rules with the same prediction power as the complete class association rule set.*

Proof. For simplicity, let R_c stand for $R_c(\sigma, \psi)$ and R_o for $R_o(\sigma, \psi)$.

First, from the previous definitions we have that $R_c \geq R_o$ and $R_o \geq R_c$, so the optimal class association rule set has the same prediction power as the complete class association rule set has.

Secondly, we prove the minimum property of optimal class association rule set. Suppose that we leave out rule r from the optimal class association rule set R_o, $R'_o = R_o - r$, and R'_o has the same prediction power as R_c has. From the definition, we know that there is no rule being stronger than rule r, so $R_o \geq R'_o$, but $R'_o \ngeq R_o$. As a result, R'_o cannot be the same prediction power as R_c is, leading to contradiction. Hence, R_o is the minimum rule set with the property of same prediction power as the complete class association rule set has.

The fact that the optimal class association rule set has the same prediction power as the complete class association rule set is because it contains all strong rules. Even though the class association rule set is usually much larger than the optimal class association rule set, it contains many weak rules that cannot provide more prediction power than their strong rules do. In other words, the optimal class association rule set is totally equivalent to the complete class association rule set in terms of prediction power. Thus, it is not necessary to keep a rule set that is larger than the optimal class association rule set, and we can find all predictive association rules from the optimal class association rule set.

In the next section, we will present an efficient algorithm to mine the optimal class association rule set.

3 Mining Algorithm

A straightforward method to obtain the optimal class association rule set R_o is to first generate the complete class association rule set R_c and then prune all weak rules from it. Clearly mining complete class association rule set R_c is very expensive and almost impossible when the minimum support is low. In this section, we present an efficient algorithm that can find the optimal class association rule set directly without generating R_c first.

Most efficient association rule mining algorithms use the upward closure property of infrequency of pattern: if a pattern is infrequent, so are all its super patterns. If we can find a similar property for weak rules, then we can avoid generating many weak rules, hence making the algorithm more efficient. In the following we will discuss an upward closure property for pruning weak rules

.

Let us begin with some definitions. We say that r_1 is a *general rule* of r_2 or r_2 is a *specific rule* of r_1 if $cond(r_1) \subset cond(r_2) \wedge cons(r_1) = cons(r_2)$. We define the *covered set* of rule r to be the set of records containing antecedent of the rule, denoted by $cov(r)$. Similarly, covered set of a pattern A is defined to be the set of records containing the pattern, denoted by $cov(A)$. It is clear that the covered set of a specific rule is a subset of the covered set of its general rule.

Suppose that X and Y are two patterns in database D, and XY is the abbreviation of $X \cup Y$. We have the following two properties of covered set.

Property 1. $cov(X) \subseteq cov(Y)$ iff $sup(X) = sup(XY)$.

Property 2. $cov(X) \subseteq cov(Y)$ iff $Y \subset X$.

Now we discuss an upward closure property for pruning weak rules. Given database D and a target value c in target attribute C, we have

Lemma 1. *If $cov(X\neg c) \subseteq cov(Y\neg c)$, then $XY \Rightarrow c$ and all its specific rules must be weak.*

Proof. We rewrite the confidence of rule $A \Rightarrow c$ as $\frac{sup(Ac)}{sup(Ac)+sup(A\neg c)}$. We know that function $f(u) = \frac{u}{u+v}$ is monotonically increasing with u when v is a constant. Noticing $sup(Xc) \geq sup(XYc)$ and $sup(X\neg c) = sup(XY\neg c)$, we have $conf(X \Rightarrow c) \geq conf(XY \Rightarrow c)$. Using relation $|cov(X \Rightarrow c)| \geq |cov(XY \Rightarrow c)|$, we have $acc(X \Rightarrow c) \geq acc(XY \Rightarrow c)$. As a result, $X \Rightarrow c \geq XY \Rightarrow c$

Since $cov(XZ\neg c) \subseteq cov(YZ\neg c)$ for all Z, we have $XZ \Rightarrow c > XYZ \Rightarrow c$ for all Z.

Consequently, $XY \Rightarrow c$ and all its specific rules are weak.

We can perceive the lemma as follows: adding a pattern to the conditions of a rule is to make the rule more precise (with less negative examples), and we shall omit the pattern that fails to do so.

Corollary 1. *If $cov(X) \subseteq cov(Y)$, then $XY \Rightarrow c$ and all its specific rule must be weak for all $c \in C$.*

We can understand the corollary in the following way: we cannot combine a super concept with a sub concept as the antecedent of a rule to make the rule more precise.

Lemma 1 and Corollary 1 are very helpful for searching strong rules, since we can remove a set of weak rules as soon as we find that one satisfies the above Lemma and Corollary. Hence, the searching space for strong rules is reduced.

To find those patterns satisfying Lemma 1 and Corollary 1 efficiently, we need to use properties 1 and 2. Property 1 enables us to find subset relation by comparing supports of two patterns. This is very convenient and easy to implement since we always have support information. By Property 2, we can always find that the covered set of a pattern (e.g. X) is a subset covered set of its $|X| - 1$ cardinality subpattern. So, we only need to compare the support of a k-pattern with that of its $(k-1)$-subpatterns in order to decide whether the k-pattern should be removed.

Since both Lemma and Corollary state upward closure property of weak rules, we can have an efficient algorithm to prune them.

Basic Idea of the Proposed Algorithm

We use a level-wise algorithm to mine the optimal class association rule set. We search strong rules from antecedent of 1-pattern to antecedent of k-pattern level by level. In each level, we select strong rules and prune weak rules. The efficiency of the proposed algorithm is based on fact that a number of weak rules are removed once satisfaction of the Lemma or the Corollary is found. Hence, searching space is reduced after each level's pruning. The number of phases of reading a database is bounded by the length of the longest rule in the optimal class association rule set.

Storage Structure

A prefix tree, or enumerate tree [5] is used as the storage structure. A prefix tree is an ordered and unbalanced tree, where each node is labeled by an element in a sorted base set, B, representing a set $S \subset B$ containing all labels from the root to the node. Since set S is unique in a prefix tree, we can use it as the identity of a node.

We use an extended prefix tree, named *candidate tree* in our algorithm. The base set here contains all attribute and value pairs and they are sorted in the order of their first references. A node in a candidate tree store a pattern A that is the identity of the node, a potential target set Z, and a supset of possible attribute and value pair sets $\mathcal{Q}$. Pattern A is the antecedents of a possible rule. The potential target set Z is a set of values of target attribute that may be consequences of A. For each target (e.g. z_j) in Z, there is a set of possible attribute and value pairs which may be conjunct with A to form more accurate rules, $Q_j \in \mathcal{Q}$.

Our algorithm is given as follows. One distinction between this algorithm and other prefix tree based algorithms is that our algorithm finds all class association rules with respect to all consequences from one candidate tree rather than many candidate trees.

Algorithm: Optimal Class Association Rule Set Miner

Input: Database D with specified target attribute C, minimum support σ and minimum confidence ψ.

Output: Optimal class association rule set R.

Set optimal class association rule set $R = \emptyset$
Count support of 1-patterns
Initiate candidate tree T
Select strong rules from T and include them in R
Generate new candidates as leaves of T
While (new candidate set is non-empty)
  Count support of the new candidates
  Prune the new candidate set
  Select strong rules from T and include them in R
  Generate new candidates as leaves of T
Return rule set R

In the following, we present and explain two unique functions in the proposed algorithm.

Function: Candidate Generating

This function generates candidates for strong rules. Let n_i denote a node of the candidate tree, A_i be the pattern of node n_i, $Z(A_i)$ be the potential target set of A_i, and $Q_q(A_i)$ be a set of potential attribute value pairs of A_i with respect to target z_q. We use $\mathcal{P}^p(A_k)$ to denote the set of all p-subsets of A_k.

for each node n_i at the p-th layer
  for each sibling node n_i and n_j (n_j is after n_i)
    generate a new candidate n_k as a son of n_i such that // combining
      $A_k = A_i \cup A_j$
      $Z(A_k) = Z(A_i) \cap Z(A_j)$
      $Q_q(A_k) = Q_q(A_i) \cap Q_q(A_j)$ for all $z_q \in Z(A_k)$
    for each $z \in Z(A_k)$ // testing
      if $\exists A \in \mathcal{P}^p(A_k)$ such that $sup(A \cup z) \leq \sigma$
        then $Z(A_k) = Z(A_k) - z$
  if $Z_k = \emptyset$ then remove node n_k

We generate the $(p+1)$-layer candidates from the p layer in the candidate tree. First, we combine a pair of sibling nodes and insert their combination as a new node in the next layer. We initiate the new node with the union of the two nodes. Next, if any of its p-subpatterns cannot get enough support with any of the possible targets (consequences), then we remove the target from the target set. When there is no possible target left, remove the new candidate.

Function: Pruning
This function prunes weak rules and infrequent candidates in the $(p+1)$-th layer of candidate tree. Let T_{p+1} be the $(p+1)$-layer of the candidate tree.

for each $n_i \in T_{p+1}$
  for each $A \in \mathcal{P}^p(A_i)$ //A is a p-subpattern of A_i
  if $sup(A) = sup(A_i)$ then remove node n_i //Corollary 1
    else for each $z_j \in Z(A_i)$
      if $sup(A_i \cup z_j) < \sigma$ then $Z(A_i) = Z(A_i) - z_j$
      // minimum support requirement
        else if $sup(A \cup \neg z_j) = sup(A_i \cup \neg z_j)$ then $Z(A_i) = Z(A_i) - z_j$
        // Lemma 1
    if $Z(A) = \emptyset$ then remove node n_i

This is the most important part of the algorithm, as it dominates the efficiency of the algorithm. We prune a leaf from two aspects, frequent rule requirement and strong rule requirement. Let us consider a candidate n_i in the $(p+1)$-th layer of tree. To examine satisfaction of Corollary 1, we test support of pattern A_i stored in the leaf with the support of its subpatterns by Property 1. There may be many such subpatterns when size of A_i is large. However, we only need to compare its p-subpatterns since upward closure property. Hence, the number of such comparisons is bounded by $p+1$. Once we find that the support of A_i equals to the support of any of its p subpattern A, we remove the leaf from the candidate tree. So all its super patterns will not be generated in all deeper layers. In this way, the number of removed weak rules may increase at an exponential rate. Examination of satisfaction of Lemma 1 is in the similar way, but it is with respect to a particular target. That is, we only remove a target from the potential target set in the leaf. Pruning those infrequent patterns is the same as that in other association rule mining algorithms. In our experiments, we will show the efficiency of weak rule pruning in dense databases.

4 Experiment

We have implemented the proposed algorithms and evaluated them on 6 real world databases from UCL ML Repository [6]. For those databases having continuous attributes, we use Discretizer in [11] to discretize them.

We have mined the complete class association rule sets and the optimal class association rule set of all testing databases with the minimum confidence of 0.5 and the minimum support of 0.1. Here support is specified as *local support* that is defined to be the ratio of the support of a rule to the support of the rule's consequence, since significance of a rule depends much on how much proportion of occurrences of its consequence it accounts for. We generate the complete class association rule set by the same algorithm without weak rule pruning and strong rule selecting. We restrict the maximum layer of candidate trees to 4 because of the observation that too specific rules (with many conditions) usually have very limited prediction power in practice. In fact, the proposed algorithm performs more efficiently when there is no such restriction, and this is clear from the second part of our experiment. We do so in order to present competitive results, since rule length constraint is an effective way to avoid combinatorial explosion. Similar constraints have been used in practice, for example, [12] restricts the maximum size of the found rule set.

The comparisons of rule set size and time to generate between the complete class association rule set and optimal class association rule set are listed in Figure 1. It is easy to see that the size of a optimal class association rule set is much smaller than that of the corresponding complete rule set, on the average less than $\frac{1}{17}$ of that. Because the optimal class association rule set has the same prediction power as the complete class association rule set has, so this rule set size reduction is very impressive. Similarly, the time for generating rules is much shorter as well. We have obtained more than $\frac{3}{4}$ reduction of mining time on average. Moreover, using a smaller optimal class association rule set instead of a lager complete class association rule set as the input for finding predictive association rules, we will have more efficiency improvement for other data mining tasks too.

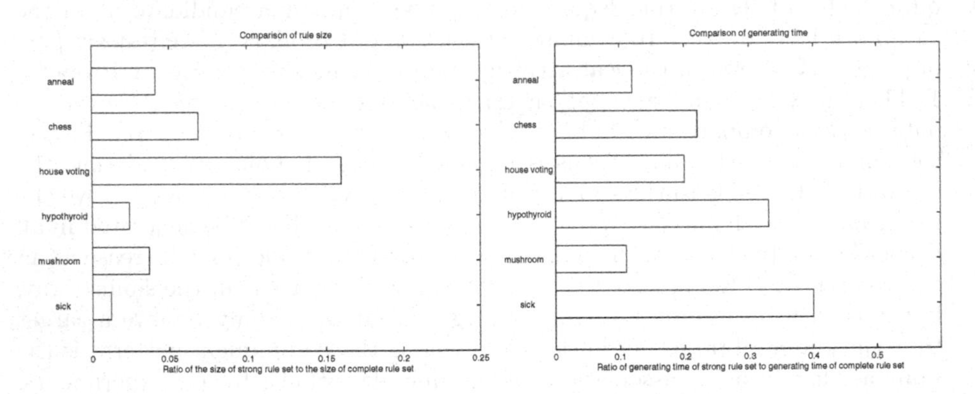

Fig. 1. Overall comparisons of rule size and generating time between R_o and R_c (in the ratio of R_o to R_c)

The core of our proposed algorithm is to prune weak rules. To demonstrate the efficiency of pruning stated in Lemma 1 and Corollary 1 on dense databases, we have illustrated the number of nodes in each layer of the candidate trees of two databases in Figure 2. In this experiment, we lift the restriction of maximum number of layers. We can see that the tree nodes explode at a sharp exponential rate without weak rule pruning. In contrast, tree nodes increase slowly with weak rule pruning, reach a low maximum quickly, and then decrease gradually.

When a pruning tree (weak rule pruning) stops growing, its corresponding unpruned tree just passes its maximum. In the deep tree level, after 4 in our case, the nodes being pruned are more than 99%. This shows how much redundancy we have eliminated. In our experiment, more than 95% time is used for such redundant computing when there is no maximum layer restriction. Considering that how much time it will take if we compute strong rules after obtaining all class association rules, we can see how effective our proposed weak rule pruning criterion is. Besides, from this detailed illustration of candidate tree growing without length restriction, we can understand that the proposed algorithm will perform more efficiently when there is no maximum layer number restriction in comparison with mining the complete class association sets.

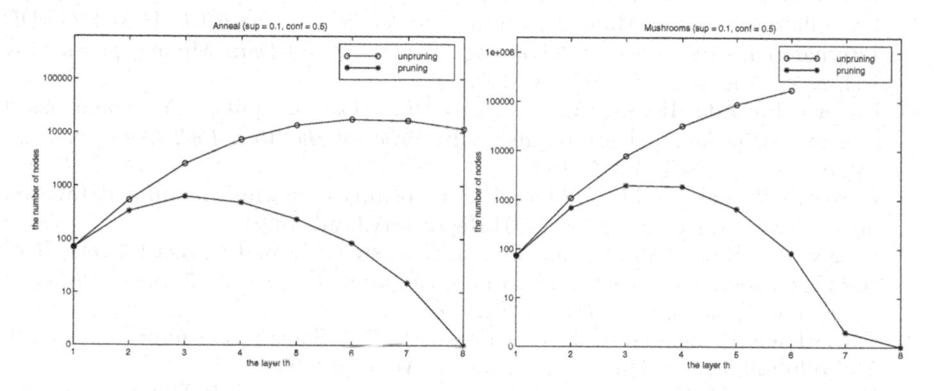

Fig. 2. Comparison of the number of candidates before and after weak rule pruning

5 Conclusion

In the paper, we studied an important problem of efficiently mining predictive association rules. We defined the optimal class association rule set, which preserves all prediction power of the complete class association rule set and hence can be used as a replacement of the complete class association rule set for finding predictive association rules. We developed a criterion to prune weak rules before they are actually generated, and presented an efficient algorithm to mine the optimal class association rule set. Our algorithm avoids redundant computation required in mining the complete class association rule set, and hence improves efficiency of the mining process significantly. We implemented the proposed algorithm and evaluated it on some real world databases. Our experimental results show that the optimal class association rule set has a much smaller size and requires much less time to generate than the complete class association rule set. It was also shown that the proposed criterion is very effective for pruning weak rules in dense databases.

References

1. R. Agrawal, T. Imielinski, and A. Swami. Mining associations between sets of items in massive database s. In *Proc. of the ACM SIGMOD Int'l Conference on Management of Data*, 1993.
2. R. Agrawal, H. Mannila, R. Srikant, H. Toivonen, and A. I. Verkamo. Fast discovery of association rules. In *Fayyad U. and et al, editors, Advances in Knowledge Discovery and Data Mining*. MIT Press, 1996.
3. Kamal Ali, Stefanos Manganaris, and Ramakrishnan Srikant. Partial classification using association rules. In David Heckerman, Heikki Mannila, Daryl Pregibon, and Ramasamy Uthurusamy, editors, *Proceedings of the Third International Conference on Knowledge Discovery and Data Mining (KDD-97)*, page 115. AAAI Press, 1997.
4. Roberto Bayardo and Rakesh Agrawal. Mining the most interesting rules. In Surajit Chaudhuri and David Madigan, editors, *Proceedings of the Fifth ACM SIGKDD International Conference on Knowledge Discovery and Data Mining*, pages 145–154, N.Y., August 1518 1999. ACM Press.
5. Roberto Bayardo, Rakesh Agrawal, and Dimitrios Gunopulos. Constraint-based rule mining in large, dense database. In *Proc. of the 15th Int'l Conf. on Data Engineering*, pages 188–197, 1999.
6. E. Keogh C. Blake and C. J. Merz. UCI repository of machine learning databases, http://www.ics.uci.edu/~mlearn/MLRepository.html, 1998.
7. Sergey Brin, Rajeev Motwani, and Craig Silverstein. Beyond market baskets: Generalizing association rules to correlations. *SIGMOD Record (ACM Special Interest Group on Management of Data)*, 26(2):265.
8. P. Clark and R. Boswell. Rule induction with CN2: Some recent improvements. In *Y. Kodratoff, editor, Machine Learning - EWSL-91*, 1991.
9. H. Toivonene M. Klemettinen P RonKainen K Hatonen and H Mannila. Pruning and grouping discovered association rules. Technical report, Department of Computer Science, University of Helsinki, Finland (`ftp.cs.helsinki.fi/pub/Reports/by_Project/PMDM/`), 1998.
10. M. Houtsma and A. Swami. Set-oriented mining of association rules in relational databases. In *11th Intl. Conf. data Engineering*, 1995.
11. Ron Kohavi, Dan Sommerfield, and James Dougherty. Data mining using MLC++: A machine learning library in C++. In *Tools with Artificial Intelligence*, pages 234–245. IEEE Computer Society Press, 1996. Received the best paper award.
12. Bing Liu, Wynne Hsu, and Yiming Ma. Integrating classification and association rule mining. In *SIGKDD 98*, pages 80–86, 1998.
13. Bing Liu, Wynne Hsu, and Yiming Ma. Pruning and summarizing the discovered associations. In *SIGKDD 99*, 1999.
14. H. Mannila, H. Toivonen, and I. Verkamo. EÆcient algorithms for discovering association rules. In *AAAI Wkshp. Knowledge Discovery in Databases*, July 1994.
15. Tom M. Mitchell. *Machine Learning*. McGraw-Hill, 1997.
16. Raymond T. Ng, Laks V. S. Lakshmanan, Jiawei Han, and Alex Pang. Exploratory mining and pruning optimizations of constrained associations rules. In *Proceedings of the ACM SIGMOD International Conference on Management of Data (SIGMOD-98)*, volume 27,2 of *ACM SIGMOD Record*, pages 13–24, New York, June 1–4 1998. ACM Press.
17. J. S. Park, M. Chen, and P. S. Yu. An effective hash based algorithm for mining association rules. In *ACM SIGMOD Intl. Conf. Management of Data*, May 1995.
18. J. R. Quinlan. *C4. 5: Programs for Machine Learning*. MK, San Mateo, CA, 1993.

Generating Frequent Patterns with the Frequent Pattern List

Fan-Chen Tseng and Ching-Chi Hsu

Department of Computer Science and Information Engineering
National Taiwan University, Taipei, Taiwan, 106
cchsu@csie.ntu.edu.tw

Abstract. The generation of frequent patterns (or frequent itemsets) has been studied in various areas of data mining. Most of the studies take the Apriori-based generation-and-test approach, which is computationally costly in the generation of candidate frequent patterns. Methods like frequent pattern trees has been utilized to avoid candidate set generation, but they work with more complicated data structures. In this paper, we propose another approach to mining frequent patterns without candidate generation. Our approach uses a simple linear list called ***Frequent Pattern List (FPL)***. By performing simple operations on FPLs, we can discover frequent patterns easily. Two algorithms, ***FPL-Construction*** and ***FPL-Mining***, are proposed to construct the FPL and generate frequent patterns from the FPL, respectively.

1 Introduction

Mining frequent patterns finds a wide range of applications in data mining. Examples include mining association rules [1], correlations [4], sequential patterns [2], and so on. Agrawal and Srikant [1] pioneered the research by proposing the Apriori-algorithm, and there were many improvements on their original works [3, 5, 8, 9]. These methods adopted the generation-and-test approach. That is, they iteratively generate the set of candidate frequent patterns of length (k+1) from the set of frequent patterns of length k, and then check their support counts in the database. However, there are two fundamental drawbacks [6, 7] with the Apriori-like generation-and-test approach. First, the generation of a huge number of candidate sets is costly. Second, the repeated scanning of the database and the testing of candidates by pattern matching is time consuming.

Han, Pei, and Yin invent a novel data structure to mine frequent patterns without candidate generation: the frequent pattern tree (FP-tree) [6], which is an extension of prefix-tree structure. The transactions in the database are encoded into FP-tree in such a way that each transaction corresponds to a path from the root to a transaction node in the tree. Based on the FP-tree, along with the associated header table, they develop a pattern fragment growth mining method to perform mining tasks recursively with such a tree.

Although novel and efficient when compared with the Apriori-like methods, the FP-tree approach still has something to be improved. First, for the same frequent item, there are duplicated tree nodes on different branches of the tree. Second, the FP-tree

D. Cheung, G.J. Williams, and Q. Li (Eds.): PAKDD 2001, LNAI 2035, pp. 376-386, 2001.

structure is rather complicated, and the recursive construction of conditional FP-trees is a nontrivial task.

In this paper, we present another approach to mining frequent patterns without candidate generation: using a simpler and more straightforward structure called the **frequent pattern list** (**FPL**). The **FPL** has the following features: (1) No duplication for the same frequent item. (2) The data structure is simple: linear lists, which can be implemented by dynamic arrays. (3) The transaction database can be partitioned neatly when mining frequent patterns, resulting in easier management of main memory.

The remaining of the paper is organized as follows. Section 2 describes the construction of the FPL. Section 3 details the algorithm for mining frequent patterns based on the FPL. Section 4 discusses our approach with other related works. Section 5 gives the conclusion.

2 Frequent Pattern List Construction (FPL-Construction)

The construction of the FPL is achieved by algorithm **FPL-Construction**. Before going into the details, let's describe the problem as follows.

2.1 Problem Description [6, 7]

Let I = {a1, a2, …, am} be **a set of items**, and a **transaction database** DB={T1, T2, …, Tn}, where Ti (i = 1, 2, ..., n) is a transaction which contains a set of items in I. The **support** (or **frequency**) of a **pattern** P, which is a set of items in I, is the number of transactions containing P in DB. A pattern P, is a **frequent pattern** if P's support is larger than or equal to a predefined *minimum support threshold t*.

Given a transaction database DB and a minimum support threshold t, the problem of finding the complete set of frequent patterns is called the **frequent pattern-mining problem**.

Table 1. The example transaction database, DB.

Transaction ID	Frequent Items
T1	f, c, a, m, p
T2	f, c, a, b, m
T3	f, b
T4	c, b, p
T5	f, c, a, m, p

2.2 Algorithm for Frequent Pattern List Construction

In this section, we describe the procedures to construct the frequent pattern list. The example transaction database (borrowed from reference [6]) shown in table 1 is used for illustration. For convenience, only frequent items are shown, and the frequent items in each transaction are listed in the order of descending frequency.

Algorithm 1 (FPL Construction: Frequent Pattern List construction)
Input: a transaction database DB and a minimum support threshold t.
Output: its frequent pattern list, FPL.
Steps:
{

1. Scan the database DB. Find the frequent items and their corresponding frequencies. Create a linear list of *item nodes* of frequent items in order of descending frequency, with the item labels and their frequencies (counts) stored in the item node. The result of step 1 is shown in Figure 1 for our example.

Item node 1	Item node 2	Item node 3	Item node 4	Item node 5	Item node 6
f: 4	c: 4	a: 3	b: 3	m : 3	p: 3

Fig. 1. The frequent pattern list (FPL) after step 1.

2. For each transaction Tx in DB, do the following:
 1) Select and sort the frequent items according to the order in the FPL.
 2) Starting from the root, traverse the FPL and compare the items in the FPL item nodes with the items in Tx. A bit string, called *transaction signature*, is formed from left to right to indicate the existence and absence of frequent items in Tx as follows: At an item node, If there is a corresponding item in Tx, set the bit to 1; otherwise, set it to 0. When all the frequent items in Tx are examined, a *transaction node (T-node)* containing the transaction signature is attached to the item node that corresponds to the rightmost item in Tx (i.e., the item with least frequency in Tx). This resultant list, shown in Figure 2, is called *frequent pattern list (FPL)*.

}

Item node 1	Item node 2	Item node 3	Item node 4	Item node 5	Item node 6
f: 4	c: 4	a: 3	b: 3	m : 3	p: 3
Empty transaction set	Empty transaction set	Empty transaction set	T3 1001	T2 11111	T1 111011
					T4 010101
					T5 111011

Fig. 2. The complete (global) FPL constructed from DB of table 1.

2.3 Properties of FPL

From the structure of FPL, we observe the following properties:

1. The transactions under item node k must contain item k, may contain items to the left of item k (items with frequencies no less than that of item k), and will not contain items to the right of item k (items with frequencies no more than that of item k). Since item k is the rightmost item contained in the transactions under node k, the LSBs (least significant bits) of their signatures must be 1.
2. The transaction database can be partitioned by the FPL according to the above criteria. All the transactions under the same item node belong to the same partition.
3. All the bit strings of the transactions under item node k have the same length k, which is also the *path length* from the root to item node k.
4. Since FPL is built in descending order of item frequencies, transactions containing less frequent items will have longer bit strings. That is, these transactions should be attached to the item nodes farther away from the root. But the number of these transactions must be small, since the rightmost items contained in these transactions are less frequent items with smaller support count.
5. Likewise, transactions containing only more frequent items will have shorter bit strings. That is, these transactions should be attached to the item nodes closer to the root. Although the number of these transactions must be large, they will not consume much memory because of the short lengths of their bit strings.

3 Mining Frequent Patterns with the FPL

After constructing the FPL, we are able to discover frequent patterns by performing simple operations on it. For this purpose, we devise algorithm **FPL Mining**, which can generate frequent patterns in a very straightforward way. An example is also given to illustrate the complete mining process.

3.1 Basic Operations

Frequent patterns can be discovered from the FPL by performing simple operations on the transaction signatures associated with its rightmost item node, which corresponds to the item with smallest frequency. These operations are described as follows:

Bit counting: for each bit position, count the number of 1-bits.

Signature trimming: since the last bit (LSB) of each signature must be 1 (refer to section 2.3, property 1), it can be removed without losing information. After this, the trailing 0-bits of the signature are also removed.

Signature migration: from the least significant 1-bit of the trimmed signature, find the corresponding item node, and migrate this trimmed signature to that item node in the FPL.

3.2 Algorithm for Mining Frequent Patterns with FPL

In this section, we give the algorithm for mining frequent patterns on the FPL. The example in section 2 is used here for explanation.

Algorithm 2 (FPL mining: **mining frequent patterns with FPL**)

Input: FPL constructed based on Algorithm 1, using DB and a minimum support threshold t.

Output: the complete set of frequent patterns.

Steps: Procedure **FP-mining** (*FPL,* t)

{

1). **Bit counting**

Visit the item node at the end of the FPL. Generate a pattern whose label and count are the same as those stored in this item node. For all the transaction signatures under this item node, conduct **bit counting** on all bit positions other than the least significant bit (LSB) position. Ignore the bit positions whose bit counts are below the minimum support threshold. The LSB position, which corresponds to the last item node, is also ignored. Figure 3-1 shows the details for the example, assuming the minimum support threshold is 3.

Item / Transaction	Bit 5 f	Bit 4 c	Bit 3 a	Bit 2 b	Bit 1 m	Bit 0 p
T1	1	1	1	0	1	1
T4	0	1	0	1	0	1
T5	1	1	1	0	1	1
Bit Count	2	3	2	1	2	3

Fig. 3-1. The result of bit counting on the last partition (item node p) for figure 2. Pattern generated: (p: 3). Remaining bit: bit 4 (item c).

2). **Output more patterns directly or by recursive calls**

For the remaining bits:

If all their bit counts are equal to the count in the end item node

Then Produce all combinations of the items corresponding to these remaining bits. **Generate patterns** by **concatenating** each of the combinations with the item of the last item node, with pattern counts being equal to the count in the end item node;

Else

(i) For each transaction signature associated with the end item node:

The least significant bit (which is a 1-bit) is removed. The signature can be discarded if it contains no more 1-bits.

(ii) All the remaining signatures, with their LSBs removed, are then used as input to **Algorithm 1** to construct a sub FPL FPL_{sub} (a FPL one order lower than its parent). A **recursive call** to **FP-mining** (FPL_{sub}, ζ) is then made to generate frequent patterns. The patterns generated from this recursive call must be **concatenated** with the last item of the parent FPL to form the final frequent patterns. The count of the final frequent pattern is the same as the count of the pattern produced by the recursive call.

For our example, there is only one remaining bit, and we get **pattern (cp: 3).**

No recursive call is made in this case.

3). **Signature trimming and migration**

For all the transaction signatures associated with the end item node:

Bit 0 (the LSB corresponding to the last item in FPL) is trimmed.

Find the next nonzero least significant bit, and migrate to the item node corresponding to this bit. All the trailing zero bits can be trimmed. The last item node of the FPL is removed. For our example, the resulting FPL, $FPL_{trimmed}$, is shown in Figure 3-2.

Item node 1	Item node 2	Item node 3	Item node 4	Item node 5
f: 4	c: 4	a: 3	b: 3	m : 3
Empty transaction set	Empty transaction set	Empty transaction set	T3 1001	T2 11111
			T4 0101	T1 11101
				T5 11101

Fig. 3-2. The resulting FPL after signature trimming and migration from the FPL in Figure 1. Item node p (item node 6) is removed.

4). **If** there is only node in the trimmed FPL (that is, the starting node becomes the final remaining node in the FPL)

Then the mining process stops by generating the final **pattern** whose label and count are the same as those stored in this item node;

Else go to step 1 with the trimmed FPL as input.

}

3.3. Details of the Mining Process

Remaining details include figure 3-3 to figure 3-11. The mining process repeats the cycle of bit counting, signature trimming and migration. Recursive calls are made, if necessary.

Item / Transaction	f	c	a	b	m
T2	1	1	1	1	1
T1	1	1	1	0	1
T5	1	1	1	0	1
Bit Count	3	3	3	1	3

Fig. 3-3. The result of bit counting for the FPL in Figure 3-2. All the surviving bits (for items f, c, a) have the same count: 3. Patterns generated**:** (m: 3), (am: 3), (fm: 3), (cm: 3), (cam: 3), (fam: 3), (fcm: 3), (fcam: 3).

Item node 1	Item node 2	Item node 3	Item node 4
f: 4	c: 4	a: 3	b: 3
Empty transaction set	Empty transaction set	T1 111	T3 1001
		T5 111	T4 0101
			T2 1111

Fig. 3-4. The result of signature trimming and migration for figure 3-2. Item node m is removed.

Item / Transaction	f	c	a	b
T3	1	0	0	1
T4	0	1	0	1
T2	1	1	1	1
Bit Count	2	2	2	3

Fig. 3-5. The result of bit counting for figure 3-4. All the remaining bits (for items f, c, a) have count less than 3. Patterns generated: only (b: 3)..

Item node 1	Item node 2	Item node 3
f: 4	c: 4	a: 3
T3 1	T4 01	T1 111
		T5 111
		T2 111

Fig. 3-6. The result of signature trimming and migration for figure 3-4. Item node b is removed.

Item / Transaction	f	c	a
T1	1	1	1
T5	1	1	1
T2	1	1	1
Bit Count	3	3	3

Fig. 3-7. The result of bit counting for figure 3-6. All the remaining bits (for items f, c) have the same count: 3. Patterns generated: (a: 3), (ca: 3), (fa: 3), (fca: 3).

Item node 1	Item node 2
f: 4	c: 4
T3 1	T4 01
	T1 11
	T5 11
	T2 11

Fig. 3-8. The result of signature trimming and migration for the FPL in Figure 3-6. Item node a is removed.

	f	c
T4	0	1
T1	1	1
T5	1	1
T2	1	1
Bit Count	3	4

Fig. 3-9. The result of bit counting for figure 3-8. The remaining bit (item f) has the count 3, but the count of the last item node (item c) has a count of 4. Pattern generated: (c: 4). T4 is discarded since it ends at item c. Signatures T1, T5, and T2, with their LSBs removed, are used to construct a sub FPL. A recursive call is made using this sub FPL as input to find frequent patterns.

Item node 1
f: 3
T1 1
T5 1
T2 1

Fig. 3-10. The resulting sub-FPL from the recursive call of figure 3-9. Pattern generated: (f: 3), which must be concatenated with item c, the last item of the parent FPL. The final pattern generated: (fc: 3).

Item node 1
f: 4
T3 1
T1 1
T5 1
T2 1

Fig. 3-11. The result of signature trimming and migration for Figure 3-8. Only one pattern is generated: (f: 4). The mining process stops.

4 Discussions

In this section we compare our **FPL** approach with other previous works: the **Apriori-like** algorithms and the **FP-tree** algorithms.

As for the issue of recursive structure of the algorithms, both FP-tree and FPL are recursive by nature. The Apriori-like algorithms, although iterative in appearance, still rely on recursive structures when checking the candidate frequent patterns.

About the order of generating frequent patterns, the Apriori-like algorithms use "size order"; that is, itemsets of smaller sizes are generated, and then itemsets of larger sizes are derived from them. This makes the **frequent pattern-mining problem** a holistic one: to check the validity of a frequent pattern, the entire database has to be scanned. The database cannot be segmented for the mining task. The FP-tree and FPL approaches, on the other hand, use the "frequency order." Patterns containing less frequent items are generated first, and then patterns with only more frequent items are generated. Using the well-defined structures of FP-tree and FPL, this mining order can be realized by partitioning the database, resulting in a divide-and-conquer methodology. For FPL, only the transaction signatures under the rightmost item node are required for mining frequent patterns at each mining stage.

When the database is huge and complicated, our algorithm allows the partition of the database (see property 2 in section 2.3). Moreover, since our FPL algorithm does not generate candidate patterns, the computation time scales linearly with the size of the database, as does FP-tree proposed in [6], rather than scales exponentially with the size of the database when Apriori-like algorithms are used.

5 Conclusions

In this paper we proposed a simple structure, the **frequent pattern list** (**FPL**), for storing information about frequent patterns, discussed the properties of FPL, and developed simple operations on FPL for mining frequent patterns.

Features of FPL are: (1) No duplication for the same frequent item. (2) The data structure is simple: linear lists. (3) The operations are simple: bit operations (*bit counting* and *signature trimming*), and signature removal and appending (called *signature migration*). Therefore, dynamic arrays can be used to implement the FPL structure. (4) The transaction database can be partitioned neatly when mining frequent patterns, resulting in easier management of main memory.

There are several issues related to FPL-based mining. For example, more efficient algorithms for performance improvement should be studied. Also, the FPL approach can be applied to other applications like the mining of user profiles.

References

[1] R. Agrawal and R. Srikant, "Fast Algorithms for Mining Association Rules in Large Databases," Proc. 20th Int'l Conf. Very Large Data Bases, pp. 478-499, Sept. 1994

[2] R. Agrawal and R. Srikant: "Mining Sequential Patterns," Proc. of the Int'l Conference on Data Engineering (ICDE), Taipei, Taiwan, March 1995

[3] S. Brin, etc, "Dynamic Itemset Counting and Implication rules for Market Basket Analysis", Proc. ACM SIGMOD Int'l Conf. Management of Data, pp. 255-264, 1997

[4] S. Brin, etc, "Beyond Market Basket: Generalizing Association Rules to Correlations", Proc. ACM SIGMOD Int'l Conf. Management of Data, pp. 265-276, 1997

[5] J.-S. Park, M.-S. Chen, and P.S. Yu, "Using A Hash-Based Method with Transaction Trimming for Mining Association Rules," IEEE Trans. On Knowledge and Data Eng., vol. 9, no. 5, pp. 813-825, Sept./Oct. 1997

[6] J. Han, J. Pei, and Y. Yin, "Mining Frequent Patterns without Candidate Generation," Proc. ACM SIGMOD, pp. 1-12, 2000

[7] J. Han and M. Kamber, *Data Mining: Concepts and Techniques.* Morgan Kaufmann Publishers, 2000.

[8] Savasere, D. Omiecinski, and S. Navathe, "An efficient Algorithm for Mining Association Rules in Large Databases", Proc. Int'l Conf. Very Large Data Bases, pp. 432-443, Sept. 1995
[9] H. Toivonen, "Sampling Large Databases for Association Rules", Proc. Int'l Conf. Very Large Data Bases, pp. 134-145, Sept. 1996

User-Defined Association Mining

Ke Wang[1] and Yu He[2]

[1] Simon Fraser University
wangk@cs.sfu.ca
http://www.cs.sfu.ca/~wangk
[2] National University of Singapore
hey@comp.nus.edu.sg
http://www.comp.nus.edu.sg/~hey

Abstract. Discovering interesting associations of events is an important data mining task. In many real applications, the notion of association, which defines how events are associated, often depends on the particular application and user requirements. This motivates the need for a general framework that allows the user to specify the notion of association of his/her own choices. In this paper we present such a framework, called the *UDA mining* (*User-Defined Association Mining*). The approach is to define a language for specifying a broad class of associations and yet efficient to be implemented. We show that (1) existing notions of association mining are instances of the UDA mining, and (2) many new ad-hoc association mining tasks can be defined in the UDA mining framework.

1 Introduction

Interesting association patterns could occur in diverse forms. Early work has defined and mined associations of different notions in separate frameworks. For example, association rules are defined by confidence/support and are searched based on the Apriori pruning [1]; correlation rules are defined by the χ^2 statistics test and are searched based on the upward-closed property of correlation [4]; causal relationships are defined and searched by using CCC and CCU rules [14]; emerging patterns are defined by the growth ratio of support [6]. With such an "one-framework-per-notion" paradigm, it is difficult to compare different notions and identify commonalities among them. More importantly, the user may not find such pre-determined frameworks suitable for his/her specific needs. For example, at one time the user likes to find all pairs $< p, c >$ such that p is some above mentioned association pattern and c is a condition under which p occurs; at another time the user likes to know all triples $< p, c_1, c_2 >$ such that association pattern p occurs in the special case c_1 but not in the general case c_2; at yet another time the user wants something else. Even for this simple example, it is not clear how the above existing frameworks can be extended to such "ad-hoc" mining. The topic of this paper is to address this extendibility.

Our approach is to propose a *language* in which the user himself/herself can *define* a new notion of association (vs. *choose* a pre-determined notion).

D. Cheung, G.J. Williams, and Q. Li (Eds.): PAKDD 2001, LNAI 2035, pp. 387–399, 2001.

In spirit, this is similar to database querying in DBMS, in that it does not predict the mining tasks that the user might require to perform; it is the expressive power of the language that determines the class of associations specifiable in this approach. The key is the notion of "user-defined associations" and its specification language. Informally, a user-defined association has two components, *events* and their *relationship*. An event is a conjunction of atomic descriptors, called items, for transactions in the database. For example, event $FEMALE \wedge YOUNG \wedge MANAGER$ is a statement about individuals, where items $FEMALE$, $YOUNG$, and $MANAGER$ are atomic descriptors of individuals [1]. A relationship is a statement about how events are associated. We illustrate the notion of "user-defined associations" through several examples.

Example I. A liberate notion of association between two events X and Y can occur in the form that X *causes* Y. As pointed out by [14], such causal relationships, which state the *nature* of the relationships, cannot be derived from the classic association rules $X \rightarrow Y$. [14] has considered the problem of mining causal relationships among 1-item events, i.e., events containing a single item. In the interrelated world, causal relationships occur more often among multi-item events than among 1-item events. For example, 2-item event $MALE \wedge POSTGRAD$ more likely causes $HIGH\_INCOME$ than each of the 1-item events $MALE$ and $POSTGRAD$ does. Though the concept of causal relationships remains unchanged for multi-item events, the search for such causal relationships turns out to be more challenging because it is unknown in advance which items form a meaningful multi-item event in a causal relationship. In our approach, such general causal relationships are modeled as a special case of user-defined associations.

Example II. The user likes to know all three events Z_1, Z_2, X such that X is more "associated" with Z_1 than with Z_2, where the notion of association between X and Z_i could be any user-defined associations. For example, if $X = HIGH\_INCOME$ is more correlated with $Z_1 = POSTGRAD \wedge MALE$ than with $Z_2 = POSTGRAD \wedge FEMALE$, the user could use it as an evidence of gender discrimination because the same education does not give woman the same pay as man. Again, multi-item events like $POSTGRAD \wedge MALE$ are essential for discovering such associations.

Example III: Sometimes, the user likes to know all combinations of events $Z_1, Z_2, X_1, \ldots, X_k$ such that the association of k events $X_1, \ldots, X_k$, in whatever notion, has sufficiently changed when the condition changes from Z_1 to Z_2. For example, $X_1 = BEER$ and $X_2 = CHIPS$ could be sold together primarily during $Z_1 = [6PM, 9PM] \wedge WEEKDAY$. Here, Z_2 is implicitly taken as $\emptyset$, representing the most general condition.

This list can go on, but several points have emerged and are summarized below.

1. *User-defined associations.* A powerful concept in user-defined association is that the user defines a class of associations by "composing" existing user-

[1] A better term for things like $FEMALE$, $YOUNG$, and $MANAGER$ is perhaps "feature" or "variable". We shall use the term "items" to be consistent with [1].

defined association. The basic building blocks in this specification, such as support, confidence, correlation, conditional correlation, etc., may not be new and, in fact, are well understood. What is new is to provide the user with a mechanism for constructing a new notion of association using such building blocks.
2. *Unified specification and mining.* A friendly system should provide a *single* framework for specifying and mining a broad class of notions of association. We do not expect a single framework to cover all possible notions of association, just as we do not expect SQL to express all possible database queries. What we expect is that the framework is able to cover most important and typical notions of association. We will elaborate on this point in Section 3.
3. *Completeness of answers.* The association mining aims to find *all* associations of a specified notion. In contrast, most work in statistics and machine learning, e.g., model search [18] and Bayesian network learning [7], is primarily concerned with finding some but not all associations. To search for such complete answers, those approaches are too expensive for data sets with thousands of variables (i.e., items) as we consider here.
4. *Unspecified event space.* The event space is not fixed in advance and must be discovered in the search of associations. This feature is different from [3,4,14] where only 1-item events are considered. Given thousands of items and that any combination of items is potentially an event, it is a non-trivial task to determine what items make a meaningful event in the association with other events. This task is further compounded by the fact that any combination of events is potentially an association.

In the rest of this paper, we present a unified framework for specifying and mining user-defined associations. The framework must be expressive enough for specifying a broad class of associations. In Section 2 and Section 3, we propose such a framework and examine its expressive power. Equally important, the mining algorithm must have an efficient implementation. We consider the implementation in Section 4. We review related work in Section 5 and conclude the paper in Section 6.

2 User-Defined Association

2.1 Definitions

The database is a collection of *transactions*. Each transaction is represented by a set of Boolean descriptors called *items* that hold on the transaction. An *event* is a conjunction of items, often treated as a set of items. We do not consider disjunction in this paper. $\emptyset$, called the *empty event*, denotes the Boolean constant TRUE or the empty set. Given the transaction database, the *support* of an event X, denoted $P(X)$, is the fraction of the transactions on which event X holds, or of which X is a subset. An event X is *large* if $P(X) \geq mini\_sup$ for the user-specified minimum support $mini\_sup$. Events that are not large occur too infrequently, therefore, do not have statistical significance. The set of large events is downward-closed with respect to the set containment[1]: if X is large and X' is a subset of X, X' is also large. For events X and Y, XY is the shorthand for event $X \wedge Y$ or $X \cup Y$, and $P(X|Y)$ for $P(XY)/P(Y)$. Thus, $X_1, \ldots, X_k$ represents k events whereas $X_1 \ldots X_k$ represents one event $X_1 \wedge \ldots \wedge X_k$. The notion of support can be extended to absence of events. For example, $P(X \neg Y \neg Z)$ denotes the fraction of the transactions on which X holds but neither Y nor Z does.

A user-defined association is written as $Z_1, \ldots, Z_p \rightarrow X_1, \ldots, X_k$. $X_1, \ldots, X_k$ are called *subject events*, whose association is of the primary concern. $Z_1, \ldots, Z_p$ are called *context events*, which provide p different conditions for comparing the association of subject events. Context events are always ordered because the order of affecting the association is of interest. The notion of user-defined association $Z_1, \ldots, Z_p \rightarrow X_1, \ldots, X_k$ is defined by the support filter and the strength filter defined below.

- *Support_Filter*. It states that events $X_1, \ldots, X_k, Z_i$ must occur together frequently: if $p > 0$, $P(X_1 \ldots X_k Z_i) \geq mini\_sup$ for $1 \leq i \leq p$; or if $p = 0$, $P(X_1 \ldots X_k) \geq mini\_sup$. In other words, $X_1 \ldots X_k Z_i$, or $X_1 \ldots X_k$ if $p = 0$, is required to be a large event. If this requirement is not satisfied, the co-occurrence of X_i under condition Z_i does not have statistical significance. This condition is called the *support filter* and is written as $Support\_Filter(z_1, \ldots, z_p \rightarrow x_1, \ldots, x_k)$, where z_i and x_i are variables representing the events Z_i and X_i in a user-defined association.
- *Strength_Filter*: It states that events $Z_1, \ldots, Z_p, X_1, \ldots, X_k$ must hold the relationship specified by a conjunction of one or more formulas of the form $\psi_i \geq mini\_str_i$. Each ψ_i measures some *strength* of the relationship and $mini\_str_i$ is the threshold value on the strength. This conjunction is called the *strength filter* and is written as $Strength\_Filter(z_1, \ldots, z_p \rightarrow x_1, \ldots, x_k)$, where z_i and x_i are variables representing the events Z_i and X_i.

In the above filters, variables x_i and z_i can be instantiated by events X_i and Z_i, and the instantiation is represented by $Support\_Filter(Z_1, \ldots, Z_p \rightarrow X_1, \ldots, X_k)$ and $Strength\_Filter\ (Z_1, \ldots, Z_p \rightarrow X_1, \ldots, X_k)$. Observe that $Support\_Filter(Z_1, \ldots, Z_p \rightarrow X_1, \ldots, X_k)$ implies that each X_i and Z_i is a large event because of the downward-closed property mentioned earlier. It remains to choose a language for specifying ψ_i, which will determine the class of associations specified and the efficiency of the mining algorithm. We will study this issue shortly. For now, we assume that such a language is chosen. As a convention, we use lower case letters $z_1, \ldots, z_p, x_1, \ldots, x_k$ for event variables and use upper case letters $Z_1, \ldots, Z_p, X_1, \ldots, X_k$ for events.

Definition 1 (The UDA specification). A *user-defined association specification (UDA specification)*, written as $UDA(z_1, \ldots, z_p \rightarrow x_1, \ldots, x_k)$, $k > 0$ and $p \geq 0$, has the form $Strength\_Filter(z_1, \ldots, z_p \rightarrow x_1, \ldots, x_k) \wedge$ $Support\_Filter$ $(z_1, \ldots, z_p \rightarrow x_1, \ldots, x_k)$. (The End)

We say that a UDA specification is *symmetric* if variables x_i's are symmetric in $Strength\_Filter$ (note that variables x_i's are always symmetric in

Support_Filter); otherwise, it is *asymmetric.* A symmetric specification is desirable if the order of subject events does not matter, such as correlation. Otherwise, an asymmetric UDA specification is desirable. For example, an asymmetric UDA is that whenever events $Z_1, \ldots, Z_p$ occur, X_1 occurs but not X_2. We consider only symmetric specification, though the work can be extended to asymmetric specification.

Definition 2 (The UDA problem). Assume that $UDA(z_1, \ldots, z_p \rightarrow x_1, \ldots, x_k)$ is given for $0 < k \leq k'$, where $p(\geq 0)$ and $k'(> 0)$ are specified by the user. Consider distinct events $Z_1, \ldots, Z_p, X_1, \ldots, X_k$. We say that $Z_1, \ldots, Z_p \rightarrow X_1, \ldots, X_k$ is a *UDA* if the following conditions hold:

1. $X_i \cap X_j = \emptyset$, $i \neq j$, and
2. $X_i \cap Z_j = \emptyset$, $i \neq j$, and
3. $UDA(Z_1, \ldots, Z_p \rightarrow X_1, \ldots, X_k)$ is true.

k is called the *size* of the UDA. $Z_1, \ldots, Z_p \rightarrow X_1, \ldots, X_k$ is *minimal* if for any proper subset $\{X_{i_1}, \ldots, X_{i_q}\}$ of $\{X_1, \ldots, X_k\}$, $Z_1, \ldots, Z_p \rightarrow X_{i_1}, \ldots, X_{i_q}$ is not a UDA. The *UDA problem* is to find all UDAs of the specified sizes $0 < k \leq k'$. The *minimal UDA problem* is to find all minimal UDAs of the specified sizes k. (The End)

Several points about Definition 2 are worth noting.

First, the number of context events, p, in a UDA $Z_1, \ldots, Z_p \rightarrow X_1, \ldots, X_k$ is fixed whereas the number of subject events, k, is allowed up to a specified maximum size k_m. This distinction comes from the different roles of these events: for subject events we do not know a prior how many of them may participate in an association, but we often examine a fixed number of conditions for each association. It is possible to allow the number of conditions p up to some maximum number, but we have not found useful applications that require this extension.

Second, context events Z_i's are not necessarily pairwise disjoint. In fact, it is often desirable to examine two context events Z_1 and Z_2 such that Z_1 is a proper superset, thereby a specialization, of Z_2. Then we could specify UDAs $Z_1, Z_2 \rightarrow X_1, \ldots, X_k$ such that the association of X_i's holds under the specialized condition Z_1 but not under the general condition Z_2. Other useful syntax constraints could be the requirement on the presence or absence of some specified items in an event, a certain partitioning of the items for context events and subject events, the maximum or minimum number of items in an event, etc. In the same spirit, the disjointness in condition 2 can be removed to express certain overlapping constraints. Constraints have been exploited to prune search space for mining association rules [16,13]. A natural generalization is to exploit syntax constraints for mining general UDAs. In this paper, however, we focus on the basic form in Definition 2.

2.2 Examples

In this section, we intend to achieve two goals through considering several examples of UDA specification: to show that disparate notions of association can be specified in the UDA framework, and to readily convey the basic idea that underly more complex specification. Once these are understood, the user can define any notion of association of his/her own choice, in the given specification language. We shall focus on specifying *Strength_Filter* because specifying *Support_Filter* is straightforward. In all examples, lower-case letters z_i and x_i represent event variables and upper-case letters Z_i and X_i represent events.

Example 1 (Association rules). Association rules $Z \to X$ *introduced in [1] can be specified by*

$$Support\_Filter(z \to x) : P(zx) \geq mini\_sup$$
$$Strength\_Filter(z \to x) : \psi(z,x) \geq mini\_conf,$$

where $\psi(z,x) = P(x|z) = P(xz)/P(z)$ *is the* confidence *of rule* $Z \to X$ *[1]. To factor in both "generality" and "predictivity" of rules in a single measure, the following* $\psi(z,x)$*, called the* J-measure *[15], can be used:*

$$P(z)[P(x|z)log_2\frac{P(x|z)}{P(x)} + (1 - P(x|z))log_2\frac{1 - P(x|z)}{1 - P(x)}].$$

Here, $P(z)$ *weighs the generality of the rule and the term in the square bracket weighs the "discrimination power" of* z *on* x*. This example shows how easy it is to adopt a different definition of association in the UDA framework. (The End)*

In the above specification, the most basic constructs are the supports $P(Z)$, $P(X)$, $P(ZX)$, $P(\neg X)$, $P(Z\neg X)$. Since *Support_Filter*$(Z \to X)$ implies that each of Z, X, ZX is a large event, these supports are readily available from mining large events (note that $P(\neg X) = 1 - P(X)$ and $P(Z\neg X) = P(Z) - P(ZX)$).

Example 2 (Multiway correlation). The notion of correlation is a special case of UDAs without context event. In particular, events $X_1, \ldots, X_k$ *are* correlated *if they occur together more often than expected when they are independent. This notion can be specified by the* χ^2 *statistic test,* $\chi^2(x_1, \ldots, x_k) \geq \chi^2_\alpha$ *[4]. Let* $R = \{x_1, \neg x_1\} \times \ldots \times \{x_k, \neg x_k\}$ *and* $r = r_1 \ldots r_k \in R$*. Let* $E(r) = N * P(r_1) * \ldots * P(r_k)$*, where* N *is the total number of transactions.* $\chi^2(x_1, \ldots, x_k)$ *is defined by:*

$$\Sigma_{r \in R} \frac{(N * P(r) - E(r))^2}{E(r)}.$$

The threshold value χ^2_α *for a user-specified significance level* α*, usually 5%, can be obtained from statistic tables for the* χ^2 *distribution. If* $X_1, \ldots, X_k$ *passes the test,* $X_1, \ldots, X_k$ *are correlated with probability* $1 - \alpha$*. The* uncorrelation *of* $X_1, \ldots, X_k$ *can be specified by Strength_Filter of the form* $1/\chi^2(x_1, \ldots, x_k) \geq 1/\chi^2_\alpha$*, where* α *is usually 95%. If* $X_1, \ldots, X_k$ *passes the filter,* $X_1, \ldots, X_k$ *are uncorrelated with probability* α*. (The End)*

The problem of mining correlation among single-item events was studied in [3,4]. One difference of correlation specified as UDAs is that each event X_i can involve multiple items, rather than a single item. One such example is the correlation of $X_1 = INTERNET$ and $X_2 = YOUNG \wedge MALE$, where $YOUNG \wedge MALE$ is a 2-item event. This generalization is highly desirable because single-item events like $X_1 = INTERNET$ and $X_2 = YOUNG$ or $X_1 = INTERNET$ and $X_2 = MALE$ may not be strongly corrected. It is not clear how the mining algorithms in [3,4] can be extended to multi-item events. A more profound difference, however, is that, as UDAs, we can model "ad-hoc" extension of correlation. The subsequent examples shows this point.

Example 3 (Conditional association). In conditional association $Z \rightarrow X_1, \ldots, X_k$*, subject events* $X_1, \ldots, X_k$ *are associated when conditioned on* Z*. For example,*

$$INT'L \wedge BUSINESS\_TRIP \rightarrow CEO, FIRST\_CLASS$$

says that $X_1 = CEO$ *and* $X_2 = FIRST\_CLASS$ *(flights) are associated for* $Z = INT'L \wedge BUSINESS\_TRIP$ *(international business trips). For example, if the association of* $X_1, \ldots, X_k$ *is taken as the correlation, we have* conditional correlation *defined by* $Strength\_Filter(z \rightarrow x_1, \ldots, x_k)$*:*

$$\frac{P(x_1 \ldots x_k|z)}{P(x_1|z) * \ldots * P(x_k|z)} \geq mini\_str \tag{1}$$

or alternatively, by the χ^2 *statistic test after replacing* $P(r)$ *and* $P(r_i)$ *in* $\chi^2(X_1, \ldots, X_k) \geq \chi^2_\alpha$ *with* $P(r|z)$ *and* $P(r_i|z)$*. (The End)*

Example 4 (Comparison association). In comparison association $Z_1, Z_2 \rightarrow X$*, subject event* X *is associated* differently *with context events* Z_1 *and* Z_2*. For example,*

$$INT'L \wedge BUSINESS\_TRIP, PRIVATE\_TRIP \rightarrow CEO \wedge FIRST\_CLASS$$

says that $X = CEO \wedge FIRST\_CLASS$ *is more associated with* $Z_1 = INT'L \wedge BUSINESS\_TRIP$ *than with* $Z_2 = PRIVATE\_TRIP$*. To compare two associations for difference, we can compare their corresponding strength* ψ_j *in* $Strength\_Filter$*. In particular, suppose that* $UDA(\rightarrow z_i, x)$ *specifies the association of* Z_i *and* X*,* $i = 1, 2$*. For each* $\psi_j(z_i, x)$ *in* $UDA(\rightarrow z_i, x)$*,* $Strength\_Filter$ $(z_1, z_2 \rightarrow x)$ *for the comparison association contains the formula:*

$$Dist(\psi_j(z_1, x), \psi_j(z_2, x)) \geq mini\_str_j. \tag{2}$$

Here, $Dist(s_1, s_2)$ *measures the distance between two strengths* s_1 *and* s_2*. Typical distance measures are* $Dist(s_1, s_2) = s_1/s_2$ *or* $Dist(s_1, s_2) = s_1 - s_2$*. (The End)*

Example 5 (Emerging association). In emerging association $Z_1, Z_2 \rightarrow X_1, \ldots, X_k$*, the association of* $X_1, \ldots, X_k$ *has changed sufficiently when the condition changes from* Z_1 *to* Z_2*. Suppose that* $UDA(z_i \rightarrow x_1, \ldots, x_k)$ *specifies the conditional association* $Z_i \rightarrow X_1, \ldots, X_k$*,* $i = 1, 2$*, as in Example 3. Then, for* *each strength function* ψ_j *in* $UDA(z_i \to x_1, \ldots, x_k)$, $Strength\_Filter(z_1, z_2 \to x_1, \ldots, x_k)$ *for emerging association contains the formula:*

$$Dist(\psi_j(z_1, x_1, \ldots, x_k), \psi_j(z_2, x_1, \ldots, x_k)) \geq mini\_str_j. \quad (3)$$

The notion of emerging association is useful for identifying trends and changes. For example, the notion of emerging patterns *[6] is a special case of emerging associations of the form* $Z_1, Z_2 \to X$*, where* Z_1 *and* Z_2 *are identifiers for the two originating databases of the transactions. An emerging pattern* $Z_1, Z_2 \to X$ *says that the ratio of the support of* X *in the two databases identified by* Z_1 *and* Z_2 *is above some specified threshold. To specify emerging patterns, we first merge the two databases into a single database by adding item* Z_i *to every transaction coming from database* i, $i = 1, 2$*, and specify* $\psi_j(z_i, x) = P(z_i x)/P(z_i)$*, where* z_i *are variables for* Z_i*, and* $Dist(s_1, s_2) = s_1/s_2$ *in Equation 3. With the general notion of emerging association, however, we can capture a context* Z_i *as an arbitrary event (not just a database identifier) and the participation of more than one subject event. (The End)*

In Examples 3, 4 and 5, the "output" UDAs (i.e., conditional association, comparison association, emerging association) are defined in terms of "input" UDAs. These input UDAs are of the form $\to X_1, \ldots, X_k$ in Example 3, $\to Z_i, X$ in Example 4, and $Z_i \to X_1, \ldots, X_k$ in Example 5, which themselves can be defined in terms of their own input UDAs. The output UDAs can be the input UDAs for defining other UDAs. In general, new UDAs are defined by "composing" existing UDAs. It is such a composition that provides the extendibility for defining ad-hoc mining tasks. We further demonstrate this extendibility by specifying causal relationships.

Example 6 (Causal association). Information about statistical correlation and uncorrelation can be used to constrain possible causal relationships. For example, if events A *and* B *are uncorrelated, it is clear that there is no causal relationship between them. Following this line, [14] identified several rules for inferring causal relationships, one of which is the so-called CCC rule: if events* Z, X_1, X_2 *are pairwise correlated, and if* X_1 *and* X_2 *are uncorrelated when conditioned on* Z*, one of the following causal relationships exists:*

$$X_1 \Leftarrow Z \Rightarrow X_2 \qquad X_1 \Rightarrow Z \Rightarrow X_2 \qquad X_1 \Leftarrow Z \Leftarrow X_2,$$

where $\Leftarrow$ *means "is caused by" and* $\Rightarrow$ *means "causes". For a detailed account of this rule, please refer to [14]. We can specify the condition of the CCC rule by* $Strength\_Filter(z \to x_1, x_2)$*:*

$$\psi_1(\emptyset, x_1, x_2) \geq mini\_str_1 \wedge \psi_1(\emptyset, x_1, z) \geq mini\_str_1 \wedge$$
$$\psi_1(\emptyset, x_2, z) \geq mini\_str_1 \wedge \psi_2(z, x_1, x_2) \geq mini\_str_2.$$

Here, $\psi_1(w, u, v) \geq mini\_str_1$ *tests the correlation of* u *and* v *conditioned on* w*, and* $\psi_2(w, u, v) \geq mini\_str_2$ *tests the uncorrelation of* u *and* v *conditioned on* w*. These tests were discussed in Examples 2 and 3. (The End)*

In all the above examples, the basic constructs used by the specification are the support of the form $P(v)$, where v is a conjunction of terms $Z_i, \neg Z_i, X_i, \neg X_i$. This syntax of v completely defines the language for *Strength_Filter* because we make no restriction on how $P(v)$ should be used in the specification. In the next section, we define the exact syntax for v.

3 The Specification Language

The term "language" is more concerned with what it can do than how it is presented. There are two considerations in choosing the language for strength functions ψ_i. First, the language should specify a wide class of association. Second, the associations specified should have an efficient mining algorithm. We start with the efficiency consideration. To specify UDAs $Z_1, \ldots, Z_p \rightarrow X_1, \ldots, X_k$, we require each strength ψ_i to be above some minimum threshold, where ψ_i is a function of $P(v)$ and v is a conjunction of X_i and Z_j. The support filter $P(X_1 \ldots X_k Z_j) \geq mini\_sup$ is used to constrain the number of candidate Z_j and X_i. The support filter implies that a conjunction v consisting of any number of subject events X_i and zero or one context event Z_j is large. Therefore, supports $P(v)$ for such v are available from mining large events if we keep the support for each large events.

The question is whether it is too restrictive to allow at most one Z_j in each v. It turns out that this is a desirability not a restriction. In fact, each Z_j serves as an individual context for the association of $X_1, \ldots, X_k$ and there is no need to consider more than one Z_j at a time. Another question is that, if absences of events, i.e., $\neg X_i$ and $\neg Z_j$, are desirable in v, as in the examples in Section 2.2, can $P(v)$ be computed efficiently? The next theorem, which is essentially a variation of the well known "inclusion-exclusion" theorem, shows that such $P(v)$ can be computed by the supports involving no absence of events.

Theorem 1. Let $V = \{V_1, \ldots, V_q\}$ be q events of the form X_i or Z_i. Let U be a conjunction of events that do not occur in V. Then

$$P(U \neg V_1 \ldots \neg V_q) = \Sigma_{W \subseteq V} (-1)^{|W|} P(UW),$$

where $|W|$ denotes the number of V_i's in W. (The End)

For example, assume that $V = \{X_2, Z_1\}$ and $U = \{X_1\}$, we have $P(X_1 \neg X_2 \neg Z_1) = P(X_1) - P(X_1 X_2) - P(X_1 Z_1) + P(X_1 X_2 Z_1)$. This rewriting conveys two important points: (1) the right-hand side contains no absence, (2) if $X_1 X_2 Z_1$ is large (as required by the support filter), the right-hand side contains only supports of large events, thus, is computable by mining large events. containing no absence. Based on these observations, we are now ready to define the syntax of v for supports $P(v)$ that appear in a strength function.

Definition 3 (Individual-context assumption). Let Z_j be a context event and X_i be a subject event. v satisfies the *ICA (Individual-Context Assumption)* if v is a conjunction of zero or more terms of the form X_i and $\neg X_i$, and zero or one term of the form Z_j and $\neg Z_j$. A support $P(v)$ satisfies the *ICA* if v satisfies the ICA. A strength function ψ satisfies the *ICA* if it is defined using only supports satisfying the ICA. A strength filter satisfies the *ICA* if it uses only strength functions satisfying the *ICA*. The *ICA-language* consists of all UDAs defined by the support filter in Section 2.1 and the strength filter satisfying the ICA. (The End)

For example, X_1X_2, $\neg X_1X_2$, $X_1X_2Z_1$, and $X_1X_2\neg Z_1$ all satisfy the ICA because each contains at most one term for context events, but $X_1X_2Z_1Z_2$ and $X_1X_2Z_1\neg Z_2$ do not. We like to point out that the ICA is a language on support $P(v)$, not a language on how to use $P(v)$ in defining a strength function ψ. This total freedom on using such $P(v)$ allows the user to define new UDAs by composing existing UDAs in any way he/she wants, a very powerful concept illustrated by the examples in Section 2.2. In fact, one can verify that all the strength functions ψ in Section 2.2 are specified in the ICA-language.

We close this section by making an observation on the "computability" of the ICA-language. The support filter implies that any absence-free v satisfying the ICA is a large event. The rewriting by Theorem 1 preserves the ICA because it only eliminates absences. Consequently, the ICA-language ensures that all allowable supports can be computed from mining large events. This addresses the computational aspect of the language.

4 Implementation

We are interested in a unified implementation for mining UDAs. Given that any combination of items can be an event and any combination of events can be a UDA, it is only feasible to rely on effective pruning strategies to reduce the search space of UDAs. Due to the space limitation, we sketch only the main ideas. Assume that the items in an event are represented in the lexicographical order and that the subject events in a UDA are represented in the lexicographical order (we consider only symmetric UDA specifications). We consider $p > 0$ context events; the case of $p = 0$ is more straightforward. Our strategy is to exploit the constraints specified by *Support_Filter* and *Strength_Filter* as earlier as possible in the search of UDAs. The first observation is that *Support_Filter* implies that all subject events X_i and context events Z_i are large. Thus, as the first step we find all large events, say by applying Apriori [1] or its variants. We assume that the mined large events are stored in a hash-tree [1] or a hash table so that the membership and support of large events can be checked efficiently.

The second step is to construct UDAs using large events. For each UDA $Z_1, \ldots, Z_p \rightarrow X_1, \ldots, X_k$, *Support_Filter* requires that $X_1 \ldots X_k Z_i$ be large for $1 \leq i \leq p$. Therefore, it suffices to consider only the *k-tuples* of the form $(X_1, \ldots, X_k, Z_i)$, where X_i's are in the lexicographical order and $X_1 \ldots X_k Z_i$ makes a large event for all $1 \leq i \leq p$. We can generate such k-tuples and UDAs of size k in a level-wise manner like Apriori by treating events as items: In the

kth iteration, a k-tuple $(X_1, \ldots, X_k, Z_i)$ is generated *only if* $(X_1, \ldots X_{k-1}, X_k)$ and $(X_1, \ldots, X_{k-1}, Z_i)$ were generated in the $(k-1)$th iteration and $X_1 \ldots X_k Z_i$ is large. The largeness of $X_1 \ldots X_k Z_i$ can be checked by looking up the hash-tree or hash table for storing large events. Also, the disjointness of X_k and Z_i, required by $Strength\_Filter$, and the lexicographical ordering of $X_1, \ldots, X_k$, can be checked before generating tuple $(X_1, \ldots, X_k, Z_i)$. After generating all k-tuples in the current iteration, we construct a candidate UDA $Z_1, \ldots, Z_p \rightarrow X_1, \ldots, X_k$ using p distinct tuples of the form $(X_1, \ldots, X_k, Z_i)$, $i = 1, \ldots, p$, that share the same prefix $X_1, \ldots, X_k$. Any further syntax constraints on Z_i, as discussed in Section 2.1, can be checked here. A candidate $Z_1, \ldots, Z_p \rightarrow X_1, \ldots, X_k$ is a UDA if $UDA(Z_1, \ldots, Z_p \rightarrow X_1, \ldots, X_k)$ holds. The above is repeated until some iteration k for which no k-tuple is generated.

For mining minimal UDAs, a straightforward algorithm is to first generate all UDAs and then remove non-minimal UDAs. A more efficient algorithm is finding all minimal UDAs without generating non-minimal UDAs. The strategy is to consider subsets of $\{X_1, \ldots, X_k\}$ for subject events X_i's before considering $\{X_1, \ldots, X_k\}$ itself, and prune all supersets from consideration if any subset is found to be a UDA. Since this is essentially a modification of the above algorithm, we omit the detail in the interest of space.

We have conducted several experiments to mine the classes of UDAs considered in Section 2.2 from the census data set used in [14], which contains 63 items and 126,229 transactions. The result is highly encouraging: it discovers several very interesting associations that cannot be found by existing approaches. For example, some strong causal associations were found among general k-item events, as discussed in Example I, but were not found in [14] because only 1-item events are considered there. This fact re-enforces our claim that the uniform mining approach does not simply unify several existing approaches; it also extends beyond them by allowing the user to define *new* notions of association. We omit the detail of the experiments due to the space limit.

5 Related Work

In [8], a language for specifying several pre-determined rules is considered, but no mechanism is provided for the user to specify new notions of association. In [10], it is suggested to query mined rules through an application programming interface. In [12,17], some SQL-like languages are adopted for mining association rules. The expressiveness of these approaches is limited by the extended SQL. For example, they cannot specify most of the UDAs in Section 1 and 2. In [11,9], a generic data mining task is defined as finding all patterns from a given pattern class that satisfy some interestingness filters, but no concrete language is proposed for pattern classes and interestingness filters. Finding causal relationships is studied in [5,14]. None of these works considers the extendibility where the user can define a new mining task.

6 Conclusion

This paper introduces the notion of user-defined association mining, i.e., the UDA mining, and proposes a specification framework and implementation. The purpose is to move towards a unified data mining where the user can mine a database with the same ease as querying a database. For the proposed approach to work in practice, however, further studies are needed in several areas. Our current work has considered only limited syntax constraints on events, and it is important to exploit broader classes of syntax constraints to reduce the search space. Also, a unified mining algorithm may be inferior to specialized algorithms targeted at specific classes of UDAs. It is important to study various optimization strategies for typical and expensive building blocks of UDAs. In this paper, we have mainly focused on the semantics and "computability" (in no theoretic sense) of the specification language. A user specification interface, especially merged with SQL, is an interesting topic.

References

1. R. Agrawal, T. Imielinski, and A. Swami: Mining Association Rules between Sets of Items in Large Datasets. SIGMOD 1993, 207-216
2. R. Agrawal and R. Srikant: Fast Algorithm for Mining Association Rules. VLDB 1994, 487-499
3. C.C. Aggarwal and P.S. Yu: A New Framework for Itemset Generation. PODS 1998, 18-24
4. S. Brin, R. Motwani, and C. Silverstein: Beyond Market Baskets: Generalizing Association Rules to Correlations. SIGMOD 1997, 265-276
5. G.F. Cooper: A Simple Constraint-based Algorithm for Efficiently Mining Observational Databases for Causal Relationships. Data Mining and Knowledge Discovery, No. 1, 203-224, 1997, Kluwer Academic Publishers
6. G. Dong, J. Li: Efficient Mining of Emerging Patterns: Discovering Trends and Differences. SIGKDD 1999, 43-52
7. D. Heckerman: Bayesian Networks for Data Mining. Data Mining and Knowledge Discovery, Vol. 1, 1997, 79-119
8. J. Han, Y. Fu, W. Wang, K. Koperski, O. Zaiane: DMQL: A Data Mining Query Language for Relational Databases. SIGMOD Workshop on Research Issues on Data Mining and Knowledge Discovery, 1996, 27-34
9. T. Imielinski and H. Mannila: A Database Perspective on Knowledge Discovery. Communications of ACM, 39(11), 58-64, 1996
10. T. Imielinski, A. Virmani, and A. Abdulghani: DataMine: Application Programming Interface and Query Language for Database Mining. KDD 1996, 256-261
11. H. Mannila: Methods and Problems in Data Mining. International Conference on Database Theory, 1997, 41-55, Springer-Verlag
12. R. Meo, G. Psaila, S. Ceri: A New SQL-like Operator for Mining Association Rules. VLDB 1996, 122-133
13. R.T. Ng, L. V.S. Lakshmanan, J. Han, A Pang: Exploratory Mining and Pruning Optimizations of Constrained Associations Rules. SIGMOD 1998, 13-24
14. C. Silverstein, S. Brin, R. Motwani, J. Ullman: Scalable Techniques for Mining Causal Structures. VLDB 1998, 594-605

15. P. Smyth, R.M. Goodman: An Information Theoretic Approach to Rule Induction from Databases. IEEE Transactions on Knowledge and Data Engineering, Vol. 4, No. 4, 301-316, August 1992.
16. R. Srikant, Q. Vu, and R. Agrawal: Mining Association Rules with Item Constraints. KDD 1997, 67-73
17. D. Tsur, J. Ullman, S. Abiteboul, C. Clifton, R. Motwani, S. Nestorov, A. Rosenthal: Query Flocks: a Generalization of Association-Rule Mining. SIGMOD 1998, 1-12
18. T.D. Wickens: Multiway Contingency Tables Analysis for the Social Sciences. Lawrence Brlbaum Associates, 1989.

Direct and Incremental Computing of Maximal Covering Rules

Marzena Kryszkiewicz

Institute of Computer Science, Warsaw University of Technology
Nowowiejska 15/19, 00-665 Warsaw, Poland
mkr@ii.pw.edu.pl

Abstract. In the paper we consider the knowledge in the form of association rules. The consequents derivable from the given set of association rules constitute the theory for this rule set. We apply maximal covering rules as a concise representation of the theory. We prove that maximal covering rules have precisely computable values of support and confidence, though the theory can contain rules for which these values can be only estimated. Efficient methods of direct and incremental computation of maximal covering rules are offered.

1 Introduction

The problem of discovery of strong association rules was introduced in [1] for sales transaction database. The association rules identify sets of items that are purchased together with other sets of items. In the paper we consider a specific problem of mining around rules rather than mining of rules in a database. Let us assume a user is not allowed to access the database and can deal only with the restricted number of rules provided by somebody else. Still, the user hopes to find new interesting relationships. On the other hand, the rules provider should be certain no secret patterns will be discovered by the user. Therefore, it is important for the provider to be aware of the consequents derivable from the delivered rule set.

The problem of inducing knowledge from the rule set was addressed first in [5]. We offered there how to use the *cover operator* and *extension operator* in order to augment the given knowledge. Unlike, the extension operator, the cover operator does not require any information on statistical importance (support) of rules. The induced rules are of the same or higher quality than the original ones. Additionally, it was introduced in [5] the notion of *maximal covering rules* that represent (subsume) the set of given and induced rules.

It was shown in [6] how to induce all knowledge (*theory*) derivable from the given rule set. Theory can contain more rules than those derived by means of cover and extension operators. The algorithms for inducing theory as well as for deriving maximal covering rules for theory were offered there. Additionally, it was shown how to test the consistency of the provided rule set and how to extract its consistent subset.

In this paper we investigate properties of maximal covering rules in order to propose more efficient method of their computing than the one proposed in [6]. In addition to efficient direct algorithm of computing maximal covering rules we propose an incremental approach.

D. Cheung, G.J. Williams, and Q. Li (Eds.): PAKDD 2001, LNAI 2035, pp. 400-405, 2001.

2 Association Rules, Rule Cover, and Maximal Covering Rules

Let I be a set of *items*. In general, any set of items is called an *itemset*. The itemset consisting of k items is called *k-itemset*. Let D be a set of transactions, where each transaction T is a subset of I. An *association rule* is an implication $X \Rightarrow Y$, where $\emptyset \neq X,Y \subset I$ and $X \cap Y = \emptyset$. *Support* of an itemset X is denoted by $sup(X)$ and defined as the number (or the percentage) of transactions in D that contain X. *Support* of the association rule $X \Rightarrow Y$ is denoted by $sup(X \Rightarrow Y)$ and defined as $sup(X \cup Y)$. *Confidence* of $X \Rightarrow Y$ is denoted by $conf(X \Rightarrow Y)$ and defined as $sup(X \cup Y) / sup(X)$. The problem of mining association rules is to generate all rules that have sufficient support and confidence. In the sequel, the set of all association rules whose support is greater than s and confidence is greater than c will be denoted by $AR(s,c)$.

A notion of a *cover operator* was introduced in [3] for deriving a set of rules from a given association rule without accessing a database. The *cover* C of the rule $X \Rightarrow Y$, was defined as follows: $C(X \Rightarrow Y) = \{X \cup Z \Rightarrow V | Z,V \subseteq Y \wedge Z \cap V = \emptyset \wedge V \neq \emptyset\}$.

Property 1 [3]. Let r: $(X \Rightarrow Y)$ and r': $(X' \Rightarrow Y')$ be association rules.

$$r' \in C(r) \text{ iff } X' \cup Y' \subseteq X \cup Y \wedge X' \supseteq X.$$

Property 2 [4]. Let r': $(X' \Rightarrow Y')$ belongs to $AR(s,c)$.

$$\exists r \in AR(s,c),\ r \neq r' \wedge r' \in C(r) \text{ iff } \exists (X \Rightarrow Y) \in AR(s,c),\ (X'=X \wedge X' \cup Y' \subset X \cup Y) \vee (X' \supset X \wedge X' \cup Y' = X \cup Y).$$

Property 3 [3]. Let $r,r' \in AR(s,c)$. If $r' \in C(r)$, then $sup(r') \geq sup(r) \wedge conf(r') \geq conf(r)$.

Example 1. Let $T_1 = \{A,B,C,D,E\}$, $T_2 = \{A,B,C,D,E,F\}$, $T_3 = \{A,B,C,D,E,H,I\}$, $T_4 = \{A,B,E\}$ and $T_5 = \{B,C,D,E,H,I\}$ are the only transactions in the database D. Let r: $(B \Rightarrow DE)$. Then, $C(r) = \{B \Rightarrow DE$ (4,80%), $B \Rightarrow D$ (4,80%), $B \Rightarrow E$ (5,100%), $BD \Rightarrow E$ (4,100%), $BE \Rightarrow D$ (4,80%)$\}$. Clearly, the support and confidence of rules in $C(r)$ are not less than the support and confidence of r.

Maximal covering rules (*MCR*) for the set of rules R were defined in [5] as follows: $MCR(R) = \{r \in R | \nexists\ r' \in R, r' \neq r \wedge r \in C(r')\}$. Whatever can be induced from R by the cover operator will be also induced from its subset $MCR(R)$.

Example 2. Let $R = \{B \Rightarrow DE,\ B \Rightarrow D,\ B \Rightarrow E,\ BD \Rightarrow E,\ BE \Rightarrow D,\ B \Rightarrow CDE,\ B \Rightarrow CD,\ B \Rightarrow CE,\ BC \Rightarrow DE\}$ be the set of association rules. This set of rules can be derived from just one maximal covering rule, namely: $MCR(R) = \{B \Rightarrow CDE\}$.

3 Inducing Theory

In this section we recollect after [6] how to induce the knowledge derivable from R. In order to augment the initial knowledge R, it will be used the information on supports of itemsets which is available in R. In the sequel, we assume the supports and confidences of all rules in R are known. By this assumption, the supports of itemsets of these rules as well as the supports of itemsets of the antecedents of these rules are also known. The support of the antecedent of a rule $r \in R$ is equal to

$sup(r) / conf(r)$. Applying this simple observation, the notion of *known itemsets* for R ($KIS(R)$) was defined in [6] as follows: $KIS(R) = \{X \cup Y | X \Rightarrow Y \in R\} \cup \{X | X \Rightarrow Y \in R\}$.

One can easily note that for any itemset X such that there are $Y,Z \in KIS(R)$, $Y \subseteq X \subseteq Z$, the support of X can be estimated as follows: $sup(Y) \geq sup(X) \geq sup(Z)$. The itemsets that can be assessed by employing the knowledge on R are called *derivable itemsets* for R ($DIS(R)$) and are defined as follows: $DIS(R) = \{X | \exists Y,Z \in KIS(R), Y \subseteq X \subseteq Z\}$. Obviously, $DIS(R) \supseteq KIS(R)$. *Pessimistic support* ($pSup$) and *optimistic support* ($oSup$) of an itemset $X \in DIS(R)$ wrt. R are defined as follows: $pSup(X,R) = \max\{sup(Z) | Z \in KIS(R) \wedge X \subseteq Z\}$, $oSup(X,R) = \min\{sup(Y) | Y \in KIS(R) \wedge Y \subseteq X\}$. Clearly, the real support of $X \in DIS(R)$ belongs to $[pSup(X,R), oSup(X,R)]$. In addition, $sup(X) = pSup(X,R) = oSup(X,R)$ for $X \in KIS(R)$.

Knowing $DIS(R)$ one can induce (approximate) rules $X \Rightarrow Y$ provided $X \cup Y \in DIS(R)$ and $X \in DIS(R)$. The *pessimistic confidence* ($pConf$) of induced rules is defined as follows: $pConf(X \Rightarrow Y,R) = pSup(X \cup Y,R) / oSup(X,R)$. The knowledge derivable from R is called *theory* for R ($T(R)$) and defined as follows: $T(R) = \{X \Rightarrow Y | X \cup Y \in DIS(R) \wedge X \in DIS(R)\}$. It is guaranteed for every rule $r \in T(R)$ that its support is not lower than $pSup(r,R)$ and its confidence is not lower than $pConf(r,R)$. By $T(R,s,c)$ we denote $T(R) \cap AR(s,c)$. In particular, $T(R,0,0)$ equals to $T(R)$.

4 Direct Generation of Maximal Covering Rules for Theory

In this section we consider generation of maximal covering rules for the theory $T(R,s,c)$, where R is a rule set, s is a minimum rule support, and c is a minimum rule confidence. Let us start with the property of rules generated from $DIS(R)$:

Property 4. Let $X,Z \in DIS(R)$, $X',Z' \in KIS(R)$, $Z' \supseteq Z \supset X \supseteq X' \neq \emptyset$, $pSup(Z,R) = sup(Z')$ and $oSup(X,R) = sup(X')$.

a) $pSup(X \Rightarrow Z \backslash X,R) = pSup(X' \Rightarrow Z' \backslash X',R) = sup(Z')$,
b) $pConf(X \Rightarrow Z \backslash X,R) = pConf(X' \Rightarrow Z' \backslash X',R) = conf(X' \Rightarrow Z' \backslash X')$,
c) $X \Rightarrow Z \backslash X \in T(R,s,c)$ iff $X' \Rightarrow Z' \backslash X' \in T(R,s,c)$,
d) $X \Rightarrow Z \backslash X \in C(X' \Rightarrow Z' \backslash X')$.

Proof: Ad. b) $pConf(X \Rightarrow Z \backslash X,R) = pSup(X \Rightarrow Z \backslash X,R) / oSup(X,R) =$ /* by Property 4a */
$= pSup(X' \Rightarrow Z' \backslash X',R) / sup(X') = pSup(X' \Rightarrow Z' \backslash X',R) / oSup(X',R)$
$= pConf(X' \Rightarrow Z' \backslash X',R) = sup(X' \Rightarrow Z' \backslash X') / sup(X') = conf(X' \Rightarrow Z' \backslash X')$.

Observations:

O1. It follows from the definition of $DIS(R)$ and $pSup$ that for every Z in $DIS(R)$ there is Z', $Z' \supseteq Z$, in $KIS(R)$ such that $pSup(Z,R) = sup(Z')$. Similarly, by definition of $DIS(R)$ and $oSup$, for every X in $DIS(R)$ there is X', $X' \subseteq X$, in $KIS(R)$ such that $oSup(X,R) = sup(X')$. Hence, and from Property 4 it follows that for every rule in $T(R,s,c)$ built from a derivable itemset with the antecedent being a derivable itemset there is a covering rule in $T(R,s,c)$ built from a known itemset with the antecedent being a known itemset. This implies that no rule built from an itemset in $DIS(R) \backslash KIS(R)$ or having antecedent built from an itemset in $DIS(R) \backslash KIS(R)$ is a maximal covering rule. Thus, generation of candidate MCR can be restricted to generating rules from itemsets in $KIS(R)$ whose antecedents are also built from itemsets in *KIS*(*R*). Since a maximal covering rule is built from a known itemset as well as its antecedent, then its support and confidence are also known.

O2. Property 4d implies that no rule $X \Rightarrow Z \backslash X$ built from $Z \in KIS(R)$ is maximal covering if there is a proper superset $Z' \in KIS(R)$ of Z having the same pessimistic support as Z.

Observations O1 and O2 are used in the *FastGenMCR* algorithm that computes *MCR*(*T*(*R*,*s*,*c*)). Our algorithm is a modification of *FastGenMaxCoveringRules*, we proposed in [6]. The difference between the two algorithm is as follows: *FastGenMaxCoveringRules* builds candidate rules $X \Rightarrow Z/X$ assuming $Z \in KIS(R)$ and $X \in DIS(R)$. *FastGenMCR* generates candidate rules $X \Rightarrow Z/X$ assuming not only Z but also X belongs to *KIS*(*R*). Both algorithms return the same *MCR*.

```
Algorithm. FastGenMCR(known itemsets with support > s: KIS, min. conf.: c);
{ MCR = ∅;
 forall k-itemsets Z ∈ KIS, k ≥ 2, do {
  Z.maxSup = max({Z'.sup| Z⊂Z'∈KIS} ∪ {0});
  if Z.sup ≠ Z.maxSup then {                    // Observation O2
   A1 = {{X}| X∈Z};                             // create 1-item antecedents
   for (i = 1; (Ai ≠ ∅) and (i < k); i++) do {
    forall itemsets X ∈ Ai ∩ KIS do {
     conf = Z.sup / X.sup;
     if conf > c then {
      /* X ⇒ Z\X is an association rule */
      if (Z.maxSup / X.sup ≤ c) then
       /* There is no longer assoc. rule X⇒Z'\X, Z'⊃Z, that covers X⇒Z\X */
       add X ⇒ Z\X to MCR;                      // Property 2 & Observation O1
      /* Antecedents of association rules are not extended */
      Ai = Ai \ {X}; }; };
   Ai+1 = AprioriGen(Ai); }; }; };     // compute (i+1)-item antecedents (see [2])
 return(MCR); }
```

5 Incremental Generation of Maximal Covering Rules for Theory

In this section we consider the issue of updating maximal covering rules for the theory when additional rules are provided. Let *MCR* be the maximal covering rules for *T*(*R*,*s*,*c*), where *R* is an initial rule set, *s* is a minimum required rule support, and *c* is a minimum required rule confidence. Let r: $X \Rightarrow Z \backslash X$ be the provided rule whose support and confidence is known. Then, the set of known itemsets augments by itemsets *X* and *Z*, which can be used for the construction of new association rules. It may happen that some new association rules will be maximal covering. On the other hand, the new association rules may invalidate some of maximal covering rules that were previously found. Hence, the update of maximal covering rules will consist from two steps: 1) generation of new maximal covering rules and 2) validation of old ones. The *IncrGenMCR* algorithm we propose shows how to update maximal covering rules *MCR* for theory *T*(*R*,*s*,*c*) for each new known itemset with sufficiently high support.

```
Algorithm. IncrGenMCR(var maximal covering rules for T(R,s,c): MCR,
var known itemsets with support > s: KIS, new known itemset: X, min. conf.: c);
{ if X∉KIS then {
  /* Step 1: add new maximal covering rules */
  ΔMCR = AddNewMCR(KIS, X, c);
  /* Step 2: remove invalidated maximal covering rules */
  RemoveInvalidatedMCR(MCR, X, c);
  add ΔMCR to MCR;
  add X to KIS;} }
```

5.1 Adding New Maximal Covering Rules

```
Algorithm. AddNewMCR(known itemsets: KIS, new known itemset: X, min. conf.: c);
{ ΔMCR = ∅;
  SuperKIS = {proper supersets of X in KIS};
  SubKIS = {proper subsets of X in KIS};
  X.maxSup = max({Z'.sup| Z'∈SuperKIS } ∪ {0});
  X.minSup = min({Z'.sup | Z'∈SubKIS} ∪ {∞});
/* new itemset X will be used as an antecedent of candidate rules */
/* KIS will be used as generators of candidate rules             */
  forall itemsets Z ∈ SuperKIS do {
  /* update minSup of superset Z of X for future use */
    Z.minSup = min(Z.minSup, X.sup);
    if Z.sup ≠ Z.maxSup then                          // Observation O2
      if IsMCR(X ⇒ Z\X, c) = true then
        add X ⇒ Z\X to ΔMCR; }
/* X will be used as a generator of candidate rules    */
/* KIS will be used as antecedents of candidate rules */
  if X.sup ≠ X.maxSup then                            // Observation O2
    forall itemsets Z ∈ SubKIS, do {
/* update maxSup of subset Z of X for future use */
      Z.maxSup = max(Z.maxSup, X.sup);
      if IsMCR(Z ⇒ X\Z, c) = true then
        add Z ⇒ X\Z to ΔMCR; }
  return(ΔMCR); }
```

Let X be a new known itemset. The *AddNewMCR* algorithm works under the assumption that for every known itemset Z there is kept an additional information on *maxSup* and *minSup*, where *maxSup* is the maximum from the supports of known itemsets that are proper supersets of X and *minSup* is the minimum from the supports of known itemsets that are proper subsets of X. At first, *AddNewMCR* determines subsets and supersets of X and computes *maxSup* and *minSup* for X. Next, each proper superset Z of X is considered as a generator of the candidate rule $X \Rightarrow Z \backslash X$. (According to Observation O2 it makes sense to limit generators to those satisfying the condition $Z.sup \neq Z.maxSup$.) Similarly, each proper subset Z of X is considered as an antecedent of the candidate rule $Z \Rightarrow X \backslash Z$. (According to Observation O2 it makes sense to consider such candidates if $X.sup \neq X.maxSup$.) The *IsMCR* function validates candidate rules as being maximal covering in $T(R,s,c)$ or not.

```
function IsMCR(candidate rule: X ⇒ Z\X, min. conf.: c);
{ conf = Z.sup / X.sup;
  if conf > c then
    /* X ⇒ Z\X is an association rule */
    if (Z.maxSup / X.sup ≤ c)
    /* There is no longer assoc. rule X⇒Z'\X, Z'⊃Z, that covers X⇒Z\X */
    and (Z.sup / X.minSup ≤ c) then
    /*There is no assoc. rule X'⇒Z\X' with antecedent X'⊂X that covers X⇒Z\X */
      return(true);
      /* X ⇒ Z\X is MCR by Property 2 & Observation O1 */
  return(false); }
```

Let $X \Rightarrow Z \backslash X$ be a candidate rule. According to Property 2, the rule is not covered if there is no association rule of the form $X \Rightarrow Z' \backslash X$, $Z' \supset Z$, and there is no association rule of the form $X' \Rightarrow Z \backslash X'$, $X \supset X'$. Clearly, if $Z.maxSup / X.sup \leq c$ then there is no association rule of the form $X \Rightarrow Z' \backslash X$, $Z' \supset Z$, and if $Z.sup / X.minSup \leq c$ then there is no association rule of the form $X' \Rightarrow Z \backslash X'$, $X \supset X'$. In such a case, the candidate rule $X \Rightarrow Z \backslash X$ is not covered by any association rules and hence it is maximal covering.

5.2 Removing Invalidated Maximal Covering Rules

```
Algorithm. RemoveInvalidatedMCR(var maximal covering rules: MCR,
new known itemset: X, min. conf.: c);
{ forall rules Y⇒Z\Y ∈ MCR, where the rule generator Z⊂X do {
   if (X.sup / Y.sup > c) then
   /* Y⇒X\Y is an association rule and covers Y⇒Z\Y - see Property 2*/
     remove Y⇒Z\Y from MCR; }
  forall rules Y⇒Z\Y ∈ MCR, where the rule antecedent Y⊃X do {
   if (Z.sup / X.sup > c) then
   /* X⇒Z\X is an association rule and covers Y⇒Z\Y - see Property 2*/
     remove Y⇒Z\Y from MCR; } }
```

The *RemoveInvalidatedMCR* algorithm applies Property 2 in order to eliminate all rules from *MCR* covered by association rules which can be built from *X*.

5.3 Computational Complexity

The two steps of updating of maximal covering rules: generating of new maximal rules (*AddNewMCR*) and validating old ones (*RemoveInvalidatedMCR*) can be performed independently. The former one is linear wrt. the number of known itemsets *KIS* and the latter one is linear wrt. the number of maximal covering rules for $T(R,s,c)$.

6 Conclusions

In the paper we proved the important property of maximal covering rules for the theory that states that both generators and antecedents of maximal covering rules can be built only from known itemsets. Hence, we know that support and confidence of maximal covering rules can be precisely computed. In order to find maximal covering rules for theory we proposed *FastGenMCR*. In addition, we proposed an incremental approach to computing maximal covering rules when the information on new rules/itemsets is provided. The offered *IncrGenMCR* algorithm is linear wrt. the number of known itemsets and the number of maximal covering rules.

References

1. Agrawal, R., Imielinski, T., Swami, A.: Mining Associations Rules between Sets of Items in Large Databases. In: Proc. of the ACM SIGMOD Conference on Management of Data. Washington, D.C. (1993) 207-216
2. Agrawal, R., Mannila, H., Srikant, R., Toivonen, H., Verkamo, A.I.: Fast Discovery of Association Rules. In: Fayyad, U.M., Piatetsky-Shapiro, G., Smyth, P., Uthurusamy, R. (eds.): Advances in Knowledge Discovery and Data Mining. AAAI, Menlo Park, California (1996) 307-328
3. Kryszkiewicz, M.: Representative Association Rules. In: Proc. of PAKDD '98. Melbourne, Australia. LNAI 1394. Springer-Verlag (1998) 198-209
4. Kryszkiewicz, M.: Representative Association Rules and Minimum Condition Maximum Consequence Association Rules. In: Proc. of PKDD '98. Nantes, France. LNAI 1510. Springer-Verlag (1998) 361-369
5. Kryszkiewicz, M.: Mining with Cover and Extension Operators. In: Proc. of PKDD '00. Lyon, France. LNAI 1910. Springer-Verlag (2000) 476-482
6. Kryszkiewicz, M.: Inducing Theory for the Rule Set. In: Proc. of RSCTC '00. Banff, Canada (2000) 353-360. To appear as a Springer-Verlag LNAI volume

Towards Efficient Data Re-mining (DRM)

Jiming Liu[1] and Jian Yin[2]

[1]Department of Computer Science, Hong Kong Baptist University
Kowloon Tong, Hong Kong jiming@comp.hkbu.edu.hk
[2]Department of Computer Science, Zhongshan University
Guangzhou, 510275, P. R. China issjyin@zsu.edu.cn

Abstract. The problem that we tackle here is a practical one: When users interactively mine association rules, it is often the case that they have to continuously tune two thresholds: minimum support and minimum confidence, which describe the users' changing requirements. In this paper, we present an efficient data re-mining (DRM) technique for updating previously discovered association rules in light of threshold changes.

1 Introduction

Various algorithms have been proposed [1,2,4,6] to discover frequent item-sets. Generally speaking, these algorithms first construct a candidate set of frequent item-sets based on certain heuristics, and then discover the subset that indeed contains frequent item-sets. This process can be done iteratively in the sense that the frequent item-sets discovered at one iteration will be used as the basis for generating the candidate set for the next iteration. For example, in [2], at the kth iteration, all frequent item-sets containing k items, referred to as frequent k-item-sets, are generated. In the next iteration, to construct a candidate set of frequent (k+1)-item-sets, a heuristic is used to expand some frequent k-item-sets into a (k+1)-item-set, if certain constraints are satisfied.

Among all the algorithms proposed, Apriori (and its variants) [2] and DHP [3] algorithms are most commonly applied. They both run a number of iterations and compute the frequent item-sets of the same size at each iteration, starting from the size-one item-sets. At each iteration, they first construct a set of candidate item-sets and then scan the database to count the number of transactions that contain each candidate set. The key to optimization lies in the techniques used to create the candidate sets. The smaller the number of candidate sets is, the faster the algorithms would be.

However, very little work has been done on the second problem mentioned earlier. A method of handling incremental database updates for the rules discovered by the generalization-based approach was briefly discussed in [5]. As related to this problem, Lee and Cheung have done some work [7], which focuses primarily on how to update association rules when a database is incrementally changed. As in real-world applications, users are often unsure about their requirements on the minimum support and confidence in the first place. This can be due to the lack of knowledge about the application domains or the outcomes resulting from different threshold settings. As a result,

D. Cheung, G.J. Williams, and Q. Li (Eds.): PAKDD 2001, LNAI 2035, pp. 406-412, 2001.

they may be repeatedly unsatisfied with the association rules discovered, and hence need to re-execute the mining procedure many times with varied thresholds. In the cases where large databases are involved, this could be a time-consuming, trial-and-error process, since all the computation done initially in finding the old frequent item-sets is wasted and all frequent item-sets have to be re-computed again from scratch. In order to deal with this situation, it is both desirable and imperative to develop an efficient means for **re-mining a database under different thresholds** in order to obtain an acceptable set of association rules. In this paper, we will explicitly address this problem and present an efficient algorithm, called Posteriori, for computing the frequent item-sets under the varied thresholds.

2 Problem Statement

2.1 Mining Association Rules

Let $I=\{i_1,i_2,\ldots,i_m\}$ be a set of literals, called items. Let D be a set of transactions, where each transaction T is a set of items such that $T\subseteq I$. Associated with each transaction is a unique identifier, called its TID. We say that a transaction T contains X, a set of some items in I, if $X\subseteq T$. An association rule is an implication of the form $X\Rightarrow Y$, where $X\subset I$, $Y\subset I$, and $X\cap Y=\varnothing$. The rule $X\Rightarrow Y$ holds in the transaction set D with confidence c if c% of transactions in D that contain X also contain Y. The rule $X\Rightarrow Y$ has support s in the transaction set D if s% of transaction in D contain $X\cup Y$.

2.2 Re-mining Association Rules

Let L be the set of frequent item-sets in D, s be the minimum support, and $|D|$ be the number of transactions in D. Assume that for each $X\in L$, its support count, X.support, which is the number of transactions in D containing X, is available.

After users have found some association rules, they may be unsatisfied with the results and want to try out new results with certain changes on thresholds, such as minsup from s to s′.

Thus, the essence of the problem of re-mining association rules is to find the set L′ of frequent item-sets under the new thresholds. Note that a frequent item-set in L may not be a frequent item-set in L′. On the other hand, an item-set X not in L may become a frequent item-set in L′.

3 Algorithm Posteriori

The following notations are used in the rest of the paper. L_k is the set of all size-k frequent items in D under the support of s%, and L_k' is the set of all frequent k-items in D under the support of s′%. C_k is the set of of size-k candidate sets in the k-th iteration of Posteriori. Moreover, X.support represents the support counts of an item-set X in D.

When minsup is changed, two cases may happen:

1. s′>s, in this case, some original frequent item-sets will become losers, i.e., they are no longer frequent under the new threshold minsup.
2. s′<s, in this case, some original frequent item-sets will become winners, i.e., they are included in the frequent L′ under the new threshold minsup.

In the first case, the updating of frequent items is simple and intuitive. Using the item support, the algorithm can be stated as follows:

```
Algorithm Posteriori_A:
Input: (1) L_k: the set of all frequent k-items in D, where k=1,...,r
       (2) s': where s'>s
Output: L ': the set of all frequent item-sets in D.
  L'=∅  /*L': initialized */
  for (k=1; k<r; k++) do begin
  L_k'={X∈L_k|X.support≥ s'} /* put winners in L_k' */
  L'=L'∪L_k'
  return L'
```

The correctness of algorithm `Posteriori_A` can be guaranteed by the following lemma:

Lemma 1: A k-item-set X is in the frequent item-set L_k' of database D under s′ only if X is in the frequent item-set L_k under s, where s′>s

Proof. Suppose that X is not in the frequent item-set L_k, then X.support < s ×|D|, since s′>s, so X.support< s′ ×|D|. That is, X is not in the frequent item-set L_k'.

Now, let us concentrate on the second case mentioned above. We will propose an efficient algorithm, called `Posteriori_B`. The framework of `Posteriori_B` is similar to that of Apriori. It contains a number of iterations. The iteration starts from the size-one item-sets, and at each iteration, all the frequent item-sets of the same size are found. Moreover, the candidate sets at each iteration are generated based on the frequent item-sets found at the previous iteration. The features of `Posteriori_B` that distinguish it from Apriori are listed as follows:

1. At each iteration, the size-k frequent item-sets in L are updated against the increment of support to add the winners.
2. While generating the candidates, a set of candidate sets, C_k, is divided into three parts. Each of them is generated separately.

These features combined together form the core in the design of Posteriori-B and make Posteriori a much faster algorithm in comparison with the re-running of Apriori on the database.

The following is a detailed description of the algorithm `Posteriori_B`. The first iteration of `Posteriori_B` is described, which is followed by the discussion of the remaining iterations.

3.1 First Iteration: Finding Size-One Winners and Generating Candidate Sets

The following properties are useful in the derivation of the frequent 1-item-sets for the updated s′.

Lemma 2: A k-item-set X in the frequent item-set L_k of database D under s is also in the frequent item-set L_k under s′, where s′<s.

Proof. Since X is in the frequent item-set L_k, X.support ≥ s×D, so X.support>s′×D. That is, X is in the frequent item-set L_k'.

Based on lemma 2, the finding of frequent 1-item-set L_1' is only to scan the item-set that is not in L_1. If we store the result C_1 of algorithm Apriori, we need only scan C_1-L_1.

So, the first iteration is very simple, that is to scan C_1-L_1 by checking the condition X.support > s′×D. As all new winners are found, we call them l_1, thus $L_1'=L_1 \cup l_1$.

3.2 Second Iteration and Beyond: Dividing Candidate Sets and Finding Re-maining Winners

The following properties are useful in the derivation of the frequent k-item-sets (where k>1) for the updated s′.

Lemma 3: All the subsets of a frequent item-set must also be frequent.

Proof. This is the basic property of frequent item-sets, as proven in [2].

Lemma 4: A k-item-set $\{X_1, X_2,\ldots, X_k\}$ not in the original frequent k-item-set L_k can become a winner under the updated s′, where s′<s, only if $\{X_1, X_2,\ldots, X_m\}$.support ≥ s′×D.

Proof. The lemma is derivable from the definitions of minimum support and frequent k-item-set.

At the first iteration, as shown above, all the frequent 1-items are divided into two non-intersecting sets: L_1 and l_1. Based on Lemma 3, all the subsets of a frequent item-set must also be frequent, so, any frequent 1-item-set corresponds to a single item of k-item-sets must be an element of L_1 or l_1. According to this, we can divide all frequent k-item-sets under the new minsup s′ into three classes:

1: for each frequent k-item-set $\{X_1, X_2,\ldots, X_k\}$, $\forall i(1 \le i \le k)$, $\{X_i\} \in L_1$;

2: for each frequent k-item-set $\{X_1, X_2,\ldots, X_k\}$, $\forall i(1 \le i \le k)$, $\{X_i\} \in l_1$;

3: for each frequent k-item-set $\{X_1, X_2,\ldots, X_k\}$, $\forall i(1 \le i \le k)$, we have two non-empty subsets x_1 and x_2, $x_1 \cup x_2=\{X_1, X_2,\ldots, X_k\}$, $x_1 \cap x_2=\varnothing$, and $x_1 \in L_1$, $x_2 \in l_1$.

So, we have obtained three mutually intersecting subsets, and we call them $L_k[1]$, $L_k[2]$, and $L_k[3]$, respectively. Moreover, we can filter original L_k from $L_k[1]$, and call remaining set l_k, then we have $L_k[1]=L_k \cup l_k$, and also we have $L_k'=L_k[1] \cup L_k[2] \cup L_k[3]$, $L_k[1] \cap L_k[2]=\varnothing$, $L_k[1] \cap L_k[3]=\varnothing$, and $L_k[2] \cap L_k[3]=\varnothing$.

Based on the above-mentioned division, we can decompose the generation of candidate sets into three parts, i.e., generating $C_k[1]$, $C_k[2]$, and $C_k[3]$, respectively.

For $C_k[1]$ and $C_k[2]$, this is simple; we just apply the Apriori_gen function in algorithm Apriori:

$C_k[1]$=Apriori_gen($L_{k-1}[1]$)-L_k

$C_k[2]$=Apriori_gen($L_{k-1}[2]$)

If we store the result of C_k, then $C_k[1]=C_k-L_k$. The key problem is how to generate $C_k[3]$. From the above discussion, we know that items in $C_k[3]$ are composed of items in $L_i[1]$ and $L_{k-i}[2]$. Thus, we can modify the Apriori_gen function into Posteriori_gen function. The Posteriori_gen function takes as argument $L_i[1]$ and $L_{k-i}[2]$, the set of all 1 and 2-classes frequent i-item-sets and (k-i)-item-sets. It returns a superset of the set of all frequent 3-class k-item-sets. The function works as follows: First, in the join step, we join $L_i[1]$ and $L_{k-i}[2]$:

insert into $C_k[3]$
select p.item$_1$, p.item$_2$,…, p.item$_i$, q.item$_1$,q.item$_2$,…, q.item$_{k-i}$
from $L_i[1]$,$L_{k-i}[2]$q

Next, in the prune step, we delete all item-sets $c\in C_k[3]$ such that some (k-1)-subset of c is not in L_{k-1}' :

for all item-sets $c\in C_k[3]$ do
  for all (k-1)-subsets s of c do
    if ($s\notin L_{k-1}'$) then
      delete c from $C_k[3]$

Correctness

As far as the correctness of our algorithm is concerned, we need to show that $C_k'\supseteq L_k'$. Since $C_k'=C_k[1]\cup C_k[2]\cup C_k[3]$, $L_k'=L_k[1]\cup L_k[2]\cup L_k[3]$, and clearly, $C_k[1]\supseteq L_k[1]$,$C_k[2]\supseteq L_k[2]$ (as proven in [2]), we only need to show $C_k[3]\supseteq L_k[3]$. As defined earlier, $C_k[3]$ is composed of two nonempty subsets x_1 and x_2, $x_1\cup x_2=\{X_1, X_2,\ldots, X_k\}$, $x_1\cap x_2=\emptyset$, and $x_1\in L_1$, $x_2\in l_1$. Because any subset of a frequent item-set must also be a frequent item-set, if we extend each item-set in x_1, x_2 with all possible items and then delete all those whose (k-1)-subsets are not in L_{k-1}', we would be left with a superset of the item-sets in $L_k[3]$.

The join is equivalent to extending x_1, x_2 with each item in the database. Thus, after the join step, we can have $C_k[3]\supseteq L_k[3]$. By similar reasoning, for the prune step, we can delete from $C_k[3]$ all item-sets whose (k-1)-subsets are not in L_{k-1}', but do not delete any item-set that could be in $C_k[3]$.

3.3 The `Posteriori_B` Algorithm

Based on the above discussions, we can now formally state the `Posteriori_B` algorithm as follows:

Algorithm `Posteriori_B`: ***An efficient algorithm for re-mining of association rules upon support changes.***
Input: (1) L_k: the set of all frequent k-items in D under s, where k=1,…,r
(2) s′: where s′<s

```
    (3) C_k: the set of all candidate k-items in D under s, where k=1,...,r
Output: L': the set of all frequent item-set in D under s'.
l_1={new frequent 1-item-sets}
L_1'= L_1 ∪ l_1
for (k=2; L_{k-1} '≠ ∅;k++) do begin
  C_k[1]=C_k- L_k              /* 1-class k candidates */
  C_k[2]=Apriori_gen(L_{k-2}[2])    /* 2-class k candidates */
  C_k[3]=∅
  for (j=1; j≤k-1;j++) do
    C_k[3]=C_k[3]∪Posteriori_gen(L_i[1],L_{k-i}[2])  /* 3-class k candidates */
  for all transactions t∈D do begin
    C_t[1] =subset(C_k[1],t)  /* candidates contained in t */
    C_t[2]=subset(C_k[2],t)
    C_t[3]=subset(C_k[3],,t)
    for all candidates c∈C_t[1]∪C_t[2]∪C_t[3] do
      c.support++
  end
  l_k={c∈ C_k[1]|c.support≥s'}
  L_k[1]=L_k ∪ l_k
  L_k[2]={c∈C_k[2]|c.support≥s'}
  L_k[3]={c∈C_k[3]|c.support≥s'}
  L_k '= L_k[1]∪L_k[2]∪L_k[3]
end
Answer=∪_k L_k '
```

4 Concluding Remarks

In this paper, we presented an efficient data re-mining (DRM) method for discovering association rules. In order to assess the efficiency and effectiveness of the Posteriori algorithm, we have conducted several experiments and compared its performance with that of Apriori. Our experiments were performed on a Pentium III PC. The obtained results have shown that Posteriori is much faster than the presently most popular mining algorithm. Furthermore, Posteriori performs 2~6 times faster than Apriori for a moderate size database of 100,000 transactions.

References

1. Agrawal, R., Imielinski, T., and Swami, T.: Mining association rules between sets of items in large databases. Proceedings of the ACM-SIGMOD International Conference on Management of Data (1993) 207-216.
2. Agrawal, R. and Srikant, R.: Fast algorithm for mining association rules. Proceedings of the International Conference on Very Large Data Bases (1994) 487-499.
3. Srikant, R. and Agrawal, R.: Mining generalized association rules. Proceedings of the International Conference on Very Large Data Bases (1995) 407-419.
4. Park, J. S., Chen, M., and Yu, P. S.: An effective hash-based algorithm for mining association rules. Proceedings of the ACM-SIGMOD International Conference Management of Data (1995) 175-186.
5. Han, J., Cai, Y., and Cercone, N.: Data-driven discovery of quantitative rules in relational databases. IEEE Transactions on Knowledge and Data Engineering, 5(1):29-40, (1993).

6. Pei, J., Han, J., and Mao, R.: CLOSET: An efficient algorithm for mining frequent closed itemsets. Proceedings of the ACM-SIGMOD International Workshop on Data Mining and Knowledge Discovery (DMKD'00) (2000).
7. Lee, S. D. and Cheung, D. W.: Maintenance of discovered association rules: When to update? Proceedings of the ACM-SIGMOD Workshop on Data Mining and Knowledge Discovery (DMKD'97) (1997).

Data Allocation Algorithm for Parallel Association Rule Discovery

Anna M. Manning[1] and John A. Keane[2]

[1] Department of Computer Science, University of Manchester,M13 9PL, UK
anna@cs.man.ac.uk
[2] Department of Computation, UMIST, Manchester, M60 1QD
jak@co.umist.ac.uk

Abstract. Association rule discovery techniques have gradually been adapt-ed to parallel systems in order to take advantage of the higher speed and greater storage capacity that they offer. The transition to a distributed memory system requires the partitioning of the database among the processors, a procedure that is generally carried out indiscriminately. However, for some techniques the nature of the database partitioning can have a pronounced impact on execution time and attention will be focused on one such algorithm, Fast Parallel Mining (FPM). A new algorithm, Data Allocation Algorithm (DAA), is presented that uses Principal Component Analysis to improve the data distribution prior to FPM.

Keywords: Data Mining, Association Rules, Parallel Algorithms, Data Partitioning, Candidate Sets

1 Introduction

The discovery of association rules [1] is an important example of data mining and we consider one particular parallel algorithm, Fast Parallel Mining (FPM) [4]. The performance of FPM improves as the inter-processor record pattern variation (data skewness, defined formally in [4]) rises as this can reduce the number of duplicate database operations that are necessary. This observation provides the opportunity for improving the performance of FPM by determining how the data should be distributed over the processors prior to its application. To achieve this, a method that provides good predictions of attribute patterns within the database must be employed. Our work investigates the application of Principal Component Analysis (PCA) [6] to guide the allocation of records in equal numbers to a set of processors in order to maximise variance between itemset supports [1,2] at each before applying FPM.

2 Proposed Method for Record Distribution

Levels of variation between itemset support counts at any one processor are particularly influential on the numbers of candidate itemsets generated [4] and

D. Cheung, G.J. Williams, and Q. Li (Eds.): PAKDD 2001, LNAI 2035, pp. 413–420, 2001.

provide the motivation for using PCA. Although the purpose of this work was to investigate the feasibility of record redistribution prior to FPM, such a technique should clearly add as little computation as possible to that of FPM. PCA [6] is well recognised as a preprocessing tool in data analysis and its computational cost is linear with the number of records in the database. It has been used efficiently to handle very large databases [7] and has been adapted to a distributed memory machine, for example with PARPACK, [7].

Data Allocation Algorithm (DAA) uses PCA to redistribute a given database as follows:

1. Find (or use statistical sampling to estimate) the variance/covariance matrix for the attribute column means in the database (referred to below as Z).
2. Apply PCA to Z, giving a matrix of weights. For example, Table 1 shows the Principal Components (PCs) for the 10 attributes of a binary database. The PCs are represented by the weights for each attribute, shown in descending order in the table columns. Each of these PCs is itself given a weight in the form of its eigenvalue - the 10 eigenvalues for the PCs in Table 1 are: 0.1751, 0.2007, 0.2707, 0.2182, 0.2374, 0.2480, 0.2508, 0.2464, 0.2407, 0.2291.

Table 1. Column PCs for sample data

	PC1	PC2	PC3	PC4	PC5	PC6	PC7	PC8	PC9	PC10
A	0.896	0.266	0.307	-0.122	-0.076	-0.079	-0.042	0.056	-0.020	-0.011
B	-0.405	0.808	0.348	-0.207	-0.113	-0.061	-0.036	0.027	-0.006	-0.034
C	-0.128	-0.459	0.378	-0.763	-0.130	-0.124	-0.070	0.063	-0.049	-0.068
D	-0.075	-0.148	0.352	0.317	-0.449	-0.080	0.029	0.022	-0.239	0.694
E	-0.047	-0.084	0.319	0.154	0.403	-0.350	-0.228	-0.101	0.687	0.214
F	-0.011	-0.049	0.260	0.023	0.056	0.807	0.058	0.417	0.306	0.066
G	0.003	0.004	0.018	-0.041	0.020	-0.247	0.928	0.193	0.195	0.017
H	-0.017	0.044	-0.267	-0.017	0.020	-0.332	-0.257	0.860	-0.038	0.096
I	0.049	0.056	-0.373	-0.152	-0.693	0.013	-0.082	-0.102	0.578	0.034
J	0.075	0.174	-0.373	-0.458	0.341	0.152	0.045	-0.145	-0.037	0.672

3. Choose the required number of processors, S, where $S \leq$ #attributes. The PCs are arranged in descending order according to the value of their eigenvalue and each processor is associated with one of the 'strongest' S PCs. If S=4 then the PCs to be considered from Table 1 (in descending order of eigenvalue size) would be 3, 7, 6 and 8.
4. Each PC will contain a group of attributes that are particularly influential in its construction; a number of methods can be used for their identification [6] and we have found graphical approaches the most useful. In order to work with all four PCs it is necessary to reach a common number of prominent attributes and the average is taken over all PCs, giving 6 in this case.

5. Each of the S processors is associated with one of the 'strongest' PCs, i.e. those with the largest eigenvalues. In order to maximise variance the prominent attributes of each PC are used to write rules that determine which records will be located at which processor. For example, the 'strongest' 6 attributes for processor 1 are the 3rd, 10th, 9th, 4th, 2nd and 5th. Intra-processor attribute variance can be maximised by concentrating records containing these attributes at one processor while at the same time attempting to eliminate the support of one or more others. The balance between attribute supports to be concentrated and those to be eliminated depends on the level of attribute supports and covariances; with high supports and low covariance the ability to eliminate records containing groups of attributes from a processor would be hindered. The choice of attributes for elimination is also an issue - the PCA process tends to be disturbed if an attribute with a relatively high weight is chosen but little impact is generally made if an attribute with a very low weight is used. Trials have shown that the prominent attribute with the lowest weight gives the best results. The identified prominent attributes for each processor are used to construct rules for record acceptance. Assuming high attribute supports and low covariance in the above example it is unlikely that the support of more than one attribute could be eliminated from each processor. The rules for processor 1 (using the 6 prominent attributes) would be as follows:
   a) add any record with a 1 for attribute 3 and a 0 for attribute 5
   b) add any record with a 1 for attribute 10 and a 0 for attribute 5
   c) add any record with a 1 for attribute 9 and a 0 for attribute 5
   d) add any record with a 1 for attribute 4 and a 0 for attribute 5
   e) add any record with a 1 for attribute 2 and a 0 for attribute 5
   f) add any record with a 0 for attribute 5
   g) add any record

   The rules for the other 3 processors are constructed in an identical manner. Sometimes clashes will occur as to which 1-itemset supports are to be eliminated; if this is the case the 'weaker' PCs are always given priority over the 'stronger'.
6. Records are taken one at a time from their centralised database and attempts are made to match them against the rules at each processor. The processor attached to the weakest of the S PCs is considered first (in this case processor 4) in order to improve load balancing. If a record satisfies rule 1 for processor 4 then it is stored at processor 4. If not, the first rule for processor 3 is considered and the record is placed at processor 3 if there is a match. Otherwise the first rule for processor 2 is considered. Once the first rules for each processor have been tried with no match the second rules are selected in the same order. The rules are chosen in rotation in this manner until there are no records left to distribute. As the last rule for each processor is always to 'add any record' each record will always have a destination. When a processor becomes full (i.e. reaches its quota of records) its rules are withdrawn from the matching process and only those from processors with space still to fill are considered.

7. During the first pass of FPM records can be selected from the centralised database and a destination processor determined for them by applying DAA. Once this is complete the support contribution of the record for that processor can be determined before the record is placed in its new location. At the end of pass 1 the data will have been redistributed and the support counts for the candidate 1-itemsets at each processor accumulated. The FPM algorithm can then proceed as before.

3 Data Preparation

With the aim of predicting the outcome of applying DAA, data was generated with two fixed parameters (data sparsity and mean-skewness), both of which can be estimated without pre-empting the results of FPM. Data sparsity refers to the percentage of 0s in the database - a dense database is a term for one with low sparsity. Mean-skewness is calculated by first measuring attribute column means over four equal-sized data partitions (as if the data had been divided in equal portions over four processors) and then applying the skewness metric in [4].

A procedure for generating datasets of fixed sparsity and mean-skewness was developed in which each attribute of each record was generated using a series of probabilities of being either 0 or 1. These probabilities were adjusted until the required levels of sparsity and mean-skewness were achieved. The datasets were given the label '(mean-skewness)_(sparsity)', e.g. 0.0_80 data refers to data with mean-skewness 0.0 and sparsity 80%. DAA was applied to each of these datasets. The highest number of candidates generated at any one processor was recorded and candidate set numbers were measured according to a selection of 10 support thresholds for each dataset

For comparison purposes each dataset was also divided across the processors in equal sized chunks (without DAA) so as to create what will be referred to as the *raw data* distributions. FPM was then applied and the candidate sets recorded in the same manner. DAA was kept separate from FPM in these experiments so that the performance of FPM on the two groups of datasets could be measured directly.

4 Results

The performance studies were carried out on an IBM SP2 with 2 'high' nodes running the AIX operation system (version 4.1). Each node contained 4 CPUs, a POWERPC 604 processor, a clock speed of 100 MHz, memory of 256MB and a 7133-500 DASD disk (5 4.5 GB drives). The Parallel Virtual Machine (PVM) system was used to view nodes as a single parallel virtual machine [5] and FPM was written in JAVA and JPVM.

The number of passes of FPM over the database vary from 2 for datasets with 85% sparsity to 10 for those with 25% and it is not easy to compare the effect of DAA. The second pass often has the highest computational cost of all passes [3] and it is therefore particularly important to focus attention on the effect of DAA over the first two passes of FPM. Execution times were taken for the first two passes over 10 support counts across 4 processors and the average percentage reduction in execution time for each pass is shown in Table 2. The average results hide significant variation; for example, improvements of up to 43.3% are made for the 0.0_80 data and up to 16.9% for the 0.0_50 data.

Table 2. Average percentage reduction in execution time

S	y												
	85%	80%	75%	70%	65%	60%	55%	50%	45%	40%	35%	30%	25%
0.0	14.6	13.5	10.9	11.2	7.4	7.8	4.2	6.2	5.2	3.0	5.1	1.6	1.5
0.02	5.6	6.2	6.7	2.8	5.2	0.0	0.0	0.0	0.0	0.0	0.0	0.0	0.0
0.04	4.2	6.2	6.0	5.8	2.5	0.0	0.0	0.0	0.0	0.0	0.0	0.0	0.0
0.06	6.2	7.5	8.2	5.9	9.3	0.0	0.0	0.0	0.0	0.0	0.0	0.0	0.0
0.08	8.0	8.6	8.7	5.9	0.7	7.0	0.0	0.0	0.0	0.0	0.0	0.0	0.0
1.0	4.2	11.2	8.5	7.4	9.5	4.4	0.0	0.0	0.0	0.0	0.0	0.0	0.0
2.0	4.0	4.1	12.3	13.8	12.8	0.0	0.0	0.0	0.0	0.0	0.0	0.0	0.0
3.0	11.2	10.1	12.7	8.3	0.0	0.0	0.0	0.0	0.0	0.0	0.0	0.0	0.0

Three main observations can be made:

1. The higher the sparsity of the dataset the better DAA performs. Lower sparsity results in higher covariance levels between attributes which hinders the ability of DAA to distribute records.
2. The performance of DAA deteriorates after a sparsity level of 25% and this relates to the rule construction of DAA. Each set of rules aims to maximise the occurrence of a set of attributes within a specific data partition and to minimise the occurrence of one or more others. No direct attention is given to any other attribute, the behaviour being left to the PCA mechanism. An attribute that is unregulated at one processor but is maximised at another will only have a negative impact on DAA if it has significantly more support than is possible to store at one data partition. Average levels of support can be estimated by considering the sparsity level of the data and the number of processors. For example, with 25% sparsity and 4 processors the average support level will be equal to the number of records at any given data partition - if data sparsity rises higher than this level then DAA will have increasing difficulty in controlling the support count patterns across the processors.
3. The performance for DAA within each fixed band of sparsity varies considerably. This can be explained by looking at the means and the variance/covariance matrix of the dataset in question. If a dataset has a few attributes with relatively high support which have relatively high covariance levels between each other then this will severely impede the ability of DAA to distribute the data efficiently. The execution results for the first line

of Table 2 are particularly high and each dataset has a very uniform variance/covariance matrix - if the relationship between attributes is uniform it is far easier for DAA to place records so that itemset generation is balanced across the processors.

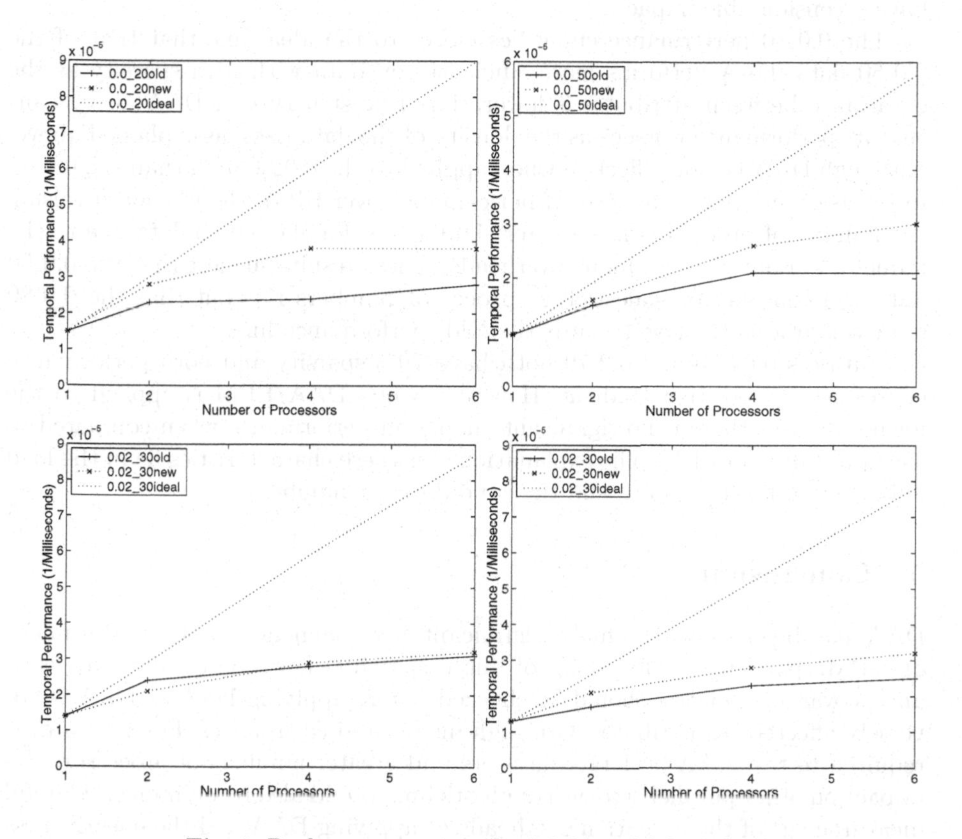

Fig. 1. Performance curves for 0.0_80 and 0.0_50 data

Execution time was also measured over 2 and 6 processors and Figure 1 shows performance curves for datasets 0.0_80, 0.0_50, 0.02_70 and 0.2_70. Performance using p processors is defined as $1/T_p$ where T_p is the time to execute the parallel code with p processors. A reasonable measure of good performance is one where performance increases in accordance with the number of processors being used, i.e. p/T_1, and the straight line in Figure 1 represents the (naive) ideal situation.

DAA/FPM outperforms FPM for the 0.0_50 and the 0.2_70 data and scales well as the number of processors rise. The time saved by scanning database partitions in parallel has more impact as the datasets become denser and more heavily populated with 1s. With the 0.0_80 data no impact is made by raising the number of processors from 4 to 6 whereas the performance of the 0.0_50 data scales well for all processor levels considered.

The benefits of faster data scanning need to be weighed up against the extra cost in message passing brought about by the addition of further processors. Messages for FPM are $O(n^2)$, where n is the number of processors. If very little improvement is gained in the performance of data scanning by partitioning the data over larger numbers of processors the increased numbers of messages will have a considerable impact.

The 0.0_80 performance curve lies closer to the ideal line that that of the 0.0_50 data. DAA performs more efficiently on data with high sparsity as the covariance between attributes is lower. The rule structure of DAA also means that its performance worsens as the density of the data rises, as explained above. Although DAA is more effective when applied to the 0.0_80 data than the 0.0_50 in terms of increasing the level of performance over FPM, the impact of raising the number of processors has greater significance for the 0.0_50 data due to the reduction in database scanning overheads. These results suggest that the 0.0_50 data will continue to scale well as processor numbers rise but that the 0.0_80 data will gradually move towards the 'old' performance line.

Datasets 0.02_70 and 0.2_70 both have 70% sparsity and both performance curves have a positive gradient. However, when DAA/FPM is applied to the former dataset there is no significant change in performance when compared to the application of FPM. The explanations for these characteristics lie in the load balancing of both the data and the candidate set numbers.

5 Conclusion

DAA has been shown to make significant improvements to the performance of FPM, particularly for data of high sparsity. For dense data the variance/covariance matrix should be checked before applying DAA as DAA is adversely affected by attributes with high means and covariances. Further work is required to test DAA on larger datasets and greater numbers of processors. Its impact on other parallel association algorithms would also be of interest. Careful measurement of the execution overheads of applying DAA and the effectiveness of sampling for PCA need to be rigorously investigated.

References

1. R. Agrawal, T. Imielinski, and A. Swami. Mining association rules between sets of items in large databases. In *proceedings of the 1993 International Conference on Management of Data (SIGMOD 93)*, pages 207-216, May 1993.
2. R. Agrawal, H. Mannila, R. Srikant, H.Toivonen, and A. Verkamo. Fast discovery of association rules. In *Advances in Knowledge Discovery and Data Mining, U. Fayyad, G. Piatetsky-Shapiro, P. Smyth, R. Uthurusamy, Eds. The AAAI Press, Menlo Park*, pages 307-328, 1996.
3. D. W. Cheung, K. Hu, and S. Xia. Asynchronous parallel algorithm for mining asso- ciation rules on a shared-memory multi-processors'. In *proceedings of the 10th Annual ACM Symposium on Parallel Algorithms and Architectures (SPAA)*, June 1998.

4. D. W. Cheung and Y. Xiao. Effect of data skewness in parallel mining of association rules. In *proceedings of the 2nd Pacific Asia Conference on Knowledge Discovery and Data Mining, (PAKDD-98), Melbourne, Australia,* pages 48-60, April 1998.
5. A. Geist, A. Beguelin, J. Dongard, W. Jiang, R. Manchek, and V. Sunderam. *PVM: Parallel Virtual Machine.* MIT Press, 1994.
6. I. Joliffe. *Principal Component Analysis.* New York: Springer Verlag, 1986.
7. R. B. Lehoucq and D. C. Sorensen. Deflation techniques for an implicitly restarted Arnoldi iteration. *SIAM Journal on Matrix Analysis and Applications,* 17(4):789-821, 1996.

Direct Domain Knowledge Inclusion in the PA3 Rule Induction Algorithm

Pedro de Almeida

CISUC – Centro de Informática e Sistemas da Universidade de Coimbra,
Polo II da Universidade de Coimbra, 3030 Coimbra, Portugal
Physics Department, Universidade da Beira Interior, 6200 Covilhã, Portugal
nop00997@mail.telepac.pt

Abstract. Inclusion of domain knowledge in a process of knowledge discovery in databases is a complex but very important part of successful knowledge discovery solutions. In real-life data mining development, non-structured domain knowledge involvement in the data preparation phase and in the final interpretation/evaluation phase tends to dominate. This paper presents an experiment of direct domain knowledge integration in the algorithm that will search for interesting patterns in the data. In the context of stock market prediction work, a recent rule induction algorithm, PA3, was adapted to include domain theories directly in the internal rule development. Tests performed over several Portuguese stocks show a significant increase in prediction performance over the same process using the standard version of PA3. We believe that a similar methodology can be applied to other symbolic induction algorithms and in other working domains to improve the efficiency of prediction (or classification) in knowledge-intensive data mining tasks.

1 Introduction

In most cases, the availability and the efficient use of Domain Knowledge (DK) during the development process of a Knowledge Discovery in Databases (KDD) system is essential for successful knowledge discovery. In fact, DK is needed for almost any practical knowledge discovery task, independently of the domain or of the data mining techniques used, since, at least, some form of DK must be involved in the problem definition, in the data preparation and in the results evaluation and utilization phases. Sometimes, however, the involvement of DK in the process does not result in all the advantages it could bring. In fact, in some real-life situations where KDD could be useful, the available formally specified DK is restricted to description or definition of data and other forms of DK (for example theories about the way domain variables interact) exist only in informal, sometimes uncertain, non-structured forms. This kind of limitation of previously existing DK, together with a somewhat scarce theoretical work on the topic, usually results in no deliberate involvement of existing DK in the specific data mining phase of many real-life KDD processes.

DK involvement in the data mining step of a KDD process always implies a conditioning of the search of hypotheses conducted by the data mining algorithm. This conditioning can operate through an "initialization bias" (introducing starting conditions for the search), or through a "search bias" (distorting the search space, or the evaluation of hypotheses) [14], [15].

D. Cheung, G.J. Williams, and Q. Li (Eds.): PAKDD 2001, LNAI 2035, pp. 421-432, 2001.

DK can be included in the data mining phase through direct integration (implicit or explicit) in the data mining algorithm, or through an associated knowledge base. In the first case, specific changes to the core data mining algorithm must be performed, in order to directly represent the involved domain knowledge through a biasing of the search. In the latter case, a very tight coupling between the domain theory description in the knowledge base and the bias representation language accepted by the learner is need, eventually involving an intermediate knowledge "translator" [3]. Anyway, both of these forms of DK integration tend to need software specifically adapted for each application case, since different kinds of domain knowledge usually involve different representations, and most data mining algorithms (and commercial data mining programs) don't allow the integration any form of DK not contained in the data.

Direct integration of DK in data mining software generally intends to direct and focus the pattern search that takes place at that KDD step. This can raise another potential limitation of this technique: If badly directed, the focused search can miss some of the potentially interesting patterns that an unbiased search could find in the data [4]. However, in spite of the limitations and potential problems, we believe that, in some cases, careful DK integration in the data mining step of a KDD process can produce significant improvements in the overall efficiency of the process.

This paper presents an experiment that integrates two domain theories directly in a rule induction data mining algorithm. The domain is short-term stock market prediction, and the two theories bias the algorithm, during rule search, against a specific class of rules, and towards another. The theories are tested over five data sets that correspond to multivariate information based on daily quotes of five of the most significant stocks in the Portuguese BVLP stock exchange. The base rule induction algorithm used, PA3 [1], is a recent general-purpose sequential cover algorithm that combines general-to-specific and specific-to-general search to develop each rule.

2 Domain Knowledge

Adopting a restrictive DK definition, we will be interested only in domain theories that explain or predict future behavior of stocks on the basis of known data. This kind of domain theory is extremely uncertain in stock market prediction. There are, basically, three different positions: Those who believe that the markets are highly efficient and, as a result, essentially unpredictable, those who advocate "fundamental analysis" of the business results of the quoted companies, and those who believe that "technical analysis" (the analysis of historical stock quotes data, isolated of other known facts) is enough to predict the future behavior of those stocks [6].

The "efficient market" hypothesis, at least in its weakest form, has been traditionally accepted in some academic circles as basically correct, and if that were really the case, any effort to predict future behavior of listed stocks would be futile. However, besides the firm belief of those who really invest in stock markets (most of the investors and all the speculators), there is a growing body of published research indicating that at least some markets exhibit imperfections (which translate to a degree of predictivity) [7], [16], [11].

Classic "fundamental analysis" has solid background theory but even when successful in the long term, is not very useful to predict short-term movements of stock values [7]. A marginal aspect related to fundamental analysis that can be linked to very important fast movements of stock prices is the announcement of surprising fundamental company information (or surprising macroeconomic information, relevant for the whole market). However, this kind of fast readjustment of fundamental expectations will not be explicitly integrated in the analysis conducted in this paper, since it does not seem relevant for the paper's objectives and it requires very complex base data, and very demanding data preparation.

The theory behind present "technical analysis" is abundant. Unfortunately it is also fragmented and many times of dubious quality, most of it corresponding to unproved, sometimes untested, hypotheses. Moreover, the fact that technical analysis theory is still not seriously established can hide a fundamental problem: Even if technical analysis is realistically possible, perhaps it cannot be generalized for different markets, or for different stocks and different time frames of a market.

3 The Problem and the Data

The work we are involved in aims to predict the future behavior of five stocks listed in the Portuguese BVL stock exchange, utilizing historical data and DK.

This paper describes work done on direct domain theory integration in a rule induction algorithm used for the prediction of the next day behavior of each stock (binary prediction of rise or fall). This kind of next-day prediction is not enough to develop an operational trading strategy, but it is frequently found in the literature [2], [9], [11], and seems adequate to test the validity of the two domain theories involved.

For this very short-term prediction task, we simplified the base data by omitting fundamental information (and by not accounting for dividend payments), and used only historical stock quotes, transaction volumes and index values. It should be noticed that this base data has low information content for the prediction task, and could never result in very high accuracy rates, even with ideal data preparation and data mining steps. This situation is similar to having very noisy data both for learning and testing, and tends to present overfitting problems during the data mining process. With this problem in mind, we selected the domain theories to integrate in the rule-induction software aiming to reduce overfitting of the training data.

The five companies chosen for prediction are among those more actively traded in the BVL stock exchange: BCP, Brisa, Cimpor, EDP and PT. For each of the 4 companies excluding Brisa, daily data from 3-Nov-1997 to 29-Oct-1999 were available. For Brisa, quotation in BVL only started in 25-Nov-1997, and so available data starts in 25-Nov-1997 and also ends in 29-Oct-1999. Each of the resulting 495 records (479 for Brisa) includes the day's date, the closing value of the stock exchange main index (BVL30), the number of shares traded, and the opening, maximum, minimum and closing values of the stock.

From each companies' base data we constructed 15 daily-based "technical indicators" to be used as features to mine. These features are functions of the base data variables and summarize relations extracted from the previous 10 days of base data. As an example, one of the features expresses the relation between the 10-day and 3-day weighted moving averages of daily "reference values" (average of maximum, minimum and closing prices). Some of these features are categorical, while the others have integer or real values. However, the data mining algorithm requires discrete values, so we converted the original values of the features to discrete integer values ranging from 1 to 5 – the categorical features resulting in unordered sets of these values, and the numerical features resulting in ordered sets. As an example, the described relation between the 10-day and 3-day moving averages results in an ordered-value feature that is discretized the following way:

If (0.96 > (MA(10-day)/MA(3-day))) then the feature value is 1;
If (0.99 > (MA(10-day)/MA(3-day)) ≥ 0.96) then the feature value is 2;
If (1.01 > (MA(10-day)/MA(3-day)) ≥ 0.99) then the feature value is 3;
If (1.04 > (MA(10-day)/MA(3-day)) ≥ 1.01) then the feature value is 4;
If ((MA(10-day)/MA(3-day)) ≥ 1.04) then the feature value is 5.

The developed features were then subjected to a selection process to reduce their number to 10. This limitation on the number of features is introduced to help to reduce overfitting problems due to the scarce number of examples available in relation to the "descriptive power" of the full set of features. To select the 10 features to retain we applied (over the learning examples) a combination of methods including (with a heavier weight) Hong's feature selection method [8] and also (with reduced weights) a measure of correlation between the feature value and the result to predict, and the simple information gain of the feature.

The final format of each prepared example consists of 10 decision features with 5 discrete values (classified as ordered or unordered) and one binary result attribute. The result attribute indicates, for each example, if the described "reference value" of the stock raises or falls in the next trading day. The total number of examples available for each stock is 478 (462 for Brisa). This number is smaller than the number of days in the original data mainly because several of the first days must be used to construct some of the features of the first example.

4 The PA3 Rule Induction Algorithm

The rule induction algorithm we used, called PA3, is a recent general-purpose sequential cover algorithm [1]. The main features of PA3 include:

- A rule evaluation function that integrates explicit evaluations for rule accuracy, coverage and simplicity
- A rule generalization step that is run immediately after each rule is developed in an initial general-to-specific development phase
- A last rule filtering step that allows a choice of the tradeoff level between the accuracy and the global coverage of the final rule list.

The rule evaluation function is

$$v = a^{\beta} \times c^{1-\beta} + \chi s \,,$$

where v is the rule value, a is the rule accuracy over the learning examples, c is the rule coverage, s is the rule simplicity and β and χ are constants that must be chosen according to the learning data characteristics (β regulates the relative importance of rule coverage and rule accuracy and χ regulates the importance of rule simplicity).

This evaluation function is used to direct the search and to choose among alternative rules during the initial general-to-specific rule development and also, in the following rule generalization step, to evaluate and choose possible generalizations of the rules that result from the initial general-to-specific development. In this generalization step the evaluation function of the standard PA3 is used with the same parameter values used in the general-to-specific rule development. This way, the algorithm only replaces a rule previously found by a more general version of that same rule if the latter is better according to the same evaluation measure.

PA3 induces an ordered list of "if…then…" rules. Each rule has the form "if <complex> then predict <class>", where <complex> is a conjunct of feature tests, the "selectors". In PA3, each selector implies testing a feature to see if its value is included in a specified range of values. So, each selector indicates the feature to be tested and the (inclusive) upper and lower limits of the range of values it has to be tested against. The postcondition of a PA3 rule is a single Boolean value that specifies the class that rule predicts for the cases that comply with all the selectors. It should be noted that, while a single PA3 rule includes a simple conjunction of tests, the final rule set is equivalent to a DNF formula.

PA3's last step uses a simple rule evaluation metric (different from the one used in the rule learning process) to filter the complete list of the induced rules, retaining only a reduced number of stronger rules. Since the rules learned by this algorithm form an ordered list, this rule filtering has to retain a set of the first contiguous rules (also maintaining the order of those rules). This filtering process is controlled by a user-defined parameter that must be set between 0 (to accept all the discovered rules) and close to 1 (to accept only the first, stronger, rules). Globally, this rule filtering method allows the user to choose the tradeoff level between a more complete case-space coverage and a reduced coverage using only the stronger rules (and therefore with greater accuracy).

5 Domain Knowledge Inclusion in PA3

Our global KDD process allows the integration and testing of domain theories of the "technical analysis" kind through a very simple process: They can be represented by the features generated from the original data. With this in mind, the theories that seem more useful when integrated at the rule induction algorithm level are "meta-theories" that can be globally applicable to the rules (in fact, combinations of "technical indicators") created by the rule induction algorithm from the data features. Since, in our domain, the relevant information present in the base data is almost completely "drowned" in noise, and overfitting tends to occur, we felt that the "meta-theories" to test should preferably be chosen to reduce overfitting.

One of the two theories we decided to test biases the learner against the selection of rules belonging to a particular class, while the other intends to promote rule generalization for another class of "marginal" rules. More specifically, the first theory states that a good rule should not include a test over an ordered-value feature that only accepts its middle value (3, since the range of possible values is 1 to 5), since that kind of "neutral" value for an ordered-value feature probably does not point strongly to clear changes in the stock value. To integrate this theory in the PA3 rule induction algorithm, we altered the evaluation of the basic (still unexpanded) rules: When, during the rule induction procedure, a rule has a selector involving an ordered-value feature with a value of 3, the standard evaluation result for that rule is multiplied by a constant (named *mod1*) with a positive real value smaller than 1, thus reducing the rule evaluation result. The second theory states that if a rule includes a test over a feature that has ordered values, and a value of 2 or 4 is accepted for that feature, then the corresponding "extreme value" (1 or 5 respectively) should also be accepted. The reasoning is that if a "strong" (high or low) value for a technical indicator seems to be predictive for the future behavior of a stock, then an even stronger (in the same direction) value for the that indicator should, most of the time, also point to the same prediction. To integrate this theory in the PA3 algorithm, we altered the evaluation of the rule expansions: When, during the expansion procedure, a rule has a selector (involving an ordered-value feature) that is expanded from a value of 2 or 4 to include (respectively) the extreme values of 1 or 5, the standard evaluation result is increased through multiplication by constant (named *mod2*) with a real value greater than 1.

The general idea behind this use of uncertain DK at the rule induction level is that if the theories are globally true, then the rules that do not agree with them have a greater probability of corresponding to statistic fluctuations found in the learning data, and not to stable patterns useful for out of sample prediction. This problem is originated by the noisy data and small learning set sizes and by the very large domain space searched. Introducing a small handicap in the evaluations of key rule classes ensures that the rules belonging to these classes that are present in the final rule list must correspond to patterns in the learning data with above-average "strength". Of course, if the theories are globally true, they should increase the out-of-sample accuracy of the predictions. If they are globally wrong, the out-of-sample predictions should present a reduced accuracy.

It is clear that increasing the number of learning examples reduces the advantages of integrating this kind of DK to focus the search, since, with a greater number of learning examples, the real patterns in the training data tend to be less obscured by noise. A marginal point to notice is that this biasing of the search will, of course, always reduce accuracy over the learning data.

6 Tests

Testing the integration of the domain theories over the available examples is not straightforward, because some characteristics of the domain and of the data limit the direct use of normal bootstrap or resampling methods.

In fact, the time series we intend to predict are far from deterministic, and their behavior can be expected to change over time due to changes in the underlying domain mechanics. This way, a prediction model that proves accurate during a certain time span can be expected to (progressively or suddenly) loose prediction accuracy in the future. This means that maintaining the temporal order of the examples is important if each test example prediction is expected to represent the real prediction setup at the time of that example. (As an example, consider the use of training examples immediately posterior to the test example being predicted: That corresponds to the use of context information that could not be available if the prediction of that example was required in a realistic situation, and can be expected to adjust the prediction model to the near-future domain behavior, artificially increasing the prediction accuracy).

This way, since the examples are "time stamped" and the domain behavior is expected to vary over time, we opted for the standard time-sequenced division of the examples, instead of a classic bootstrap or resampling method. To ensure unbiased test results, the available examples were divided into separate learning/validation and test sets. We used the first 300 examples (284 for Brisa) from each stock for learning and parameter selection and kept apart the last 178 examples from each set for testing.

To determine the best values for *mod1* and *mod2*, the first 200 of the 300 examples (184 of 284 for Brisa) were used for learning with different values for *mod1* and *mod2*, and the resulting rule sets were tested on the remaining 100 examples from the learning sets. The test results were averaged over the five stocks, and the best global values for *mod1* and *mod2* were selected.

Those values were then used to develop rule lists from the complete sets of 300 learning examples (284 for Brisa), and the prediction accuracy of those rule lists (over the test sets of 178 examples) was compared with the one achieved by rule lists obtained using the standard, unbiased, PA3 (*mod1* = *mod2* = 1).

PA3 uses 3 internal parameters:

- β and χ are used to regulate rule evaluation during the general-to-specific and specific-to-general rule development phases, and must be set considering the domain characteristics
- The final rule filtering parameter must be chosen according to the users desired tradeoff level between prediction precision and model coverage.

Since our aim with these tests is not to achieve the best possible prediction results, but to compare the results with and without the integration of the domain theories, we chose to simplify our test procedure setting, from the start, the β and χ parameters to the "standard" values of 0.8 and 0.01 [1], instead of optimizing them through tests over the training/validation data. Also to simplify the test procedure, the final rule filtering parameter was set to prevent any rule filtering, and a default rule was added to the end of each learned (ordered) rule set. This way, every learned model is guaranteed to produce a prediction for every possible test case.

During the initial test phase, to determine the best values for the two theories parameters, 7 values were tried for the *mod1* parameter (0.4, 0.5, 0.6, …,0.9, 1.0) and 11 values were tried for *mod2* (1.0, 1.1, 1.2, …, 1.9, 2.0). The results for each *mod1* value were obtained as an average over the *mod2* values and vice-versa. In all, 11 runs of the induction algorithm (over each of the 5 examples sets) are averaged to obtain each of the accuracy values for *mod1* and 7 runs (also over each of the 5 examples sets) are done for each of the accuracy values for *mod2*.

This test procedure does not try to optimize the *mod1* and *mod2* parameters for each stock. Naturally, each of the tested theories can present a different behavior over each stock involved in the study, and better final accuracy values could be expected if individual *mod1* and *mod2* values were used for each stock. However, considering the small number of examples available for each stock, we opted to use accuracy values averaged over the five sets, in order to obtain more robust values for *mod1* and *mod2*: This way, the chosen values are those that resulted in the global best results across the 5 stocks.

The average accuracy (as tested over the last 100 learning examples) of the rule sets learned over the first 200 (184 for Brisa) examples of each stock set is shown in percentage in Figure 1 for the tested values of *mod1* and *mod2*.

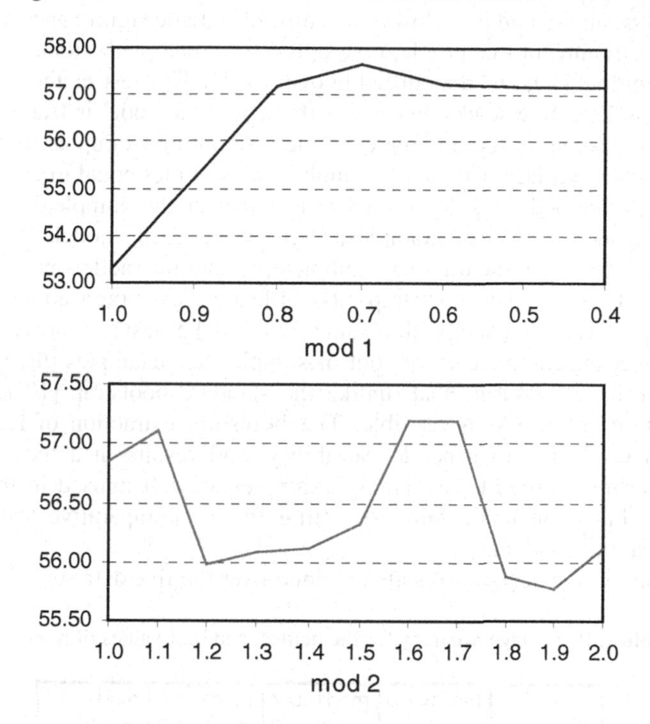

Fig. 1. Accuracy (in %) over the last 100 learning examples (averaged over the 5 stock sets)

As can be seen from the charts in Figure 1, for some of the tested modifier values both theories produce an improvement over the standard PA3 (*mod1*=*mod2*=1). However, the lack of regularity of the second theory chart contrasts with the "well behaved" first theory chart. In fact, in this first test, the second theory achieves an improvement for some of the *mod2* values, but several of the *mod2* values tested produce worse results than the basis value of 1.0. However, since these very simple first tests were based on relatively few examples and only intended to assist in choosing the values for *mod1* and *mod2* to be used in more extensive comparative tests, neither the stable behavior of the first theory nor the much less stable results for the second theory can be seem as very relevant.

Among the values tried for *mod1* and *mod2*, the best results were obtained for *mod1* = 0.7 and for *mod2* = 1.7. Those best values for *mod1* and *mod2* were then tested with rule sets developed over the sets of 300 learning examples (284 for Brisa), and applied over the five sets of 178 testing examples.

In these tests, a more complex procedure was used to try to achieve more stable results. Due to the method used to choose the "best" values for the two theory modifiers, and to the non-stationary nature of the time series involved, we wanted to keep a separation between the training and test sets, based on a strict time frontier: All the examples before that point are seen as training examples with a known outcome, and all the examples after that point are regarded as previously unseen test examples. That would lead to a simple holdout testing method that, due to the reduced number of available examples and to the small number of individual tests, would not produce reliable results, and would not allow a meaningful statistic significance analysis.

To try to circumvent this problem we opted for a test methodology that combines the simple holdout [10] and a modified bootstrap [5]. This test methodology uses 100 tests for each of the five stocks. In each of those tests, a model is learned on the basis of a bootstrap sample of the training examples (sampling examples from the original training set, using replacement, until a number of examples equal to the number in the original set is attained) and that model is tested over the complete, original, set of previously unseen 178 test examples. This way, each model is learned from approximately 63.2% of the training examples [5], and the models present exactly the same variability of standard bootstrap models learned over the training examples (in fact, they are learned exactly the same way). The tests, however, are always performed over the complete set of "out-of-sample" test examples (the best set of test examples we have), assuring that (unlike the standard bootstrap [10]) no optimistic "contamination" of results is possible. The bootstrap extraction of learning sets of examples is used only to generate variability, and results in a reduced prediction accuracy (because some of the training examples are left unused in the learning of each model) but maintains a fair test setting for the comparative tests of modifier values we want to conduct.

Table 1 shows the accuracy results obtained over the five data sets.

Table 1. Percentage accuracy for the neutral and best values of mod1 and mod2

	mod1=1.0 mod2=1.0	mod1=0.7 mod2=1.0	mod1=1.0 mod2=1.7	mod1=0.7 mod2=1.7
BCP	55.88	55.81	55.99	56.82
Brisa	52.02	52.83	52.46	53.20
Cimpor	52.66	53.09	52.36	53.53
EDP	51.56	52.26	52.03	52.31
PT	57.19	56.95	57.97	58.15
Average	53.86	54.19	54.16	54.80

Comparing the results of Table 1 (accuracy values close to 54%) and those indicated in Figure 1 (values close to 56%), a global accuracy decrease is clear. This decrease is mainly due to a very different behavior of the BVL stock exchange during the period corresponding to the learning examples (high volatility with a strong global raise) and during the period used to generate the test examples (a steady drop in the quote values). In those conditions, being able to achieve, over the test examples, global results clearly above the 50% level seems a strong indication that valid prediction patterns were in fact extracted from the training examples (both using the standard version of PA3 and using the versions with integrated DK). A secondary reason for the reduced accuracy is the test methodology that only uses about 63.2% of the available training examples to generate the prediction models, but this factor is partially offset by the increased number of available training examples in these tests.

The average results in Table 1 show that both theories, used in isolation, produce small accuracy improvements and that, used together, they result in a clearly greater improvement.

The results obtained for each stock show that when the integration of one of the theories in isolation results in a decreased accuracy (in only 3 out of 10 tests), the decrease if very small. The two theories combined always result in improved accuracy (in 5 out of 5 tests). This behavior seems to indicate that the average results can be regarded as relatively stable.

As previously referred, the prediction of this kind of financial time series would be impossible if the markets involved were theoretically efficient. That does not seem to be the case of most markets, and specifically of the market we are studying. However, even when stock markets are not theoretically efficient, that hypothesis does not seem to be very far from being true, and the predictability of stock quotes time series is always marginal. This way, in our binary prediction setting, a prediction accuracy close to 50% should be expected and any global percent accuracy improvement based in better data mining techniques must be marginal. This tends to result in a difficult setting for the analysis of the statistic significance of any data mining improvements. This global problem is compounded by the time-based sequential nature of the problems, by the relatively small number of available examples and by a large variability effect that can be associated with the noisy data [12].

To conduct a meaningful significance test of the theories integration, we used our bootstrap-based setup to analyze the number of times the altered algorithms achieved better, equal or worse results than the non-altered version (with mod1 = mod2 = 1.0). The involved test methodology allows us to conduct any desired number of tests with models that exhibit a bootstrap-like variance and still are tested in a strict holdout setup, allowing the tightening of the confidence intervals [13].

Table 2 shows the test results for the 100 runs for each of the five stocks that also produced the accuracy results of Table 1.

Table 2. Number of better, equal and worse results in relation to the basic, unmodified algorithm

	mod1=0.7 mod2=1.0			mod1=1.0 mod2=1.7			mod1=0.7 mod2=1.7		
	B	E	W	B	E	W	B	E	W
BCP	50	6	44	49	5	46	54	8	38
Brisa	52	5	43	47	9	44	56	1	43
Cimpor	53	6	41	43	6	51	58	5	37
EDP	51	5	44	51	4	45	49	5	46
PT	47	4	49	52	3	45	54	4	42
Average	50.6	5.2	44.2	48.4	5.4	46.2	54.2	4.6	41.2

One of the points that can be noticed in the results shown in Table 2 is the relatively large number of equal results. This is basically due to the fact that, in some runs, the mined data (that, in these tests, includes a number of repeated examples) is stable enough to generate exactly the same rule sets, in spite of the introduced search bias. Another interesting point is that in these results, only 2 of the 10 tests that compare the isolated theories with the unmodified algorithm produce more worse than better results (the slightly worse result of the isolated first theory in the accuracy results of the BCP stock is now inverted). This (average) worse accuracy result (see Table 1) is due to a small number of very bad results in some of the 100 accuracy tests (results that can be considered outliers).

The global results for each theory and for the two theories combined are consistent with the accuracy results shown in Table 1: In isolation, both theories produce a small but clear improvement over the unmodified algorithm, and the first theory produces a greater improvement then the second. When used together, the two theories produce a considerably greater improvement.

Applying a traditional significance analysis (single-sided paired *t* tests [13]) to the results in Table 2, the same general effects are detected: The average results for the first theory prove to be better than those of the unmodified algorithm version with 95% significance. The average results for the second theory are better than those of the unmodified algorithm version, but only with 68% significance. The average results for the two theories combined are better than those of the unmodified algorithm version with 98% significance. This last result seems to correspond to a meaningful prediction improvement in the difficult domain involved.

It should be pointed out that the test methodology we used expands the number of available examples by using simultaneous data from different stocks instead of a longer time frame of the same stock. This is common in stock time series data mining, but implies that the examples from each stock are not correlated which, of course, is not entirely true.

7 Conclusions

The work described in this paper reinforced our belief that direct use of DK in the core data mining phase of a KDD process can improve the overall efficiency of some knowledge discovery processes. In particular, changing the rule evaluation in order to introduce domain specific deformations in what would otherwise be an unbiased setting seems a promising way of integrating domain specific knowledge in the data mining phase of KDD processes that use rule induction algorithms.

As further work, we intend to test the present theories over more extensive stock market data. We also intend to evaluate, over the same domain, other globally applicable theories in the line of those involved in the present tests.

Acknowledgements. The author would like to thank L. Torgo and the anonymous reviewers of the paper submission for PAKDD-2001 for the helpful comments and corrections suggested.

This work was partially supported by the Portuguese Ministry of Science and Technology (MCT), through the Foundation for Science and Technology (FCT).

References

1. Almeida, P. and Bento, C.: Sequential Cover Rule Induction with PA3. To appear in Proceedings of the 10th International Conference on Computing and Information (ICCI'2000), Kuwait. Springer-Verlag (2000)
2. Choey, M. and Weigend, A.: Nonlinear Trading Models Through Sharpe Ratio Maximization. In Decision Technologies for Financial Engineering: Proceedings of the Fourth International Conference on Neural Networks in the Capital Markets (NNCM-96). World Scientific (1997)
3. Cohen, W.: Compiling Prior Knowledge Into an Explicit Bias. In Proceedings of the Ninth International Conference on Machine Learning. Morgan Kaufmann (1992)
4. Cook, D., Holder, L. and Djoko S.: Scalable Discovery of Informative Structural Concepts Using Domain Knowledge. IEEE Expert/Intelligent Systems & Their Applications, 11 (5) (1996)
5. Efron, B. and Tibshirani, R.: An Introduction to the Bootstrap. Chapman & Hall (1993)
6. Fama, E.: Efficient Capital Markets: A Review of Theory and Empirical Work. Journal of Finance, May (1970)
7. Herbst, A.: Analyzing and Forecasting Futures Prices. John Wiley & Sons (1992)
8. Hong S.: Use of Contextual Information for Feature Ranking and Discretization. IEEE Transactions on Knowledge and Data Engineering, 9 (5) (1997)
9. Hutchinson, J.: A Radial Basis Function Approach to Financial Time Series Analysis. Ph.D. Dissertation, Department of Electrical Engineering and Computer Science, Massachusetts Institute of Technology (1994)
10. Kohavi, R.: A Study of Cross-Validation and Bootstrap for Accuracy Estimation and Model Selection. In Proceedings of the 14th International Joint Conference on Artificial Intelligence (IJCAI-95). Morgan Kaufmann (1995)
11. Lawrence, S., Tsoi, A. and Giles C.: Noisy Time Series Prediction using Symbolic Representation and Recurrent Neural Network Grammatical Inference. Technical Report UMIACS-TR-96-27 and CS-TR-3625, Institute for Advanced Computer Studies, University of Maryland, College Park, MD (1996)
12. LeBaron, B. and Weigend, A.: A Bootstrap Evaluation of the Effect of Data Splitting on Financial Time Series. IEEE Transactions on Neural Networks, 9 (1) (1998)
13. Mitchell, T.: Machine Learning. McGraw-Hill (1997)
14. O'Sullivan, J.: Integrating Initialization Bias and Search Bias in Neural Network Learning. Unpublished research paper from April 1996, available in: http://www.cs.cmu.edu/~josullvn/research.html
15. Pazzani, M.: When Prior Knowledge Hinders Learning. In Proceedings of the AAAI Workshop on Constraining Learning with Prior Knowledge. San Jose, CA (1992)
16. Weigend, A., Abu-Mostafa and Refenes A.-P. (eds): Decision Technologies for Financial Engineering (Proceedings of the Fourth International Conference on Neural Networks in the Capital Markets, NNCM-96). World Scientific (1997)

Hierarchical Classification of Documents with Error Control

Chun-hung Cheng[1], Jian Tang[2], Ada Wai-chee Fu[1], and Irwin King[1]

[1] Department of Computer Science and Engineering
The Chinese University of Hong Kong, Shatin, Hong Kong
{chcheng,adafu,king}@cse.cuhk.edu.hk

[2] Department of Computer Science
Memorial University of Newfoundland, St. John's, NF, A1B 3X5 Canada
jian@cs.mun.ca

Abstract. Classification is a function that matches a new object with one of the predefined classes. Document classification is characterized by the large number of attributes involved in the objects (documents). The traditional method of building a single classifier to do all the classification work would incur a high overhead. Hierarchical classification is a more efficient method — instead of a single classifier, we use a set of classifiers distributed over a *class taxonomy*, one for each internal node. However, once a misclassification occurs at a high level class, it may result in a class that is far apart from the correct one. An existing approach to coping with this problem requires terms also to be arranged hierarchically. In this paper, instead of overhauling the classifier itself, we propose mechanisms to detect misclassification and take appropriate actions. We then discuss an alternative that masks the misclassification based on a well known software fault tolerance technique. Our experiments show our algorithms represent a good trade-off between speed and accuracy in most applications.

Keywords: Hierarchical document classification, naive Bayesian classifier, error control, class taxonomy, parallel algorithm

1 Introduction

Classification is a function that matches a new object with one of the predefined classes. A special kind of classification, *document classification*, has recently caught researchers' attention [4,12,20]. A document classifier categorizes the documents into the classes based on their content. This problem is characterized by the large number of attributes involved in the objects (documents). While a few hundred attributes are considered as very big for a traditional classifier, documents often contain thousands or even tens of thousands of terms. The traditional method of building a single classifier for all the classification work, known as *flat classification*, would incur a high overhead.

Koller and Sahami [12] propose the use of hierarchical classification in this context. Instead of a single classifier, a set of classifiers distributed over a *class*

D. Cheung, G.J. Williams, and Q. Li (Eds.): PAKDD 2001, LNAI 2035, pp. 433–443, 2001.

taxonomy are used, one for each node. A document is classified in a top-down fashion from the root to the leaf. For each current node (i.e. class), the child of maximum likelihood is selected. Thus, by decomposing a job into smaller jobs like this and some other techniques (e.g. feature selection), the amount of work can be maintained at a manageable level. This method is called *simple hierarchical classification* in this paper. However, once a misclassification occurs at a high level node, there is little chance to accommodate it at the low levels. The deeper the classification goes, the further it drifts away from the correct one. A variation of simple hierarchical classification, known as *TAPER*, is proposed in [4]. To avoid misclassification, it attempts to search for a global optimal probability by assigning the probability to the edge of the taxonomy graph in some ways that would transform the search into a least-cost path problem.

Weiss and Kulikowski [20] propose a different scheme of which one of the main goals is to remedy the misclassification problem. They utilize a single classifier over 'global' terms. The classifier is actually a set of special kind of association rules whose right sides are class labels, but only a portion of the rules are selected for the classification. A problem with this scheme is that in addition to class hierarchy, a term hierarchy is also required, which does not always exist. Also it is not clear if the selection rule can adequately reduce the number of association rules to make the job by the lone classifier manageable in the general case.

In this paper, we attack this misclassification problem from a different angle. We adopt hierarchical classification model due to its efficiency, but instead of trying to reduce the misclassification rate by overhauling the classifier itself, we develop mechanisms to detect the misclassification as early as possible and then take appropriate actions. We also discuss an alternative that masks the misclassification using a well known software fault tolerance technique.

The rest of this paper is organized as follows. In Section 2, we present a general model for document classification using hierarchical classifiers. In Section 3, the two error control schemes are introduced. We move to the experimental results in Section 4 and finally, we conclude this paper in Section 5.

2 Document Classification

Informally, a document is a pattern which consists of a number of terms and is attached with a class value (topic). Each term can occur multiple times in a document. The dependencies between the class values and the terms follow certain probabilistic distribution.

More specifically, we adopt a naive Bayesian model from [4]. Each class c is associated with a multinomial term-variable V_c. V_c can take values i, $1 \leq i \leq n_c$, with probability $p_{i,c}$ where each i denotes a term and n_c is the total number of different terms. A document in a class is then modeled as a collection of values (duplicates allowed) that the associated variable V_c generates successively. Let d be a given document in class c, h_d be its length, $z_{i,d}$ be the number of occurrences of value i in d, and $z_c = \sum_{d \in c, j=1,2,\cdots,n_c} z_{j,d}$. Let $P(d \mid c)$ be the probability that a randomly chosen document is d given that it is in c. Then we have

$$P(d \mid c) = \frac{h_d!}{z_{1,d}! z_{2,d}! \cdots z_{n_c,d}!} \Pi_{j=1}^{n_c} p_{j,c}^{z_{j,d}}. \tag{1}$$

where the value of $p_{i,c}$ can be estimated as $\frac{(\sum_{d \in c} z_{i,d})+1}{z_c + n_c}$. It is not the more intuitive value $\frac{\sum_{d \in c} z_{i,d}}{z_c}$. See [18] for a justification.

Let T be the class taxonomy, c be an internal node and c_i where $1 \leq i \leq q$ be the ith child of c. Given a document d, the classifier at node c classifies it into one of $c_1, \cdots, c_q$ by choosing c_i that maximizes $P(c_i \mid c, d)$, the probability of d belonging to c given it belongs to c'.

$$P(c_i \mid c, d) = \frac{P(c_i, d)}{P(c, d)} = \frac{P(c_i)P(d \mid c_i)}{\Sigma_{j=1}^{q} P(c_j)P(d \mid c_j)} \tag{2}$$

where $P(c_i, d)$ is the probability that we are given a document d and d belongs to c_i; $P(d \mid c_j)$ is estimated as stated above and $P(c_k)$ can be estimated as the fraction of the number of the documents that belongs to class c_k.

Since documents can contain a large number of terms, we must perform *feature selection* to reduce the cost. In addition, it can separate unindicative terms, or *noise*, from feature terms and increase accuracy of the classifier, since too many features may cause overfitting and loss of generality. As described in [12], features are context sensitive, meaning that we have different features at different splits in the taxonomy. Thus feature selection should be carried out at each split in the taxonomy.

One way to do feature selection is to use *Fisher Index* [4]. Let t_k be the kth term, $w(t_k, d)$ be the relative frequency of term t_k in document d, and $aw(t_k, c)$ $= \frac{1}{|c|} \Sigma_{d \in c} w(t_k, d)$. Thus, the Fisher Index of t_k for class c is:

$$Fisher(t_k, c) = \frac{\Sigma_{i=1}^{\ell} \mid c_i \mid (aw(t_k, c_i) - aw(t_k, c))^2}{\Sigma_{i=1}^{\ell} (\frac{1}{|c_i|} \Sigma_{d \in c_i} (w(t_k, d) - aw(t_k, c_i))^2)} \tag{3}$$

The idea is that a smaller value for the denominator implies a closer distance along dimension t_k among the points within each class, and a larger value for the numerator signifies a larger distance between any class and c. Thus a larger Fisher Index indicates a larger discriminative power of a term for a class. Let L be the list of terms in the descending order of their Fisher Indexes for c. We pick up a prefix F of L, and use F for the classification for c. Since F leaves out most noise terms, it reduces misclassification. The number of terms in F, known as the *feature length*, is a choice by users.

3 Error Control Schemes

Since simple hierarchical classification is problematic when a misclassification occurs at an early level, our approach is to incorporate error control mechanisms into the algorithm. We propose two schemes, namely recovery oriented error handling and error masking. The latter is a parallel algorithm and should run on a multi-processor machine.

3.1 Recovery Oriented Error Handling

The recovery oriented error handling approach is inspired by the way a transactional database is recovered upon failure. When a failure occurs in a transactional database, a previous consistent state is reconstructed, and an appropriate recovery action is taken based on that state. To bring the idea of database recovery to document classification, a consistent state here means an ancestor class node to which a document is classified with high confidence. We call it a *High Confidence Ancestor* (HCA). When a document is misclassified into a wrong path in the class taxonomy, we can restart from the HCA and then select another path. However, from our empirical studies rollback and reclassification are very time consuming. To simulate the effects of recovery, we try to identify the *wrong paths* first and avoid them during the classification.

To detect the wrong paths, we associate each document with a value called *closeness indicator* (CI) to indicate how close the document is to a given topic. Once a document is misclassified, the more it descends along the selected path, the further it would drift away from the distribution represented by the nodes in the path. When CI drops below a certain threshold, we may conclude that we are on the wrong path. For example, consider the class taxonomy depicted in Fig. 1. Assume a document is about 'folk dance', but has been misclassified into 'Business'. While this may seem not entirely unacceptable, it would be less acceptable to classify it into either 'Financial' or 'Insurance'. Suppose it is classified into 'Financial' by the classifier at 'Business' node. Then it faces the choices of 'Investment in stock market' and 'Portfolio arrangement of mutual funds'. Neither of these is remotely related to 'folk dance', so CI would fall to a small value and the path would be rejected.

Clearly, CI should be calculated without referring to the probabilities we used in the classification. Therefore, instead of the one-step probability, the probability of d belongs to c given the HCA is used as CI. Let c be the class that document d has been classified into. Let c' be the HCA of c for d. The CI of d with respect to c under c' is computed as:

$$CI(d, c \mid c') = P(c \mid d, c') = \frac{P(c, d \mid c')}{P(d \mid c')} = \frac{P(c \mid c')P(d \mid c)}{\Sigma_{i \in c'} P(i \mid c')P(d \mid i)}. \quad (4)$$

A simple way to determine the threshold is to use $1/N$, where N is the total number of classes at the same level as c and leaf classes at some level above c[1], in the subtree rooted at c'.

We maintain a moving window of l levels where l is a user parameter. The top and the bottom of the window correspond respectively to the levels of the current HCA and the class into which the document is being classified. Initially, the HCA is the root. The window moves downwards one level when the class at the bottom edge passes the test by CI, resulting in a new HCA at one level lower than it was prior to the move of the window.

[1] The class taxonomy can be an unbalanced tree.

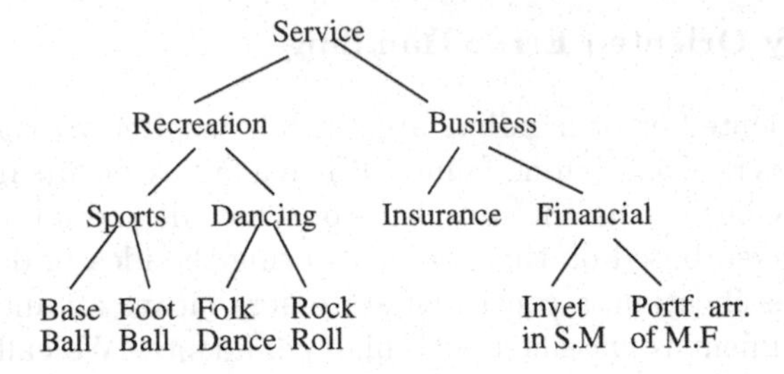

Fig. 1. A class taxonomy

```
Algorithm hc_recovery_oriented(T, d, l)
// T: class taxonomy
// d: document to be classified
// l: difference of the level of HCA and that of the current node
1. HCA ← root(T)
2. Loop
3.     CI_list ← find_CI_list(HCA, l)
4.     If no_of_element(CI_list) = 1 then
5.         result_class ← only element of CI_list
6.     Else
7.         result_class ← arg max_{c∈CI_list} {local_prob(HCA, c) }
8.     Endif
9.     If result_class is leaf Then return result_class
10.    HCA ← child of HCA who is an ancestor of result_class
11. Until forever
```

Fig. 2. Pseudo code for recovery oriented error control

Fig. 2 shows the pseudo code of the recovery oriented scheme. Before we do the real classification, the CI of nodes l levels ahead are calculated so the list of classes that pass the CI test is known. The algorithm will select the optimal path with maximum local_prob(HCA, c) (defined below) among all such classes. If there is only one class passing the CI test, we jump to that class directly without further calculations. The functions used are listed below:

find_CI_list(c, l) Suppose the level of c is i. Return a list of classes at level $l+i$ that passes CI test and any leaf classes between level $i+1$ and level $i+l-1$ that passes CI test.

local_prob(r_0, r_n) Suppose r_0 is an ancestor of r_n and the path along r_0 to r_n is $r_0 \rightarrow r_1 \rightarrow \cdots \rightarrow r_n$. Return $p(r_1|r_0)\ p(r_2|r_1) \cdots p(r_n|r_{n-1})$. In plain words, we are multiplying the one-step probabilities along r_0 to r_n together to get $p(r_n|r_0)$.

3.2 Error Masking

The error masking scheme is based on the idea behind software fault tolerance. Instead of detecting error and then performing recovery, we use multiple programs employing different designs. Among the outputs generated by these pro-

```
Algorithm hc_error_mask(T,d,l,f,f')
// T: class taxonomy
// d: document to be classified
// l: difference of the level of HCA and that of the current node
// f, f': two feature lengths for use with two O-classifiers
1.  n ← root(T) //level zero
2.  level ← l + 1
3.  (c1, c2, c3) ← (n, n, n)
4.  While c1 is not leaf do
5.      Start three threads:
6.          (i)     c1 ← O-classifier(c1, T, d, level, f)
7.          (ii)    c2 ← N-classifier(n, T, d, level)
8.          (iii)   c3 ← O-classifier(c1, T, d, level, f')
9.      Wait the finish of all threads
10.     If not (c1 = c2 = c3) Then
11.         c1 ← majority of c1, c2 and c3 (take a predefined action, e.g. using c2, if no majority)
12.         level ← level + 1
13.         n ← child of n who is an ancestor of c1
14.     Else
15.         level ← level + l
16.         n ← c1
17.     Endif
18. Endwhile
19. Return c1
```

Fig. 3. Pseudo code for error masking scheme

grams, the one generated by a majority is considered correct. More programs will generate more reliable results, but consume more resource.

We adopt a moderate approach. We run three classification methods in parallel. The first and third classifications are hierarchical classifications of traditional sense. The second classification is performed by dynamically skipping some levels in the class taxonomy. For example, to classify a document based on the taxonomy in Fig. 1, we can perform an additional classification by first skipping level 1 (i.e. {Recreation, Business}). Say the three classifications end up with class 'Dancing'. We then classify it at node 'Recreation'. But this time we skip the left part of level 2, i.e., {Sports, Dancing}. In the following discussion, we use the terms 'N-classifier' and 'O-classifier' respectively to refer to the classifiers with and without skipping the levels. The third classification is to employ O-classifier again but with a different feature length. A majority voting scheme is used to decide the overall output.

Skipping some levels has the effect of (partially) globalizing the information for the classification, and therefore can possibly reducing the misclassification rate. The more levels skipped, the more likely it is to reduce misclassification rate. In the extreme, if all but the leaf and root levels are skipped, we have a flat classifier. However, skipping a large number of levels beats one of the main motivations for using a hierarchical classifier, i.e., handling the complexity involved in the document classification. Thus a trade-off must be made. How to make such a trade-off is application dependent and is determined by users. In general, more levels can be skipped if the taxonomy has a large height but a small width than the other way around.

Fig. 3 shows the pseudo code for the error masking scheme. At line 11, if there is no majority formed by the classifiers, a user-defined action should be taken. For our experiments, this action is to use c_2, because usually, this is the most accurate (and slowest) classifier. Line 15-16 are some optimization codes. If c_1, c_2 and c_3 all match, we are confident that it is on the right track and so we can make a bold move — Instead of advancing one level, our algorithm moves l levels ahead. Some functions used are defined below:

O-classifier(c, T, d, k, f) To classify the document d using O-classifier in the taxonomy T from class c to reach a class in level k or a leaf class at a level higher than k. The feature length to use in classification is f.

N-classifier(c, T, d, k) To classify the document d using N-classifier in the taxonomy T from class c to reach a class in level k or a leaf class at a level higher than k.

4 Performance Evaluation

In this section, we study the performance of the algorithms. We implemented five document classification algorithms in C++. Our algorithms, namely the recovery oriented and the error masking schemes, are compared against simple hierarchical classification, flat classification and TAPER [4]. We run the experiments on a Sun Enterprise E4500 machine with 12 processors. Response time, rather than total CPU time, is measured so the error masking scheme can take advantage of parallelism.

We are interested in data sets with reasonably large class taxonomies, because the advantages of skipping levels can only be fully exploited in such data sets. We have chosen the data set of US patents[2] because they are organized in a large taxonomy. Three sets of data are collected from the US patent database. For convenience, we name them *Data_388*, *Data_TAPER* and *Data_Four*. Data_388 is the top-level class numbered 388 (motor control system) on the patent database. The class taxonomy is formed by all the 98 subclasses under the class 388. In each subclass, we download at most 20 patents, resulting in 901 patents. Data_TAPER highly resembles a data set used in [4]. The taxonomy of this data set is shown at Fig. 4(a). There are 12 leaf classes, each of which is a top-level class in the patent database. There are 500 training patents and 300 validation patents picked randomly from each leaf. However, since Data_TAPER is a three level data set, it is insufficient to demonstrate all the features of our algorithms while Data_388 only consists of a small number of patents. In Data_Four, we expand Data_TAPER by introducing more classes from the US patent database and grow the taxonomy by one level. The resulting class taxonomy is shown at Fig. 4(b).

Fig. 5, 6 and 7 show the accuracy and performance of the different algorithms on the three data sets. First of all, we achieve 65-70% accuracy in Data_TAPER,

[2] Available at several places on Internet, e.g. Delphion Intellectual Property Network (http://www.delphion.com/).

```
Patent: Communication:  329 Demodulator
                        332 Modulator
                        343 Antenna
                        379 Telephony
        Electricity:    307 Transmission
                        318 Motive
                        323 Regulator
                        219 Heating
        Electronics:    330 Amplifier
                        331 Oscillator
                        338 Resistor
                        361 System
```

(a) Class taxonomy for Data_TAPER

```
Patent: Physics and   Physics:        131 Fluid handling
        Chemistry:                    261 Contact apparatus
                                      096 Gas separation appratus
                                      095 Gas separation process
                      Chemistry:      422 Chemistry apparatus etc
                                      423 Inorganic compounds
                                      071 Fertilizers
                                      585 Hydrocarbon compounds
        Engineering:  Communication:  329 Demodulator
                                      332 Modulator
                                      343 Antenna
                                      379 Telephony
                      Electricity:    307 Transmission
                                      318 Motive
                                      323 Regulator
                                      219 Heating
                      Electronics:    330 Amplifier
                                      331 Oscillator
                                      338 Resistor
                                      361 System
```

(b) Class taxonomy for Data_Four

Fig. 4. Class taxonomies for some data sets used in the experiments

which is similar to the result in [4]. Among all the experiments, simple hierarchical classification is always the fastest algorithm and therefore it is the baseline of our comparison. Generally speaking, it is not justified to use a more complicated classification scheme unless it is more accurate than the fastest algorithm.

As simple hierarchical classification classifies the document in a greedy manner but TAPER searches the whole tree for the maximum overall probability, TAPER guarantees at least as good accuracy as simple hierarchical classification, although more time is required for the extra search. From the experiments, however, TAPER gives almost the same accuracy as the simple hierarchical classification. This result suggests the greedy approach of the simple hierarchical algorithm is close to optimal. Exhaustive search does not help to boost the accuracy. If we are to increase the accuracy, there must be a different approach to classify the documents. This is another reason to skip levels.

Flat classification gives the best accuracy in most cases[3]. However, from our experiments, it is clear that this is also the most time consuming algorithm except in the smallest taxonomy (Data_TAPER). Our algorithms stand on a middle ground between speed and accuracy. Our algorithms consistently beat TAPER and simple hierarchical algorithms in terms of accuracy. The recovery oriented scheme even slightly suppresses the flat classification in accuracy on Data_TAPER. However, it does not run fast since the recovery oriented scheme is essentially doing both a simple hierarchical and flat classification in a three level data set. In a bigger taxonomy (Data_388 and Data_Four), the recovery oriented scheme is clearly faster than flat classification, and at the same time more accurate than TAPER and simple hierarchical classification.

Like the recovery oriented scheme, the error masking scheme is also faster than flat classification and more accurate than TAPER and simple hierarchical classification in a large taxonomy. When comparing between the two error control schemes, it is found that there are some cases that either scheme is faster than the other. Due to parallelism, it is easy to understand why the error masking scheme is faster. However, the optimization adopted in our implementation also

[3] There seems to be no theoretical support for this in the general case. For example, the contrary is claimed in [4].

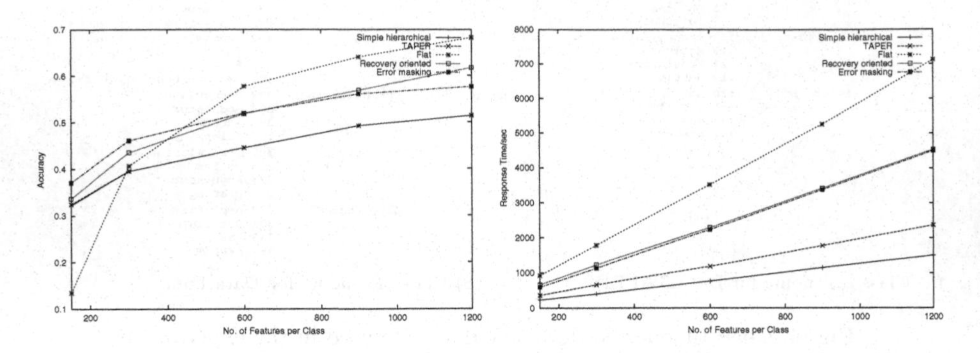

Fig. 5. Experimental result on Data_388

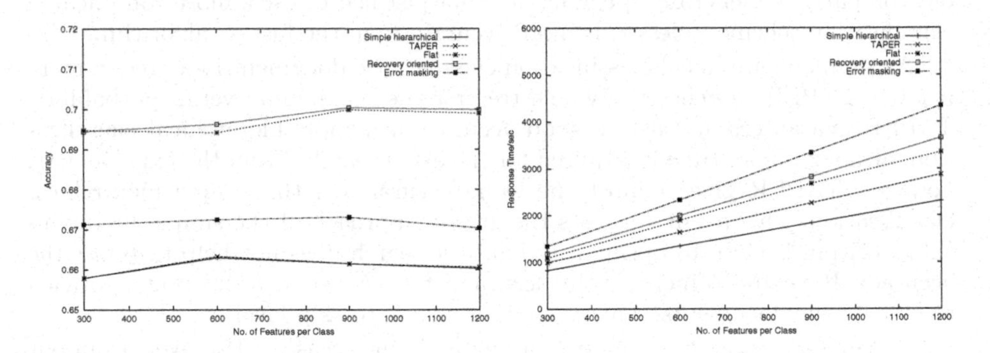

Fig. 6. Experimental result on Data_TAPER

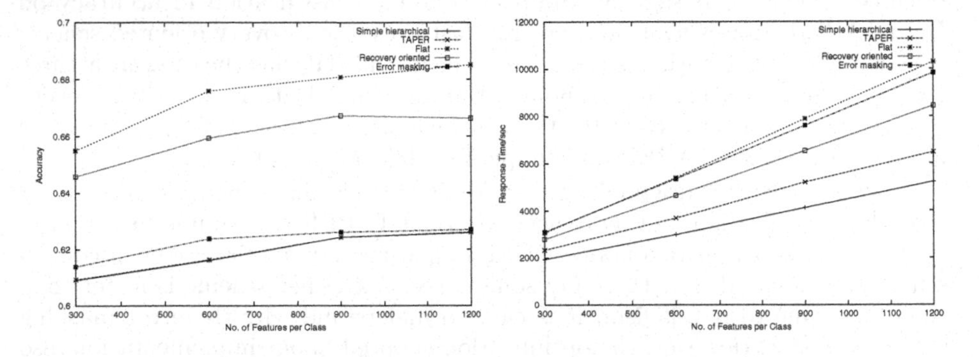

Fig. 7. Experimental result on Data_Four

gives the recovery oriented scheme an edge which explains why the recovery oriented scheme is faster in many cases. In recovery oriented scheme, if only one class is found to pass the closeness indicator, we will jump to that class directly. The one-step probability is not calculated. This is a considerable saving in running time. In contrast, the error masking scheme always classifies the document by three different classifiers and the slowest one will determine the response time. As for accuracy, the error masking scheme, while still ahead of TAPER and simple hierarchical classification, is often less accurate than the recovery oriented scheme. As the recover oriented scheme does not require a multi-processor machine, we feel that the recovery oriented scheme is preferable over the error masking scheme.

In a serious application, we expect a large class taxonomy. From the experiments, the response time difference between flat and simple hierarchical classification widens as the size of taxonomy grows. While the accuracy of simple hierarchical classification may not be satisfactory, switching to flat classification is too radical and computationally expensive. As our algorithms can be faster than flat classification at a taxonomy of as low as four levels (Data_Four), they represent a good trade-off between speed and accuracy for most applications.

5 Related Work and Conclusion

Classification has been studied extensively in the last decades [2,3,9,13,15,14, 17,21]. However, most of the work on the classification ignores the hierarchical structure of classes. In [1], the authors explore the hierarchical structure of attributes to improve the efficiency, but assume only a single level of classes. The work reported in [4,12] propose hierarchical classification based on the class taxonomy in the context of document classification. The work in [20] discusses document classification without using hierarchical classification. Bayesian network as a model for data mining has been studied in [6,5,10,7]. Feature selections are discussed in some work [11,16]. The general method is to define a measure first and then search for a subset of features that optimize this measure. Fisher Index method in [4] also follows this line, it does so however in a 'localized' manner, i.e. one term at a time. Although this local method has the weakness of not considering the fact that sometimes terms may be related, it does reduce the complexity when the number of features is very large.

In this paper, we have studied document classification using hierarchical classifiers with error control capability. We demonstrate that some well established strategies in other areas can also find a way to enhance the performance in our context. Two methods are proposed, recovery oriented and error masking. Recovery oriented method 'detects' an error and rejects it, while error masking method 'masks' an outcome under suspicion by adopting a better one. Our experiments show that both methods consistently reduce the misclassification rate against TAPER and simple hierarchical classification. The cost is extra running time, but they are faster than flat classification on a large taxonomy. Our algorithms are suitable for classifying documents into a large taxonomy where the users are willing to spend the extra time to trade for a higher accuracy.

References

1. H. Almualim, Y. Akiba, S. Kaneda, "An efficient algorithm for finding optimal gain-ratio multiple-split tests on hierarchical attributes in decision tree learning", Proc. of National Conf. on Artificial Intelligence, AAAI 1996, pp 703 - 708.
2. R. Agrawal, S. Ghosh, T. Imielinski, B. Iyer and A. Swami, "An interval classifier for database mining applications", Proc. of VLDB, 1992, pp 560 - 573.
3. L. Breiman, J. Friedman, R. Olshen and C. Stone, "Classification and regression trees", Wadsworth, Belmont, 1984.
4. S. Chakrabarti, B. Dom, R. Agrawal and P. Raghavan, "Using taxonomy, discriminants, and signatures for navigating in text databases", Proc. of the 23rd VLDB, 1997, pp 446 - 455.
5. K. Cios, W. Pedrycz and r. Swiniarski, "Data mining methods for knowledge discovery", Kluwer Academic Publishers, 1998.
6. P. Cheeseman, J. Kelly, M. Self, "AutoClass: a Bayesian classification system", Proc. of 5th Int'l Conf. on Machine Learning, Morgan Kaufman, June 1988.
7. N. Friedman and M. Goldszmidt,"Building classifiers using Bayesian networks", Proc. of AAAI, 1996, 1277 - 1284.
8. T. Fukuda, Y. Morimoto and S. Morishita, "Constructing efficient decision trees by using optimized numeric association rules", Proc. Of VLDB, 1996, pp 146 - 155.
9. J. Gehrke, R. Ramakrishnan and V. Ganti, "Rainforest - a framework for fast decision tree construction of large datasets", Proc. of VLDB, 1998, pp 416 -427.
10. D. Heckerman, "Bayesian networks for data mining", *Data Mining and Knowledge Discovery*, 1, 1997, pp 79 - 119.
11. D. Koller and M. Sahami, "Toward optimal feature selection", Proc. of Int'l. Conf. on Machine Learning, Vol. 13, Morgan-Kaufmann, 1996.
12. D. Koller and M. Sahami, "Hierarchically classifying documents using very few words", Proc. of the 14th Int'l. Conf. on Machine Learning, 1997, pp 170 - 178.
13. M. Mehta, R. Agrawal and J Rissanen, "SLIQ: a fast scalable classifier for data mining", Proc. of fifth Int'l Conf. on EDBT, March 1996
14. J. Quinlan, "Induction of decision trees", Machine Learning, 1986, pp 81 - 106.
15. J. Quinlan, "C4.5: programs for machine learning", Morgan Kaufman, 1993.
16. G. Salton, "Automatic text processing, the transformation analysis and retrieval of information by computer", Addison - Wesley, 1989.
17. J. Shafer, R. Agrawal and M. Mehta, "Sprint: a scalable parallel classifier for data mining", Proc. of the 22nd VLDB, 1996, pp 544 - 555.
18. E. S. Ristad, "A natural law of succession", Research report CS-TR-495-95, Princeton University, July 1995.
19. S. Weiss, and C. Kulikowski, "Computer systems that learn: Classification and prediction methods from statistics, neural nets, machine learning and expert systems", Morgan Faufman, 1991.
20. K. Wang, S. Zhou and S. C. Liew, "Building hierarchical classifiers using class proximity", Proc. of the 25th VLDB, 1999, pp 363 - 374.
21. Y. Morimoto, T. Fukuda, H. Matsuzawa, T. Tokuyama and K. Yoda, "Algorithms for mining association rules for binary segmentations of huge categorical databases", Proc. of VLDB, 1998.

An Efficient Data Compression Approach to the Classification Task

Claudia Diamantini and Maurizio Panti

Computer Science Institute, University of Ancona, via Brecce Bianche, 60131 Ancona, Italy
{Diamanti, Panti}@inform.unian.it

Abstract. The paper illustrates a data compression approach to classification, based on a stochastic gradient algorithm for the minimization of the average misclassification risk performed by a Labeled Vector Quantizer. The main properties of the approach can be summarized in terms of both the efficiency of the learning process, and the efficiency and accuracy of the classification process. The approach is compared with the strictly related nearest neighbor rule, and with two data reduction algorithms, SVM and IB2, on a set of real data experiments taken from the UCI repository.

1 Introduction

Data mining can be viewed as a *model induction* task, that is the task of building a descriptive or predictive model of a phenomenon starting from a set of instances of the phenomenon itself. In particular, the scope of this paper is on predictive model induction. This problem has been undertaken for a long time in disciplines like statistics, pattern recognition and machine learning. In pattern recognition, non-parametric classification methods has been developed, such as the nearest neighbor classifier [2]. This method is often considered for data mining tasks, for its conceptual simplicity, associated to good classification performance, which often turns out to compete with those of other, more sophisticated, approaches [5,6]. However, it presents also a severe limit for data mining, namely the fact that the entire training set has to be processed in order to classify a new datum, making infeasible its application to huge databases.

To reduce the classification cost of the nearest neighbor classifier, *data reduction* techniques were introduced [1,8,13,14]. The aim of data reduction is to select, from the whole set of data, the subset which allows the minimum degradation in performance, introducing in this way an accuracy vs efficiency tradeoff. It also introduces a cost for "learning" (i.e. the running of the reduction algorithm) which, very often, turns out to be itself too expensive to be applied to large databases, so an important research topic in data mining is the development of techniques in order to improve algorithm scalability [9,12,15].

A complementary approach to data reduction is *data compression* [10,11,17]. In this approach, the aim of learning is to compress the information contained

D. Cheung, G.J. Williams, and Q. Li (Eds.): PAKDD 2001, LNAI 2035, pp. 444–454, 2001.

in the original data in a new, reduced, set of elements. The main advantages of compression over reduction techniques can be summarized in the fact that compression units are free to move in the feature space, allowing in principle to reach a more accurate classifier design, and in the fact that the complexity of the design is independent of the size of the training set. Of course, compression, as well as data reduction, works fine only if the learning criterion is appropriate. As a matter of fact, in data reduction, learning algorithms were naturally concerned with the classification problem, while data compression is historically related to the problem of data reproduction, hence to the minimum mean squared distortion criterion [17,11]. Kohonen [10] introduced for the first time some data compression algorithms complying with the classification task, but mainly on an intuitive basis.

In this paper, we want to bring to the attention of researchers the features of a data compression approach to classification, based on a stochastic gradient algorithm for the minimization of the *average misclassification risk* performed by a Labeled Vector Quantizer (LVQ). The approach has the following advantages:

- Minimization of the average misclassification risk allows to guarantee that the learning guides the LVQ towards (local) optimal classification performances. That is, it guarantees the effectiveness of the learning process;
- The use of a stochastic gradient algorithm guarantees the efficiency of the learning process. In particular, the use of one sample per iteration allows to keep data on hard-disks during the learning, with no accuracy vs efficiency tradeoff;
- The particular quantization architecture adopted allows to design a nearest neighbor classification rule, that is a very simple rule, which outperform the classical nearest neighbor classifier;
- LVQ architectures allows to strongly compress the information contained in the training set. Thus the classification is based on a very small number of elements with respect to the training set size.

In the following, such advantages will be experimented on real data sets taken from the UCI Machine Learning repository, comparing the method with the classical Nearest Neighbor (1-NN) classifier, Support Vector Machines (SVM) [16] and the IB2 algorithm of the Instance Based Learning family [1].

2 The Bayes Vector Quantizer

In the statistical approach to classification, data are described by a continuous random vector $\mathbf{x} \in \mathcal{R}^n$ (feature vector) and classes by a discrete random variable $\mathbf{c} \in \mathcal{C} = \{c_1, c_2, \ldots, c_C\}$. Each class c_i is described in terms of the conditional probability density function (cpdf) $p_{\mathbf{x}|\mathbf{c}}(x|c_i)$ and the a priori probability $P_{\mathbf{c}}(c_i)$. The predictive accuracy of a classification rule $\Phi : \mathcal{R}^n \rightarrow \mathcal{C}$ is evaluated by the *average misclassification risk*

$$R(\Phi) = \int R(\Phi(x)|x) p_{\mathbf{x}}(x) dV_x \,, \tag{1}$$

where $p_{\mathbf{x}}(x) = \sum_{i=1}^{C} P_{\mathbf{c}}(c_i) p_{\mathbf{x}|\mathbf{c}}(x|c)$, dV_x denotes the differential volume in the x space, and $R(c_j|x)$ is the risk in deciding for class c_j when a particular x is observed. $R(c_j|x)$ is defined as

$$R(c_j|x) = \sum_{i=1}^{C} b(c_i, c_j) P_{\mathbf{c}|\mathbf{x}}(c_i|x) . \tag{2}$$

In (2), $b(c_i, c_j) \geq 0$ expresses the cost of an erroneous classification, i.e. the cost of deciding in favor of class c_j when the true class is c_i, with $b(c_i, c_i) = 0\ \forall i$. If $b(c_i, c_j) = 1\ \forall i, j, i \neq j$ the average misclassification risk turns to the simpler error probability. $P_{\mathbf{c}|\mathbf{x}}(c_i|x)$ can be derived from $P_{\mathbf{x}|\mathbf{c}}(x|c_i)$ by the Bayes theorem. A well known result is that the best possible classification of a feature vector consists in mapping it to the class with the minimum conditional risk (2) (Bayes rule):

$$\Phi_B(x) = \min_{c \in \mathcal{C}}{}^{-1} \{R(c|x)\} . \tag{3}$$

The development of non-parametric methods and learning algorithms for classification, arise from the attempt to overcome the limits of applicability of this optimal rule, related to the the fact that cpdfs involved in are in general unknown. Thus, their ultimate goal is to obtain an estimate of such functions of x on the basis of the training set.

In this paper we take a different approach, based on the observation that a classification rule $\Phi : \mathcal{R}^n \rightarrow \mathcal{C}$, being a total and surjective function, induces a partition of the feature space $\mathcal{R}^n$ into C regions $R_1, \ldots, R_C$ (decision regions), where R_i is the set points which are pre-images of class c_i in $\mathcal{R}^n$. Starting from this observation, we propose an algorithm to adapt an initial labeled partition, where labels represent classes, towards the optimal partition induced by the Bayes rule. Notice that, in this way, the function of x we try to approximate is directly $\Phi_B(x)$, which, under the hypothesis of piecewise continuity of cpdf, is a piecewise constant function. We encode a labeled partition by a Labeled nearest neighbor Vector Quantizer.

A *nearest neighbor* Vector Quantizer (VQ) of size M is a mapping

$$\Omega : \mathcal{R}^n \rightarrow \mathcal{M} ,$$

where $\mathcal{M} = \{m_1, m_2, \ldots, m_M\}$, $m_i \in \mathcal{R}^n, m_i \neq m_j$, which defines a partition of $\mathcal{R}^n$ into M regions $\mathcal{V}_1, \mathcal{V}_2, \ldots, \mathcal{V}_M$, such that

$$\mathcal{V}_i = \{x \in \mathcal{R}^n : d(x, m_i) < d(x, m_j),\ j \neq i\} .$$

Basically, a VQ performs data compression since it represents each point of a region $\mathcal{V}_i$ by one point: m_i. $\mathcal{V}_i$ is the *Voronoi region* of *code vector* m_i and d is some distance measure. If the Euclidean measure is adopted, then Voronoi region boundaries turns out to be piecewise linear. In particular, the boundary $\mathcal{S}_{i,j}$ between two regions $\mathcal{V}_i$ and $\mathcal{V}_j$ is a piece of the hyperplane equidistant from m_i and m_j (see Figure 1(a)). In the following we will always refer to nearest neighbor VQs with Euclidean distance.

The VQ is extended with a further mapping $\Lambda:\mathcal{M} \to \mathcal{C}$, which assigns a label from $\mathcal{C}$ to each code vector. We will call this extended VQ a Labeled VQ (LVQ). An LVQ can be used to define a classification rule: let $l_i = \Lambda(m_i) \in \mathcal{C}$ denote the label assigned to m_i. The decision taken by the LVQ when x is presented at its input is

$$\Phi_{LVQ}(x) = \Lambda \circ \Omega(x) = l_i, \text{ if } x \in \mathcal{V}_i\,. \tag{4}$$

In practice, the classification is performed by finding in $\mathcal{M}$ the code vector at minimum distance from x, and then by declaring its label. Thus an LVQ implements a simple nearest neighbor rule, and each point of a Voronoi region $\mathcal{V}_i$ is implicitly labeled with the same label of m_i. Figure 1 (b) shows an example of labeled partition induced by an LVQ. Notice that, even if each Voronoi region

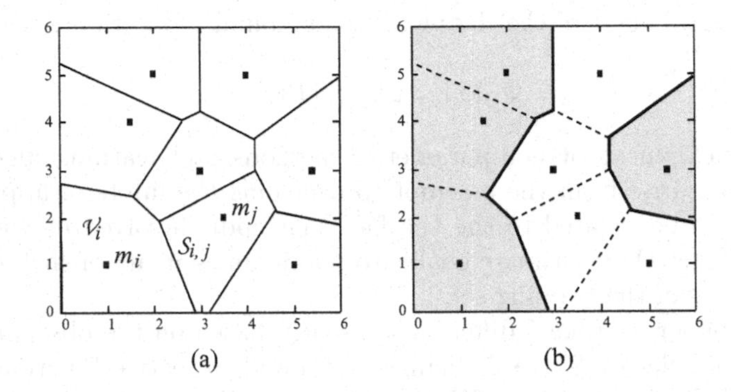

Fig. 1. (a) A nearest neighbor VQ of size 8 in $\mathcal{R}^2$ and (b) A possible labeled partition induced by labeling the VQ.

is convex, we can construct non-convex and non connected decision regions as well. Notice also that boundaries between two Voronoi regions with the same label (the dashed lines in Figure 1(b)) do not contribute to the definition of decision region boundaries (decision boundaries). The adoption of the simple Euclidean distance limits us to piecewise linear decision boundaries. However, with other distance measures, non linear boundaries could be obtained as well [7, §10.4].

In order to develop an algorithm to find an optimal approximation of the Bayes partition, a crucial observation is that the average risk of Φ_{LVQ} depends only on the labeling function Λ and on the mutual position of code vectors m_i, which determines the form of the integration regions. Thus, keeping the labeling function Λ fixed, and under the continuity hypothesis for cpdfs, average risk is differentiable w.r.t. $\mathcal{M}$. The gradient of $R(\mathcal{M})$ w.r.t. the generic m_i has been derived for the first time in [4], and has the form

$$\nabla_i R(\mathcal{M}) = \sum_{j=1}^{C} \sum_{q=1,q\neq i}^{M} \frac{b(c_j,l_q) - b(c_j,l_i)}{\| m_i - m_q \|} P_{\mathbf{c}}(c_j) \int_{\mathcal{S}_{i,q}} (m_i - x) p_{\mathbf{x}|\mathbf{c}}(x|c_j) dS_x\,, \tag{5}$$

where dS_x denotes the differential surface in the x space.

Almost surprisingly, the variation of the risk w.r.t. code vector m_i depends only on what happens on boundary surfaces between $\mathcal{V}_i$ and each neighbor region $\mathcal{V}_q$ (obviously $\mathcal{S}_{i,q}$ vanishes if m_i and m_q are not neighbors), and only in the case that $b(c_j, l_q) \neq b(c_j, l_i)$. In the case of error probability this means simply that $l_i \neq l_q$, i.e. that the boundary surface between $\mathcal{V}_i$ and $\mathcal{V}_q$ actually represents a part of the decision boundary. This result formalizes and generalizes the original intuition of Hart [8], elaborated also in the more recent, so called, "boundary hunting" methods [1,16,10], that all the relevant information about a classification problem is found in those samples falling near the decision border.

The use of a stochastic Parzen estimate for $p_{\mathbf{x}|\mathbf{c}}(x|c_j)$, and some approximations introduced for the sake of simplicity, for which we refer the interested readers to [4], leads to a class of stochastic gradient algorithms for the minimization of $R(\mathcal{M})$. We consider the algorithm, here called Bayes VQ (BVQ), obtained when a uniform window of side Δ is adopted as the Parzen window.

Let us assume that the labeled training set $\mathcal{L} = \{(t_1, u_1), \ldots, (t_T, u_T)\}$ is given, where $t_i \in \mathcal{R}^n$ is the feature vector and $u_i \in \mathcal{C}$ is its class. At each iteration, the algorithm considers a labeled sample randomly picked from the training set. If the sample turns out to fall near the decision boundary, then the position of the two code vectors determining the boundary is updated, moving the code vector with the same label of the sample towards the sample itself, and moving away that with a different label. More precisely, the k-th iteration of the BVQ algorithm is:

BVQ Algorithm - k-th iteration

1. randomly pick a training pair $(t^{(k)}, u^{(k)})$ from $\mathcal{L}$;
2. find the two code vectors $m_i^{(k)}$ and $m_j^{(k)}$ nearest to $t^{(k)}$;
   /* note: certainly such vectors are neighbors! */
3. $m_t^{(k+1)} = m_t^{(k)} \quad for \quad t \neq i, j;$
4. compute $t_{i,j}^{(k)}$, the projection of $t^{(k)}$ on $\mathcal{S}_{i,j}^{(k)}$[1];
5. if $\| t^{(k)} - t_{i,j}^{(k)} \| \leq \Delta/2$ then

$$m_i^{(k+1)} = m_i^{(k)} - \gamma^{(k)} \frac{b(u^{(k)}, l_j) - b(u^{(k)}, l_i)}{\| m_i - m_j \|} (m_i^{(k)} - t_{i,j}^{(k)});$$

$$m_j^{(k+1)} = m_j^{(k)} + \gamma^{(k)} \frac{b(u^{(k)}, l_j) - b(u^{(k)}, l_i)}{\| m_i - m_j \|} (m_j^{(k)} - t_{i,j}^{(k)});$$

else $m_t^{(k+1)} = m_t^{(k)} \quad for \quad t = i, j.$

Figure 2 illustrates the behavior of BVQ, considering two equiprobable classes (called black (B) and white (W) class) and error probability as the performance measure. In this case, the point P, located where cpdfs coincide is, by definition, the optimal Bayes decision boundary. Figure 2(a) shows a set of samples of the W and B classes represented by small white and black dots respectively,

[1] Such projection is a function of the two code vectors and of $t^{(k)}$ only.

whose distribution in the feature space follows class statistics. In Figure 2(b) is depicted an LVQ of size three, with two code vectors labeled as white and one code vector labeled as black. Voronoi region boundaries are $\mathcal{S}_{1,2} = \frac{m_1+m_2}{2}$ and $\mathcal{S}_{2,3} = \frac{m_2+m_3}{2}$. The decision regions induced by the LVQ are graphically represented by the dashed and solid lines. Samples falling near $\mathcal{S}_{1,2}$ are not used

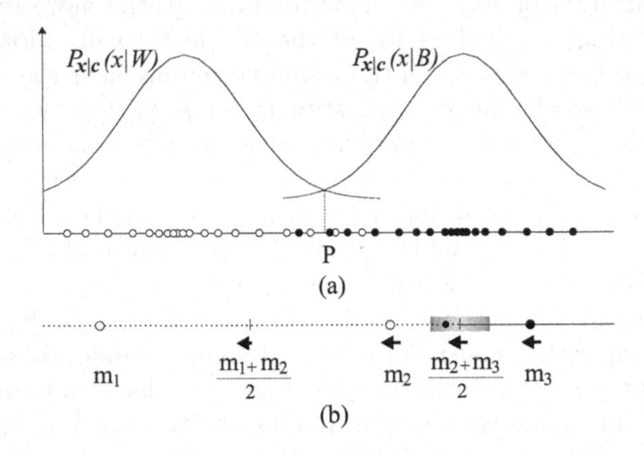

Fig. 2. A graphical representation of the adaptation step performed by BVQ.

to update code vectors. This is consistent with the fact that perturbations of $\mathcal{S}_{1,2}$ do not change the decisions taken in its around, which is always in favor of the white class (note however that the updating of m_2 and m_3 indirectly influences also $\mathcal{S}_{1,2}$). Samples around $\mathcal{S}_{2,3}$ are used only if the sample falls inside the window of side Δ, represented in the Figure by the shaded area. If such a sample is of the B class, then $b(u^{(k)}, l_3) = 0$ and $b(u^{(k)}, l_2) = 1$, so $b(u^{(k)}, l_2) - b(u^{(k)}, l_3) = 1$. Hence m_3, which, with this setting, plays the role of m_i in the algorithm, is moved toward the sample, while m_2 is moved away. As a result, both code vectors (and $\mathcal{S}_{2,3}$ as well) will be moved toward left. Vice-versa, if the sample is of the W class, both code vectors will be moved toward right. Repeated iterations of the BVQ algorithm moves $\mathcal{S}_{2,3}$ from its initial position toward left, since at the beginning black samples are more frequent than white samples. This drifting continues until a point is reached where the frequency of black and white samples falling inside the window is the same. In the asymptotic case, the number of samples is arbitrarily large, and the optimal value of Δ is zero, so this point is exactly the point P where cpdfs coincide. In the finite case, P can be only approached, since $\Delta > 0$. However, more samples we have, smaller values of Δ can be set, more accurate approximations of the Bayes decision boundary can be found. The comparison of BVQ with other VQ based approaches, performed in [4], enlightens the advantages of a learning formally based on average risk minimization.

The computational cost of a single iteration of the BVQ can be divided into: the cost for finding the two code vectors nearest to the training vector, the cost for calculating the projection of the training vector on the decision border, and the cost for code vectors updating. It is simple to see that the dominating factor is the first one, corresponding to a total of $M \cdot n$ multiplications. This cost is thus independent of the training set size. This result compares favorably with the costs of other learning algorithms used in Data Mining, in particular SVM. As to the number of iterations which are necessary to approach Bayes risk, preliminary experimental results in [3] suggest that it does not depend on the training set size as well, but only on the initial position of code vectors and on the initial value of the step size γ.

3 Experimental Evidences

In order to show the advantages of the BVQ algorithm, we compare it to 1-NN, SVM and IB2 on a set of real data taken from the UCI Machine Learning repository[2]. Table 1 gives the database size and dimension n of the feature space for each experiment.

Table 1. The data sets from the UCI Repository used in the experiments.

Data set	Size	Dimension
Australian	690	14
Diabetes	768	8
German	1000	24
Ionosphere	351	34
Liver-Disorders	345	6
Mushroom	8124	22

Table 2 shows the error probabilities achieved by SVM, 1-NN and IB2, together with the size of each classifier in brackets. This size is expressed by the number of vectors used in the respective classification rules. Since 1-NN does not discard any training sample, the size of this classifier correspond to the training set size. The results are taken from [14]. They correspond to average measures obtained by a 10-fold cross validation method. Thus, in order to compare the results, the same experimental procedure was adopted for the BVQ.

Before applying the BVQ, a normalization of data is performed, in such a way that each feature ranges in the same interval. This is an invertible transformation of data which allows to give equal importance to each vector component during the learning. The problem of local minima suffered by "greedy" methods afflicts BVQ as well. This can be alleviated by a proper initialization of code vectors, but in this case some domain knowledge should be given. In the experiments the LVQs are initialized simply by the first training vectors of each training set. The decreasing law for the step size was set to $\gamma^{(k)} = \gamma^{(0)} j_k^{-0.51}$, where j_k denotes the number of non-null updatings until step k and $\gamma^{(0)}$ is an initial value experimentally determined in the range $[0.1, 0.005]$. In Table 3, for each data set,

[2] http://www.ics.uci.edu/~mlearn/MLRepository.html

Table 2. Error probability and number of vectors for SVM, 1-NN and IB2 on the UCI repository data sets.

Data set	SVM	1-NN	IB2
Australian	0.1537 (203.9)	0.185 (603)	0.2642 (151.5)
Diabetes	0.2292 (401.7)	0.3048 (691.2)	0.3505 (253.5)
German	0.249 (487)	0.331 (900)	0.388 (338.7)
Ionosphere	0.0543 (167.1)	0.1543 (315.9)	0.1314 (54.7)
Liver	0.3132 (209.7)	0.3765 (310.5)	0.4201 (121.3)
Mushroom	0.0 (437.3)	0.0 (7311)	0.0041 (25.6)

the optimal value of the parameter Δ and the best error performances achieved by BVQ with LVQs of varying size are reported.

Table 3. Error probability vs number of code vectors for the BVQ on the UCI repository data sets. Values in boldface denote the best error probability for each experiment.

No. Code Vectors	*Australian* $\Delta = 0.346$	*Diabetes* $\Delta = 0.49$	*German* $\Delta = 1.549$	*Ionosphere* $\Delta = 1.897$	*Liver* $\Delta = 0.154$	*Mushroom* $\Delta = 1.897$
2	0.1493	0.2396	0.291	0.1454	0.3338	0.2254
4	0.1478	0.2358	0.290	0.1398	0.3219	0.2129
8	0.1449	**0.2265**	0.277	0.1343	**0.286**	0.100
16	0.1493	0.2358	0.270	0.1231	0.3038	0.0242
32	**0.1435**	0.2422	**0.253**	**0.112**	0.3033	**0.0138**

Some comments on this Table are in order. First, we have to report a phenomenon typical of VQ architectures, called the "dead neuron problem" by Kohonen. In practice, it can happens that a code vector is never used to classify input data. The removal of such code vectors would not modify the error probability for the data at hand. Thus the number of code vectors reported is always greater than or equal to the true number of used code vectors. Second, we can observe a non monotonic trend of error probability vs number of code vectors for the australian, diabetes and liver data sets. This somewhat counter-intuitive result can be explained by the intrinsic "jitter" of stochastic methods, which make the system to wander around the optimum, by the dead neuron problem and by the dependency of the result from initialization and labeling of code vectors.

Turning to the comparison of results in Tables 2 and 3, we can observe that the BVQ is always the worst on the mushroom experiment. This bad performance can be explained by noticing that, in this experiment, samples are described by purely categorical features, hence cpdfs are not piecewise continuous, and the basic assumption for the applicability of the method is not satisfied. As a consequence, we can report a long number of iterations needed to converge, and a great sensitivity of the algorithm to the initialization of code vectors and to the value of the parameter $\gamma^{(0)}$. Vice-versa, the algorithm proves to work nicely if at least some feature turns out to be continuous, as it is the case for the other experiments. Here, the BVQ clearly outperforms both 1-NN and IB2, while its performance can be considered at least comparable to that of SVM. In particular, BVQ performance is about the 7%, 1% and 10% better than SVM on the Australian, Diabetes and Liver data sets respectively, while it is about the 2% and 50% worse than SVM on the German and Ionosphere data sets respectively. For this experiment, we failed to find a number of code vectors for which a comparable error probability could be obtained. This fact is likely to be related to the small training set size, especially compared with the dimension of the feature space. In this case, the large value of the window introduces a distortion in the estimate of (5) and of the cost function, and the BVQ can find only a poor suboptimum. It is nevertheless better than the one found by 1-NN and IB2. Since the latter result is likely to be due to the small training set size, it is not very serious, in the perspective of very large databases.

On the other hand, the advantage of BVQ in memory requirements is striking. As we can see, two code vectors are sufficient to BVQ to obtain the best of all error probability on the Australian data set, they are sufficient to reach a lower error probability than 1-NN in all the experiments and a lower error probability than IB2 in all but the ionosphere experiment. This fact, together with the fact that, using 2 code vectors, BVQ already achieves as almost good results as using 32 code vectors, gives some insights on these problems, allowing the hypothesis that the optimal decision boundary can be quite accurately approximated by a linear decision boundary.

The highest memory requirements for the BVQ are on the german data set, where 32 code vectors are needed, to reach an error probability comparable to that of SVM, which needs 487 vectors out of 900 training samples.

These results assume greater relevance in the perspective of very large databases, in the light of the fact that, with BVQ, the classifier size turns out to be related only to the geometry of the problem, that is to the number of classes and shape of the decision boundaries, while with the other methods the classifier size grows with the training set size. Preliminary results supporting this statement can be found in [3].

The size of BVQ, IB2 and 1-NN classifiers also allows to directly compare their computational cost, as they all use the same decision function (4). For instance, on the german data set, the BVQ allows to classify a sample by calculating only 2 distances, against the 900 distance calculation of the 1-NN and the 339 distance calculation of IB2. The decision function of SVM takes a different form. SVM can directly manage only two class problems[3]. Assuming that class labels are encoded by integers 1 and -1, SVM decision function is

$$\Phi_{SVM}(x) = sgn[\sum_{i=1}^{S} u_i \alpha_i K(x \cdot s_i) + b] , \tag{6}$$

where, $\{(s_1, u_1), \ldots, (s_S, u_S)\} \subseteq \mathcal{L}$ is the set of Support Vectors and S is the size of the classifier. $K(\cdot)$ is either a linear or a non linear kernel (typical kernels are

[3] problems with $C > 2$ classes have to be managed by designing C different classifiers, each separating one class from the rest [13].

the gaussian and sigmoid functions). If K is a linear kernel, then this decision function requires approximately the same number of multiplications of (4). In fact, we can rearrange the last one as follows

$$\Phi_{LVQ}(x) = \Lambda(\min_{m_i}{}^{-1}\{\| m_i - x \|^2\}) = \Lambda(\min_{m_i}{}^{-1}\{\| m_i \|^2 - x \cdot m_i\}) .$$

Hence, if we store the squared norm of code vectors, which equals the cost of storing the weights α_i in SVM, both the decision functions require a number of inner products equal to the size of the classifier. In the general case, however, the SVM decision function is computationally heavier than (4), since we have to add the cost of the non linear kernels calculus. In the above experiments, in order to obtain the reported accuracy, the adopted kernel is always gaussian, except for the mushroom experiment where it is linear. Thus, the comparison of the size of BVQ and SVM classifiers allows to establish an even greater computational advantage of the former over the latter than the advantage observed over 1-NN and IB2.

4 Conclusions

In the paper we presented a data compression approach to the classification task, based on the stochastic gradient algorithm BVQ. The main properties of the approach can be summarized in terms of the efficiency of the learning process, and efficiency and accuracy of the classification processes. Efficiency of the learning process is due both to the use of a stochastic gradient algorithm, which exploits only one training sample per iteration, and to the light computational cost of each iteration. Efficiency of the classification process is due to the use of nearest neighbor vector quantizer architectures, which allows to implement a simple nearest neighbor rule, based on a very small number of elements with respect to the training set size. Finally, both efficiency and effectiveness of the classification process gains from the use of the average misclassification risk as a learning criterion, which allows to design the VQ in such a way that the (locally) optimal linear approximation of the Bayes decision border is found, with the given number of code vectors. Furthermore, although in the experiments we focus on error probability as the performance measure, the BVQ is a general algorithm for the minimization of the average risk. Thus the introduction of misclassification cost matrices in the formulation of the classification problem can be supported as well. This feature is important for practical applications, where some classification errors are often considered more serious than others (for instance, evaluating a client reliable for a loan when it is unreliable can be more dangerous for a bank than evaluating a reliable client unreliable). The use of general cost matrices in real applications will be the scope of future research. Other directions of research include the study of techniques to improve BVQ performance, by finding better initialization strategies of code vectors and by developing non greedy versions of the algorithm to escape local minima. Also of interest is the study on the exploitation of geometric characteristics of VQs in order to extract symbolic classification rules from it.

References

1. D. W. Aha, D. Kibler, and M. K. Albert. Instance-based learning algorithms. *Machine Learning*, 6(1):37–66, 1991.
2. T. M. Cover and P. E. Hart. Nearest neighbor Pattern Classification. *IEEE Trans. on Information Theory*, 13(1):21–27, Jan. 1967.
3. C. Diamantini and M. Panti. An efficient and scalable data compression approach to classification. *ACM SIGKDD Explorations*, 2(2):54–60, 2000.
4. C. Diamantini and A. Spalvieri. Quantizing for Minimum Average Misclassification Risk. *IEEE Trans. on Neural Networks*, 9(1):174–182, Jan. 1998.
5. P. Domingos and M. Pazzani. On the Optimality of the Simple Bayesian Classifier under Zero-One Loss. *Machine Learning*, 29:103–130, 1997.
6. J. H. Friedman. On Bias, Variance, 0/1-Loss, and the Curse-of-Dimensionality. *Data Mining and Knowledge Discovery*, 1(1):55–77, 1997.
7. A. Gersho and R. M. Gray. *Vector Quantization and Signal Compression.* Kluwer Academic Publishers, 1992.
8. P. E. Hart. The condensed nearest neighbor rule. *IEEE Trans. Information Theory*, 4:515–516, May 1968.
9. T. Joachims. Making Large-Scale SVM Learning Practical. In B. Scholkopf, C. J. Burges, A. J. Smola, editor, *Advances in Kernel Methods - Support Vector Learning.* MIT Press, 1999.
10. T. Kohonen. The self organizing map. *Proc. of the IEEE*, 78(9):1464–1480, Sept. 1990.
11. G. F. McLean. Vector quantization for texture classification. *IEEE Trans. on Systems, Man, and Cybernetics*, 23:1637–649, May. 1993.
12. F. J. Provost and V. Kolluri. A survey of methods for scaling up inductive learning algorithms. Tech. Rep. ISL-97-3, Intelligent System Lab., Dept. of Comp. Science, Univ. Pittsburg, 1997.
13. B. Schoelkopf, C. Burges, and V. Vapnik. Extracting Support Data for a Given Task. In U. Fayyad and R. Uthurusamy, editors, *Proc. 1st Int. Conf. on Knowledge Discovery and Data Mining*, Menlo Park, CA, 1995. AAAI Press.
14. N. A. Syed, H. Liu, and K. K. Sung. A Study of Support Vectors on Model Independent Example Selection. In S. Chaudhuri and D. Madigan, editors, *Proc. 5th ACM SIGKDD Conf. on Knowledge Discovery and Data Mining*, pages 272–276, New York, 1999. ACM Press.
15. N. A. Syed, H. Liu, and K. K. Sung. Handling Concept Drift in Incremental Learning with Support Vector Machines. In S. Chaudhuri and D. Madigan, editors, *Proc. 5th ACM SIGKDD Conf. on Knowledge Discovery and Data Mining*, pages 317–321, New York, 1999. ACM Press.
16. V. Vapnik. *Statistical Learning Theory.* J. Wiley and Sons, New York, 1998.
17. Q. Xie, C. A. Laszlo, and R. K. Ward. Vector quantization technique for non-parametric classifier design. *IEEE Trans. on Pattern Analysis and Machine Intelligence*, 15(12):1326–1330, Dec. 1993.

Combining the Strength of Pattern Frequency and Distance for Classification

Jinyan Li[1], Kotagiri Ramamohanarao[1], and Guozhu Dong[2]

[1] Dept of CSSE, The University of Melbourne, Vic. 3010, Australia.
{jyli, rao}@cs.mu.oz.au
[2] Dept of CSE, Wright State University, USA. gdong@cs.wright.edu

Abstract. Supervised classification involves many heuristics, including the ideas of decision tree, k-nearest neighbour (k-NN), pattern frequency, neural network, and Bayesian rule, to base induction algorithms. In this paper, we propose a new instance-based induction algorithm which combines the strength of pattern frequency and distance. We define a neighbourhood of a test instance. If the neighbourhood contains training data, we use k-NN to make decisions. Otherwise, we examine the support (frequency) of certain types of subsets of the test instance, and calculate support summations for prediction. This scheme is intended to deal with outliers: when no training data is near to a test instance, then the distance measure is not a proper predictor for classification. We present an effective method to choose an "optimal" neighbourhood factor for a given data set by using a guidance from a partial training data. In this work, we find that our algorithm maintains (sometimes exceeds) the outstanding accuracy of k-NN on data sets containing pure continuous attributes, and that our algorithm greatly improves the accuracy of k-NN on data sets containing a mixture of continuous and categorical attributes. In general, our method is much superior to C5.0.

Keywords: classification, neighbourhood, emerging patterns, outlier.

1 Introduction

Supervised classification, where prediction is performed after training instances are provided, has been intensively studied in the machine learning and pattern recognition communities over a long period of time. Instance-based induction [2], contrast to eager-learning based classification (as exemplified by C4.5 [17]), is an important approach to classification. A typical example of instance-based induction algorithms is the k-nearest neighbour (k-NN) rule [4]. Given a test instance T and a database of training instances, the k-NN rule finds k training instances which are the nearest to T according to some kind of distance, and chooses the class label prevailing among these training instances.

Recently, with the advances in data mining, supervised classification also becomes an interesting topic in the KDD field [8]. Liu et al [15] have proposed the CBA classifier based on the idea of association rules [1]. Meretakis and Wuthrich [16] have explored the use of frequent and long patterns in optimising posterior

D. Cheung, G.J. Williams, and Q. Li (Eds.): PAKDD 2001, LNAI 2035, pp. 455–466, 2001.

probabilities which are used for classification. Dong, Li, Ramamohanarao et al have proposed the CAEP [6], JEP-C [14], and DeEPs [13] classifiers based on the concept of emerging patterns [5]. A common aspect of the above classifiers is that they make use of the support (frequency) of the interesting patterns as a basis to construct discriminating power, rather than the distance as used in the classic k-NN rule. In this paper, we investigate how to combine the strength of pattern frequency and distance to solve supervised classification tasks.

Suppose an instance T is to be classified. The basic idea of the approach proposed in this paper is to utilize distance as a measure to predict the class of T when a local area of T contains some training instances; otherwise (when there exists no training instance in the local area) to make use of the support of some subsets of T. The constrained use of distance reflects our belief that

- when a test instance is an outlier, i.e., its local area contains no training instance, the k-NN rule may not properly predict the class of the outlier.

In this case,

- we examine the significant support change between classes of some subsets of the outlier, and summarize those changes to make a decision.

The idea of compactly aggregating support changes for classification originated in our previous work DeEPs [13], an instance-based classifier. In brief, the idea behind DeEPs is to discover those subsets of a given test instance whose support changes significantly from one class to another, and then base decisions on the supports of the discovered subsets. In this paper, we improve the support aggregation method of DeEPs. As a result, the improved DeEPs can properly handle the data sets with a very unbalanced class distribution.

Without any prior knowledge, it is difficult to define an outlier: to define a point's neighbourhood within which there is no other points. We propose here a method to determine an appropriate neighbourhood of a test instance by using partial training data as a guide. Basically, we initially set three neighbourhoods, and we choose one of them to be applied to all test instances if the selected one is the best (in terms of accuracy) for the partial training data.

The remainder of the paper is organized as follows: Section 2 begins with a set of basic term definitions, followed by a brief description of k-NN and DeEPs which are closely related to the current work. Our main contributions are also described in this section. Section 3 details our methods including selection of a neighbourhood, summation of pattern supports, and combination of k-NN and DeEPs. Section 4 reports our experimental results on 30 widely used data sets. The results show that the proposed approach is generally superior to the performance of k-NN and C5.0, a commercial version of C4.5 [17]. Section 5 concludes this paper.

2 Related Work and Our Contributions

In this section we provide basic definitions and background materials. We also describe a related work including k-NN and our previous work DeEPs, each followed with our new contributions made to them. These two instance-based classifiers are combined into a new approach. We present the new approach in Section 3.

2.1 Background

An **attribute** is one of the most elementary terms in relational (including binary transactional) databases. For example, SHAPE can be an attribute in some databases. Usually, there exist at least two values for an attribute. The SHAPE attribute can have such **categorical** values as `square`, `circle`, and `diamond`. Another type of attribute value are called **continuous** values. For example, HEIGHT can have continuous values ranging from 0 to 3 meters.

Item is an important notion in data mining. An item is defined as such a pair as attribute-value (e.g., SHAPE-`square`), or attribute-interval (e.g., HEIGHT-$[0.1, 1.2)$), respectively with regard to categorical or continuous attributes. An **itemset** is a set of items. The process of partitioning a value range of a continuous attribute into a number of intervals is referred to as discretization [9,11, 18]. Many classification algorithms could not be applied to real-world classification tasks unless the continuous attributes are first discretized [7]. However, our methods do not need to pre-discretize data. This is one of our advantages over other methods.

An **instance** is also defined as a set of items. Usually, different instances within a data set have the same number of attributes. When an instance is labelled with a class, the instance is called a **training** instance. When an instance's class is unknown or is assumed unknown, then it is referred to as a **test** instance. The **test accuracy** of a classifier is the percentage of test instances which are correctly classified.

The support of an itemset is used to measure the occurrence (or frequency) of the itemset in a data set. Given a database $\mathcal{D}$ and an itemset X, the **support** of X in $\mathcal{D}$, denoted $supp_{\mathcal{D}}(X)$, is the percentage of instances in $\mathcal{D}$ containing X. An itemset X is **contained** (or occurred) in an instance Y if $X \subseteq Y$.

2.2 The k-Nearest Neighbour Rule

With extensive theoretical analysis and rigorous empirical evaluation, the k-NN rule [4] has widely attracted the attention of many researchers since its conception in 1951 [10]. The idea of k-NN is straightforward. Given a test instance and for $k = 1$, one finds the stored training instance which is nearest, notes the class of the retrieved case, and predicts the new instance will have the same class [12]. In spite of its simplicity, k-NN is powerful in solving those classification tasks where all attributes are continuous. For example, on the letter-recognition dataset [1], the k-NN rule can reach a test accuracy of 95.58%; in comparison, the notable decision tree based classifier C5.0 obtains a test accuracy of only 88.06%,

[1] All data sets used in this paper were taken from the UCI Machine Learning Repository [3].

7.5% lower than k-NN. However, for handling those data sets which contain a mixture of categorical and continuous attributes, k-NN loses its power. For example, on the australian data set, k-NN and C5.0 reach 66.69% and 85.94% accuracy respectively. The degradation in accuracy of k-NN on those data sets is mainly caused by the confusing distance contributions between categorical and continuous attributes. So, one of the extensively studied issues on k-NN is how to justify a balance of the distance contributions by the two types of attributes.

In this paper, our proposed method can appropriately deal with both continuous and categorical attributes. When continuous attributes are present in a data set, we scale the training values of every attribute into the range of $[0, 1]$, and use the same parameters to scale their values in test instances. On the other hand, when categorical attributes are present, we transform the categorical values into continuous values: transform the value of any categorical attribute of a test instance into 1, and transform its different values (respectively, the same value) in training instances into 0 (respectively, 1). We then calculate the Euclidean distance. The experimental results show that our method maintains the outstanding accuracy of k-NN on data sets containing pure continuous attributes, and that our method greatly improves the accuracy of k-NN particularly on data sets containing a mixture of the two types of attributes.

2.3 Brief Description of DeEPs

DeEPs is a recently proposed instance-based classifier [13]. It makes decisions based on the supports of certain types of subsets of a test instance. Assume we are given a set $\mathcal{D}_p$ of positive instances and a set of $\mathcal{D}_n$ of negative instances, and a test instance T. DeEPs selects two special collections of subsets of T: one consists of subsets which only occur in $\mathcal{D}_p$ but not in $\mathcal{D}_n$; the other of those which occur in $\mathcal{D}_n$ but not in $\mathcal{D}_p$. Then, DeEPs calculates a support summation over each of the two collections. DeEPs assigns T the class where a larger summation is obtained.

There are a series data reduction and concise knowledge representation techniques used in DeEPs. The significant reduction is achieved by removing from training data the irrelevant values to a test instance: If an attribute is categorical, DeEPs removes those values of this attribute in the training data which are different from the value of the test instance; If an attribute is continuous, DeEPs removes those values which are beyond a neighbourhood of the value of the test instance. Table 1 demonstrates these points.

In this paper, we improve DeEPs by proposing a new method to summarize the supports. This new method is specially designed to handle those data sets containing extremely unbalanced class distributions. For example, in the lym data set, the distribution is allocated as: 2 instances for class 1, 81 for class 2, 61 for class 3, and 4 for class 4. Let A and B be two subsets of a test instance. Suppose A occurs only once in class 1, but is not present in any other classes. Suppose B occurs 27 times in class 2, but does not in any other classes. None of the other subsets has any occurrence in the four classes. Then, $supp_1(A) = 50\%$, and $supp_2(B) = 33.3\%$. Therefore, the original DeEPs would choose class 1.

Table 1. Original training data are transformed into binary data after removing values which are irrelevant to the test instance $T = \{\texttt{square}, 0.30, 0.25\}$. A chosen neighbourhood of 0.30 is $[0.28, 0.32]$ and a chosen neighbourhood of 0.25 is $[0.23, 0.27]$.

original training data				binary data		
circle	0.31	0.24	→	0	1	1
square	0.80	0.70	→	1	0	0
diamond	0.48	0.12	→	0	0	0

$$T = \{\texttt{square}, 0.30, 0.25\}$$

Apparently, this decision is unreasonable as the occurrence of B in class 2 is much more frequent than A's occurrence in class 1. In this work, we add a weight to scale down the supports contributed by small classes. This adaption is detailed in Section 3.3.

As mentioned in the Introduction, a more important contribution of this paper is to combine the use of pattern frequency and distance to strengthen a classifier's discriminating power. We discuss when and how the system uses k-NN rules or the system uses the DeEPs ideas. These proposals are provided in the following section.

3 Our Methods

An important preliminary step of our method is to normalise all training values of every continuous attribute into the range of $[0, 1]$. For each continuous attribute, we used the formula $\frac{x-min}{max-min}$, where x is an original training value, max and min are the maximum and minimum value respectively in the training data. The normalisation parameters max and min are stored and will be used to scale the values in any test instances.

3.1 Main Steps

Suppose we are given a classification problem in which a data set $\mathcal{D}$ contains at least C ($C \geq 2$) classes of data. Our methods consist of the following two main steps, when a test instance T is given to classify.

(a) If a chosen neighbourhood of T covers some training instances, we apply 3-NN to classify T. (On special situations where only two or one instance is covered, we apply 1-NN.)

(b) Otherwise, when the neighbourhood does not contain any training instances, we apply DeEPs. Note that we consider T as an outlier in this case.

These basic ideas are illustrated in Figure 1.

In the subsequent two subsections, we describe our methods to select proper neighbourhoods, and to improve DeEPs' support summation process.

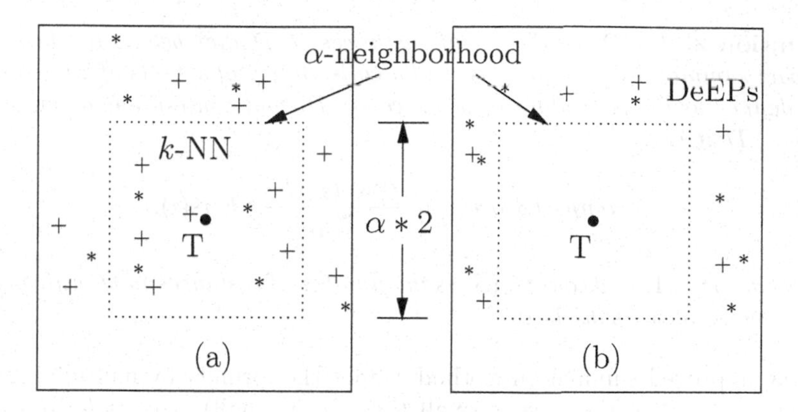

Fig. 1. Our classification algorithm deals with two cases when a test instance T is required to classify. The signs "*" and "+" represent instances from two different classes.

3.2 Selecting a Proper Neighbourhood

We first describe a definition of a neighbourhood of $T = \{a_1, a_2, \cdots, a_n\}$.

Definition 1. *An α-neighbourhood of T is defined as the area:*

$$\{X \mid a_i - \alpha \leq x_i \leq a_i + \alpha \;\; \text{if} \;\; a_i \;\text{is continuous, and}\; x_j = a_j \;\text{if}\;\; a_j \;\text{is categorical}\},$$

where $X = \{x_1, x_2, \cdots, x_n\}, 1 \leq i \leq n, 1 \leq j \leq n.$

Depending on the property and complexity of training data, one would see that the parameter α should not be uniform for different data sets. We propose here a heuristic to optimize α for a given data set. Recall that one of the primary objectives of classification algorithms is to accurately predict the class of test instances. The proposed heuristic is closely related to this objective.

We initially set three values for α: 0.05, 0.10, and 0.20. We then randomly choose 10% of training data and view them as "test" instances. Therefore, three accuracies, by using our classification method (as described in Section 3.1), on this special collection of "test" instances, can be obtained. Then, we apply the value of α, by which the highest accuracy is obtained, to the real test instances. This heuristic emphasizes the guidance role played by a partial training data. Sometimes, such a process is referred to as tuning by training data.

3.3 An Improved Support Summation Method

In the original DeEPs algorithms [13], the supports of some subsets of a test instance is aggregated by the compact summation method. We review the compact summation using the following definition.

Definition 2. *Let $\mathcal{D}_i$ be the set of instances in $\mathcal{D}$ that belong to class i. The compact summation score for class i of a collection $\mathcal{S}$ of subsets of a test instance T is defined as the percentage of instances in $\mathcal{D}_i$ that contain one or more of the subsets. That is:*

$$compactScore(i) = \frac{count_{\mathcal{D}_i}(\mathcal{S})}{|\mathcal{D}_i|} * bias(i),$$

where $bias(i) = 1$, and $count_{\mathcal{D}_i}(\mathcal{S})$ is the number of instances in $\mathcal{D}_i$ which contain one or more of the subsets in $\mathcal{S}$.

The improved summation method refines the formula by multiplying a bias weight ($\neq 1$) when the class is small (i.e., $\mathcal{D}_i$ is small). We set $bias(i)$ as $\frac{2}{3}$ if the number of instances in $\mathcal{D}_i$ is less than 20 and the numbers of other classes data are at least three times larger than 20. This amendment can help DeEPs to adjust the compact scores contributed by unbalanced classes.

Many other options for bias are possible. For example, the selection of the weight can rely on Bayesian rules. This is one of our future investigations.

4 Experimental Results

We in this section report a performance of our method in comparison to the performance conducted by k-NN and C5.0. We used 30 data sets for experimental evaluation. For each dataset-algorithm combination, the test accuracies were measured by a ten-fold stratified cross validation. Each of the exclusive ten folds test instances were randomly selected from the original data sets. The same splits of the data were used for all the three classification algorithms.

4.1 Accuracy Comparison

Table 2 provides the data set names and the number and type of the attributes. (See columns 1, 2, and 3.) Columns 4, 5, and 6 show the test accuracies achieved by our approach, C5.0, and k-NN respectively. Along the vertical direction, Table 2 is organized into three groups according to the performance differences between our method and C5.0: significant differences ($\geq$ 2.00%) are in the top and bottom groups, while slight differences ($<$ 2.00%) in the middle.

We observed the following interesting points from Table 2.

- Among the 30 data sets, our method is significantly superior to C5.0 on 16 data sets. The accuracy gaps can reach up to 14.93% (in sonar), half of them are around 6.5%.
- Our method is not always better than C5.0. We lose on four data sets, particularly on auto.
- On average over the 30 data sets, our accuracy is 2.18% higher than C5.0, and 7.27% higher than 3-NN.

Table 2. Accuracy comparison among our algorithm, C5.0, and k-NN.

Data Sets	Numbers of Attributes		Accuracy (%)			Difference
	cont.	categ.	3-NN	C5.0	Ours	ours vs. C5.0
australian	6	8	66.69	85.94	88.41	+2.47
cleve	5	8	62.64	77.16	83.18	+6.02
crx	6	9	66.64	83.91	86.37	+2.46
german	7	13	63.1	71.3	74.40	+3.10
heart	6	7	64.07	77.06	81.11	+4.05
hepatitis	6	13	70.29	74.70	82.56	+7.86
iris	4	0	96.00	94.00	96.00	+2.00
letter-recog.	16	0	95.58	88.06	95.51	+7.45
labor-neg	8	8	93.00	83.99	91.67	+7.68
lym	3	15	74.79	74.86	84.10	+9.24
pendigits	16	0	99.35	96.67	98.81	+2.14
satimage	36	0	91.11	86.74	90.82	+4.08
sonar	60	0	82.69	70.20	85.13	+14.93
soybean-s	35	0	100.0	98.00	100.0	+2.00
waveform	21	0	80.86	76.5	83.78	+7.28
wine	13	0	72.94	93.35	95.55	+2.20
anneal	6	32	89.70	93.59	95.11	+1.52
breast-w	10	0	96.85	95.43	96.28	+0.85
diabetes	8	0	69.14	73.03	73.17	+0.14
horse-colic	7	15	66.31	84.81	85.05	+0.24
hypothyroid	7	18	98.26	99.32	98.17	-1.15
ionosphere	34	0	83.96	91.92	91.08	-0.84
pima	8	0	69.14	73.03	73.17	+0.14
segment	19	0	95.58	97.28	96.62	-0.66
shuttle-s	9	0	99.54	99.65	99.74	+0.09
yeast	8	0	54.39	56.14	54.62	-1.52
auto	15	10	40.86	83.18	74.04	-9.14
glass	9	0	67.70	70.01	67.98	-2.03
sick	7	22	93.00	98.78	96.55	-2.23
vehicle	18	0	65.25	73.68	68.71	-4.97
Average			78.98	84.07	86.25	+2.18

– Our method maintains (and sometimes exceeds) the outstanding discriminating power of k-NN when coping with the data sets containing pure continuous attributes. Moreover, the accuracy is greatly improved from k-NN to our approach when data sets contain a mixture of continuous and categorical attributes such as australian, cleve, crx, and so on.

The data sets in this paper do not include those which contain pure *categorical* attributes. Assuming no duplicate instances in a data set which contains pure categorical attributes, then the current classification algorithm is equivalent to the original DeEPs [13] except that the current support summation method is refined. Our previous experimental results show that the accuracy of DeEPs is on average comparable to C5.0 when handling data sets containing pure categorical attributes. The accuracy is sometimes better (e.g., on the tic-tac-toe data set), but sometimes worse (e.g., on the splice data set).

4.2 Accuracy Variation among Folds

The next set of experimental results are used to demonstrate the accuracy variations among the ten folds. We chose the australian, anneal, and sick data sets as examples which are respectively from the three data set groups in Table 2. The results are summarized in Table 3, where "Maximum difference" represents the difference between the minimum and the maximum accuracy in the ten folds.

Table 3. Ten folds accuracy variations in our method, C5.0, and 3-NN.

Data sets	Algorithms	Accuracies(%)			Standard	Maximum
		min	average	max	deviation	difference
australian	ours	81.16	88.41	94.29	3.82	13.13
	C5.0	77.9	85.94	92.8	4.91	14.9
	3-NN	60.00	66.69	74.29	4.91	14.29
anneal	ours	91.09	95.11	99.00	2.27	7.91
	C5.0	90.1	93.59	96.00	1.66	5.9
	3-NN	87.13	89.70	93.00	1.83	5.87
sick	ours	95.76	96.55	97.63	0.63	1.87
	C5.0	98.10	98.78	99.5	0.43	1.40
	3-NN	92.31	93.00	93.90	0.58	1.59

From Table 3, we observe that a better classification algorithm did not remarkably change the standard deviation over the other's, and did not always reduce the standard deviation. This indicates that a better algorithm evenly increases its accuracy on every fold. This point can be also seen from the min and max accuracy change trends from one algorithm to others.

4.3 Effects of Randomisation Process and Neighbourhood Factor

We have also conducted experiments to examine the effect of the ten-fold randomisation process on the accuracy of our method. We set five different randomisation seeds. Accordingly, the five splits over the original data sets should be different. Then, we applied our method to obtain five ten-fold average accuracies based on the different data splits. For the australian data set, the average accuracy over the five splits is 87.11%, and the standard deviation is 0.71. Similarly, for the anneal data set, the average accuracy is 95.32%, and the deviation is 0.15; for the sick data set, the average accuracy is 96.50%, and the deviation is 0.15. Observe that different splits can indeed produce different accuracy (though slightly). Therefore, when comparing the accuracy performance of two classification algorithms, one should strictly apply the two algorithms to the same split of the original data as done in the current work.

Our final round of experiments were conducted to show the effect of the neighbourhood factor α on the accuracy of our classification algorithm. We ran our algorithm when α varied from 0.02, 0.05, 0.08, 0.10, 0.15, to 0.20. The corresponding accuracies on the australian and german data sets are plotted in Figure 2. Note that the two curves in this figure are shaped in a different manner. The left curve reaches its summit when $\alpha = 0.05$, and then goes down with increasing α values. However, the right curve starts with a decline till $\alpha = 0.1$, and then climbs to its summit at $\alpha = 0.2$. These facts strongly indicate that data sets have different properties. They also confirm the usefulness of our idea of selecting an "optimal" neighbourhood for a given data sets by using a guidance role played by a partial training data.

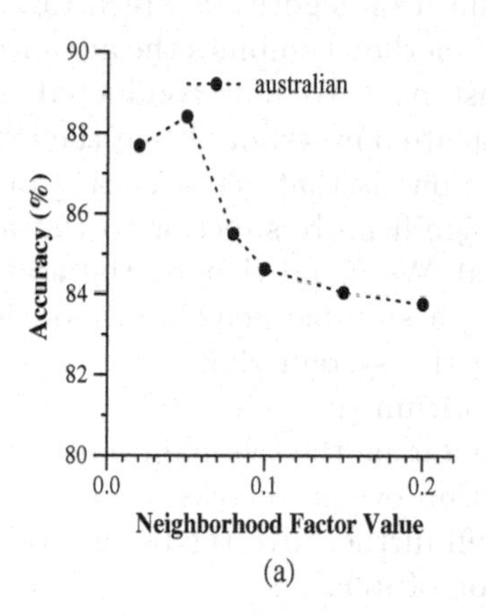

(a)

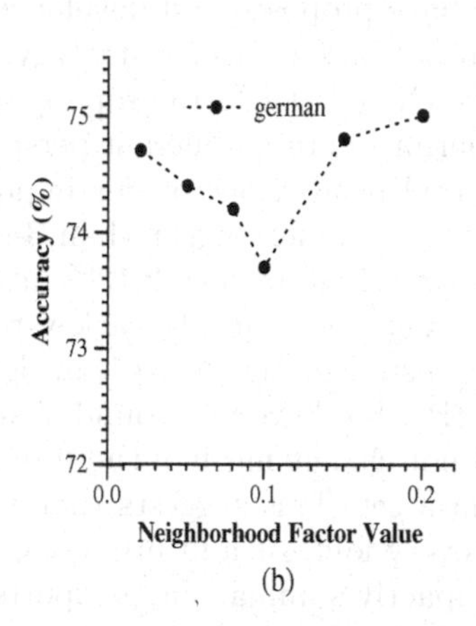

(b)

Fig. 2. Different neighbourhoods of a test instance can produce different accuracies by our method. Partial training data can be used to select proper neighbourhoods of test instances to improve performance of our classifier. (a) Accuracy varies in the range of [84.20%, 88.50%] on the australian data set. (b) Accuracy varies in the range of [73.70%, 75.00%] on the german data set.

4.4 Discussions on Speed

In our experiments, we found that the speed of our method was not much worse (on average, 1.1 times slower) than the original DeEPs. We originally thought that the speed of our method should be faster than the original DeEPs, and even better than k-NN because

1. k-NN calculates the distance of a test instance to all training points, and sorts the distance values. However, our method limits the distance calculation only in the region of a neighbourhood of the test instance. Usually, the points

in such a neighbourhood constitutes a very small percentage of the whole training data.
2. It is very efficient to locate those training data which are covered by the neighbourhood.
3. The number of outliers should be small.

In reality, with the computation of selecting an "optimal" neighbourhood factor and the computation required by DeEPs to make decisions for outliers, the speed of the proposed classifiers was not improved over the original DeEPs. However, the speed was generally improved over (on average, 1.2 times faster than) the original DeEPs when a neighbourhood was fixed.

5 Conclusion

We have proposed and developed a new classification algorithm which takes an instance-based learning strategy. The proposed method combines the advantages of k-NN and DeEPs to properly treat outlier instances. We have conducted many experiments from different perspectives to evaluate our system. From the experimental results, one of our important observations is that the accuracy of our method is much higher than 3-NN and also significantly superior to (on average over 30 data sets, 2.18% higher than) C5.0. We found that partial training data guidance can play a key role in selecting a suitable neighbourhood for a test instance, and hence can determine when the system should use k-NN or DeEPs. We have also found that a better algorithm (in ours, k-NN, and C5.0) did not change much of the standard deviation among the ten fold accuracies on a data set. This suggests that a better algorithm evenly increases its accuracy on every fold. As a future research issue, we will further investigate methods for compactly summarising supports of a collection of itemsets.

References

1. R. Agrawal, T. Imielinski, and A. Swami. Mining association rules between sets of items in large databases. In *Proceedings of the 1993 ACM-SIGMOD International Conference on Management of Data*, pages 207–216, Washington, D.C., May 1993. ACM Press.
2. D. W. Aha, D. Kibler, and M. K. Albert. Instance-based learning algorithms. *Machine Learning*, 6:37–66, 1991.
3. C.L. Blake and P.M. Murphy. The UCI machine learning repository. [`http://www.cs.uci.edu/~mlearn/MLRepository.html`]. In *Irvine, CA: University of California, Department of Information and Computer Science*, 1998.
4. T. M. Cover and P. E. Hart. Nearest neighbour pattern classification. *IEEE Transactions on Information Theory*, 13:21–27, 1967.
5. Guozhu Dong and Jinyan Li. Efficient mining of emerging patterns: Discovering trends and differences. In *Proceedings of the Fifth ACM SIGKDD International Conference on Knowledge Discovery and Data Mining*, pages 43–52, San Diego, CA, 1999. ACM Press.

6. Guozhu Dong, Xiuzhen Zhang, Limsoon Wong, and Jinyan Li. CAEP: Classification by aggregating emerging patterns. In *Proceedings of the Second International Conference on Discovery Science, Tokyo, Japan*, pages 30–42. Springer-Verlag, December 1999.
7. James Dougherty, Ron Kohavi, and Mehran Sahami. Supervised and unsupervised discretization of continuous features. In *Proceedings of the Twelfth International Conference on Machine Learning*, pages 94–202. Morgan Kaufmann, 1995.
8. U. M. Fayyad, G. Piatetsky-Shapiro, and P. Smyth. From data mining to knowledge discovery: An overview. In U.M. Fayyad, G. Piatetsky-Shapiro, P. Smyth, and R. Uthurusamy, editors, *Advances in Knowledge Discovery and Data Mining*, pages 1–34. AAAI/MIT Press, 1996.
9. U.M. Fayyad and K.B. Irani. Multi-interval discretization of continuous-valued attributes for classification learning. In *Proceedings of the 13th International Joint Conference on Artificial Intelligence*, pages 1022–1029. Morgan Kaufmann, 1993.
10. E. Fix and J. Hodges. Discriminatory analysis, non-parametric discrimination, consistency properties. Technical Report Technical Report 4, Project Number 21-49-004, USAF School of Aviation Medicine, Randolph Field, TX, 1951.
11. R. Kohavi, G. John, R. Long, D. Manley, and K. Pfleger. MLC++: A machine learning library in C++. In *Tools with artificial intelligence*, pages 740 – 743, 1994.
12. Pat Langley and Wayne Iba. Average-case analysis of a nearest neighbour algorithm. In *Proceedings of the Thirteenth International Joint Conference on Artificial Intelligence*, pages 889 –894, Chambery, France, 1993.
13. Jinyan Li, Guozhu Dong, and Kotagiri Ramamohanarao. Instance-based classification by emerging patterns. In *Proceedings of the Fourth European Conference on Principles and Practice of Knowledge Discovery in Databases*, pages 191–200, Lyon, France, September 2000. Springer-Verlag.
14. Jinyan Li, Guozhu Dong, and Kotagiri Ramamohanarao. Making use of the most expressive jumping emerging patterns for classification. In *Knowledge and Information Systems: An International Journal*, to appear.
15. Bing Liu, Wynne Hsu, and Yiming Ma. Integrating classification and association rule mining. In *Proceedings of the Fourth International Conference on Knowledge Discovery and Data Mining*, pages 80–86, New York, USA, August 1998. AAAI Press.
16. Dimitris Meretakis and Beat Wuthrich. Extending naive bayes classifiers using long itemsets. In *Proceedings of the Fifth ACM SIGKDD International Conference on Knowledge Discovery and Data Mining*, pages 165–174, San Diego, CA, 1999. ACM Press.
17. J. R. Quinlan. *C4.5: Programs for Machine Learning*. Morgan Kaufmann, San Mateo, CA, 1993.
18. J. R. Quinlan. Improved use of continuous attributes in C4.5. *Journal of Artificial Intelligence Research*, 4:77–90, 1996.

A Scalable Algorithm for Rule Post-pruning of Large Decision Trees

Trong Dung Nguyen, Tu Bao Ho, and Hiroshi Shimodaira

Japan Advanced Institute of Science and Technology
Tatsunokuchi, Ishikawa, 923-1292 JAPAN
{nguyen, bao, sim}@jaist.ac.jp

Abstract. Decision tree learning has become a popular and practical method in data mining because of its high predictive accuracy and ease of use. However, a set of if-then rules generated from large trees may be preferred in many cases because of at least three reasons: (i) large decision trees are difficult to understand as we may not see their hierarchical structure or get lost in navigating them, (ii) the tree structure may cause individual subconcepts to be fragmented (this is sometimes known as the "replicated subtree" problem), (iii) it is easier to combine new discovered rules with existing knowledge in a given domain. To fulfill that need, the popular decision tree learning system C4.5 applies a rule post-pruning algorithm to transform a decision tree into a rule set. However, by using a global optimization strategy, C4.5rules functions extremely slow on large datasets. On the other hand, rule post-pruning algorithms that learn a set of rules by the separate-and-conquer strategy such as CN2, IREP, or RIPPER can be scalable to large datasets, but they suffer from the crucial problem of overpruning, and do not often achieve a high accuracy as C4.5. This paper proposes a scalable algorithm for rule post-pruning of large decision trees that employs incremental pruning with improvements in order to overcome the overpruning problem. Experiments show that the new algorithm can produce rule sets that are as accurate as those generated by C4.5 and is scalable for large datasets.

1 Introduction

Data mining algorithms have usually to deal with very large databases. For the prediction data mining task, in addition to the requirements of high accurate and understandability of discovered knowledge, the mining algorithms must be scalable, i. e., given a fixed amount of main memory, their runtime increases linearly with the number of records in the input database.

Decision tree learning has become a popular and practical method in data mining because of its significant advantages: the generated decision trees usually have acceptable predictive accuracy; the hierarchical structure of generated trees makes them are quite easy to understand if trees are not large; and especially the learning algorithms, which employ the *divide-and-conquer* (or simultaneous covering) strategy to generate decision trees, do not require complex processes

D. Cheung, G.J. Williams, and Q. Li (Eds.): PAKDD 2001, LNAI 2035, pp. 467–476, 2001.

of computation. However, it happens that in certain domains the comprehensibility and predictive accuracy of decision trees decrease considerably because of the problem known as *subtree replication* [10] (when the subtree replication occurs, identical subtrees can be found at several different places in the same tree structure).

The solution to the problem of subtree replication in the most well-known decision tree learning system C4.5 [11] is to convert a generated decision tree into a set of rules using a *post-pruning* strategy [8]. The conversion of trees into rules is not only an effective way to avoid the subtree replication problem but also offers other significant advantages: while large trees generated from large datasets are difficult to understand, discovered knowledge in form of rules is much easier to understand. Also, in our practical experience domain experts often feel more comfortable to analyze and validate rules than trees if trees become large. Moreover, it appears that the generated rule sets usually have equal or higher predictive accuracy than the original decision tree. However, the C4.5rules algorithm is not scalable to large databases as the simulated annealing, which is employed to achieve an optimal generalization, requires $O(n^3)$ time complexity where n is the number of records in the input database [4].

The *separate-and-conquer* (or simultaneous covering) strategy is an alternative approach to learn rules directly from databases. The most well-known separate-and-conquer algorithms include CN2 [2], REP [1], IREP [6], RIPPER [3], PART [5]. Among them, CN2 and REP also require a computation with high complexity, and therefore cannot be applicable to large data bases. IREP and RIPPER solve the problem of complexity by using a scheme called *incremental pruning*. The result is that they can run very fast and generate small rule sets with acceptable predictive accuracy. However, incremental pruning may lead to the problem of *overpruning* (or hasty generalization) that reduces the accuracy of the algorithms in many cases. PART [5] is an attempt to combine divide-and-conquer and separate-and-conquer strategies, and was claimed to be effective and efficient.

This paper concerns with scalable algorithms for rule-post pruning from large decision trees. In particular it proposes a solution to the problem of high complexity in C4.5rules by using a scheme similar to incremental pruning. The essence of the proposed algorithm is to avoid the problem of overpruning by appropriate improvements in incremental pruning. Experiments show that the proposed algorithm produces rule sets that as accurate as those generated by C4.5 and is scalable for very large data sets.

2 Related Works

A variety of approaches to learning rules have been investigated. One is to begin by generating a decision tree, then to transform it into a rule set, and finally to simplify the rules (the *divide-and-conquer* strategy as used in the system C4.5 [11]). Another is to use the *separate-and-conquer* strategy [10] to generate and an initial rule set, then applying a rule pruning algorithm.

2.1 Rule Post-pruning in C4.5

The rule learner in C4.5 does not employ a separate-and-conquer method to generate a set of rules—it achieves this by simplifying an unpruned decision tree using the decision tree inducer included in the C4.5 software. Then it transforms each leaf of the decision tree into a rule. This initial rule set will usually be very large because no pruning has been done. Therefore C4.5 proceeds to prune it using various heuristics.

First, each rule is simplified separately by greedily deleting conditions in order to minimize the rule's estimated error rate. Following that, the rules for each class in turn are considered and a "good" subset is sought, guided by a criterion based on the minimum description length principle. The next step ranks the subsets for the different classes with respect to each other to avoid conflicts, and determines a default class. Finally, rules are greedily deleted from th whole rule set one by one, so long as this decreases the rule set's error on the training data.

Unfortunately, the global optimization process is rather lengthy and time-consuming. Cohen [3] shows that C4.5 can scale with the cube of the number of examples on noisy datasets.

2.2 Other Related Rule Pruning Algorithms

The earliest approaches to pruning rule sets are based on *global optimization*. These approaches build a full, unpruned rule set using a separate-and-conquer strategy. Then they simplify these rules by deleting conditions from some of them, or by discarding entire rules. The simplification procedure is guided by a pruning criterion that the algorithm seeks to optimize [1]. The optimized solution can only be found via exhaustive search. In practice, some heuristic searches are applied, but they are still quite time consuming.

There is a faster approach to rule pruning called incremental pruning that is introduced first in IREP [7], and also used in RIPPER and RIPPERk [3]. The key idea is to prune a rule immmediately after it has been built, before any new rules are generated in subsequent steps of the separate-and-conquer algorithm. By integrating pruning into each step of the separate-and-conquer algorithm, this approach can avoid the high complexity of a global optimization process.

2.3 The Problem of Overpruning

Although the incremental pruning used in IREP (and its variants) avoids the high complexity process of global optimization of C4.5rules, it may suffer the problem of overpruning or hasty generalization [5]. By using pre-pruning approach, the algorithm does not know about potential new rules when it consider a rule to prune; the pruning decisions are based on the accuracy estimation of the current rule only. In other words, the algorithm cannot estimate how the pruning decisions on a single rule will effect the accuracy of the whole final rule set. Therefore, it may happen that the pruning decisions increase the accuracy

Table 1. Potential rules in a hypothetical dataset

Rule	Coverage			
	Growing Set		Pruning Set	
1: $A = true \rightarrow yes$	600	60	200	20
2: $A = false \wedge B = true \rightarrow yes$	1200	60	400	20
3: $A = false \wedge B = false \rightarrow no$	0	30	0	10

of the current rule but may in fact decrease the accuracy of the potential final rule set on the same estimation.

Table 1 shows a simple example of overpruning. The example is taken from [5] with a modification to make it easier to calculate. Consider a binary dataset with two attributes A and B; examples can belong to either class *yes* or *no*. There are three potential rules on the dataset as shown on the table. Assume that the algorithm generates first rule

$$A = true \rightarrow yes. \tag{1}$$

Now consider whether the rule should be further pruned. Its error rate on the pruning set is $1/10$, and the pruned rule

$$\rightarrow yes \tag{2}$$

has an error rate of $1/13$, which is smaller, thus the rule will be pruned to that null rule. As the null rule covers all the data the algorithm stops and satisfies with a final rule set consisting only of that trivial rule. But the found rule set actually has a greater error rate compare comparing to $4/65$ which is the error rate of the set three rules showed in the table. Note that this happens because the algorithm concentrates on the accuracy of rule 1 when pruning—it does not make any guesses about the benefits of including further rules in the classier.

3 A Scalable Algorithm for Rule Post-pruning

As an attempt to solve the high complexity problem of C4.5rules we have developed a new algorithm that adopts a scheme similar to incremental pruning used in IREP, we have named it CABROrule as it is integrated in our decision tree learning CABRO [9]. Similar to C4.5rules, CABROrule uses a bottom-up search instead of IREP's top-down approach: The final rule set is found by repeatedly removing conditions and rules from an input unpruned rules rather than adding new rules to an initial empty set. In the other words, CABROrule uses post-pruning approach in contrast to pre-pruning one used in IREP. By taking the advantage of working with a full grown rule set throughout the pruning process, we can improve the incremental pruning scheme in CABROrule to avoid the problem of hasty generalization.

Table 2. The Main Procedure of CABROrule

```
procedure CABROrule(UnprunedSet, Data)

   PrunedSet ← ∅
   while (Data ≠ ∅)
      Rule ← SelectRule(UnprunedSet, Data)
      UnprunedSet ← UnprunedSet \ {Rule}
      PrunedRule ← PruneRule(Rule, UnprunedSet, Data)
      PrunedSet ← PrunedSet ∪ {PrunedRule}
      Data ← Data \ Match(PrunedRule, Data)
   return PrunedSet
```

3.1 Description of the Algorithm

Similar to C4.5rules, CABROrule begins with a set of unpruned rules. The rule set is taken directly from an unpruned decision tree where each rule corresponds to a path from the tree root and a leave node. To prune the rule set, CABROrule follows a separate-and-conquer strategy: first choosing one rule to prune at a time, then removing the covered examples and repeating the process on the remaining examples. Table 2 shows the main procedure of CABROrule, the procedure to prune a single rule is in Table 3.

The efficiency of CABROrule comes from the avoidance of the process of searching for an "optimized" subset of rules such as the one in C4.5rule. We will analyze the reason why C4.5rules requires a global optimization but CABROrule does not. In C4.5rules, each individual rule is pruned with respect to the *all* training data. Deleting conditions from a rule—and thereby increasing its coverage—ultimately may result in a rule set with many overlaps. The optimized exclusive subset of rules can only be found via exhaustive search. In practise, exhaustive search is infeasible and C4.5rules apply some heuristic approximations (two alternatives in C4.5rules are greedy search and simulated annealing), but even these approximate algorithms are quite time consuming. In contrast, each single rule in CABROrule is pruned with respect to the *remaining* training data after removing all examples covered by previous pruned rules. The exhaustion of training data serves as a stop condition; and the final rule set has no overlap. It is noticeable that while there is no natural order for rules generated by C4.5rules, CABROrule generates ordered rule sets those sometimes are known as decision lists [12].

To calculate the complexity of CABROrule, we assume that the data set consists of n examples described by a attributes. To choose a condition to prune the procedure *PruneRule* needs to examine all the conditions of the considered rule each require n tests on the training examples. Therefore complexity of pruning a condition is $O(an)$. Suppose the length of an unpruned rule is a, then pruning a rule requires $O(a^2n)$. If we assume that the size of the final theory is constant

Table 3. Pruning a Single Rule

```
procedure PruneRule(Rule, UnprunedSet, Data)

  repeat
    Accuracy ← EstimateAccuracy({Rule} ∪ UnprunedSet, Data)
    DeltaAccuracyA ← 0
    for each (Condition ∈ Rule) do
      NewRule ← Rule \ Condition
      NewAccuracy ← EstimateAccuracy({NewRule} ∪ UnprunedSet, Data)
      NewDeltaAccuracy ← NewAccuracy − Accuracy
      if (NewDeltaAccuracy > DeltaAccuracy)
        DeltaAccuracy ← NewDeltaAccuracy
        BestCondition ← Condition
    if (DeltaAccuracy > 0)
      Rule ← Rule \ Condition
  until (DeltaAccuracy < 0)
  return Rule
```

[6], the complexity will be linear to n. Because a decision tree can be built in time $O(nlogn)$, the overall cost to build a rule set from data is $O(nlogn)$. The complexity is the same as that of PART [5] and better than $O(nlog^2n)$ of IREP or RIPPER, or $O(n^3)$ of C4.5rules.

Before going to the next subsection which addresses the problem of overpruning we discuss briefly about the procedure *SelectRule* in CABROrule. There is only a number of input rules those have their chance to be considered to prune and add to the final rule set. Certainly, we want as many "significant" rules having that chance as possible. A measure is necessary to judge the "significance" of a rule. In general, there are several existing measures those may be suitable for that purpose such as *relative frequency*, *m-estimate of accuracy*, or *entropy* [8]. However, when we apply CABROrule on an unpruned decision tree resulting from C4.5 we use the coverage of a rule as a criterion to select which rule will be prune first. That because when growing a decision tree, C4.5 already optimized each path (corresponding to an unpruned rule) by information gain. Therefore, it is reasonable that rules with larger coverage may be more important and need to be considered first in the pruning process. If CABROrule is applied on rule sets that grown by other algorithms, other criteria for selecting rules can be better choices.

3.2 Avoiding Overpruning

CABROrule uses a greedy search algorithm for pruning a single rule. At a time, the algorithm searches for a condition to prune. The pruning continues until the accuracy estimation cannot be improved anymore. A description of the algorithm for pruning a single rule is in Table 3.

To overcome the problem of overpruning in the original algorithm of incremental pruning used in IREP and its variants, CABROrule takes a different approach to estimate the accuracy when making pruning decisions. Instead of estimating the accuracy only on the rule under consideration, the procedure *EstimateAccuracy* does estimation on that rule together with all remaining unpruned rules. As we have stated in the previous section, the overpruning occurs when pruning decisions on a rule improving the accuracy estimation of that rule, but in fact potentially reducing the accuracy of the final rule set. By taking into account of remaining unpruned rules when pruning a single rule, we can make sure that a condition will be pruned if that potentially improves the accuracy on the whole final rule set not only on that single rule locally.

We return to the example of overpruning in section 2 to illustrate the new approach. Assume that the CABROrule considers pruning rule 1 back to a null rule. Instead of estimating the accuracy of only rule 1 before and after pruning it, the algorithm does estimation on all three rules to make that pruning decision. From Table 2, we can see that after pruning rule 1, the accuracy estimation is 1/13 which less than 4/65 before the pruning decision. Therefore the algorithm cancels that pruning decision and avoids a case of overpruning.

The problem of overpruning or hasty generation is not restricted to a particular method of accuracy estimation [5], and our solution to the problem does not depend on estimation methods. The estimation of reduce error pruning is used in the example only to make calculations easier. In CABROrule we uses pessimistic estimation [11] similar to the one used in C4.5 to estimate the accuracy of a rule set. The estimation is done by calculating the rule accuracy over the training data, then calculating the standard deviation in this estimated accuracy assuming a binomial distribution. For a given confidence level (we used 95% in our experiments), the lower-bound estimate is then taken as the measure of rule performance. The accuracy estimate of a rule set is the average of the estimates over its members with respect to their coverage on the data set.

4 Experimental Results

In order to evaluate the performance of CABROrule we designed two experiments. The first experiment evaluates the predictive accuracy of CABROrule comparing to C4.5 and C4.5rules, the second evaluates the run-time of CABROrule comparing to C4.5rules.

For the first experiment we used 31 standard datasets from UCI collection. The datasets and their characteristics, together with experimental results are listed in Table 4. We performed 10-fold cross-validation on these datasets with C4.5, C4.5rules and CABROrule. The same folds were used for each program. A numbers in the result columns is the average of error rates or size of rule sets over ten times of running. A symbol "•" in the last column indicates that CABROrule has an error rate lower than both C4.5 and C4.5rules on that dataset, while a

Table 4. Experimental Results

Dataset	#Exam	NumAtt	NomAt	Class	C4.5 size	C4.5 error	C4.5rules size	C4.5rules error	CABROrule size	CABROrule error	
anneal	898	6	32	5	60.0	4.3	11.3	4.4	12.0	3.1	•
audiology	226	0	69	24	49.0	9.3	21.0	8.8	22.0	8.8	
australian	690	6	9	2	34.5	15.2	13.2	16.2	9.6	14.5	•
auto	205	15	10	6	68.7	19.5	22.2	18.4	20.8	20.4	×
balance-scale	625	4	0	3	82.0	22.1	37.0	21.1	28.7	22.2	×
breast	699	9	0	2	27.4	5.1	9.0	4.6	8.2	4.9	×
breast-cancer	286	0	9	2	12.1	25.9	7.8	29.7	3.0	26.2	◦
german	100	7	13	2	86.0	29.7	19.7	29.6	14.6	29.1	•
glass	214	9	0	6	44.0	32.7	13.8	30.8	13.6	30.8	
glass2	163	9	0	2	23.4	21.9	8.1	20.2	8.0	20.2	
heart	303	6	7	2	24.0	12.2	8.8	14.4	7.2	11.5	•
hepatitis	155	6	13	2	18.6	25.2	8.4	20.1	6.7	21.4	×
horse-colic	168	7	15	2	8.2	15.2	5.9	16.0	4.1	14.7	•
hypothyroid	3772	7	22	4	12.2	0.6	6.0	0.6	5.1	0.6	
ionosphere	351	34	0	2	25.0	9.4	9.1	8.8	9.2	8.8	
iris	150	4	0	3	8.8	5.3	4.1	4.6	4.0	4.6	
labor-neg	57	8	8	2	5.7	19.3	4.0	21.0	2.6	19.3	◦
lymphography	148	3	15	4	26.9	22.8	9.6	22.8	9.2	22.9	×
mushroom	8124	0	22	2	29.7	0.0	17.0	0.0	16.9	0.0	
pima	768	8	0	2	45.2	25.7	10.7	26.3	9.9	25.7	◦
primary-tumor	339	0	17	21	77.8	59.3	17.1	60.2	13.9	59.9	◦
segment	2310	19	0	7	87.0	2.8	28.2	3.7	27.6	3.7	
sick-euthyroid	372	7	22	2	24.6	2.2	12.0	2.4	9.2	2.2	◦
sonar	208	60	0	2	27.2	28.9	8.7	30.3	8.7	30.3	
soybean-large	683	0	25	19	94.9	7.8	35.8	7.0	3.1	6.7	•
splice	3190	0	61	3	220.2	5.9	74.1	6.5	60.0	6.0	◦
vehicle	840	18	0	4	135.8	28.7	26.6	27.1	25.9	26.8	•
vote	435	0	16	2	13.0	6.0	6.4	5.3	5.1	6.2	×
waveform-21	301	21	0	3	542.2	23.2	68.1	22.4	67.6	22.3	•
waveform-40	5002	34	0	3	584.6	24.9	66.1	23.4	68.0	23.0	•
zoo	101	1	15	7	17.4	7.6	7.8	7.6	7.8	7.6	

"◦" indicates that CABROrule has an error rate lower than C4.5rules, and a "×" indicates that C4.5rules has an error rate lower than that of CABROrule.

We can observe from Table 4 that CABROrule outperforms C4.5rules on 15 over 31 datasets, among them there are 9 datasets CABROrule outperforms both C4.5rules and C4.5, whereas C4.5rules has a lower error rate comparing to CABROrule on 6 datasets (totally C4.5rules has lower error than C4.5rules on 15 datasets, while is with higher error rate on 4 datasets). In some datasets the differences between error rates are too small to say that they are significant, but this experiment showed that CABROrule at least as good as C4.5rules if not better in the predictive accuracy.

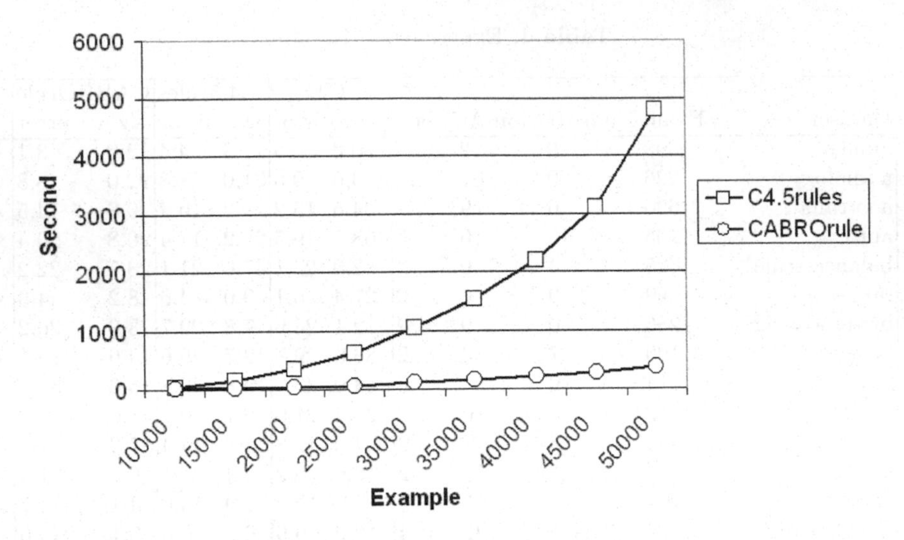

Fig. 1. Comparison of Running Time

About the size of rule sets, CABROrule generated smaller rule sets on a major number of datasets comparing to C4.5rules, and both reduce the number of rules comparing to C4.5 substantially. That reduction, in many case, will increase the understandability of result models, and this experiment reconfirms the advantage of transforming decision trees to rules.

In order to evaluate the efficiency of CABROrule, the second experiment is done with the census-income dataset. We began with 10000 examples and repeatedly ran CABROrule and C4.5rules, each time with a bigger number of examples, to learn in what order the run-time increases according to the size of data. Figure 1 is the graph drawn from the experiment results. This graph confirms and illustartes that the run-time of C4.5rules is higher than $O(n^2)$, while the run-time of CABROrule is about $O(nlogn)$ that confirms our calculation about the algorithm complexity in the previous section.

Some significant conclusions can be drawn from these two experiments:

- By using incremental pruning approach to post-pruning problem, CABROrule can reduce the run-time substantially in comparion to C4.5rules. It allows us to apply the algorithm to large datasets those are very common in data mining.
- There is no lost in criteria of predictive accuracy and model size. In fact, there is some gain in accuracy, and CABROrule usually generates smaller rule sets in comparion to C4.5rules.
- Transferring decision trees to rules may increase both understandability and predictive accuracy of models.

5 Conclusion

This paper has presented a new algorithm for rule post-pruning of decision trees. It can be considered an alternative algorithm for C4.5rules when the input data become very large. The problem of high complexity in C4.5 is solved by adopting an incremental pruning scheme. However the algorithm does not suffer the problem of hasty generalization such as in the original incremental pruning approach. Experiments have shown that the new algorithm generates rule sets as accuracy as those of C4.5 but with far less time of computation.

References

1. Brunk, C. A. and Pazzani, M. J.: An Investigation of Noise-Tolerant Relational Concept Learning Algorithms. *Proceedings of the 8th International Workshop on Machine Learning* (1991), 389–393.
2. Clark, P., Niblett, T.: The CN2 Induction Algorithm. *Machine Learning*, 3 (1989), 261–283.
3. Cohen, W. W.: Fast Effective Rule Induction. *Proceeding of the 12th International Conference on Machine Learning*, Morgan Kaufmann (1995) 115-123.
4. Cohen, W. W.: Efficient Pruning Methods for Separate-and-Conquer Rule Learning Systems. *Proceeding of the 13th International Joint Conference on Artificial Intelligence*, Chambery, France (1993), 988–995.
5. Frank, E., Witten, I.H.: Generating Accurate Rule Sets Without Global Optimization. *Proceeding of the 15th International Conference on Machine Learning*, Morgan Kaufmann (1998), 144–151.
6. Furnkranz, J.: Pruning Algorithms for Rule Learning. *Machine Learning*, 27 (1997), 139–171.
7. Furnkranz, J. and Widmer, G.: Incremental Reduced Error Pruning. *Proceeding of the 11th International Conference on Machine Learning*, Morgan Kaufmann (1994), 70–77.
8. Mitchell, T.M.: *Machine Learning*, McGraw-Hill (1997).
9. Nguyen, T.D. and Ho, T.B.: An Interactive-Graphic System for Decision Tree Induction. *Journal of Japanese Society for Artificial Intelligence*, Vol. 14, N. 1 (1999), 131–138.
10. Pagallo, G. and Haussler, D.: Boolean Feature Discovery in Empirical Learning. *Machine Learning*, 5 (1990), 71–99.
11. Quinlan, J. R.: *C4.5: Programs for Machine Learning*, Morgan Kaufmann (1993).
12. Rivest, R. L.: Learning Decision Lists. *Machine Learning*, 2 (1987), 229–246.

Optimizing the Induction of Alternating Decision Trees

Bernhard Pfahringer, Geoffrey Holmes, and Richard Kirkby

University of Waikato, Hamilton, New Zealand,
{bernhard, geoff, rbk1}@cs.waikato.ac.nz,
http://www.cs.waikato.ac.nz/~ml

Abstract. The alternating decision tree brings comprehensibility to the performance enhancing capabilities of boosting. A single interpretable tree is induced wherein knowledge is distributed across the nodes and multiple paths are traversed to form predictions. The complexity of the algorithm is quadratic in the number of boosting iterations and this makes it unsuitable for larger knowledge discovery in database tasks. In this paper we explore various heuristic methods for reducing this complexity while maintaining the performance characteristics of the original algorithm. In experiments using standard, artificial and knowledge discovery datasets we show that a range of heuristic methods with log linear complexity are capable of achieving similar performance to the original method. Of these methods, the random walk heuristic is seen to outperform all others as the number of boosting iterations increases. The average case complexity of this method is linear.

1 Introduction

Highly accurate classifiers can be found using the boosting procedure but as [6] discovered for standard decision trees the combination of each classifier produced at each iteration into a single classifier is multiplicative. Freund and Mason [4] introduced a method capable of inducing a single classifier without the exponential growth in tree size by changing the representation of the underlying tree.

Standard decision trees have interior nodes that perform tests on the data and leaf nodes labelled with class values. Classification is achieved by following the unique path from the root to a leaf for a given unknown instance. The alternating decision tree introduces a new node called a predictor node which can be either an interior or a leaf node. The tree has a predictor node at its root and then alternates between test node and further predictor nodes, hence the name. Classification is achieved by summing the contributions from the predictor nodes of all paths that an instance successfully traverses. A positive sum implies membership of one class and a negative sum membership of the other. While the original algorithm was restricted to two class problems it appears that the algorithm can be extended to multiclass problems by using the framework of [7].

Each boosting iteration adds a test (weak hypothesis) and two predictor nodes to the tree. The test chosen to extend the tree is the one that minimizes a

D. Cheung, G.J. Williams, and Q. Li (Eds.): PAKDD 2001, LNAI 2035, pp. 477–487, 2001.

function that measures the "impurity" of the test. The tree can be extended from any of its existing predictor nodes which means that for each boosting iteration the minimization function must be computed for each possible test, i.e. the algorithm is quadratic in the number of boosting iterations. This paper explores heuristic methods for restricting the number of predictor nodes that need to be examined for the possible addition of new test nodes. By maintaining the performance levels of the original algorithm and reducing its complexity we aim to demonstrate that it is possible to produce a practical form of the alternating decision tree that can be applied to larger knowledge discovery tasks.

The paper is organized as follows. In the next section we outline our interpretations of the original algorithm (some aspects of the induction of alternating decision trees were not clearly defined in the original paper). Section 3 looks at ways in which the original algorithm can be made more efficient with no loss of performance. While these techniques improve the algorithm they do not have any effect on the overall complexity, and so Section 4 introduces three heuristic search mechanisms for constructing useful paths in the tree without exploring all tests at each predictor node. Each of these methods is log-linear in the worst case. Section 5 outlines an experiment to determine the efficacy of the heuristic methods compared to the original. Accuracy, runtime, and "shallowness" of the tree are measured for a range of standard, artificial and knowledge discovery in database datasets. Shallowness is measured as the number of leaves in the tree. This measure gives a picture of the effect the heuristics have on the overall shape of the trees that they prune. Section 6 provides a discussion of the results and outlines some avenues for further work.

2 Inducing Alternating Decision Trees

Alternating decision trees provide a mechanism for combining the weak hypotheses generated during boosting into a single representation. Keeping faith with the original implementation, we use inequality conditions that compare a single feature with a constant as the weak hypotheses generated during each boosting iteration. In [4] some typographical errors and omissions make the algorithm difficult to implement so we include below a more complete description of our implementation.

At each boosting iteration t the algorithm maintains two sets, a set of preconditions and a set of rules, denoted $\mathcal{P}_t$ and $\mathcal{R}_t$, respectively. A further set $\mathcal{C}$ of weak hypotheses is generated at each boosting iteration.

Initialize. Set the weights associated with each training instance to 1. Set the first rule $\mathcal{R}_1$ to have a precondition and condition which are both true. Calculate the prediction value for this rule as $a = \frac{1}{2} \ln \frac{W_+(c)}{W_-(c)}$ where $W_+(c)$, $W_-(c)$ are the total weights of the positive and negative instances that satisfy condition c in the training data. The initial value of c is simply True.

Pre-adjustment. Reweight the training instances using the formula

$$w_{i,1} = w_{i,0} e^{-a y_t}$$

(for two class problems, the value of y_t is either +1 or -1).

Do for t = 1, 2, ..., T

1. Generate the set $\mathcal{C}$ of weak hypotheses using the weights associated with each training instance $w_{i,t}$

2. For each base precondition $c_1 \in \mathcal{P}_t$ and each condition $c_2 \in \mathcal{C}$ calculate

$$Z_t(c_1, c_2) = 2\left(\sqrt{W_+(c_1 \wedge c_2) W_-(c_1 \wedge c_2)} + \sqrt{W_+(c_1 \wedge \neg c_2) W_-(c_1 \wedge \neg c_2)} \right) + W(\neg c_1)$$

3. Select c_1, c_2 which minimize $Z(c_1, c_2)$ and set $\mathcal{R}_{t+1}$ to be $\mathcal{R}_t$ with the addition of the rule r_t whose precondition is c_1, condition is c_2 and two prediction values are:

$$a = \frac{1}{2} \ln \frac{W_+(c_1 \wedge c_2) + 1}{W_-(c_1 \wedge c_2) + 1}, \qquad b = \frac{1}{2} \ln \frac{W_+(c_1 \wedge \neg c_2) + 1}{W_-(c_1 \wedge \neg c_2) + 1}$$

4. Set $\mathcal{P}_{t+1}$ to be $\mathcal{P}_t$ with the addition of $c_1 \wedge c_2$ and $c_1 \wedge \neg c_2$.

5. Update the weights of each training example according to the equation

$$w_{i,t+1} = w_{i,t} e^{-r_t(x_i) y_t}$$

Output the classification rule that is the sign of the sum of all the base rules in R_{T+1}:

$$class(x) = sign\left(\sum_{t=1}^{T} r_t(x) \right)$$

The best value of T for stopping the boosting process is still an open research question. In [4] the value is decided by cross-validation. In this paper we look at the effects of heuristics on fixed values for T.

Figure 1 depicts a sample alternating decision tree. A hypothetical example with attribute values $A1 = true$ and $A2 = false$ would be classified according to the following sum derived by going down all appropriate paths in that tree collecting all prediction values encountered: $0.5 + -1.2 + -3.4 + 0.2 = -3.9$ (indicated by horizontal arrows in the figure).

The changes to the original algorithm are fairly minor - the pre-adjustment phase may have been "implicitly" defined, and the change in the Z_t formula to represent all instances that do not satisfy the precondition must be typographical as is the missing minus sign in the updating phase. The formulas for the newly generated predictor nodes in stage 3 have a unit value added to avoid zero-frequency problems [8].

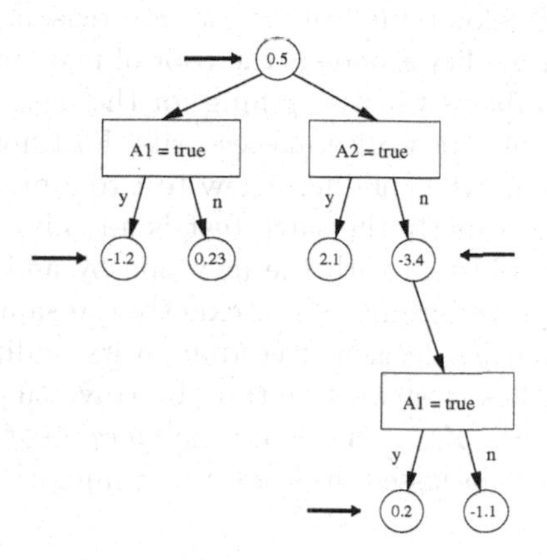

Fig. 1. A sample ADTree. The horizontal arrows indicate all predictor nodes encountered when classifying an example with $A1 = true$ and $A2 = false$.

3 Optimizing the Original Algorithm

The algorithm described in Section 2 is quadratic in the number of boosting iterations because the calculation of the Z-value for each of the set of $\mathcal{C}$ weak hypotheses is performed at every predictor node in the tree. While this complexity is unavoidable, there are ways to avoid performing the Z-value calculation unnecessarily. We call this value $Z_{pure}(c)$. It is the best possible Z-value that would result from a pure split of the training instances under consideration.

$$Z_{pure}(c) = 2(\sqrt{W_+(c)} + \sqrt{W_-(c)}) + W(\neg c_1)$$

Straightforwardly using the formula for Z given in the previous section would have yielded a lower bound of $Z_{pure} = W(\neg c_1)$ only. But as we adjust all weight sums in a way reminiscent of the Laplace correction, we are able to derive this more stringent lower bound. Z_{pure} is a lower bound on the Z-value a good test could possibly achieve [1]. So if Z_{pure} for some predictor node in the tree is worse

[1] We omit the proof here, but basically one has to show that the following inequality holds for all $a, b, c, d \geq 0$: $\sqrt{a+b+1} + \sqrt{c+d+1} \leq \sqrt{(a+1)(c+1)} + \sqrt{(b+1)(d+1)}$.

than the best test found so far, we do not need to evaluate any test at this node. Furthermore, one can show that all possible tests at all successor nodes of this node are also bounded by the same Z_{pure}, as they involve subsets of the current set of examples only. Therefore we can omit evaluating the complete subtree rooted at a node cutoff by Z_{pure}.

Duplicated tests have been identified as another source of unnecessary inefficiency. Especially with larger numbers of boosting iterations (100 and 200 in our experiments reported below) duplicated tests are reasonably common to justify special attention. If a predictor node is the root of two identical tests, both tests will induce identical subsets when searching for the next best test to add to the tree. Thus we will duplicate work unnecessarily. Fortunately, there is a simple remedy for the problem: when adding a new test to a predictor node, we simply need to check whether exactly the same test is already present at this node. If so, we just merge the old test with the new one by adding the respective prediction values. This procedure results in exactly the same predictive behaviour of the induced alternating decision tree due to its additive nature. But when determining the next best test we save time by traversing a smaller tree.

The effects of both the Z_{pure} cutoff and the *merging of tests* have been studied experimentally and are discussed in Section 5. Summary results are depicted in Figure 3.

4 Heuristic Search Variants

Even though both methods described in the previous section do improve the efficiency of the algorithm, they do not alter its quadratic nature in general. Determining the next best test to add still involves looking at (almost all) current predictor nodes and for each of those evaluating all possible tests. With every new test, i.e. at every boosting iteration where we are not able to merge tests, we add two more predictor nodes to the tree. A way of reducing the total complexity is to limit the search to just a subset of all predictor nodes, hopefully including the node that would have yielded the next best test using the exhaustive search of the original induction algorithm.

Figure 2 demonstrates the heuristic we have chosen to investigate here. Instead of recursively exploring the complete tree, we limit search at each boosting iteration to just one path down the tree. Obviously this must reduce the complexity, as now we will only be exploring a logarithmically-sized subset of all predictor nodes. Additionally, this procedure seems to yield more shallow trees on average, thus improving efficiency further. On the other hand, such heuristically induced trees are different from the original trees, so we will have to explore whether we are trading off gains in efficiency for worse predictive error or less comprehensible trees, or even both.

The next section will empirically explore these questions, but first we need to define the heuristics used for determining the particular paths to be explored. Basically, we would like to have a good chance of including the node with the best test. In order to achieve this we invented the first two of the following three

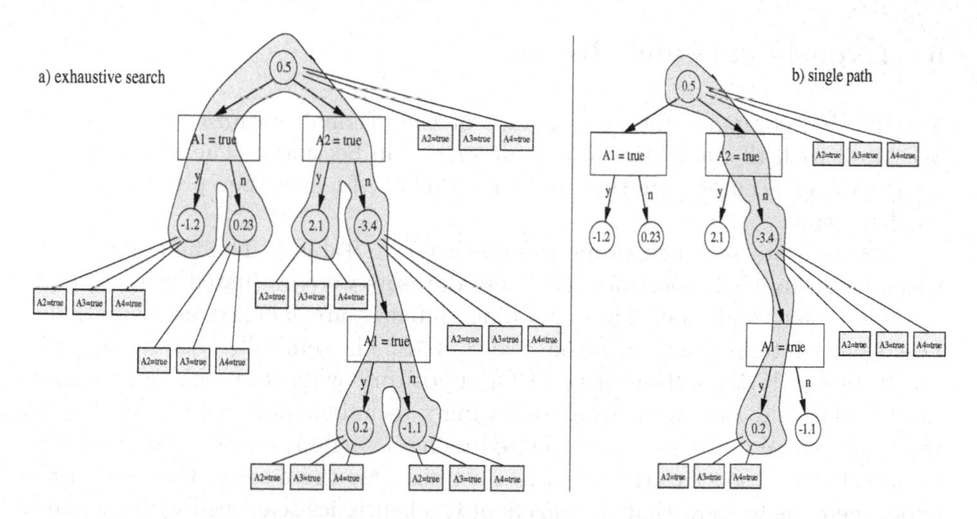

Fig. 2. The exhaustive method (a) has to evaluate all possible additional tests for all predictor nodes. Going down just one path (b) considerably reduces the number of tests to evaluate.

heuristics. The third heuristic was initially added to simply function as a bottom line for comparisons, but turned out to perform pretty well in practise, too.

1. Heaviest path: looking at the formula for Z we see that larger sets of more important examples, i.e. "heavier" sets can lead to a larger reduction, provided we find a test that separates both classes reasonably well. Therefore this heuristic always follows the path of the heaviest[2] subset of examples.
2. Best possible Z_{pure}: reflecting on the previous heuristic we see that it sometimes might lead us astray. The heaviest subset could consist of large a number of examples of just one class, so every conceivable split would still not result in a particularly good Z value. Consequently, this heuristic chooses to follow down the path of the subset with the smallest possible value for Z_{pure}. Clearly, it too cannot provide any guarantees on whether we will be able to find such a split performing as well as theoretically possible.
3. Random walk: this heuristic is a bottom line for comparison and it involves the least computational effort of all. Interestingly, as its choices are purely random, it will explore all paths with equal probability. So every single path has a fair chance of being chosen for evaluation at some boosting iteration or the other.

No matter which method we choose for selecting a single path, we will always be exploring considerably less predictor nodes, which should at least result in considerable savings in terms of time needed for induction. We will try to quantify these savings empirically in the next section.

[2] Heaviest is literally correct here as we sum the weights of the subset of examples at a node to determine how heavy it is.

5 Experiments and Results

This section compares the performance of the original optimized algorithm of Section 3 with the heuristic variants described in Section 4. The methods were compared for accuracy, runtime and the number of leaves they produced in the resulting tree.

The datasets and their properties are listed in Table 1. The first sixteen are taken from the UCI repository [1]. These datasets were evaluated using a single ten-fold cross validation. The remaining datasets are labelled as "Knowledge Discovery" datasets and come from two sources. The sets called `adult` and `coil` are from the KDD section of the UCI repository while the artificial datasets `art1`, `art2`, and `art3` were generated using a technique described in [5]. Due to their size, these datasets were evaluated using a single train and test split. The table lists the respective train and test set sizes for these cases. One aim of the experiments is to show that the effects of the heuristics scale well with the data.

Figure 3a shows the effect on average relative runtimes of optimizing the original algorithm by merging common branches and employing the Z_{pure} cutoff across all the UCI datasets in Table 1. The figures for the four variations are

Table 1. Datasets used for the experiments.

Dataset	Instances	Missing values (%)	Numeric attributes	Nominal
UCI Datasets				
breast-cancer	699	0.2	9	0
cleveland	303	0.2	6	7
credit	690	0.6	6	9
diabetes	768	0.0	8	0
hepatitis	155	5.4	6	13
hypothyroid	3772	5.4	7	22
ionosphere	351	0.0	34	0
kr-vs-kp	3196	0.0	0	36
labor	57	33.6	8	8
mushroom	8124	1.3	0	22
promoters	106	0.0	0	57
sick-euthyroid	3163	6.5	7	18
sonar	208	0.0	60	0
splice	3190	0.0	0	61
vote	435	5.3	0	16
vote1[3]	435	5.5	0	15
KDD Datasets				
coil	5822/4000	0.0	85	0
adult	32561/16281	0.2	6	8
art1	50000/50000	0.0	0	50
art2	50000/50000	0.0	25	25
art3	50000/50000	0.0	50	0

shown relative to the original algorithm runtime at 10 iterations. The variations are, the original algorithm with no optimization, the original merging common branches only, the original employing the Z_{pure} cutoff only and finally the original using both optimizations. For numbers of boosting iterations up to 50 there is little to be gained by these methods, but beyond 50 significant gains can be made, particularly by merging. The biggest reduction occurs at 200 iterations when both optimizations are used, making an approximate average runtime saving of around one third.

Figure 3b charts the average relative runtimes across the same datasets comparing the original optimized algorithm with the three heuristic methods described in Section 4. The figures for heuristic improvements are relative to "random search" at 10 iterations. The relative differences in performance are only negligible at 10 iterations. Beyond this value the heuristic methods are clearly superior. The random walk method especially is twice as fast as the other two heuristic methods at all iterations, and an order of magnitude faster than the original algorithm at 100 and 200 iterations. In general, the heaviest path and Z_{pure} heuristics have a rather similar runtime behaviour, sometimes they even induced identical trees. The random walk method follows a runtime curve which we suspect is linear. A possible explanation for this surprising average case behaviour is given in the next section.

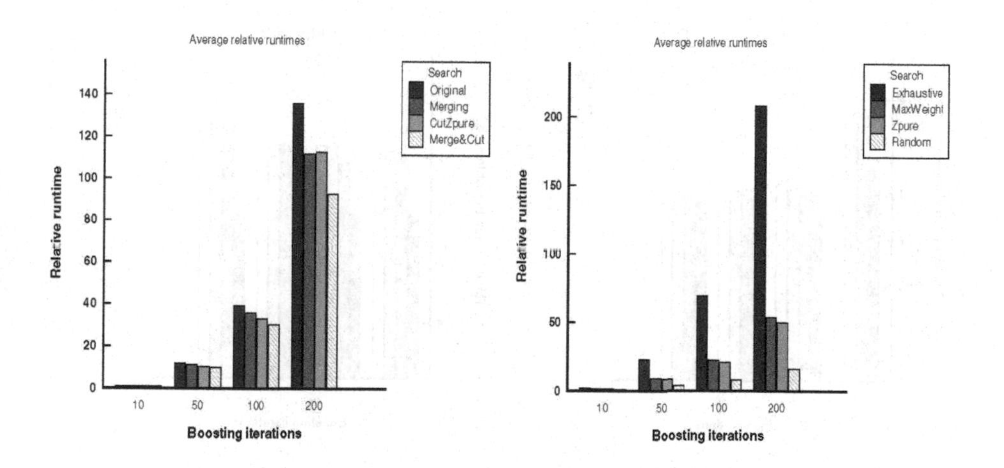

Fig. 3. Average relative runtimes for (a) variations on the original algorithm and (b) the various heuristic search methods.

The runtime performance of the heuristic variants is only relevant if there is no appreciable degradation in predictive accuracy for these methods when compared to the original. Figure 4a shows their performance relative to the original. All error figures are shown relative to the original at 10 iterations. It can be seen that for 10 iterations the heuristic methods fail to produce the same performance as the original, particularly the random walk heuristic which is up to 20% worse. At 50 iterations the gap has closed considerably and beyond 50 the random walk method actually outperforms the original algorithm. The other two heuristics again have the same relative performance which is consistently slightly worse than the original. One explanation for the superior performance of the random walk method is that it may avoid overfitting due to "natural" pruning, but this is only an hypothesis.

The number of leaves (predictor nodes) produced by the various methods give some indication of the shape of the trees being produced by the heuristic methods. Figure 4b shows these leaf figures relative to the original algorithm at the respective number of iterations. As can be seen, the heuristic methods have significant numbers of additional predictor nodes relative to the original method. This is a clear indication that these trees are more shallow, i.e. that they contain more but shorter paths on average. To understand this result we need to visualize the possible shapes of an alternating decision tree. For a fixed total number N of predictor nodes the minimum number of leaves $\frac{N}{2}$ is achieved by a perfectly binary tree. The maximum number $N-1$ is achieved by a flat list of tests; such a totally flat alternating decision tree is actually equivalent to an ensemble of boosted decision stumps. Thus a higher number of leaves indicates a more decision stump-like tree shape.

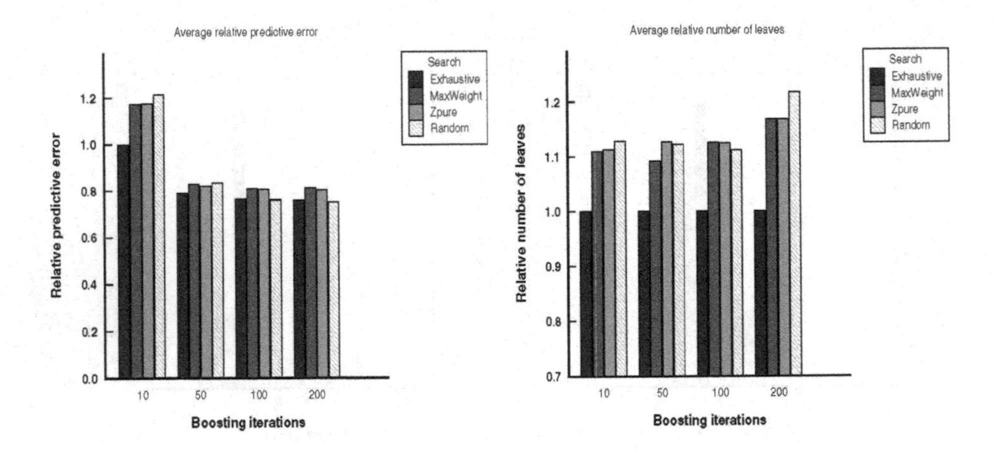

Fig. 4. Average relative accuracies and number of leaves for the various heuristic search methods.

6 Conclusions and Further Work

This paper has presented an improved version of the original alternating decision tree algorithm of Freund and Mason. This improved method is still quadratic in the number of boosting iterations and as such is not particularly useful for knowledge discovery. The use of heuristics to speed up the algorithm was investigated and we have shown that it is possible to achieve results similar, and occasionally better than the original, particularly for large numbers of boosting iterations. In terms of runtime, all heuristic methods were superior. Informal analysis would indicate that all heuristic methods have $O(n \log n)$ worst case complexity, but that they enjoy $O(n)$ average case complexity thanks to the shallowness of the trees they are inducing.

This perceived shallowness also has some impact on comprehensibility. The heuristic methods need to induce larger trees to be competitive with respect to predictive accuracy. Obviously, larger trees are harder to read. But due to the additive nature of alternating decision trees, they can be understood as the sum of all paths. Consequently, we can look at single paths in isolation to understand their respective contribution to the final prediction. Luckily, in a shallow tree most of these paths are rather short, thus they will be relatively easy to comprehend.

In future work we will investigate other approaches on speeding up the original algorithm, which will be based on adaptive caching of some of the statistics that are currently recomputed over and over again. Furthermore, the alternating decision tree algorithm can be extended in a variety of ways. The first and most important is to produce a version of the algorithm capable of handling multiple classes. It would also make sense to apply the trees to regression and cost-sensitive classification problems. Unlike standard decision trees where combining is multiplicative, combining alternating trees is linear which opens up the possibility of being able to bag [2] them to hopefully perform well. Especially in the presence of noise which is problematic for boosting algorithms in general [3], such a bagging approach might alleviate boosting's tendency to overfit the noise.

The improved ADTree induction algorithm as well as the artificial dataset generator described above will both be included into the next version of the WEKA machine learning workbench [9], which is available[4] under the Gnu Public License.

References

1. Blake, C. L., Keogh, E., Merz, C.J.: UCI Repository of Machine Learning DataBases. Irvine, CA: University of California, Department of Information and Computer Science. [http://www.ics.uci.edu/ mlearn/MLRepository.html] (1998).
2. Breiman L.: Bagging Predictors, Machine Learning, 24(2), 1996.
3. Dietterich T.G.: An Experimental Comparison of Three Methods for Constructing Ensembles of Decision Trees: Bagging, Boosting, and Randomization, Machine Learning, 40(2), 139-158, 2000.
4. Freund, Y., Mason, L.: The alternating decision tree learning algorithm. Proceedings of the Sixteenth International Conference on Machine Learning, Bled, Slovenia, (1999) 124-133.

[4] WEKA can be downloaded from `http://www.cs.waikato.ac.nz/~ml`

5. Pfahringer, B., Bensusan, H., Giraud-Carrier, C.: Meta-Learning by Landmarking Various Learning Algorithms. Proceedings of the Seventeenth International Conference on Machine Learning, Stanford University, California, USA (2000) 743-750.
6. Quinlan, J.R.: MiniBoosting Decision Trees. Draft (1999) (available at `http://www.cse.unsw.EDU.AU/~quinlan/miniboost.ps`).
7. Schapire, R.E., Singer, Y.: Improved boosting algorithms using confidence-rated predictions. Machine Learning **37** (3) (1999) 297-336.
8. Witten, I.H., Bell, T.C.: The zero-frequency problem: estimating the probabilities of novel events in adaptive text compression. IEEE Transactions on Information Theory 37(4) (1991) 1085-1094.
9. Witten, I.H., Frank, E.: Data Mining: Practical Machine Learning Tools and Techniques with Java Implementations. Morgan Kaufmann Publishers, San Francisco, California (2000).

Building Behaviour Knowledge Space to Make Classification Decision

Xiuzhen Zhang[1], Guozhu Dong[2], and Kotagiri Ramamohanarao[1]

[1] Department of CSSE, The University of Melbourne, Victoria 3010, Australia
{xzhang, rao}@cs.mu.oz.au
[2] Department of CSE, Wright State University, Ohio 45435, USA
gdong@cs.wright.edu

Abstract. CAEP, namely *Classification by Aggregating Emerging Patterns*, builds classifiers from *Emerging Patterns* (EPs). EPs mined from the training data of a class are distinguishing features of the class. To classify a test instance t, the scores by aggregating EPs in t measures the weight we put on each class; direct comparison of scores decides t's class. However the skewed distribution of EPs among classes and intricate relationship between EPs sometimes make the decision by directly comparing scores unreliable. In this paper, we propose to build *Score Behaviour Knowledge Space* (SBKS) to record the behaviour of training data on scores; classification decision is drawn from SBKS from a statistical point of view. Extensive experiments on real-world datasets show that SBKS frequently improves CAEP classifiers, especially on datasets where they have relatively poor performance. The improved CAEP classifiers outperform the start-of-the-art decision tree classifier C5.0.

1 Introduction

The recently proposed classification model CAEP, [1] namely *Classification by Aggregating Emerging Patterns* [2,9], builds classifiers from *Emerging Patterns* (EPs)[1]. EPs mined from the training data of a class are distinguishing features of the class. Functions have been proposed to measure the aggregate contribution of EPs that appear in a test instance t, resulting in *scores*; [2] t is labelled the class with the highest score. Although the order of scores is generally a good indication of t's label, the unbalanced distribution of EPs among classes and intricate relationship between EPs sometimes make it unreliable. In this paper, rather than relying on the absolute order of scores, we propose to consider the *behaviour* of training data on scores to make classification decision. Specifically, we build *Score Behaviour Knowledge Space* (SBKS) for training data and derive the final classification decision from a statistical point of view. Experiments on 28 datasets from the UCI machine learning repository (`http://www.ics.uci.edu/~mlearn/MLRepository.html`)

[1] CAEP also refers to a specific classifier(§ 2), which will be clear from context.
[2] Scores refer to both the aggregate score of CAEP and encoding cost of iCAEP(§ 2).

D. Cheung, G.J. Williams, and Q. Li (Eds.): PAKDD 2001, LNAI 2035, pp. 488–494, 2001.

show that SBKS frequently improves the accuracy of CAEP classifiers, especially on datasets where they are relatively weak.

Behaviour knowledge space [3] was first proposed in pattern recognition for combining multiple experts (CME): When K classifiers give their individual decisions $e(1)$, ..., $e(K)$ about the identity of an input x, what is the combination function $E(e(1), ..., e(K))$ which can produce the best final decision? Behaviour knowledge space consists of units where the decision of classifiers are recorded and CME decision is drawn. We incorporate the idea of recording the behaviour of training data on scores and develop elaborate scheme constructing SBKS and efficient algorithms searching SBKS for reliable classification decision.

CAEP maps training instances into points in an Euclidean space for scores. Models like decision tree or nearest neighbour can be used to classify such space. Decision tree [4] divides the space into regions by considering dimensions *sequentially*. In contrast, SBKS *concurrently* divides score space into fine-grained "hypercubes". K-nearest neighbour [7] classifies a test point by voting on its k nearest points, where a dimension may be dominant in the distance measure. With SBKS, each dimension is of equal weight and our classification algorithm searches subspace of SBKS for decision.

2 Classification by Aggregating Emerging Patterns

Items are `(continuous_attr, interval)` or `(discrete_attr, value)` pairs. An itemset is a set of items; an instance defined by n attributes is an itemset of n items. The *support* of itemset x in dataset D, $supp_D(x)$, is $\frac{|\{t \in D | x \subseteq t\}|}{|D|}$. Given D' and D'', the growth rate of an itemset x from D' to D'' is $GR(x) = \frac{supp_{D''}(x)}{supp_{D'}(x)}$ ($\frac{0}{0} = 0$ and $\frac{>0}{0} = \infty$); EPs from D' to D'', or simply EPs of D'', are itemsets with growth rate greater than a threshold $minrate$ ($minrate > 1$).

Example 1. e is an EP from the Malignant (M) to Benign (B) class of Breast cancer(Wisc): e=`{(Bare-Nuclei,1),(Bland-Chromatin,3),(Normal-Nucleoli,1),(Mitoses,1)}`,. $supp_M(e) = 0.41\%$, $supp_B(e) = 20.31\%$ and $GR(e) = 49.54$. e has high predictive power: With odds of 98%, instances containing e are benign.

For training dataset D of m classes, $D = D_1 \cup D_2 \cup ... \cup D_m$, where D_i consists of training instances for class C_i, EPs of all classes, $E = E_1 \cup E_2 \cup ... \cup E_m$, is the model for D; E_i is the EP set for C_i, consisting of EPs from $D - D_i$ to D_i. In Example 1, if an instance t only contains e, we tend to assign t "Benign". However, if t contains EPs of both the Benign and Malignant, as will be discussed next, we classify t by aggregating the EPs appearing in t.

CAEP [2] aggregates EPs from a probabilistic perspective. The combined power of *EPs* of C_i that appear in t is t's aggregate score (or score) for C_i, where the first item computes the odds that t belongs to C_i:

$$score(t, C_i) = \sum_{e \subseteq t, e \in E_i} \frac{GR(e)}{GR(e) + 1} * supp_{D_i}(e)$$

Naturally, one tend to assign t the class C_i where $score(t, C_i)$ is the maximum. It turns out, however, classes present different features on EPs and direct comparison of scores often leads to inaccurate decision. *Normalization* is proposed to overcome the problem. Instead of letting the class with the highest raw score win, CAEP lets the class with the highest normalized score win: $norm\_score(t, C_i) = \frac{score(t,C_i)}{base\_score(C_i)}$, where $base\_score(C_i)$ is got at a percentile (50%-85%) when the scores are in decreasing order.

iCAEP [9] aggregates EPs in an information-based approach. According to the minimum message length theory, t should be labelled C_i where the total cost of encoding C_i and of encoding t under C_i is the minimum. E is the model for D, and each class have the same encoding length under E. However, the encoding length of t under different classes is different: EPs of E_j, with high support in D_j and low support in $D_i (i \neq j)$, have the smallest encoding cost in C_j. E^t, the representative EP set to encode t, consists of long EPs of all classes [9]:

$$E^t = \bigcup_{j=1}^{m} E_j^t,\ E_j^t = \{e_k \in E_j | k = 1..p_j\}\ is\ a\ partition\ of\ t$$

Under the encoding scheme that a message of probability P incurs an encoding cost of $log_2(P)$ bits, the encoding cost of t under C_i, $L(t||C_i)$, is

$$L(t||C_i) = -\sum_{k=1}^{p} log_2 P(e_k|C_i),\ e_k \in E^t$$

t is assigned the class C_i where $L(t||C_i)$ is the minimum. Experiments show that normalization can not notably improve classification accuracy.

ConsEPMiner[8] is employed to mine EPs. Given support threshold *minsupp*, growth rate threshold *minrate*, growth-rate improvement threshold *minrateimp*, (the growth-rate improvement of an EP e, $rateimp(e)$, is $min(\forall e' \subset e, GR(e) - GR(e'))$,) using all constraints to effectively control the blow-up of candidate EPs, ConsEPMiner successfully mines EPs from large high-dimensional datasets.

3 Building Score Behaviour Knowledge Space to Make Classification Decision

To further solve the problem of unreliable decision by comparing scores, we propose to build score behaviour knowledge space to make classification decision.

3.1 Score Behaviour Knowledge Space

Given a training dataset of m class labels, a *Score Behaviour Knowledge Space* (SBKS) is an m-dimensional space where each dimension corresponds to the score for a class. As will be discussed in Section 3.2, for each dimension i, $1 \leq i \leq m$, the score range $[0, \infty)$ is first divided into K_i intervals numbered as 1, 2, 3, ..., K_i. A *unit* is an m-dimensional hypercube defined by m intervals, denoted as $u = (u_1, u_2, ..., u_m)$; SBKS then consists of $K_1 * K_2 * ... * K_m$ units.

A subspace of SBKS, $S = [[u_{11}, u_{12}], ..., [u_{m_1}, u_{m_2}]]$, consists of the units falling into the range defined by S. For example, in a 2-dimensional SBKS, the subspace $S = [[1, 2], [5, 6]]$ consists of 4 units, namely, $S = \{(1,5), (1,6), (2,5), (2,6)\}$.

For a training instance t where $score(t) = \langle x_1, ..., x_m \rangle$, t is associated with the unique unit $u = (u_1, ..., u_m)$, where x_i falls into the interval $u_i = [s_{i_1}, s_{i_2})$. In SBKS, each unit records the number of incoming (training) instances for each class. Given a unit u, let $n_u(ll)$ denote the number of instances of class ll in u and T_u denote the total number of instances in u, $T_u = \sum_{i=1}^{m} n_u(i)$. For subspace S, $n_S(i) = \sum_{u \in S} n_u(i)$, $T_S = \sum_{u \in S} T_u$. The semantics of SBKS is clear: for a subspace S, the probability that an input falling into S belongs to class i is $\frac{n_S(i)}{T_S}$ ($T_S > 0$). The class with the largest probability is called the *representative class* of S, denoted as $E(S)$. When no training instances fall into S or when classes have the same probability, the representative class of S is nil.

$$E(S) = \begin{cases} ll & if\ T_S > 0,\ \frac{n_S(ll)}{T_S}\ is\ the\ maximum \\ nil & otherwise \end{cases}$$

Example 2. In CAEP, for 100 randomly selected training instances from Horse colic (2 classes), dividing score dimensions into 5 intervals, we get the SBKS of Table 1, where for a unit u we record $n_u(C_1)/n_u(C_2)$. With unit (4, 2), a total of 9 instances fall into this unit; 8 are C_1 instances and 1 is a C_2 instance. For an instance t where $score(t, C_1) \in [17.6, 26.3)$, and $score(t, C_2) \in [0.02, 42.5)$, with probability of $\frac{8}{8+1} = 88.89\%$, t belongs to C_1; with probability of $\frac{1}{8+1} = 11.11\%$, t belongs to C_2. $E((4,2)) = C_1$.

Table 1. The SBKS for 100 Horse colic Instances

$C_1 \setminus C_2$	$1: [0, 0.02)$	$2: [0.02, 42.5)$	$3: [42.5, 85)$	$4: [85, 127.4)$	$5: [127.4, \infty)$
$1: [0, 0.04)$	2/0	0/0	0/2	0/1	0/12
$2: [0.04, 8.8)$	4/0	7/6	3/3	1/3	0/7
$3: [8.8, 17.6)$	1/0	12/1	4/0	0/0	0/0
$4: [17.6, 26.3)$	1/0	8/1	1/0	0/0	0/0
$5: [26.3, \infty)$	2/0	18/0	0/0	0/0	0/0

3.2 Building SBKS to Make Classification Decision

To build SBKS, we need to first divide $[0..\infty)$, the score range for training and test instances, into intervals. Sorting the scores of training instances into increasing order, we get the range $[S_{i_1}, S_{i_2})$, where 80% of training instances fall into. By equally dividing $[S_{i_1}, S_{i_2})$ into K intervals, $K+2$ intervals are formed: $[0, S_{i_1}), [S_{i_1}, S_{i_1}+a), ..., [S_{i_1}+(K-1)*a, S_{i_2}), [S_{i_2}, \infty)$, where $a = (S_{i_2} - S_{i_1})/K$.

Given an m-dimensional SBKS, to classify an input t where $score(t) = \langle x_1, \ldots x_m \rangle$, suppose that t falls into unit u, if $E(u) \neq nil$, t is assigned $E(u)$; otherwise, larger subspace centered at u is searched. As shown in Fig. 1, at line 6, function S.increment() forms a larger subspace centered at S: Suppose $S = [[u_{11}, u_{12}], \ldots, [u_{m_1}, u_{m_2}]]$, after S.increment(), $S = [[u_{11} - 1, u_{11} + 1], \ldots, [u_{m_1} - 1, u_{m_2} + 1]]$, which takes in all the units surrounding S. From lines 3 to 6, S keeps growing, until t's label is decided or the whole SBKS is searched.

```
classify(SBKS B, test instance t)
;; decide t's label from B, score(t) = ⟨x1, x2, ..., xm⟩, xi is the score for Ci
1)  ll ← nil;
    ;; for 1 ≤ i ≤ m, ui = [si1, si2) and si1 ≤ xi < si2
2)  S ← [ [u1, u1], ..., [um, um] ];
3)  while ll = nil do
4)      ll ← E(S);
5)      if ll ≠ nil then return ll;
6)      S.increment();
```

Fig. 1. Classify(): Classify a test instance by SBKS

Suppose that each dimension consists of K intervals, to decide the label of an input t, Classify() searches from a minimum of 1 unit to a maximum of K^m units. In implementing the algorithm, to speed up searching, SBKS is stored as a hash tree, where units are hashed on the first two dimensions. Usually the algorithm can finish within one iteration. When several iterations are necessary, as m is typically very small ($m < 10$), the algorithm is still efficient.

Example 3. Fig. 2 plots the SBKS of Table 1, with the representative class for each unit. Given an input t, $score(t) = \langle 16.8, 86.7 \rangle$, t's label can not be determined from (3,4), for $T_{(3,4)} = 0$ and $E((3,4)) = nil$. A larger subspace centered at (3,4), the square area with bold borders is searched: $S = [\ [2,4], [3,5]\ ] = \{$(2,3), (2,4), (2,5), (3,3), (3,4), (3,5), (4,3), (4,4), (4,5) $\}$. From Table 1, $n_S(C_1) = 9$ and $n_S(C_2) = 13$, and thus $E(S) = C_2$. t is labelled C_2. Apparently SBKS can build complex classification models, even when the space is not linearly separable.

4 Experimental Results

Applying SBKS to CAEP and iCAEP, CAEP$^\sharp$ and iCAEP$^\sharp$ are produced, where for CAEP$^\sharp$, SBKS is constructed from the original scores of CAEP before normalization. Experiments were done on 28 UCI datasets with the following settings: Classification accuracy was measured with 10-fold cross validation. For ConsEPMiner, $minsupp = 1\%$ or a count of 5, whichever is larger, $minrate = 5$, $minrateimp = 0.01$, and the EP set for a class is limited to 100,000 EPs. For

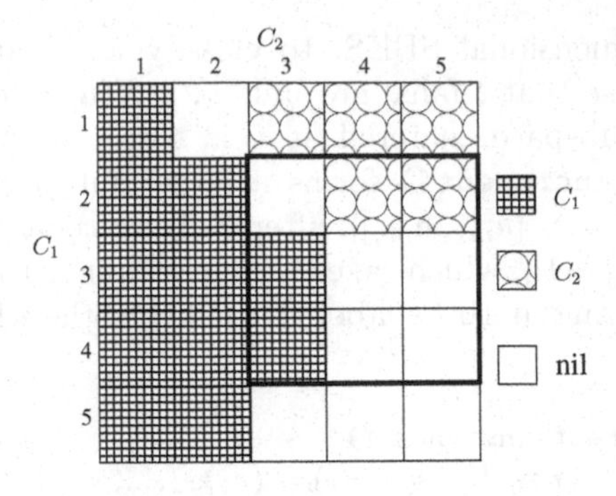

Fig. 2. Classification decsion from the SBKS of Table 1

CAEP, the base score is chosen at 85%. For CAEP$^\sharp$ and iCAEP$^\sharp$, SBKS was built with 10 intervals for each score dimension.

Fig. 3 plots the accuracy ratio of CAEP$^\sharp$/CAEP and iCAEP$^\sharp$/iCAEP on 28 datasets. Points above the line at the ratio 1.0 indicate that SBKS produces improvement. Several conclusions can be drawn from the figure: (1) Obviously SBKS improves the accuracy of both CAEP and iCAEP, by an average of 1.6% and 1.3% respectively. With the accuracy ratio of CAEP$^\sharp$/CAEP ranging from 0.945 to 1.128, and that of iCAEP$^\sharp$/iCAEP ranging from 0.949 to 1.119, while SBKS produces accuracy improvement up to 12.8%, it never significantly degrades the performance of CAEP or iCAEP. (2) Compared with CAEP, CAEP$^\sharp$ improves accuracy on 17, ties on one and loses on 10 datasets; compared with iCAEP, iCAEP$^\sharp$ improves accuracy on 17 and loses on 11 datasets. More importantly, SBKS produces significant improvement on datasets where the performance of CAEP and iCAEP is relatively poor. We also compared CAEP classifiers with C5.0 [5]. The average accuracy of iCAEP$^\sharp$, CAEP$^\sharp$ and C5.0 are 88.08%, 87.44% and 87.07 respectively; both iCAEP$^\sharp$ and CAEP$^\sharp$ outperform C5.0. Further experiments are underway comparing the performance of CAEP classifiers with that of C5.0 under boosting [6].

5 Conclusions

We have presented a behaviour knowledge-based approach for making classification decision, and have evaluated the method on CAEP— *Classification by Aggregating Emerging Patterns.* In the original CAEP classifiers, aggregate contribution of EPs for each class is quantified as scores and classification decision is reached by comparing scores. With *Score Behaviour Knowledge Space* (SBKS), we record the behaviour of training data on scores, and thus can take into account the varying features of EPs for each class. Experiments on 28 UCI datasets show that SBKS improves the performance of CAEP classifiers and the resulting CAEP classifiers outperform C5.0, the most advanced decision tree classifier.

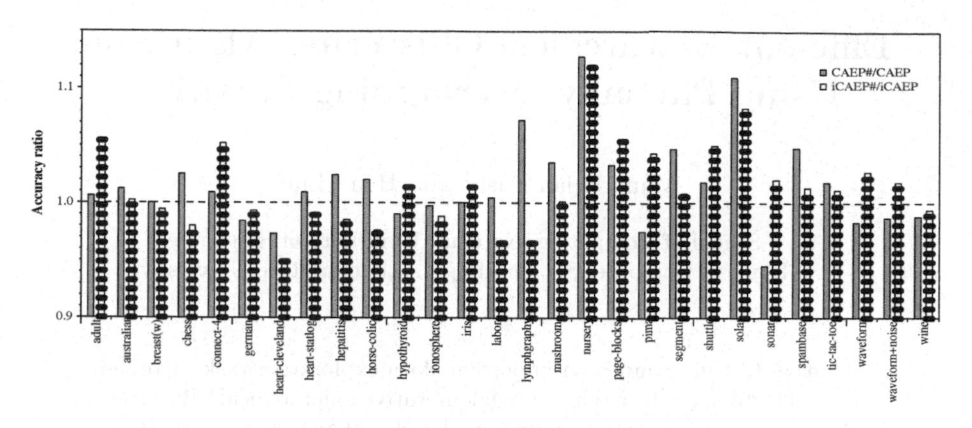

Fig. 3. Accuracy ratio: $\text{CAEP}^\sharp$ vs. CAEP and $\text{iCAEP}^\sharp$ vs. iCAEP

References

1. G Dong and J Li. Efficient mining of emerging patterns: Discovering trends and differences. In *Proc. 1999 ACM SIGKDD Conf.*, pages 15–18, USA, Aug. 1999.
2. G Dong, X Zhang, L Wong, and J Li. CAEP: Classification by aggregating emerging patterns. In *Proc. DS'99*, LNAI 1721, Tokyo, Japan, Dec. 1999.
3. Y S Huang and C Y Suen. A method of combining multiple experts for the recognition of unconstrained handwritten numerals. *IEEE Transactions on Pattern Recognition and Machine Intelligence*, 17(1):90–94, Jan. 1995.
4. J R Quinlan. *C4.5: Programs for Machine Learning.* Morgan Kaufmann, 1993.
5. Rulequest Research Pty Ltd. See5/C5.0. http://www.rulequest.com, 1999.
6. R E Schapire. The strength of weak learnability. *Machine Learning*, 5(2):197–227, 1990.
7. C Stanfill and D Waltz. Toward memory-based reasoning. *Communications of ACM*, 29:1213–1228, 1986.
8. X Zhang, G Dong, and K Ramamohanarao. Exploring constraints to efficiently mine emerging patterns from large high-dimensional datasets. In *Proc 2000 ACM SIGKDD Conf.*, pages 310–314, Boston, USA, Aug. 2000.
9. X Zhang, G Dong, and K Ramamohanarao. Information-based classification by aggregating emerging patterns. In *Proc IDEAL'2000*, LNCS 1983, HK, 2000.

Efficient Hierarchical Clustering Algorithms Using Partially Overlapping Partitions

Manoranjan Dash[1] and Huan Liu[2]

[1] School of Computing, National University of Singapore
[2] Dept. of Computer Sci. and Engg., Arizona State University

Abstract. Clustering is an important data exploration task. A prominent clustering algorithm is agglomerative hierarchical clustering. Roughly, in each iteration, it merges the closest pair of clusters. It was first proposed way back in 1951, and since then there have been numerous modifications. Some of its good features are: a natural, simple, and non-parametric grouping of similar objects which is capable of finding clusters of different shape such as spherical and arbitrary. But large CPU time and high memory requirement limit its use for large data. In this paper we show that geometric metric (centroid, median, and minimum variance) algorithms obey a 90-10 relationship where roughly the first 90iterations are spent on merging clusters with distance less than 10the maximum merging distance. This characteristic is exploited by partially overlapping partitioning. It is shown with experiments and analyses that different types of existing algorithms benefit excellently by drastically reducing CPU time and memory. Other contributions of this paper include comparison study of multi-dimensional vis-a-vis single-dimensional partitioning, and analytical and experimental discussions on setting of parameters such as number of partitions and dimensions for partitioning.

1 Introduction

Clustering is an important data exploration task. It is applied in different areas including data mining. Surveys on clustering algorithms can be found in [1, 9]. Among the more prominent clustering algorithms hierarchical clustering is one. It was first proposed in 1951 [6], and since then there have been numerous modifications including the recent ones [8,10]. The fact that it has been a prominent algorithm for last half a century shows that it has stood the test of time. Among its various good features it is a non-parametric (assumes very little in the way of data characteristics), natural and simple way of grouping objects, and capable of finding clusters of different shapes such as spherical, arbitrary. But its large CPU time and high memory requirement make it unsuitable for large data. Efficient techniques to handle large data can be roughly classified as sampling, summarizing, and partitioning. Sampling has been used in several algorithms and recently in CURE [8]. Basically, clustering is done over a small sample and results are extended to the whole data set. Although it is criticized for missing out small clusters, sampling usually retains the underlying cluster structure. Summarizing algorithms are based on the fact that data points that

D. Cheung, G.J. Williams, and Q. Li (Eds.): PAKDD 2001, LNAI 2035, pp. 495–506, 2001.

are very close to each other can be merged and their summary information can be used for efficient clustering [5,18]. Partitioning is used in [8] and in parallel algorithms in [12] where data is partitioned and each partition is then clustered and later the results are consolidated to determine a global set of clusters. These techniques help in reducing the size of the data while trying to retain the original cluster structure but they still require to run the traditional algorithms on the reduced data – a sample, a summary, or a partition. Improving the traditional algorithms will have complementary effect on these algorithms. This paper focuses on reducing the time and memory requirement of hierarchical algorithms.

2 Background

It is assumed that data contains N data points with each point x described by an M-tuple $x = (x_1, ..., x_M)$ of real numbers where M is the number of attributes or dimensions [1]. Distance between two objects is usually calculated using *Minkowski* metrics where for fixed d and for any two objects x and y, $L_d(x, y) = \{\sum_{j=1}^{M} |x_j - y_j|^d\}^{\frac{1}{d}}$. This family includes the *Manhattan* metric L_1, *Euclidean* metric L_2, and *Chebychev* metric L_∞. For ease of explanation it is assumed that distance is measured in *Euclidean* metric.

Hierarchical algorithms can be broadly classified as divisive and agglomerative. In divisive type, starting with one cluster that contains all data points, a cluster is divided in each iteration until each data point belongs to a distinct cluster. In agglomerative type, starting with each point in a distinct cluster, the closest pair of clusters are merged in each iteration until there remains only one cluster containing all data points. The output of hierarchical clustering is a dendrogram which is a hierarchy of divisions or agglomerations with the top level or root representing all points in one cluster, and the bottom level or leaves representing each point in a distinct cluster. The computation needed in an agglomerative algorithm to go from one level to another is usually simpler than a divisive algorithm. In this paper we are concerned with agglomerative algorithms. They can be classified into two categories depending on the type of similarity measure used, namely graph and geometric metrics. Algorithms that use graph metrics are single link, complete link, and average link, and those using geometric metrics are centroid, median and minimum variance. A basic difference between the two types is that in graph metrics each point is a representative while in geometric metrics each cluster has only one representative, eg. centroid. This difference is important from the point of view of performance and shape of clusters. For example, graph metric algorithm (eg. single link) can discover arbitrarily shaped clusters while geometric metric algorithm (eg. centroid method) is more suitable for clusters of spherical shape.

It is observed that for geometric metric algorithms, except for the last portion of the iterations, the size (i.e. number of points) and the merging distance of the closest pair of clusters is very small compared to their maximums respectively. It is further observed that initial iterations are typically much costlier

[1] Hierarchical clustering over binary and nominal data types are discussed in [7].

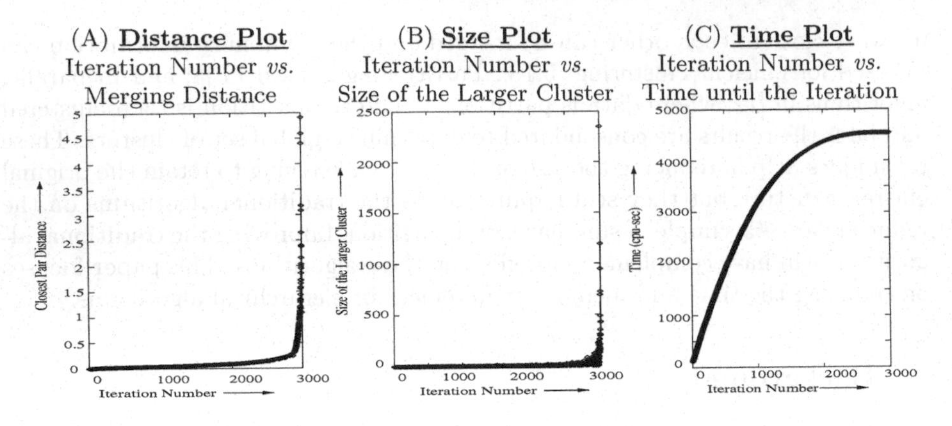

Fig. 1. Important observations for geometric metric algorithms

than those towards the end. See Figure 1 for plots showing these observations pictorially. The plots are based on the results obtained for a 2-D data set containing 3000 points in 100 clusters, each a Gaussian ball, distributed randomly in the feature space with some noise. Each dimension has a range [0.0 – 10.0]. This data is a simulated version of the data set DS2 reported in BIRCH [18]. All experiments here and later in the paper are run on Digital 8400 5/350 with 350 MHz processor and 1 GByte of ECC main memory. The results are for centroid type algorithm. In Figure 1(A) the merging distance is plotted against iteration number. Although the merging distance is not monotonically increasing, i.e. in a later iteration it can be larger than that in an earlier iteration, but still, roughly only in the last 10% or so iterations it is more than 10% or so of the maximum merging distance. This behavior is seen for other geometric metric algorithms and for different data including high-dimensional, multi-resolution, and skewed distribution. But it may not be observed for graph metrics. The reason is in geometric metric (eg. centroid type), a cluster is represented by single point (i.e. centroid); after merging the closest pair of clusters, the distance of the new cluster from most of the other clusters (except for those clusters very close to the bisecting hyper-plane between the two merging centroids) is larger than the smaller of the two distances before merging, whereas in graph metric (eg. single link) distance after merging will be equal to the smaller of the two distances before merging [2]. We will call these observations *90-10 relationship* which means *roughly 90% or so iterations from the beginning merge clusters that are separated by less than 10% or so of the maximum merging distance.*

Centroid and median types of geometric metric algorithms merge clusters whose centroids are the closest. Centroid type is known as unweighted as it treats each point in a cluster equally whereas median type is known as weighted as it weights all clusters the same, so points in small clusters are weighted more heavily than points in large clusters. The minimum variance Ward's method merges clusters that results in minimum change in square error [16]. These algo-

[2] Because of this characteristic single link algorithm follows reducibility property [11].

rithms observe the 90-10 relationship. In the rest of the paper centroid method will be discussed in detail except when otherwise specified. The first centroid algorithm proposed uses a similarity matrix. In each iteration the similarity matrix is searched to find the closest pair of clusters. Let us call this '*step 1*'. For n ($n = N...2$) clusters in an iteration, it requires $O(n^2)$ time to search through the similarity matrix, and for $N-1$ iterations this step takes $O(N^3)$ time. After merging the closest pair the similarity matrix is updated by deleting the column entries for the pair that merged and by creating a new row for the merged cluster by determining distance from other clusters. Let us call this '*step 2*'. For n clusters in an iteration this step takes $O(n)$ time to update, and so for $N-1$ iterations it takes $O(N^2)$ time. Memory required is $O(N^2)$ because of the similarity matrix. This simple algorithm can be improved by maintaining a nearest neighbor array that stores nearest neighbor for each cluster. This way *step 1* requires only to find the minimum of the nearest neighbor distances from the nearest neighbor array in $O(n)$ time for n clusters in any iteration. But *step 2* requires $O(n^2)$ time in each iteration if naive nearest neighbor algorithm of $O(n)$ time complexity is used. In 1984, Day and Edelsbrunner [2] suggested two ways to improve the time complexity of this algorithm: Type (1) – Obtain an improved bound on the required number of nearest neighbor updates – i.e. improve the *step 1*; Type (2) – Obtain an improved bound on the time required for each update – i.e. improve the *step 2*. For type (1) they suggested to use a heap-based priority queue that requires $O(\log n)$ time to find nearest neighbor giving an over-all time complexity of $O(N^2 \log N)$. For type (2) algorithm if there are α number of clusters to be updated after each iteration then the over-all complexity becomes $O(\alpha N^2)$. Using geometric preliminaries they proposed an upper bound for α as $2(3^M - 2)$ where M is the number of dimensions.

Anderberg classified hierarchical algorithms into 'stored matrix' and 'stored data' [1]. Stored matrix algorithms maintain a similarity matrix whereas stored data algorithms do not but instead calculate the similarities as required. A major distinction is that stored matrix methods are preferred when memory is sufficient to store the similarity matrix of size $O(N^2)$, otherwise stored data is the way out. Type (1) algorithm of previous paragraph that uses priority queues is a stored matrix method while type (2) algorithm is more suitable as stored data if M is not large and if memory is not enough to store $O(N^2)$ similarity matrix. Recent algorithms on hierarchical clustering [8,10,12] uses priority queues. The algorithms presented here improves both stored matrix and stored data algorithms by reducing their CPU time significantly, and furthermore it reduces the memory requirement substantially for stored matrix algorithm.

3 Proposed Algorithms

In the previous section we discussed the existence of a 90-10 relationship between the stage of algorithm and the merging distance. We also observed that initial iterations are very costly. In this section we propose algorithms using *partially overlapping partitioning* that exploits these properties. The following figure pictorially shows a single-dimensional partially overlapping partitioning approach.

Data is divided into p number of partitions or cells by dividing a dimension range (in this case A) into p number of smaller ranges. Each cell is sandwiched between two δ-regions which are typically much smaller than the cell. The first and the last cells have only one adjacent δ-region. Henceforth, by definition, a *cell* is inclusive of its adjacent δ-regions. Note that every δ-region is included in two cells. The value of δ can be mentioned as a percentage of the range of the attribute or as an absolute value. Using this data structure we propose two algorithms: *2-phase* and *nested*. In the rest of the section we use priority queue algorithm to explain although complexity analysis is done for other algorithms.

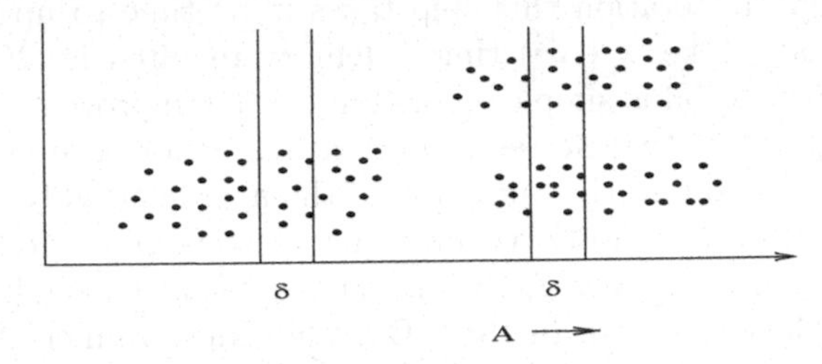

2-Phase Algorithm. The basic idea is instead of creating priority queues for all data, now there will be priority queues for each cell separately. In each iteration closest pair is found for each cell, and from those the over-all closest pair is found. If the over-all closest pair has distance less than δ then they are merged and the priority queues of only the corresponding cell are updated. If any of the merging cluster centers or merged cluster center is in any δ-region then the priority queue of the affected cell is also updated. *This procedure is repeated until the merging distance is larger than* δ. This is the *phase 1* of the algorithm. In *phase 2* traditional priority queue algorithm is employed over the remaining clusters to complete the dendrogram. Note that the same δ is used to partition the data as well as to stop the algorithm. This way large number of small sized clusters are merged in the *phase 1* that uses partitioning and only small number of larger clusters are merged in *phase 2* that uses traditional algorithm. Partially overlapping partitions are employed earlier in [13,17] in order to speed-up nearest neighbor search. In nearest neighbor algorithm δ is ideally set to a value slightly larger than the minimum distance between all points so that actual nearest neighbor is located in the same cell as the candidate point and optimum performance is obtained. But an ideal δ for clustering purposes is the distance corresponding to the point at which distance plot takes a sharp up-turn as shown in Figure 1. More on this is discussed later. Efficient nearest neighbor algorithm cannot benefit the priority queue clustering algorithm where nearest neighbors of each cluster is already in the top of each queue. Instead it requires to be maintained efficiently as the algorithm progresses. Nearest neighbor algorithms gain in time but not in memory, whereas, as will be evident soon, clustering algorithms gain significantly in both.

Table 1 exhibits the 2-phase partitioning algorithm using priority queues with time and memory complexities. Steps 1–10 is *phase 1* and 11-12 is *phase 2*. *Notations*: $|\delta|$ is number of points in a δ-region, p is number of cells, k' is the remaining number of clusters after *phase 1*. Complexity analysis is done by assuming all cells to be of equal size.

Complexity Analysis: The 2-phase algorithm in Table 1 uses priority queues

Table 1. An example of 2-phase partitioning approach over the priority queue algorithm

Algorithm	Time Complexity	Memory Complexity
Input: Data (N,M), p, δ		
Output: Dendrogram		
1. choose an attribute to partition the data	$O(1)$	
2. divide the data into p number of cells where each cell has two adjacent regions of width δ (except first and last cells)	$O(N)$	$O(N + p * \lvert\delta\rvert)$
3. Create priority queues P for each cell	$p * O((\frac{N}{p} + \lvert\delta\rvert)^2)$	$p * O((\frac{N}{p} + \lvert\delta\rvert)^2)$
4. while merging distance $< \delta$		
5. for each cell find the closest pair of clusters, C_1 and C_2	$p * O(\frac{n}{p} + \lvert\delta\rvert)$	
6. find the over-all closest pair	$O(p)$	
7. merge the over-all closest pair	$O(1)$	
8. update corresponding P	$O((\frac{n}{p} + \lvert\delta\rvert)\log(\frac{n}{p} + \lvert\delta\rvert))$	
9. if any of the pair or merged cluster in δ-region then update P of affected cell	$O((\frac{n}{p} + \lvert\delta\rvert)\log(\frac{n}{p} + \lvert\delta\rvert))$	
10. remove the duplicate copies of clusters in δ-regions, if any	$p * O(k')$	
11. cluster the remaining k' clusters	$O(k'^2 \log k')$	$O(k'^2)$
12. return dendrogram		
Over-All Complexity	$(N - k')*$ $O((\frac{N}{p} + \lvert\delta\rvert)\log(\frac{N}{p} + \lvert\delta\rvert))$ $+O(k'^2 \log k')$	$p * O((\frac{N}{p} + \lvert\delta\rvert)^2)$ or $O(k'^2)$

first proposed in [2]. The original algorithm has an over-all time complexity $O(N^2 \log N)$ and memory complexity $O(N^2)$. On the other hand, the 2-phase algorithm has an over-all time complexity $(N - k') * O((\frac{N}{p} + |\delta|)\log(\frac{N}{p} + |\delta|)) + O(k'^2 \log k')$ assuming $\log(\frac{N}{p} + |\delta|)$ to be greater than p. Memory complexity is $p * O((\frac{N}{p} + |\delta|)^2)$ or $O(k'^2)$ whichever is larger. If $|\delta|$ and k' are small and if they are assumed negligible then the complexities simplify significantly giving time complexity $N * O(\frac{N}{p} \log \frac{N}{p})$, i.e. $O(\frac{N^2}{p} \log \frac{N}{p})$, and memory complexity $p * O(\frac{N^2}{p^2})$, i.e. $O(\frac{N^2}{p})$.

In Table 2 we give time and memory complexities of 'stored matrix' and 'stored data' types of algorithms for before and after simplification. Under 'stored matrix' category, in addition to 'priority queue' type there is 'similarity matrix' algorithm that uses a similarity matrix in place of priority queues.

Correctness: The overlapping partitioning data structure guarantees a correct dendrogram. Note that number of correct dendrogram can be more than one due to the existence, if any, of ties between merging distances.

Nested Algorithm. Time taken by a 2-phase algorithm depends on the total time taken by *phase 1* and *2*. As *phase 2* is the traditional algorithm whose time complexity directly depends on the remaining number of clusters (k') after *phase 1*, for good performance *phase 1* should take small time and k' should be small

Table 2. Time and Memory Complexity Comparison: * – Simplification is done by assuming $|\delta|$ and k' negligible

Type of Algorithm	Traditional Algorithm	2-Phase Algorithm: Before* Simplification	2-Phase Algorithm: After* Simplification
	Time Complexity		
Stored Matrix			
– Similarity Matrix	$O(N^3)$	$(N-k') * p * O((\frac{N}{p}+\lvert\delta\rvert)^2)$ $+O(k'^2 \log k')$	$O(\frac{N^3}{p})$
– Priority Queues	$O(N^2 \log N)$	$(N-k') * O((\frac{N}{p}+\lvert\delta\rvert)*$ $\log(\frac{N}{p}+\lvert\delta\rvert)) + O(k'^2 \log k')$	$O(\frac{N^2}{p} \log \frac{N}{p})$
Stored Data	$O(\alpha * N^2)$	$(N-k') * O(\alpha * (\frac{N}{p}+\lvert\delta\rvert))$ $+O(\alpha * k'^2)$	$O(\alpha * \frac{N^2}{p})$ if $\alpha > p$ or, $O(N^2)$ if $p > \alpha$
	Memory Complexity		
Stored Matrix			
– Similarity Matrix	$O(N^2)$	$p * O((\frac{N}{p}+\lvert\delta\rvert)^2)$	$O(\frac{N^2}{p})$
– Priority Queues	$O(N^2)$	$p * O((\frac{N}{p}+\lvert\delta\rvert)^2)$	$O(\frac{N^2}{p})$
Stored Data	$O(N)$	$p * O(\frac{N}{p}+\lvert\delta\rvert)$	$O(N)$

as well. Experiments show that it may not be easy to detect the point at which the total time is the least. Experiments are conducted over DS2 data set (used earlier for distance plots in Figure 1) with N=3k in 2-D. Number of cells is fixed to 10 while δ varies from 0.25 to 9.0. Figure 2(A) shows the time taken by *phases 1, 2* and their total time. We redraw the distance plot of Figure 1(A) with some

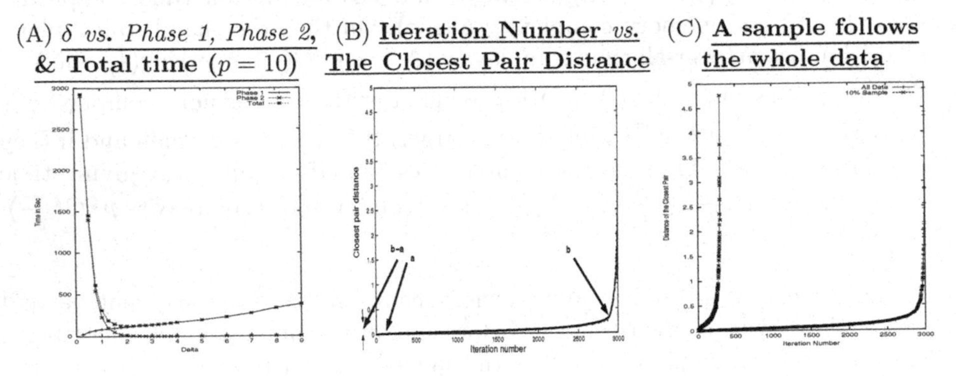

Fig. 2. Results and Analysis of Two-Phase Algorithm

labels in Figure 2(B). Label b points to a proper value of δ, label a points to the minimum merging distance. The reason why b is considered as an ideal δ value is that while on one hand it is very small, on the other hand it is able to capture most of the iterations leading to good performance. For δ smaller than b *phase 2* takes long time as remaining number of clusters is still large, whereas for δ larger than b *phase 1* takes more time than the optimal as larger δ means larger overlapping between cells. So we see that b is an ideal value for δ for efficient clustering. Contrast this with the ideal δ for nearest neighbor search methods discussed in [13,14] that employs partially overlapping partitioning. Hence, we see that a is an ideal δ for nearest neighbor search while b is ideal for clustering.

It may not be easy to guess b value properly though. The following approach may be helpful. Take a random sample of points and obtain its distance plot. Figure 2(C) shows that a sample follows the whole data where the b values for the sample and the whole data almost match. The main reason is a sample will typically increase the smallest distance among *points* in a cluster but will retain the larger distances among the *clusters* themselves. This is also the reason why the sampling approach has been efficient for clustering. Manual observation can find a proper δ value. A window function to detect the point at which distance increases significantly can lead to automatic detection.

We propose a nested algorithm that does not require a manual setting of δ. It tries to make parameter setting even easier while gaining in performance in comparison to 2-phase algorithm. It is similar to the 2-phase algorithm except that it does not switch to *phase 2* immediately. Instead it performs *phase 1* with reduced number of clusters until a negligible number of clusters remain (eg. 2). The algorithm starts typically with a large value for p and a small value for δ. With each iteration of *phase 1*, p is gradually reduced while δ is increased. Arguably this is easier than to set δ to b from the start itself as in 2-phase algorithm. Nested partitioning attempts to divide the difference (b-a) (shown in Figure 2(B)) to gain in performance over 2-phase algorithm. More discussion on how to break the range (b-a) is given in the next section. *Complexity Analysis*: Let us assume that the nested sequence is specified by $<p_j,\delta_j>$ for j=1...s where p_j and δ_j are p and δ values in j^{th} nested iteration respectively, and s is the number of iterations in the nested sequence. Let n_j be the number of clusters remaining after nested iteration $j-1$. Initially n_1 is N and finally n_{s+1} is k'. Time complexity of nested algorithm is given as $\sum_{j=1}^{s}((n_j - n_{j+1})O((\frac{n_j}{p_j} + |\delta_j|)\log(\frac{n_j}{p_j} + |\delta_j|))) + O(k'^2 \log k')$. In this case k' is made negligible by specifying sufficient number of iterations in the nested sequence so that the last term $O(k'^2 \log k')$ can be discarded from the complexity. Memory complexity of nested algorithm is $\max_{j=1...s}(p_j * O((\frac{n_j}{p_j} + |\delta_j|)^2))$ because after each iteration of the nested sequence the priority queues are freed.

Overlapping Partitioning for Minimum Variance Ward's Method. Partially overlapping partitions can be applied with some modification to Ward's minimum variance method [16]. Like centroid method, it represents a cluster by the centroid. But unlike centroid method, in each iteration the pair of clusters with the least increase in the sum of square error are merged. Square-error for cluster k is the sum of squared distances to the centroid for all points in cluster k. Mathematically, $e_k^2 = \sum_{i=1}^{n_k}\sum_{j=1}^{M}[x_{ij}^{(k)} - \mu_j^{(k)}]^2$ where $x_{ij}^{(k)}$ is the jth dimension value of point x_i of centroid k, $\mu_j^{(k)}$ is the jth dimension value of cluster k and $j = 1...M$. If cluster r and s merge in an iteration to create cluster t, then the change in square errors is given as: $\Delta E_{rs}^2 = e_t^2 - e_r^2 - e_s^2$. By replacing the expressions for square error and after simplification, we obtain that: $\Delta E_{rs}^2 = \frac{n_r n_s}{n_r+n_s} \sum_{j=1}^{M} [\mu_j^{(r)} - \mu_j^{(s)}]^2$. Note that the second term of the right hand side is the squared distance of the centroids of clusters r and s. The first term $\frac{n_r n_s}{n_r+n_s}$ can take values less than, equal to, or greater than 1. It is less than 1 when at least one of the clusters has size 1. This term makes direct application of the partially overlapping partitioning unsuitable although it shows 90-10 relationship. The reason is there is no guarantee that the closest pair is merged in each iteration as they may be in different cells. But note that the minimum value of this first term is $\frac{1}{2}$ for the case where size of both the merging clusters r and s is 1. The minimum possible ΔE_{rs}^2 with distance δ is $\frac{1}{2}\delta^2$. Hence to guarantee that the two clusters merged belong to at least one common partition the following condition must hold: *the ΔE_{rs}^2 of the closest pair r and s must be less than $\frac{1}{2}\delta^2$ so that distance between them will be less than δ.*

4 Performance

In this section we discuss different performance issues, and then perform experiments using the suggested parameter settings. *Data Sets* used for experiments include synthetic and benchmark data. Synthetic data sets are: DS2 used in BIRCH [18]; t4.8k, t5.8k, t8.8k, t7.10k used in CHAMELEON [10] and CURE [8]; SEQUOIA benchmark data [15] used in DBSCAN [4]. We generated DS2 data in different dimensions with uniform (DS2_U) and Gaussian (DS2_G) distribution within clusters . The rationale behind experimenting over these data is to compare the partitioning algorithm with traditional algorithm for different distribution of data and different shape of clusters in varying dimension and noise.

Performance Issues. From the complexity analyses in the previous section, it is apparent that performance of the partitioning algorithm depends largely on number of cells p, and δ. Other important factors are (a) number (m) and choice of dimensions to partition, (b) how to change δ and p values in nested partitioning.
How Many Dimensions to Partition (m): It is affected by these factors: (1) the larger the m (possibly) the more uniform the distribution of points across cells leading to CPU time reduction, (2) given p and δ, the larger the m the smaller the number of δ regions leading to less CPU time and memory. These two factors indicate that larger m is preferable. But this may not be true as the larger the m the larger the number of cross-sections of δ-regions. As the number of cells required to update if a point is in the cross-section of m dimensions is 2^m, larger m will require more updating when a point is in cross-section. We conducted experiments by running phase 1 only for different m over DS2_U and DS2_G data sets, and for different p and δ. Our findings can be summarized as follows: multi-dimensional outperforms single-dimensional partitioning; the change in memory requirement is insignificant; for $m > 1$ the CPU time depends on different factors but a noticeable thing is there is no consistent trend and it is very close in general.
Choice of Dimensions can affect performance. Some of the factors affecting it

are variance, range and outliers. If the range is not affected by outliers, then attributes with larger range and variance are usually preferred. Other possible approaches include principal components analysis where data is projected to the first m principal components.
Setting δ & p for Nested Partitioning In nested partitioning a sequence of δ, p is given as the input. Experiments suggest to initialize δ to a very small value and then increment it gradually by small values until there remains negligible number of clusters. Value of p can be set by first setting a probable number (n') of clusters in each cell and then p is set to $\frac{N}{n'}$. Experiment with different data sets show that optimum performance is obtained for n' between 5 and 20.

Experimental Results. Experiments are conducted on data sets described before. Nested algorithm is used where $n' = 7$ (i.e. $p = \frac{N}{7}$), number of dimensions for partitioning $m = 2$ each with $\lfloor\sqrt{p}\rfloor$ divisions, δ is initialized to 0.25% and is incremented by 0.25% for each nested iteration. Experiments are conducted to find speed-up and memory scale-down for varying N, M, and number of clusters.
Varying N: For this experiment N varies from 0.5k to 3k. As the traditional algorithm exceeds the memory allocated in our server we could not test for N larger than 3k. Figure 3(A) shows the speed-up of the nested multi-dimensional parti-

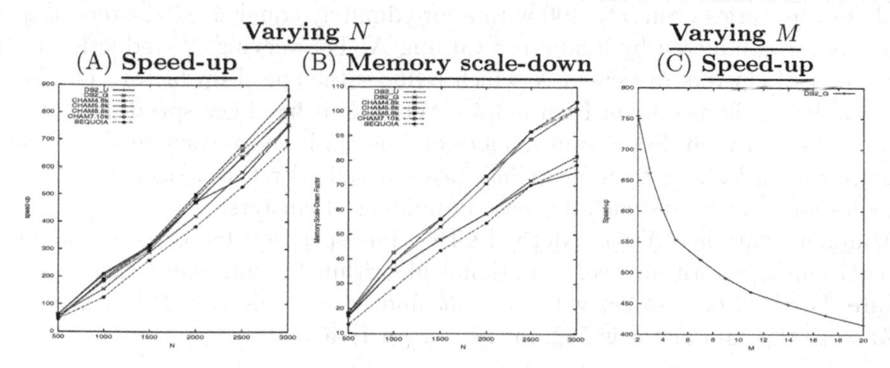

Fig. 3. Speed-up and Memory Scale-Down for 2-D data sets (A and B) and speed-up for varying M (C)

tioning algorithm over the traditional priority queue algorithm. The maximum speed-up is 862 for t8.8k with N=3k where traditional algorithm took 4717.7 CPU-sec while nested algorithm took only 5.47 CPU-sec. Note that speed-up factor increases with N. The reason is the complexity of jth iteration of the nested sequence is given as $(n_j - n_{j+1})O((\frac{n_j}{p_j} + |\delta_j|)\log(\frac{n_j}{p_j} + |\delta_j|))$. Note that $\frac{n_j}{p_j}$ is set to 7 in our experiments. So the second term of the expression is mostly indifferent to increase in N except for the increase in $|\delta|$. In fact the time difference for partitioning algorithm for data set DS2_G for N=0.5k and 3k is less than 6 CPU-sec. But, as increase in N significantly affects CPU time of the traditional algorithm, hence the gain factor increases with N. Variation in speed-up for different data sets can be attributed to different distributions. Figure 3(B) shows the memory scale-down of the nested method over the traditional. For both algorithms the reported memory is the maximum during the complete duration of program where memory requirement of the programs themselves is negligible. The maximum scale-down is 105 for t8.8k data set where for N=3k, the traditional algorithm requires 385M and the nested algorithm requires only 5.1M. The increase in scale-down factor with N is due to the same reason as explained for CPU time speed-up. The variation in the scale-down factor among different data sets is more than that for the CPU time speed-up because the distribution has a direct effect over the memory requirement.

Varying M: Experiments are conducted over DS2_G data set with N=3k, M=2 to 20. Number of dimensions to partition m is set to 2 for all experiments. As seen in Figure 3(C) the gain factor for CPU time reduces with increasing M for the following reason. Increasing M does not affect the traditional algorithm much because the dominant time factor of maintaining the heap-based priority queues due to the insertions and deletions is not affected by increasing M. Hence, the small increase in time due to the larger M for nested algorithm reduces the gain factor, while on the other hand, memory complexity is affected very little as priority queues are the dominant factor which do not change with dimensions.

Varying Number of Clusters: DS2_G data with N=3k in 2-D is used. Number of clusters varies from 5 to 100 with approximately equal sized clusters. Experiments are conducted by fixing and varying N. For varying N, reduction in the number of clusters decreases N which reduces the speed-up factor. This result is similar to the results in Figure 3(A). When N is fixed the speed-up factor reduces as well with decrease in number of clusters because each cluster contains more points leading to more points in each cell. For this reason the memory requirement increases with decrease in number of clusters.

Minimum Variance Ward's Method CPU time speed-up factor is measured for partitioning algorithm over traditional algorithm for minimum variance measure. For DS2_G data set with $N = 3k$ and $M = 2$ the gain factor is 835; for $M = 10$ the gain factor is 532; for $M = 20$ it is 460.

5 Conclusion and Future Work

In this paper we studied an important behavior of geometric metric hierarchical clustering algorithms that show a 90-10 relationship where roughly 90% or so iterations from the beginning merge clusters that are separated by less than 10% or so of the maximum merging distance. We proposed two algorithms using partially overlapping partitions that exploit this behavior. These algorithms are suitable for both stored data and stored matrix type of algorithms. Complexity analysis and experiments show significant reduction in time and memory. It is found that gain factor increases with increase in number of points in the data. Increase in dimensions reduces the gain factor but still the gain is significant for high-dimensional data. We introduced a modified method to apply this partitioning technique to minimum variance Ward's method.

A future work is to parallelize this technique. In [12] parallel algorithms employ naive partitioning where all N priority queues are distributed among various processors. But partially overlapping partitioning is foreseen to be very suitable for parallelization as a processor will be in charge of priority queues for clusters in only one or more number of cells which should reduce memory and CPU time. Other future work includes testing the suitability of other more sophisticated partitioning [3].

References

1. M. R. Anderberg. *Cluster Analysis for Applications.* Academic Press, NY, 1973.
2. W. H. E. Day and H. Edlesbrunner. Efficient algorithms for agglomerative hierarchical clustering methods. *Journal of Classification*, 1(1):7–24, 1984.
3. W. DuMouchel, C. Volinsky, T. Johnson, C. Cortes, and D. Pregibon. Squashing at files atter. In *Proceedings of KDD'99*, pages 6–15, 1999.
4. M. Ester, H. P. Kriegel, J. Sander, and X. Xu. A density-based algorithm for discovering clusters in large spatial databases with noise. In *Proceedings of KDD'96*, pages 226–231, 1996.
5. U. Fayyad, C. Reina, and P. S. Bradley. Initialization of iterative refinement clustering algorithms. In *Proceedings of KDD'98*, pages 194–198, 1998.
6. K. Florek, J. Lukaszewicz, J. Perkal, H. Steinhaus, and S. Zubrzycki. Sur la liason et la division des points d'un ensemble fini. *Colloq. Math.*, 2:282–285, 1951.
7. S. Guha, R. Rastogi, and S. Kyuseok. ROCK: A robust clustering algorithm for categorical attributes. In *Proceedings of ICDE'99*, pages 512–521, 1999.
8. S. Guha, R. Rastogi, and K. Shim. CURE: An efficient clustering algorithm for large databases. In *Proceedings of ACM SIGMOD'98*, pages 73–84, 1998.
9. A. K. Jain and R. C. Dubes. *Algorithm for Clustering Data*, chapter Clustering Methods and Algorithms. Prentice-Hall Advanced Reference Series, 1988.
10. G. Karypis, E-H. Han, and V. Kumar. CHAMELEON: A hierarchical clustering algorithm using dynamic modeling. *IEEE Computer*, 32:68–75, 1999.
11. F. Murtagh. A survey of recent advances in hierarchical clustering algorithms. *The Computer Journal*, 26:354–359, 1983.
12. C. F. Olson. Parallel algorithms for hierarchical clustering. *Parallel Computing*, 21:1313–1325, 1995.
13. M. O. Rabin. Probabilistic algorithms. In J. F. Traub, editor, *Algorithms and Complexity*, pages 21–39. Academic Press, New York, 1976.
14. F. J. Rohlf. Computation efficiency of agglomerative clustering algorithms. Technical Report Report RC 6831, IBM T. J. Watson Research Center, NY, 1977.
15. M. Stonebraker, J. Frew, K. Gardels, and J. Meredith. The SEQUOIA 2000 storage benchmark. In *Proceedings of ACM SIGMOD*, pages 2–11, 1993.
16. Jr. J. H. Ward. Hierarchical grouping to optimize an objective function. *Journal of the American Statistical Association*, 58:236–244, 1963.
17. G. Yuval. Finding nearest neighbors. *Information Processing Letters*, 5:63–65, 1976.
18. T. Zhang, R. Ramakrishnan, and M. Livny. BIRCH: An efficient data clustering method for very large databases. In *Proceedings of ACM SIGMOD* pages 103–114, 1996.

A Rough Set-Based Clustering Method with Modification of Equivalence Relations

Shoji Hirano[1], Tomohiro Okuzaki[1], Yutaka Hata[1], Shusaku Tsumoto[2], and Kouhei Tsumoto[3]

[1] Department of Computer Engineering, Himeji Institute of Technology 2167 Shosha, Himeji, 671-2201 Japan
hirano@ieee.org
[2] Department of Medical Informatics, Shimane Medical University, School of Medicine, 89-1 Enya-cho, Izumo-city, Shimane, 693-8501 Japan
[3] Department of Bimolecular Engineering, Graduate School of Engineering, Tohoku University, Aoba-yama07, Sendai, 980-8579 Japan

Abstract. This paper presents a clustering method for nominal and numerical data based on rough set theory. We represent relative similarity between objects as a weighted sum of two types of distances: the Hamming distance for nominal data and the Mahalanobis distance for numerical data. On assigning initial equivalence relations to every object, modification of slightly different equivalence relations is performed to suppress excessive generation of categories. The optimal clustering result can be obtained by evaluating the cluster validity over all clusters generated with various values of similarity thresholds. After classification has been performed, features of each class are extracted based on the concept of value reduct. Experimental results on artificial data and amino acid data show that this method can deal well with both types of attributes.

1 Introduction

Recent databases store a large amount of data composed of both nominal and numerical attributes. Clustering has been receiving considerable attention as one of the most promising approaches for revealing underlying structure in such databases. However, the well-known clustering methods, K-means [1] and Fuzzy C-Means (FCM) [2], have difficulty in handling nominal data since they require distance between objects that is represented on a ratio scale. Although the agglomerative hierarchical clustering method [3] can deal with nominal data by using relative similarity, it still has a problem that the clustering result strongly depends on the order of processing objects.

This paper presents a rough set-based clustering method for data containing both nominal and numerical attributes. Rough sets, proposed by Pawlak [4], have been receiving considerable attention in the field of knowledge discovery since they provide tools to mathematically treat roughness of the knowledge. Rough sets can easily handle nominal data since their basic properties are related to

D. Cheung, G.J. Williams, and Q. Li (Eds.): PAKDD 2001, LNAI 2035, pp. 507–512, 2001.

the indiscernibility relation. In our method, we first form equivalence relations among objects based on their relative similarity, and classify them into some categories according to the relations. Similarity between objects is determined as a weighted sum of their Hamming/Mahalanobis distances. Similar equivalence relations will be modified so that they represent the same, more simple knowledge which generates adequate number of categories. The optimal clustering result can be obtained by evaluating the cluster validity, defined using upper and lower approximations of a cluster, over all clusters generated with various values of similarity thresholds. After classification has been performed, features of each class are extracted based on the concept of value reduct. Experimental results on artificial data and amino acid data show that this method can deal well with both attributes and produce good clustering results.

2 Clustering Method

Clustering is performed according to indiscernibility relations defined on the basis of relative similarity between objects. The overall procedure is summarized as follows.

Step1) For every object, assign an initial equivalence relation using a similarity threshold Th_1.
Step2) Modify similar equivalence relations using a threshold Th_2.
Step3) Iterate Steps 1-2 using various values of Th_1 and $Th2$, and obtain the best clustering result that yields maximum validity.
Step4) Extract features of each class based on the concept of value reduct.

2.1 Initial Equivalence Relation

The first procedure is to assign an initial equivalence relation to every object. Let $U = \{x_1, x_2, ..., x_n\}$ be the entire set of objects we are interested in. Each object has p attributes represented by nominal or numerical values.

[Definition 1] Equivalence relation

An equivalence relation R_i for object x_i is defined by

$$R_i = \{\{x_j| \ s(x_i, x_j) \geq Th_1\}, \{x_j|others\}\}, \quad \text{for all } j(1 \leq j \leq n),$$

where $s(x_i, x_j)$ denotes similarity between objects x_i and x_j, and Th_1 denotes a threshold value of similarity. Obviously, $R_i = \{\{[x_i]_{R_i}\}, \{\overline{[x_i]_{R_i}}\}\}$, $[x_i]_{R_i} \cap \overline{[x_i]_{R_i}} = \phi$, and $[x_i]_{R_i} \cup \overline{[x_i]_{R_i}} = U$ hold. The equivalence relation R_i classifies U into two categories: one containing objects similar to x_i and another containing objects dissimilar to x_i. When $s(x_i, x_j)$ is larger than Th_1, object x_j is considered to be indiscernible to x_i. Similarity $s(x_i, x_j)$ is calculated as a weighted sum of the Hamming distance $d_H(x_i, x_j)$ of nominal attributes and the Mahalanobis distance $d_M(x_i, x_j)$ of numerical attributes as follows:

$$s(x_i, x_j) = \frac{p_d}{p} \times (1 - d_H(x_i, x_j)/p_d) + \frac{p_c}{p} \times (1 - d_M(x_i, x_j)),$$

where p_d and p_c denote the numbers of nominal and numerical attributes, respectively. □

2.2 Modification of Equivalence Relations

Objects should be classified into the same category when most of equivalence relations commonly regard them indiscernible. However, depending on the value of Th_1, there could be some equivalence relations which classify these similar objects into different categories. Such equivalence relations will cause to generate unpreferable clustering result consists of small categories. Following is an example of the case.

Let $U = \{x_1, \ldots, x_4\}$ be the entire set of objects and let $\mathbf{R} = \{R_1, \ldots, R_4\}$ be a set of equivalence relations over U. Suppose that $\mathbf{R}$ classifies U as follows:

$$\begin{aligned} U/R_1 &= \{\{x_1, x_2, x_3\}, \{x_4\}\}, \\ U/R_2 &= \{\{x_1, x_2, x_3\}, \{x_4\}\}, \\ U/R_3 &= \{x_1, x_2, x_3, x_4\}, \\ U/R_4 &= \{\{x_3, x_4\}, \{x_1, x_2\}\}, \\ U/IND(\mathbf{R}) &= \{\{x_1, x_2\}, \{x_3\}, \{x_4\}\}. \end{aligned}$$

In this case, equivalence relation R_4 classifies objects x_2 and x_3 into different categories although other three relations classify them into the same category. Consequently, three fine categories are obtained. To avoid excessive generation of categories, we modify similar equivalence relations so that they represent the same, more simplified knowledge. First, we define subordination degree, $\gamma(R_i, R_j)$, of two equivalence relations (R_i, R_j) as follows.

[Definition 2] Subordination degree of equivalence relations

$$\gamma(R_i, R_j) = \delta_{ij} \frac{\#(([x_i]_{R_i} \cap [x_j]_{R_j}) \cup (\overline{[x_i]_{R_i}} \cap \overline{[x_j]_{R_j}}))}{\#(U)},$$

$$\delta_{ij} = \left\{ \begin{array}{l} 1, \text{ if } [x_i]_{R_i} \cap [x_j]_{R_j} \neq \phi \\ 0, \text{ if } [x_i]_{R_i} \cap [x_j]_{R_j} = \phi. \end{array} \right\},$$

where $\#(Y)$ denotes cardinality of a set Y. □

[Definition 3] Modification of equivalence relations
Let $R_i, R_j \in \mathbf{R}$ be initial equivalence relations and let $R'_i, R'_j \in \mathbf{R}'$ be equivalence relations after modification. For an initial equivalence relation R_i, a modified equivalence relation R'_i is defined as

$$R'_i = \{\{x_j | x_j \in P_i\}, \{x_j | others\}\}, \quad \text{for all } j (1 \leq j \leq n),$$

where P_i denotes a subset of objects represented by

$$P_i = \bigcup_{1 \leq j \leq n} \{x_j | \gamma(R_i, R_j) \geq Th_2\}.$$

The value Th_2 denotes the lower threshold value to regard R_i and R_j as the same relations. □

2.3 Evaluation of Validity

Depending on the values of two thresholds Th_1 and Th_2, a varicty of sets of equivalence relations can be obtained in the preceding steps. We then evaluate validity of their clustering results based on the following criteria and obtain the best set of equivalence relations which yields maximum validity.

[Definition 4] Validity of clustering result

Let U denote the entire set of objects, $\mathbf{R}$ denote an initial set of equivalence relations, and $\mathbf{R}'$ denote the modified set of $\mathbf{R}$, respectively. Suppose that $\mathbf{R}'$ classifies U into l categories, $U/IND(\mathbf{R}') = \{\mathbf{C}_1, \mathbf{C}_2, ..., \mathbf{C}_l\}$.

Validity of the clustering result, $V(\mathbf{R}')$, obtained using $\mathbf{R}'$ is defined by

$$V(\mathbf{R}') = \frac{1}{l}\sum_{k=1}^{l}\left(\frac{\#(\underline{\mathbf{R}}\mathbf{C}_k)}{\#(\overline{\mathbf{R}}\mathbf{C}_k)} \times \#(\mathbf{C}_k)\right),$$

where $\#(\mathbf{C}_k)$ denotes the number of objects in the k-th category, $\underline{\mathbf{R}}\mathbf{C}_k$ and $\overline{\mathbf{R}}\mathbf{C}_k$ denote $\mathbf{R}$-lower and $\mathbf{R}$-upper approximations of $\mathbf{C}_k$ given below:

$$\underline{\mathbf{R}}\mathbf{C}_k = \bigcup_{x_i \in \mathbf{C}_k} \{[x_i]_{R_i} | [x_i]_{R_i} \subseteq \mathbf{C}_k\},$$

$$\overline{\mathbf{R}}\mathbf{C}_k = \bigcup_{x_i \in \mathbf{C}_k} \{[x_i]_{R_i} | [x_i]_{R_i} \cap \mathbf{C}_k \neq \phi\}.$$

□

2.4 Feature Extraction by Value Reduct

After classification is performed, we examine features of classified objects based on the concept of value reduct. Note that we here regard 'value reduct' as 'a set of attributes which are essential to specify an object' and define it as follows.

[Definition 5] Reduct

Let $\mathbf{R}$, $\mathbf{P}$ and $\mathbf{Q}$ denote a set of equivalence relations, a subset of $\mathbf{R}$ and a proper subset of $\mathbf{P}$, respectively. Here we define a reduct r_i of object x_i as

$$r_i = \{A_i(\mathbf{P}) \mid [x_i]_\mathbf{P} \subseteq [x_i]_\mathbf{R},\ [x_i]_\mathbf{Q} \not\subseteq [x_i]_\mathbf{R}\},\ \text{ for all } \mathbf{P} \subseteq \mathbf{R}!\ \mathbf{Q} \subset \mathbf{P},$$

where $A_i(\mathbf{P})$ denotes a set of attribute values of x_i associated with a set of relations $\mathbf{P}$.
□

3 Experimental Results

The method was first applied to the BALLOON database [5]. This database contained 20 objects and each data had 5 nominal attributes. Table 1 shows the clustering result. Here, the row 'μ' denotes rough membership degree of each object to its corresponding class. Objects 1-8 certainly ($\mu = 1.0$) belonged to class 1, and objects 9-16 belonged to class 2 or 3 also with high membership

Table 1. Clustering result on the BALLOON database.

Obj	Color	Size	Act	Age	Infl	Cls	μ
1	YEL	SMA	STR	ADL	T	1	1.0
2	YEL	SMA	STR	CHI	T	1	1.0
3	YEL	SMA	DIP	ADL	T	1	1.0
4	YEL	SMA	DIP	CHI	T	1	1.0
5	YEL	SMA	STR	ADL	T	1	1.0
6	YEL	SMA	STR	CHI	T	1	1.0
7	YEL	SMA	DIP	ADL	T	1	1.0
8	YEL	SMA	DIP	CHI	T	1	1.0
9	YEL	LAR	STR	ADL	F	2	0.75
10	YEL	LAR	STR	CHI	F	2	0.75
11	YEL	LAR	DIP	ADL	F	2	0.75
12	YEL	LAR	DIP	CHI	F	2	0.75
13	PUR	SMA	STR	ADL	F	3	0.75
14	PUR	SMA	STR	CHI	F	3	0.75
15	PUR	SMA	DIP	ADL	F	3	0.75
16	PUR	SMA	DIP	CHI	F	3	0.75
17	PUR	LAR	STR	ADL	F	4	0.2
18	PUR	LAR	STR	CHI	F	5	0.2
19	PUR	LAR	DIP	ADL	F	6	0.2
20	PUR	LAR	DIP	CHI	F	7	0.2

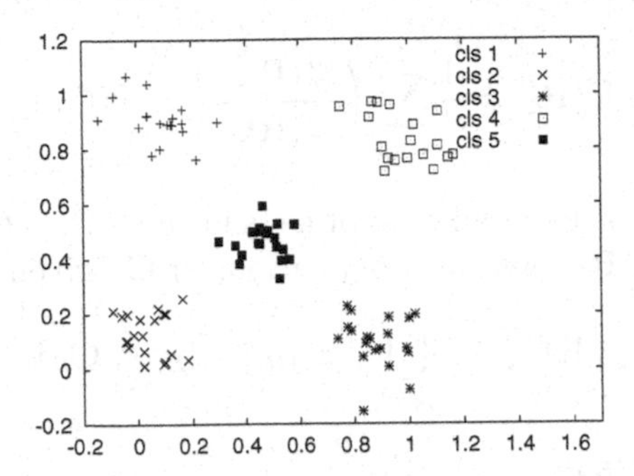

Fig. 1. Clustering result on the numerical data.

grades. Objects 17-20 were divided into four classes since merging these classes into other class required excessive modification of equivalence relations. Figure 1 shows the clustering result of numerical data generated using Neyman-Scott's method [6]. The data consisted of 98 objects. As shown in Figure 1, the objects were clearly divided into the expected clusters.

As a practical application, this method was applied to the analysis of partially mutated anti-lysozyme antibody HyHEL-10 [7]. The data contained 35 objects. Each object had 16 nominal attributes (selected amino-acid sequence) and a numerical attribute (measured value; combining coefficients K_a). Relations between amino acids and K_a were examined as follows: (1) Classify antibodies according to K_a (Sect. 2.1-2.3). (2) Extract amino acid residues that characterize each antibody (Sect. 2.4). (3) Evaluate relations between the classification result and extracted features of antibodies. Table 2 shows the clustering result. Reduct of each antibody is denoted by []. A remarkable feature was found on antibody #23, in which all residues were marked as reduct. This implies that antibody #23 is a base antibody for mutation. Class 11 contained antibodies #5! #9! #20 and #34 that lost affinity for antigen (represented by ND). In these antibodies, one of the residues in VH33, VH50, VH98, VL92 had been mutated to Ala. Since other antibodies which had mutation in these sites but to other amino acids (for example, #6 (VH33=Leu), #8 (VH33=Trp), #10 (VH50=Leu) and #11 (VH50=Phe)) did not lose affinity, mutation in these sites to Ala may be the reason to reduce affinity of the antibody to the antigen.

Table 2. Clustering result on the amino-acid data.

Class	Acid	VH									VL							Ka
		31	32	33	50	53	56	58	98	99	31	32	50	53	91	92	96	
1	23	[s]	[d]	[y]	[y]	[y]	[s]	[y]	[w]	[d]	[n]	[n]	[y]	[q]	[s]	[n]	[y]	42.0
	24	s	d	y	y	y	s	y	w	d	[a]	n	y	q	s	n	y	34.0
2	19	s	d	y	y	y	s	[f]	w	d	n	n	y	q	s	n	y	23.0
	4	s	[n]	y	y	y	s	y	w	d	n	n	y	q	s	n	y	22.0
	21	s	[d]	y	y	y	s	y	w	[a]	n	n	y	q	s	n	y	21.0
3	3	s	[e]	y	y	y	s	y	w	d	n	n	y	q	s	n	y	16.9
	32	s	d	y	y	y	s	y	w	d	n	n	y	q	[a]	n	y	15.6
	14	s	d	y	y	[p]	s	y	w	d	n	n	y	q	s	n	y	15.0
	33	s	d	y	y	y	s	y	w	d	n	n	y	q	s	[d]	y	14.0
	15	s	d	y	y	[w]	s	y	w	d	n	n	y	q	s	n	y	13.0
	16	s	d	y	y	y	[a]	y	w	d	n	n	y	q	s	n	y	12.0
	35	s	d	y	y	y	s	y	w	d	n	n	y	q	s	n	[f]	12.0
	13	s	d	y	y	[l]	s	y	w	d	n	n	y	q	s	n	y	11.0
4	2	s	[a]	y	y	y	s	y	w	[d]	n	n	y	q	s	n	y	10.0
	1	[a]	d	y	y	y	s	y	w	d	n	n	y	q	s	n	y	9.60
	29	s	d	y	y	y	s	y	w	d	n	n	[f]	q	s	n	y	9.20
	25	s	d	y	y	y	s	y	w	d	[d]	n	y	q	s	n	y	8.80
	30	s	d	y	y	y	s	y	w	d	n	n	y	[a]	s	n	y	7.80

Class	Acid	VH									Vl							Ka
		31	32	33	50	53	56	58	98	99	31	32	50	53	91	92	96	
4	7	s	d	[f]	y	y	s	y	w	d	n	n	y	q	s	n	y	7.10
	31	s	d	y	y	y	s	y	w	d	n	n	y	[e]	s	n	y	6.90
5	22	s	[a]	y	y	y	s	y	w	[a]	n	n	y	q	s	n	y	4.00
	18	s	d	y	y	y	s	[l]	w	d	n	n	y	q	s	n	y	3.40
	11	s	d	y	[f]	y	s	y	w	d	n	n	y	q	s	n	y	2.90
	17	s	d	y	y	y	s	[a]	w	d	n	n	y	q	s	n	y	2.60
	8	s	d	[w]	y	y	s	y	w	d	n	n	y	q	s	n	y	2.30
6	6	s	d	[l]	y	y	s	y	w	d	n	n	y	q	s	n	y	1.50
7	27	s	d	y	y	y	s	y	w	d	n	[d]	y	q	s	n	y	0.97
	28	s	d	y	y	y	s	y	w	d	n	n	[a]	q	s	n	y	0.64
8	10	s	d	y	[l]	y	s	y	w	d	n	n	y	q	s	n	y	0.37
9	12	s	d	y	y	[a]	s	y	w	d	n	n	y	q	s	n	y	0.19
10	26	s	d	y	y	y	s	y	w	d	n	[a]	y	q	s	n	y	0.04
11	5	s	d	[a]	y	y	s	y	w	d	n	n	y	q	s	n	y	ND
	9	s	d	y	[a]	y	s	y	w	d	n	n	y	q	s	n	y	ND
	20	s	d	y	y	y	s	y	[a]	d	n	n	y	q	s	n	y	ND
	34	s	d	y	y	y	s	y	w	d	n	n	y	q	s	[a]	y	ND

4 Conclusions

This paper has proposed the clustering method based on rough sets. In the experiments on amino-acid data analysis, 35 objects were classified into 11 clusters with adequate steps of attribute values among the classes. This indicates that integration of equivalence relations successfully suppressed excessive generation of clusters. Besides, the results on artificial data showed that this method can handle both nominal and numerical attributes and produce good clustering results. It remains as a future work to investigate behavior of the method when thresholds are independently assigned to each object.

References

1. S. Z. Selim and M. A. Ismail, "K-means-type Algorithms: A Generalized Convergence Theorem and Characterizatin of Local Optimality," *IEEE Trans. Pattern Anal. Mach. Intell.*, **6**(1), 81-87, 1984.
2. J. C. Bezdek, *Pattern Recognition with Fuzzy Objective Function Algorithm,* Plenum Press, 1981.
3. M. R. Anderberg, *Cluster Analysis for Applications,* Academic Press, 1973.
4. Z. Pawlak, *Rough Sets, Theoretical Aspects of Reasoning about Data,* Kluwer Academic Publishers, 1991.
5. URL: http://www.ics.uci.edu/pub/machine-learning-databases/
6. J. Neyman and E. L. Scott., "Statistical Approach to Problems of Cosmology," *Journal of the Royal Statistical Society,* Series B20, 1-43, 1958.
7. URL: http://www.shimane-med.ac.jp/med_info/kdd/kbsamino.csv

Importance of Individual Variables in the k-Means Algorithm

Juha Vesanto

Neural Networks Research Centre,
Helsinki University of Technology,
P.O.Box 5400, 02015 HUT, Finland
Juha.Vesanto@hut.fi

Abstract. In this paper, quantization errors of individual variables in k-means quantization algorithm are investigated with respect to scaling factors, variable dependency, and distribution characteristics. It is observed that Z-norm standardation limits average quantization errors per variable to unit range. Two measures, quantization quality and effective number of quantization points are proposed for evaluating the goodness of quantization of individual variables. Both measures are invariant with respect to scaling/variances of variables. By comparing these measures between variables, a sense of the relative importance of variables is gained.

Keywords: k-means, quantization, scaling, normalization, standardation

1 Introduction

Unsupervised clustering algorithms are important tools in exploratory data analysis. Because clustering criteria is usually based on some distance measure between individual data vectors, they are highly sensitive to the scale, or dispersion, of the variables. It is easy to come up with examples where the clustering result can be considerably changed by a simple linear rescaling of the variables (see e.g. [5] p.5). Therefore, apart from the case when the original values of the variables are somehow meaningful with respect to each other, some kind of rescaling or standardation procedures are normally recommended prior to the clustering [8,5,1].

The most common standardation procedure is to treat all variables independently and transform each to so-called Z-scores by substracting the mean and dividing by the standard deviation of each variable. Another widely used method is to normalize the range of the variable to unit interval. Also other nonlinear, multidimensional, and even local standardation operations are possible [6,5,8,4]. However, what these more complex may gain in flexibility, they loose in interpretative power.

In this paper, vector quantization of numerical data sets is investigated with respect to the quality of quantization of individual components. The aim is to derive easily understandable measures of quantization quality to be used as part of a data understanding framework.

D. Cheung, G.J. Williams, and Q. Li (Eds.): PAKDD 2001, LNAI 2035, pp. 513–518, 2001.

2 Definitions

Let D be a $n \times d$ sized numerical matrix such that each row of the matrix corresponds to one data sample $\mathbf{x}_i$, and each column to one variable $\mathbf{v}_j$. Without loss of generality, let each $\mathbf{v}_j$ be Z-normed to zero mean and unit variance. Scaling D with a set of scaling factors $\{f_j\}$ is a simple multiplication operation: $\hat{\mathbf{v}}_j = f_j \mathbf{v}_j$, which results in a new data set $\hat{D}$. The variances of the new variables are equal to squared scaling factors $\hat{\sigma}_j^2 = f_j^2 \sigma_j^2 = f_j^2$.

The scaled data set $\hat{D}$ is quantized (or clustered). In this paper, the batch k-means algorithm is used for quantization [3,7]. The algorithm finds a set of k prototype vectors $\hat{\mathbf{m}}_\mathbf{l}$ which minimize the representation error:

$$E = \frac{1}{n}\sum_{i=1}^{n} ||\hat{\mathbf{x}}_i - \hat{\mathbf{m}}_{b_i}||^2 = \sum_{j=1}^{d} \frac{1}{n}\sum_{i=1}^{n} |\hat{x}_{ij} - \hat{m}_{b_i j}|^2 = \sum_{j=1}^{d} E_{\hat{\mathbf{v}}_j} = \sum_{j=1}^{d} \hat{E}_j, \quad (1)$$

where $b_i = \arg_l \min(||\hat{\mathbf{x}}_i - \hat{\mathbf{m}}_l||)$, $||\cdot||$ is euclidean distance metric, and $\hat{E}_j$ is the average quantization error of scaled variable $\hat{\mathbf{v}}_j$.

The variable-wise quantization errors $\hat{E}_j$ can be represented as functions of the number of effective number of quantization points k_j, ie. the number of quantization points which are needed to get the same error when only $\mathbf{v}_j$ is quantized. The k-means algorithm finds a local minima of E in the space defined by k_j. The derivative of E gives insight to what happens when the importance of a variable increases:

$$\frac{\delta E}{\delta k_{j'}} = \sum_{j=1}^{d} \hat{\sigma}_j^2 \frac{\delta E_j}{\delta k_{j'}} = \sum_{j=1}^{d} \hat{\sigma}_j^2 \frac{\delta E_j}{\delta k_j}\frac{\delta k_j}{\delta k_{j'}}. \quad (2)$$

This shows that the allocation of quantization points is dependent on three factors: the variance of each variable $\hat{\sigma}_j^2$, distribution characteristics of the variable $\frac{\delta E_j}{\delta k_j}$ and dependencies between variables: $\frac{\delta k_j}{\delta k_{j'}}$. In general, since the total supply of quantization points k is limited, the partial derivatives $\frac{\delta k_j}{\delta k_{j'}}$, $j \neq j'$ are negative. On the other hand, for those variables which are (highly) dependent on variable j', $\frac{\delta k_j}{\delta k_{j'}}$ is positive, and thus increased $k_{j'}$ actually benefits them.

The function $\hat{E}_j(k)$ can be estimated directly from data by making a number of quantizations of each variable with varying values of k. This is relatively light operation compared to quantization of the whole data space. For uniform distribution, the function is also easy to derive analytically: $\hat{E}_j(k) = \hat{\sigma}_j^2 k^{-2}$, in which case $k_j = (\hat{\sigma}_j^2/\hat{E}_j)^{0.5}$. A similar formula can also be used for other continuous distributions, for example for gaussian distribution $\hat{E}_j(k) \approx \hat{\sigma}_j^2 k^{-1.7}$.

For each variable-wise quantization error $\hat{E}_j$, the minima is reached when $k_j = k$, and maxima when $k_j = 1$. In the latter case the data is quantized using just its mean, in which case $\hat{E}_j = \hat{\sigma}_j^2$. Thus, the quantization errors $\hat{E}_j$ are limited by:

$$\hat{E}_j(k) \leq \hat{E}_j \leq \hat{\sigma}_j^2. \quad (3)$$

Since $\hat{E}_j = f_j^2 E_j$ and $f_j = \hat{\sigma}_j$ the limits can also be defined with respect to the original unscaled variables:

$$E_j(k) \leq E_j \leq 1. \tag{4}$$

The quantization quality of a variable can be estimated as how close E_j is to the minimum possible error $E_j(k)$:

$$q_j = \frac{E_j - E_j(k)}{1 - E_j(k)} = \frac{\hat{E}_j - \hat{E}_j(k)}{\hat{\sigma}_j^2 - \hat{E}_j(k)} \tag{5}$$

With sufficiently large k, $\hat{E}_j(k) = 0$ and thus $q_j \approx \hat{E}_j/\hat{\sigma}_j^2$. Since quantization quality is intricately linked with variable importance — the more important variable, the better it is quantized — q_j acts as a measure of variable importance.

3 Experiments

To get further insight to how the proposed methods work in practice, some tests were made using artificial data sets. The number of data points was 1000, and they were quantized using 100 quantization points. To ensure good quantization, ten k-means runs with 50 epochs were made and the best one was utilized. The data points were from four different kinds of distributions: gaussian, uniform, exponential and "2-spikes", which was formed as a mixture of two supergaussian distributions with equal prior probabilities.

Figure 1 studies the effect of scaling factors in a 10-dimensional case. Instead of a steady increase in quantization error $\hat{E}_j$ of the scaled variable, which might be expected, there is a transfer area for scaling factors in range [1, 10], where the error of the scaled variable is about equal to the quantization error of all the other variables. The behaviour is due to the limits imposed on $\hat{E}_j$ by the possible values of k_j. With small scaling factors, $k_1 \approx 1$, while for large scaling factors $k_1 \approx 100$.

The 2-spikes distribution (Figure 1 top right corner) has a sudden decrease in quantization error for $1 < f_1 < 5$. This decrease shows the effect of increasing k_1 over the threshold of 2: at this point, the quantization points are divided to two groups, one for either spike of the variable.

Both importance measures q_j and k_j work very well in all cases showing how the importance of the first variable increases with increasing scaling factor in the significant range of scaling factors, and levels off when the scaling factor does not really matter.

Figure 2 studies the effect of variable dependencies. In the test, 10-dimensional data sets with 1 to 9 identical variables were quantized. The quantization errors behave exactly as if there were actually 10 to 2 variables, respectively: the sum of errors of the dependent variables is equal to the errors of each of the independent variables.

As an example of the usage of the proposed indicators, the IRIS data set [2] was quantized using 15 prototypes, see Table 1. Both measures q_j and k_j indicate

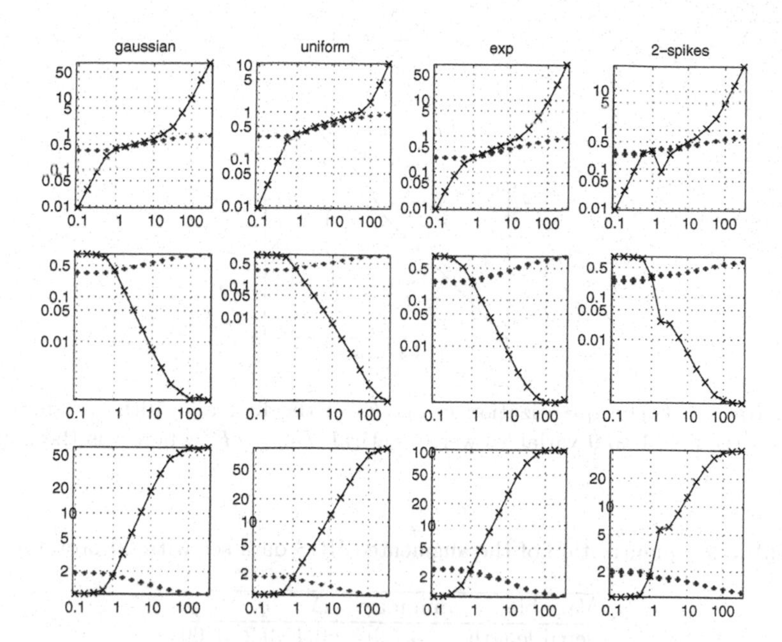

Fig. 1. Quality of quantization. $\hat{E}_j$ on top row, q_j in the middle, and k_j on the bottom. The data consists of 10 similar variables (gaussians on the left, uniform and exponential on the middle, and 2-spikes distributions on the right) the first of which is scaled by a factor of $f_1 \in [0.1, 316]$. The solid line corresponds to the first variable, and the separate markers (actually: boxplots) to the other 9 variables. Note that all axes are logarithmic.

that petal length variable was the most important variable in this quantization, although its descaled quantization error (Δ column) is the biggest of the four variables.

Table 2 shows results for a modified version of IRIS data set. Four new variables have been added: one discreet variable which indicates the class information of the sample (values 1, 2 and 3 for the three different Iris subspecies) and three random uniformly distributed variables. The random variables are clearly the least important.

Table 1. Quantization of the IRIS data set with 15 prototypes. Both min, max and Δ values are in original value range in order to faciliate interpretation by domain experts.

Variable	[min,max]	Δ	q_j	k_j
sepal length	[4.3,7.9]	±0.2	0.063	5.07
sepal width	[2.0,4.4]	±0.1	0.096	4.29
petal length	[1.0,6.9]	±0.3	0.018	5.62
petal width	[0.1,2.5]	±0.1	0.033	3.87

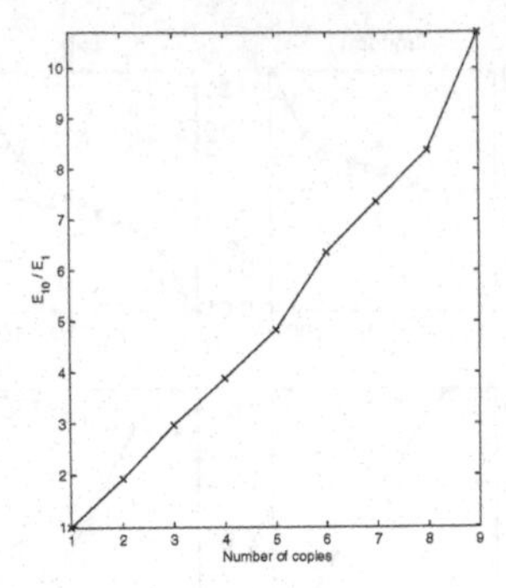

Fig. 2. Ratio between quantization errors $\hat{E}_j$ of the first and 10th variables (out of 10) when the first 1 to 9 variables were identical: $\hat{E}_{10} = c\hat{E}_1$ where c is the number of copies.

Table 2. Quantization of the augmented IRIS data set with 20 prototypes.

Variable	[min,max]	Δ	q_j	k_j
sepal length	[4.3,7.9]	±0.4	0.2	2.69
sepal width	[2.0,4.4]	±0.3	0.33	2.29
petal length	[1.0,6.9]	±0.3	0.036	3.60
petal width	[0.1,2.5]	±0.2	0.065	2.94
iris species	[1.0,3.0]	±0.2	0.062	2.75
random 1	[0.0,1.0]	±0.1	0.23	2.06
random 2	[0.0,1.0]	±0.2	0.33	1.86
random 3	[0.0,1.0]	±0.2	0.3	1.89

4 Discussion

Various studies have investigated and compared different kinds of standardation/scaling methods in clustering problems. For example in [6] several standardation procedures were compared to each other in an artificial clustering problem. Standardation based on range was often found to be superior to the Z-norm standardation. This is understandable since clustered distributions, for example dicreet variables, retain more of their variance than continuous variables in scaling by range. Thus they have, in the view of the results in this paper, bigger inherent scaling factors. However, Z-norm provides a more uniform starting point in quantization, since the maximum quantization errors are equal for all variables.

The importance of a variable can be viewed as the gain — decrease in the quantization error — the quantization algorithm achieves through increasing the effective number of quantization points k_j of some variables, and (therefore) decreasing k_j of the others. The allocation of k_j seems to depend primarily on three factors: scaling of the variables, their distribution characteristics, and their dependency on the other variables (see Eq. 2). Of these scaling has quite straightforward effect, and the effect of distribution characteristics can be assessed by calculating the 1-dimensional quantization errors to a range of k-values. The third factor is the most problematic, and also the most interesting, because it appears to allow a way to investigate variable dependencies through vector quantization.

Variable importance and quantization quality are important pieces of information when analysing and interpreting a quantization or a clustering result. The final quantization error of a variable — even when compared to errors of the other variables — does not by itself give very clear picture of the quantization quality of the variable. In this paper, two measures q_j and k_j have been proposed which are well suited for evaluating the quantization quality of single variables.

Acknowledgments. This work has been carried out in the "KÄD" project in Helsinki University of Technology. The financing of TEKES and Jaakko Pöyry Consulting is gratefully acknowledged.

References

1. Michael R. Anderberg. *Cluster Analysis For Applications.* Academic Press, 1973.
2. E. Anderson. The irises of the gaspe peninsula. *Bulletin of American Iris Society*, 1935.
3. Robert M. Gray. Vector quantization. *IEEE ASSP Magazine*, pages 4–29, April 1984.
4. Jari A. Kangas, Teuvo K. Kohonen, and Jorma T. Laaksonen. Variants of Self-Organizing Maps. *IEEE Transactions on Neural Networks*, 1(1):93–99, March 1990.
5. Leonard Kaufman and Peter J. Rousseeuw. *Finding Groups in Data: and Introduction to Cluster Analysis.* John Wiley & Sons, Inc., 1990.
6. Glenn W. Milligan and Martha C. Cooper. A study of standardation of variables in cluster analysis. *Journal of Classification*, 5:181–204, 1988.
7. John Moody and Christian J. Darken. Fast Learning in Networks of Locally-Tuned Processing Units. *Neural Computation*, 1(2):281–294, 1989.
8. Dorian Pyle. *Data Preparation for Data Mining.* Morgan Kaufmann Publishers, 1999.

A Hybrid Approach to Clustering in Very Large Databases[1]

Aoying Zhou, Weining Qian[2], Hailei Qian, Jin Wen, Shuigeng Zhou, and Ye Fan

Department of Computer Science, Fudan University, 200433, P.R.China
{ayzhou, wnqian, hlqian}@fudan.edu.cn

Abstract. Current clustering methods always have such problems: 1) High I/O cost and expensive maintenance; 2) Pre-specifying the uncertain parameter k; 3) Lacking good efficiency in treating arbitrary shape under very large data set environment. In this paper, we first present a hybrid-clustering algorithm to solve these problems. It combines both distance and density strategies, and makes full use of statistics information while keeping good cluster quality. The experimental results show that our algorithm outperforms other popular algorithms in terms of efficiency, cost, and even get much more speedup as the data size scales up.

1 Introduction

Current clustering algorithms often consider some single criterion and take fixed strategy alone. Because of such limitations, these methods always have advantages in some aspects but weak in other aspects. Moreover, in very large databases the already existed information is not fully utilized. Another problem is the pre-specified k, which is unreasonable to determine before moving forward to the final goal. We think that the following requirements for clustering algorithms are necessary: to achieve good time efficiency under very large datasets, to identify arbitrarily shaped clusters, to remove noise or outliers effectively and to cluster without any pre-specified k.

In this paper, a new clustering algorithm is proposed. It works on a hierarchical framework and takes hybrid criterion based on both distances between clusters and density within each cluster. This hybrid method can easily identify arbitrarily shaped clusters and can be scaled up to very large databases efficiently.

The rest of the paper is organized as follows. We first generalize the related work in 1.1. In Section 2, a new clustering algorithm is presented. Section 3 discusses its enhancement behavior. In Section 4, we show the experimental evaluation of the algorithm. Finally in Section 5,concluding remarks are offered.

---

[1] This work is partially supported by the National Key Fundamental Research Program (G1998030414) and NSFC (60003016)

[2] The author is supported by Microsoft Fellowship.

D. Cheung, G.J. Williams, and Q. Li (Eds.): PAKDD 2001, LNAI 2035, pp. 519-524, 2001.

1.1 Related Work

An agglomerative algorithm for hierarchical clustering starts with the separate set of clusters, which is each data point under initial case. Pairs of sub-clusters are then successively merged until the distance between clusters satisfies the minimum requirements. CURE [1], a hierarchical algorithm, uses multi-representative data points in order to control arbitrary shape well.

DBSCAN [2] relies on a density-based notion of clustering. They are designed to discover clusters of arbitrary shapes. DBSCAN uses R*-tree to achieve better performance. But when the data size is very large, DBSCAN needs frequent I/O swap to load data into memory. And it also needs an uncertain choice of optimal cut point.

Wang etc. gave STING [3], which divides the spatial data into rectangular cells using a hierarchical structure with statistical information stored together. However, in hierarchy, the parent cell may not be built up correctly for the reason of statistical numbers, although high efficiency are obtained. When the agglomerative procedure moves on, the cells cannot represent the precise information they originally have.

2 Hybrid Clustering Algorithm Based on Distance and Density

2.1 Hybrid Clustering Algorithm

The hybrid algorithm needs three parameters: M-DISTANCE, M-DENSITY and M-DIAMETER. M-DIAMETER will be introduced later.

Definition 1: **M-DISTANCE** is the minimum distance between two clusters.

Definition 2: **M-DENSITY** is the minimum value among each **density**, which is the number of data in a cell belonging to corresponding cluster.

The main clustering algorithm starts from original sub-clusters (including units or data points). It is detailed as below:

```
1. CLUSTERING (M-DISTANCE, M-DENSITY)
2. {  sort the sub-clusters in heap;
3.    for each sub-cluster i with minimum distance
        between i and i.closest
4.    {  if (distance (i, i.closest) < M-DISTANCE)
5.          merge (i, i.closest);
6.       else
7.          if (CONNECTIVITY (i, i.closest, M-DENSITY)
              == TRUE)
8.             merge (i, i.closest);
9.          else
10.            note i & i.closest aren't connected;}}
```

The procedure obtains sub-cluster with minimum distance between itself and the closest sub-cluster to it. If this distance is smaller than M-DISTANCE, then two sub-clusters must belong to one cluster. There exists such situation that some sub-clusters that should be merged, but the distance between them is large. We test the connectivity of such sub-clusters. Two connected sub-clusters must belong to one cluster. If these two sub-clusters could not connect to each other, they must belong to two clusters unless the distance between them is smaller than M-DISTANCE.

We make use of statistics information to test the connectivity of sub-clusters.

Definition 3: A cluster's **diameter** is the maximum distance of two data points in it.

Definition 4: **M-DIAMETER** is the minimum diameter clusters may have.

Definition 5: A **cell** is a small data grid, the length of whose diagonal distance is smaller than min {1/2*M-DISTANCE, M-DIAMETER}.

Then some properties about cell can be presented. Since the lack of space, the proof is omitted here.

Theorem 1: There will not be any cell that contains data points belonging to two different clusters, and there will not be any cell that contains a whole cluster either.

Theorem 2: If cell i belongs to cluster A, cell j is a neighbor cell of i and density(j)>M-DENSITY, then cell j must belong to cluster A too.

Definition 6: **Noises** are the data points in the cell whose density is smaller than M-DENSITY.

Finally, the following procedure is to judge the connectivity of two sub-clusters.

```
1. CONNECTIVITY (cluster_i, cluster_j, M-DENSITY)
2. {   QUEUE q;
3.     for each cell k in cluster_i
4.        q.ADD (k);
5.     for each neighbor cell l of cells in q AND l do
          not in q
6.     { if (density(l) > M-DENSITY)
7.           if (l.belongto==cluster_j)
8.              return TRUE;
9.           else
10.          {  q.ADD (l);
11.             merge (cluster_i, l.belongto);
12.             for each cell m in l.belongto
13.                q.ADD (m);}}
14.    return FALSE;}
```

2.2 Scale Up to Very Large Databases

To handle very large databases, the algorithm constructs units instead of original sub-clusters while using sampling. To obtain these units, first, partitioning work is done to make the data points into cells. Some statistics information is also obtained in each dimension of every cell. Then we test to see if a cell forms a unit. The definition of unit is as below:

Definition 7: A **unit** is a cell whose data points belong to a certain cluster.

The density of units is bigger than that of other cells. Therefore, we determine if a cell is a unit by density. If the density of a cell is M (M•1) times the average density of all the data points, we regard such cell as a unit. There may be some cells with low density but are in fact parts of final clusters, they may not be included into units. We treat such data points as separated sub-clusters. This method will greatly reduce the time complexity, as will be shown in the experiments.

2.3 Labeling Data in Databases

The clusters identified in this algorithm are denoted by the representative points. In most cases, user needs to know the detailed information about the clusters and the data points included. Therefore, we need not test every data point which cluster it belongs to. Instead, we can only find the cluster a cell belonging to, then all data points in it belong to such cluster. This process can also improve the speed of the whole clustering process.

2.4 Time and Space Complexity

The time complexity of our clustering algorithm can be O (m^2*log$m$) in upper case. Here, m is the number of sub-clusters in the beginning. In general, when the data distributes in well proportioned dense area, the m will be small. Because the cell and data are stored in linear space, the space complexity needed in our method is O (n).

3 Enhancements for Different Data Environment

3.1 Handling Noises and Outliers

Noises are random disturbance that reduces the clarity of clusters. In our algorithm, we can easily find noises and wipe off them by finding the cells with very low density and eliminate the data points in them precisely. This method can reduce the influence of the noises both on efficiency and on time.

Unlike the noises, outliers are not well proportioned. Outliers are data points that are away from the clusters and have smaller scale compared to clusters. So outliers will not be merged to any cluster. When the algorithm finishes, the sub-clusters that have rather small scale are outliers. Our algorithm can determine this scale by the parameter M-DIAMETER, which denotes the minimum diameter a cluster may have.

3.2 Handling Data Sets Having Clusters with Arbitrary Shape

By using multi-representatives technique and distance-plus-density strategy, the hybrid algorithm can accurately identify most arbitrarily-shaped clusters which are difficult to be processed by other methods, as shown in Fig. 1.

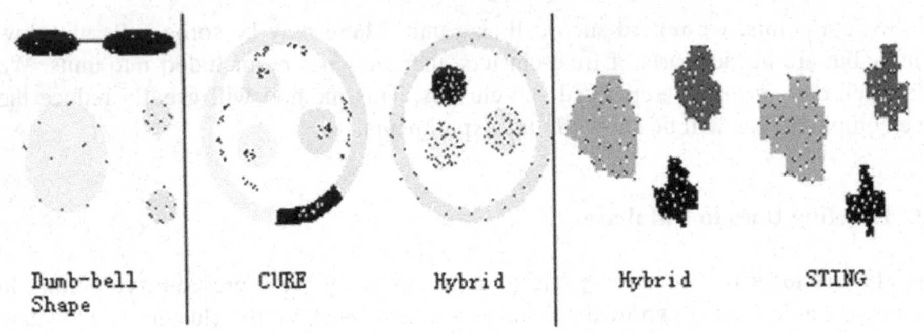

Fig. 1. Data sets with clusters in dumb shape

4 Performance Evaluation

The experiments were run under such environment as: Microsoft Windows NT 4.0, Intel Pentium II 350 × 2, and 512M RAM. We research the performance of our hybrid clustering algorithm to see its effectiveness and efficiency for clustering compared to DBSCAN and CURE.

Fig. 2 illustrates the performance of the hybrid-clustering algorithm and CURE as the number of sample size from 1000 to 6000. It shows that our algorithm far outperforms CURE while keeping the good clustering quality

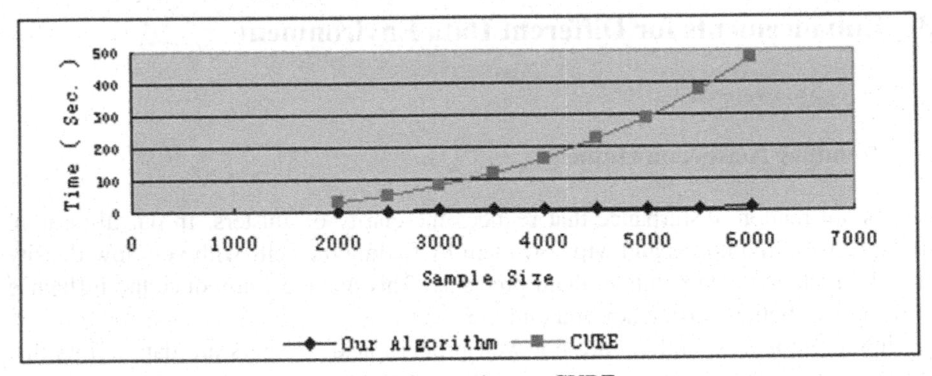

Fig. 2. Comparison to CURE

The hybrid-clustering algorithm can successfully handle arbitrarily large number of data points. Fig. 3 illustrates the performance of our algorithm, DBSCAN, and CURE, as the number of data size from 30,000 to 1,000,000.

5 Conclusions

In this paper, we present a hybrid-clustering algorithm. This algorithm identifies the clusters both by distance between clusters and density of within clusters. The algorithm can easily identify arbitrarily shaped clusters with good quality, and user need not pre-specify the number of clusters. With the help of statistics information, it greatly reduces the computational cost of the clustering process. Our experimental results demonstrate that it can outperform other popular clustering methods.

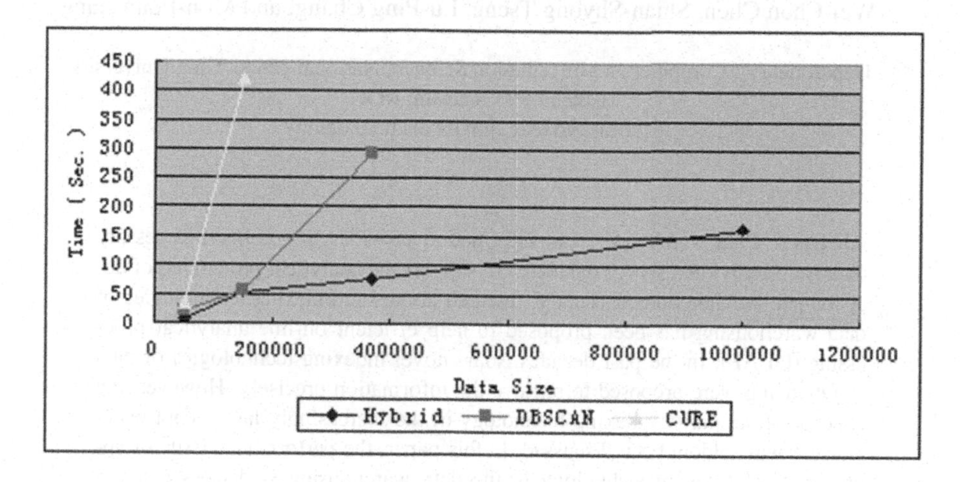

Fig. 3. Scale-up Experiments

Acknowledgements. We thank Dr. Joerg Sander for providing source code of DBSCAN, and Prof. Jiawei Han gave valuable critical comments to our algorithm.

References

[1] S. Guha, R. Rastogi and K.Shim, CURE: An Efficient Clustering Algorithm for Large Databases, In *Proc. of ACM SIGMOD'98 Conf.*, 1998.

[2] M. Ester, H-P. Kriegel, J. Sander, and X. Xu, A Density-Based Algorithm for Discovering Clusters in Large Spatial Database with Noise. In *Proc. of the 2nd Int'l Conf. on KDD*, 1996.

[3] W. Wang, J. Yang, and R. Muntz. STING: A Statistical Information Grid Approach to Spatial Data Mining. In *Proc. of the 23rd Int'l Conf. on VLDB*, 1997.

A Similarity Indexing Method for the Data Warehousing – Bit-Wise Indexing Method[1]

Wei-Chou Chen, Shian-Shyong Tseng, Lu-Ping Chang, and Mon-Fong Jiang

Department of Computer and Information Science, National Chiao Tung University, Hsinchu 300, Taiwan, ROC
{sirius, sstseng, jmf}@cis.nctu.edu.tw

Abstract. Data warehouse is an information provider that collects necessary data from individual source databases to support the analytical processing of decision-support functions. Recently, research about the indexing technologies of data warehousing has been proposed to help efficient on-line analytical processing (OLAP). In the past decades, some novel indexing technologies of data warehousing were proposed to retrieve the information precisely. However, the concept of similarity indexing technology in the increasingly larger data warehousing was seldom been discussed. In this paper, the performance issue of approximation indexing technology in the data warehousing is discussed and a new similarity indexing method, called bit-wise indexing method, and the corresponding efficient algorithms are proposed for retrieving the similar cases of a case-based reasoning system using a data warehouse to be the storage space. Some experiments are made for comparing the performance with two other methods and the results show the efficiency of the proposed method.

1 Introduction

Data warehouse is an information provider that collects necessary data from individual source databases to support the analytical processing of decision-support functions[14]. Recently, research about the indexing technologies of a data warehouse has been proposed to help efficient on-line analytical processing (OLAP). A critical issue of performance for data warehousing is to retrieve necessary tuples according to the query statements. Many researchers have proposed the useful indexing technologies to retrieve records precisely.

Case-based reasoning (CBR) is a methodology of problem solving in AI [3]. Just like human being, CBR uses prior cases to find out suitable solution for the new problems. The method of CBR uses useful prior cases to solve the new problems. It has been successfully applied in many areas [2][4][8][10][13]. The major tasks of CBR can be divided into five phases, including Case Representation, Indexing, Matching,

---

[1] This work was supported by Ministry of Education and National Science Council of the Republic of China under Grand No. 89-E-FA04-1-4, High Confidence Information Systems.

D. Cheung, G.J. Williams, and Q. Li (Eds.): PAKDD 2001, LNAI 2035, pp. 525-537, 2001.

Adaptation and Storage. A critical task of CBR is to retrieve similar prior cases accurately and many researchers have proposed some useful technologies to handle such problem [7][11][14]. However, performance of retrieving similar cases was seldom been discussed. When the number of cases in the case base is large, the processing time for retrieving similar cases increases rapidly. Therefore, the process of retrieving similar cases becomes an overhead task of CBR. Thus the retrieving time should be taken into consideration in a successful CBR. In this work, we propose a novel indexing method with suitable similarity-measuring function. The corresponding case retrieving algorithm of CBR using a data warehouse to be the storage space has also proposed to highly accelerate the performance of cases indexing and retrieving. The new indexing method and the corresponding algorithm are easy to be parallelized and thus improve the performance substantially in case retrieving and similarity measuring in the data warehousing. Finally, the experiments and the comparison with the traditional relational database and bitmap index method of the data warehousing have been made. The results show the correctness and efficiency of the proposed method.

2 Related Works

In this section, some related topics would be discussed, including data warehousing, case-base reasoning and bitmap indexing method.

The concept of data warehousing was first proposed by Inmon [14] in 1993. A data warehouse contains information collected from individual data sources and integrated into a common repository for efficient querying and analysis. When the data sources are distributed over several locations, a data warehouse is responsible for collecting the necessary data and saving it in appropriate forms.

Case-based reasoning (CBR) is a methodology of problem solving in AI. The method of CBR reuses past cases to solve the new problems. The success of a CBR system mainly depends on an effective retrieval on similar cases for the problem; therefore, the indexing and matching thus become the important tasks in CBR [6][7][11]. In the similarity-based matching function, the weights provide a surrogate method of representing the complex interrelationships in similarity measurement, and represent the degree of importance of a feature toward the goal of the solving problem. A critical task of CBR is to retrieve similar cases accurately and many researchers have proposed some useful technologies to handle such problem[5]. However, performance of retrieving similar cases has seldom been discussed. Retrieving similar cases needs more time when the matching function becomes more complex or the number of cases in the case base becomes very large. Therefore, how to quickly retrieve similar cases becomes an important issue in CBR and the retrieving time should be taken into consideration in a successful CBR. Since retrieving cases in attributed based CBR is similar to retrieving records in DWs, bitmap indexing technology seems to be able to be directly applied to indexing and retrieval phrases in CBR[9]. However, there are still some problems should be solved:

1. Prior cases are most likely not the same as the new case in CBR. Rather than exact matching that can be used to retrieve records in DWs, partial matching may be required to retrieve prior cases in CBR.
2. Some extra computation of the similarity between the new case and prior cases is required.

In other words, it is unsuitable to straightly apply the bitmap indexing method to the indexing phase of CBR. It needs some adaptation and we will discuss the details of our new indexing method.

3 Bit-Wise Indexing CBR

3.1 Architecture of Bit-Wise Indexing CBR

As described above, we propose a novel indexing method with suitable similarity-measuring function to speed up retrieving similar cases in CBR. The architecture of Bit Wise Indexing CBR (BWI-CBR) is shown in Fig. 1. Most parts of BWI-CBR are the same with that of general CBR, except the following:

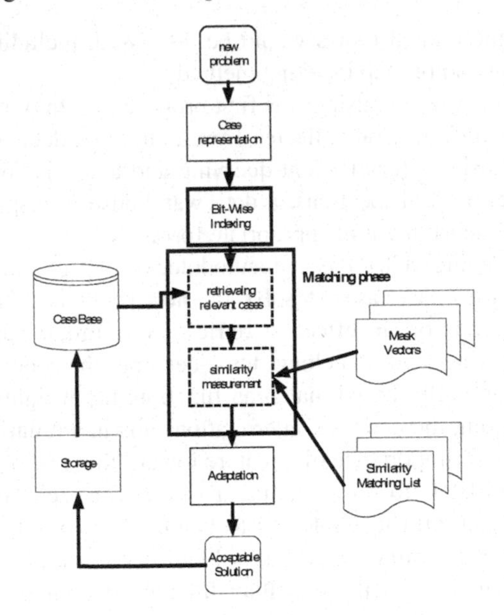

Fig. 1. The architecture of BWI-CBR

1. Bit-wise indexing phase: replace the indexing method in traditional CBR with bit-wise indexing method. It can highly speed up retrieving time in the Matching phase.
2. Matching phase: **a**. *Retrieving relevant cases phase*: To match bit-wise indexes between the new case and prior cases, we can select relevant prior cases and filter out irrelevant cases. Moreover, the matching result of bit-wise indexing can be used

to calculate the similarity degree of prior cases directly in Similarity measurement phase. **b.** *Similarity measurement phase*: Computing the similarity between the relevant cases and the new arrival case is to find the similar cases in case base. The similarities between relevant prior cases and the new cases cannot be computed without knowing which attributes with the same values. To solve the problem, we use the Mask Vector. We can pre-compute all-possible similarities and construct the Similarity Mapping List. Accordingly, the similarity of each prior cases and new arrival case can be quickly found out by seeking the Similarity Mapping List. Therefore, the computing overhead can thus be largely reduced.

In our approach, we first use the bit-wise indexing method to replace the traditional indexing method in CBR. Then the bit-wise operations can be used to select the relevant prior cases in retrieving relevant cases phase. By this way, the irrelevant cases can be filtered out quickly and the number of prior cases, which are needed to compute the similarities with the new case, can be reduced. Therefore, the similarities between relevant prior cases and the new case can be quickly measured in similarity measurement phase.

3.2 Some Definitions of Bit-Wise Indexing CBR

Assume there is a set of cases, called *examples*, denoted as $C=<C_1, C_2,\ldots, C_i>$ where $i\geq 1$, needs to be stored in the CBR in a specific domain, denoted as *DOM*, for reasoning. Assume A is the set of attributes and all cases of C in the domain *DOM* can be abstracted into ∂ attributes, denoted as $A=<A_1,A_2,\ldots,A_\partial>$. For each attribute A_k of case C_j, its *attribute value* is denote as $V_k(j)$ and $V_k(j)\neq null$. Moreover, Denote $V_*(j)=<V_1(j), V_2(j),\ldots, V_\partial(j)>$ is the attribute value set of case C_j. Denote $V_i=<V_{i1}, V_{i2},\ldots, V_{i\alpha(i)}>$ where V_{ij} is a possible attribute value of A_i, that is, $V_{ij}=V_k(x)$ for some C_x, $C_x\in C$, and $V_{ij}\neq V_{ik}$ for $j\neq k$. $\alpha(i)$ is the number of values in A_i, all elements of V_i is the collection of all attribute values of attribute A_i of C, called *attribute value domain* of attribute A_i of C. In a CBR system of domain *DOM*, the cases of C need to be stored in the CBR system for solving the new arrival case. A, the set of its significance attributes, acts like indexes to books in a library, helps the CBR system to select cases likely to fulfill the needs of the problems for new arrival case. When retrieving cases, matching function is used to retrieve cases based on a weighted sum of matched attributes in the input cases. Attributes can be viewed as indexing features of a design case or as the decision variables relevant to the original design situation. An index of case can be formally defined as follows.

DEFINITION 1 (Index of Case in CBR system for Domain DOM): The index IND_k of a case C_k in CBR system for domain *DOM* is defined as $IND_k=\{A_1=V_1(k), A_2=V_2(k),\ldots, A_\partial=V_\partial(k)\}$.

Example 3.1: As shown in Fig. 2(a), there are five cases, each of which has three attributes OS, PL and DB. Therefore, ∂ is equal to 3. Attribute values domain of OS, PL and DB are V_1=<WinNT, OS2, Linux, Mac, Solaris>, V_2=<C, Basic, Java, Pascal> and V_3=<SQL-Server, ORCALE, SYBASE>. The index of Case 1 is written as {OS=WinNT, PL=C, DB=SQL-Server}.

In CBR, cases represent an experienced situation. This experienced situation constructs knowledge which can be used in future. When a similar situation arises, the knowledge that goes into making them provide a starting point for interpreting the new situation or solving the problem it poses.

Formally, let Cv be the set of the case's contents including case description representing the problem needed to be solved at that time, the stated or derived solution to the problem specified in the problem description, and the resulting state when the solution was carried out. A case in CBR can be formally defined as follows.

DEFINITION 2 (Case in CBR system for Domain DOM): The case C_k in CBR system for domain *DOM* is a twin $\{IND_k, cv_k\}$, where $cv_k \in Cv$ and $C_k \in C$.

Example 3.2: We assume that a software company wants to design software, which allows multi-user access the data stored in SQL database. The Case 1 of Fig. 2(a) can be transformed as follows:

Case1 {Index IND_1: Operation System=WinNT, Program language=C language and Database=SQL-server. Contents of the case C_1: It adopts: WinNT as the OS, C language as developing tools and SQL-server as database. Result : good performance.}

(a) An example Case base

	OS	*PL*	*DB*
Case1	WinNT	C	SQL-Server
Case2	OS2	Basic	ORCALE
Case3	Linux	Java	SYBASE
Case4	Mac	Java	ORCALE
Case5	Solaris	Pascal	SQL-Server

(b) The bitmap indexing for (a)

B_{WinNT}	B_{OS2}	B_{Linux}	B_{Mac}	$B_{Solaris}$	B_C	B_{Basic}	B_{Java}	B_{Pascal}	B_{SQL}	B_{Orcale}	B_{Sybase}
1	0	0	0	0	1	0	0	0	1	0	0
0	1	0	0	0	0	1	0	0	0	1	0
0	0	1	0	0	0	0	1	0	0	0	1
0	0	0	1	0	0	0	1	0	0	1	0
0	0	0	0	1	0	0	0	1	1	0	0

Fig. 2. An example case base and corresponding bitmap indexes

A case of Case-based reasoning using the proposed indexing method, called *bit-wise indexing method*, is formally defined as follows:

DEFINITION 3 (the i-th attribute bit-wise indexing vector of the case): The bit-wise indexing vector B_i of the *i*-th attribute for the case C_k in C is a bit string. $B_i = b_{i1}b_{i2}...b_{i\alpha(i)}$, $b_{ij}=1$ if $V_i(k)=V_{ij}$ otherwise $b_{ij}=0$, where $0<i\leq\partial$.

DEFINITION 4 (bit-wise indexing vector of case): A bit-wise indexing vector BWI_k of case C_k is a concatenation of the attribute bit-wise indexing vectors. That is, $BWI_k=B_1B_2...B_\partial$ for ∂ attributes.

Example 3.3: According to the DEFINITION 3 and 4, the bit-wise indexing vector of attributes OS, PL and DB for Case 1 in Fig. 2(a) are B_1="10000", B_2="1000" and B_3="100", respectively. We have the bit-wise indexing vector BWI_1 of Case 1 is $B_1B_2B_3$="***100001000100***".

DEFINITION 5 (Matrix of bit-wise indexes for case-based reasoning): A matrix of bit-wise indexing for CBR T_{BWI} is written as $\begin{bmatrix} BWI_1 \\ BWI_2 \\ \vdots \\ BWI_{|C|} \end{bmatrix}$.

Example 3.4: According to DEFINITION5, the T_{BWI} of the cases of Fig. 2(a) is shown as below:

BWI_1	10000	1000	100
BWI_2	01000	0100	010
BWI_3	00100	0010	001
BWI_4	00010	0010	010
BWI_5	00001	0001	100

4 The Phases of Bit-Wise Indexing CBR

4.1 Indexing Phase in Bit-Wise Indexing CBR

Unlike the bitmap indexing method which takes each attribute value as a bit vector to constructs indexes, Bit-Wise Indexing method views each case as a bit vector. For the cases shown in Fig. 2(a), Fig. 2(b) is the bitmap indexes. When a casc (OS=WinNT; PL=Java; DB=ORCALE) arrives, some partial matched cases may be obtained by checking B_{WinNT}, B_{Java}, B_{ORCALE}, (B_{WinNT} AND B_{Java}), (B_{WinNT} AND B_{Java}), (B_{WinNT} AND B_{ORCALE}), and (B_{WinNT} AND B_{Java} AND B_{ORCALE}) vectors for bitmap indexing method. To compute the similarity of the *i*th case and the new case, we need to check the *i*th bit position in B_{WinNT}, B_{Java}, B_{ORCALE}, (B_{WinNT} AND B_{Java}), (B_{WinNT} AND B_{Java}), (B_{WinNT} AND B_{ORCALE}), and (B_{WinNT} AND B_{Java} AND B_{ORCALE}) vectors. Once the attributes of cases or the number of cases are large, the computing time of scanning to the *i*-th position in vectors is dramatically increasing. Therefore, we propose a new indexing method, which is suitable for similarity computing. Let the length of bit-wise indexes l= $\sum$. The bit-wise indexes creation algorithm and the matrix of bit-wise indexes creation algorithm are shown as follows:

***Algorithm* 4.1 (*Bit-wise indexes creation algorithm*):**
Input : C_i of C.
Output : The BWI_i of T_{BWI}.
Step 1:Create bit-wise vector BWI_i of case C_i with length l.
Step 2:For each bit b_{jk} of BWI_i, if $V_j(i)=V_{jk}$, b_{jk}=1; otherwise, b_{jk}=0.
Step 3:Return BWI_i.

***Algorithm* 4.2 (*Matrix of bit-wise indexes creation algorithm*):**
Input : C of CBR.
Output : The T_{BWI} of the CBR.
Step 1:Create an empty matrix T_{BWI} and set counter *i* to 1.
Step 2:For each case C_i in CBR, do the following sub-steps.

Step 2.1: Use *Bit-wise indexes creation algorithm* to get the BWI_i of C_i.
Step 2.2: Add the BWI_i into T_{BWI}.
Step 2.2: if $i=|C|$, exit sub-procedure, otherwise, set $i=i+1$ and go to Step 2.1.
Step 3:Return T_{BWI}.

After the bit-wise indexing matrix is built, the bit-wise operations can be easily applied on it. According to the characteristics of bit-wise indexing method, we will discuss our similarity measurement algorithms in following sections.

4.2 Matching Phase in Bit-Wise Indexing CBR

For the obtained bit-wise indexing matrix, the bit-wise operations can be used to retrieve similar cases in the CBR easily. However, computing the similarity among all cases using matching function is a time-consuming task. A two-phase *Similar Cases Seeking algorithm*, including *relevant cases retrieving* phase and *similarity computing* phase, is proposed. Since the bit-wise operations are quite fast, the major concern of our algorithm is to reduce the computing time by filtering all irrelevant cases before calculating the similarities. If all irrelevant cases can be filtered out first, the time of retrieving useful prior cases can then be decreased largely. Therefore, in the *relevant cases retrieving* phase, all irrelevant cases will be filtered out quickly and the similarities of other cases will then be computed efficiently in the *similarity computing* phase.

***Algorithm* 4.3 (*Similar Cases Seeking Algorithm*):**
Input : The T_{BWI} and a new case C_N.
Output : A set of similar cases *Rc* and a set of its corresponding similarity degree *Rs*.
Step 1:Use *Bit-wise indexes creation algorithm* to get BWI_N of case C_N with length *l*.
Step 2:Initialize the counter *j* to 1 and let *Rc* and *Rs,* be empty.
Step 3:For each BWI_j in T_{BWI}, do the following sub-steps (where $1<j\leq|C|$):
Step 3.1: Call *Search-relevant Algorithm* to compare the relevant degree rdi_j between BWI_j and BWI_N.
Step 3.2: If $rdi_j=0$, the rdi_j is dropped and go to Step 3.4.
Step 3.3: Call *Similarity Computing Algorithm* to compute sim_j, and then add sim_j and the $case_j$ into *Rs* and *Rc*.
Step 3.4: Add 1 to *j*.
Step 4:Sort *Rc* in descending order according to its corresponding similarity degree in *Rs*.
Step 5:Output *Rc* and *Rs*.

The Retrieving Relevant Prior Cases of Bit Wise Indexing CBR

For one prior case, if it is relevant to the new case, at least one of its attributes has the same value as that of the same attribute of the new case. That is, the bits in the corresponding positions of the same attributes should be set as "1" in their bit vectors and can be found by using the ***AND*** bit-wise operation to compare these two bit vectors. It means the two cases have the same attribute-value for the corresponding attribute. In other words, these two cases are similar in some degree. The *Search-relevant algorithm* is described in the following:

Algorithm 4.4 **(*Search-relevant algorithm*):**
Input: The bit-wise indexing vector BWI_N of a new arrival case C_N and BWI_j of case C_j in C.
Output: the relevant degree rdi_j.
Step 1: Use ***AND*** bit-wise operation on BWI_N and BWI_j, and then store the result into rdi_j which is also a bit-wise indexing vector with length l.
Step 2: Return rdi_j.

Since the ***AND*** bit-wise operation takes one instruction execution time, the *Search-relevant algorithm* can be used to select the relevant prior cases quickly. In the real implementation, some integers are used to represent the prior case's indexes, the new arrival case's indexes and *rdi*. If *rdi* is zero, then the bit vector of the prior case will be filtered. By this way, all irrelevant prior cases can be filtered out efficiently and precisely.

The Similarity Computing of Bit-Wise Indexing CBR

After all relevant prior cases have been retrieved, in order to select the useful prior cases, the similarities between these relevant prior cases and the new case need to be computed. As discussed above, we use the matching function to retrieve cases. The matching function is based on a weighted sum of attributes for the input case that matches the prior cases in the case base. Each attribute has its own weights. The similarities between relevant prior cases and the new cases can not be computed without knowing which attributes with the same values. Since each attribute of every case has only one attribute value, at most one bit of *rdi* is set after executing the *Search-relevant algorithm*. Accordingly, we propose a special bit-wise vector, call *Mask Vector*, to solve the bottleneck of *similarity computing* phase. Denote $<1>_i=b_{i1}b_{i2}...b_{i\alpha(i)}$, $\forall b_{ij}=1$ is the 1-vector of length $\alpha(i)$ and $<0>_i=b_{i1}b_{i2}...b_{i\alpha(i)}$, $\forall b_{ij}=0$ is the 0-vector of length $\alpha(i)$. The definition of *Mask Vector* is shown as below:

DEFINITION 6 (Mask Vector): A mask bit-wise indexing vector *Mask* is a set of $Mask_k$, where $0<k\leq\partial$. The $Mask_k$, the mask vector of attribute A_k, is a concatenation of bit string S_i where $Mask_k=S_1S_2...S_\partial$ for $S_k=<1>_k$ and $\forall i\neq k$, $S_i=<0>_i$.

Example 4.1: Continuing the Example 3.4 in Section 3.2, the Mask vector $Mask_1$ $Mask_2$ and $Mask_3$ of attribute OS can be generated to {11111 0000 000} {00000 1111 000} and {00000 0000 111} respectively.

By applying the 'AND' operation on *Mask vector* and the bit-wise vector of the results generated from *search-relevant* phase(*rdi*), called *Mask-Vector processing*. Based upon these results, the similarities of each attribute for new arrival cases and prior cases can be computed easily.

As discussed above, we proposed a suitable similarity-measuring function for BWI-CBR to compute similarity according to the result of *Mask-Vector processing*. The function is:

$$SIM(Case_i) = \frac{\sum_{j=1}^{\partial}(PC_{ij} \times W_j)}{\sum_{j=1}^{\partial} W_j} \quad (1)$$

Where $SIM(Case_i)$ is the similarity between i-th prior case and new case, W_j is the weight of the attribute j. If the result of performing AND bit-wise operation on the relevant degree rdi_i and $Mask_j$ is 0 then set PC_{ij} as 0; set PC_{ij} as 1, otherwise.

Example 4.2: Assume that the weights of attributes OS, PL and DB, denoted as $Weight_1$, $Weight_2$ and $Weight_3$, are set to 1, respectively. Also, assume that the result of *Mask-Vector processing* Cases 1, 2, 3 and 4 are "001","001", "011", and "100", respectively. Therefore, the similarity of these cases 0.333, 0.333, 0.667 and 0.333, respectively.

Since the computing results (similar degree) of Case 1 and Case 2 are the same, however, it can been computed only once by pre-computing all-possible similarities and storing them into the *Similarity Mapping List*. The computing overhead can be largely reduced and the processing time can thus be eliminated.

DEFINITION 7 (Similarity Mapping List)*:* Let L be the Similarity Mapping List. L_i is the element in L with the index value i, where $i=b_{i1}b_{i2}...b_{i\partial}$, and $b_{i1}b_{i2}...b_{i\partial}$ is the binary representation of i, where $1 \leq i \leq 2^{|\partial|}-1$.

$$L_i = \frac{\sum_{j=1}^{\partial} b_{ij} \times W_j}{\sum_{j=1}^{\partial} W_j} \qquad (2)$$

***Algorithm* 4.5 (*Similarity Mapping List Creation Algorithm*):**

Input: Weight for indexes of CBR.

Output: The similarity mapping list L.

Step 1: Initialize the counter k to 1 and let List L be empty.

Step 2: For each k, do the following sub-steps

Step 2.1: Encode k into a binary string $k=<b_{i1}b_{i2}...b_{i\partial}>$.

Step 2.2: Calculate the similarity degree L_k by Formula 1 in Definition 8.

Step 2.3: Add L_k into L.

Step 2.4: If $k = 2^{|\partial|}-1$, then exit the processing of sub-steps; Otherwise, let $k = k + 1$ and repeat the sub-steps of Step 2.

Step 3: Return L.

After the *Similarity* Mapping *List* had been built, the similarity of each prior cases and new arrival case can be quickly found out by the following algorithm:

***Algorithm* 4.6 (*Similarity Computing Algorithm*):**

Input: The relevant degree rdi_j, the *Mask Vector* and the *Similarity Mapping List* L.

Output: The similarity of $case_j$.

Step 1: Initialize k to be a binary string with length ∂.

Step 2: For each i, set the i-th position of k to 1 if the result of using AND bit-wise operation on the $Mask_i$ and rdi_j is not all 0; otherwise, set it to 0.

Step 3: Transform k into an integer j, set L_j to sim_j.

Step 4: Return sim_j.

During the pre-processing step, we had constructed the *Similarity Mapping List* and *Mask Vector*. In *Similarity computing algorithm*, only the 'AND' bit-wise operation needs to be done on *Mask Vector* and bit-wise vectors of relevant case. Therefore, we can use the Similarity *Mapping List* to find out the similarities between relevant prior cases and the new case quickly and easily. Since the redundancy of similarity measure is avoided, the computing overhead and time computing similarity for each relevant prior case can be largely reduced.

Example 4.3: In this Example, we will show our *Similar Cases Seeking Algorithm* in detail. Following the Example 3.4, for the new arrival case C_N: {OS=Solaris, PL=Java, DB=ORCALE}, we have BWI_N = {00001 0010 010}. In Step 2 of *Similar Cases Seeking Algorithm*, Initial state:(j=1) and set *Rc* and *Rs* to empty. For each BWI_j in T_{BWI}, do the following sub-steps (where $1<j\leq|C|$):

- For BWI_1 : The *Search-relevant Algorithm* is used to compare the relevant degree rdi_1 between BWI_1 and BWI_N and the return value rdi_1 = {00000 0000 000}. Because the bit of rdi_1 are all "0", Case 1 is filtered out.
- For BWI_2 : Similarly, we have rdi_2 = {00000 0000 010}. Since one of the bit in rdi_2 is equal to "1", the *Similarity Computing Algorithm* is called to compute sim_2=0.333 and then add sim_2 and the $case_2$ into *Rs* and *Rc*, respectively ($sim_2=L_k=L_{001(2)}=L_1$=0.333).
- For BWI_3 : Similarly, we have rdi_3 = {00000 0010 000}. Since one of the bit in rdi_3 is equal to "1", sim_3=0.333 and then add sim_3 and the $case_3$ into *Rs* and *Rc*.
- For BWI_4 : Similarly, we have rdi_4 = {00000 0010 010}. Since one of the bit in rdi_4 is equal to "1", sim_4=0.667 and then add sim_4 and the $case_4$ into *Rs* and *Rc*.

For BWI_5 : Similarly, we have rdi_5 = {00001 0000 000}.Since one of the *bit* in rdi_5 is equal to "1", sim_5=0.333 and then add sim_5 and the $case_5$ into *Rs* and *Rc*. After sorting the element pairs of *Rc* and *Rs* in decreasing order, new *Rs* and new *Rc* become:

Rc:	**Case 4**	**Case 2**	**Case 3**	**Case 5**
Rs:	**0.667**	**0.333**	**0.333**	**0.333**

5 Experiments and Discussions

To evaluate the performance of BWI-CBR, we compare BWI-CBR with two other indexing methods, including SQL-CBR, which uses index of the relational database to be the index method of the CBR, and Bitmap-CBR, which uses bitmap indexing method to construct the indexes of cases. Our target machine is a Pentium-166 dual processors system, running the Microsoft Windows NT multithreaded OS. The system includes 512K L2 cache and 128MB shared-memory.

Compare BWI-CBR with SQL-CBR

In this comparison, the SQL-CBR uses Microsoft SQL server as the case base. The result of comparing BWI-CBR and SQL-CBR is shown in Fig. 3.

According to the result, we can see the BWI-CBR is much more efficient than SQL-CBR. The reasons are:

- In *relevant cases* retrieving phase, BWI-CBR transfers the indexes to bit-wise indexes and uses bit operation to retrieve relevant prior cases. Because bit operation is quite fast, our retrieving relevant prior cases algorithm can filter out irrelevant cases efficiently. In the SQL-CBR, the target SQL statement needs to be transformed into several SQL statements with different WHERE clauses for handling the partial matching in retrieving relevant cases. Therefore, the BWI-CBR is much faster than SQL-CBR in this phase.
- In *similarity computing* phase, the BWI-CBR uses bit operation to compare the new case and prior case in case base, and uses the Mask vectors to check the similarity between the new case and prior cases. That is, it doesn't need to check the contents

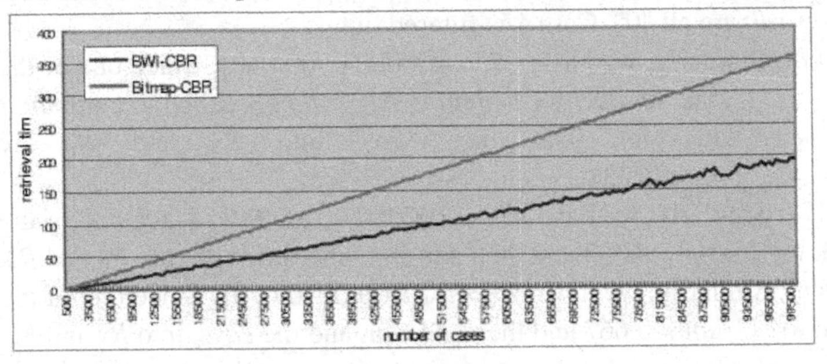

Fig. 3. BWI-CBR vs SQL-CBR

one by one for getting the similarities. However, the SQL-CBR must checks the detail of each feature's contents for computing the similarity. Therefore, this task in BWI-CBR is still much faster than that in SQL-CBR.

Fig. 4. BWI-CBR vs Bitmap-CBR

Compare BWI-CBR with Bitmap-CBR

The result of comparing the BWI-CBR with the CBR-Bitmap is shown in Fig. 4. We can see that BWI-CBR is faster than CBR-bitmap, The reasons are:

- In *relevant cases retrieving* phase, the Bitmap indexing technology is not suitable for retrieving similar cases. It needs to check the more vectors than the BWI-CBR. So it needs more time than BWI-CBR.
- In *similarity computing* phase, the Bitmap-CBR needs to check the *i*th-bit position in B_{WinNT}, B_{Java}, B_{ORCALE}, (B_{WinNT} AND B_{Java}), (B_{WinNT} AND B_{ORCALE}), (B_{Java} AND B_{ORCALE}) and (B_{WinNT} AND B_{Java} AND B_{ORCALE}) vectors for case c_i. Scanning to the *i*th position in the bitmap vectors needs some extra time, especially when the number of attribute of cases or the number of cases in the case base are large. The waste time is lengthy and unbearable. Therefore, the performance of similarity computing in BWI-CBR is much faster than that in Bitmap-CBR.

6 Conclusion and Future Work

In addition to accuracy, performance issue should also be taken into consideration in retrieving similar cases in CBR, especially when the number of cases in CBR is increasingly large. In this paper, the performance issue of large-scale CBR that using a data warehouse to be the storage space is discussed and a new indexing method, called *bit-wise indexing method*, has been proposed. Also, the correspondingly algorithms, including index creation and case retrieving algorithms, are proposed. Finally, some experiments are made for comparing the performance with two other methods, including traditional indexing method and the bitmap indexing method of data warehousing, and the results show the performance of proposed method is admirable. In the future, we will attempt to apply the indexing method and corresponding retrieving algorithm to CBR with multi-processor data warehousing system.

References

1. Gonzalez, A. J., Xu, L., Gupta, U. M.: Validation Techniques for Case-Based Reasoning Systems. IEEE Transactions on Systems, Man, and Cybernetic-Part A: Systems And Humans **28**(4) (1998) 465-477
2. Gardingen, D., Watson, I.: A web based CBR system for heating ventilation and air conditioning systems sales support. Knowledge-Based Systems **12** 1999 207-214.
3. Waston, I.: Case-based reasoning is a methodology not a technology. Knowledge-Based Systems **12** (1999) 303-308
4. Daengdej, J., Lukpse, D., Tsui, E., Beinat, P., Prophet, L.: Combining case-based reasoning and statistical method for proposing solution in RICAD. Knowledge-Based Systems, **1** (1997).153-159
5. Wu, K. L., Yu, P. S.: Range-Based Bitmap Indexing for High Cardinality Attributes with Skew. Proceedings of 22nd Annual International Conference on Computer Software and Applications (1998) 61-66.
6. Gupta, K. M., Montazemi, A. R.: Empirical Evaluation of Retrieval in Case-Based Reasoning Systems Using Modified Cosine Matching Function. IEEE Transactions on Systems, Man, and cybernetics-Part A: Systems and Humans **27**(5) (1997) 601-612
7. Shin, K. S., Han, I.: Case-based reasoning supported by genetic algorithms for corporate bond rating. Expert Systems with applications 16 (1999) 85-95
8. Li, L. L. X.: Knowledge-based problem solving: an approach to health assessment. Expert Systems with Application **16** (1999) 33-42
9. Wu M. C., Buchmann, A. P.: Encoded Bitmap Indexing for Data Warehouses. Proceedings of IEEE ICDE (1998) 220 -230
10. Suh, M. S., Jhee, W. C., Ko, Y. K., Lee, A.: A case-based expert system approach for quality design. Expert Systems With Applications **15** (1998) 181-190
11. Cercone, N., Aijun, A., Chan, C.: Rule-Induction and Case-Based Reasoning: Hybrid Architectures Appear Advantageous. IEEE Transactions. On Knowledge and Data Engineering **11**(1) (1999), 166-174
12. O'Neil, P., Quass, D.: Improved Query Performance with Variant Indexes. Proceedings of ACM SIGMOD (1997)

13. Dutta, S., Wierenga B., Dalebout, A.: Case-Based Reasoning Systems: From Automation to Decision-Aiding and Stimulation. IEEE Transactions. On Knowledge and Data Engineering **9**(6) 1997 911-922
14. Inmon, W. H., Kelley, C.: Rdb/VMS: Developing The Data Warehouse. QED Publishing Group, Boston, Massachusetts (1993)

Rule Reduction over Numerical Attributes in Decision Trees Using Multilayer Perceptron

DaeEun Kim[1] and Jaeho Lee[2]

[1] Division of Informatics,
University of Edinburgh, Edinburgh, EH1 2QL
United Kingdom
daeeun@dai.ed.ac.uk
[2] Department of Electrical Engineering
University of Seoul, Tongdaemoon-ku,
Seoul, 151-011, Korea
jaeho@uofs.ac.kr

Abstract. Many data sets show significant correlations between input variables, and much useful information is hidden in the data in a non-linear format. It has been shown that a neural network is better than a direct application of induction trees in modeling nonlinear characteristics of sample data. We have extracted a compact set of rules to support data with input variable relations over continuous-valued attributes. Those relations as a set of linear classifiers can be obtained from neural network modeling based on back-propagation. It is shown in this paper that variable thresholds play an important role in constructing linear classifier rules when we use a decision tree over linear classifiers extracted from a multilayer perceptron. We have tested this scheme over several data sets to compare it with the decision tree results.

1 Introduction

The discovery of decision rules and recognition of patterns from data examples is one of the most challenging problems in machine learning. If data points contain numerical attributes, induction tree methods need the continuous-valued attributes to be made discrete with threshold values. Induction tree algorithms such as C4.5 build decision trees by recursively partitioning the input attribute space [17]. The tree traversal from the root node to each leaf leads to one conjunctive rule. Each internal node in the decision tree has a splitting criterion or threshold for continuous-valued attributes to partition some part of the input space, and each leaf represents a class related to the conditions of each internal node.

Approaches based on decision trees involve making the continuous-valued attributes discrete in input space, creating many rectangular divisions. As a result, they may have the inability to detect data trends or desirable classification surfaces. Even in the case of multivariate methods of discretion which search in parallel for threshold values for more than one continuous attribute [5,15], the decision rules may not reflect data trends or the decision tree may build many

D. Cheung, G.J. Williams, and Q. Li (Eds.): PAKDD 2001, LNAI 2035, pp. 538–549, 2001.

rules with the support of a small number of examples or ignore some data points by dismissing them as noisy.

A possible process is suggested to grasp the trend of the data. It first tries to fit it with a given data set for the relationship between data points, using a statistical technique. It generates many data points on the response surface of the fitted curve, and then induces rules with a decision tree. This method was introduced as an alternative measure regarding the problem of direct application of the induction tree to raw data [12,13]. However, it still has the problem of requiring many induction rules to reflect the response surface.

In this paper we use a hybrid technique to combine neural networks and decision trees for data classification [14]. It has been shown that neural networks are better than direct application of induction trees in modeling nonlinear characteristics of sample data [4,16,18,6,20]. Neural networks have the advantage of being able to deal with noisy, inconsistent and incomplete data. A method to extract symbolic rules from neural networks has been proposed to increase the performance of the decision process [1,21,8,22,18]. The KT algorithm developed by Fu [7] extracts rules from subsets of connected weights with high activation in a trained network. The M of N algorithm clusters weights of the trained network and removes insignificant clusters with low active weights. Then the rules are extracted from the weights [22].

A simple rule extraction algorithm that uses discrete activations over continuous hidden units is presented in by Setiono and Taha [18,21]. They used in sequence a weight-decay back-propagation over a three-layer feed-forward network, a pruning process to remove irrelevant connection weights, a clustering of hidden unit activations, and extraction of rules from discrete unit activations. They derived symbolic rules from neural networks that include oblique decision hyperplanes instead of general input attribute relations [19]. Also the direct conversion from neural networks to rules has an exponential complexity when using search-based algorithm over incoming weights for each unit [8,22]. Most of the rule extraction algorithms are used to derive rules from neuron weights and neuron activations in the hidden layer as a search-based method. An instance-based rule extraction method is suggested to reduce computation time by escaping search-based methods [14]. After training two hidden layer neural networks, the first hidden layer weight parameters are treated as linear classifiers. These linear differentiated functions are chosen by decision tree methods to determine decision boundaries after re-organizing the training set in terms of the new linear classifier attributes.

Our approach is to train a neural network with sigmoid functions and to use decision classifiers based on weight parameters of neural networks. Then an induction tree selects the desirable input variable relations for data classification. Decision tree applications have the ability to determine proper subintervals over continuous attributes by a discretion process. This discretion process will cover oblique hyperplanes mentioned in Setiono's papers. In this paper, we have tested linear classifiers with variable thresholds and fixed thresholds. The methods are tested on various types of data and compared with the method based on the decision tree alone.

2 Problem Statement

Induction trees are useful for a large number of examples, and they enable us to obtain proper rules from examples rapidly [17]. However, they have the difficulty in inferring relations between data points and cannot handle noisy data.

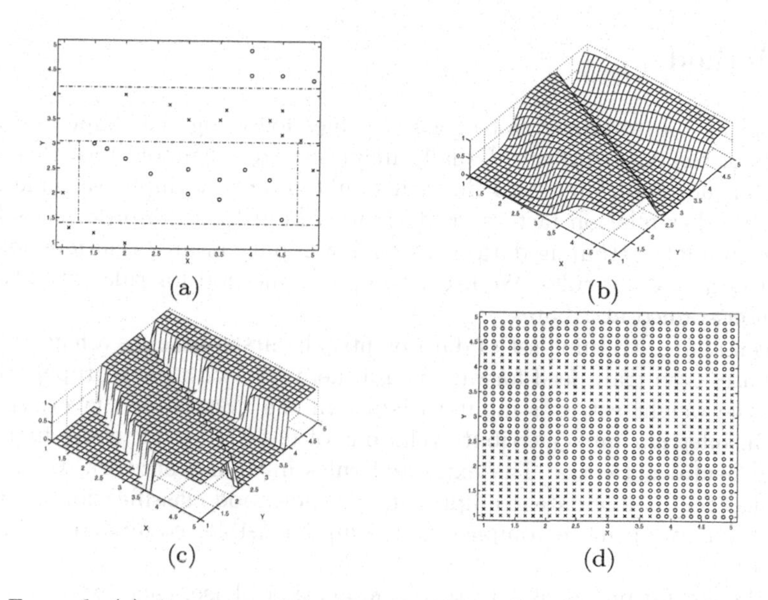

Fig. 1. Example (a) data set and decision boundary (O : class 1, X : class 0) (b)-(c) neural network fitting (d) data set with 900 points

We can see a simple example of undesirable rule extraction discovered in the induction tree application. Fig.1(a) displays a set of 29 original sample data with two classes. It appears that the set has four sections that have the boundaries of direction from upper-left to lower-right. A set of the dotted boundary lines is the result of multivariate classification by the induction tree. It has six rules to classify data points. Even in C4.5 run, it has four rules with 6.9 % error, making divisions with attribute y. The rules do not catch data clustering completely in this example. Fig.1(b)-(c) shows neural network fitting with the back-propagation method. In Fig.1(b)-(c) neural network nodes have slopes $alpha = 1.5, 4.0$ for sigmoids, respectively. After curve fitting, 900 points were generated uniformly on the response surface for the mapping from input space to class, and the response values of the neural network were calculated as shown in Fig.1(d). The result of C4.5 to those 900 points followed the classification curves, but produced 55 rules. The production of many rules results from the fact that decision tree makes piecewise rectangular divisions for each rule. This happens in spite of the fact that the response surface for data clustering has a correlation between the input variables.

As shown above, the decision tree has a problem of over-generalization for a small number of data and an over-specialization problem for a large number of data. A possible suggestion is to consider or derive relations between input variables as another attribute for rule extraction. However, it is difficult to find input variable relations for classification directly in supervised learning, while unsupervised methods can use statistical methods such as principal component analysis [9].

3 Method

The goal for our approach is to generate rules following the shape and characteristics of response surfaces. Usually induction trees cannot trace the trend of data, and they determine data clustering only in terms of input variables, unless we apply other relation factors or attributes. In order to improve classification rules from a large training data set, we allow input variable relations for multi-attributes in a set of rules. We use a two-phase method for rule extraction over continuous-valued attributes.

Given a large training set of data points, the first phase, as a feature extraction phase, is to train feed-forward neural networks with back-propagation and collect the weight set over input variables in the first hidden layer. A feature useful in inferring multi-attribute relations of data is found in the first hidden layer of neural networks. The extracted rules involving network weight values will reflect features of data examples and provide good classification boundaries. Also they may be more compact and comprehensible, compared to induction tree rules.

In the second phase, as a feature combination phase, each extracted feature for a linear classification boundary is combined together using Boolean logic gates. In this paper, we use an induction tree to combine each linear classifier.

The highly nonlinear property of neural networks makes it difficult to describe how they reach predictions. Although their predictive accuracy is satisfactory for many applications, they have long been considered as a complex model in terms of analysis. By using expert rules derived from neural networks, the neural network representation can be more understandable.

It has been shown that a particular set of functions can be obtained with arbitrary accuracy by at most two hidden layers given enough nodes per layer [3]. Also one hidden layer is sufficient to represent any Boolean function [10]. Our neural network structure has two hidden layers, where the first hidden layer makes a local feature selection with linear classifiers and the second layer receives Boolean logic values from the first layer and maps any Boolean function. The second hidden layer and output layer can be thought of as a sum of the product of Boolean logic gates. The n-th output of neural networks for a set of data is $F_n = f(\sum_k^{N_2} W_{kn}^2 f(\sum_j^{N_1} W_{jk}^1 f(\sum_i^{N_0} W_{ij}^0 a_i)))$ After training data patterns with a neural network by back-propagation, we can have linear classifiers in the first hidden layer.

For a node in the first hidden layer, the activation is defined as $H_j = f(\sum_i^{N_0} a_i W_{ij})$ for the j-th node where N_0 is the number of input attributes, a_i is an input, and $f(x) = 1.0/(1.0 + e^{-\alpha x})$ is a sigmoid function. When we train neural networks with the back-propagation method, α, the slope of the sigmoid function is increased as iteration continues. If we have a high value of α, the activation of each neuron is close to the property of digital logic gates, which has a binary value of 0 or 1.

Except for the first hidden layer, we can replace each neuron by logic gates if we assume we have a high slope for the sigmoid function. Input to each neuron in the first hidden layer is represented as a linear combination of input attributes and weights, $\sum_i^N a_i W_{ik}$. This forms linear classifiers for data classification as a feature extraction over data distribution. When Fig.1(a) data is trained, we can introduce new attributes $aX + bY$, where a, b are constants. We use two hidden layers with 4 nodes and 3 nodes, respectively, where every neuron node has a high sigmoid slope to guarantee desirable linear classifiers as shown in Fig.1(c).

We transformed 900 data points in Fig.1(d) into four linear classifier data points, and then we added the classifier attributes to the original attributes x, y. Induction tree algorithm used those six attributes for its input attributes. Then we could obtain only four rules with C4.5, while a simple application of C4.5 for those data generated 55 rules. The rules are given as follows:

$$\begin{aligned}
&\text{rule 1}: \text{ if } (1.44x + 1.73y <= 5.98), \text{ then class 0}\\
&\text{rule 2}: \text{ if } (1.44x + 1.73y > 5.98)\\
&\qquad \text{and } (1.18x + 2.81y <= 12.37) \text{ then class 1}\\
&\text{rule 3}: \text{if}(1.44x + 1.73y > 5.98)\\
&\qquad \text{and } (1.18x + 2.81y > 12.37)\\
&\qquad \text{and } (0.53x + 2.94y < 14.11), \text{ then class 0}\\
&\text{rule 4}: \text{if}(1.44x + 1.73y > 5.98)\\
&\qquad \text{and } (1.18x + 2.81y > 12.37)\\
&\qquad \text{and } (0.53x + 2.94y > 14.11), \text{ then class 1}
\end{aligned}$$

These linear classifiers exactly match with the boundaries shown in Fig.1(c), and they are more dominant for classification in terms of entropy minimization than a set of original input attributes itself. Even if we include input attributes, the entropy measurement leads to a rule set with boundary equations. These rules are more meaningful than those of direct C4.5 application to raw data since their divisions show the trend of data clustering and how each attribute is correlated.

Our approach can be applied to the data set that has both discrete and continuous values. If there is a set of input attributes, $Y = \{D_1, ..., D_m, C_1, ..., C_n\}$, then D_i is a discrete attribute, C_j is a continuous-valued attribute, m is the number of discrete attributes, and n is the number of continuous-valued attributes. Any discrete attribute D_x has a finite set of values available. For example, if there is a value set $\{d_{x1}, d_{x2}, d_{x3}, ..., d_{xp}\}$ for a discrete attribute D_x, we can have a Boolean logic value for each discrete attribute, using the conditional equation $D_x = d_{xj}$, for $j = 1, .., p$. We can put this state as a node in the first hidden layer, and then one of linear classifiers obtained with neural network is $L_k = \sum_i^{m+n} a_i W_{ik}^0 = \sum_i^n c_i W_{ik}^0 + \sum_i^m d_i W_{ik}^0$ where a_i is an instance of data in the form of the set Y, c_i is an instance of numeric attributes, and d_i is an instance of discrete attributes. The second term $\sum_i^m d_i W_{ik}^0$ can be treated as a threshold for the linear classifier L_k.

Since we have no interest in the relation of discrete attributes whose numeric conditions and coefficient values are not meaningful in this model, the discrete attributes can be taken as variable thresholds. As a result, the value of linear classifier L_k only depends on a linear combination of continuous attributes and weights. The choice of discrete attributes in rules can be handled using induction tree algorithms more properly, without interfering the relations of continuous-valued attributes.

Induction trees can split any continuous value by selecting thresholds for given attributes, while it cannot derive the relation of input attributes directly. In our method, we can add to the data set of the induction tree, new attributes $L_k = \sum_i^n c_i W_{ik}$ for $k = 1, .., r$, where r is the number of nodes in the first hidden layer for continuous-valued attributes. The new set of attributes for the induction tree is $Y' = \{D_1, D_2, ..., D_m, C_1, C_2, ..., C_n, L_1, L_2, ..., L_r\}$. The entropy measurement will try to find out a significant classification over the new set of attributes. We can have another attribute set which consists of only linear classifiers generated by neural network as follows:

$$Y'' = \{D_1, D_2, ..., D_m, L_1, L_2, ..., L_r\}$$

$\{C + L\}$ linear classifiers, including both original input attributes and neural network linear classifiers together, were tested with some data sets in the UCI depository [2] to compare it with L-linear classifier method which only includes neural network linear classifiers [14]. It is believed that a compact set of attributes to represent the data set shows a better performance. Adding original input attributes does not improve the result, but it makes its performance worse in most cases. C4.5 has a difficulty in selecting properly the most significant attributes for a given set of data, because it chooses attributes with local entropy measurement and the method is not a global optimization of entropy. Also, especially when only linear classifiers from neural network are used, it is quite effective in reducing the number of rules [14].

Generally when we give many feature attributes to the induction rule generator based on C4.5, it has a tendency to worsen performance. This is because the induction tree is based on a locally optimal entropy search. In this paper, a compact L-linear classifier method was tested. We compared L-linear classifiers with fixed thresholds and variable thresholds ; fixed thresholds are chosen by neural network and variable thresholds are selected by induction tree algorithm.

All instances in the training data can be converted into Boolean logic values and then they are applied to the induction tree algorithm C4.5. This method uses given thresholds determined by neural network training, and each hidden node activation is taken as a Boolean logic value. It is equivalent to logic circuit minimization problem that finds a simple form of Boolean circuits. As a result, it classifies data with newly constructed attributes. Decision trees can be seen as a kind of heuristic method for logic circuit minimization. We collect a set of Boolean logic instances depending on the activation of each node in the first hidden layer of the neural network and then it is given to the C4.5 induction tree to construct logical functions over linear classifiers. C4.5 can also prune rules over Boolean logic instances, even when it sees inconsistent class mappings from input Boolean instances to output classes.

Another method is to use a set of linear classifiers as continuous-valued attributes. The C4.5 application over instances of linear classifiers will try to find the best splitting thresholds for discretion over each linear classifier attribute to classify training data. In this case, each linear classifier attribute may have multivariate thresholds, different from thresholds obtained in neural network training. This will show the difference between neural networks and induction trees in handling marginal boundaries of linear classifiers.

4 Experiments

Our method has been tested on several sets of data in the UCI depository [2]. Fig.2 and Table 1 show average classification error rates for neural networks and the C4.5 [17] algorithm, and Fig.3 and Table 2 show error rates in our two methods. We show the results of linear classifier methods, the pure C4.5 method, and neural networks. The error rates were estimated by running the complete 10-fold cross-validation ten times, and the average and the standard deviation for ten runs were given in the table.

Our methods using linear classifiers are better than C4.5 in some sets and worse in other data sets such as *glass* and *pima* which are hard to predict even in neural network. The result supports the fact that the methods greatly depend on neural network training. If neural network fitting is not correct, then the fitting errors may mislead the result of linear classifier methods. Normally, the C4.5 application shows the error rate is very high for training data in Table 1. The neural network can improve training performance by increasing the number of nodes in the hidden layers as shown in Table 1. However, it does not mean that it improves test set performance. In many cases, reducing errors in a training set tends to increase the error rate in a test set by overfitting.

The error rate difference between a neural network and linear classifiers explains that some data points are located on marginal boundaries of classifiers. It is due to the fact that our neural network model uses sigmoid functions with high slopes instead of step functions. When activation is near 0.5, the weighted sum of activations may lead to different output classes. If the number of nodes in the first hidden layer is increased, this marginal effect becomes larger as observed in Table 1 and Table 2. Fig.3(b) and Table 3 shows that the number of rules using our method is significantly smaller than that using conventional C4.5 in all the data sets. To reduce the number of rules, linear classifiers with the Boolean circuit model greatly depend on the number of nodes in the first hidden layer. It decreases the number of rules when the number of nodes decreases in the first hidden layer, while the error rate performance is similar within some limit, regardless of the number of nodes. The linear classifier method with variable thresholds also depends on the number of nodes. The reason why the number of rules is proportional to the number of nodes is related to the search space of Boolean logic circuits. The linear classifier method with the Boolean circuit model often tends to generate rules that have a small number of support exam-

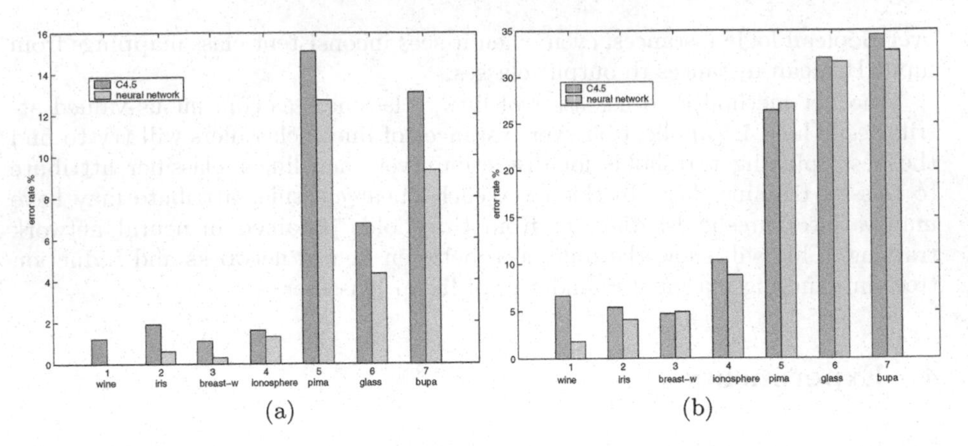

Fig. 2. Difference between C4.5 and neural network (a) average error rate in training data (b) average error rate in test data

Table 1. Data classification error rate result in neural network and C4.5

	neural network					*C4.5*	
data	pat / attr	train (%)	test (%)	nodes	train (%)	test (%)
wine	178 / 13	0 ± 0	1.8 ± 0.6	13-8-5-3	1.2 ± 0.1	6.6 ± 1.2
	178 / 13	0 ± 0	1.9 ± 0.6	13-5-5-3	1.2 ± 0.1	6.6 ± 1.2
iris	150 / 4	0.6 ± 0.1	4.1 ± 1.3	4-5-4-3	1.9 ± 0.1	5.4 ± 0.7
breastw	683 / 9	0.3 ± 0.1	4.9 ± 0.7	9-8-5-3	1.1 ± 0.1	4.7 ± 0.5
ion	351 / 34	1.3 ± 0.2	8.8 ± 1.3	34-10-7-2	1.6 ± 0.2	10.4 ± 1.1
pima	768 / 8	4.9 ± 0.4	27.8 ± 0.7	8-15-9-2	15.1 ± 0.8	26.4 ± 0.9
	768 / 8	8.8 ± 0.7	27.8 ± 0.7	8-10-7-2	15.1 ± 0.8	26.4 ± 0.9
	768 / 8	12.7 ± 0.4	27.1 ± 0.9	8-7-7-2	15.1 ± 0.8	26.4 ± 0.9
glass	214 / 9	2.3 ± 0.4	32.1 ± 2.6	9-15-12-7	6.7 ± 0.4	32.0 ± 1.5
	214 / 9	4.3 ± 0.9	31.6 ± 1.2	9-15-8-7	6.7 ± 0.4	32.0 ± 1.5
	214 / 9	5.0 ± 0.7	32.1 ± 1.9	9-10-8-7	6.7 ± 0.4	32.0 ± 1.5
bupa	345 / 6	7.3 ± 1.4	32.7 ± 2.2	6-10-7-2	13.1 ± 0.7	34.5 ± 1.8
	345 / 6	9.3 ± 0.7	32.1 ± 1.9	6-8-6-2	13.1 ± 0.7	34.5 ± 1.8
	345 / 6	15.2 ± 0.9	32.8 ± 1.9	6-5-5-2	13.1 ± 0.7	34.5 ± 1.8

ples, while variable threshold model prunes rules by adjusting splitting thresholds in the decision tree.

Most of the data sets in the UCI depository have a small number of data examples relative to the number of attributes. The significant difference between a simple C4.5 application and a combination of C4.5 application and a neural network is not seen distinctively in UCI data in terms of error rate unlike the synthetic data in Fig.1. Information of data trend or input relations can be more definitely described when given many data examples relative to the number of attributes.

Table 1 and 2 show that neural network classification is better than linear classifier applications. Even though linear classifier methods are good approxi-

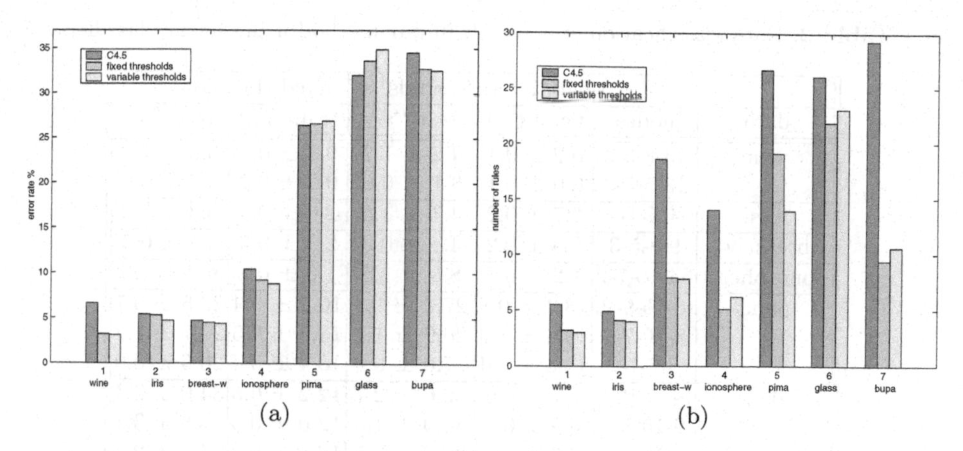

(a) (b)

Fig. 3. Difference between C4.5 and linear classifier methods (a) test data average error rate for C4.5, linear classifiers with fixed and variable thresholds (b) the number of rules for C4.5, linear classifiers with fixed and variable thresholds

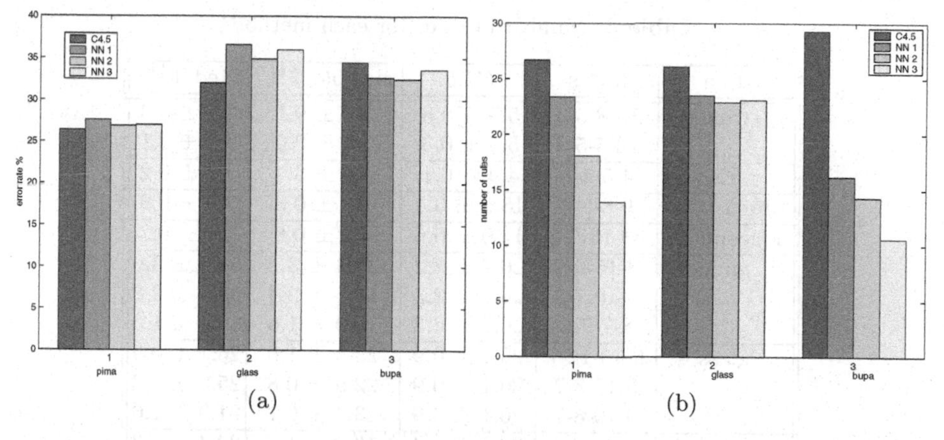

(a) (b)

Fig. 4. Linear classifier methods with several neural networks (a) test data average error rate for C4.5 and linear classifiers with 3 different neural networks (see Table2) (b) the number of rules for C4.5 and linear classifiers with 3 different neural networks (see Table3)

mations to nonlinear neural network modeling in the experiments, we still need to reduce the gap between neural network training and linear classifier models. Also it may be necessary to prove that linear classifiers with variable thresholds may be a close approximation to neural network modeling theoretically. There is a trade-off between the number of rules and error rate performance. We need to explain what is the optimal number of rules for a given data set for future study.

Table 2. Data classification error result in our method using linear classifiers

		variable thresholds$^{\{L\}}$		fixed thresholds$^{\{L\}}$	
data	nodes	train (%)	test (%)	train (%)	test (%)
wine	13-8-5-3	0.2 ± 0.1	4.2 ± 1.2	0.3 ± 0.1	3.6 ± 1.3
	13-5-5-3	0.0 ± 0.1	3.1 ± 0.6	0.2 ± 0.2	3.2 ± 0.8
iris	4-5-4-3	0.7 ± 0.2	4.7 ± 1.5	2.0 ± 0.4	5.3 ± 1.2
breast-w	9-8-5-3	0.9 ± 0.2	4.4 ± 0.3	2.2 ± 0.3	4.5 ± 0.3
ionosphere	34-10-7-2	1.2 ± 0.2	8.8 ± 1.5	1.6 ± 0.2	9.2 ± 1.6
pima	8-15-9-2	13.4 ± 0.8	27.6 ± 1.3	10.4 ± 0.4	28.5 ± 0.7
	8-10-7-2	14.6 ± 1.0	26.9 ± 1.1	13.5 ± 0.5	27.7 ± 1.2
	8-7-7-2	15.8 ± 0.4	27.0 ± 0.9	16.4 ± 0.4	26.6 ± 0.9
glass	9-15-12-7	6.7 ± 0.6	36.6 ± 2.7	12.2 ± 0.5	34.0 ± 2.4
	9-15-8-7	6.6 ± 0.8	34.9 ± 1.6	12.0 ± 0.6	33.6 ± 3.1
	9-10-8-7	7.6 ± 1.1	36.0 ± 2.5	13.8 ± 0.5	34.4 ± 2.3
bupa	6-10-7-2	15.2 ± 1.3	32.7 ± 2.9	17.3 ± 0.7	32.7 ± 1.6
	6-8-6-2	13.6 ± 1.5	32.5 ± 1.7	15.1 ± 0.6	34.0 ± 2.4
	6-5-5-2	17.5 ± 1.0	33.6 ± 2.3	22.2 ± 1.1	34.3 ± 2.3

Table 3. Number of rules for each method

data	nodes	$C4.5$	variable T$^{\{L\}}$	fixed T$^{\{L\}}$
wine	13-8-5-3	5.5 ± 0.3	3.1 ± 0.1	3.4 ± 0.3
	13-5-5-3	5.5 ± 0.3	3.0 ± 0.0	3.2 ± 0.1
iris	4-5-4-3	4.9 ± 0.1	4.0 ± 0.4	4.1 ± 0.2
breast-w	9-8-5-3	18.6 ± 0.7	7.8 ± 0.8	7.9 ± 0.9
ionosphere	33-10-7-2	14.0 ± 0.6	6.2 ± 0.8	5.1 ± 0.3
pima	8-15-9-2	26.7 ± 2.3	23.4 ± 3.0	58.9 ± 3.0
	8-10-7-2	26.7 ± 2.3	18.1 ± 2.4	33.9 ± 2.8
	8-7-7-2	26.7 ± 2.3	13.9 ± 1.0	19.1 ± 1.0
glass	9-15-12-7	26.1 ± 0.9	23.5 ± 1.0	26.1 ± 0.9
	9-15-8-7	26.1 ± 0.9	22.9 ± 0.8	25.0 ± 1.3
	9-10-8-7	26.1 ± 0.9	23.1 ± 0.7	21.9 ± 1.0
bupa	6-10-7-2	29.3 ± 1.4	16.2 ± 1.7	24.4 ± 2.3
	6-8-6-2	29.3 ± 1.4	14.3 ± 2.1	19.5 ± 1.3
	6-5-5-2	29.3 ± 1.4	10.6 ± 1.3	9.4 ± 0.9

5 Conclusions

This paper presents a hybrid method for constructing a decision tree from neural networks. Our method uses neural network modeling to find unseen data points and then an induction tree is applied to data points for symbolic rules, using features from the neural network. The combination of neural networks and induction trees will compensate for the disadvantages of one approach alone. This method has advantages over a simple decision tree method. First, we can obtain good features for a classification boundary from neural networks by training input patterns. Second, because of feature extractions about input variable relations, we can obtain a compact set of rules to reflect input patterns.

We still have much work ahead, such as applying the minimum description length principle to reduce the number of rules and error rate, and finding the optimal number of linear classifiers.

Acknowledgments. We would like to thank Yuval Marom for his comments. Also we especially thank anonymous reviewers of the original manuscript. The first author is supported by Manna Information System.

References

1. R. Andrews, J. Diederich, and A.B. Tickle. A survey and critique of techniques for extracting rules from trained artificial neural networks. *Knowledge-Based Systems*, 8(6), 1996.
2. C. Blake, E. Keogh, and C.J. Merz. *UCI* repository of machine learning databases. In *Preceedings of the Fifth International Conference on Machine Learning*, http://www.ics.uci.edu/~mlearn, 1998.
3. G. Cybenko. Continuous valued neural networks with two hidden layers are sufficient. Technical report, Technical Report, Department of Computer Science, Tufts University, Medford, MA, 1988.
4. T.G. Dietterich, H. Hild, and G. Bakiri. A comparative study of id3 and backpropagation for english text-to-speech mapping. In *Proceedings of the 1990 Machine Learning Conference*, pages 24–31. Austin, TX, 1990.
5. U.M. Fayyad and K.B. Irani. Multi-interval discretization of continuous-valued attributes for classification learning. In *Proceedings of IJCAI'93*, pages 1022–1027. Morgan Kaufmann, 1993.
6. D.H. Fisher and K.B. McKusick. An empirical comparison of id3 and backpropagation. In *Proceedings of 11th International Joint Conference on AI*, pages 788–793, 1989.
7. L. Fu. Rule learning by searching on adaptive nets. In *Preceedings of the 9th National Conference on Artificial Intelligence*, pages 590–595, 1991.
8. L. Fu. *Neural Networks in Computer Intelligence.* McGraw-Hill, New York, 1994.
9. S.S. Haykin. *Neural networks : a comprehensive foundation.* Prentice Hall, Upper Saddle River, N.J., 2nd edition, 1999.
10. J. Hertz, R.G. Palmer, and A.S. Krogh. *Introduction to the Theory of Neural Computation.* Addision Wesley, Redwood City, Calif., 1991.
11. K. Hornik, M. Stinchrombe, and H. White. Multilayer feedforward networks are universal approximators. *Neural Networks*, 2(5):359–366, 1989.
12. K.B. Irani and Qian Z. Karsm : A combined response surface / knowledge acquisition approach for deriving rules for expert system. In *TECHCON'90 Conference*, pages 209–212, 1990.
13. D. Kim. Knowledge acquisition based on neural network modeling. Technical report, Directed Study, The University of Michigan, Ann Arbor, 1991.
14. D. Kim and J. Lee. Handling continuous-valued attributes in decision tree using neural network modeling. In *European Conference on Machine Learning, Lecture Notes in Artificial Intelligence 1810*, pages 211–219. Springer Verlag, 2000.
15. W. Kweldo and M. Kretowski. An evolutionary algorithm using multivariate discretization for decision rule induction. In *Proceedings of the Third European Conference on Principles of Data Mining and Knowledge Discovery*, pages 392–397. Springer, 1999.

16. J.R. Quinlan. Comparing connectionist and symbolic learning methods. In *Computational Learning Theory and Natural Learning Systems*, pages 445–456. MIT Press, 1994.
17. J.R. Quinlan. Improved use of continuous attributes in c4.5. *Journal of Artificial Intelligence Approach*, (4):77–90, 1996.
18. R. Setiono and Huan Lie. Symbolic representation of neural networks. *Computer*, 29(3):71–77, 1996.
19. R. Setiono and H. Liu. Neurolinear: A system for extracting oblique decision rules from neural networks. In *European Conference on Machine Learning*, pages 221–233. Springer Verlag, 1997.
20. J.W. Shavlik, R.J. Mooney, and G.G. Towell. Symbolic and neural learning algorithms: An experimental comparison. *Machine Learning*, 6(2):111–143, 1991.
21. Ismail A. Taha and Joydeep Ghosh. Symbolic interpretation of artificial neural networks. *IEEE Transactions on Knowledge and Data Engineering*, 11(3):448–463, 1999.
22. G.G. Towell and J.W. Shavlik. Extracting refined rules from knowledge-based neural networks. *Machine Learning*, 13(1):71–101, Oct. 1993.

Knowledge Acquisition from Both Human Expert and Data

Takuya Wada, Hiroshi Motoda, and Takashi Washio

Institute of Scientific and Industrial Research,
Osaka University
Mihogaoka, Ibaraki, Osaka 567-0047, JAPAN

Abstract. A Knowledge Acquisition method "Ripple Down Rules" can directly acquire and encode knowledge from human experts. It is an incremental acquisition method and each new piece of knowledge is added as an exception to the existing knowledge base. This knowledge base takes the form of a binary tree. There is another type of knowledge acquisition method that learns directly from data. Induction of decision tree is one such representative example. Noting that more data are stored in the database in this digital era, use of both expertise of humans and these stored data becomes even more important. In this paper, we attempt to integrate inductive learning and knowledge acquisition. We show that using the minimum description length principle, the knowledge base of Ripple Down Rules is automatically and incrementally constructed from data and thus, making it possible to switch between manual acquisition by a human expert and automatic induction from data at any point of knowledge acquisition. Experiments are carefully designed and tested to verify that the proposed method indeed works for many data sets having different natures.

1 Introduction

We pay attention to the Ripple Down Rule Method (RDR) [2,8] as a promising approach to constructing a knowledge-based system in an environment in which the rapid innovation in technology makes existing knowledge being out-of-date in a very short time and requires frequent updates [9,15]. In RDR, Knowledge Acquisition (KA) is regarded as a continuous refinement of existing knowledge. It is interactive and there is no distinction between knowledge acquisition and maintenance.

Since RDR is primarily a method to capture knowledge from a human expert, it heavily relies on human expert's judgment. Although it is known that a human expert is good at explaining why a particular instance is misclassified and justifying what kind of remedy needs to be made, she is by no means almighty. Humans make mistakes. Recent advancement of machine learning makes it possible to induce a classifier from data quite efficiently, *e.g.* [10]. Further, it is often the case that there has already been a large quantity of data on databases. There is no reason not to use these data in building knowledge-based systems.

D. Cheung, G.J. Williams, and Q. Li (Eds.): PAKDD 2001, LNAI 2035, pp. 550–561, 2001.

In this paper, we explore possibility of integrating the inductive learning method and the standard RDR method in a unified way. We propose to use the Minimum Description Length Principle [12] (MDLP) as an underlying principle for this integration. Inducing the RDR knowledge base from data has been studied by Gains and Compton [5,6]. They used Induct [3,4] in which the basic algorithm is to search for the premise for a given conclusion that is least likely to predict that conclusion by chance. Their method, although it produces the RDR knowledge base, does not seem to have a common strategy by which a knowledge acquisition approach and a machine learning approach are integrated.

We examine the proposed methods against 25 benchmark data sets [1, 14] and show that they work as expected. This suggests that it is feasible to develop a flexible knowledge-based system: the knowledge base is constructed by a human expert at an earlier stage of the development where there is not many data available, and then it is refined by using an induction technique at a later stage where there are enough data available.

2 Ripple Down Rules Revisited

The basis of RDR is the maintenance and retrieval of cases. When a case is incorrectly retrieved by an RDR system, the KA (maintenance) process requires the expert to identify how a case stored in a knowledge-based system differs from the present case.

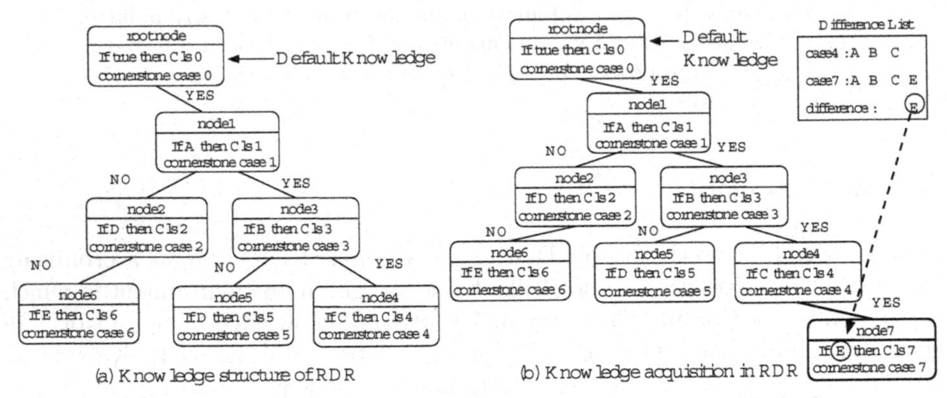

Fig. 1. Knowledge structure of the Ripple Down Rules Method

The tree structure of an RDR knowledge base is shown in Fig. 1.(a). Each node in the binary tree is a rule with a desired conclusion. Each node has a "cornerstone case" associated with it, that is, the case that prompted the inclusion of the rule. An inference process for an incoming case starts from the root node of the binary tree and continues until there is no branch to move on. The conclusion for the case is the conclusion part of the "last satisfied node". If the class is different from the class which a human expert judges the case to be, knowledge (new rule) is acquired from the human expert. The KA process in

RDR is illustrated in Fig. 1.(b). When the expert wants to add a new rule, there must be a case that is misclassified by a rule in RDR. The system asks the expert to select conditions from the "difference list" between the misclassified case and the cornerstone case. Then the misclassified case is stored as the refinement case (new cornerstone case) with the new rule whose condition part distinguishes these cases. Depending on whether the last satisfied node is the same as the end node, the new rule and its cornerstone case are added at the end of YES or NO branch of the end node. Knowledge is never removed or changed, simply modified by the addition of exception rules. The tree structure of the knowledge and the keeping of cornerstone cases ensure that the knowledge is always used in the same context when it was added.

3 The Minimum Description Length Principle

Occam's razor, or minimizing the description length, is the normal practice for selecting the most plausible one of many alternatives. Occam's razor prefers the simplest hypothesis that explains the data. Simplicity of hypothesis can be measured by description length (DL), originally proposed by Rissanen [12]. DL can express the complexity of specifying a hypothesis, and the value of DL is calculated as the sum of two encoding costs: one for the hypothesis and the other for cases misclassified with the hypothesis. The MDLP has been used as a criterion to select a good model in machine learning, *e.g.* in decision trees [11], neural networks [7] and Bayesian networks [13].

Table 1. Examples of cases

ID No.	Att. Swim	Att. Breath	Att. Legs	Class
1	can	lung	2legs	Dog
2	can	lung	4legs	Penguin
3	can	skin	2legs	Monkey
4	can	skin	4legs	Dog
5	can_not	lung	2legs	Dog
6	can_not	lung	4legs	Monkey
7	can_not	gill	2legs	Penguin
8	can_not	gill	4legs	Dog
9	can_not	skin	2legs	Dog
10	can_not	skin	4legs	Monkey

We illustrate the concept of the MDLP, using a communication problem. Let us suppose that both a sender A and a receiver B have the same list of Table 1 except that B does not know the class information. A communication problem is to send the class information from A to B through a communication path with as few bits as possible. Knowledge-base models such as decision trees or binary trees in RDR can be thought to be composed of a splitting method and a representative class for each split subset. Thus, the calculation of DL consists of the following 4 steps.

Step 1 : Split cases in the list into several subsets according to the attribute values on the basis of some splitting method (a model).

Step 2 : Encode the model on the basis of a certain coding method, and send the resultant bit sequence to B.

Step 3 : Encode the representative class of each subset, and send the respective bit sequence to B.

Step 4 : Encode the true class of each misclassified case, which is different from the representative class, and send the respective bit sequence to B.

If the splitting method used in **Step 1** is complex, most likely the DLs in **Step 2** and in **Step 3** are large, and the DL in **Step 4** is small. In reverse, the DL in **Step 4** will be larger if the splitting method is simpler. Therefore, there is a trade-off relation between 1) the DL for the splitting method (**Step 2**) and the DL for the representative classes (**Step 3**) and 2) the DL for the class information of the misclassified cases (**Step 4**). The total DL (the DL for a knowledge-base model + the DL for misclassified cases in the subsets) can be calculated by assuming a certain encoding method. The MDLP says that a knowledge-base model with the smallest total DL predicts the classes of unseen cases well. In the case of RDR, it is to say that a binary tree with the smallest total DL has the lowest error rate for unseen cases.

4 Calculation of the Total DL

In this section, we briefly explain how we encode the DL. Note that this is not the only way to encode the DL. There are other ways to do so, but the experimental results show that our encoding method is reasonable.

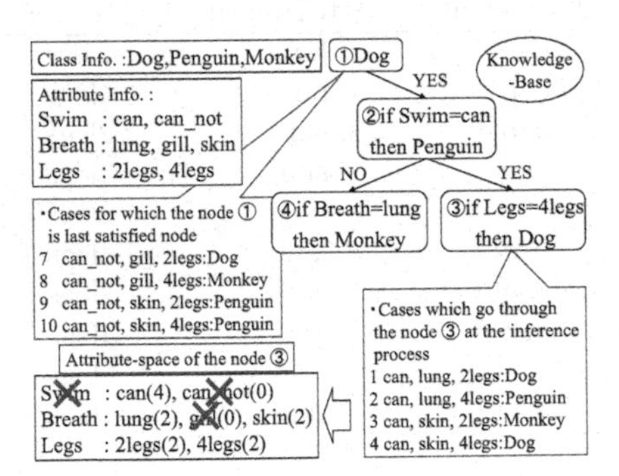

Fig. 2. An example knowledge base to calculate the DL

The total DL is the sum of the DL for the binary tree and the DL for cases misclassified by the tree. We explain the way to calculate the DL of a binary tree in subsection 4.1 and that of misclassified cases by this tree in subsection 4.2. Explanation is general but we instantiate it by using the RDR knowledge base given by Fig. 2 and the set of cases given by Table 1.

4.1 The DL of the Binary Tree

First, inference is made for all the cases in the list to obtain the attribute-space of each node in the binary tree. Assume that k cases go through a node C in the tree. From those k cases, we obtain the frequency distribution of each attribute-value in the node C. The corresponding attribute-space consists of a set of attributes each having at least 2 attribute-values with its frequency of at least 1 case. The attribute-space of node No.3 is {Att.Breath:lung,skin Att.Legs:2legs,4legs}, which is depicted in the lower left-hand side of Fig. 2.

There are two kinds of information to be encoded for the node C: one for the branches below the node and the other for an If-Then rule that is stored in the node as knowledge itself. The branch-information of the node No.3 is {YES-branch:no NO-branch:no} and the rule-information of the node is {If Legs=4legs then Dog}. There are four branch-combinations for all nodes except for the root node, that is, the number of candidates is 4. Therefore, $\log_2 {}_4C_1$ bits are necessary to encode this information for the node C. The DL for the branch-information of node No.3 is also $\log_2 {}_4C_1$ bits. The information needed to describe the If-Then rule consists of 4 components, (**1**) {the number of attributes used in the condition part}, (**2**) {attributes used in the condition part}, (**3**) {the attribute value for each used attribute} and (**4**) {the class used in the conclusion part}. In the case of node No.3, (**1**){the number of attributes:1},(**2**){attributes:Legs},(**3**) {the value of Att.Legs:4legs},(**4**){the class:Dog}. $\log_2 {}_nC_1$ bits are necessary to encode the information of (**1**) because there are n candidates, {1,2,...,n}. In the case of node No.3 it is $\log_2 {}_2C_1$ because the attribute-space has two attributes. Let the number of attributes used in the condition part be t. The information of (**2**) can be encoded by $\log_2 {}_nC_t$ bits because the number of combinations of having t 1's and n-t 0's is given by ${}_nC_t$. In the case of node No.3, it is $\log_2 {}_2C_1$ bits. Next, $\log_2 {}_{m_i}C_1 + \log_2 {}_2C_1$ bits are necessary for each used attribute to encode the information of (**3**). The second term is the DL necessary to specify whether it is negated or not. In the case of node No.3, it is $\log_2 {}_2C_1 + \log_2 {}_2C_1$ bits because the attribute Legs is the only one used. Finally $\log_2 {}_{class\_num-1}C_1$ bits are necessary to encode the information of (**4**) because the number of classes that are possible to use as the conclusion in each node except for the root node is $class\_num - 1$. Here, $class\_num$ is the number of classes in the problem domain. In the case of node No.3, it is $\log_2 {}_2C_1$ bits because the candidates are Dog and Monkey. The sum of DLs for the information of (**1**), (**2**), (**3**) and (**4**) is the DL necessary to encode the If-Then rule in the node C.

The sum of the DL for the branch-information, which was mentioned in the beginning, and the one for the rule-information is the DL for the node C. In the case of node No.3, it is $5 \log_2 {}_2C_1 + \log_2 {}_4C_1$ bits. If the sender A encodes the information of all nodes in the tree and send it to the receiver B, B can decode it and obtain the used splitting method and the representative classes (the binary tree itself).

4.2 The DL for Misclassified Cases

We next explain how to calculate the DL that is necessary to encode the true class information for the cases misclassified by the given binary tree. This DL can be calculated at each node in the tree.

Assume that there is a node D in the binary tree which is the last satisfied node for r cases as a results of running all cases in the list. That is, these r cases form a subset that has the conclusion of the node D as its representative class. In the case of node No.1 in Fig. 2, they are {7,8,9,10}. Further assume that k cases out of r are misclassified. In the case of node No.1, 3 cases out of 4 are misclassified and have different classes from the representative class: Dog. First, it is necessary to encode the number of classes that are different from the representative one in r cases. The DL is $\log_2 {}_{class\_num}C_1$ because the candidate number is *class_num*, that is, $\{0, 1, ..., class\_num - 1\}$. In the case of node No.1 it is $\log_2 {}_3C_1$ bits. Next, we calculate the DL which is necessary to specify which cases are which classes. Let the number of cases with the i-th different class be $p_i(i = 1, 2, ..., s)$ (s: the number of classes different from the representative one). The different classes are ordered to satisfy $p_s \geq p_{s-1} \geq ... \geq p_2 \geq p_1$. In the case of node No.1, it is $p_2 = 2$ (Penguin), $p_1 = 1$ (Monkey). With this preparation, the DL is calculated by the algorithm shown in Fig. 3.

```
function DescriptionLength: real;
variable dl: real;
         case, i: integer;
begin
  dl := 0;
  case := r; # number of cases
  i := s;    # number of different classes
  repeat
    dl := dl + log2 (i); # specifying the i-th class
    if root node then
      begin
        dl := dl + log2 (case − i + 1); # specifying the value of p_i
      end
    else
      begin
        dl := dl + log2 (case − i); # specifying the value of p_i
      end
    dl := dl + log2 (_case C_p_i); # specifying p_i cases
    case := case − p_i;      # number of remaining cases
    i := i − 1;              # number of remaining classes
  until i = 0;
  DescriptionLength= dl;
end
```

Fig. 3. Algorithm to calculate the DL to specify p_i cases $(i = 1, 2, ..., s)$

It is now possible to find the true classes of k cases by decoding the encoded signal of "*DescriptionLength*" bit long. All the remaining $r - k$ cases have the representative class. In the case of node No.1, the 8th case is Monkey, and the 9th and the 10th cases are Penguin. If the sender A encodes the information for the misclassified cases in each subset, the receiver B can get the true classes of the cases misclassified by the tree.

4.3 The Total DL

Finally, by sending the encoded signal with the total DL (the sum of the DLs mentioned in subsection 4.1 and subsection 4.2), the receiver B is able to find the class information of all cases in the list.

It is empirically known that many encoding methods used for the construction of knowledge-based systems based on the MDLP, including the one mentioned here, tend to overestimate the DL necessary to encode the knowledge base, compared with the one to encode true classes of the misclassified cases. Therefore, it is common to use a weighted sum of the two.

$$\begin{aligned} \textit{the total DL} &= (\textit{the DL for classes of misclassified cases}) \\ &+ W \times (\textit{the DL for the knowledge base}) \end{aligned} \tag{1}$$

Here, W is a coefficient, which is less than 1, and in this paper we empirically found that 0.3 is a good value for W.

5 The MDLP-Based Knowledge Acquisition Methods

We propose two kinds of knowledge acquisition methods that are based on the MDLP in RDR. One is for constructing a binary tree by using data alone and the other is for constructing the tree by using both data and human experts.

5.1 A Method That Uses Data Alone

In the standard RDR, a set of elements selected from the difference list by a human expert becomes the condition part of a new node to be added to the so far grown binary tree. However, based on the MDLP, we want to select a set of elements among the possible sets in the difference list that gives the minimum total DL for the whole tree. Let us assume that a problem domain has n attributes $\{A_i | i = 1, ..., n\}$ and m_i attribute-values $\{v_{i,j} | j = 1, ..., m_i\}$. Let a current case misclassified by the so far grown RDR tree be defined as *case* $A : \{v_{1,a}, v_{2,a}, ..., v_{n,a}(v_{i,a} \in A_i)\}$, and a cornerstone case whose node has derived the false conclusion (the last satisfied node) be defined as *case* B : $\{v_{1,b}, v_{2,b}, ..., v_{n,b}(v_{i,b} \in A_i)\}$. Figure 4 is an example in which the *case* A is $\{v_{1,2}, v_{2,1}, v_{3,2}\}$ and the *case* B is $\{v_{1,1}, v_{2,1}, v_{3,1}\}$. Details of the search algorithm is omitted due to the space limitation, but it is a greedy search as shown in Fig. 4. The search starts with a condition which is specialized to *case* A[1], and while expanding its search space, it finds a condition that falls in a (local) minimum of the total DL. This method enables to construct an RDR knowledge base by using data alone[2] without the help of human experts.

[1] It is natural to start with this condition because this is the only evidence that is against the cornerstone case. Further note that any choice of an element in the difference list differentiates between the the cornerstone case and the misclassfied case.

[2] Implicit assumption is that the data are labeled.

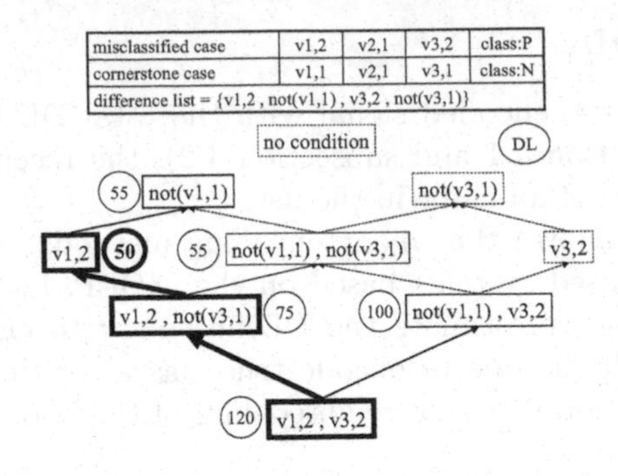

Fig. 4. Search space that uses data alone

5.2 A Method That Uses Both Data and Human Experts

The greedy search described in subsection 5.1 starts with a condition specialized to a current misclassified case, and this is not necessarily a good starting point. Another possibility is to start with the condition that an expert has selected if such an advice is available as in the case of the standard RDR. The advantage of this method is that it can find a better condition even when the expert fails to select the best correct conditions from the difference list, and in general the better the expert's guess is, the smaller the found DL is.

6 Experiments

Before examining the effectiveness of the methods we proposed in previous section, we need to ascertain that the MDLP holds for binary trees of the standard RDR. After we confirm this, we examine whether the method proposed in subsection 5.2 can indeed construct more accurate binary trees than the standard RDR binary trees and whether the method proposed in subsection 5.1 can construct binary trees which are as accurate as the induction rule sets obtained by C4.5 [10], which we select as a standard machine learning method.

Databases used for experiments: We have selected 24 databases from University of California Irvine Data Repository [1] and 1 database from University of Toronto Data Repository [14]. Of these, 13 databases have only nominal attributes, 9 databases only numerical attributes and 3 databases mixed attributes We discretized the numeric attributes beforehand. We ran C4.5 to build a decision tree using the whole data for each database that has numeric attributes and used the same discretization thresholds that C4.5 found by information gain ratio criterion.

Simulated expert: We use a Simulated Expert (SE) (machine generated expert) instead of a human expert for the reproduction of experiments and consistent performance estimation. The SE has a set of If-Then rules derived from a decision tree constructed by C4.5 using a whole training data (explained in subsection 6.1). A set of elements selected from the difference list by the SE is defined as the intersection between the list and the condition part of the If-Then rule in the SE which predicts correctly the case misclassified by the RDR system.

6.1 Experiment 1 to Examine Whether the MDLP Holds for the Standard RDR

We randomly selected 75% cases from a database as a "training data set" and used them as incoming cases to the RDR system, and remaining 25% cases as a "test data set" for testing the accuracy of binary trees. The SE is constructed using the identical training data set. The same set of 75% cases is also used as the data set for calculating total DLs for binary trees. Then, using both the SE and training data set, a binary tree is constructed in the standard RDR. We plot the total DL and the number of misclassified cases of the test data each time a new case comes in and a new node is added. Because the results vary and depend on the order of incoming cases, we randomly generate 10 different orderings and plot all of the 10 trials in the same two-dimensional plane.

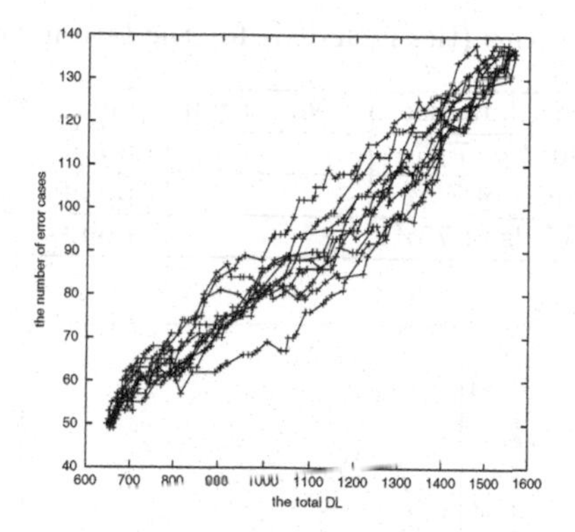

Fig. 5. Result for the database "Car Evaluation" for the default class "unacceptable"

A typical result is given in Fig. 5. This is the result obtained for the database "Car Evaluation". The horizontal axis is the value of the total DL, and the vertical one is the number of misclassified cases. We see from this figure that the fewer misclassified cases for the test data set are, the smaller the total DL is. This tendency is seen in many other databases.

6.2 Experiment 2 to Compare the Effect of Using Both Data and Human Expert with the Standard RDR

75% cases are selected from a database for a training data set, and the remaining 25% cases used for a test data set. However, we use only the two thirds of 75% cases to construct the SE. This is to simulate that the SE does not have long experience in the field and to examine how the data can support constructing a good binary tree. The data available to construct the tree (and to calculate the total DLs) was assumed to be one third, two thirds and three thirds of the 75% cases. Three kinds of binary trees are constructed by the method proposed in subsection 5.2 using the SE and the respective data set as far as the total DL continues to decrease. These results are also compared with those corresponding to standard RDR trees that are constructed using the same SE and the same three data sets. Note that ten different knowledge bases are constructed for each default class of each data set changing the order of the training data at random. By taking the average of these 10 runs we have 94 points as the total for the whole 25 databases.

Table 2 summarizes the number of wins (ties included) of the proposed method for the three different training data of different sizes. Fig. 6 shows the plots in the case of one thirds data set.

Table 2. Number of wins (ties included) for the training data of different sizes

training data size	No. of wins (out of 94 points)
one third of 75%	65 points
two thirds of 75%	74 points
three thirds of 75%	80 points

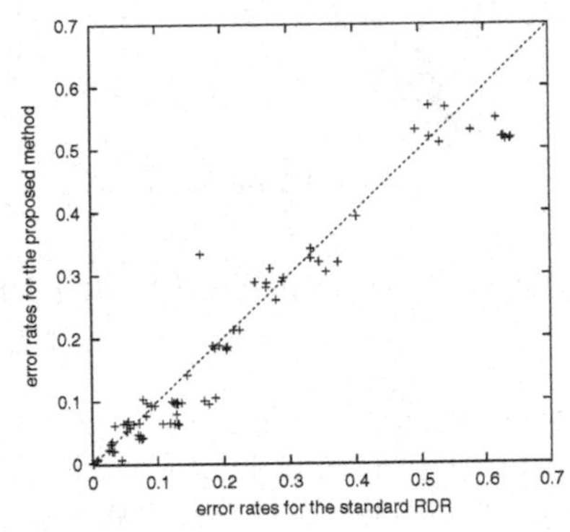

Fig. 6. Comparison of the proposed method and the standard method

From these results, it is experimentally shown that the more data are available, the more accurate knowledge bases can be constructed by the proposed method based on the MDLP than by the standard RDR, even when the SE has not enough expertise. Data can complement the lack of expertise.

6.3 Experiment 3 to Compare the Effect of Using Data with C4.5

We consider constructing accurate knowledge bases from data alone without relying on the SE. Using only 75% cases and nothing else, we construct the knowledge base by the method proposed in subsection 5.1 as far as the total DL continues to decrease. Because a different ordering of incoming cases and a different default class result in a different knowledge base of RDR, we individually change these parameters, and construct a total of *(the number of classes)* × 10 knowledge bases. Then we select the knowledge base with the smallest total DL. This knowledge base is expected to be the best one that can be constructed by the proposed method.

We compare the accuracy of this knowledge base with that of induction rule set obtained by C4.5 (This set is called "c4.5rules" below). The results for the 25 databases are shown in Fig. 7 where the horizontal axis is the error rates of c4.5rules and the vertical axis is the error rates of the selected knowledge bases.

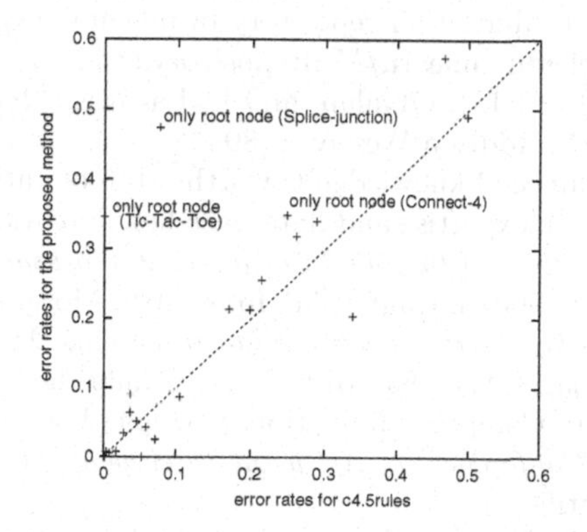

Fig. 7. Comparison of the proposed method with the c4.5rules

There are 11 databases out of 25 for which the knowledge bases constructed by the proposed method have equal or smaller error rates compared with those of c4.5rules. However, there are a few databases where the proposed method gives very high error rates. Especially, for those points noted as "only root node" in Fig. 7, no new nodes are added to the starting root nodes and the resultant knowledge bases consist of the root nodes alone. This may be due to the way the starting point of the search is selected (see subsection 5.1). It would have been possible to construct knowledge bases as accurate as c4.5rules if changes were made to the starting conditions. Excluding these points with "only root node", the accuracy is about the same as that of c4.5rules.

7 Conclusion

We explored possibility of integrating inductive learning (to extract knowledge from data) into the Ripple Down Rules Method (KA method from human experts), and proposed to use the MDLP as an underlying principle.

It is experimentally shown that data can complement the lack of expertise, i.e. the more data are available, the more accurate knowledge bases can be constructed by the proposed method than by the standard RDR, even when a human expert has not enough expertise. On the other hand, it is also found that there are situations where no growth of knowledge bases is made and the predictive accuracy is much worse than c4.5rules, but this is rather rare. Overall, with our proposed method we can construct a knowledge base which is equivalent to c4.5rules when the same amount of data is allowed to be used. The datasets used in the experiments are all artificial although taken from standard benchmark repositories. However, we prospect the sucess of the proposed approach for a real world dataset based on the known success of RDR approoach.

We, thus, conclude that the proposed method enables to make effective use of both human expertise and accumulated data as separate sources of knowledge, and to switch between manual acquisition by a human expert and automatic induction from data at any point of knowledge acquisition.

References

1. C.L. Blake and C.J. Merz. UCI repository of machine learning databases, 1998. http://www.ics.uci.edu/ mlearn/MLRepository.html.
2. P. Compton, K. Horn, J.R. Quinlan, and L. Lazarus. *Maintaining an Expert System*, pages 366–385. Addison Wesley, 1989.
3. B.R. Gaines. An ounce of knowledge is worth a ton of data: Quantiative studies of the trade-off between expertise and data based on statistically well-founded empirical induction. In *Proc. of the 6th International Workshop on Machine Learning*, pages 156–159, San Mateo, California, June 1989. Morgan Kaufmann.
4. B.R. Gaines. *The Trade-Off Between Knowledge and Data In Knowledge Acquisition*, chapter 29, pages 491–505. MIT Press, Cambridge, Mass, 1991.
5. B.R. Gaines and P. Compton. Induction of ripple-down rules. In *Proc. of the 5th Australian Joint Conference on Artificial Intelligence*, pages 349–354, Singapore, 1992. World Scientific.
6. B.R. Gaines and P. Compton. Induction of ripple-down rules applied to modeling large databases. *Journal of Intelligent Information Systems*, 5(2):211–228, 1995.
7. D.K. Gary and J.H. Trevor. Optimal network construction by minimum description length. *Neural Computation*, pages 210–212, 1993.
8. B.H. Kang. *Validating Knowledge Acquisition: Multiple Classification Ripple Down Rules.* PhD thesis, Dept. of Electrical Engineering, University of New South Wales, 1996.
9. K. Morik, S. Wrobel, J. Kietz, and W. Emde, editors. *Knowledge Acquisition and Machine Learning: Theory, Methods, and Applications.* Academic Press, 1993.
10. J.R. Quinlan, editor. *C4.5: Programs for Machine Learning.* Morgan Kaufmann, 1993.
11. J.R. Quinlan and R.L. Rivest. Inferring decision trees using the minimum description length principle. *Information and Computation* pages 227–248, 1989.
12. J. Rissanen. Modeling by shortest data description. *Automatica*, pages 465–471, 1978.
13. J. Suzuki. A construction of bayesian networks from databases on an mdl principle. In *Proc. of the 9th Conference on Uncertainty in Artificial Intelligence*, 1993.
14. The University of Toronto. Data for evaluating learning valid experiments, 1998. http://www.cs.utoronto.ca/~delve/data/datasets.html.
15. S. Wrobel, editor. *Concept Formation and Knowledge Revision.* Kluwer Academic Publishers, 1994.

Neighborhood Dependencies for Prediction

Renaud Bassée and Jef Wijsen

Université de Mons-Hainaut, Institut de Mathématique et d'Informatique,
Avenue du Champ de Mars 6, B-7000 Mons, Belgium
{renaud.bassee, jef.wijsen}@umh.ac.be

Abstract. We introduce the construct of *neighborhood dependency* (ND) to express regularities like: "Families with similar size and income, tend to own cars of similar size." Arguably, the discovery of such regularities is useful for prediction purposes. We have implemented and tested an algorithm for mining NDs. The discovered NDs are then used in the *P-neighborhood method* to predict unknown values.

Keywords: instance-based learning, k-nearest-neighbor, neighborhood dependency, P-neighborhood.

1 Introductory Example

Most data mining tasks involve predicting a target variable based on a number of predictor variables [3,7]. For example, predicting into what class a case falls (classification), or predicting what number value a variable will have (regression). In instance-based learning, one starts from a number of existing cases with known values for target and predictor variables. Each new case is then compared with existing ones using a distance metric on the predictor variables, and the closest existing cases are used to compute the value for the target variable. Obviously, the distance metric to determine closeness is crucial in this approach, and should satisfy the following intuitive property:

If two cases are close w.r.t. the predictor variables, then the two cases should have similar values for the target variable.

Without this property, close cases could have quite dissimilar values for the target variable, and hence closeness would be meaningless for prediction purposes. To make this intuitive property rigorous, we introduce the construct of *neighborhood dependency* (ND), which is exemplified in the next paragraph. But first we note that a database jargon will be used in the remainder of this paper; that is, variables will be denoted as "attributes," and cases as "tuples."

For the example, suppose a mini telephone poll of 23 families has resulted in a relation over *Tel*, *FSize*, *Income*, *Monovol*. The attribute *FSize* is the number of persons in the family. *Monovol* is the target attribute, and indicates whether the family owns a monovolume car. That is, the main question is: can one predict whether a new family would be interested in buying a monovolume car, given its telephone number, size, and income. Part of the dataset is shown in Fig. 1.

D. Cheung, G.J. Williams, and Q. Li (Eds.): PAKDD 2001, LNAI 2035, pp. 562–567, 2001.

POLL	predictor attributes			target	
	Tel	*FSize*	*Income*	*Monovol*	
	053/664842	6	250K	Y	(t_1)
	016/260660	5	262.5K	Y	(t_2)
	116/296597	3	150K	N	(t_3)
			...		
			(23 tuples in total)		

Fig. 1. Example relation.

We equip each attribute A with a *closeness function* $\theta_A(\cdot,\cdot)$, which takes as its input two attribute values, and outputs a number between 0 and 1. The nearer the output is to 1, the closer the input values. For example, the closeness function θ_{FSize} on family sizes yields 1 if the two sizes are equal, 0.5 if they are one unit apart, and 0 otherwise. The closeness function $\theta_{Monovol}$ for the attribute *Monovol* yields 1 if the two arguments are equal (both Y or both N), and 0 otherwise.

A *neighborhood predicate* (NP) maps attributes to thresholds numbers between 0 and 1. An example is:

$$FSize^{0.5}Income^{0.8} ,$$

fixing a threshold of 0.5 for *FSize*, and 0.8 for *Income*. Two tuples t_1 and t_2 are *neighbors* under this NP if the closeness function θ_{FSize} applied on the *FSize*-values in both tuples, i.e., on $t_1(FSize)$ and $t_2(FSize)$, yields a number ≥ 0.5 and θ_{Income} applied on the *Income*-values yields a number ≥ 0.8. Fig. 2 visualizes this for two-dimensional space. Given a tuple t, the NP under consideration gives rise to a rectangle around t that distinguishes neighbors of t under $FSize^{0.5}Income^{0.8}$ from non-neighbors; the figure depicts the neighborhoods around t_1, t_2, and t_3. Note that such visualization is generally impossible, as attribute values may be nominal or the closeness function may not reflect Euclidean distance.

An example *neighborhood dependency* (ND) is the expression:

$$FSize^{0.5}Income^{0.8} \rightarrow Monovol^{1.0} ,$$

expressing the hypothesis that neighbors under $FSize^{0.5}Income^{0.8}$ are also neighbors under $Monovol^{1.0}$. Put in simple words, families of similar size and income, behave similar w.r.t. the ownership of monovolume cars. The strength of this hypothesis is measured by the notion of *confidence*, i.e., the probability that two tuples that are neighbors under $FSize^{0.5}Income^{0.8}$ are also neighbors under $Monovol^{1.0}$. In Fig. 2, the filled-circle denoted t_1 represents a family with a monovolume car; among the seven families in the rectangle around t_1, five own a monovolume car, just like t_1, but two don't. We say that the confidence of the above ND in t_1 is 5/7. The confidence in t_3 is 2/3, since 2 of the 3 neighbors of t_3 are like t_3: they don't own a monovolume car. Another measure, called *support*, is introduced to reflect the number of neighbors of a tuple (7 in t_1, 5 in t_2, and 3 in t_3). The foregoing confidence and support measures are defined relative to a given tuple in the data set. Additionally, weighted average support and confidence measures over all tuples are introduced, where the weight factor for a given tuple is proportional to its number of neighbors.

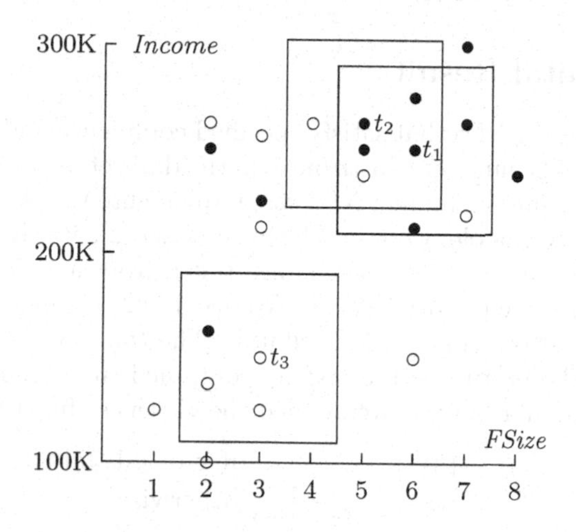

Fig. 2. Filled-circles represent tuples t for which $t(Monovol)$ = Y; open-circles represent tuples t for which $t(Monovol)$ = N.

One data mining task is as follows: given a target right-hand side NP, like $Monovol^{1.0}$, find a predictor left-hand side NP such that the resulting ND has sufficient support and confidence. This involves not only determining the attributes for the left-hand side, but also their associated threshold numbers (the width and the height of the rectangles in Fig. 2). If the thresholds are lowered, the neighborhood of each tuple—and hence its support— increases, but its confidence may decrease.

Next, if a "good" left-hand side neighborhood predicate P is discovered, it is used for prediction purposes in what we call the *P-neighborhood method*. Clearly, an ND by itself does not tell us how to compute a target value when values for the predictor attributes are given. In this respect, it is different from, for example, decision trees and regression models, which provide effective procedures to compute a target value from predictor values. The proposed *P-neighborhood method* therefore relies on the existing tuples in the database, and hence can be categorized as *instance-based learning*. In particular, the P-neighborhood method predicts the target value of a new tuple based on all existing neighbors of the tuple under P. For example, if the ND $FSize^{0.5} Income^{0.8} \rightarrow Monovol^{1.0}$ turns out to have high support and confidence, then, given the values for *FSize* and *Income* of a new tuple t, the value $t(Monovolume)$ is predicted by taking into consideration all neighbors of t under $FSize^{0.5} Income^{0.8}$ that exist in the database.

The important contribution of our work is the introduction of a simple yet semantically meaningful notion of "neighborhood" that can be used to make predictions in a way that can be easily understood. It is more intuitive than, for example, the factor k or the distance metric used in k-nearest-neighbor, where there is no preset definition of what constitutes a "good" distance metric or k.

2 Experimental Results

The problem of finding NDs that satisfy specified confidence and support thresholds, is **NP**-hard if the number of attributes in the dataset is used as a complexity measure. An algorithm for mining NDs was implemented in ANSI C.

We tested our methodology on the El Niño dataset [2], which contains oceanographic and surface meteorological readings taken from a series of buoys positioned throughout the equatorial Pacific. We used 782 readings from 23 May 98 to 5 June 98; the dataset contains 9 attributes. The *training set* consisted of 587 (75%) randomly chosen tuples; the *test set* contained the remaining 195 tuples. For each numerical attribute A, we applied the closeness function:

$$\theta_A(x,y) = \begin{cases} 0 & \text{if } x \text{ or } y \text{ is missing} \\ 1 - \frac{|x-y|}{\max_A - \min_A} & \text{otherwise} \end{cases}$$

where $\max_A$ and $\min_A$ denote the maximal and minimal values for A found in the dataset under consideration. For each categorical attribute A, the closeness function was:

$$\theta_A(x,y) = \begin{cases} 1 \text{ if both } x \text{ and } y \text{ are non-missing and } x = y \\ 0 \text{ otherwise} \end{cases}$$

Note that these closeness functions account for the many missing values in the dataset. We discovered the "strong" ND:

$$Buoy\#^{1.0} AirT^{0.8} \rightarrow SeaSurfaceT^{0.9} \quad ,$$

indicating that the buoy number and the air temperature together determine the sea surface temperature. It is interesting to add that the confidence *gain* w.r.t. the neighborhood dependency $\{\} \rightarrow SeaSurfaceT^{0.9}$, with an empty left-hand, was significant.

The P-neighborhood method with $P = Buoy\#^{1.0} AirT^{0.8}$ was then applied to predict $SeaSurfaceT$-values for each tuple in the test set. That is, for each tuple t in the *test* set, the value $t(SeaSurfaceT)$ was predicted by averaging the $SeaSurfaceT$-values of all neighbors of t under $Buoy\#^{1.0} AirT^{0.8}$ in the *training* set.

In order to assess the prediction quality, the distance between observed and predicted values was characterized by the Root Mean Square (RMS) error of the predictions converted to a percentage of the mean, called coefficient of RMS error (CRMS):

$$\mathrm{CRMS} = 100 \times \frac{\sqrt{\frac{\sum (O-E)^2}{N}}}{\frac{\sum O}{N}} \quad ,$$

where O are the observed values in the test set, E are the expected values predicted by P-neighborhood, N is the cardinality of the test set, and the summation is over all tuples of the test set. The P-neighborhood method yielded a CRMS of 1%, indicating a high-quality prediction.

For these data, a linear relationship between *AirT* and *SeaSurfaceT* is known. Predicting target values in the test set by a linear regression model, yielded a CRMS of 1.9%, and hence was less precise than the P-neighborhood method.

Next, in the same training set, the ND:

$$\mathit{Buoy\#}^{1.0} \rightarrow \mathit{Latitude}^{0.9}$$

had a confidence of 1.0. It expresses that the latitude of a buoy changes very little. Predicting *Latitude*-values in the test set by the P-neighborhood method with $P = \mathit{Buoy\#}^{1.0}$, yielded a CRMS of 2%. As *Buoy#* is in fact a non-numeric attribute, classical linear regression is not applicable here.

Recently, we also tested our approach on the Meningoencephalitis Diagnosis dataset, donated by Dr. Shusaku Tsumoto (Shimane Medical University, Japan). The results are promising.

3 Related Work

NDs generalize functional dependencies (FDs) [1] by comparing attribute values for *similarity* instead of *equality*. For example, the FD *Buoy#* → *Latitude* would express that the same buoy stays at the same latitude. As buoys move around to different locations, this FD does not hold for the El Niño dataset. Unlike NDs, FDs cannot express that the latitude of a buoy can change little from the average latitude.

NDs are different from quantitative association rules (QARs) [5], which associate intervals to attributes. For example,

$$\mathit{AirT} : 27..28 \rightarrow \mathit{SeaSurfaceT} : 28..29 \ ,$$

stating that the sea surface temperature is between 28 and 29 if the air temperature is between 27 and 28. An ND, on the other hand, reveals an overall relationship between *AirT* and *SeaSurfaceT*. One can expect the existence of strong QARs in the presence of strong NDs (the opposite is not necessarily true). Significantly, a QAR expresses a relationship between values *within* tuples, whereas NDs also express relationships *among* tuples.

One of the incentives of our work was to overcome certain semantic difficulties with roll-up dependencies (RUDs) [6]. In that work, attribute values are compared for equality after “rolling up” the values to a higher level of abstraction. For example, the *cities* Brussels and Mons are equal at the *country* level (both roll up to Belgium). The problem lies in the treatment of numeric attributes, where the roll-up is usually determined by fixed intervals. For the attribute *Income*, these intervals may be [0K..99K], [100K..199K], [200K..299K],... Then 100K and 199K are equal at the interval level, but 199K and 200K are not, and the roll-up will treat 199K as being closer to 100K than to 200K, which is counterintuitive. A similar problem has been raised for QARs [4]. NDs avoid these problems by using a closeness function instead of intervals: 199K will be much closer to 200K than to 100K.

The P-neighborhood method is inspired by, but different from, k-nearest-neighbor: to predict the value for a (numeric of categorical) target attribute A of a new tuple t, we rely upon all existing P-neighbors of t—rather than on the k "nearest neighbors" of t. The neighborhood predicate P itself is the result of a preceding data mining task: P results from the discovery of a rule $P \rightarrow A^r$ with high support and confidence, and r close to 1.

4 Conclusion

We proposed the P-neighborhood method for predicting a numeric or categorical target. It is an instance-based learning method that predicts the target of a new tuple based on its existing P-neighbors in the database. The construct of P-neighbor has natural semantics. The actual value for P is established during a preceding data mining task. Tests on real-life datasets show that the method is easy to apply (for example, it can treat numeric as well as categorical attributes; it can deal with missing values; default closeness functions can be used), and that the quality of the predictions is very promising.

References

1. S. Abiteboul, R. Hull, and V. Vianu. *Foundations of Databases.* Addison-Wesley, 1995.
2. C.L. Blake and C.J. Merz. UCI repository of machine learning databases, 1998.
3. J. Han and M. Kamber. *Data Mining: Concepts and Techniques.* Morgan Kaufmann, 2000.
4. R.J. Miller and Y. Yang. Association rules over interval data. In *Proc. ACM SIGMOD Int. Conf. Management of Data*, pages 452–461, Tucson, AZ, 1997.
5. R. Srikant and R. Agrawal. Mining quantitative association rules in large relational tables. In *Proc. ACM SIGMOD Int. Conf. Management of Data*, pages 1–12, Montreal, Canada, 1996.
6. J. Wijsen, R.T. Ng, and T. Calders. Discovering roll-up dependencies. In *Proc. ACM SIGKDD Int. Conf. Knowledge Discovery and Data Mining*, pages 213–222, San Diego, CA, 1999.
7. I.H. Witten and E. Frank. *Data Mining. Practical Machine Learning Tools and Techniques with Java Implementations.* Morgan Kaufmann, 2000.

Learning Bayesian Networks with Hidden Variables Using the Combination of EM and Evolutionary Algorithms*

Fengzhan Tian, Yuchang Lu, and Chunyi Shi

The State Key Laboratory of Intelligent Technology and System,
The Department of Computer Science and Technology,
Tsinghua University, Beijing, China 100084
tfz@263.net lyc@tsinghua.edu.cn scy@est4.cs.tsinghua.edu.cn

Abstract. In this paper, a new method, called EM-EA, is put forward for learning Bayesian network structures from incomplete data. This method combines the EM algorithm with an evolutionary algorithm (EA) and transforms the incomplete data to complete data using EM algorithm and then evolve network structures using the evolutionary algorithm with the complete data. In order to learn Bayesian networks with hidden variables, a new mutation operator has been introduced and the function of the crossover has been correspondingly expanded. The results of the experiments show that EM-EA is more accurate and practical than other network structure learning algorithms that deal with the incomplete data.

1 Introduction

A Bayesian network is applied more and more widely, and has become a main method to deal with the uncertainty in the field of artificial intelligence[5]. Particularly, in recent years there has been a growing interest in learning Bayesian networks from data[2][5][6] . At present, there have been effective methods for structure learning and parameter learning from complete data and good methods for parameter learning from incomplete data under fixing network structure. However, there are few effective and efficient methods for learning the network structures from incomplete data. Further, it is an especially difficult problem to learn network structures with hidden variables.

In 1998, Friedman put forward structural expectation-maximization algorithm, which he named MS-EM[4]. In his method, EM algorithm[8] and greedy search algorithm are employed. But when the search space is very large and multimodal landscape, the greedy search algorithm will stop at the local optimal model.

In 1996, Larranaga et al discussed learning network structure using an evolutionary algorithm[7]. The results of their experiments show that their method can learn good network structures and avoid getting into the local maxima with complete data. But for incomplete data, the results are not ideal.

* This research has been supported by Natural Science Foundation of China, National 973 Fundamental Research Program and 985 Program of Tsinghua University.

D. Cheung, G.J. Williams, and Q. Li (Eds.): PAKDD 2001, LNAI 2035, pp. 568-574, 2001.

In 1999, W. Myers et al improved Larranaga's work to make it adapt to incomplete data[9]. Their method not only evolved network structures but also evolved missing data to complete the incomplete data using generic operations.

While their method met the efficiency problem due to the enlarged search space and the convergence problem caused by the strong randomness of the genetic operators for the missing data.

In this paper, we present a new method called EM-EA. Compared to the work before, our method makes two improvements: (1) combines the EM algorithm with evolutionary algorithm organically, solves effectively the network structure learning problem from incomplete data and the problem of getting into local maxima; (2) expands EA of W. Myers et al to learn Bayesian networks with hidden variables.

2 Evolutionary Algorithm

The very large, multi-dimensional, multi-modal landscape immediately suggests the use of evolutionary algorithms. A Bayesian network can be broken down into local structures——a variable and all its parents that can be considered genes. Then the whole network structure can be represented as a chromosome. Furthermore, We can use the MDL score as the fitness function evaluating the network structure.

Formally, the structure, S, can be represented as an adjacency list, see Figure 1, where each row represents a variable v_i and the parents of v_i, π_{V_i}. The adjacency list can be thought of as a chromosome, where each row is a gene and the π_{V_i} are the alleles. This representation is convenient because the log form of MDL is the summation of scores for each variable. Because of this, each gene can be scored separately and added to generate the fitness score for the entire structure. Of course, this assumes the MDL is closed which is the case for complete data.

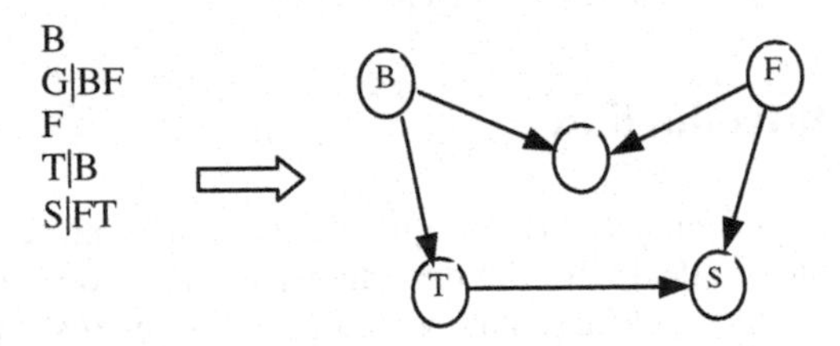

Fig. 1. Structure Mapping Genotype to Phenotype

However, there is no closed form expression for evaluating structures when the data are incomplete. In literature [9], W. Myers et al turn the incomplete data problem into a complete data problem by evolving the missing data and imputing these values into the data. So, they evolve not only the network structures but also the missing values. They represent each cell from the dataset that has a missing value as a gene. The gene takes on sampled values from the set of values of the corresponding variable. The chromosome is a string of missing values.

For the missing data chromosomes, W. Myers et al chose uniform parameterized crossover[3][10]. As for the mutation operator, they randomly select a value from the remaining possible values of the corresponding variable. They also use uniform parameterized crossover for the structure chromosome. They employed three basic mutation operators for the network structure chromosome. Two of them are adding and deleting a node to a gene. They have the effect in the phenotype of adding and deleting arcs respectively. The third one is reversing an arc, which is implemented in the phenotype by deleting the parent-child arc and adding the child-parent arc.

3 EM-EA Algorithm

The EA of W. Myers et al can avoid getting into the local maxima, but it has also two disadvantages. One is that it exponentially enlarges the search space (the number of missing data×network structure). When the number of missing data is big, the search space is so large that the efficiency of the algorithm will be very low and it is difficult to get satisfactory results. The more important is that the completion from incomplete data to complete data achieved by the generic operators has strong randomness and can not reflect the probability distribution that the missing data actually follow. So, it is difficult for their method to assure its convergence.

As for the disadvantages of the EA of W. Myers et al, we combine the EM algorithm with evolutionary algorithms organically, handle the incomplete data with EM algorithm, and learn Bayesian network structures with evolutionary algorithms. In addition, in order to make our method be able to learn the network structure with hidden variables, we improve the EA of W. Myers by introducing a new mutation operator and expanding the function of the crossover operator.

The mutation operator that we introduced can add some new vertices and arcs to the network and delete some arcs from the network. However, we can not add vertices and arcs arbitrarily, and we must follow some criteria while employing this operator. Our criterion is when finding some vertices depend on each other and connect thickly, we then add a vertex representing a hidden variable to the network. The parents of the vertex added are the common parent nodes of those vertices depending on each other, while the parent nodes of those vertices depending on each other are replaced by the vertex added. So the interdependent relationship among those vertices is represented by a hidden variable. And thus simplify the network structure. The experiments show that when there are three variables whose parent sets have a common subset, this mutation operator can be used for evolutionary calculation.

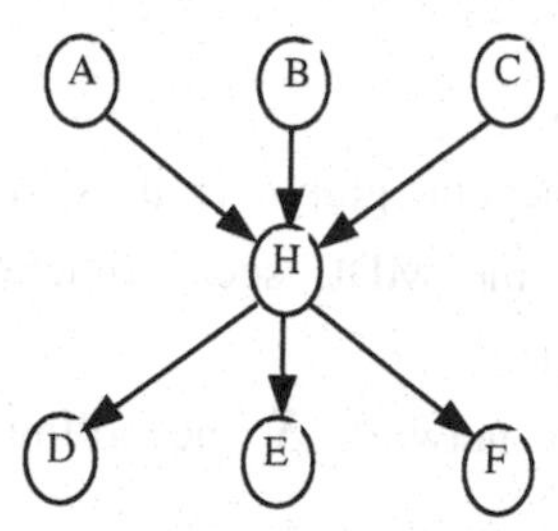

Fig. 2. a) An example of a network

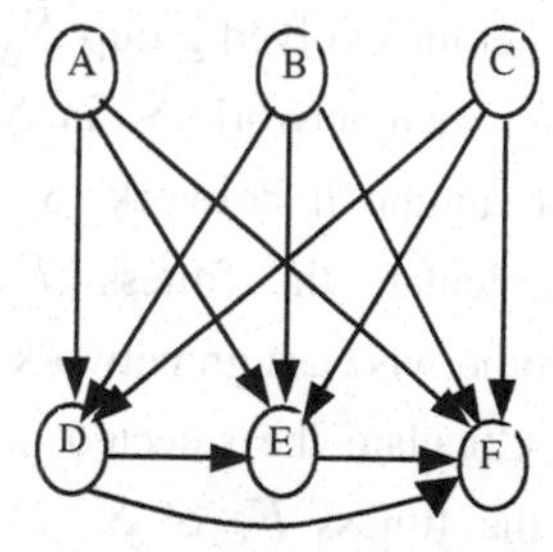

(b) The simplest network that can capture the same distribution without using the hidden variable

A simple example, originally given by Binder et al[1], is shown in Figure 2. In figure 2, the network structure (b) can be evolved to network structure (a) using our mutation operator. The corresponding adjacency lists for network structures in figure 2 are shown in Figure 3. The concrete process is as follows: by analyzing the adjacency list in Figure 3b, we can find that vertices A, B and C appear together most frequently in the alleles and the corresponding vertices whose alleles include A, B and C, are D, E and F. Therefor, we add a new gene corresponded with a hidden variable H in the adjacency list whose allele is ABC, and replace the alleles of D, E and F with H. Thus the adjacency list shown in Figure 3a is formed whose corresponding network structure is (a) in Figure 2.

(a)	(b)
A	A
B	B
C	C
D|H	D|ABC
E|H	E|ABCD
F|H	F|ABCD
H|ABC	

Fig. 3. The adjacency list corresponded with the networks in Figure 2

But the introduction of our mutation operator also raises a new problem. The adding of the gene has changed the length of the chromosome after we apply this new mutation operator and thus brought difficulties for applying the crossover operator. So we have to expand the crossover operation. The concrete means is to add a corresponding virtual gene that is correspondent with the hidden variables to the shorter chromosome to make the lengths of the two chromosomes making crossover operation same. The so-called virtual gene means its allele is empty and its corresponding variable does not appear in the alleles of other genes. In fact, adding a virtual gene equals to adding an isolated vertex in the network. After adding virtual genes, the two chromosomes can make the usual crossover operation.

After expanding the evolutionary algorithm mentioned above, EM-EA algorithm can find and evolve network structures with hidden variables. The whole process of the EM-EA method is as follows:

(1) Complete the incomplete dataset D using the current network S_c and EM algorithm, and get the complete dataset D_c;

(2) As for the original group S_Δ, make crossover or mutation operations, and get the evolved group $S_{\Delta'}$.

(3) As for each network S in $S_{\Delta'}$, do as follows:

a) Examine if network S is a directed acyclic graph. If it is, then calculate the fitness F_S according to the MDL score function; otherwise assign network S a small value.

b) Calculate the selected probability P_S of network S according to the fitness F_S of S.

(4) Choose λ individuals having the highest selected probabilities from S_{Δ}, to form the next generation. Where λ represents the size of the evolutionary group.

(5) Select $S^{'}$, make $F_{S'} = \max\arg(F_S)$. If $F_{S'} > F_{S_c}$, then $S_c = S^{'}$.

(6) Judge if the terminative condition of the algorithm is satisfied. If satisfied, then quit; otherwise, go to (1) and continue the above process.

4 Experiment Results

In order to validate our method, we compare the EM-EA algorithm with the EA of W. Myers et al and Friedman's MS-EM algorithm respectively.

While comparing our method with the EA of W. Myers et al, for convenience, we also use the Bayesian network known as ASIA, which has been used by W. Myers[9]. Furthermore, we use the same experiment process as that of W. Myers. In addition, the original evolutionary group is obtained with the following method: from the 1000 samples that include missing data mentioned above, we use a computer program to create a supposed complete dataset, and choose some better network structure as the original group based on the complete dataset. The current network can be selected at random from the original group. In the experiments, mutation and crossover probability are set as 0.05 and 0.5 respectively; the group size is set as 40.

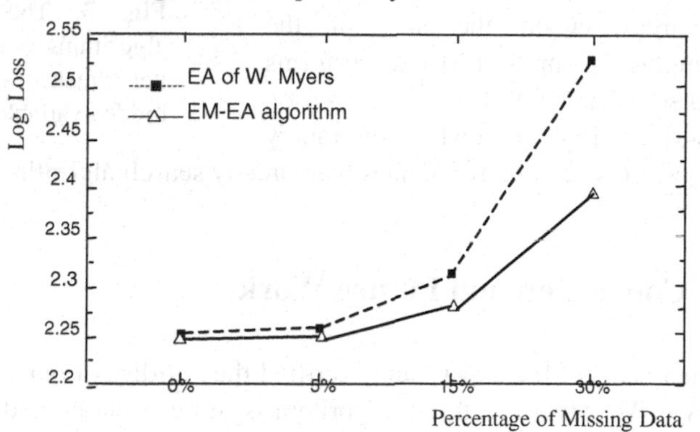

Fig. 4. The comparison of the two algorithms in terms of Log Loss

Figure 4 shows the log loss score of the two algorithms for each level of missing data. As can be seen from the figure, both EA and EM-EA could find good predictive networks at 0%, 5%, and 15% missing data. While at 30% the predictive accuracy of the two algorithms degrades sharply. However, the performance of the EM-EA is better than that of the EA of W. Myers, especially at the 30% missing data.

While comparing our method with Friedman's MS-EM, we use the same experimental conditions as that of MS-EM[4]. The stopping criterion for the algorithm is set at 1000 generations. Except that the mutation rate is set as 0.1 and the group size is set as 50, the selection of the original group, current network and crossover probability are the same as that in the above experiment. We tested the average log-loss of our algorithm on a separate test set. The results are summarized in Figure 5.

From Figure 5, we can see that in terms of the comparison of these two algorithms, the performance of the EM-EA is comparable with that of the MS-EM for the large samples, but the former is more robust as the hidden variables varies. And for the middle-sized samples, the EM-EA works better than MS-EM, especially when the hidden variables increase. Whereas, when the size of the samples is small, EM-EA performs worse than EM-EA. The reason possibly lies in that evolutionary algorithms need more samples than greedy search algorithms to some extent.

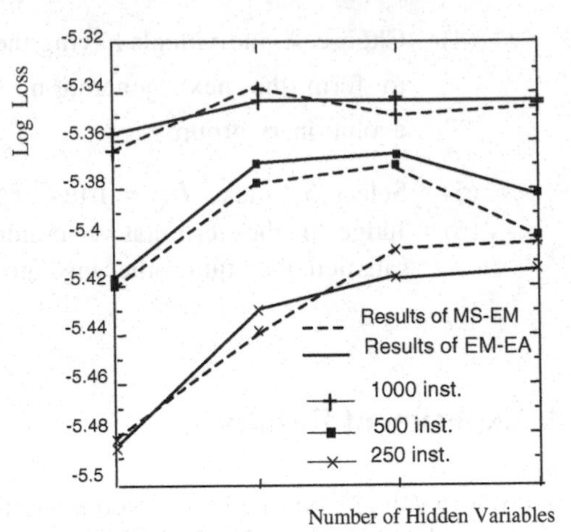

Fig. 5. The comparison of the two algorithms in terms of learning performance of network structures with hidden variables

5 Conclusion and Future Work

The results of the experiments verified the validity of our method. Compared with the EA of W. Myers et al, our algorithm is more accurate and efficient. And in terms of learning network structures with hidden variables, our algorithm is comparable with MS-EM. However, MS-EM starts with a given set of hidden variables and attempts to find a model that includes them. While our algorithm could create hidden variables on as-needed basis during the learning process, and so is more flexible and practical than MS-EM.

Next, we will test our method further with BDe score. In addition, we are planning to explore application of this method. In particular, we will try to put this method into use for the macroeconomics prediction.

References

1. Binder, J., Koller, D., Russell, S., Kanazawa, K.: Adaptive probabilistic networks with hidden variables. Machine Learning 29 (1997) 213—244
2. Cooper, G., Herskovits, E.: A Bayesian method for the induction of probabilistic networks from data. Machine Learning 9 (1992) 309—347
3. DeJong, K. A., Spears, W. M.: An Analysis of the Interacting Roles of Population Size and Crossover in Genetic Algorithms. Proceedings of the First International Conference on Parallel Problem Solving from Nature (1990), Dortmund, Germany

4. Friedman, N.: Learning Belief Networks in the Presence of Missing Values and Hidden Variables. Fourteenth International Conference on Machine Learning (ICML'97) (1998), Vanderbilt University, Morgan Kaufmann Publishers
5. Heckerman, D.: A tutorial on learning Bayesian network. Technical Report MSR-TR-95-06, Microsoft research (1995)
6. Lam, W., Bacchus, F.: Learning Bayesian belief networks: An approach based on the MDL principle. Computational Intelligence 10 (1994) 269—293
7. Larranaga, P., Poza, M. et al: Structure Learning of Bayesian Networks by Genetic Algorithms: A Performance Analysis of Control Parameters. IEEE Journal on Pattern Analysis and Machine Intelligence 18(9) (1996) 912—926
8. Lauritzen, S. L.: The EM algorithm for graphical association models with missing data. Computational Statistics and Data Analysis 19 (1995) 191—201
9. Myers J. W., Laskey K. B., DeJong K. A.: Learning Bayesian networks from incomplete data using evolutionary algorithms. In GECCO'99 (1999)
10. Syswerda, G.: Uniform Crossover in Genetic Algorithms. Proceedings of the 3rd International Conference on Genetic Algorithms (1989), Morgan Kaufman

Interactive Construction of Decision Trees

Jianchao Han and Nick Cercone

Department of Computer Science, University of Waterloo
Waterloo, Ontario, N2L 3G1, Canada
Email: {j2han, ncercone}@math.uwaterloo.ca

Abstract. We introduce an interactive decision tree construction system, DTViz, which consists of five components and maintains two interaction windows, and attempts to integrate the user's preference and domain knowledge into the construction process.

1 Introduction

There are several visualization systems for constructing decision trees. Most of them visualize the decision tree structure only [6, 7]. The PBC system developed in [1, 2] is very close to our work and attempts to interactively construct decision trees. However, it uses *circle segments* technique to visualize the original data, and some visualization space is wasted and the number of tuples in the training data is limited. Secondly, the decision tree being constructed is not visually displayed so that the user is not able to clearly see what's going on and make decision at critical moments, though [1] improves. Finally, the PBC system doesn't provide any approaches for cleaning the raw data.

We present a novel approach and develop a visualization system, DTViz, for interactively building decision trees based on our RuleViz model [3, 4] by visualizing the entire process. DTViz is a fully interactive system. During the decision tree construction, the user can integrate the domain knowledge, see the intermediate decision trees, evaluate tree nodes, and feedback his/her perception.

2 The DTViz System

The DTViz system consists of the following components: data tuple visualization, data reduction, decision tree node construction node evaluation, and decision tree visualization, shown in Fig. 1.

DTViz maintains two interaction windows, *data window* and *tree window*. The training data are visualized in the data window in order for the user to see the data distribution vertically and horizontally. The data window is often limited and can not accommodate large amount of tuples. If scrollable windows are used, the user is not capable of observing the entire data set at the first glance and thus the user's attention is distracted. Thus, large data sets must be reduced. The selected attributes and randomly sampled tuples (if necessary) are used as the training examples for constructing decision trees.

D. Cheung, G.J. Williams, and Q. Li (Eds.): PAKDD 2001, LNAI 2035, pp. 575–580, 2001.

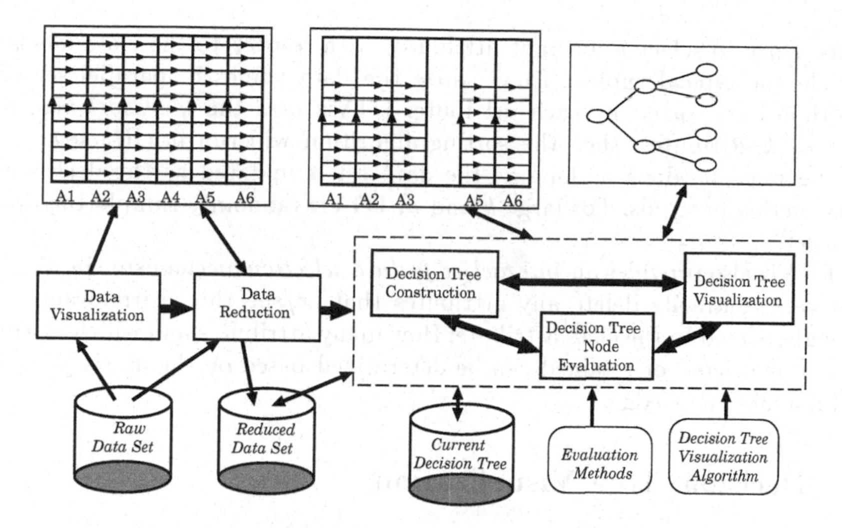

Fig. 1. The DTViz System

Decision trees are interactively constructed by iteratively performing the last three components and switching between the two windows. The data window visualizes the data tuples contained in the current node. Initially, it visualizes the entire data set corresponding to the decision tree root. The construction of decision trees is an interactive process with feedback from the user.

The current decision tree is visualized in a separate window, and is drawn from left to right with the root at the middle of left side. The nodes in the same level of the tree are uniformly arranged.

3 Data Visualization and Reduction

We combine two techniques, *Circle segments* [2] and *parallel coordinates* [3] to develop a new visualization technique, called *parallel segments*. Fig. 1 illustrates this technique, where the left-top snapshot demonstrates the visualization of a six dimensional data set.

The visualization area is divided into d equal sized segments for d-dimensional data set with each segment corresponding to an attribute. Within each segment, the pixels are arranged to start from the left bottom and end at the right top in a row-by-row and bottom-up fashion.

To visualize the training tuples, we map the attribute values occurring in the data set to the pixels in the attribute columns. The attribute values are sorted by each attribute and mapped to the pixels in the arrangement order. The pixels are rendered as the color determined by the class label of tuples to which attribute values belong. The color scale for class labels used in DTViz is derived from the PBC system [2] which is based on the HSI color model.

The size of data that can be visualized in *parallel segments* is determined by the data window size and the number of attributes. For large data sets, it is necessary to select important attributes with respect to the class labels and sample the typical tuples. To visualize the data values in parallel segments, the tuples are sorted by each attribute. DTViz uses the *quicksort* algorithm. Assume k attributes, then the sorting algorithm will run in $O(kn \log n)$. The sorting must be done as long as the data set is updated as the decision tree construction proceeds. For large k and n, DTViz randomly sample the original data set.

DTViz also provides an *interactive feature selection* mechanism, in which the user can arbitrarily delete any attributes that he/she thinks irrelevant or not strongly related to the class attribute. How many attributes and which attributes should be deleted or retained can be determined based on the user's perception and domain knowledge.

4 Decision Tree Visualization

The intermediate trees and the final decision tree are visualized in the *tree window*. The root is at the middle of the first column. The children of the root are evenly distributed in the second column. Generally, all the tree nodes at the i-th level are evenly arranged in the i-th column ($i \geq 1$), as shown in Fig. 3.

There are two types of tree nodes, *labeled* and *unlabeled*. Labeled nodes represent the leaf nodes of the final tree that can not be split further and are labeled with the most frequent class labels respect with to the nodes. The labeled nodes are drawn as rounded rectangles, and rendered in terms of the node evaluation and the HSI color model. Assuming that the prediction accuracy and support of a labeled node are a and s, respectively. Then the node color is calculated as follows: hue $= 0.5 + 2 \times a$, intensity $= 0.5 + 2 \times \frac{s}{n}$, saturation $= 1.0$, where n is the size of the test data set. Moreover, the class label and the $(support, accuracy)$ pair are displayed in the labeled node with rounded rectangles. The unlabeled nodes are drawn as rectangles, which are filled with the split attributes. If an unlabeled node is under construction, then it is left blank and ready to be split.

5 Interactive Construction of Decision trees

The decision tree is constructed from the training set. DTViz provides five interaction operations for the user to interactively build decision trees based on his/her perception and node evaluation. Fig. 2 depicts the interaction model developed in DTViz.

Initially, the decision tree contains only the *root* node, that covers the entire training set. The tree window displays the blank root. At this moment, the root is the current node for split.

As the decision tree construction proceeds, the user can arbitrarily select interaction operations to view the state of tree nodes and to control the growth and shrink of the decision tree. The left four operations in Fig. 2 change the current decision tree. The data window may need updated to reflect the current tree node. The final decision tree is obtained when all leaf nodes are labeled.

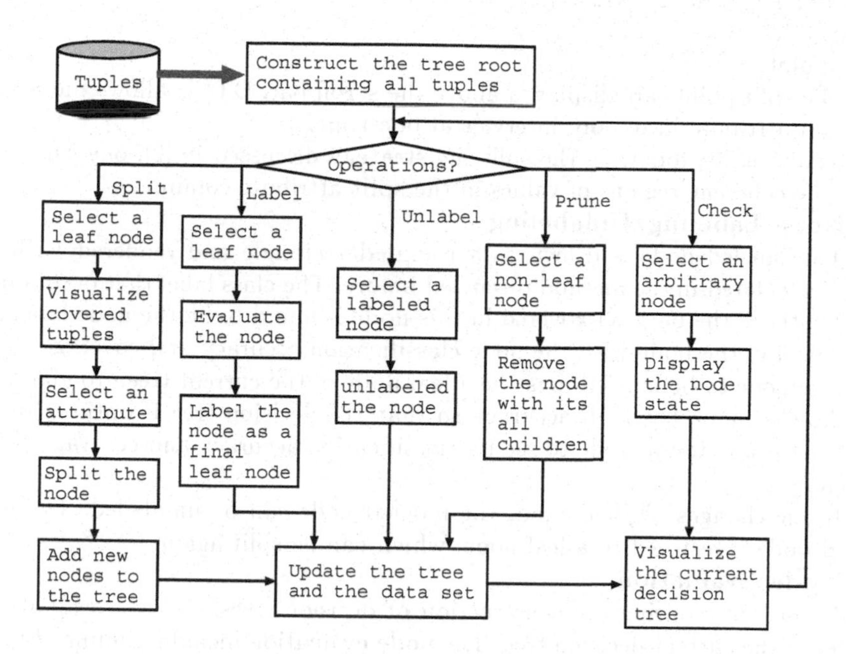

Fig. 2. The interaction model of constructing decision trees used in DTViz

Node Split

The *node split* includes three steps, selecting a split node, selecting a split attribute, and selecting split points of the attribute. The first step is performed in the tree window. The labeled nodes can not be selected for split.

The second and third steps are performed in the data window. When a node is specified to be split, the data window only visualizes the data tuples covered by this node, and the segments corresponding to the attributes that appear in the nodes on the path from the root to the current node are left blank.

The split attribute can be interactively specified by the user. Attribute selection follows two strategies: (1) *the more clear the clusters in a parallel segment, the better the corresponding attribute for splitting*; and (2) *The more approximate the size of clusters and non-clusters, the better the corresponding attribute.*

To select a split point for the split attribute, one just needs to click upon the pixel that separates clusters. Note that the separation of two different colors is not the only criteria for determining the exact split point because the same attribute values may belong to tuples of different classes. To solve this problem and help the user identify the reasonable split points, the DTViz system provides feedback to the user in the following ways: (1) *the attribute value of the pixel at the position of the mouse pointer is displayed in the status area at the bottom line of the data window when the mouse is moving*; and (2) *the scroll bar provides another means of viewing attribute values and class labels.* First, point the mouse around the boundary of two differently colored clusters to get a rough range in which the possible split point is because the attribute values are sorted. Then slightly move the slider around the range to see if the boundary is actually a split point.

The split points are displayed above the scroll bar. DTViz allows the user to split an attribute into more intervals at one time.

Additionally, following the splitting strategy discussed in [2], one can partition the coherent regions of values in the split attribute column.

Node Labeling/Unlabeling

The labeled node is drawn as a rounded rectangle and rendered with the color calculated in the method discussed before. The class label that occurs most frequently in the data set covered in this node is found to be the node label and displayed in the node with the node classification accuracy and coverage.

The node to be labeled must be a leaf node of the current tree. To guarantee the labeled node has high accuracy so that the decision tree is optimized, one can first evaluate the node to see its classification accuracy and coverage before labeling it.

If one changes his/her mind, the labeled node can be unlabeled. The unlabeled node is restored to a leaf node, which can be split again.

Node Evaluation

In any time during the construction of decision trees, one can evaluate any nodes in the current decision tree. The node evaluation includes finding *the node class label* that most frequently occurs in the node tuples; *the node support*, which is the number of node tuples; and *the node classification accuracy*, which is calculated as the occurring frequency of the node label in the set of tuples covered by the node.

Decision Tree Pruning

The decision tree pruning is needed in the following cases: (1) the user is not satisfied with the structure of the current decision tree; (2) the user feels hard to split some unlabeled leaf nodes to get high node evaluation; or (3) the final tree is too large. Note that only non-leaf node can be pruned, and the pruned node is not removed, while its all descendants are removed. The pruned node can be re-split or just labeled to be a final leaf node.

6 DTViz Implementation and Experiment

We implement the approach described in this chapter with Visual J++ 6.0 on Windows 98. We experiment the DTViz system with data sets from the UCI repository [5], including Adult, Iris, Car, Flag, Breast-Cancer, etc. Fig. 3 illustrates the decision tree construction with the Adult data set.

The Adult database consists of 14 condition attributes with 6 continuous and 8 nominal attributes. The class attribute *Salary* has two values, $> 50K$ and $<= 50K$. Due to the data window size, 5000 instances are randomly sampled and visualized. In Fig. 3, the left window shows a decision tree under construction. Four leaf nodes are labeled, and four nodes are unlabeled, three of which are split and one is to be labeled or split, depending on the node evaluation and the user's decision. The right window visualizes the data set contained in a tree node in the third level since two attributes have already been split.

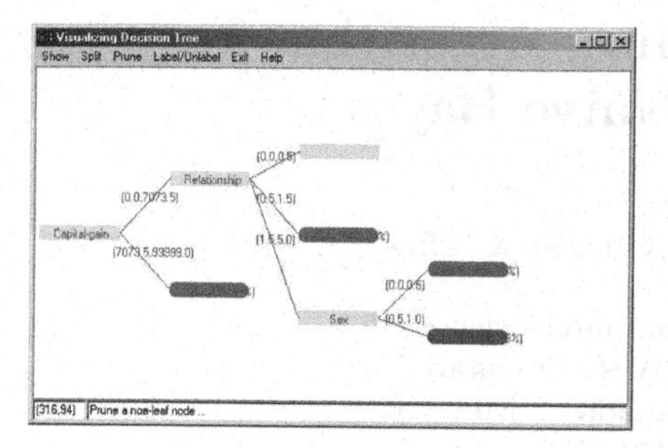

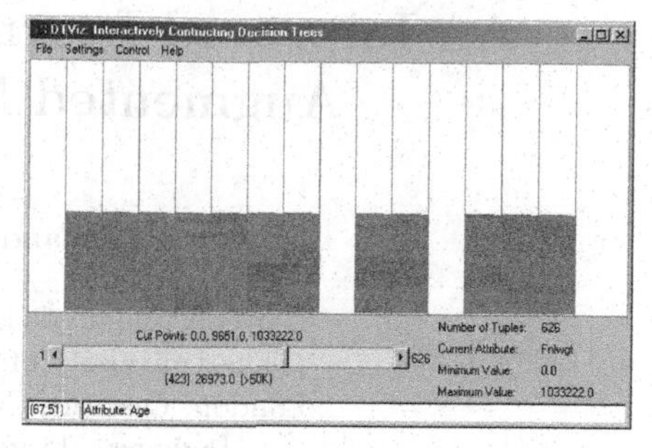

Fig. 3. The tree window and the data window

7 Conclusion

We presented an visualization system, DTViz, for interactively constructing decision trees. DTViz consists of five components and two interaction windows, To visualize the training data, a pixel-oriented visualization technique, *parallel segments*, is developed. The strategies for selecting split attributes and split points are discussed. The five interactive operations help the user grow, prune, and revise the decision tree iteratively until the final result is satisfying. The characteristics of DTViz include easy to use, uncertain results, varying accuracy, understandable decision tree structure, and on-demand node and attribute discretization.

Acknowledgments: The authors are members of the Institute for Robotics and Intelligent Systems (IRIS) and wish to acknowledge the support of the Networks of Centers of Excellence Program of the Government of Canada, the Natural Sciences and Engineering Research Council, and the participation of PRECARN Associates Inc. We also wish to thank Glaxo Wellcome Corp. of Mississagua, Canada for their financial contribution.

References

1. M. Ankerst, M. Ester, and H. P. Kriegel, Towards an Effective Cooperation of the User and the Computer for Classification, KDD-2000, 179-188.
2. M. Ankerst, C. Elsen, M. Ester, and H. P. Kriegel, Visual Classification: An Interactive Approach to Decision Tree Construction, KDD-99, 392-397.
3. J. Han and N. Cercone, Visualizing the Process of Knowledge Discovery, *Journal of Electronic Imaging*, SPIE 9(4), 404-420, 2000.
4. J. Han and N. Cercone, RuleViz: A Model for Visualizing Knowledge Discovery Process, KDD-2000, 223-242.
5. P. M. Murphy and D. W. Aho, UCI Repository of Machine Learning Databases, URL: *http://www.ics.uci.edu/~mlearn/MLRepository.html*, 1996.
6. T. D. Nguyen, T. B. Ho, and H. Shimodaira, Interactive Visualization in Mining Large Decision Trees, PAKDD-2000, 345-348.
7. J. S. Rao and W. J. E. Potts, Visualizing Bagged Decision Trees, KDD-97, 243-246.

An Improved Learning Algorithm for Augmented Naive Bayes

Huajie Zhang and Charles X. Ling

Department of Computer Science
The University of Western Ontario
London, Ontario, Canada N6A 5B7
{hzhang, ling}@csd.uwo.ca

Abstract. Data mining applications require learning algorithms to have high predictive accuracy, scale up to large datasets, and produce comprehensible outcomes. Naive Bayes classifier has received extensive attention due to its efficiency, reasonable predictive accuracy, and simplicity. However, the assumption of attribute dependency given class of Naive Bayes is often violated, producing incorrect probability that can affect the success of data mining applications. We extend Naive Bayes classifier to allow certain dependency relations among attributes. Comparing to previous extensions of Naive Bayes, our algorithm is more efficient (more so in problems with a large number of attributes), and produces simpler dependency relation for better comprehensibility, while maintaining very similar predictive accuracy.

1 Introduction

Learning Bayesian classifier is a process of constructing a special Bayesian network from a given set of pre-classified examples, each of which is represented by a vector of attribute values. Assume A_1, A_2,..., A_n are n attributes which take values $a_1, a_2, , ..., a_n$ respectively. Those attributes will be used collectively to predict the value c of another attribute C, called class label.

According to the Bayesian rule, the probability of an example E being in class c is:

$$p(C = c|a_1, a_2, ..., a_n) = \frac{p(a_1, a_2, ..., a_n|C = c)p(C = c)}{p(a_1, a_2, ..., a_n)}$$

The classification is taken as the C's value with the largest probability.

Assume all attributes are independent given the class. That is:

$$p(a_1, a_2, ..., a_n|c) = p(a_1|c)p(a_2|c)...p(a_n|c)$$

The resulting Bayesian classifier is called the *Naive Bayesian classifier*.

Where strong dependent relations do exist among attributes (such as $A_i = A_j$), the probability of Naive Bayes will not be correct. Friedman (1997) presented his work on learning tree-like Bayesian networks, TAN (Tree-Augmented

D. Cheung, G.J. Williams, and Q. Li (Eds.): PAKDD 2001, LNAI 2035, pp. 581–586, 2001.

Naive Bayes), in which non-classification attributes can form a tree structure. A TAN is a compromised representation between a full Bayesian network and Naive Bayes.

The basic idea is to approximate the underlying probability distribution by conditional mutual information. Its time complexity is $O(n^2)$, where n is the number of the attributes. [1] Keogh and Pazzani (1999) present another approach to learn TANs, which searches heuristically for a TAN guided by the predictive accuracy. Their algorithm, called *SuperParent*, also has the time complexity $O(n^2)$. They show that their algorithm consistently predicts more accurately than Friedman (1997)'s TAN, which in turn, predicts more accurately than the Naive Bayes. *SuperParent* consists of two major steps. The first step searches for a best super parent that improves the predictive accuracy the most. A super parent is a node with arcs pointing to all other nodes without a parent (not counting the class label). The second step determines one best child for the super parent chosen at the first step, again based on the predictive accuracy. After this iteration of the two steps, one arc is added on the TAN, and this process repeats until no improvement is achieved, or $n-1$ arcs are added into the tree. Obviously, *SuperParent* is a greedy algorithm with complexity $O(n^2)$. Our algorithm, called *StumpNetwork*, is constant-factor faster than *SuperParent*, while maintaining very similar predictive accuracy. The constant becomes larger in domains with a large number of attributes. This speed-up is important for data mining problems with a large number of attributes.

2 An Improved Algorithm for Learning TANs

We extend Keogh and Pazzani (1999) by proposing a more efficient algorithm, called *StumpNetwork*, to construct a special class of TAN. The motivation of *StumpNetwork* derives from the observation that the dependence among attributes tends to cluster into groups in many read-world domains with a large number of attributes. Attributes in each cluster form a simple dependency relation: a one-level tree structure, called the tree stump.[2] For example, in the customer database of a commercial bank, the amount of deposit, credit, and debt may be dependent on customer's total income, while the house price, utility bills, and neighbourhood are dependent on the postal code. But those tree stumps may not be totally independent, so some simple dependency relations are allowed among tree stumps. That it, we search for a special class of the tree structure as the topology of our Bayesian network.

Our algorithm constructs simple tree stumps first, and then construct links among the tree stumps. Similar to SuperParent (Keogh and Pazzani, 1999), search in *StumpNetwork* is guided by the predictive accuracy. That is, our criterion of constructing tree stumps and links between tree stumps is solely to improve the predictive accuracy of the Augmented Naive Bayesian classifier.

[1] The time complexity is also linear to the size of the training set. But this term is same for all Augmented Naive Bayesian classifiers in our discussion, so we omit it in our paper.

[2] Holte (1993) studied tree stumps as decision tree classifiers.

2.1 *StumpNetwork*

We first define the kind of tree structures as our hypothesis space, and then describe greedy search strategies for finding such tree structures.

Definition 1: A tree $T(r, N)$, where r is an attribute and N is a set of attributes, is called a *tree stump*, if the tree is of root r and of height 1, and there is an arc from r to each node in N.

Definition 2: A set of tree stumps is called Stump Network if the intersection of any two tree stumps is empty, and tree stumps may be connected in such a way that the root of one tree stump is pointed by at most a leaf node of another tree stump.

Figure 1 shows an example of a Stump Network. Clearly Stump Network is a special form of the tree topology where attribute dependencies tend to form clusters.

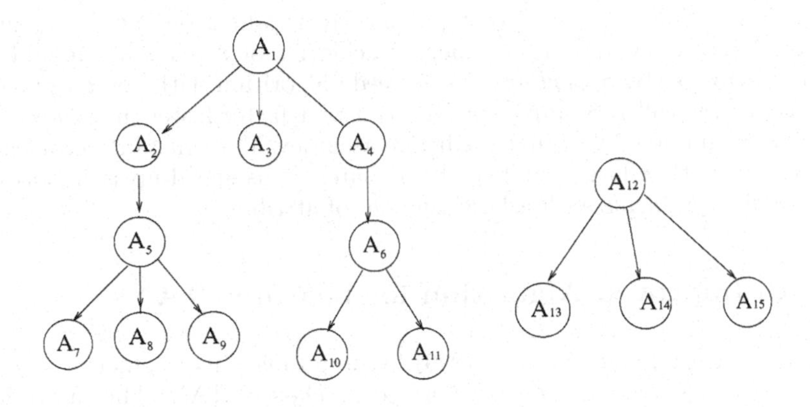

Fig. 1. An example of a Stump Network, a special class of TAN. Note that classification node C and all links from C to all A_i for all i are omitted for simplicity.

Our algorithm for searching Stump Network heuristically consists of two main steps. In the first step, it finds out a set of tree stumps, if adding such tree stumps into the Naive Bayes improves the the predictive accuracy on a testing set. The resulting set of tree stumps are sorted by the improvement of the predictive accuracy. The second step goes through the sorted list of the tree stumps once, to see if each tree stump can be pointed by a leaf of the previous tree stumps in the sorted list that will result in an improved predictive accuracy. If so, such a link will be remained in the Stump Network. The algorithm called *StumpNetwork* is presented below.

1. Read the data D
2. Initialize B to be Naive Bayes and calculate its predictive accuracy
3. Let node set N be all attributes (except C)and *TreeStump* queue be empty
4. Make a tree stump for each node in N. Let T_s be the tree stump with the highest improvement on the predictive accuracy

5. For each arc on T_s, if the predictive accuracy does not decrease after deleting the arc, remove it from T_s
6. Put T_s in *TreeStump* queue
7. Remove all nodes of T_s from N. If N is not empty, go to 4
8. Go over *TreeStump* queue once, for each tree stump in the queue, add a link from a leaf of the previous tree stumps in the queue to the root of this tree stump, if the predictive accuracy increases

Similar to *SuperParent*, our algorithm finds one super parent that links to all other possible attributes with the highest improvement on the predictive accuracy (Step 4). However, unlike *SuperParent* which keep just one child of the parent node found in the previous step (thus *SuperParent* needs to loop n times), we keep all the child nodes they do not result in a decreased predictive accuracy (Step 5). The resulting one-level tree forms a tree stump. Tree stumps may be linked together in Step 8.

2.2 Theoretical Comparison of *StumpNetwork* to *SuperParent*

From the description of the *StumpNetwork* algorithm above, the time for the creation of the tree stumps (from *step 4* to *step 7*) is $O(n^2)$ and the time for *step 8* is $O(n)$, where n is the number of attributes. Therefore, we consider the time complexity of constructing tree stumps as the time complexity of *StumpNetwork* T_c. We have:

$$T_c(n) = 2n + 2(n - m_1) + 2(n - m_1 - m_2) + ... + 2(n - \sum_{i=1}^{N-1} m_i) \qquad (1)$$

where N is the number of the tree stumps, and m_i is the size of the tree stump i. Let $k = \sum_{i=1}^{N} m_i$. We have:

$$T_c(n) = \frac{n(n+k)}{k} \qquad (2)$$

Because the tree stump found in Step 4 improves the predictive accuracy, and deleting edges in Step 5 will not decrease the predictive accuracy, the resulting tree cluster must have its size greater than 1. That is, $k \geq 2$.

It is easy to obtain the time complexity of *SuperParent*, T_s, as (Keogh and Pazzani 1999):

$$T_s(n) = 2n + 2(n-1) + ... + 1 = n(n+1) \qquad (3)$$

We omit the items of $O(n)$ in Equation 2 and Equation 3, then we have:

$$\frac{T_c(n)}{T_s(n)} = \frac{1}{k} \qquad (4)$$

Since $k \geq 2$, $T_c(n) \leq 1/2T_s(n)$. The greater the k, the less $T_c(n)$ than $T_s(n)$.

In many data mining applications, the number of attributes is often quite large (hundreds to thousands). In addition, those attributes can be clustered into several large groups. For these problems, our algorithm will be many times more efficient than *SuperParent*. The greater of the number of the attributes in each group, the more efficient our algorithm will be.

The empirical results shown in the next section verify this conclusion. However, little is lost with a simplified TAN and an improved efficiency: the predictive accuracies of the two algorithms will be shown to be very similar in the next subsection.

2.3 Experimental Results of *StumpNetwork*

We compare *StumpNetwork* with *SuperParent* using several datasets from the UCI repository (Merz, 1997), and one real-world dataset from a bank that we worked on in a previous project. Table 1 lists the properties of the datasets we used in our experiment.

Table 1. Descriptions of domains used in our experiments.

Dataset	Attributes	Class	Instances
Ecoli	7	8	336
Vote	16	2	435
Pima	8	2	768
Australia	14	2	690
Breast	10	2	683
Segment	19	7	1540
Vehicle	18	4	846
Bank	20	2	1162

Our experiments follow the procedure below:

1. The continuous attributes in the dataset are discretized by Fayyad(1993)'s entropy-based method.
2. Calculate the average predictive accuracy of *SuperParent* and *StumpNetwork* respectively, with 5 fold cross validation.

Table 2 shows the experimental results. As we can see, *StumpNetwork* achieves essentially the same testing accuracy as *SuperParent*. On the other hand, the time used in constructing Augmented Naive Bayes is quite different. On average over the datasets we compared, *StumpNetwork* is 4.8 times faster than *SuperParent*. There is a clear trend of larger saving with larger datasets and datasets with a larger number of attributes.

3 Conclusions

In this paper we present a new algorithm for constructing a special kind of tree augmented Naive Bayes based on work by Keogh and Pazzani (1999). Our

Table 2. Comparison of our *StumpNetwork* and *SuperParent*.

Dataset	*StumpNetwork*		*SuperParent*	
	Accuracy	**Time**	**Accuracy**	**Time**
Ecoli	84.43±1.07	4.08	83.11±1.16	7.74
Vote	96.07±1.99	16.33	94.02±2.23	39.63
Pima	78.98±2.28	2.75	79.39±1.83	6.36
Australia	85.43±1.83	17.97	86.07±1.47	53.59
Breast	96.38±0.70	37.17	96.56±0.91	97.61
Segment	93.37±0.96	1675.80	94.55±0.78	8275.60
Vehicle	69.10±1.86	152.39	68.91±1.13	668.55
Bank	53.22± 1.62	348.37	53.58±1.94	1726.20
Average	81.87	281.9	82.02	1359.4

algorithm works best in domains with a large number of attributes, and attributes tend to form large clusters with simple dependency relations inside clusters. Both experimental and theoretical analyses show that our algorithm is constant-factor faster than Keogh and Pazzani (1999)'s algorithm, and the constant is proportional to the size of the clusters. Our algorithm also produces simpler tree structure than an arbitrary tree, thus producing more comprehensible results. Empirical comparisons demonstrate that both algorithms have very similar predictive accuracies.

In our future research, we will study other efficient and specialized Augmented Naive Bayes suitable for domains possessing certain properties commonly occurring in real-world applications.

Acknowledgements. This research is based on the work of Mr. Keogh and Pazzani and Dr. Pazzani. They kindly provided us with the source codes of *SuperParent*, which is a great help to us.

References

[1997] Friedman, N. : Bayesian network classifier. Machine Learning, **29** (1997) 131–161.

[1993] Holte, R. : Very simple classification rules perform well on most commonly used datasets. Machine Learning, **3** (1993), 63–91.

[1999] Keogh and Pazzani , E. , Pazzani, M. : Learning augmented bayesian classifiers. Proceedings of Seventh International Workshop on AI and Statistics. (1999) Ft. Lauderdale.

[1997] Merz, C.: Uci repository of machine learning databases. Dept. of Information and Computer Science, University of California, (1997) Irvine.

Generalised RBF Networks Trained Using an IBL Algorithm for Mining Symbolic Data

Liviu Vladutu[1], Stergios Papadimitriou[1,2],
Severina Mavroudi[1], and Anastasios Bezerianos[1]

[1] University of Patras, Department of Medical Physics Patras 26500, Greece
liviu@heart.med.upatras.gr
[2] University of Patras, Department of Computer Engineering and Informatics, Greece

Abstract. The application of neural networks to domains involving prediction and classification of symbolic data requires a reconsideration and a careful definition of the concept of distance between patterns. Traditional distances are inadequate to access the differences between the symbolic patterns. This work proposes the utilization of a statistically extracted distance measure in the context of Generalized Radial Basis Function (GRBF) networks. The main properties of the GRBF networks are retained in the new metric space. The regularization potential of these networks can be realized with this type of distance. Furthermore, the recent engineering of neural networks offers effective solutions for learning smooth functionals that lie on high dimensional spaces.

1 Introduction

The emergence of neural network (NN) technology [1] offers valuable solutions to solve complicated data mining problems. Patterns arising both from commercial databases and from many engineering databases (as those that describe biosequences) involve data defined over a space that lacks the fundamental properties of distance metric spaces. This work constructs a proper distance metric for expressing the distance between values of features in symbolic domains. This metric owns some geometric properties that make it effective in the context of the regularization formulation of the Generalized Radial Basis Function (GRBF) networks. Regularization techniques impose the learning of a smooth functional from the network [2,4]. Therefore, it is justifiable to expect from the network to be able of learning the underlying smooth dependence of the outcomes on the attributes, even in the presence of noise that induces the perturbation. The potential of this distance metric to regularize the solution of the GRBF networks is the theoretical justification of the improved performance related to the simple nearest neighbor schemes. The paper proceeds as follows: Section 2 presents the proposed Statistical Distance Metric (SDM). Section 3 discusses how the statistical distance is fitted in the context of GRBF networks. Section 4 introduces the SDM within the framework of the GRBFs. Section 5 discusses the heuristic

D. Cheung, G.J. Williams, and Q. Li (Eds.): PAKDD 2001, LNAI 2035, pp. 587–593, 2001.

instance based parsing of the training set in order to improve the GRBF parameters (i.e. the selection of centers and their spreads). In the last section are presented the conclusions of the present work.

2 The Statistical Distance Metric (SDM)

The key problem for applications involving symbolic features is the definition of the distance metric. In domains where features are numeric, it is straightforward to compute the distance between two points in the pattern space in terms of a geometric distance. Indeed, the traditional RBF learning algorithms have been formulated and operate effectively in numeric domains with such distances. However, when the features are symbolic (as is usually in data mining applications using databases from bioinformatics or characteristic to a certain type of disease), the utilization of the traditional types of distances yields inadequate performance. There are two common approaches for handling symbolic information: one is the overlap method and the second the orthogonal representation [3, 5], both of them yielding poor performance in case of symbolic data. In order to be able to obtain an effective formulation of the distances between patterns with symbolic feature values we have adapted the distance measure proposed in [5]. This statistical distance measure takes into account the overall similarity of classification of all instances for each possible value of each feature. The method extracts with a statistical approach from the training set, a matrix that defines the distances between all possible values of a given feature. Therefore, a separate matrix for each feature is obtained. The distance measure for a specific feature is defined according to the following equation:

$$d(V_A, V_B) = \sum_{i=1}^{N} \left| \frac{C_{A_i}}{C_A} - \frac{C_{B_i}}{C_B} \right|^k \quad (1)$$

In the equation above, V_A and V_B denote two possible values for the feature, e.g. for the DNA promoter data they will be two nucleotides. The distance between the values is the sum over all the N classes. For example, for the DNA promoter example (discussed below) there are two classes, either the sequence is a promoter (i.e. a sequence that initiates a process called transcription) or not. The number of patterns for which the value V_A (V_B) is classified to class i , is denoted by C_{A_i} (C_{B_i}). Also, the total number of patterns of class A (B) is denoted by C_A (C_B), and k is a constant usually set to 1. These counts are computed over all patterns of the training set. It becomes easily evident that the more correlated are the classifications of patterns pertaining to two values for a feature the smallest is their statistical distance computed with equation (1). Therefore for feature values belonging to training set patterns with similar classifications a small statistical distance will be computed. The distance between two patterns is obtained by a weighted sum of distances between the values of the individual features of these patterns:

$$D(X,Y) = \sum_{i=1}^{F} w_{fi} d(V_{X_i}, V_{Y_i})^r \tag{2}$$

where F is the number of features, w_{f_i} accounts for the weight assigned to feature f_i reflecting its significance and r is a parameter that controls how distances between individual features scale for the computation of the total pattern distance (usually r=1 or 2). Also, V_{Xi} and V_{Yi} denote the values for the ith feature of X and Y.

3 Generalized RBFs with the Statistical Distance Metric (SDM)

The Generalized Radial Basis Functions networks explore the Tikhonov's regularization theory for obtaining a good generalization performance, as described in [1,2]. One prerequisite for the application of SDM distance type is to have enough training data for the accurate construction of the SDM space. However, the training sets of size large enough for providing the essential information for generalization, provide also the necessary information for the computation of an effective distance matrix. In contrast to example based nearest neighbor learning schemes, the GRBF learns a smooth functional that weights the contribution of each example subject to the requirements imposed by the regularizing term for the smoothness of the solution. This fact is the theoretical explanation for the superior performance of GRBF networks related to the Instance Based Learning (IBL) schemes. A parameter of particular importance is the region of influence of the GRBF kernels that is determined by their spread parameter σ. This problem becomes more complicated within the domain of statistical distances and the heuristic suggestions of [1] to compute σ as

$$\sigma = \frac{d_{max}}{\sqrt{2m}} \tag{3}$$

where d_{max} is the maximum distance and m the number of RBF centers has not been proved effective in practice. In order to obtain an effective setting for the spread parameter, a sensible approach is to obtain at the first step an estimate of the average distance d_{av}, of patterns within the space defined with the SDM. Then the region of influence of the RBF kernels is designed by requiring that at a particular distance *Spread* from the RBF center expressed in units of d_{av} , the attenuation of influence is decreased by a. Mathematically, this requirement is formulated as: $exp(-DF \cdot d_{av} \cdot Spread) = a$ and therefore the required parameter DF is derived as:

$$DF = \frac{-log(a)}{Spread \cdot d_{av}} \tag{4}$$

Values of these parameters that realize good results are for example $Spread =$ 5 and $a = 0.01$ meaning that at a distance from an RBF center 5 times larger than the average distance between patterns, the influence of the RBF function attenuates with a factor of 0.01. The RBF centers own an influence at a distance x from their center expressed by: $exp(-DF \cdot x)$ However, since the above scheme trains globally the spreads of RBF centers the peculiarities and irregularities of the state space are ignored. An additional instance based learning step that is described in the following section can estimate the relative importance of each RBF center and therefore can improve the performance of the designed RBF solution.

4 Instance-Based Learning for the Determination of the Parameters of the RBF Networks

It is highly desirable to exploit the reliable examples as centers of the RBF network. Also, the more reliable an example is, larger should its region of influence be when the example is used as an RBF center. The extent of the region of influence is expressed with the spreading parameter σ of the RBF center.A heuristically driven learning strategy is adopted for the determination of the examples that should be used as RBF centers and of their widths. The proposed GRBF training approach consists of two steps. At the first step, the Instance Based Learning (IBL) step, successive learning steps evaluate the potential of each example for serving as an RBF center, i.e. how representative the example is. This step is of a heuristic type and it tries to discover the reliability and the importance of the training examples with an instance based learning scheme that resembles the functionality of PEBLS [5]. This solution can be implemented with nearest neighbor schemes and if it is viewed as an input-output mapping it tends to create many class boundaries and discontinuous "islands" of misclassified regions placed near erroneously classified examples. The structure of the decision boundaries is smoothed and most of the regions with artifacts are extracted to reject the influence of noisy examples at the designed classification system. These examples do not yield satisfactory performance at the initial IBL step, so they are not selected as RBF centers. The second learning step constructs the Green's matrix with the estimated spreads of the Gaussian kernels estimated from the heuristically driven first step. During the first step, an empirical approximation to the solution is constructed. There are three basic approaches that can be exploited at the first heuristic learning pass.
1) *The one pass approach* is an exemplar weighting method that is used in conjunction with the nearest neighbor parameter. The learning is accomplished with only one pass through the training examples. At this training step, for each training instance its k nearest neighbors are found from among the remaining training set. If j neighbors have a matching class then the weight is assigned to the current instance according to the simple formula: $weigth = 1 + k - j$ Therefore, the more the class of the exemplar is reinforced by its neighbors, the less the weight (i.e. the more reliable the exemplar is). Algorithmically, *the one pass* instance based learning algorithm takes the form:

for each pattern P of the training set **do begin**
1 detect the k nearest neighbors to P from the training set according to the SDM;
2 Let j=number of nearest neighbors with the same class label as the class of P;
3 Set the weight parameter that quantifies the reliability of the exemplar as weight= 1+k-j
end;

The other two approaches that we tested for the weighting were 2) *the used correct* and, respectively, 3) *the increment* method.

5 Applications

We have applied the GRBF based solutions to a variety of data mining problems both from the engineering domain and from the commercial databases domain. Below we describe shortly one application from bioinformatics, one from data mining of commercial databases and finally some examples using databases from the UCI Machine Learning Repository, only from the medical field (http://www.ics.uci.edu/ mlearn/MLRepository.html). The first application concerns the prediction of promoter sequence [3]. This task involves predicting whether or not a given subsequence of a DNA sequence is a promoter, i.e. a sequence of genes that initiates a process called transcription. The data set contains 106 examples, 53 of which were positive examples (promoters) and the rest negative ones. A training pattern consists of a sequence of 57 nucleotides (features) from the alphabet a, c, g and t with the respective classification (promoter or not promoter). Since the available number of patterns were small the classification performance was tested with the leave-one-out methodology, i.e. repeatedly trials have been performed by training on 105 examples and testing on the remaining one. The computed performance was 2/106 (i.e. an average of 2 errors over 106 trials) versus 4/106 for a competitive experiment that used the KBANN neural network model [6].

We can observe from the Table 1 that the utilization of IBL within the framework of GRBFs improves the generalization performances obtained with the classic PEBLS algorithm. However, we cannot easily conclude that a particular IBL learning approach (from those described in the previous section) is better.

6 Conclusions

Neural network algorithms for learning are very effective in domains in which all features have numeric values. At these domains, the examples are treated as points and distance metrics obeys to standard definitions. However the usual domain of data mining applications is the symbolic domain. The utilization of

Table 1. Performances of the proposed $GRBF + IBL$ data mining algorithm

Database	**PEBLS**	**GRBF**	**GRBF+IBL** *One − pass*	**GRBF+IBL** *Used − correct*	**GRBF+IBL** *Increment*
Hypothyroid	97.90	98.04	98.33	**98.34**	98.29
Breast cancer	94.23	95.8	96.01	**96.12**	96.08
Iris	94.62	95.2	**96.22**	96.2	95.09
Hepatitis	76.59	78.23	79.45	**84.31**	81.29
Liver Disorders	63.45	62.98	65.9	72.5	**74.56**
Heart Disease	81.90	82.34	82.28	83.20	**85**
Audiology	77.90	78.91	**81.06**	81.03	79.44

the traditional distance metrics for data mining with neural networks usually results in modest results. The paper has adapted a SDM for application in the context of the GRBF neural networks. This distance metric extends the area of effectiveness of GRBF neural networks to the symbolic world. The results indicate that the generalization potential of neural networks can be utilised for patterns with symbolic features when the learning and evaluation algorithms are designed with the statistically extracted distance metric.
Future work to upgrade further the proposed GRBF and IBL hybrid data mining algorithms can proceed along many different directions, such as: finding optimal multisplits (for numerical attributes), or using simulated annealing algorithm (for the discretization of the continuous attributes).

Acknowledgements. The authors want to express their gratitude to the Research Committee of the University of Patras, Greece, for the partial support of this research, with a KARATHEODORIS contract, No. 2454.

References

[1] Simon Haykin, Neural Networks, MacMillan College Publishing Company, Second Edition,1999.
[2] T. Poggio and F. Girosi. Regularization algorithms for learning that are equivalent to multilayer perceptrons, Science 247 (1990):978-982.
[3] Pierre Baldi, Soren Brunak, Bioinformatics, MIT Press, 1998.
[4] Federico Girosi, "An Equivalence Between Sparse Approximation and Support Vector Machines", Neural Computation, 10:6, (1998) , pp. 1455-1480.
[5] C. Stanfill, D. Waltz, "Toward memory-based reasoning", Communications of the ACM, 29:12 (1986) , pp.1213-1228.
[6] G. Towell, J. Shavlik, M. Noordewier, "Refinement of aproximate domain theories by knowledge-based neural networks", Proceedings Eight National Conference on Artificial Intelligence, pp. 861-866, Menlo Park, CA:AAAI Press.

[7] Justin C. W. Debuse and Victor Jayward-Smith, "Discretisation of Continuous Commercial Database Features for a Simulated Annealing Data Mining Algorithm" Applied Intelligence 11, pp. 285-295 (1999).
[8] D. Randall Wilson, Tony R. Martinez, "Improved Heterogenous Distance Functions", Journal of Artificial Intelligence Research 6 (1997), p. 1-34.
[9] Christopher M. Bishop, Neural Networks for Pattern Recognition Clarendon Press-Oxford, 1996).

Author Index